Springer Proceedings in Mathematics

Volume 11

Erwin Bolthausen (with kind permission of Alexander Drewitz)

For further volumes:
http://www.springer.com/series/8806

Jean-Dominique Deuschel • Barbara Gentz
Wolfgang König • Max von Renesse
Michael Scheutzow • Uwe Schmock
Editors

Probability in Complex Physical Systems

In Honour of Erwin Bolthausen and Jürgen Gärtner

Springer

Editors
Jean-Dominique Deuschel
Max von Renesse
Michael Scheutzow
TU Berlin
Institute for Mathematics
Berlin
Germany

Wolfgang König
Weierstrass Institute for
Applied Analysis and Stochastics
Berlin
Germany

Barbara Gentz
University of Bielefeld
Faculty of Mathematics
Bielefeld
Germany

Uwe Schmock
Vienna University of Technology
Institute for Mathematical
Methods in Economics
Vienna
Austria

ISSN 2190-5614
ISBN 978-3-642-23810-9 e-ISBN 978-3-642-23811-6
DOI 10.1007/978-3-642-23811-6
Springer Heidelberg Dordrecht London New York

Library of Congress Control Number: 2012930493

Mathematical Subject Classification (2010): 60FXX, 60GXX, 82BXX, 82CXX, 82DXX

Printed on acid-free paper

Springer is part of Springer Science+Business Media (www.springer.com)

Preface

Probabilistic approaches have played a prominent role in the study of complex physical systems for more than 30 years. Two outstanding protagonists of this approach are Jürgen Gärtner and Erwin Bolthausen, to whom this volume is dedicated. Each of them was honored with a workshop in 2010; these took place at Technische Universität Berlin, where they both worked for decades. The conferences were devoted to the most important aspects of their interests: 'Random media' and 'Probabilistic techniques in complex physical systems'. They were organized by the DFG Research Unit FOR718 *Analysis and Stochastics in Complex Physical Systems* on the occasion of Jürgen's 60th birthday and Erwin's 65th birthday.

Jürgen and Erwin have been recognized for decades as outstanding experts in the probabilistic treatment, spiced with a dash of analysis, of problems in statistical mechanics and related fields. Their high esteem and profound impact are reflected by their great number of students and collaborators and by their large number of invitations to conferences, editorships, etc. over the years.

Erwin started his career with various distributional limit results of central limit and martingale type, but soon turned to problems coming from large-deviation analysis, like Laplace approximations and the maximum entropy principle. One of the main types of problems that accompanied his career for decades are intricate questions about the extremal behavior of the volume of the path of a random walk or a Wiener sausage and of the intersection of two independent such objects. Here he has derived a number of striking and deep results over the years. Another core area of his research, which is closely related, is the description of paths under the influence of a self-attracting or self-repellent force, partially motivated by the polaron problem. In particular, Erwin derived several fundamental properties of polymers with various kinds of interactions. His results also had a strong influence on the understanding of interface models with gradient-type interactions. Some of his favorite subjects in recent years have been random walks in random environments, and spin glasses and the little-understood phenomenon of ultrametricity.

Jürgen was educated within the Russian school in the 1970s, pioneering the application of large deviation analysis to various models in statistical mechanics.

One of the fundamental tools, the Gärtner–Ellis theorem, is a side-result of his thesis. Later he built up a theory of large deviations for projective limits. Also his contributions to the McKean–Vlasov equation remain a vital element of the theory. Over the last two decades, he has been one of the most active researchers on the parabolic Anderson model, the Cauchy problem for the heat equation with random potential.

Many of the above results were derived in close collaboration with students, colleagues, and friends, many of whom also presented talks on the occasion of the two 2010 workshops. The present volume collects 20 research and review papers by participants in the fields in which Jürgen and Erwin are best known for their contributions. Most of these papers are, in some way or another, influenced by Jürgen's and Erwin's work, and all of them present state-of-the-art results in topics that accompanied the two for decades and received significant impacts from them over the years. All papers have been peer-refereed according to highest standards.

Almost half of the contributions to this volume are devoted to the parabolic Anderson model, one of the most active research fields of Jürgen. For more than 20 years, Jürgen has formed and extended this subject like nobody else. Jürgen's co-authors and students and their students and colleagues give an impressive account on some of the latest developments for the parabolic Anderson model, among which there are results on the long-time behavior for various time-dependent and time-independent potentials, and novel aspects like several moving catalysts, acceleration/deceleration, and front propagation.

Another main topic covered by this volume is random polymers interacting with random and nonrandom environment and their critical behavior, a topic that received much attention from Erwin and his coauthors. Furthermore, special aspects of branching processes and interacting measure-valued processes are considered, topics that Jürgen studied many years ago. Finally, this volume offers a choice of results on various models that Erwin worked on or was interested in for many years, like Parisi's formulas for the generalized random energy model, metastability, hydrodynamic limits for gradient models and dimers.

In total, the collection of 20 papers in this volume presents important contributions to and surveys on research areas that are of current interest and have been strongly influenced by these two eminent mathematicians. It is not too much to say that these fields have benefited tremendously from their work.

Berlin
June 2011

Jean-Dominique Deuschel
Barbara Gentz
Wolfgang König
Max von Renesse
Michael Scheutzow
Uwe Schmock

Workshop on

Probabilistic Techniques in Statistical Mechanics
Celebrating the 65th Birthday of Erwin Bolthausen

Organized by:
Jean-Dominique Deuschel, Wolfgang König, Max von Renesse, Michael Scheutzow and the DFG Research Unit FOR718 *Analysis and Stochastics in Complex Physical Systems*

Venue:
Technische Universität Berlin, Institute for Mathematics, Str. des 17. Juni 136, 10623 Berlin, Germany, Room MA043

Period:
October 14–16, 2010

Speaker/Title:

Gérard Ben Arous (Courant Institute New York)
Extreme gaps in the spectrum of random matrices

Michiel van den Berg (University of Bristol)
Minimization of Dirichlet eigenvalues with geometric constraints

Anton Bovier (University of Bonn)
Almost sure ageing

David Brydges (University of British Columbia, Vancouver)
The strong interaction limit of continuous-time weakly self-avoiding walk

Francesco Caravenna (University of Padova)
The weak coupling limit of disordered copolymer models

Amir Dembo (Stanford University)
Low temperature expansion for matrix models

Tadahisa Funaki (University of Tokyo)
Hydrodynamic limit for two- and three-dimensional Young diagrams

Giambattista Giacomin (University Paris Diderot – Paris 7)
Impurities, defects and critical phenomena

Ilya Goldsheid (Queen Mary College)
Simple random walks in one-dimensional random environment: limiting behaviour in the sub-diffusive regimes

Alice Guionnet (ENS Lyon)
Potts model on random graphs

Frank den Hollander (Leiden University and EURANDOM)
Variational approach to copolymers near linear interfaces

Dmitry Ioffe (Technion)
Stretched polymers in random environment

Nicola Kistler (University of Bonn)
Traveling waves through the spin glass

Alain-Sol Sznitman (ETH Zurich)
Random walks and random interlacements

Béatrice de Tilière (University Pierre et Marie Curie, Paris)
The critical Z-invariant Ising model via dimers: local statistics and combinatorics

Fabio Toninelli (ENS Lyon)
On the zero temperature dynamics of the three-dimensional Ising model
(joint work with P. Caputo, F. Martinelli and F. Simenhaus)

Bálint Tóth (Technical University Budapest)
Superdiffusive lower bound for self-repelling processes in the critical dimension

Yvan Velenik (University of Geneva)
A new approach to the Aizenman–Higuchi theorem

Wendelin Werner (University Paris-Sud 11 in Orsay and ENS Paris)
Self-interacting random walks with finite spatial interaction range

Ofer Zeitouni (University of Minnesota and Weizmann Institute of Science)
Fluctuations of the (discrete) Gaussian free field, and branching random walks

Workshop on

Random Media
Celebrating the 60th Birthday of Jürgen Gärtner

Organized by:
Jean-Dominique Deuschel, Wolfgang König, Max von Renesse, Michael Scheutzow and the DFG Research Unit FOR718 *Analysis and Stochastics in Complex Physical Systems*

Venue:
Technische Universität Berlin, Institute for Mathematics, Str. des 17. Juni 136, 10623 Berlin, Germany, Room MA043

Period:
April 8–10, 2010

Speaker/Title:

Gérard Ben Arous (Courant Institute New York)
Random matrices and Morse theory in many dimensions: the case of mixtures of spherical spin glasses and the (possibility of) full replica symmetry breaking

Marek Biskup (UCLA and University of South Bohemia)
Eigenvalue order statistics for the random Schrödinger operator

Erwin Bolthausen (University of Zurich)
Kac-type interactions in a one-dimensional system with a continuous symmetry

Anton Bovier (University of Bonn)
Metastability in Ginzburg–Landau type stochastic differential equations

Donald Dawson (University of Ottawa)
McKean–Vlasov mutation–selection dynamics

Klaus Fleischmann (Berlin)
Recent properties of states of super-α-stable motion with branching of index $1 + \beta$

Mark Freidlin (University of Maryland)
Perturbation theory for systems with many invariant measures

Markus Heydenreich (Free University of Amsterdam)
Random walk on high-dimensional incipient infinite cluster

Frank den Hollander (Leiden University and EURANDOM)
The mathematical work of Jürgen Gärtner

Grégory Maillard (University of Provence, Marseille)
Parabolic Anderson model with a finite number of moving catalysts

Peter Mörters (University of Bath)
Geometric approaches to intermittency in the parabolic Anderson model

Tom Mountford (EPF Lausanne)
Parabolic Anderson model with voter model noise

Alejandro Ramírez (Pontifical Catholic University of Chile)
Ballisticity conditions for random walk in random environment

Vladas Sidoravicius (CWI Amsterdam)
On random growth and random walks in dynamically evolving random environment

Rongfeng Sun (University of Singapore)
Annealed versus quenched asymptotics for the parabolic Anderson model with moving catalysts or traps

Alain-Sol Sznitman (ETH Zurich)
Disconnecting discrete cylinders

Contents

Laudatio: The Mathematical Work of Jürgen Gärtner

Frank den Hollander

Abstract Over the past 35 years, Jürgen Gärtner has made seminal contributions to probability theory and analysis. In this brief laudatio, I describe what I consider to be his five most important lines of research: (1) Gärtner-Ellis large deviation principle; (2) Kolmogorov–Petrovskii–Piskunov equation; (3) Dawson–Gärtner projective limit large deviation principle; (4) McKean–Vlasov equation; (5) Parabolic Anderson model. Each of these lines is placed in its proper context, but no attempt is made to fully trace the literature. What characterizes the papers of Jürgen is that they all deal with hard fundamental problems requiring a delicate combination of probabilistic and analytic techniques. A red thread through his work is the symbiosis of large deviation theory and potential theory, which he masterfully combines to reach powerful and elegant solutions.

1 Gärtner–Ellis Large Deviation Principle

In 1977, Jürgen proved what is nowadays considered to be the *most general form of Cramér's theorem in large deviation theory* [21, 22]. This work, which was suggested to him by Mark Freidlin, took place while the architectural foundations of large deviation theory were being laid. As such, Jürgen's theorem belongs to the very heart of the field, as developed in the 1970s by Freidlin and Wentzell [20] and Donsker and Varadhan [16]. In 1984, the assumptions under which Jürgen had proved his theorem were weakened by Richard Ellis [17].

F. den Hollander (✉)
Mathematical Institute, Leiden University, P.O. Box 9512, 2300 RA Leiden, The Netherlands
EURANDOM, P.O. Box 513, 5600 MB Eindhoven, The Netherlands
e-mail: denholla@math.leidenuniv.nl

J.-D. Deuschel et al. (eds.), *Probability in Complex Physical Systems*,
Springer Proceedings in Mathematics 11, DOI 10.1007/978-3-642-23811-6_1,

Theorem 1.1. *Given a sequence* $(X_n)_{n\in\mathbb{N}}$ *of random variables taking values in* $\mathbb{R}^d$*, let*

$$\phi_n(t) = E(e^{\langle t, X_n \rangle}), \qquad t \in \mathbb{R}^d,$$

denote their moment generating functions (where $\langle\cdot,\cdot\rangle$ *is the standard inner product on* $\mathbb{R}^d$*). Suppose that*

$$\lim_{n\to\infty} \frac{1}{n} \log \phi_n(nt) = \Phi(t) \qquad \text{exists for all } t \in \mathbb{R}^d,$$

and is everywhere finite and differentiable. Then the family $(P_n)_{n\in\mathbb{N}}$ *with* $P_n(\cdot) = P(X_n \in \cdot)$ *satisfies the large deviation principle (LDP) on* $\mathbb{R}^d$ *with rate function* $I\colon \mathbb{R}^d \to [0,\infty]$ *given by the Legendre transform*

$$I(x) = \sup_{t\in\mathbb{R}^d} [\langle t, x\rangle - \Phi(t)], \qquad x \in \mathbb{R}^d.$$

Theorem 1.1 says that

$$\liminf_{n\to\infty} \frac{1}{n} \log P(X_n \in O) \geq - \inf_{x\in O} I(x) \qquad \forall\, O \subset \mathbb{R}^d \text{ open},$$

$$\limsup_{n\to\infty} \frac{1}{n} \log P(X_n \in C) \leq - \inf_{x\in C} I(x) \qquad \forall\, C \subset \mathbb{R}^d \text{ closed},$$

which informally reads as

$$P(X_n \approx x) \approx e^{-nI(x)} \qquad \forall\, x \in \mathbb{R}^d, \qquad n \to \infty.$$

Hence Theorem 1.1 gives full control of the deviations of the random variable X_n away from its typical values for large n.

For the special case where

$$X_n = \frac{1}{n}(Y_1 + \cdots + Y_n), \quad n \in \mathbb{N}, \quad (Y_i)_{i\in\mathbb{N}} \text{ i.i.d.},$$

we have $\phi_n(nt) = [\phi(t)]^n$ with $\phi(t) = E(e^{\langle t, Y_1\rangle})$ the moment-generating function of Y_1, and Theorem 1.1 reduces to Cramér's theorem for the empirical mean of i.i.d. random sequences. However, in its full generality, the theorem is applicable far beyond the i.i.d. setting, including Markov sequences, Gibbs random fields, and random processes in random media.

Over the years, the Gärtner–Ellis LDP has become *one of the workhorses of large deviation theory*. Due to its simplicity, generality, and flexibility, it appears in every textbook on large deviations. It has been and is being applied in a great many different contexts. For refinements as well as additional background, see the monographs by Varadhan [39] and Dembo and Zeitouni [15].

2 Kolmogorov–Petrovskii–Piskunov Equation

In 1982, Jürgen wrote a seminal paper [23] on the famous semi-linear diffusion equation introduced in 1937 by Kolmogorov, Petrovskii, and Piskunov [32]:

$$\frac{\partial u}{\partial t}(x,t) = \frac{1}{2}\Delta u(x,t) + f(u(x,t)), \quad x \in \mathbb{R}^d,\, t \geq 0.$$

Here, $f\colon [0,1] \to [0,\infty)$ is assumed to be once continuously differentiable with $f(0) = f(1) = 0$ and $0 < f(u)/u \leq f'(0)$ for $u \in (0,1)$. The initial condition is taken to be

$$u(x,0) = g(x), \qquad x \in \mathbb{R}^d,$$

for some appropriate $g\colon \mathbb{R}^d \to [0,1]$ that is strictly positive near $x = 0$ and tends to zero rapidly at infinity. The KPP equation describes a system of particles that diffuse and that split into two at a rate that depends on their local density via the function f, both in the continuum limit of many particles with small mass. This is why it is referred to as a *reaction-diffusion equation*. The KPP equation plays a key role in the understanding of wave front propagation phenomena, occurring, e.g., in combustion processes.

Theorem 2.1. *Abbreviate* $v^* = [2f'(0)]^{1/2}$, *and define*

$$h(z) = \sup_{u \in (0,z]} [f'(0) - f(u)/u], \qquad z \in (0,1].$$

Suppose that

$$\int_0^1 h(z)\, z^{-1} \log^2(1/z)\, \mathrm{d}z < \infty,$$

and that $g(x) = \bar{g}(\|x\|)$ *with*

$$\limsup_{r\to\infty} r^{-1/2} \log[e^{v^* r}\bar{g}(r)] < \infty.$$

Then, for every $\epsilon \in (0,\frac{1}{2})$, *there exists a* $\rho(\epsilon) \in (0,\infty)$ *such that, for all t sufficiently large,*

$$\begin{aligned}&\big\{x \in \mathbb{R}^d\colon \epsilon < u(x,t) < 1-\epsilon\big\}\\ &\qquad \subset \big\{x \in \mathbb{R}^d\colon m(t) - \rho(\epsilon) < \|x\| < m(t) + \rho(\epsilon)\big\},\end{aligned}$$

where

$$m(t) = v^* t - \frac{d+2}{2v^*}\log t + \frac{1}{v^*}\log \int_0^\infty r^{(d+1)/2}\, e^{v^* r}\, \bar{g}(r)\, \mathrm{d}r.$$

The first condition in Theorem 2.1 controls the behavior of f near zero, and the second condition controls the behavior of g near infinity. The result identifies the location of the *expanding wave front* around which u drops from $u \approx 1$ to $u \approx 0$: this wave front is an annulus of finite width around the surface of the ball of radius $m(t)$. The leading term in $m(t)$ says that the speed of the wave front is v^*, the correction terms in $m(t)$ are computed up to and including order 1.

Earlier work by McKean [33,34], Aronson and Weinberger [1], Bramson [5] and Uchiyama [38] had fallen short of identifying the constant in $m(t)$ and had required more severe restrictions on g, such as compact support. Part of this work was for $d = 1$ only.

The proof of Theorem 2.1 centers around a delicate estimate of the first-exit time distribution for a Brownian motion in a time-dependent domain. In later work, Jürgen extended Theorem 2.1 to a much broader class of reaction-diffusion equations. This work was subsequently picked up and pushed further by others. See Freidlin [18] for a survey.

3 Dawson–Gärtner Projective Limit Large Deviation Principle

In 1987, Jürgen and Don Dawson proved a theorem that considers a *nested sequence* of LDPs and obtains from this a new LDP via a *projective limit* [8]. This theorem is a powerful tool, because it allows to first derive an LDP in a simple setting (e.g., on a finite or a compact space) and then draw from that an LDP in a more difficult setting (e.g., on an infinite or a noncompact space). Over the years, also the Dawson–Gärtner projective limit LDP has become *one of the workhorses of large deviation theory*.

Theorem 3.1. *Let $(P_n)_{n\in\mathbb{N}}$ be a family of probability measures on a Hausdorff topological space χ. Let $(\pi^N)_{N\in\mathbb{N}}$ be a nested family of projections acting on χ, and let*

$$\chi^N = \pi^N \chi, \qquad P_n^N = P_n \circ (\pi^N)^{-1}, \qquad N \in \mathbb{N}.$$

If, for each $N \in \mathbb{N}$, the family $(P_n^N)_{n\in\mathbb{N}}$ satisfies the LDP on χ^N with rate function $I^N\colon \chi^N \to [0,\infty]$, then the family $(P_n)_{n\in\mathbb{N}}$ satisfies the LDP on χ with rate function $I\colon \chi \to [0,\infty]$ given by

$$I(x) = \sup_{N\in\mathbb{N}} I^N(\pi^N x), \qquad x \in \chi.$$

The π_N's can for instance be discretizations or truncations,

$$\chi = \mathbb{R},\ \chi_N = 2^{-N}\mathbb{Z} \quad \text{or} \quad \chi = \mathbb{Z},\ \chi_N = \mathbb{Z} \cap [-N, N],$$

and the P_N's can for instance be probability distributions of the empirical means of a sequence of random variables. The supremum defining I is monotone in N and can often be computed explicitly. Apart from the nestling condition, the result in Theorem 3.1 is again simple, general, and flexible. For more background, see, e.g., the monograph by Dembo and Zeitouni [15].

4 McKean–Vlasov Equation

In the period 1987–1989, Jürgen and Don Dawson wrote a series of papers on the McKean–Vlasov equation [8–10, 24]. Their main result reads as follows. Let $H_N\colon \mathbb{R}^N \to \mathbb{R}$ be the N-particle *mean-field* Hamiltonian

$$H_N(x) = \frac{1}{2N} \sum_{i,j=1}^{N} f(x_j - x_i) + \sum_{i=1}^{N} g(x_i), \quad x = (x_1, \ldots, x_N),$$

with f even and f, g both twice continuously differentiable (f is a pair interaction, g is an external field). For $T > 0$, which plays the role of a time horizon, let $(X(t))_{[0,T]} = ((X_1(t), \ldots, X_N(t)))_{[0,T]}$ evolve according to the system of N *coupled diffusion equations*

$$\mathrm{d}X_i(t) = \frac{\partial H_N}{\partial x_i}(X(t))\,\mathrm{d}t + \mathrm{d}B_i(t), \quad i = 1, \ldots, N,$$

where $(B(t))_{[0,T]} = ((B_1(t), \ldots, B_N(t)))_{t\in[0,T]}$ are i.i.d. standard Brownian motions. This system defines a stochastic dynamics that is reversible w.r.t. the Gibbs measure with Hamiltonian H_N. A typical initial condition is where $X(0)$ has distribution λ^N for some probability measure on $\mathbb{R}$.

Define the *empirical path measure*

$$L_N = \frac{1}{N} \sum_{i=1}^{N} \delta_{(X_i(t))_{t\in[0,T]}},$$

which is an element of $M_1(C[0,T])$, the space of probability measures on the set of continuous functions from $[0,T]$ to $\mathbb{R}$.

Theorem 4.1. *The family $(P_N)_{N\in\mathbb{N}}$ with $P_N(\cdot) = P(L_N \in \cdot)$ satisfies the LDP on $M_1(C[0,T])$ with rate function $I\colon M_1(C[0,T]) \to [0,\infty]$ given by*

$$I(Q) = \begin{cases} \int \log\left(\frac{\mathrm{d}Q}{\mathrm{d}P^Q}\right) \mathrm{d}Q, & \text{if } Q \ll P^Q, \\ \infty, & \text{otherwise}, \end{cases}$$

where P^Q is the law of a single diffusion with self-interaction.

Formally, P^Q is the law of the unique strong solution of the one-dimensional Itô stochastic differential equation

$$\mathrm{d}x(t) = \beta^{\pi_t Q}(x(t))\,\mathrm{d}t + \mathrm{d}b(t), \qquad t \in [0,T],$$

where $x(0)$ has probability distribution λ, $(b(t))_{[0,T]}$ is a standard Brownian motion on $\mathbb{R}$, $\pi_t Q$ is the evaluation of Q at time t, and

$$\beta^q(x) = -\int_{\mathbb{R}} f'(y-x)\,q(\mathrm{d}y) - g'(x), \qquad x \in \mathbb{R},\ q \in M_1(\mathbb{R}).$$

Theorem 4.1 describes the large deviation properties of the paths of the interacting diffusions. The rate function I has a unique zero solving the equation

$$Q = P^Q.$$

The solution of this equation determines the law of $(x(t))_{[0,T]}$ via the successive time evaluations of Q. The resulting process, which is called the *McKean–Vlasov process*, is a diffusion with a time-inhomogenous drift that is to be determined from *self-consistency*. This self-consistency is typical for mean-field models. In terms of the McKean–Vlasov process, I can be written as an action functional, in the spirit of Freidlin–Wentzell theory.

Related work was done by Sznitman [35, 36] and by Ben Arous and Brunaud [2]. The results were later extended to random mean-field interactions by Dai Pra and den Hollander [7] and to spin-glass mean-field interactions by Ben Arous and Guionnet [3, 4] and by Jürgen's student Malte Grunwald [30].

In the period 1991–1997, while extending their work on the McKean–Vlasov equation, Jürgen and Don Dawson introduced the notion of *multi-level large deviations*, describing the large deviation behavior of multi-array families of dependent random variables [11, 12]. This work in turn gave rise to the Dawson–Greven *renormalization program for hierarchically interacting diffusions* [13, 14], introduced in 1993 and since then pursued by various groups. For an overview on the latter, see den Hollander [31].

5 Parabolic Anderson Model

In 1990, Jürgen wrote a seminal paper with Stas Molchanov on *intermittency* in the Parabolic Anderson Model [26]. A lot of earlier work had been done in the physics and in the chemistry literature, but this was the first paper that put the model on a firm mathematical basis and provided a new way of looking at intermittency via the study of *Lyapunov exponents*. A follow-up paper in 1998 [27] pushed the subject further. Since then Jürgen has been working intensively on the PAM with several colleagues, both senior and junior. There are two versions of the model: *static* and *dynamic*.

The PAM is the partial differential equation:

$$\frac{\partial u}{\partial t}(x,t) = \kappa \Delta u(x,t) + \xi(x,t)\, u(x,t), \quad x \in \mathbb{Z}^d,\ t \geq 0,$$

where Δ is the discrete Laplacian, $\kappa \in (0,\infty)$ is the diffusion constant, and $\xi(x,t)$ is a space–time random medium that drives the equation. Typical initial conditions are:

$$u(x,0) = 1 \quad \text{or} \quad u(x,0) = \delta_0(x).$$

The solution of the PAM describes the behavior of a *reactant* u under the influence of a *catalyst* ξ.

The key objects of interest are the Lyapunov exponents

$$\lambda_p = \lim_{t\to\infty} \frac{1}{pt} \log E([u(0,t)]^p), \quad p > 0,$$

$$\lambda_0 = \lim_{t\to\infty} \frac{1}{t} \log u(0,t), \quad \xi\text{-a.s.},$$

where E denotes expectation over the ξ-field. The λ_p's are referred to as the *annealed* Lyapunov exponents, λ_0 as the *quenched* Lyapunov exponent. The PAM is said to be *intermittent* when

$$p \mapsto \lambda_p \text{ is strictly increasing.}$$

The geometric interpretation behind this property is that the u-field develops *sparse high peaks*, with λ_p being dominated by different classes of peaks for different p (see Gärtner, König, and Molchanov [28]). This is the reason why λ_p and λ_0 provide insight into the behavior of u in space and time.

A key tool to study the PAM is the *Feynman–Kac formula*

$$u(x,t) = E_x\left(\exp\left[\int_0^t \xi(X^\kappa(s), t-s)\, ds\right] u(X^\kappa(t),0)\right),$$

where $(X^\kappa(t))_{t\geq 0}$ is simple random walk jumping at rate $2d\kappa$, and E_x denotes expectation given that $X^\kappa(0) = x$. This shows that understanding the PAM is equivalent to understanding the large deviation properties of *a random walk in a random scenery*.

- *Static version:* For the case where ξ is time-independent, i.e., $\xi(x,t) = \xi(x,0) = \xi(x)$, the PAM is by now fairly well understood. The typical case is where $\xi(x)$, $x \in \mathbb{Z}^d$, are i.i.d., in which case there are *four subclasses* of distributions of $\xi(0)$ leading to qualitatively different behavior. A detailed description has been obtained for the location and the height of the peaks in the u-field, which tend to concentrate around the peaks in the ξ-field. The peaks in the u-field tend to live on *sparse islands*, whose locations and sizes change over time. For an overview, see Gärtner and König [25]. The development of the static PAM took place parallel to the work by Alain-Sol Sznitman on *Brownian motion among Poissonian obstacles* [37]. Both have substantially enriched our understanding of random processes in random media.

- *Dynamic version:* For the case where ξ is time-dependent, work is still in progress. Early work was done by Carmona and Molchanov [6] when ξ consists of i.i.d. Brownian noises. Since then the focus has been on a number of choices where ξ evolves like an interacting particle system:

  1. Independent random walks
  2. Exclusion process
  3. Voter model

  It turns out that the behavior of λ_p as a function of d and κ is extremely rich. For instance, there is a critical dimension d_c such that λ_p is constant in κ for $d < d_c$ and nonconstant in κ for $d \geq d_c$, with a delicate asymptotics for $\kappa \to \infty$ at $d = d_c$. For an overview, see Gärtner, den Hollander, and Maillard [29].

The main collaborators of Jürgen on the PAM have been S. Molchanov, F. den Hollander, W. König, and G. Maillard. Many others have made important contributions, including:

- *Jürgen's colleagues:* M. Biskup, F. Castell, M. Cranston, O. Gün, R. van der Hofstad, H. Kesten, H.-Y. Kim, L. Koralov, H. Lacoin, P. Mörters, T. Mountford, M. Ortgiese, A. Ramirez, T. Shiga, V. Sidoravicius, N. Sidorova, R. Sun, F. Viens, A. Vizcarra.
- *Jürgen's students:* A. Drewitz, J. Hähnel, M. Heydenreich, A. Schnitzler, A. Vosz, T. Wolff.

The present Festschrift contains several papers on the PAM, which include many references to the literature.

The PAM has been the main focus of Jürgen's work in the past decade. He has been the leader in the field and has shown to his colleagues what challenges the PAM is offering. The above list of names shows that he has made school.

6 Personal Remarks

Over the years, three collaborators of Jürgen have been a major inspiration to him:

- Mark Freidlin
- Don Dawson
- Stas Molchanov

Each of them has played an important role in his career: Mark as his thesis advisor, later co-authoring Jürgen's most cited paper (on wave propagation in random media [19]) and following him ever since, Don as a long-term collaborator pursuing a variety of different themes over more than a decade, and Stas as the person who brought him to the PAM, which became Jürgen's main focus in later years. Each of them has drawn Jürgen into exciting new areas of research, which he has subsequently pursued with all his force. Without them, Jürgen's mathematical itinerary would no doubt have been quite different.

For me, personally, it has been a wonderful experience to work with Jürgen. Our discussions over the past 20 years have covered a vast area. Most of what we spoke about was never written up, but part did make it to the literature: we wrote 7 papers together, and number 8 appears in the present Festschrift. What I value most in Jürgen, apart from his mastery of probability theory and analysis, is his ability to look far ahead, his constant search for elegance, his unwavering computational skills, his humor and scepticism, as well as his friendship and loyalty.

Jürgen holds the record as the most frequent visitor at EURANDOM. I trust that he will continue to push up this record in the years to come!

References

1. Aronson, D.G., Weinberger, H.F.: Multidimensional nonlinear diffusion arising in population genetics. Adv. Math. **30**, 33–76 (1978)
2. Ben Arous, G., Brunaud, M.: Methode de Laplace: étude variationnelle des fluctuations de diffusions de type champ moyen. Stochastics **31–32**, 79–144 (1990)
3. Ben Arous, G., Guionnet, A.: Large deviations for Langevin spin glass dynamics. Probab. Theory Relat. Fields **102**, 455–509 (1995)
4. Ben Arous, G., Guionnet, A.: Symmetric Langevin spin glass dynamics. Ann. Probab. **25**, 1367–1422 (1997)
5. Bramson, M.D: Maximal displacement of branching Brownian motion. Comm. Pure Appl. Math. **31**, 531–582 (1978)
6. Carmona, R.A., Molchanov, S.A.: Parabolic Anderson Problem and Intermittency, AMS Memoir 518. American Mathematical Society, Providence RI (1994)
7. Dai Pra, P., den Hollander, F.: McKean–Vlasov limit for interacting random processes in random media. J. Stat. Phys. **84**, 735–772 (1996)
8. Dawson, D.A., Gärtner, J.: Large deviations from the McKean–Vlasov limit for weakly interacting diffusions. Stochastics **20**, 247–308 (1987)
9. Dawson, D.A., Gärtner, J.: On the McKean–Vlasov limit for interacting diffusions. Math. Nachr. **137**, 197–248 (1988)
10. Dawson, D.A., Gärtner, J.: Large deviations, free energy functional and quasi-potential for a mean-field model of interacting diffusions. Mem. Amer. Math. Soc. **78** (398), iv+94 (1989)
11. Dawson, D.A., Gärtner, J.: Multilevel large deviations and interacting diffusions. Probab. Theory Related Fields **98**, 423–487 (1994)
12. Dawson, D.A., Gärtner, J.: Analytic aspects of multilevel large deviations. In: Asymptotic Methods in Probability and Statistics, Ottawa, ON, 1997, pp. 401–440. North-Holland, Amsterdam (1998)
13. Dawson, D.A., Greven, A.: Hierarchical models of interacting diffusions: multiple time scales, phase transitions and cluster-formation. Probab. Theory Relat. Fields **96**, 435–473 (1993)
14. Dawson, D.A., Greven, A.: Multiple scale analysis of interacting diffusions. Probab. Theory Relat. Fields **95**, 467–508 (1993)
15. Dembo, A., Zeitouni, O.: Large Deviations Techniques and Applications, 2nd edn. Springer, New York (1998)

16. Donsker, M.D., Varadhan, S.R.S.: Asymptotic evaluation of certain Markov process expectations for large time. Comm. Pure Appl. Math. **28**, 1–47 (Part II) and 279–301 (Part III) (1975)
17. Ellis, R.S.: Large deviations from a general class of random vectors. Ann. Probab. **12**, 1–12 (1984)
18. Freidlin, M.I.: Wave Front Propagation for KPP Type Equations, Surveys in Applied Mathematics, vol. 2, pp. 1–62. Plenum, New York (1995)
19. Freidlin, M.I., Gärtner, J.: The propagation of concentration waves in periodic and random media. (Russian) Dokl. Akad. Nauk SSSR **249**, 521–525 (1979)
20. Freidlin, M.I., Wentzell, A.D.: Random Perturbations of Dynamical Systems, 2nd edn. Grundlehren der Mathematischen Wissenschaften 260. Springer, New York (1998)
21. Gärtner, J.: Large deviation theorems for a class of random processes. (Russian) Teor. Verojatnost. i Primenen **21**, 95–106 (1976)
22. Gärtner, J.: On large deviations from an invariant measure. (Russian) Teor. Verojatnost. i Primenen **22**, 27–42 (1977)
23. Gärtner, J.: Location of wave fronts for the multi-dimensional KPP equation and Brownian first exit densities. Math. Nachr. **105**, 317–351 (1982)
24. Gärtner, J.: On the McKean–Vlasov limit for interacting diffusions. Math. Nachr. **137**, 197–248 (1988)
25. Gärtner, J., König, W.: The parabolic Anderson model. In: Deuschel, J.-D., Greven, A. (eds.) Interacting Stochastic Systems, pp. 153–179. Springer, Berlin (2005)
26. Gärtner, J., Molchanov, S.A.: Parabolic problems for the Anderson Hamiltonian. I. Intermittency and related topics. Comm. Math. Phys. **132**, 613–655 (1990)
27. Gärtner, J., Molchanov, S.A.: Parabolic problems for the Anderson Hamiltonian. II. Second-order asymptotics and structure of high peaks. Probab. Theory Relat. Fields **111**, 17–55 (1998)
28. Gärtner, J., König, W., Molchanov, S.A.: Geometric characterization of intermittency in the parabolic Anderson model. Ann. Probab. **35**, 439–499 (2007)
29. Gärtner, J., den Hollander, F., Maillard, G.: Intermittency on catalysts. In: Blath, J., Mörters, P., Scheutzow, M. (eds.) Trends in Stochastic Analysis, London Mathematical Society Lecture Note Series 353, pp. 235–248. Cambridge University Press, Cambridge (2009)
30. Grunwald, M.: Sanov results for Glauber spin-glass dynamics. Probab. Theory Relat. Fields **106**, 187–232 (1996)
31. den Hollander, F.: Renormalization of interacting diffusions: a program and four examples. In: Operator Theory, Advances and Applications, vol. 168. Partial Differential Equations and Functional Analysis: The Philippe Clément Festschrift, pp. 123–136. Birkhäuser, Basel (2006)
32. Kolmogorov, A.N., Petrovskii, I.G., Piskunov, N.S.: Etude de l'équation de la diffusion avec croissance de la quantité de matière et son application à un problème biologique. Bulletin Université d' Etat à Moscou (Bjul. Moskowskogo Gos. Univ.), Série Internationale, Section A 1, 1–26 (1937)
33. McKean, H.P.: Application of Brownian motion to the equation of Kolmogorov-Petrovskii-Piskunov. Comm. Pure Appl. Math. **28**, 323–331 (1975)
34. McKean, H.P.: A correction to Application of Brownian motion to the equation of Kolmogorov-Petrovskii-Piskunov. Comm. Pure Appl. Math. **29**, 553–554 (1976)
35. Sznitman, A.-S.: Equations de type de Boltzmann spatialement homogenès. Z. Wahrsch. verw. Gebiete **66**, 559–592 (1984)
36. Sznitman, A.-S.: Nonlinear reflecting diffusion processes, and the propogation of chaos and fluctuations associated. J. Funct. Anal. **56**, 311–336 (1984)
37. Sznitman, A.-S.: Brownian Motion, Obstacles and Random Media. Springer, Berlin (1998)
38. Uchiyama, K.: Brownian first exit from and sojourn over one sided moving boundary and application. Z. Wahrsch. verw. Gebiete **54**, 75–116 (1980)
39. Varadhan, S.R.S.: Large Deviations and Applications. CBMS-NSF Regional Conference Series in Appl. Math., vol. 46. SIAM, Philadelphia (1984)

Part I
The Parabolic Anderson Model

The Parabolic Anderson Model with Long Range Basic Hamiltonian and Weibull Type Random Potential

Stanislav Molchanov and Hao Zhang

Abstract We study the quenched and annealed asymptotics for the solutions of the lattice parabolic Anderson problem in the situation in which the underlying random walk has long jumps and belongs to the domain of attraction of the stable process. This type of stochastic dynamics has appeared in recent work on the evolution of populations.

The i.i.d random potential in our case is unbounded from above with regular Weibull type tails. Similar models but with the local basic Hamiltonian (lattice Laplacian) were analyzed in the very first work on intermittency for the parabolic Anderson problem by J. Gärtner and S. Molchanov.

We will show that the long-range model demonstrates the new effect. The annealed (moment) and quenched (almost sure) asymptotics of the solution have the same order in contrast to the case of the local models for which these orders are essentially different.

1 Dedication and Introduction

Asymptotic analysis of the Anderson parabolic problem, the surrounding bifurcations (depending on the tail behavior of the random potential), the phenomenon of intermittency, etc., were the central topics in the research activity of Jürgen Gärtner and his school starting from the early 1990s. I am pleased to have collaborated with Jürgen in this area.

The present paper (with my US student H. Zhang) is closely related to the early stage of Jürgen's and my work and is dedicated to Jürgen on the occasion of his 60th birthday.

S. Molchanov · H. Zhang (✉)
Department of Mathematics and Statistics, University of North Carolina at Charlotte, Charlotte, NC 28223, USA
e-mail: hzhang1@uncc.edu

J.-D. Deuschel et al. (eds.), *Probability in Complex Physical Systems*, Springer Proceedings in Mathematics 11, DOI 10.1007/978-3-642-23811-6_2,

We thank the reviewer of the first version of this paper for numerous corrections and improvements. We took these remarks into account in the preparation of the final text. We also gratefully acknowledge the help of J. Whitmeyer.

The Anderson parabolic problem can appear in the following situation. Assume that at the moment $t = 0$. we have on the lattice Z^d the particle field $n(0,x)$. The r.v.s $n(0,x)$, $x \in Z^d$, are i.i.d (for instance, Poissonian) with mean value $\rho_0 = En(0,x)$. The parameter ρ_0 is the initial density of the population.

The stochastic dynamics of the system include:

(1) Division (or splitting) (independent for different particles) with rate $\nu(x)$, annihilation (or death) with rate $\mu(x)$;
(2) Transition from a given position x to $x+z$ with rate $\kappa a(z)$, $x, z \in Z^d$.

We will assume that $a(z) = a(-z) \geq 0$, $a(0) = 0$ and $\sum_{z \neq 0} a(z) = 1$, i.e., κ is the rate of the exponentially distributed time that the underlying random walk spends in each site $x \in Z^d$. The random walk $X(s), s \geq 0$ has the following infinitesimal transition probabilities:

$$\begin{aligned} &P\{X(s+\mathrm{d}s) = x+z | X(s) = x\} = \kappa a(z)\mathrm{d}s, z \neq 0. \\ &P\{X(s+\mathrm{d}s) = x | X(s) = x\} = 1 - \kappa a(z)\mathrm{d}s. \end{aligned} \tag{1}$$

We call the rate κ "diffusivity."

Let us introduce the operator $\mathcal{L}$ of this random walk and call it the basic Hamiltonian:

$$\mathcal{L}f(x) = \kappa \sum_{z \neq 0} [f(x+z) - f(x)]a(z), \tag{2}$$

where $f \in l^{\infty}(Z^d)$.

If $\mathcal{L}f(x) = \kappa \Delta f(x) = \frac{\kappa}{2d} \sum_{z \neq 0} [f(x+z) - f(x)]$ is the lattice Laplacian, then we can call the underlying random walk the diffusion with diffusivity $\kappa > 0$.

We will consider here the nonlocal random walk with long jumps:

$$\mathcal{L}f(x) = \kappa \sum_{z \neq 0} [f(x+z) - f(x)]a(z). \tag{3}$$

Conditions on $a(z)$ will be presented later in Lemma 1.2.

Clearly, the transition probabilities $p(t,x,y)$ depend on the difference $z = y - x$ (i.e., the process X_s has independent increments) and, due to the symmetry of $a(z)$, are symmetric. Furthermore, it follows easily from (1) that

$$\frac{\partial p}{\partial t} = \kappa(\mathcal{L}p)(t,0,y), \ \ p(0,0,y) = \delta_0(y). \tag{4}$$

This equation is most easily analyzed using the Fourier transform (characteristic function). To do so, we use the following lemma.

Lemma 1.1. *Define* $\hat{\mathcal{L}}(\varphi) = \kappa \sum_{z\neq 0}(1-\cos(\varphi,z))a(z)$.
Then, $\widehat{\mathcal{L}f}(\varphi) = -\hat{f}(\varphi)\hat{\mathcal{L}}(\varphi)$.

Proof. By definition,

$$\begin{aligned}\widehat{\mathcal{L}f}(\varphi) &= \kappa \sum_x \mathrm{e}^{i(\varphi,x)} \sum_{z\neq 0} a(z)(f(x+z)-f(x))\\ &= \kappa \sum_{z\neq 0} a(z)\left[\mathrm{e}^{-i(\varphi,z)}\sum_x \mathrm{e}^{i(\varphi,x+z)} f(x+z) - \sum_x \mathrm{e}^{i(\varphi,x)} f(x)\right]\\ &= \kappa \sum_{z\neq 0} a(z)(\mathrm{e}^{-i(\varphi,z)}-1)\hat{f}(\varphi) = -\kappa \hat{f}(\varphi)\sum_{z\neq 0}(1-\cos(\varphi,z))a(z)\\ &= -\hat{f}(\varphi)\hat{\mathcal{L}}(\varphi).\end{aligned}$$

□

In the subsequent discussion, we use the following notation:

$$\widehat{\mathcal{L}f}(\varphi) = \hat{f}(\varphi)\hat{\mathcal{L}}(\varphi),\ \hat{f}(\varphi) = \sum_{x\in Z^d} \mathrm{e}^{i(\varphi,x)} f(x),\ \hat{\mathcal{L}}(\varphi) = \kappa \sum_{z\neq 0}(1-\cos(\varphi,z))a(z). \tag{5}$$

Operator $\mathcal{L}$ is the bounded and self-adjoint in $L^2(Z^d)$. In the dual Fourier space $L^2(T^d, d\varphi)$, it acts as the operator of multiplication by $-\hat{\mathcal{L}}(\varphi)$.

The characteristic function of the random walk is

$$E_0 \mathrm{e}^{i(\varphi,X_t)} = \sum_{y\in Z^d} p(t,0,y)\mathrm{e}^{i(\varphi,y)} = \hat{p}(t,0,\varphi),$$

where $\varphi \in [-\pi,\pi)^2 = T^2$. From (3), we have

$$\frac{\partial \hat{p}}{\partial t} = -\kappa \hat{\mathcal{L}}(\varphi), \quad \hat{p}(0,0,\varphi) = 1, \tag{6}$$

which gives the explicit expression

$$\hat{p}(t,0,\varphi) = \mathrm{e}^{-\kappa t \hat{\mathcal{L}}(\varphi)}.$$

If $\sum_{z\neq 0} a(z)|z|^2 < \infty$, then $\hat{\mathcal{L}}(\varphi) \approx \frac{(B\varphi,\varphi)}{2}$, $|\varphi| \to 0$. For fixed $\lambda \in R^2$:

$$E\mathrm{e}^{i(\lambda \frac{X(t)}{\sqrt{t}})} \underset{t\longrightarrow\infty}{\rightarrow} E\mathrm{e}^{i\frac{\kappa(B\lambda,\lambda)}{2}}$$

i.e., $\frac{X_{ts}}{\sqrt{t}}$, $s \in [0,1]$, is asymptotically two-dimensional Brownian motion with correlation matrix κB (calculation not shown).

We are interested in the situation $\sum_{z\neq 0} a(z)|z|^2 = \infty$. As usual in the theory of stable distributions, we impose the regularity condition on $a(z)$ given by (7).

Lemma 1.2. *(See [8]) Suppose*

$$a(z) = \frac{h(\theta)}{|z|^{2+\alpha}}\left(1 + O\left(\frac{1}{|z|^2}\right)\right),\ z \neq 0 \tag{7}$$

with $0 < \alpha < 2$, $\theta = \arg \frac{z}{|z|} \in [-\pi, \pi) = T^1$, $h \in C^2(T^1)$, $h > 0$ *and so satisfies the heavy tails assumption. Then,*

$$\begin{aligned} &a)\hat{\mathcal{L}}(\varphi) = c_\alpha |\varphi|^\alpha \mathcal{H}(\gamma) + O(|\varphi|^2)\ as\ \varphi \to 0,\\ &\quad where\ \mathcal{H}(\gamma) = \int_{-\pi}^{\pi} h(\theta)|\cos(\theta - \gamma)|^\alpha \mathrm{d}\theta\ and\ c_\alpha = \int_0^\infty \frac{1-\cos t}{t^{1+\alpha}}\mathrm{d}t,\\ &b) P\{x(t) = x\} \xrightarrow[t\to\infty]{} \frac{1}{t^{d/2}} St_{\beta,\mathcal{H}(\frac{x}{t^{1/\alpha}})} (1 + o(1))\ uniformly\ in\ x \in Z^d. \end{aligned} \tag{8}$$

This lemma is the local form of the usual statement that

$$\frac{x(t)}{t^{\frac{1}{\beta}}} \xrightarrow[t\to\infty]{\text{law}} St_{\beta,\mathcal{H}(\cdot)} \Leftrightarrow P\left\{\frac{x(t)}{t^{\frac{1}{\beta}}} \in \Omega\right\} = \int_\Omega St_{\beta,\mathcal{H}(\cdot)}(z)\mathrm{d}z. \tag{9}$$

In addition, this local form indicates the absence of "large deviations." A similar "global" theorem was published recently in [18]. We will give a sketch of the proof following the idea of [18]. See [8] for the detailed proof.

Proof. We have $\hat{\mathcal{L}}(\varphi) = \sum_{z\neq 0} a(z)(1 - \cos(\varphi \cdot z))$.

Let us consider the following integral $I(\varphi)$, which will give a good approximation of $\hat{\mathcal{L}}(\varphi)$, $\varphi \in [-\pi, \pi)^2 = T^2$:

$$I(\varphi) = \sum_{\mathbf{n}\neq 0} I_{\mathbf{n}}(\varphi),\ \text{where } I_{\mathbf{n}}(\varphi) = \int_{A(n)} \frac{h\left(\frac{\mathbf{x}}{|\mathbf{x}|}\right)}{|\mathbf{x}|^{2+\alpha}}(1 - \cos(\varphi \cdot \mathbf{x}))\mathrm{d}x.$$

For large $|\mathbf{n}|$, the leading term in the expansion of $I_{\mathbf{n}}(\varphi)$ is equal to

$$\psi_{\mathbf{n}}(\varphi) := \frac{1}{|\mathbf{n}|^{2+\alpha}} h\left(\frac{\mathbf{n}}{|\mathbf{n}|}\right)(1 - \cos(\mathbf{n}\cdot\varphi))$$

while the remaining term in the expansion of $I_{\mathbf{n}}(\varphi)$ will give a contribution of order $O(|\varphi|^2)$, $|\varphi| \to 0$. With $\sum_{\mathbf{n}\neq 0} \psi_{\mathbf{n}}(\varphi) = \hat{\mathcal{L}}(\varphi)$, we have

$$\hat{\mathcal{L}}(\varphi) = I(\varphi) + O(|\varphi|^2),\ |\varphi| \to 0.$$

Some calculations about $I(\varphi)$ give

$$I(\varphi) = \int\limits_{R^2 - A(0)} \frac{h(\frac{\mathbf{x}}{|\mathbf{x}|})}{|\mathbf{x}|^{2+\alpha}}(1 - \cos(\varphi \cdot \mathbf{x}))\mathrm{d}x = c_\alpha |\varphi|^\alpha \mathcal{H}(\gamma),$$

where $\mathcal{H}(\gamma) = \int_{-\pi}^{\pi} h(\theta)|\cos(\theta - \gamma)|^\alpha \mathrm{d}\theta$, $\gamma = \arg\varphi$, $\mathcal{H}(\gamma) \in C(T^1)$, $\mathcal{H}(\gamma) > 0$ and $c_\alpha = \int_0^\infty \frac{1-\cos t}{t^{1+\alpha}}\mathrm{d}t$. Then:

$$\hat{\mathcal{L}}(\varphi) = c_\alpha |\varphi|^\alpha \mathcal{H}(\gamma) + O(|\varphi|^2),\ |\varphi| \to 0. \qquad \square$$

Corollary 1.3. *Under our condition on $a(z)$,*

$$E_0 \mathrm{e}^{i\left(\varphi \frac{X(t)}{t^{1/\beta}}\right)} \underset{t\to\infty}{\longrightarrow} \mathrm{e}^{-\kappa c_\alpha |\varphi|^\alpha \mathcal{H}(Arg\varphi)}$$

(center symmetric distribution with parameter $0 < \beta < 2$ and angular measure $\mathcal{H}(\gamma)$, $\gamma = \arg\varphi$).

Corollary 1.4. *Under our condition on $a(z)$, the random walk $X(S)$ on Z^2 is transient for any $0 < \beta < 2$.*

In fact:

$$p(t,0,0) = \frac{1}{(2\pi)^2}\int\limits_{T^2} \mathrm{e}^{-\kappa t \hat{\mathcal{L}}(\varphi)}\mathrm{d}\varphi \Rightarrow \int\limits_0^\infty p(t,0,0)\mathrm{d}t = \frac{1}{(2\pi)^2\kappa}\int\limits_{T^2}\frac{\mathrm{d}\varphi}{\hat{\mathcal{L}}(\varphi)} < \infty.$$

Random walks with long jumps appear in a recent paper [8] by Y. Feng, S. Molchanov, and J. Whitmeyer about models of stochastic dynamics that are stationary in time and space of biological populations.

Under mild conditions on the functions $\nu(x)$ and $\mu(x)$, the particle field $n(t,x)$ exists, is unique and has all statistical moments. In particular, the first moment, or the population density,

$$u(t,x) = En(t,x)$$

is the solution of the parabolic problem

$$\begin{cases} \dfrac{\partial u}{\partial t} = \mathcal{L}u + V(x,\omega_m)u \\ u(0,x) \equiv \rho_0 > 0, \end{cases} \tag{10}$$

where $V(x) = \nu(x) - \mu(x)$. We will assume that the values of $V(x,\omega_m)$, $x \in Z^d$ are i.i.d. r.v.s on the probability space $(\Omega_m, \Gamma_m, P_m)$. The points $\omega_m \in \Omega_m$ represent the realization of random media (a random environment) and P_m is the distribution on Ω_m. In other words, we consider random particle dynamics in a random environment.

Consider the case of the potential $V(x, \omega_m)$ unbounded from above, i.e, $P\{V(\cdot) > a\} > 0$ for any $a > 0$. A typical example is the i.i.d. $N(0, 1)$ r.v.s. See [17] for detailed analysis of the Gaussian case.

General study of the problem (10) in the particular case of local diffusion $\mathcal{L} = \kappa\Delta$ was started in a paper by J. Gärtner and S. Molchanov [10] and later was expanded by J. Gärtner and his students and other collaborators in many different directions. The central idea is the justification of the intermittency phenomenon: the random environment leads to a highly nonuniform population structure (see J. Gärtner and S. Molchanov [10–12], J. Gärtner, W. König and S. Molchanov [13], S. Molchanov [17], J. Gärtner and W. König [9]). Corresponding effects were known in the physics literature [20]. There are many works on random environments changing in time (in the areas in physics of turbulent flow, moving catalysts, etc.; see [2, 3, 6, 7, 14–16, 19]). Some of them are included in in this volume of proceedings. We will discuss only the stationary environment.

Let us mention one of the results from [10]. Under the condition $\langle e^{\lambda V(\cdot)} \rangle = \Psi(\lambda) < \infty, \forall(\lambda \in R^1)$, the particle field $u(t, x)$ has all statistical moments and

$$m_k(t) = \langle u^k(t, 0) \rangle = \langle u^k(t, x) \rangle, \quad k = 1, 2, \ldots.$$

Here, and in the future, the symbol $\langle\ \rangle$ means the expectation with respect to the random environment. The notation E will be used for the expectation over the random process $x(t)$ associated with $\mathcal{L}$. For instance, due to the Kac–Feynman formula for the fixed environment $V(\cdot, \omega_m)$, the quenched representation of the first moment is

$$u(t, \omega_m, x) = E_x e^{\int_0^t V(x_s, \omega_m) ds} \rho_0.$$

At the same time,

$$m_1(t, 0) = \langle E_0 e^{\int_0^t V(x_s, \omega_m) ds} \rho_0 \rangle = E_0 \langle e^{\int_0^t V(x_s, \omega_m) ds} \rho_0 \rangle$$

is the corresponding annealed first moment.

In the following content, without loss of generality we take $\rho_0 = 1$.

In this paper, we are mainly concerned with the potential $V(x, \omega_m)$ with $\mathbf{E} e^{tV(\cdot)} < \infty$ for all $t > 0$ in the following two Weibull type forms:

$$P\{V(\cdot) > x\} = \exp\{-h(x)\} = \exp\left\{-\frac{x^\alpha}{\alpha}\right\}, \tag{11}$$

and

$$P\{V(\cdot) > x\} = \exp\{-h(x)\} = \exp\left\{-\frac{x^\alpha}{\alpha} L(x)\right\} \text{ with } \alpha > 1, \tag{12}$$

where $L(x)$ is a slowly varying function with some restrictions (see below). Recall that a slow varying function is a positive measurable function f satisfying

$$\frac{f(\lambda x)}{f(x)} \to 1 \quad (x \to \infty), \quad \forall \lambda > 0,$$

and a regularly varying function of index ρ is a measurable function $f > 0$ satisfying

$$\frac{f(\lambda x)}{f(x)} \to \lambda^{\rho} \quad (x \to \infty), \quad \forall \lambda > 0,$$

alternatively written $f \in R_{\rho}$. For details and further reference see Bingham et al. [4].

The particular case when $V(x_s, \omega_m)$ is i.i.d. $N(0, \sigma^2)$ r.v.s is also essential:

$$P\{V(\cdot) > a\} = \frac{1}{\sqrt{2\pi}\sigma} \int_a^{\infty} \mathrm{e}^{-\frac{x^2}{2\sigma^2}} \mathrm{d}x \underset{a \to \infty}{\sim} \frac{\mathrm{e}^{-\frac{a^2}{2\sigma^2}}}{a\sqrt{2\pi}\sigma}.$$

It is close to the Weibull situation with $\alpha = 2$.

Let us formulate the annealed (moment) asymptotics and quenched (P_m-a.s.) asymptotics of $u(t,0)$ in the Gaussian case and the random walk with Laplacian operator

$$\mathcal{L}f(x) = \frac{\kappa}{2d} \sum_{|z-x|=1} [f(x+z) - f(x)].$$

Theorem 1.5. *(See [10, 17])*

(a) For every $p \geq 1$ and every $t \geq 0$,

$$\exp\left\{\frac{p^2 t^2}{2}\sigma^2 - pkt\right\} \leq \langle u^p(t,0)\rangle \leq \exp\left\{\frac{p^2 t^2}{2}\sigma^2\right\},$$

or

$$\lim_{t \to \infty} \frac{\ln\langle u^p(t,0)\rangle}{t^2} = \frac{p^2}{2}\sigma^2.$$

A more precise version of (a) is

$$(a') \ln\langle u^p(t,0)\rangle = \frac{p^2 t^2}{2}\sigma^2 - pkt + o(t) \text{ (see also [17])}.$$

(b) P_m-a.s.: $\lim_{t \to \infty} \frac{\ln u(t,0)}{t\sqrt{\ln t}} = \sqrt{2d}\sigma.$

Let us stress that the annealed and quenched asymptotics have completely different orders. It is a general feature of all models with the local basic Hamiltonian (see [10, 17], etc.). We will see that for the long range Hamiltonian the opposite situation obtains, that is, the annealed and quenched asymptotics have the same order.

Let us return to the formula (10) and impose some technical conditions on the function $L(x)$ in the spirit of the paper [1].

2 The Annealed and Quenched Asymptotic Properties of $u(t,0)$ with Weibull Potential $V(x,\omega_m)$: $P\{V(\cdot) > x\} = \exp\left\{-\frac{x^\alpha}{\alpha}\right\}$

For a tail probability of the form $P\{V(\cdot) > x\} = \exp\{-\frac{x^\alpha}{\alpha}\}$, we have the following annealed asymptotic result for $u(t,x)$.

Theorem 2.1. *For every $p \in N$ and every $t \geq 0$,*

$$\exp\left\{\frac{p^{\alpha'}t^{\alpha'}}{\alpha'} - p\kappa t + O(\ln t)\right\} \leq \langle u^p(t,x)\rangle \leq \exp\left\{\frac{p^{\alpha'}t^{\alpha'}}{\alpha'} + O(\ln t)\right\},$$

and

$$\lim_{t\to\infty}\frac{\ln\langle u^p(t,x)\rangle}{t^{\alpha'}} = \frac{p^{\alpha'}}{\alpha'}, \text{ where } \alpha' \text{ satisfies } \frac{1}{\alpha} + \frac{1}{\alpha'} = 1.$$

Remark 2.2. Except for the specific calculation of the Laplace transformation, this theorem is the direct representation of the corresponding general result from [10].

Proof. **(a) Lower Estimation of the Annealed Asymptotics of $m_p(t)$.** The first moment of the solution $u(t,0)$ is

$$m_1(t) = \langle u(t,0)\rangle \geq \langle e^{tV(0)}\rangle e^{-\kappa t}.$$

For the Weibull tail, we calculate the term $\langle e^{tV(0)}\rangle$ as follows:

$$\langle e^{tV(0)}\rangle = \int_0^\infty e^{tx-\frac{x^\alpha}{\alpha}} x^{\alpha-1}\,dx.$$

Changing variables by setting $x = t^\beta y$ and selecting $\beta : 1+\beta = \alpha\beta$ gives

$$\langle e^{tV(0)}\rangle = \int_0^\infty e^{t^{\alpha'}(y-\frac{y^\alpha}{\alpha})} t^{\alpha\beta} y^{\alpha-1}\,dy,$$

where $\beta = \frac{1}{\alpha-1}$, $\alpha' = \frac{\alpha}{\alpha-1} = 1+\beta$.

The term $y - \frac{y^\alpha}{\alpha}$ is maximal when $y = 1$. Then, using Laplace's method, we obtain:

$$\int_0^\infty e^{t^{\alpha'}\left(y-\frac{y^\alpha}{\alpha}\right)} t^{\alpha\beta} y^{\alpha-1} dy = e^{\frac{t^{\alpha'}}{\alpha'}+\frac{\alpha'}{2}\ln t+\frac{1}{2}\ln\left(\frac{2\pi}{\alpha-1}\right)+o(1)}.$$

Thus, $m_1(t) \geq e^{\frac{t^{\alpha'}}{\alpha'}-\kappa t+\frac{\alpha'}{2}\ln t+\frac{1}{2}\ln\left(\frac{2\pi}{\alpha-1}\right)+o(1)}$. For $p > 1$, we can consider p independent copies x_s^i, $i = 1, \dots, p$ of the random walk x_s. For the pth moment of the solution $u(t,0)$, we have

$$\begin{aligned}\langle u^p(t,x)\rangle &= \left\langle E_0\left\{\exp\left(\sum_{i=1}^p \int_0^t \xi(x_s^i)\right)\right\}\right\rangle \\ &\geq \exp\left\{\frac{p^{\alpha'} t^{\alpha'}}{\alpha'} - p\kappa t + \frac{\alpha'}{2}\ln t + \frac{1}{2}\ln\left(\frac{2\pi}{\alpha-1}\right) + o(1)\right\} \\ &= \exp\left\{\frac{p^{\alpha'} t^{\alpha'}}{\alpha'} - p\kappa t + O(\ln t)\right\}.\end{aligned}$$

(b) Upper Estimation of $m_p(t)$. We obtain the following results after applying Hölder's inequality, Jensen's inequality and Fubini's theorem.

$$\begin{aligned}\langle u^p(t,x)\rangle &= \left\langle\left(E_0 \exp\left(\int_0^t \xi(x_s)ds\right)\right)^p\right\rangle \leq \left\langle\left(E_0 \exp\left(p\int_0^t \xi(x_s)ds\right)\right)\right\rangle \\ &= E_0\left\langle\left(\exp\left(p\int_0^t \xi(x_s)ds\right)\right)\right\rangle \\ &\leq \frac{1}{t}\int_0^t dsE_0\langle \exp pt\xi(x_s)\Big)\Big\rangle = \exp\left\{\frac{p^{\alpha'} t^{\alpha'}}{\alpha'} + O(\ln t)\right\}.\end{aligned}$$

Combining the lower and upper estimations of the $m_p(t)$ of the solution $u(t,0)$, we get the result. □

Remark 2.3. Following the method in [17], the more precise upper estimation can be proved:

$$\langle \ln u^p(t,x)\rangle = \frac{p^{\alpha'} t^{\alpha'}}{\alpha'} - p\kappa t + O(\ln t).$$

For the discussion of the quenched asymptotic properties of the solution $u(t,0)$, we need the following lemma concerning the asymptotics of $\max_{|x|\leq n} V(x)$ as $n \to \infty$ for the potential $P\{V(x,\omega_m) > a\} = \exp\{-\frac{a^\alpha}{\alpha}\}$.

Lemma 2.4. *P_m-a.s.,*

$$\max_{|x|\le n} V(x) \underset{n\to\infty}{\sim} (\alpha d \ln n)^{1/\alpha}. \tag{1}$$

Proof. Using the Borel–Cantelli lemma for the event $A_x^{\pm} = \{V(x) > (1 \pm \epsilon)(\alpha d \ln |x|)^{1/\alpha}\}$, $|x| = |x_1| + \cdots + |x_k|$, straightforward calculation proves the lemma. □

Theorem 2.5. *P_m-a.s. for $t \to \infty$,*

$$\limsup_{t\to\infty} \frac{\ln u(t,0)}{t^{\alpha'}} \le \frac{1}{\alpha'},$$

$$\liminf_{t\to\infty} \frac{\ln u(t,0)}{t^{\alpha'}} \ge \frac{1}{\alpha'}\left(\frac{d}{d+\beta}\right)^{\frac{\alpha'}{\alpha}}.$$

Proof. **(a) Lower Estimation for the Quenched Asymptotics of** $u(t,0)$. To check the lower estimation, let us consider the "almost optimal" trajectory x_s, $s \ge 0$. This trajectory spends the time $t \le 1$ at the origin, then jumps to the point $x_0 = x_0(t,\omega_m)$ of the very high local maximum of $V(x,\cdot)$ and stays there until moment t (i.e., time at least $t-1$.)

Assume that $|x_0| \in [R(1-\delta'), R]$ for some R, $R \gg 1$ and $V(x_0) \ge (1-\delta)(\alpha d \ln R)^{1/\alpha}$. Then, for $R \to \infty$,

$$\begin{aligned} u(t,0) &\ge \max_R\, [\mathrm{e}^{-\kappa t} \cdot C_2 \cdot \frac{1}{R^{d+\beta}} \exp\{(t-1)(1-\delta)(\alpha d \ln R)^{1/\alpha}]\} \\ &\ge C_2 \mathrm{e}^{-\kappa t} \max_R \mathrm{e}^{-(d+\beta)\ln R + t(1-\delta'')(\alpha d \ln R)^{1/\alpha}} \left(\text{because } a(0,x_0) \ge \frac{C}{|x_0|^{d+\beta}}\right). \end{aligned}$$

Putting $x = \ln R$, we find

$$\max_x\, [-(d+\beta)x + \tilde{t}(\alpha \mathrm{d} x)^{1/\alpha}],\ \ \tilde{t} = (1-\delta'')t.$$

The equation for the critical point gives:

$$\tilde{t} d(\alpha \mathrm{d} x_0)^{1/\alpha - 1} = d + \beta \quad \Rightarrow x_0 = \frac{1}{\alpha d}\left(\frac{\tilde{t} d}{d+\beta}\right)^{\alpha'}.$$

The value at the critical point is

$$\tilde{t}^{\alpha'} \frac{1}{\alpha'}\left(\frac{d}{d+\beta}\right)^{\frac{\alpha'}{\alpha}}.$$

Because $\tilde{t} = (1-\delta'')t$ and δ'' is arbitrarily small, we have proved the lower estimation

$$\liminf_{t\to\infty}\frac{\ln u(t,0)}{t^{\alpha'}} \geq \frac{1}{\alpha'}\left(\frac{d}{d+\beta}\right)^{\frac{\alpha'}{\alpha}}.$$

(b) Upper Estimation for the Quenched Asymptotics of $u(t,0)$. It follows from the Kac–Feynman formula that $u \uparrow$ as a function of t for fixed ω_m. It means that for $n \leq t < n+1$,

$$u(n,0) \leq u(t,0) \leq u(n+1,0).$$

But $\forall(\epsilon > 0)$ and $t = n$ or $n+1$,

$$P\left\{u(t,0) > \mathrm{e}^{(1+\epsilon)\frac{t^{\alpha'}}{\alpha'}}\right\} \leq \frac{\langle u(t,0)\rangle}{\mathrm{e}^{(1+\epsilon)\frac{t^{\alpha'}}{\alpha'}}} \leq \frac{\langle \exp(tV(0))\rangle}{\mathrm{e}^{(1+\epsilon)\frac{t^{\alpha'}}{\alpha'}}}.$$

Trivial calculation gives

$$\langle \exp(tV(0))\rangle \sim \mathrm{e}^{\frac{t^{\alpha'}}{\alpha'}}$$

and, due to the Borel–Cantelli lemma, $\forall(\epsilon > 0)$

$$u(t,0) \leq \mathrm{e}^{(1+\epsilon')\frac{t^{\alpha'}}{\alpha'}}, \quad t \geq t_0(\omega),$$

first for integer t, then, due to the monotonicity of $u(t,0)$, for all $t > t_0(\omega)$.
It means that

$$\limsup_{t\to\infty}\frac{\ln u(t,0)}{t} \leq (1+\epsilon)\frac{1}{\alpha'}.$$

Because ϵ can be arbitrarily small, we have proved the upper estimation. □

3 The Annealed and Quenched Asymptotic Properties of $u(t,0)$ with Potential $V(x,\omega_m)$ of the Form $P\{V(\cdot) > x\} = \exp\{-\frac{x^\alpha}{\alpha}L(x)\}$

In the spirit of the paper [1] by G. Ben Arous, L. Bogachev, and S. Molchanov on REM model, we will introduce some essential technical developments of the Weibull case of Sect. 2. Specifically, we will study models when $P\{V(\cdot) > x\} = \exp\{-\frac{x^\alpha}{\alpha}L(x)\}$, and $L(\cdot)$ is a slowly varying function with some additional regularity assumptions. We will start from several definitions and propositions.

The following lemmas and definitions are fundamental to our paper and can be found in [4].

Lemma 3.1. *(Uniform Convergence Theorem by Karamata and Korevaar et al.) If $f(x)$ is slowly varying then $\frac{f(\lambda x)}{f(x)} \to 1 (x \to \infty)$ uniformly on each compact λ-set in $(0, \infty)$.*

It is known (see [4]) that a function $f(x) \in R_\rho$ iff $f(x)$ admits the Karamata representation

$$f(x) = c(x) \exp\left\{ \int_a^x \frac{\rho + \epsilon(u)}{u} \mathrm{d}u \right\} \quad (x \geq a) \tag{1}$$

for some $a > 0$, where $c(\cdot)$, $\epsilon(\cdot)$ are measurable functions and $c(x) \to c_0 > 0$, $\epsilon(x) \to 0$ as $x \to \infty$.

Definition 3.2. (After [4].) The function f is a normalized regularly varying function, or $f \in NR_\rho$, if it can be represented in the form (1) with $c(\cdot) =$ *constant*> 0.

One of the important properties of a normalized regularly varying function $f(x)$, $f(x) \in NR_\rho$, is provided by the following lemma (see [4, page 24]).

Lemma 3.3. *A positive measurable function f is a normalized regularly varying function, or $f(x) \in NR_\rho$, iff for every $\epsilon > 0$ $\frac{f(x)}{x^{\rho-\epsilon}}$ is ultimately increasing and $\frac{f(x)}{x^{\rho+\epsilon}}$ is ultimately decreasing.*

Another important property of a normalized regularly varying function $f(x) \in NR_\rho$ is given by the following lemma (see [4, page 15]).

Lemma 3.4. *Let f be a positive measurable function. Then $f \in NR_\rho$ iff f is differentiable (a.e.) and when $x \to \infty$,*

$$\frac{x f'(x)}{f(x)} \to \rho.$$

Kasahara-de Bruijn's Tauberian theorem (see Bingham et al. [4, page 253]) is fundamental to this paper. In the following, $f^{\leftarrow}(y)$ is the generalized inverse function of f, defined $f^{\leftarrow}(y) = \inf\{x : f(x) \geq y\}$.

Theorem 3.5. *(Kasahara-de Bruijn's Tauberian Theorem). Let μ be a measure on $(0, \infty)$ such that*

$$M(\lambda) = \int_0^\infty \mathrm{e}^{\lambda x} \mathrm{d}\mu(x) < \infty$$

for all $\lambda > 0$. If $0 < \alpha < 1$, $\phi \in R_\alpha$, put $\psi(\lambda) = \lambda/\phi(\lambda) \in R_{1-\alpha}$; then, for $B > 0$,

$$-\ln \mu(x, \infty) \sim B\phi^{\leftarrow}(x) \quad (x \to \infty) \tag{2}$$

if and only if

$$\ln M(\lambda) \sim (1-\alpha)(\alpha/B)^{\alpha/(1-\alpha)}\psi^{\leftarrow}(\lambda) \quad (\lambda \to \infty).$$

To make use of the theorem, we restrict our attention to $L(x)$ that satisfy:

Assumption 3.6. *The function $L(x)$ in (12) is slowly varying and $L(x) \in NR_0$, that is,*

$$\frac{xL'(x)}{L(x)} \to 0 \quad (x \to \infty).$$

With Assumption 3.6, we see that $\frac{xh'(x)}{h(x)} \to \alpha$ when $x \to \infty$, where $h(x) = \frac{x^{\alpha}L(x)}{\alpha}$. From Lemmas 3.3 and 3.4, $h(x)$ is a normalized regularly varying function and is ultimately increasing.

This assumption is not completely sufficient for our analysis. To prove the results similar to the Theorem 2.1 and 2.5, we need the additional technical

Assumption 3.7. *$L(x)$ in (12) satisfies*

$$L(xL^{\alpha}(x))/L(x) \to 1 \ (x \to \infty) \text{ locally uniformly in } \alpha \in R.$$

We will use the Assumption 3.7 to control the critical point x_0 in the application of Laplace method. The Assumption 3.7 is fulfilled for all "standard" slowly varying functions, for example, $L(x) = \ln^{\beta}(2+x)$, $\beta \in R^1$, $L(x) = \ln^{\gamma}\ln(x+4)$, $\gamma \in R^1$ and $L(x) = \exp(\ln^{\beta}(2+x))$, $0 < \beta < \frac{1}{2}$. However, the function $L(x) = \exp(\ln^{\beta}(2+x))$, $\frac{1}{2} < \beta < 1$, is slowly varying but elementary calculation gives $\lim\limits_{x\to\infty}\frac{L(xL(x))}{L(x)} = +\infty$, i.e., the Assumption 3.7 restricts the growth of $L(x)$.

The following lemma (see Bojanic and Seneta [5] and see Bingham et al. [4, page 77, 78]) is related to the sufficient condition for the Assumption 3.7.

Lemma 3.8. *(Bingham et al. [4], Bojanic and Seneta [5]) If $L(x)$ is nondecreasing slowly varying function, $L(x) > 1$ and $L(x)$ is continuously differentiable, then*

$$\frac{xL'(x)}{L(x)}\ln L(x) \to 0 \tag{3}$$

implies

$$L(xL^{\alpha}(x))/L(x) \to 1 \ (x \to \infty) \text{ locally uniformly in } \alpha \in R.$$

This result is very general and useful. Its main defect is the assumption that $L(x)$ is nondecreasing.

The following theorem is similar to Theorem 2.1 but in our more general setting:

Theorem 3.9. *Under Assumptions 3.6 and 3.7, for every $p \in N$ and $t \to \infty$,*

$$\langle e^{ptV(0)}\rangle e^{-\kappa pt} \le \langle u^p(t,x)\rangle \le \langle e^{ptV(0)}\rangle$$

and

$$\lim_{t\to\infty}\frac{\ln\langle u^p(t,x)\rangle}{t^{\alpha'}}L^{\frac{1}{\alpha-1}}(t^{\frac{1}{\alpha-1}})=\frac{p^{\alpha'}}{\alpha'},\ \textit{where } \alpha' \textit{ satisfies } \frac{1}{\alpha}+\frac{1}{\alpha'}=1.$$

Proof. The first moment of the solution $u(t,0)$ with tail probability $P\{V(\cdot)>x\}=\exp\{-h(x)\}$, where $h(x)=\frac{x^\alpha}{\alpha}L(x)$, is

$$m_1(t)=\langle u(t,0)\rangle\geq\langle e^{tV(0)}\rangle e^{-\kappa t}. \tag{4}$$

Using integration by parts, we get

$$\langle e^{tV(0)}\rangle=1+t\int_0^\infty e^{tx}P\{V(0)>x\}dx=1+t\int_0^\infty e^{tx-\frac{x^\alpha}{\alpha}L(x)}dx,$$

i.e., we have to evaluate asymptotically

$$I(t)=\int_0^\infty e^{tx-h(x)}dx.$$

The natural idea is to apply the Laplace method. One can do this under additional assumption that $L(x)\in C^2_{\mathrm{loc}}$ and $\frac{x^2L''(x)}{\alpha L(x)}\to 0$. However, on the level of the logarithmical asymptotics, the initial Assumption 3.7 is sufficient.

Fix the point $x_0=\left(\frac{t}{L(t^{\frac{1}{\alpha-1}})}\right)^{\frac{1}{\alpha-1}}$ (which is not exactly the extreme for $tx-h(x)$) and divide $[0,\infty)$ into the following intervals:

$$\Delta_{-1}=\left[0,\frac{1}{2}x_0\right),\ \Delta_0=\left[\frac{1}{2}x_0,2x_0\right),\ \ldots,\ \Delta_n=\left[2^nx_0,2^{n+1}x_0\right),\ \ldots$$

Since $x_0(t)\to\infty$ and L is slowly varying, there exists function $\delta=\delta(t)\to 0$, $t\to\infty$ such that for $x\in\Delta_n$, $n\geq 0$, we have

$$(1-\delta)L(2^nx_0)\leq L(x)\leq(1+\delta)L(2^nx_0).$$

Finally, the exponent $tx-h(x)$ is increasing on the $[0,\frac{1}{2}x_0)$ and decreasing on $[2x_0,\infty)$ as the function of x.

Consider the integral $I_0(t)=\int_{\Delta_0}e^{tx-h(x)}dx$, then

$$\int_{\Delta_0}e^{tx-(1+\delta)\frac{x^\alpha}{a}L(x_0)}dx\leq I_0\leq\int_{\Delta_0}e^{tx-(1-\delta)\frac{x^\alpha}{a}L(x_0)}dx.$$

The critical points here are $x_{\pm}(t) = \left(\frac{t}{L(x_0)}\right)^{\frac{1}{\alpha-1}} \frac{1}{(1\mp\delta)^{\frac{1}{\alpha-1}}}$. The usual Laplace method (L now is constant) gives

$$\ln I_0 \sim \frac{t^{\frac{\alpha}{\alpha-1}}}{L^{\frac{1}{\alpha-1}}(x_0)}\left(1-\frac{1}{\alpha}\right).$$

On Δ_{-1}, we can use a very rough estimation:

$$\begin{aligned} I_{-1}(t) &= \int\limits_{\Delta_{-1}} \mathrm{e}^{tx-h(x)}\mathrm{d}x \le |\Delta_{-1}|\cdot \mathrm{e}^{t\cdot\frac{x_0}{2}-h\left(\frac{x_0}{2}\right)} \\ &\le \frac{x_0}{2}\mathrm{e}^{\frac{t^{\alpha/(\alpha-1)}}{L^{1/(\alpha-1)}(x_0)}\left(\frac{1}{2}-\left(\frac{1}{2}\right)^{\alpha}\cdot\frac{1}{\alpha}\frac{L\left(\frac{x_0}{2}\right)}{L(x_0)}\right)} \end{aligned}$$

and $I_{-1}(t)$ is exponentially smaller than $I_0(t)$.

Similarly for $n \ge 1$,

$$\begin{aligned} I_n(t) &= \int\limits_{\Delta_n} \mathrm{e}^{tx-h(x)}\mathrm{d}x \le |\Delta_n|\cdot \mathrm{e}^{t\cdot\frac{2^n x_0}{2}-h(2^n x_0)} \\ &\le 2^n x_0\cdot \mathrm{e}^{\frac{t^{\alpha/(\alpha-1)}}{L^{1/(\alpha-1)}(x_0)}\left(2^n-\frac{2^{\alpha n}}{\alpha}\cdot\frac{L(2^n x_0)}{L(x_0)}\cdot\frac{L^{\alpha/(\alpha-1)}(x_0)}{L^{\alpha/(\alpha-1)}(t^{1/(\alpha-1)})}\right)} \qquad (5) \\ &\le 2^n x_0\cdot \mathrm{e}^{\frac{t^{\alpha/(\alpha-1)}}{L^{1/(\alpha-1)}(x_0)}\left(2^n-\frac{2^{(\alpha-\delta)n}}{\alpha}\right)}. \end{aligned}$$

In (5) we use the Assumption 3.7, which provides that $L(x_0)/L(t^{1/(\alpha-1)}) \to 1$.

Again $\sum_{n\ge1} I_n(t)$ is exponentially small than I_0. Finally,

$$\ln I(t) \underset{t\to\infty}{\sim} \frac{t^{\frac{\alpha}{\alpha-1}}}{L^{\frac{1}{\alpha-1}}(x_0)}\left(1-\frac{1}{\alpha}\right) \sim \frac{t^{\alpha'}}{\alpha'}\frac{1}{L^{\frac{1}{\alpha-1}}(t^{\frac{1}{\alpha-1}})}.$$

In the last step, we use the Assumption 3.7 again. □

For the quenched asymptotics of $u(t,x)$, $t\to\infty$, as in the previous section we need the asymptotics of $\max\limits_{|x|\le n} V(x)$ similar to Lemma 2.4.

Lemma 3.10. *If function $L(x)$ satisfies Assumption 3.7, P_m-a.s., then*

$$\max_{|x|\le n} V(x) \underset{n\to\infty}{\sim} \left(\frac{\alpha d\ln n}{L(\ln^{1/\alpha} n)}\right)^{1/\alpha}.$$

Proof. **(a) Lower Estimation of** $V(x)$. For notational convenience, we define $\beta(x) = \frac{\alpha d \ln|x|}{L(\ln^{1/\alpha}|x|)}$. Consider the events $A_{x,\delta} = \{\omega_m : V(x) > (1+\delta)\beta^{1/\alpha}(x)\}$.

Then, for any $\delta > 0$,

$$P(A_{x,\delta}) = \exp\left\{-(1+\delta')\cdot\beta(x)\cdot\frac{1}{\alpha}L(\beta^{1/\alpha}(x))\right\}$$
$$= \frac{1}{|x|^{d(1+\delta')(1+o(1))}}.$$

The last step follows from Assumption 3.7.

Because $\sum_x \frac{1}{|x|^{d(1+\delta')(1+o(1))}} < \infty$, we have that for $|x| \geq C(\delta,\omega_m)$, $V(x) \leq (1+\delta)\beta^{1/\alpha}(x)$, due to the Borel–Cantelli lemma.

(b) Upper Estimation of $V(x)$. We define $\beta(R_n) = \frac{\alpha d \ln R_n}{L(\ln^{1/\alpha} R_n)}$. Let us split Z^d into Ω_n: $\Omega_n = \{\mathbf{x} : R_n < \|\mathbf{x}\|_\infty \leq R_{n+1}\}$, where $R_n = (1+\gamma)^n$ and $\gamma > 0$, and consider the event $B_{n,\delta} = \{\omega_m : \max_{\Omega_n} V(x) < (1-\delta)\beta^{1/\alpha}(R_n)\}$. The events $B_{n,\delta}$ are independent for different n and due to independence

$$P\{B_{n,\delta}\} = P^{|\Omega_n|}\{V(x) < (1-\delta)\beta^{1/\alpha}(R_n)\}$$
$$= \left(1 - P\{V(x) \geq (1-\delta)\beta^{1/\alpha}(R_n)\}\right)^{|\Omega_n|} \sim \mathrm{e}^{-|\Omega_n| P\{V(x) > (1-\delta)\beta^{1/\alpha}(R_n)\}}.$$

But with $|\Gamma_n| \sim \gamma R_n \cdot (2R_n)^{d-1}$ and $P\{V(x) > (1-\delta)\beta^{1/\alpha}(R_n)\} \sim \frac{c}{R_n^{d(1-\delta')}}$,

$$P\{B_{n,\delta}\} \sim \mathrm{e}^{-\gamma R_n\cdot(2R_n)^{d-1}\cdot\frac{c}{R_n^{d(1-\delta')}}} \leq \mathrm{e}^{-c\gamma R_n^{\delta' d}}, \quad \sum_n P\{B_{n,\delta}\} < \infty.$$

So for $n > n_0(\omega)$, $\max_{\Omega_n} V(x) \geq (1-\delta)\beta^{1/\alpha}(R_n)$. Thus, Lemma 3.10 is proved. □

Let us give some illustrations.

Corollary 3.11. *If* $L(x) = \ln^\beta x$, P_m*-a.s.:*

$$\max_{|x|\leq n} V(x) \underset{n\to\infty}{\sim} \left(\frac{\alpha d \ln n}{\ln \ln^\beta n}\right)^{1/\alpha} \alpha^{\beta/\alpha}.$$

Corollary 3.12. *If* $L(x) = \ln\ln^\beta x$, P_m*-a.s.:*

$$\max_{|x|\leq n} V(x) \underset{n\to\infty}{\sim} \left(\frac{\alpha d \ln n}{\ln\ln\ln^\beta n}\right)^{1/\alpha}.$$

Corollary 3.13. *If* $L(x) = \mathrm{e}^{\ln^\beta x}$, $0 < \beta < \frac{1}{2}$, P_m*-a.s.:*

$$\max_{|x|\leq n} V(x) \underset{n\to\infty}{\sim} \left(\frac{\alpha d \ln n}{\mathrm{e}^{(\frac{\ln\ln n}{\alpha})^\beta}}\right)^{1/\alpha}.$$

Using Lemma 3.10, we can obtain the following theorem for the quenched asymptotics of $u(t,x)$ for potential $P\{V(x,\omega_m) > x\} = \exp\{-\frac{x^\alpha}{\alpha} L(x)\}$.

Theorem 3.14. *Under Assumptions 3.6 and 3.7, P_m-a.s., for $t \to \infty$:*

$$\limsup_{t\to\infty} \frac{\ln u(t,0)}{t^{\alpha'}} L^{\frac{1}{\alpha-1}}\left(t^{\frac{1}{\alpha-1}}\right) \le \frac{1}{\alpha'},$$

$$\liminf_{t\to\infty} \frac{\ln u(t,0)}{t^{\alpha'}} L^{\frac{1}{\alpha-1}}\left(t^{\frac{1}{\alpha-1}}\right) \ge \frac{1}{\alpha'}\left(\frac{d}{d+\beta}\right)^{\frac{\alpha'}{\alpha}}, \text{ where } \alpha' \text{ satisfies } \frac{1}{\alpha}+\frac{1}{\alpha'}=1.$$

Proof. **(a) Lower Estimation for the Quenched Asymptotics of $u(t,0)$.** Assume that $|x_0| \in [R(1-\delta'), R]$ for some R, $R \gg 1$ and $V(x_0) \ge (1-\delta)\left(\frac{\alpha d \ln R}{L(\ln^{\frac{1}{\alpha}} R)}\right)^{1/\alpha}$. Then, for $R \to \infty$,

$$\begin{aligned} u(t,0) &\ge \max_R \left[\mathrm{e}^{-\kappa t} \cdot C_2 \cdot \frac{1}{R^{d+\beta}} \exp\left\{ (t-1)(1-\delta)\left(\frac{\alpha d \ln R}{L(\ln^{\frac{1}{\alpha}} R)}\right)^{1/\alpha} \right]\right\} \\ &\ge C_2 \mathrm{e}^{-\kappa t} \max_R \mathrm{e}^{-(d+\beta)\ln R + t\left(1-\delta''\right)\left(\frac{\alpha d \ln R}{L(\ln^{\frac{1}{\alpha}} R)}\right)^{1/\alpha}}. \end{aligned}$$

Putting $x = \ln^{\frac{1}{\alpha}} R$, we find

$$\max_x \left[-(d+\beta)x^\alpha + \tilde{t}\frac{(\alpha d)^{1/\alpha} x}{L^{\frac{1}{\alpha}}(x)} \right], \quad \tilde{t} = (1-\delta'')t.$$

The equation for the critical point is:

$$\hat{t}\frac{1-\frac{x_0 L'(x_0)}{\alpha L(x_0)}}{L^{\frac{1}{\alpha}}(x_0)} - \alpha(d+\beta)x_0^{\alpha-1} = 0,$$

where $\hat{t} = \tilde{t}(\alpha d)^{1/\alpha}$. Under Assumption (3.7)

$$L^{\frac{1}{\alpha}}(x_0) = \frac{\hat{t}}{\alpha(d+\beta)x_0^{\alpha-1}},$$

or

$$x_0 = \left(\frac{\bar{t}}{L^{\frac{1}{\alpha}}(x_0)}\right)^{\frac{1}{\alpha-1}}, \tag{6}$$

where $\bar{t} = \frac{\hat{t}}{\alpha(d+\beta)}$.

We use the iterative method to solve (6) by setting the initial point $x_0 = \bar{t}^{\frac{1}{\alpha-1}}$. We obtain

$$x_0 \sim \frac{\bar{t}^{\frac{1}{\alpha-1}}}{L^{\frac{1}{\alpha}\frac{1}{\alpha-1}}\left(\bar{t}^{\frac{1}{\alpha-1}}\right)}.$$

The value at the critical point is

$$\bar{t}^{\alpha'}\frac{1}{\alpha'}\left(\frac{d}{d+\beta}\right)^{\frac{\alpha'}{\alpha}}\frac{1}{L^{\frac{1}{\alpha-1}}\left(\bar{t}^{\frac{1}{\alpha-1}}\right)}.$$

Because δ'' is arbitrarily small, we have proved the lower estimation:

$$\liminf_{t\to\infty}\frac{\ln u(t,0)}{t^{\alpha'}}L^{\frac{1}{\alpha-1}}\left(t^{\frac{1}{\alpha-1}}\right) \geq \frac{1}{\alpha'}\left(\frac{d}{d+\beta}\right)^{\frac{\alpha'}{\alpha}}.$$

(b) Upper Estimation for the Quenched Asymptotics of $u(t,0)$. Applying the upper estimation of the quenched asymptotics of $u(t,0)$, Theorem 2.5 and the result in Theorem 3.9, we have

$$\limsup_{t\to\infty}\frac{\ln u(t,0)}{t^{\alpha'}}L^{\frac{1}{\alpha-1}}\left(t^{\frac{1}{\alpha-1}}\right) \leq \frac{1}{\alpha'}.$$

Combining the lower and upper estimations of the quenched asymptotics of $u(t,0)$, we obtain the result. □

4 Concluding Remark

Let us emphasize that we did not find the exact quenched asymptotics but only the upper and lower estimations of the same order. To find the true asymptotics, we need a better understanding of the a.s. behavior of the underlying random walk $x(t), t \to \infty$ (upper and lower functions, etc.). In the 1-d case the situation is clear, but the multidimensional case is more difficult. We will return to this subject in the future.

References

1. Ben Arous, G., Bogachev, L., Molchanov, S.: Limit theorems for sums of random exponentials. Probab. Theory Relat. Fields **132**(4), 579–612 (2005)
2. Ben Arous, G., Molchanov, S., Ramírez, A.F.: Transition from the annealed to the quenched asymptotics for a random walk on random obstacles. Ann. Probab. **33**(6), 2149–2187 (2005)

3. Ben Arous, G., Molchanov, S., Ramírez, A.F.: Transition asymptotics for reaction-diffusion in random media. In: Probability and Mathematical Physics, 1–40, CRM Proc. Lecture Notes, vol. 42. Amer. Math. Soc., Providence, RI (2007)
4. Bingham, N.H., Goldie, C.M., Teugels, J.L.: Regular Variation. Cambridge University Press, Cambridge (1989)
5. Bojanic, R., Seneta, E.: Slowly varying functions and asymptotic relations. J. Math. Anal. Appl. **34**, 302–315 (1971)
6. Castell, F., Gún, O., Maillard, G.: Parabolic Anderson model with a finite number of moving catalysts. In: Probability in Complex Physical Systems, Springer Proceedings in Mathematics, Vol. 11, Springer (2012)
7. Drewitz, A., Gärtner, J., Ramírez, A.F., Sun, R.: Survival Probability of a Random Walk Among a Poisson System of Moving Traps. In: Probability in Complex Physical Systems, Springer Proceedings in Mathematics, Vol. 11, Springer (2012)
8. Feng, Y, Molchanov, S., Whitmeyer, J.: Random walks with heavy tails and limit theorems for branching processes with migration and immigration. 2010. (Preprint)
9. Gärtner, J., König, W.: Moment asymptotics for the continuous parabolic Anderson model. Ann. Appl. Probab. **10**(1), 192–217 (2000)
10. Gärtner, J., Molchanov, S.: Parabolic problems for the Anderson model. I. Intermittency and related topics. Commun. Math. Phys. **132**, 613–655 (1990)
11. Gärtner, J., Molchanov, S.: Parabolic problems for the Anderson model. II: Second-order asymptotics and structure of high peaks. Probab. Theory Relat. Fields **111**(1), 17–55 (1998)
12. Gärtner, J., Molchanov, S.: Moment asymptotics and Lifshitz tails for the parabolic Anderson model. Stochastic models, Ottawa, ON, 1998. CMS Conf. Proc., vol. 26, pp. 141–157. Amer. Math. Soc., Providence, RI (2000)
13. Gärtner, J., König, W., Molchanov, S.: Geometric characterization of intermittency in the parabolic Anderson model. Ann. Probab. **35**(2), 439–499 (2007)
14. Gärtner, J., den Hollander, F., Maillard, G.: Quenched Lyapunov exponent for the parabolic Anderson model in a dynamic random environment. In: Probability in Complex Physical Systems, Springer Proceedings in Mathematics, Vol. 11, Springer (2012)
15. König, W., Schmidt, S.: The Parabolic Anderson Model with Acceleration and Deceleration. In: Probability in Complex Physical Systems, Springer Proceedings in Mathematics, Vol. 11, Springer (2012)
16. Maillard, G., Mountford, T., Schöpfer, S.: Parabolic Anderson model with voter catalysts: dichotomy in the behavior of Lyapunov exponents. In: Probability in Complex Physical Systems, Springer Proceedings in Mathematics, Vol. 11, Springer (2012)
17. Molchanov, S.: Lectures on random media. In: Bakry, D., Gill, R.D., Molchanov, S. (eds.) Lectures on Probability Theory, Ecole d'Eté de Probabilités de Saint-Flour XXII-1992, LNM 1581, pp. 242–411. Springer, Berlin (1994)
18. Molchanov, S., Petrov, V.V., Squartini, N.: Quasicumulants and limit theorems in case of the Cauchy limiting law. Markov Process. Relat. Fields **17**(3), 597–624 (2007)
19. Schnitzler, A., Wolff, T.: Precise asymptotics for the parabolic Anderson model with a moving catalyst or trap. In: Probability in Complex Physical Systems, Springer Proceedings in Mathematics, Vol. 11, Springer (2012)
20. Zeldovich, Ya., Molchanov, S., Ruzmaikin A., Sokoloff D.: Intermittency, diffusion and generation in a non-stationary random media. Sov. Sci. Rev., Sec C **7**, 1–100 (1988)

Parabolic Anderson Model with Voter Catalysts: Dichotomy in the Behavior of Lyapunov Exponents

Grégory Maillard, Thomas Mountford, and Samuel Schöpfer

Abstract We consider the parabolic Anderson model $\partial u/\partial t = \kappa \Delta u + \gamma \xi u$ with u: $\mathbb{Z}^d \times \mathbb{R}^+ \to \mathbb{R}^+$, where $\kappa \in \mathbb{R}^+$ is the diffusion constant, Δ is the discrete Laplacian, $\gamma \in \mathbb{R}^+$ is the coupling constant, and ξ: $\mathbb{Z}^d \times \mathbb{R}^+ \to \{0, 1\}$ is the voter model starting from Bernoulli product measure ν_ρ with density $\rho \in (0, 1)$. The solution of this equation describes the evolution of a "reactant" u under the influence of a "catalyst" ξ.

In Gärtner, den Hollander and Maillard [Gärtner et al., Intermittency on catalysts: Voter model. Ann. Probab. 38, 2066–2102 (2010)] the behavior of the *annealed* Lyapunov exponents, i.e., the exponential growth rates of the successive moments of u w.r.t. ξ, was investigated. It was shown that these exponents exhibit an interesting dependence on the dimension and on the diffusion constant.

In the present paper we address some questions left open in [Gärtner et al., Intermittency on catalysts: Voter model. Ann. Probab. 38, 2066–2102 (2010)] by considering specifically when the Lyapunov exponents are the a priori maximal value in terms of strong transience of the Markov process underlying the voter model.

G. Maillard (✉)
CMI-LATP, Université de Provence, 39 rue F. Joliot-Curie, 13453 Marseille Cedex 13, France
EURANDOM, P.O. Box 513, 5600 MB Eindhoven, The Netherlands
e-mail: maillard@cmi.univ-mrs.fr

T. Mountford · S. Schöpfer
Institut de Mathématiques, École Polytechnique Fédérale, Station 8, 1015 Lausanne, Switzerland
e-mail: thomas.mountford@epfl.ch; samuel.schoepfer@epfl.ch

J.-D. Deuschel et al. (eds.), *Probability in Complex Physical Systems*,
Springer Proceedings in Mathematics 11, DOI 10.1007/978-3-642-23811-6_3,

1 Introduction

1.1 Model

The parabolic Anderson model (PAM) is the partial differential equation

$$\frac{\partial}{\partial t}u(x,t) = \kappa \Delta u(x,t) + \gamma \xi(x,t)u(x,t), \qquad x \in \mathbb{Z}^d,\ t \geq 0, \tag{1}$$

with u a $\mathbb{R}^+$-valued field, $\kappa \in \mathbb{R}^+$ a diffusion constant, Δ the discrete Laplacian, acting on u as

$$\Delta u(x,t) = \sum_{\substack{y \in \mathbb{Z}^d \\ y \sim x}} [u(y,t) - u(x,t)]$$

($y \sim x$ meaning that y is a nearest neighbor of x), $\gamma \in \mathbb{R}^+$ a coupling constant and

$$\xi = (\xi_t)_{t \geq 0} \quad \text{with} \quad \xi_t = \{\xi_t(x) := \xi(x,t) \colon x \in \mathbb{Z}^d\}$$

the *Voter Model* (VM) taking values in $\{0,1\}^{\mathbb{Z}^d \times \mathbb{R}^+}$. As initial condition, we choose

$$u(x,0) = 1, \qquad x \in \mathbb{Z}^d. \tag{2}$$

One can interpret (1) in terms of population dynamics. Consider a system of two types of particles, A "catalyst" and B "reactant," subject to:

- A-particles evolve autonomously according to the voter dynamics.
- B-particles perform independent random walks at rate $2d\kappa$ and split into two at a rate that is equal to γ times the number of A-particles present at the same location.
- The initial configuration of B-particles is one particle everywhere.

Then $u(x,t)$ can be interpreted as the average number of B-particles at site x at time t conditioned on the evolution of the A-particles.

1.2 Voter Model

The VM is the Markov process on $\{0,1\}^{\mathbb{Z}^d}$ with generator L acting on cylindrical functions f as

$$Lf(\eta) = \sum_{x \in \mathbb{Z}^d} p(x,y) \sum_{\substack{y \in \mathbb{Z}^d \\ y \sim x}} \Big(f(\eta^{x,y}) - f(\eta) \Big),$$

where $p: \mathbb{Z}^d \times \mathbb{Z}^d \to [0,1]$ is the transition kernel of an irreducible random walk and $\eta^{x,y}$ is the configuration

$$\begin{cases} \eta^{x,y}(z) = \eta(z) & \forall z \neq x, \\ \eta^{x,y}(x) = \eta(y). \end{cases}$$

In words, $\xi(x,t) = 1$ and $\xi(x,t) = 0$ mean the presence and the absence of a particle at site x at time t, respectively. Under the VM dynamics, the presence and absence of particles are imposed according to the random walk transition kernel $p(\cdot,\cdot)$.

The VM was introduced independently by Clifford and Sudbury [3] and by Holley and Liggett [8], where the basic results concerning equilibria were shown. Let $(S_t)_{t\geq 0}$ be the Markov semigroup associated with L, $p^{(s)}(x,y) = (1/2)[p(x,y)+p(y,x)]$, $x,y \in \mathbb{Z}^d$, be the *symmetrized* transition kernel associated with $p(\cdot,\cdot)$, and μ_ρ be the *equilibrium measure* with density $\rho \in (0,1)$. When $p^{(s)}(\cdot,\cdot)$ is *recurrent* all equilibria are *trivial*, i.e., of the form $\mu_\rho = (1-\rho)\delta_0 + \rho\delta_1$, while when $p^{(s)}(\cdot,\cdot)$ is *transient* there are also *nontrivial* equilibria, i.e., ergodic measures μ_ρ, different from the previous one, which are the unique shift-invariant and ergodic equilibria with density $\rho \in (0,1)$.

For both cases we have

$$\mu S_t \to \mu_\rho \qquad \text{weakly as } t \to \infty \tag{3}$$

for any starting measure μ that is stationary and ergodic with density ρ (see Liggett [11], Corollary V.1.13). This is in particular the case for our choice $\mu := \nu_\rho$, the Bernoulli product measure with density $\rho \in (0,1)$.

1.3 Lyapunov Exponents

Our focus of interest will be on the pth *annealed* Lyapunov exponent, defined by

$$\lambda_p = \lim_{t\to\infty} \frac{1}{t} \log \mathbb{E}_{\nu_\rho}\big([u(0,t)]^p\big)^{1/p}, \qquad p \in \mathbb{N}, \tag{4}$$

which represents the exponential growth rate of the pth moment of the solution of the PAM (1), where $\mathbb{E}_{\nu_\rho}$ denotes the expectation w.r.t. the ξ-process starting from Bernoulli product measure ν_ρ with density $\rho \in (0,1)$. Note that λ_p depends on the parameters κ, d, γ, and ρ with the two latter being fixed from now. If the above limit exists, then, by Hölder's inequality, $\kappa \mapsto \lambda_p(\kappa)$ satisfies

$$\lambda_p(\kappa) \in [\rho\gamma, \gamma] \qquad \forall \kappa \in [0,\infty).$$

The behavior of the annealed Lyapunov exponents with VM catalysts has already been investigated by Gärtner, den Hollander and Maillard [17], where it was shown that:

- The Lyapunov exponents defined in (4) exist and do not depend on the choice of the starting measure $\nu_\rho S_T$, $T \in [0,\infty]$, where $\nu_\rho S_\infty := \mu_\rho$ denotes the equilibrium measure of density ρ (recall (3))
- The function $\kappa \mapsto \lambda_p(\kappa)$ is globally Lipschitz outside any neighborhood of 0 and satisfies $\lambda_p(\kappa) > \rho\gamma$ for all $\kappa \in [0,\infty)$
- The Lyapunov exponents satisfy the following dichotomy (see Fig. 1):
  - When $1 \le d \le 4$, if $p(\cdot,\cdot)$ has zero mean and finite variance, then $\lambda_p(\kappa) = \gamma$ for all $\kappa \in [0,\infty)$
  - When $d \ge 5$
    - $\lim_{\kappa\to 0} \lambda_p(\kappa) = \lambda_p(0)$
    - $\lim_{\kappa\to\infty} \lambda_p(\kappa) = \rho\gamma$
    - If $p(\cdot,\cdot)$ has zero mean and finite variance, then $p \mapsto \lambda_p(\kappa)$ is strictly increasing for $\kappa \ll 1$

The following questions were left open (see [17], Sect. 1.8):

(Q1) Does $\lambda_p < \gamma$ when $d \ge 5$ if $p(\cdot,\cdot)$ has zero mean and finite variance?
(Q2) Is there a full dichotomy in the behavior of the Lyapunov exponents? Namely, $\lambda_p < \gamma$ if and only if $p^{(s)}(\cdot,\cdot)$ is strongly transient, i.e.,

$$\int_0^\infty t p_t^{(s)}(0,0)\,dt < \infty.$$

Since any transition kernel $p(\cdot,\cdot)$ in $d \ge 5$ satisfies $\int_0^\infty t\, p_t(0,0)dt < \infty$, a positive answer to (Q2) will also ensure a positive one to (Q1) in the particular case when

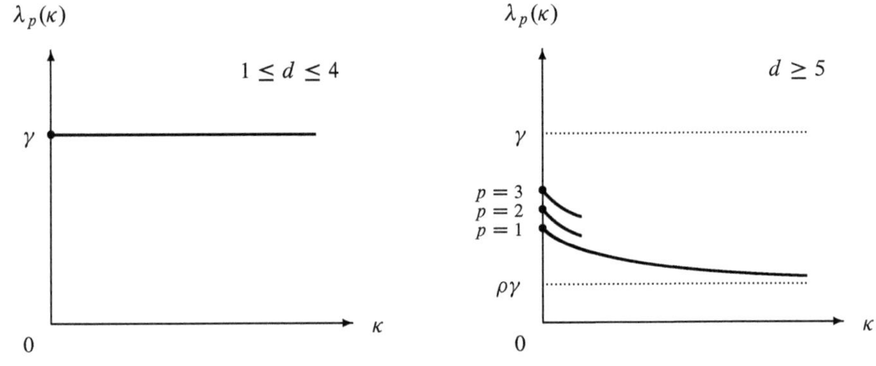

Fig. 1 Dichotomy of the behavior of $\kappa \mapsto \lambda_p(\kappa)$ when $p(\cdot,\cdot)$ has zero mean and finite variance

$p(\cdot,\cdot)$ is symmetric. Theorems 1.2 and 1.4 in Sect. 1.4 give answers to question (Q2), depending on the symmetry of $p(\cdot,\cdot)$. A positive answer to (Q1), given in Theorem 1.1, can also be deduced from our proof of Theorem 1.2.

By the Feynman–Kac formula, the solution of (1) and (2) reads

$$u(x,t) = \mathbb{E}_x\left(\exp\left[\gamma\int_0^t \xi\left(X^\kappa(s), t-s\right)\,\mathrm{d}s\right]\right),$$

where $X^\kappa = (X^\kappa(t))_{t\geq 0}$ is a simple random walk on $\mathbb{Z}^d$ with step rate $2d\kappa$ and $\mathbb{E}_x$ denotes the expectation with respect to X^κ given $X^\kappa(0) = x$. This leads to the following representation of the Lyapunov exponents:

$$\lambda_p = \lim_{t\to\infty} \Lambda_p(t)$$

with

$$\Lambda_p(t) = \frac{1}{pt}\log\left(\mathbb{E}_{\nu_\rho}\otimes\mathbb{E}_0^{\otimes p}\right)\left(\exp\left[\gamma\int_0^t\sum_{j=1}^p \xi\left(X_j^\kappa(s), t-s\right)\mathrm{d}s\right]\right),$$

where X_j^κ, $j = 1,\ldots,p$, are p independent copies of X^κ. In the above expression, the ξ and X^κ processes are evolving in time reversed directions. It is nevertheless possible to let them run in the same time evolution by using the following arguments. Let $\widetilde{\Lambda}_p(t)$ denote the ξ-*time-reversal* analogue of $\Lambda_p(t)$ defined by

$$\widetilde{\Lambda}_p(t) = \frac{1}{pt}\log\left(\mathbb{E}_{\nu_\rho}\otimes\mathbb{E}_0^{\otimes p}\right)\left(\exp\left[\gamma\int_0^t\sum_{j=1}^p \xi\left(X_j^\kappa(s), s\right)\mathrm{d}s\right]\right)$$

and denote $\underline{\Lambda}_p(t)$

$$= \frac{1}{pt}\log\max_{x\in\mathbb{Z}^d}\left(\mathbb{E}_{\nu_\rho}\otimes\mathbb{E}_0^{\otimes p}\right)\left(\exp\left[\gamma\int_0^t\sum_{j=1}^p \xi\left(X_j^\kappa(s), t-s\right)\mathrm{d}s\right]\prod_{j=1}^p \delta_x\left(X_j^\kappa(t)\right)\right)$$

$$= \frac{1}{pt}\log\max_{x\in\mathbb{Z}^d}\left(\mathbb{E}_{\nu_\rho}\otimes\mathbb{E}_0^{\otimes p}\right)\left(\exp\left[\gamma\int_0^t\sum_{j=1}^p \xi\left(X_j^\kappa(s), s\right)\mathrm{d}s\right]\prod_{j=1}^p \delta_x\left(X_j^\kappa(t)\right)\right),$$

where in the last line we reverse the time of the ξ-process by using that ν_ρ is shift-invariant and X_j^κ, $j = 1,\ldots,p$, are time-reversible. As noted in [13], Sect. 2.1, $\lim_{t\to\infty}[\Lambda_p(t) - \underline{\Lambda}_p(t)] = 0$ and, using the same argument, $\lim_{t\to\infty}[\widetilde{\Lambda}_p(t)$

$-\underline{\Lambda}_p(t)] = 0$, after which we can conclude that

$$\lambda_p(\kappa) = \lim_{t\to\infty} \frac{1}{pt} \log\left(\mathbb{E}_{\nu_\rho} \otimes \mathbb{E}_0^{\otimes p}\right)\left(\exp\left[\gamma \int_0^t \sum_{j=1}^p \xi\left(X_j^\kappa(s), s\right) \mathrm{d}s\right]\right).$$

1.4 Main Results

In what follows we give answers to questions (Q1) and (Q2) addressed in [17] concerning when the Lyapunov exponents are trivial, i.e., equal to their a priori maximal value γ.

Our first theorem gives a positive answer to (Q1). It will be proved in Sect. 2 as a consequence of the proof of Theorem 1.3.

Theorem 1.1. *If $d \geq 5$ and $p(\cdot,\cdot)$ has zero mean and finite variance, then $\lambda_p(\kappa) < \gamma$ for all $p \geq 1$ and $\kappa \in [0,\infty)$.*

Our two next theorems state that the full dichotomy in (Q2) holds in the case when $p(\cdot,\cdot)$ is symmetric (see Fig. 2). They will be proved in Sects. 2 and 3, respectively.

Theorem 1.2. *If $p(\cdot,\cdot)$ is symmetric and strongly transient, then $\lambda_p(\kappa) < \gamma$ for all $p \geq 1$ and $\kappa \in [0,\infty)$.*

Theorem 1.3. *If $p(\cdot,\cdot)$ is symmetric and not strongly transient, then $\lambda_p(\kappa) = \gamma$ for all $p \geq 1$ and $\kappa \in [0,\infty)$.*

A similar full dichotomy also holds for the case where ξ is symmetric exclusion process in equilibrium, between recurrent and transient $p(\cdot,\cdot)$ (see [6]).

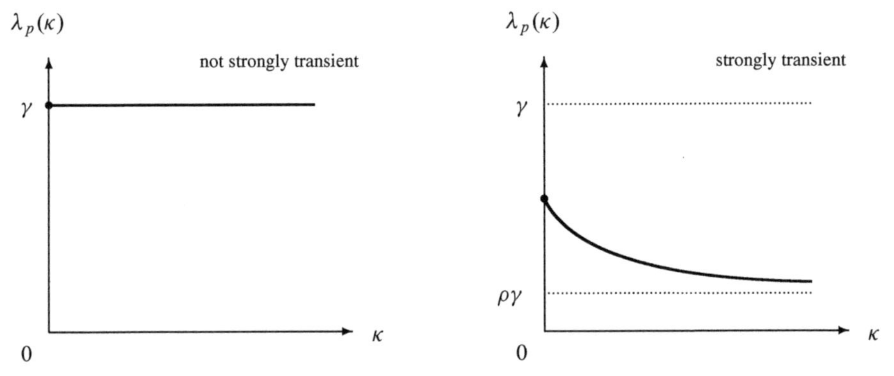

Fig. 2 Full dichotomy of the behavior of $\kappa \mapsto \lambda_p(\kappa)$ when $p(\cdot,\cdot)$ is symmetric

Our fourth theorem shows that this full dichotomy only holds for symmetric transition kernels $p(\cdot,\cdot)$, ensuring that the assertion in (Q2) is not true in its full generality.

Theorem 1.4. *There exists $p(\cdot,\cdot)$ not symmetric with $p^{(s)}(\cdot,\cdot)$ not strongly transient such that $\lambda_p(\kappa) < \gamma$ for all $p \geq 1$ and $\kappa \in [0,\infty)$.*

In the strongly transient regime, the following problems remain open:

(a) $\lim_{\kappa\to 0} \lambda_p(\kappa) = \lambda_p(0)$.
(b) $\lim_{\kappa\to\infty} \lambda_p(\kappa) = \rho\gamma$.
(c) $p \mapsto \lambda_p(\kappa)$ is strictly increasing for $\kappa \ll 1$.
(d) $\kappa \mapsto \lambda_p(\kappa)$ is convex on $[0,\infty)$.

In [17], (a) and (b) were established when $d \geq 5$, and (c) when $d \geq 5$ and $p(\cdot,\cdot)$ has zero mean and finite variance. Their extension to the case when $p(\cdot,\cdot)$ is strongly transient remains open.

In what follows, we use generic notation $\mathbb{P}$ and $\mathbb{E}$ for probability and expectation whatever the corresponding process is (even for joint processes) and denote $\xi_s(x) := \xi(s,x)$.

2 Proof of Theorems 1.1 and 1.3

We first give the proof of Theorem 1.3. Recall that the transition kernel associated to the Voter Model ξ is assumed to be symmetric. At the end of the section we will explain how to derive the proof of Theorem 1.1.

We have to show that $\lambda_p(\kappa) < \gamma$ for all $\kappa \in [0,\infty)$. In what follows we assume without loss of generality that $p = 1$, the extension to arbitrary $p \geq 1$ being straightforward. Our approach is to pick a *bad environment set* B_E associated to the ξ-process and a *bad random walk set* B_W associated to the random walk X^κ so that, for all $n \in \mathbb{N}$,

$$\mathbb{E}\left(\exp\left[\gamma\int_0^n \xi_s(X^\kappa(s))\,\mathrm{d}s\right]\right) \tag{5}$$
$$\leq \Big(\mathbb{P}(B_E)+\mathbb{P}(B_W)\Big)\mathrm{e}^{\gamma n} + \mathbb{E}\left(1\!\!1_{B_E^{\mathrm{c}}\cap B_W^{\mathrm{c}}}\exp\left[\gamma\int_0^n \xi_s(X^\kappa(s))\,\mathrm{d}s\right]\right)$$

with, for some $0<\delta<1$,

$$\mathbb{P}(B_E) \leq \mathrm{e}^{-\delta n}, \qquad \mathbb{P}(B_W) \leq \mathrm{e}^{-\delta n}, \tag{6}$$

and,

$$\int_0^n \xi_s(X^\kappa(s))\,\mathrm{d}s \le n(1-\delta) \qquad \text{on } B_E^{\mathrm{c}} \cap B_W^{\mathrm{c}}. \tag{7}$$

Since, combining (5)–(7), we obtain

$$\lim_{n\to\infty} \frac{1}{n} \log \mathbb{E}\Big(\exp\Big[\gamma \int_0^n \xi_s(X^\kappa(s))\,\mathrm{d}s\Big]\Big) < \gamma,$$

it is enough to prove (6) and (7).

The proof of (6) is given in Sects. 2.1–2.2 and (7) will be obvious from our definitions of B_E and B_W.

2.1 Coarse-Graining and Skeletons

Write $\mathbb{Z}_{\mathrm{e}}^d = 2\mathbb{Z}^d$ and $\mathbb{Z}_{\mathrm{o}}^d = 2\mathbb{Z}^d + 1$, where $1 = (1,\dots,1) \in \mathbb{Z}^d$. We are going to use a *coarse-graining representation* defined by a *space–time block partition* B_y^j and a *random walk skeleton* $(y_i)_{i\ge 0}$. To that aim, for a fixed M, consider

$$B_y^j = \prod_{k=1}^d \big[(y_k-1)M, (y_k+1)M\big) \times \big[jM, (j+1)M\big) \subset \mathbb{Z}^d \times \mathbb{R}_+,$$

where $j \in \mathbb{N}_0 := \mathbb{N} \cup \{0\}$ and

$$y \in \begin{cases} \mathbb{Z}_{\mathrm{e}}^d & \text{when } j \text{ is even,} \\ \mathbb{Z}_{\mathrm{o}}^d & \text{when } j \text{ is odd.} \end{cases}$$

Without loss of generality we can consider random walk trajectories on interval $[0,n]$ with $n \in \mathbb{N}$ multiple of M. Define the *M-skeleton set* set by

$$\varXi = \Big\{\Big(y_0,\dots,y_{\frac{n}{M}}\Big) \in (\mathbb{Z}^d)^{\frac{n}{M}+1}\colon y_{2k} \in \mathbb{Z}_{\mathrm{e}}^d,\ \ y_{2k+1} \in \mathbb{Z}_{\mathrm{o}}^d \ \ \forall k \in \mathbb{N}_0\Big\}$$

and the *M-skeleton set associated to a random walk X* by

$$\varXi(X) = \Big\{\Big(y_0,\dots,y_{\frac{n}{M}}\Big) \in \varXi\colon X(kM) \in B_{y_k}^k \ \forall k \in \{0,\dots,n/M\}\Big\}.$$

In what follows, we will consider the M-skeleton $\varXi(X^\kappa)$, but, as X^κ starts from $0 \in \mathbb{Z}^d$, the first point of our M-skeleton will always be $y_0 := 0 \in \mathbb{Z}^d$ (see Fig. 3).

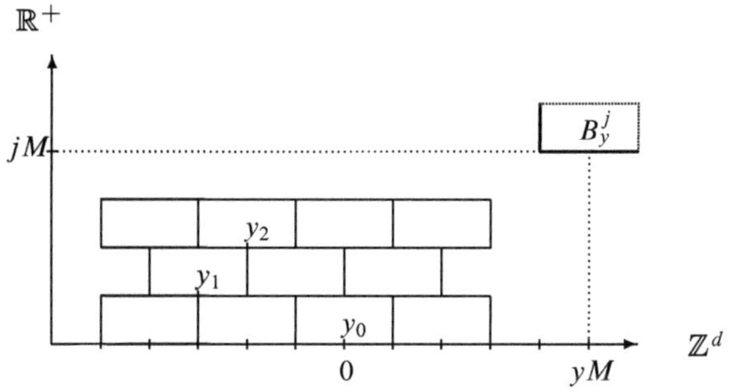

Fig. 3 Illustration of a M-skeleton $(y_0, y_1, y_2, \dots) \in \Xi$ and coarse-grained B_y^j blocks

In the next lemma we prove that the number of M-skeletons not oscillating too much is at most exponential in n/M. For that, define

$$\Xi_A = \left\{ (y_0, \dots, y_{\frac{n}{M}}) \in \Xi \colon \sum_{j=1}^{n/M} (\|y_j - y_{j-1}\|_\infty - 1) \le \frac{n}{Md} \right\}, \tag{8}$$

where $\|\cdot\|_\infty$ is the standard l_∞ norm, the set of all M-skeletons that are *appropriate*.

Lemma 2.1. *There exists some universal constant $K \in (1, \infty)$ such that, for any $n, M \in \mathbb{N}$,*

$$|\Xi_A| \le K^{n/M}.$$

Proof. For any fixed $y_1 \in \mathbb{Z}^d$ and $N \in \mathbb{N}_0$, let

$$I(N) = \left|\left\{ y_2 \in \mathbb{Z}^d \colon \|y_1 - y_2\|_\infty - 1 = N \right\}\right|$$

be the number of elements of $\mathbb{Z}^d$ on the boundary of the cube of size $2N+3$ centered at y_1. For any $N \in \mathbb{N}$, we have $I(N) = (2N+3)^d - (2N+1)^d$ and $I(0) = 3^d - 1$, therefore, for any $N \in \mathbb{N}_0$,

$$I(N) \le 3^d (N+1)^d. \tag{9}$$

Define, for any $N, k \in \mathbb{N}$,

$$I(N,k) = \left| \left\{ (y_j)_{0 \le j \le k} \in (\mathbb{Z}^d)^{k+1} \colon \sum_{j=1}^{k} (\|y_j - y_{j-1}\|_\infty - 1) = N \right\} \right| \tag{10}$$

the number of sequences in $(\mathbb{Z}^d)^{k+1}$ having size N. By (9) and (10), we have

$$\begin{aligned} I(N,k) &= \sum_{\substack{(N_1,\dots,N_k):\\ \sum_{i=1}^k N_i=N}} \left(\prod_{i=1}^k I(N_i)\right) \\ &\le 3^{dk} \sum_{\substack{(N_1,\dots,N_k):\\ \sum_{i=1}^k N_i=N}} \left(\prod_{i=1}^k (N_i+1)\right)^d \\ &\le 3^{dk} \binom{k+N+1}{N} 2^{dN}, \end{aligned} \tag{11}$$

where, in the last line, we used that

$$\max_{\substack{(N_1,\dots,N_k):\\ \sum_{i=1}^k N_i=N}} \prod_{i=1}^k (N_i+1) = 2^N.$$

Using (11) and the fact that $\Xi \subset \mathbb{Z}^d$, we obtain

$$\begin{aligned} |\Xi_A| &\le \sum_{N=0}^{\lfloor \frac{n}{Md} \rfloor} I\left(N, \frac{n}{M}\right) \le 3^{\frac{n}{M}} \sum_{N=0}^{\lfloor \frac{n}{Md} \rfloor} \binom{\frac{n}{M}+N+1}{N} 2^{dN} \\ &\le 3^{\frac{n}{M}} \sum_{N=0}^{\lfloor \frac{n}{Md} \rfloor} \binom{\frac{2n}{M}}{N} 2^{dN} \le 3^{\frac{n}{M}} \sum_{N=0}^{\frac{2n}{M}} \binom{\frac{2n}{M}}{N} 2^{dN} \\ &= 3^{\frac{n}{M}} (2^d+1)^{\frac{2n}{M}}, \end{aligned}$$

which ends the proof of the lemma. □

2.2 The Bad Environment Set B_E

This section is devoted to the proof of the leftmost part of (6) for suitable set B_E defined below.

We say that an environment ξ is *good* w.r.t. an M-skeleton $(y_0, \dots, y_{\frac{n}{M}})$ if we have

$$\left|\left\{0 \le j < \frac{n}{M} : \exists (x, jM) \in B^j_{y_j} \text{ s.t. } \xi_s(x) = 0 \ \forall s \in [jM, jM+1]\right\}\right| \ge \frac{n}{4M}.$$

Since we want the environment to be good w.r.t. all appropriate M-skeleton, we define the *bad environment set* as

$$B_E = \{\exists \text{ an } M\text{-skeleton} \in \Xi_A \text{ s.t. } \xi \text{ is not good w.r.t. it}\}.$$

In the next lemma we prove that for any fixed M-skeleton, the probability that ξ is not good w.r.t. it is at most exponentially small in n/M.

Lemma 2.2. *Take $(y_i)_{0\le i\le \frac{n}{M}} \in \Xi$ an M-skeleton. For M big enough, we have*

$$\mathbb{P}\left(\xi \text{ is not good w.r.t. } (y_i)_{0\le i\le \frac{n}{M}}\right) \le (4K)^{-n/M},$$

where K is the universal constant defined in Lemma 2.1.

Therefore, combining Lemmas 2.1 and 2.2, we get

$$\mathbb{P}(B_E) \le 4^{-n/M}$$

for M big enough, from which we obtain the leftmost part of (6).

Before proving Lemma 2.2, we first give an auxiliary lemma. In order to study the evolution of the VM, we consider, as usual, the dual process, namely, a coalescing random system that evolves backward in time. To that aim, define $(X^{x,t}(s))_{0\le s\le t}$ to be the random walk starting from 0 at time t (i.e., $X^{x,t}(0) = 0$). From the graphical representation of the VM, we can write $\xi_t(x) = \xi_0(X^{x,t}(t))$, $x \in \mathbb{Z}^d$, and therefore the the VM process can be expressed in terms of its initial configuration and a system of coalescing random walks. Two random walks $X^{x,s}$ and $X^{x',s'}$ with $s' < s$ meet if there exists $u \le s'$ such that $X^{x',s'}(s'-u) = X^{x,s}(s-u)$. It is therefore the same to say that $X^{x,s}$ and $X^{x',s'}$ with $s' < s$ meet (in some appropriate time interval) if there exists $t \ge 0$ in this interval (by letting $t = s' - u$) such that

$$X^{x',s'}(t) = X^{x,s}(t + s - s').$$

For convenience we will adopt this notation in the rest of the section.

Lemma 2.3. *Take two independent random walks $X^{x,s}$ and $X^{x',s'}$ with $s' < s$. Then the probability they ever meet is bounded above by*

$$\int_{s-s'}^{\infty} p_t(0,0)\, dt.$$

Proof. Consider the random variable

$$W = \int_0^\infty 1\left\{X^{x,s}(t) = X^{x',s'}(t + (s-s'))\right\} dt.$$

By symmetry, its expectation satisfies

$$\mathbb{E}(W) = \int_0^\infty \mathbb{P}\left(X^{0,0}(2t + s - s') = x - x'\right) \mathrm{d}t$$

$$\leq \int_0^\infty \mathbb{P}\left(X^{0,0}(2t + s - s') = 0\right) \mathrm{d}t$$

$$= \frac{1}{2} \int_{s-s'}^\infty p_t(0,0)\, \mathrm{d}t.$$

Moreover, we have

$$\mathbb{E}(W \mid W > 0) = \int_0^\infty p_{2t}(0,0)\, \mathrm{d}t \geq \frac{1}{2}$$

and then, since $\mathbb{E}(W) = \mathbb{E}(W \mid W > 0)\, \mathbb{P}(W > 0)$, it follows that

$$\mathbb{P}(W > 0) \leq 2\mathbb{E}(W) = \int_{s-s'}^\infty p_t(0,0)\, \mathrm{d}t.$$

□

We are now ready to prove Lemma 2.2.

Proof. Recall that $\xi_s(x) = \xi_0(X^{x,s}(s))$, where ξ_0 is distributed by a product Bernoulli law with density $\rho \in (0,1)$. We first consider any M-skeleton $(y_0, \ldots, y_{n/M}) \in \Xi$ (even not appropriate). For each $0 \leq j < n/M$, we choose R sites $(x_1^j, jM), \ldots, (x_R^j, jM) \in B_{y_j}^j$ such that

$$\mathbb{P}\left(\exists 0 \leq k, k' \leq R,\ \ k \neq k' \colon X^{x_k^j, jM}(s) = X^{x_{k'}^j, jM}(s) \text{ for some } s \in [0, jM]\right) \leq \epsilon \tag{12}$$

for $\epsilon \ll 1$ to be specified later (see Fig. 4). Remark that we first fix ϵ and R and then we choose M large enough so we can find these R sites. As we are in the strongly transient regime, we know that these points exist. If two such random walks hit each other, then we freeze all the random walks issuing from the corresponding block j.

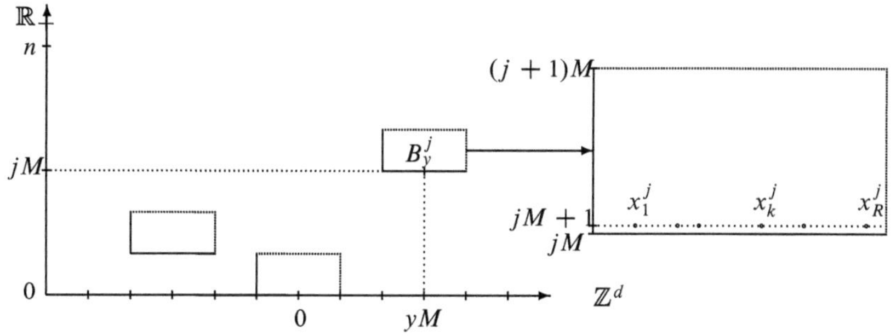

Fig. 4 Illustration of sites (x_k^j, jM), $k \in \{1, \ldots, R\}$, for a fixed M-skeleton

For any $1 \leq j < \frac{n}{M}$, $1 \leq k \leq R$ for some $R > 0$, we have

$$\mathbb{E}\left(\sum_{j'=j+1}^{\frac{n}{M}-1} \sum_{k'=1}^{R} 1\left\{X^{x_{k'}^{j'},j'M}(s+(j'-j)M) = X^{x_k^j,jM}(s) \text{ for some } s \in [0, jM]\right\}\right)$$

$$\leq R \sum_{j'=j+1}^{\frac{n}{M}-1} \int_{(j'-j)M}^{\infty} p_t(0,0)\,\mathrm{d}t$$

$$\leq R \int_{M}^{\infty} \frac{t}{M}\, p_t(0,0)\,\mathrm{d}t,$$

and therefore, summing over $1 \leq k \leq R$, we get

$$\mathbb{E}\left(\sum_{j'=j+1}^{\frac{n}{M}-1} \sum_{k,k'=1}^{R} 1\left\{X^{x_{k'}^{j'},j'M}(s+(j'-j)M) = X^{x_k^j,jM}(s) \text{ for some } s \in [0, jM]\right\}\right)$$

$$\leq \frac{R^2}{M} \int_{M}^{\infty} t\, p_t(0,0)\,\mathrm{d}t \leq \epsilon^2, \tag{13}$$

for M sufficiently large. Again, remark that we first fix ϵ and R, then we choose M large enough. For each j, we now define the filtration

$$\mathscr{F}_t^j = \sigma\left(X^{x_k^j,jM}(s) : 0 \leq s \leq t,\ 1 \leq k \leq R\right)$$

and the sub-martingale $Z^j(t) :=$

$$\mathbb{E}\left(\sum_{j'=j+1}^{\frac{n}{M}-1}\sum_{k,k'=1}^{R} 1\left\{X^{x_{k'}^{j'},j'M}(s+(j'-j)) = X^{x_k^j,jM}(s) \text{ for some } s\in[0,t]\right\}\,\middle|\,\mathscr{F}_t^j\right),$$

with the stopping time

$$\tau^j = \inf\left\{t \geq 0\colon Z^j(t) > \epsilon\right\}.$$

We freeze every random walk issuing from block j at time $jM \wedge \tau^j$. Using (13) and the Doob's inequality, we can see that

$$\mathbb{P}(\tau^j \leq jM) \leq \mathbb{P}\left(\sup_{0\leq t\leq jM} Z_t^j \geq \epsilon\right) \leq \frac{R^2}{M\epsilon}\int_M^\infty tp_t(0,0)\,\mathrm{d}t \leq \epsilon. \tag{14}$$

Since, Z^j is a continuous sub-martingale except at jump times of one of the random walks $X^{x_k^j,jM}$ and when a jump occurs, the increment is at most

$$R\sum_{j'=j+1}^{\frac{n}{M}-1} p_{(j'-j)M}(0,0) \leq \epsilon^2$$

if M is big enough. Therefore, for all $0 \leq t \leq \tau^j$, we get

$$Z^j(t) < \epsilon + \epsilon^2 \leq 2\epsilon \qquad \mathbb{P}\text{-a.s.}. \tag{15}$$

Now we say that j is *good* if

- $\tau^j > jM$
- The R random walks $X^{x_1^j,jM},\ldots,X^{x_R^j,jM}$ do not meet
- The random walks $X^{x_k^j,jM}$ do not hit any point $x_{k'}^{j'}$ during interval $[(j-j')M-1,(j-j')M]$ for $j' < j$
- The random walks $X^{x_k^j,jM}$ do not meet $X^{x_{k'}^{j'},j'M}$ for $j' < j$

By (14), we know that the probability that the first condition does not occur is smaller than ϵ. By definition of the sites $x_1^j,\ldots,x_R^j$, we know that the probability that random walks issuing from the same block j at sites x_k^j, $k = 1,\ldots,R$, hit each other is smaller than ϵ (recall (12)). Moreover, the probability that the third condition does not occur is bounded from above by

$$R\sum_{j'=1}^{j-1}\int_{(j-j')M-1}^{(j-j')M} p_t(0,0)\,\mathrm{d}t,$$

which is as small as we want for M large because we are in a transient case. We still have to compute the probability that the fourth condition does not occur. Furthermore, we can see that two random walks issuing from the same block evolve independently until they meet, provided they do not meet a previous random walk. From the above consideration, we get

$$\mathbb{P}\left(j \text{ is not good} \mid \mathscr{G}^{j-1}\right) \leq 3\epsilon + \sum_{j'=1}^{j-1} \sum_{k,k'=1}^{R} \mathbb{P}\left(X^{x_k^j, jM} \text{ meets } X^{x_{k'}^{j'}, j'M} \mid \mathscr{G}^{j'}\right),$$

where

$$\mathscr{G}^j = \sigma\left(X^{x_k^{j'}, j'M}(s)\colon 1 \leq k \leq R,\ 1 \leq j' \leq j,\ 0 \leq s \leq j'M \wedge \tau^{j'}\right).$$

Here, we recall that $X^{x_k^j, jM}$ meets $X^{x_{k'}^{j'}, j'M}$ (with $j' < j$) if we have

$$X^{x_k^j, jM}(s + (j - j')M) = X^{x_k^{j'}, j'M}(s) \text{ for some } s \in \left[0, \tau^{j'} \wedge j'M\right].$$

By (15), we have, for all j' fixed,

$$\sum_{j=j'+1}^{\frac{n}{M}-1} \sum_{k,k'=1}^{R} \mathbb{P}\left(X^{x_k^j, jM} \text{ meets } X^{x_{k'}^{j'}, j'M} \;\middle|\; \mathscr{G}^{j'}\right) \leq 2\epsilon.$$

Summing over all $1 \leq j' \leq n/M - 2$, we get

$$\sum_{j'=1}^{\frac{n}{M}-2} \sum_{j=j'+1}^{\frac{n}{M}-1} \sum_{k,k'=1}^{R} \mathbb{P}\left(X^{x_k^j, jM} \text{ meets } X^{x_{k'}^{j'}, j'M} \;\middle|\; \mathscr{G}^{j'}\right) \leq \frac{n}{M} 2\epsilon,$$

and then, interchanging the sums, we arrive at

$$\sum_{j=2}^{\frac{n}{M}-1} \sum_{j'=1}^{j-1} \sum_{k,k'=1}^{R} \mathbb{P}\left(X^{x_k^j, jM} \text{ meets } X^{x_{k'}^{j'}, j'M} \;\middle|\; \mathscr{G}^{j'}\right) \leq \frac{n}{M} 2\epsilon.$$

Thus, there are at most $\left\lfloor \frac{n}{2M} \right\rfloor$ random positions j with the property

$$\sum_{j'=1}^{j-1} \sum_{k,k'=1}^{R} \mathbb{P}\left(X^{x_k^j, jM} \text{ meets } X^{x_{k'}^{j'}, j'M} \;\middle|\; \mathscr{G}^{j'}\right) \geq 4\epsilon,$$

and so at least $\lceil \frac{n}{2M} \rceil - 2$ random positions j have the property

$$\sum_{j'=1}^{j-1} \sum_{k,k'=1}^{R} \mathbb{P}\left(X^{x_k^j, jM} \text{ meets } X^{x_{k'}^{j'}, j'M} \;\middle|\; \mathscr{G}^{j'} \right) < 4\epsilon.$$

For these random positions j, we then have

$$\mathbb{P}\big(j \text{ is good} \mid \mathscr{G}^{j-1}\big) \geq 1 - 7\epsilon.$$

Using an elementary coupling, we have at least $\frac{n}{3M}$ positions that are good with probability bounded by

$$\mathbb{P}\left(Y \geq \frac{n}{3M}\right) \geq 1 - \mathrm{e}^{-c(\epsilon)n/M}$$

for

$$Y \sim B\left(\left\lceil \frac{n}{2M} \right\rceil - 2, 1 - 7\epsilon\right)$$

and $c(\epsilon) \to \infty$ as $\epsilon \to 0$. Therefore, outside of a small probability $\mathrm{e}^{-c(\epsilon)n/M}$, for at least $\frac{n}{3M}$ positions j, we have that the random walks $X^{x_k^j, jM}$ are disjoint and so the values $\xi_0(X^{x_k^j, jM}(s))$ are independent until time $s \leq jM$. Then, using the fact that $(\xi_0(x))_{x\in\mathbb{Z}^d}$ are i.i.d. Bernoulli product with parameter ρ, we have that the number of positions j so that there exists $(x, jM) \in B_{y_j}^j$ with $\xi_s(x) = 0$ and $s \in [jM, jM+1]$ is at least $\frac{n}{4M}$ outside the probability

$$\mathbb{P}\left(Y' \geq \frac{n}{12M}\right) \leq \left(\frac{1}{4K}\right)^{n/M}$$

with $Y' \sim B\left(\frac{n}{3M}, \left((1-\mathrm{e}^{-1}(1-\rho))^R\right)\right)$, where ϵ and R are chosen small and large enough, respectively. □

The proof of the leftmost part of (6) is now completed.

2.3 The Bad Random Walk Set B_W

This section is devoted to the proof of the rightmost part of (6).

We are now interested in the random walk X^κ. We are going to prove that $(X^\kappa(s))_{0\leq s\leq n}$ has an appropriate M-skeleton and touches enough zeros outside a probability event exponentially small in n (see Lemmas 2.4 and 2.5 below, respectively). To define the bad random set B_W announced in (5), we are going to define $B_W = B_{W_1} \cup B_{W_2}$, where the bad sets B_{W_1} and B_{W_2} correspond, respectively, to random walk trajectories X^κ which do not have appropriate M-skeleton and do not touch enough sites occupied by a zero configuration of the VM. To be more precise, define

$$B_{W_1} = \{(\Xi(X^\kappa) \notin \Xi_A\}.$$

In the next lemma, we prove that the probability of B_{W_1} is exponentially small in n.

Lemma 2.4. *Take* $(X^\kappa(s))_{0\le s\le n}$ *and* $\Xi(X^\kappa) = (y_0, \dots, y_{n/M})$ *the associated* M-*skeleton. Then, there exists a constant* K' *not depending on* n *such that*

$$\mathbb{P}(B_{W_1}) \le \mathrm{e}^{-K'n}.$$

Proof. In order to have the random walk moving from one block of the skeleton to a nonadjacent one, the random walk has to make at least M steps in the same direction. Keeping that in mind, define

$$Y_j(s) = X^\kappa(jM+s) - X^\kappa(jM)$$

and let

$$\tau_1^j = \inf\{s : \|Y_j(s)\|_\infty \ge M\}, \quad \tau_i^j = \inf\{s > \tau_{i-1}^j : \|Y_j(s) - Y_j(\tau_{i-1}^j)\|_\infty \ge M\}.$$

Next, define

$$W_i^j = \mathbf{1}_{\{\tau_j^i < M\}}$$

and use an elementary coupling to have

$$\mathbb{P}(W_1^j = 1) \le \mathrm{e}^{-c/M} \quad \text{and} \quad \mathbb{P}(W_i^j = 1 | W_{i-1}^j = 1) \le \mathrm{e}^{-c(i)/M} \le \mathrm{e}^{-c/M}$$

for some constants c and $c(i)$ which verify $c(i) \ge c$. Therefore, we have that the number of jumps for the jth block is bounded above by the number of W_i^j that are equal to 1. Using a coupling we can see that this is bounded above by a geometric law with parameter $\mathrm{e}^{-c/M}$. Now if we consider all the blocks, by elementary properties of geometric random variables, we have

$$\mathbb{P}(B_{W_1}) \le \mathbb{P}\left(Y \ge \frac{n}{Md}\right) \le \mathrm{e}^{-c'n}$$

for $Y \sim B\left(\frac{n}{M}\left(1+\frac{1}{d}\right), \mathrm{e}^{-c/M}\right)$, some constant $c' > 0$, M being large and the proof is done. □

Lemma 2.4 proves the first part of the rightmost part of (6), namely the part concerned with bad set B_{W_1}. Now we look at the number of times X^κ stays on a site where the VM has zero value. For that, define

$$\tau_{i+1} = \inf\{t > \tau_i + 1 : \exists x \in \mathbb{Z}^d \text{ s.t. } \|x - X^\kappa(t)\|_\infty \le 2M, \xi_s(x) = 0 \ \forall s \in [t, t+1]\}$$

with $\tau_0 = 0$ and

$$k(M) = \mathrm{e}^{-1/2} \inf_{\|x\|_\infty \le 2M} \mathbb{P}\big(X^\kappa(1/2) = x \mid X^\kappa(0) = 0\big)$$

(remark that $k(M)$ does not depend on n and is strictly positive). Finally, we define

$$B_{W_2} = \left\{ (X^\kappa, \xi) \colon \tau_{\lfloor \frac{n}{2M} \rfloor} \le n - 1 \text{ and } \int_0^n \xi_s(X^\kappa(s))\, \mathrm{d}s \ge n\left(1 - \frac{k(M)}{8M}\right) \right\}$$

as being the bad set corresponding to random walk trajectories X^κ which do not touch enough sites occupied by a zero configuration of the VM. In the next lemma, we prove that such a set has an exponentially small probability in n.

Lemma 2.5. *There exists a constant $\delta > 0$ not depending on n, such that, for M big enough, we have*

$$\mathbb{P}(B_{W_2}) \le \mathrm{e}^{-n\delta}.$$

Proof. Take any realization of ξ and for each time τ_i, define a random variable Y_i which takes value 1 if X^κ reaches a site with value zero at time $\tau_i + \frac{1}{2}$ and stays at that point until time $\tau_i + 1$, and takes value 0 otherwise. Remark that after having fixed n, we can choose the state of ξ_t, $t > n$, as we want, for example, full of zeros. Continue until $\tau_{\lfloor n/(2M) \rfloor}$ which is finite if ξ is well chosen after time n. Using the strong Markov property, for every $k_i \in \{0, 1\}$, we see that

$$\mathbb{P}(Y_i = 1 | Y_j = k_j, j < i) = \mathbb{P}(Y_i = 1 | Y_{i-1} = k_{i-1}) \ge k(M).$$

Then, it follows that $Y := \sum_{i=1}^{\lfloor n/(2M) \rfloor} Y_i$ is stochastically greater than Y' the binomial random variable $B\left(\frac{n}{2M}, k(M)\right)$. Moreover, if $\tau_{\lfloor n/(2M) \rfloor} \le n - 1$, we have

$$\int_0^n \xi_s(X(s))\, \mathrm{d}s \le n - \frac{1}{2} Y.$$

Hence, we get

$$\begin{aligned}
\mathbb{P}(B_{W_2}) &\le \mathbb{P}\left(\tau_{\lfloor \frac{n}{2M} \rfloor} \le n - 1 \ \text{ and } \ n - \frac{1}{2} Y \ge n - \frac{nk(M)}{8M}\right) \\
&\le \mathbb{P}\left(n - \frac{Y}{2} \ge n - \frac{nk(M)}{8M}\right) \\
&= \mathbb{P}\left(Y \le \frac{nk(M)}{4M}\right) \\
&\le \mathrm{e}^{-cn}
\end{aligned}$$

for n sufficiently large and c, a positive constant not depending on n. This result being shown for any realization of ξ (up to time n), this ends the proof. □

Lemma 2.5 proves the second part of the rightmost part of (6), namely the part concerned with bad set B_{W_2}. To complete the proof of (7), it suffices to use the definition of B_E and $B_W = B_{W_1} \cup B_{W_2}$, and to remark that if B_E and B_{W_1} do not occur, then the first condition of B_{W_2}, namely $\tau_{\lfloor n/2M \rfloor} \leq n-1$, is satisfied and therefore the second must be violated.

2.4 *Proof of Theorem 1.1*

The proof of Theorem 1.1 can be deduced from the proof of Theorem 1.3. Without assuming that $p(\cdot,\cdot)$ is symmetric, it is enough to see that

- $\int_0^\infty tp_t(0,0)\,\mathrm{d}t < \infty$, by local CLT
- There is enough symmetry because there exists some $C > 0$ such that $p_t(x,0) \leq Cp_t(0,0)$ for all $x \in \mathbb{Z}^d$ and $t \in [0,\infty)$. Therefore, Lemma 2.2 can still be applied.

From these two observations, the proof of Theorem 1.1 goes through the same lines as the one of Theorem 1.3.

3 Proof of Theorem 1.4

In this section we consider the Lyapunov exponents when the random walk kernel associated to the voter model noise is symmetric and also not strongly transient, that is

$$\int_0^\infty tp_t(0,0)\,\mathrm{d}t = \infty.$$

We want to show that when $p(\cdot,\cdot)$ is symmetric and not strongly transient, then

$$\lambda_p(\kappa) \equiv \gamma \qquad \forall \kappa \in (0,\infty),\ \forall p \geq 1.$$

Since the result is easily seen for recurrent random walks, we can and will assume in the following that

$$\int_0^\infty p_t(0,0)\,\mathrm{d}t < \infty.$$

Given the reasoning of [6], Sect. 3.1, and [7], Sect. 5.1, this result will follow from Proposition 3.1. Consider, in the graphical representation associated to the VM ξ,

$$\chi(t) := \text{ number of distinct coalescing random walks produced on } \{0\} \times [0,t]$$

(This quantity is discussed in Bramson, Cox and Griffeath [1]).

Proposition 3.1. *Assume that $p(\cdot,\cdot)$ is symmetric and not strongly transient, then for any $\epsilon > 0$, we have that*

$$\lim_{t\to\infty} \mathbb{P}\big(\chi(t) \leq \epsilon t\big) = 1.$$

Before proving Proposition 3.1, we will first give the proof of Theorem 1.4.

Proof. From the graphical representation of the VM and Proposition 3.1, we can see that for all $\delta > 0$ and $M < \infty$,

$$\mathbb{P}\Big(\xi(x,s) = 1 \;\; \forall \|x\|_\infty \leq M, \;\; \forall s \in [0,t]\Big) \geq e^{-\delta t}$$

for all t sufficiently large (see [7], proof of Lemma 5.1). Thus, just as in [7], Sect. 5.1, we have for all $p \geq 1$,

$$\begin{aligned}
&\mathbb{E}([u(0,t)]^p)\\
&\geq e^{\gamma p t}\,\mathbb{P}\Big(\|X^\kappa(s)\|_\infty < M \;\; \forall s \in [0,t]\Big)^p \mathbb{P}\Big(\xi(x,s) = 1 \;\; \forall \|x\|_\infty < M, \; \forall s \in [0,t]\Big)\\
&\geq e^{t(\gamma p - \delta - c(M)p)}
\end{aligned}$$

for $c(M) \to 0$ as $M \to \infty$. From this, it is immediate that

$$\lim_{t\to\infty} \frac{1}{t} \log \mathbb{E}([u(0,t)]^p)^{1/p} = \gamma.$$

□

To prove Proposition 3.1, we consider the following system of coalescing random walks:

$$I = \{X^t : t \in \mathscr{P}\}$$

for $\mathscr{P}$ a two sided, rate one Poisson process and X^t a random walk defined on $s \in [t,\infty)$, starting at 0 at time t (we could equally well consider a system of random walks indexed by $h\mathbb{Z}$ for some constant h). The coalescence is such that for $t < t' \in \mathscr{P}$, X^t, $X^{t'}$ evolve independently until $T^{t,t'} = \inf\{s > t' : X^t(s) = X^{t'}(s)\}$, and then, for $s \geq T^{t,t'}$, $X^t(s) = X^{t'}(s)$.

We will be interested in the density or number of distinct random walks at certain times. To aid this line we will adopt a labeling procedure for the random walks, whereby effectively when two random walks meet for the first time, one of them (chosen at random) dies; in this optic the number of distinct random walks will be the number still alive. Our labeling scheme involves defining for each $t \in \mathscr{P}$, the label process l_s^t for $s \geq t-$ (it will be helpful to be able to define $l_{t-}^t = t$, though since at time t there may well be other random walks present at the origin, it will not necessarily be the case that $l_t^t = t$). These processes will be defined by the following properties:

- If for $t \neq t' \in \mathscr{P}$, $X^t(s) \neq X^{t'}(s)$, then $l_s^t \neq l_s^{t'}$.
- If $t_1, t_2, \ldots, t_r$ are elements of $\mathscr{P}$, then at $s \geq \max\{t_1, \ldots, t_r\}$, if $X^{t_1}(s) = X^{t_2}(s) = \cdots = X^{t_r}(s)$, then $l_s^{t_1} = l_s^{t_2} = \cdots = l_s^{t_r} = u$ for some $u \in \mathscr{P}$ with $X^{t_1}(s) = X^u(s)$.
- If for $t \neq t' \in \mathscr{P}$, X^t meets $X^{t'}$ for the first time at s, then independently of past and future random walks or labeling decisions $l_s^t = l_s^{t'} = l_{s-}^{t'}$ with probability $\frac{1}{2}$ and with equal probability $l_s^t = l_s^{t'} = l_{s-}^t$.
- The process l_s^t can only change at moments where X^t meets a distinct random walk for the first time.

For $t \in \mathscr{P}$, $s > t$, we say that t is *alive at time* s, if $l_s^t = t$; it *dies* at time s if $l_{s-}^t = t, l_s^t \neq t$. We say X^t, X^u *coalesce* at time s if this is the first time at which the two labels are equal. The following are easily seen:

- The events $A_s^t = \{l_s^t = t\}$ for $t \in \mathscr{P}$ are decreasing in s.
- A_s^t depends only on the random motions of the coalescing random walks and on the labeling choices involving X^t.
- For $s > 0$, the number of independent random walks $X^t(s)$, $t \in \mathscr{P} \cap [-n, 0]$ is simply equal to the number of distinct labels l_s^t, $t \in \mathscr{P} \cap [-n, 0]$.

Let

$$c_0 = \lim_{s \to \infty} \mathbb{P}^t(A_s^t) \in [0, 1], \tag{16}$$

according to palm measure, $\mathbb{P}^t$, for $t \in \mathscr{P}$. We obtain easily the following proposition.

Proposition 3.2.

$$\lim_{s \to \infty} \frac{1}{s} \left| \{\textit{distinct random walks } X^t(0) \colon t \in \mathscr{P} \cap [-s, 0)\} \right| = c_0 \qquad \textit{a.s.}$$

Proof. Using the definition of c_0 in (16) and ergodicity of the system we see that the limit is greater than c_0. Then, Lemma 3.1 gives the result.

Lemma 3.1. *For c_0 as defined in (16), for each $\epsilon > 0$, there exists $R < \infty$ so that if we consider the finite system of coalescing random walks $(X^t)_{t \in (-R,0] \cap \mathscr{P}}$, then with probability at least $1 - \epsilon$ at time R there are less than $(c_0 + \epsilon)R$ distinct random walks labels.*

Proof. By definition of c_0, for all $\epsilon > 0$ there exists a T_0 so that

$$\mathbb{P}^0\big(\text{label } 0 \text{ is alive at time } s\big) < c_0 + \frac{\epsilon}{100} \qquad \forall s \geq T_0.$$

Now pick R_1 so that

$$\mathbb{P}\big(\|X^0(s)\|_\infty \leq R_1 \ \ \forall s \in (0, T_0)\big) \geq 1 - \frac{\epsilon}{100}.$$

Therefore,

$$\mathbb{P}^0\big(\text{label } 0 \text{ is not alive at time } s \geq T_0 \text{ and } \|X^0(s)\|_\infty \leq R_1 \ \ \forall s \in (0, T_0)\big) \\ \geq 1 - c_0 - \frac{2\epsilon}{100}.$$

We then pick T_1 so that

$$\mathbb{P}\big(\exists t \in \mathscr{P} \cap [-T_1, T_1]^{\mathrm{c}} \colon \big\|X^t(s)\big\|_\infty \leq R_1 \ \text{ for some } s \in (0, T_0)\big) < \frac{\epsilon}{100}.$$

Thus

$$\mathbb{P}^0\big(\exists t \in \mathscr{P} \cap [-T_1, T_1] \setminus \{0\} \colon l^0_{R_1} = t\big) \geq 1 - c_0 - \frac{\epsilon}{30}.$$

From the translation invariant property of the system and ergodicity if

$$\lambda_s := \Big|\Big\{t \in [-s, s] \cap \mathscr{P} \colon X^t \text{ loses its label to a random walk } X^{t'} \text{ with } |t - t'| \leq T_1\Big\}\Big|,$$

then

$$\liminf_{s \to \infty} \frac{\lambda_s}{2s} \geq 1 - c_0 - \frac{\epsilon}{30} \qquad \text{a.s.}$$

The result now follows easily. □

Proposition 3.1 will be proven by showing the following proposition.

Proposition 3.3. *If $p(\cdot, \cdot)$ is symmetric and not strongly transient, then* $c_0 = 0$.

The proof of Proposition 3.3 will work for any Poisson process rate, in particular for $\mathscr{P}$ having rate $M \gg 1$. The distinct random walks treated in Proposition 3.1 can be divided into those coalesced with a random walk from the system derived from $\mathscr{P}$ (and so by Proposition 3.3 of small "density") and those uncoalesced (also of small "density" if M is large). Thus Proposition 3.1 follows almost immediately from Proposition 3.3.

The argument for Proposition 3.3 is low level and intuitive. We argue by contradiction and suppose that $c_0 > 0$. From this we can deduce, loosely speaking, that after a certain time either a random walk has lost its original label, or it will keep it forever. We then introduce coupling on these random walks so that we may regard these random walks as essentially independent random walks starting at 0 (at different times). We then introduce convenient comparison systems so that we can analyze subsequent coalescences. We will use automatically, without reference, the following "obvious" result.

Lemma 3.2. *Consider two collections of coalescing random walks $\{Y^i\}$ and $\{(Y')^i\}$ for i in some index set. If the coalescence rule is weaker for the $\{(Y')^i\}$ system in that if two walks $(Y')^i$ and $(Y')^j$ are permitted to coalesce at time t, then so are Y^i and Y^j, then there is a coupling of the two systems so that the weaker contains the stronger.*

We now fix $\epsilon > 0$ so that $\epsilon \ll c_0$ (by hypothesis $c_0 > 0$). We choose R according to Lemma 3.1 and divide up time into intervals $I_j = [jR, (j+1)R)$. We first consider the coalescing system where random walks X^t, $X^{t'}$, $t, t' \in \mathscr{P}$, can only "coalesce" (or destroy a label t or t') if t, t' are in the same I_j interval. Thus we have a system of random walks that is invariant to time shifts by integer multiples of R. We now introduce a system of random walks Y^t, $t \in V := \cup_j \left\{[jR, (j+1)R) \cap jR + \frac{1}{c_0}\mathbb{Z}\right\}$. The random walks Y^t, $t \in \left[jR, (j+1)R\right)$ are not permitted to coalesce up until time $(j+1)R$ (at least) and will evolve independently of the system $(X^t)_{t\in\mathscr{P}}$ until time $(j+1)R$. We will match up the points in $V \cap I_j$, with those in $\mathscr{P} \cap I_j \cap K$ in a maximal measurable way for $K = \{t \in \mathscr{P}\colon \text{label } t \text{ survives to time } (j+1)R\}$.

Lemma 3.3. *Unmatched points in $\cup_j \mathscr{P} \cap I_j \cap K$ and in V have density less than 2ϵ for R fixed sufficiently large.*

Remark: the system is not translation invariant with respect to all shifts but it possesses enough invariance for us to speak of densities.

We similarly have the following lemma.

Lemma 3.4. *Unmatched Y particles have density less than 2ϵ for R fixed sufficiently large.*

It is elementary that two random walks X, Z can be coupled so that for t sufficiently large $X(t) = Z(t)$. For given $\epsilon > 0$ we choose M_0 and then M_1 so that

$$\mathbb{P}\left(\sup_{t\le R} \|X(t)\|_\infty \ge M_0\right) < \frac{\epsilon}{10} \tag{17}$$

and

$$\sup_{|z|\le 2M_0} \mathbb{P}\left(X_0, Z_z \text{ not coupled by time } M_1\right) < \frac{\epsilon}{10}, \tag{18}$$

where X_0 and Z_z denote that the random walks X and Z start at point 0 and point z, respectively.

We then (on interval I_j) couple systems Y and X by letting married pairs Y^t, $X^{t'}, t \in V, t' \in \mathscr{P} \cap K$, evolve independently of other Y, X random walks so that they couple by time $M_1 + (j+1)R$ with probability at least $1 - \frac{3\epsilon}{10}$.

Thus we have two types of random walk labels, l^t, for the X system which are equal to t at time $t + R + M_1$: those for which the associated random walk was paired with a Y random walk and such that the random walks have coupled by time $t + R + M_1$ said to be *coupled* and the others, said to be *decoupled*. This is similar for the points in V associated to Y random walks. We note that the foregoing implies that the density of uncoupled labels is bounded by $2\epsilon + \frac{3\epsilon}{10} \le 3\epsilon$. The point is that modulo this small density, we have an identification of the coalescing X random walks and the Y random walks.

We now try to show that enough Y particles will coalesce in a subsequent time interval to imply that there will be a significant decrease in surviving labels for the X system. To do this we must bear in mind that, essentially, it will be sufficient to show a decrease in the density of Y random walk labels definitely greater than ϵ. Secondly, as already noted, we will adopt a coalescence scheme that is a little complicated namely $Y^i, Y^{i'}$ in V can only "coalesce" at time $t \ge \max\{i, i'\}$ if

- $(t - i', t - i)$ are in some time set to be specified
- For $i < i'$, $\frac{t-i'}{i'-i} \in \left(\frac{9}{10}, \frac{11}{10}\right)$

We now begin to specify our coalescence rules for the random walk system $\{Y^i\}_{i \in V}$. The objective here will be to facilitate the necessary calculations. A first objective is to have coalescence of $Y^i, Y^{i'}$ at times $t > \max\{i, i'\}$ so that $p_t(0,0)$ is well behaved around $t - i$, $t - i'$. It follows from symmetry of the random walk that $t \mapsto p_t(0,0)$ is decreasing. The problem we address is that it is not immediate how to achieve bounds in the opposite direction. This is the purpose of the next result.

Lemma 3.5. *Consider positive $\{a_n\}_{n\ge0}$ so that $\sum_{n=0}^{\infty} a_n = \infty$. For all $r \in \mathbb{Z}_+$, there exists a subsequence $\{a_{n_i}\}_{i\ge0}$ so that*
(i) $\sum a_{n_i} = \infty$
(ii) $a_{n_i} > \frac{1}{2} a_{n_i - r}$, $\forall i \ge 0$

Proof. If $r > 1$ we may consider the r subsequences $\{a_{ri+j}\}_{i\ge0}$ for $j \in \{0, 1, \ldots, r-1\}$. At least one of these must satisfy $\sum a_{ri+j} = \infty$, so, without loss of generality, we take $r = 1$.

Now we classify i as good or bad according to whether $a_i > a_{i-1}/2$ or not. This decomposes $\mathbb{Z}$ into intervals of bad sites, alternating with intervals of good sites. By geometric bounds, the sum of bad sites is bounded by the sum of the good a_i for which i is the right end point of a good interval. Thus we have

$$\sum_{i \text{ good}} a_i = \infty,$$

from which the result is immediate. □

Corollary 3.1. *For our symmetric kernel $p_t(0,0)$ we can find $n_i \uparrow \infty$ so that*
(i) $\sum_i \int_{2^{n_i}}^{2^{n_i+1}} tp_t(0,0)\mathrm{d}t = \infty$
(ii) $p_{2^{n_i-1}}(0,0) \le 2^{12} p_{2^{n_i+3}}(0,0)$

Proof. In Lemma 3.5 take

$$a_n = \int_{2^{n+3}}^{2^{n+4}} tp_t(0,0)\mathrm{d}t$$

and take $r = 5$. Then by the monotonicity of $t \to p_t(0,0)$, we have

$$2^{2n_i+7} p_{2^{n_i+3}}(0,0) \ge a_{n_i} \ge \frac{1}{2} a_{n_i-5} \ge \frac{1}{2} 2^{2n_i-4} p_{2^{n_i-1}}(0,0).$$

□

We fix such a sequence $\{n_j\}_{j\ge 1}$ once and for all.

We assume, as we may, that $n_j < n_{j+1} - 4$ for all $j \ge 1$ and also assume again, as we may, that

$$\int_{2^{n_j}}^{2^{n_j+1}} tp_t(0,0)\mathrm{d}t < \frac{\epsilon}{100}$$

for all $j \ge 1$. We are now ready to consider our coalescence rules. We choose $\epsilon \ll \alpha \ll 1$ (we will fully specify α later on but we feel it is more natural to defer the technical relations). We then choose k_0 so that $2^{n_{k_0}} > R + M_1$ with R as in Lemmas 3.1 and 3.3 and M_1 as in (18), and

$$k_1 := \inf\left\{k > k_0 \colon \sum_{j=k_0}^{k} \int_{2^{n_j}}^{2^{n_j+1}} tp_t(0,0)\,\mathrm{d}t > \alpha\right\}. \tag{19}$$

We have coalescence between Y^i and $Y^{i'}$, for $i < i'$ only at $t \in [i'+2^{n_j}, i'+2^{n_j+1}]$, $j \in [k_0, k_1]$ if

(a) $\frac{t-i'}{i'-i} \in (9/10, 11/10)$
(b) The interval of $t \in [i' + 2^{n_j}, i' + 2^{n_j+1}]$ satisfying (a) is of length at least 2

We say (i,i') and (i',i) are in j and write $(i,i') \in j$ if the above relations hold. To show that sufficient coalescence occurs, we essentially use Bonferroni inequalities (see, e.g., [4], p. 21).

To aid our argument we introduce a family of independent (noncoalescing) random walks $\{(Z^i(s))_{s\ge i}\}_{i\in V}$ such that for each $i \in V$, $Y^i(s) = Z^i(s)$ for $s \ge i$ such that $l_s^i = i$. In the following we will deal with random walks Y^0, Z^0, but lack of total translation invariance notwithstanding, it will be easy to see that all bounds obtained for these random walks remain valid for more general random walks Y^i, Z^i. For a given random walk Y^0, say, the probability that Y^0 is killed by Y^i (with i possible in the sense of the above rules) is in principle a complicated event given the whole system of coalescing random walks. Certainly the event

$$\{Z^0 \text{ meets } Z^i \text{ in appropriate time interval after first having met } Z^k\}$$

is easier to deal with than the corresponding Y event. From this point on we will shorten our phraseology by taking "Z^i hits Z^k" to mean that Z^i meets Z^k at a time t satisfying the conditions (a) and (b) with respect to i, k.

For $(Z^0(s))_{s\geq 0}$ and $(Z^i(s))_{s\geq i}$ independent random walks each beginning at 0, we first estimate

$$\sum_i \mathbb{P}\big(Z^0 \text{ hits } Z^i\big).$$

This of course decomposes as

$$\sum_j \sum_{(0,i)\in j} \mathbb{P}\big(Z^0 \text{ hits } Z^i\big).$$

We fix j and consider $i > 0$ so that $(0,i) \in j$ (the case $i < 0$ is similar). That is the interval of times s with

$$(s-i)/i \in (9/10, 11/10), \quad s \in [i + 2^{n_j}, i + 2^{n_j+1}]$$

is at least 2 in length: we note that for each $i \in \left(\frac{5}{4}2^{n_j}, \frac{7}{4}2^{n_j}\right)$ the relevant interval $\frac{19}{10}i \leq s \leq \frac{21}{10}i$ is an interval of length greater than $\frac{5}{4}2^{n_j}\frac{1}{5} = \frac{2^{n_j}}{4}$.

Lemma 3.6. *There exists $c_2 \in (0,\infty)$ so that for any interval I of length at least 1 contained in $(1,\infty)$,*

$$\frac{1}{c_2}\int_I p_t(0,0)\,\mathrm{d}t \leq \mathbb{P}\big(X^0(t) = 0 \textit{ for some } t \in I\big) \leq c_2 \int_I p_t(0,0)\,\mathrm{d}t$$

Proof. Consider random variable $W = \int_a^{b+1} 1_{\{X^0(s)=0\}}\,\mathrm{d}s$ for $I = [a,b]$. Then

$$\mathbb{E}(W) = \int_a^{b+1} \mathbb{P}(X^0(s) = 0)\,\mathrm{d}s = \int_a^{b+1} p_s(0,0)\,\mathrm{d}s \leq 2\int_I p_s(0,0)\,\mathrm{d}s,$$

by monotonicity of $p_s(0,0)$ and the fact that $b - a \geq 1$. But for $\tau := \inf\{s \in I : X^0(s) = 0\}$ we have $\mathbb{E}(W|\mathscr{F}_\tau) \geq \mathrm{e}^{-1}$ on $\{\tau < \infty\}$ so

$$\mathbb{P}(\tau < \infty) = \mathbb{P}\big(X^0(t) = 0 \text{ for some } t \in I\big) \leq \mathbb{E}(W)e \leq 2e\int_I p_s(0,0)\mathrm{d}s.$$

Equally for $W' = \int_a^b 1_{\{X^0(s)=0\}}\,\mathrm{d}s$, we have

$$\mathbb{E}(W' \mid \mathscr{F}_\tau) \le \gamma = \int_0^\infty p_s(0,0)\,\mathrm{d}s \quad \text{on } \{\tau < \infty\}$$

and so

$$\mathbb{P}(\tau < \infty) \ge \frac{\mathbb{E}(W')}{\gamma} = \frac{1}{\gamma}\int_a^b p_s(0,0)\,\mathrm{d}s.$$

□

Proposition 3.4. *For some universal* $c_3 \in (0,\infty)$,

$$c_3^{-1} 2^{2n_j} p_{2^{n_j}}(0,0) \le \sum_{(0,i)\in j} \mathbb{P}\big(Z^0 \textit{ hits } Z^i\big) \le c_3 2^{2n_j} p_{2^{n_j}}(0,0).$$

Proof. We consider first the upper bound. There are less than 2^{n_j} relevant i. For such an i,

$$\mathbb{P}\big(Z^0 \text{ hits } Z^i\big) \le \mathbb{P}\big(X^0(t) \text{ hits } 0 \text{ for some } t \in \big[a + 2^{n_j}, a + 3\cdot 2^{n_j}\big]\big)$$

for some $a \ge 0$. By monotonicity of $t \to p_t(0,0)$ and using Lemma 3.6, this is bounded by

$$c_2 \int_{2^{n_j}}^{3\cdot 2^{n_j}} p_s(0,0)\,\mathrm{d}s \le c_3 2^{n_j} p_{2^{n_j}}(0,0)$$

for some $c_3 > 0$. On the other side the number of $i \in \big(\frac{5}{4}2^{n_j}, \frac{7}{4}2^{n_j}\big)$ is greater than $c_1 \frac{1}{3} 2^{n_j}$ if R was fixed sufficiently large and for each such i, $(\frac{9}{10}i, \frac{11}{10}i) \subset [2^{n_j}, 2^{n_j+1}]$. Moreover, we have

$$\mathbb{P}\big(Z^0 \text{ hits } Z^i\big) \ge \frac{1}{c_2}\int_{\frac{28}{10}i}^{\frac{32}{10}i} p_s(0,0)\,\mathrm{d}s \ge \frac{1}{c_2}\frac{4}{10}\, i\, p_{2^{n_j+3}}(0,0) \ge \frac{1}{c_2} 2^{n_j-1} p_{2^{n_j+3}}(0,0)$$

$$\ge \frac{2^{-13}}{c_2} 2^{n_j} p_{2^{n_j-1}}(0,0) \ge c_3^{-1}\, 2^{n_j} p_{2^{n_j}}(0,0),$$

because of Lemma 3.1, Corollary 3.1 (by our choice of j), monotonicity of $t \to p_t(0,0)$ and possibly after increasing c_3. □

Thus, using that $j \in [k_0, k_1]$ (recall 19), we have a universal c_4 such that

$$c_4\alpha \geq \sum_j \sum_{(0,i)\in j} \mathbb{P}\big(Z^0 \text{ hits } Z^i\big) \geq \frac{\alpha}{c_4}.$$

There are two issues to address to show that

(a)

$$\mathbb{P}\big(\exists j, \exists i \text{ so that } (0,i) \in j, \;\; Z^0 \text{ hits } Z^i\big)$$

is of the order α

(b) (a) holds with Z^0, Z^i replaced by our coalescing random walks Y^0, Y^i

In fact both parts are resolved by the same calculation.

We consider the probability that random walk Z^0 is involved in a "3-way" collision with Z^i and $Z^{i'}$ either due to Z^0 hitting Z^i in the appropriate time interval and then hitting $Z^{i'}$, or Z^0 hitting Z^i and, subsequently Z^i hitting $Z^{i'}$. The first case is important to bound so that one can use simple Bonferroni bounds to get a lower bound on $\mathbb{P}(\exists i \text{ so that } Z^i \text{ hits } Z^0)$. The second is to take account of the fact that we are interested in the future coalescence of a given random walk Y^0. As already noted, we can couple the systems in the usual way so that for all t, $\cup_i\{Y_t^i\} \subseteq \cup_i\{Z_t^i\}$. The problem is that if for some i, Z^i hits Z^0 due to coalescence this need not imply that Y^i hits Y^0: if the Y^i particles coalesced with a $Y^{i'}$ before Z^i hits Z^0. Fortunately this event is contained in the union of events above over i, i'.

Proposition 3.5. *There exists universal constant K so that for all i and i' with $(0,i') \in j'$*

$$\mathbb{P}\big(Z^0 \textit{ hits } Z^i \textit{ and then } Z^{i'}\big) \leq K 2^{n_{j'}} p_{2^{n_{j'}}}(0,0)\mathbb{P}\big(Z^0 \textit{ hits } Z^i\big).$$

Proof. There are several cases to consider: $i < 0 < i'$, $i < i' < 0$, $i' < i < 0$, $i' < 0 < i$, $0 < i < i'$ and $0 < i' < i$. All are essentially the same so we consider explicitly $0 < i < i'$. We leave the reader to verify that the other cases are analogous. We choose j, j' so that $(0,i) \in j$ and $(0,i') \in j'$ (so necessarily $j' \geq j$). We condition on T_j, $Z^i(T_j)(= Z^0(T_j))$, for

$$T_j := \inf\left\{s \in \left(\frac{19i}{10}, \frac{21i}{10}\right) \cap \left[i + 2^{n_j}, i + 2^{n_j+1}\right] : Z^i(s) = Z^0(s)\right\} < \infty.$$

With $x = Z^0(T_j)$ we have

$$\begin{aligned}
&\mathbb{P}\left(\exists s' \geq T_j \in \left(\frac{19i'}{10}, \frac{21i'}{10}\right) \cap \left[i' + 2^{n_{j'}}, i' + 2^{n_{j'}+1}\right] : Z^{i'}(s') = Z^0(s') \;\Big|\; G^{0,i}\right) \\
&\quad = \mathbb{P}\big(Z^0(t) = x \text{ for some } t \in I_j\big),
\end{aligned}$$

where I_j is the image of the interval

$$\left[(i' + 2^{n_{j'}}) \vee T_j \vee \frac{19i'}{10}, i' + 2^{n_{j'}+1} \wedge \frac{21i'}{10}\right],$$

by the function $t \mapsto 2t - T_j - i'$, for $G^{0,i} = \sigma(Z^0(s), Z^i(s): s \leq T_j)$. By elementary algebra this is less than

$$\mathbb{P}\left(Z^0(t) = x \text{ for } t \in \left(\frac{9i'}{10}, \frac{16i'}{5}\right)\right),$$

but by arguing as in Lemma 3.6, this is bounded by

$$c_2 \int_{\frac{9i'}{10}}^{\frac{16i'}{5}} p_s(0,x)\,\mathrm{d}s \leq c_2 \int_{\frac{9}{10}i'}^{\frac{16}{5}i'} p_s(0,0)\,\mathrm{d}s \leq c_2 \frac{23}{10} i' p_{2^{n_{j'}-1}}(0,0)$$

$$\leq c_2 \frac{23}{9} 2^{13} 2^{n_{j'}} p_{2^{n_{j'}+3}}(0,0) \leq c' 2^{n_{j'}} p_{2^{n_{j'}}}(0,0)$$

for some universal constant c', where we use symmetry and monotonicity of $p_s(\cdot,\cdot)$ and Corollary 3.1, by the choice of our $n_{j'}$. So given that

$$\mathbb{P}(T_i < \infty) = \mathbb{P}\big(Z^0 \text{ hits } Z^i\big),$$

the desired bound is achieved. □

Corollary 3.2. *For α sufficiently small*

$$\mathbb{P}\big(\exists i: Z^0 \text{ hits } Z^i\big) \geq \frac{\alpha}{2}.$$

Proof. By Bonferroni, the desired probability is superior to

$$\sum_i \mathbb{P}\big(Z^0 \text{ hits } Z^i\big) - \sum_{i,i'} \mathbb{P}\big(Z^0 \text{ hits } Z^i \text{ and then } Z^{i'}\big) \geq \alpha - K c_3^2 \alpha^2 \geq \frac{\alpha}{2}$$

if $\alpha \leq 1/(2Kc_3^2)$. □

We similarly show the following proposition.

Proposition 3.6. *There exists universal constant K so that for all i' and i with $(0,i) \in j$,*

$$\mathbb{P}\big(Z^0 \text{ hits } Z^i \text{ after } Z^i \text{ hits } Z^{i'}\big) \leq K 2^{n_j} p_{2^{n_j}}(0,0) \mathbb{P}\big(Z^i \text{ hits } Z^{i'}\big).$$

This gives the following corollary.

Corollary 3.3. *For the coalescing system $\{Y^i\}_{i\in V}$ provided α is sufficiently small,*

$$\mathbb{P}\big(Y^i \text{ dies after time } R\big) \geq \frac{\alpha}{5}.$$

Proof. We have of course from the labeling scheme

$$\begin{aligned}
&\mathbb{P}\big(Y^i \text{ dies after time } R\big)\\
&\quad\geq \frac{1}{2}\mathbb{P}\big(Y^i \text{ hits } Y^{i'} \text{ in appropriate time interval for some } i'\big)\\
&\quad\geq \frac{1}{2}\mathbb{P}\big(Z^i \text{ hits } Z^{i'} \text{ in appropriate time interval for some } i'\big)\\
&\quad-\frac{1}{2}\mathbb{P}\big(Z^i \text{ hits } Z^{i'} \text{ in appropriate time interval for some } i' \text{ so that}\\
&\qquad\qquad Z^{i'} \text{ hits some } Z^{i''} \text{ previously}\big)\\
&\quad\geq \frac{\alpha}{4} - Kc_3^2\alpha^2 \geq \frac{\alpha}{5}
\end{aligned}$$

for $\alpha \leq 1/(20Kc_3^2)$. □

We can now complete the proof of Proposition 3.3 and hence that of Proposition 3.1. If we have $c_0 > 0$, then we can find $0 < \epsilon < \alpha/200$ and α so small that the relevant results above hold, in particular Corollary 3.3. Thus the density of Y's is reduced by at least $\alpha/5$. But by our choice of ϵ and Lemma 3.3, the density of X's is reduced by at least $\alpha/5 - 6\epsilon \geq \alpha/6 \geq 3\epsilon$ which is a contradiction with Proposition 3.2, because it would entail the density falling strictly below c_0.

4 Proof of Theorem 1.4

In what follows we assume, as in Sect. 2, that $p = 1$, the extension to arbitrary $p \geq 1$ being straightforward.

We begin by specifying the random walk $(X(t))_{t\geq 0}$ on $\mathbb{Z}^4$ defined by

$$X(t) = S(t) + e_1 N(t)$$

with $(S(t))_{t\geq 0}$ denoting a simple random walk on $\mathbb{Z}^4$, $(N(t))_{t\geq 0}$ a rate 1 Poisson process and $e_1 = (1,0,0,0)$ the first unit vector in $\mathbb{Z}^4$.

Thus our random walk $(X(t))_{t\geq 0}$ is highly transient but its symmetrization is a mean zero random walk and by the local central limit theorem, we have

$$\int_0^\infty t p_t^{(s)}(0,0)\,\mathrm{d}t = \infty,$$

where $p_t(\cdot,\cdot)$ is the semigroup associated to $(X(t))_{t\geq 0}$.

It remains to show that $\lambda_p(\kappa) < \gamma$ for all $\kappa \in [0, \infty)$. Our approach is modeled on the proof of the first part of Theorem 1.3. We wish again to pick *bad environment set* B_E associated to the ξ-process and *bad random walk set* B_W associated to the random walk X^κ so that

$$\mathbb{E}\left(\exp\left[\gamma\int_0^n \xi_s(X^\kappa(s))\,\mathrm{d}s\right]\right) \tag{20}$$

$$\leq (\mathbb{P}(B_E)+\mathbb{P}(B_W))\,\mathrm{e}^{\gamma n} + \mathbb{E}\left(\mathbb{1}_{B_E^{\mathrm{c}}\cap B_W^{\mathrm{c}}}\exp\left[\gamma\int_0^n \xi_s(X^\kappa(s))\,\mathrm{d}s\right]\right) \tag{21}$$

with, for some $0 < \delta < 1$,

$$\mathbb{P}(B_E) \leq \mathrm{e}^{-\delta n}, \qquad \mathbb{P}(B_W) \leq \mathrm{e}^{-\delta n}, \tag{22}$$

and, automatically from the definition of B_E and B_W,

$$\int_0^n \xi_s(X^\kappa(s))\,\mathrm{d}s \leq n(1-\delta) \qquad \text{on } B_E^{\mathrm{c}}\cap B_W^{\mathrm{c}} \tag{23}$$

(as in the proof of Theorem 1.3). Since, combining (20)–(23), we obtain

$$\lim_{n\to\infty}\frac{1}{n}\log\mathbb{E}\left(\exp\left[\gamma\int_0^n \xi_s(X^\kappa(s))\,\mathrm{d}s\right]\right) < \gamma,$$

it is enough to prove (22). All of this has been done in the proof of Theorem 1.3 in a different situation. The major difference is that we need to modify the collection of skeletons used.

Lemma 4.1. *Let $X^\kappa(\cdot)$ be a speed κ simple random walk in four dimensions. Fix $M \in \mathbb{N}\setminus\{1\}$. There exists $c > 0$ so that for M large and all n, outside of an e^{-cn} probability event, there exists $0 \leq i_1 < i_2 < \cdots < i_{n/2M} \leq n$ so that*

$$X^\kappa_{(1)}(i_jM + kM) - X^\kappa_{(1)}(i_jM) > -\frac{kM}{2}, \qquad j \in \{1,\ldots,n/(2M)\},\ k \geq 0, \tag{24}$$

where $(X^\kappa_{(1)}(t))_{t\geq 0}$ denotes the first coordinate of $(X^\kappa(t))_{t\geq 0}$.

Proof. Define

$$\sigma_1 = \inf\left\{kM > 0\colon X^\kappa_{(1)}(kM) \leq -kM/2\right\}$$

and recursively

$$\sigma_{i+1} = \inf\big\{kM > \sigma_i \colon X^{\kappa}_{(1)}(kM) - X^{\kappa}_{(1)}(\sigma_i) \le -(kM - \sigma_i)/2\big\}.$$

Since the event

$$\{rM \le \sigma_1 < \infty\} \subset \cup_{k=r}^{\infty}\{X^{\kappa}_{(1)}(kM) \le -kM/2\},$$

we have easily that, for all r,

$$\mathbb{P}(rM \le \sigma_1 < \infty) \ \le \ \mathrm{e}^{-rMc}$$

for $c > 0$ not depending on n or M. If we now define

$$\tau_1 = \inf\big\{kM > 0 \colon X^{\kappa}_{(1)}(jM) - X^{\kappa}_{(1)}(kM) > -(j-k)M/2 \ \ \forall j > k\big\}$$

and recursively

$$\tau_{i+1} = \inf\big\{kM > \tau_i \colon X^{\kappa}_{(1)}(jM) - X^{\kappa}_{(1)}(kM) > -(j-k)M/2 \ \ \forall j > k\big\},$$

it is easily seen that

$$\begin{aligned}
\mathbb{P}(\tau_1 \ge rM) &\le \mathbb{P}(\exists 1 \le k \le r \colon rM \le \sigma_k < \infty)\\
&\le \sum_{k=1}^{r-1} \sum_{0<x_1<\dots<x_k<r} \mathbb{P}\big(\sigma_i = x_i M \ \ \forall i \le k,\ rM \le \sigma_{k+1} < \infty\big)\\
&\le \sum_{k=1}^{r-1} \sum_{0<x_1<\dots<x_k<r} \mathrm{e}^{-rMc} \le \mathrm{e}^{-rMc} 2^r
\end{aligned}$$

which is less than $\mathrm{e}^{-rMc/2}$ if M is fixed sufficiently large. We have

- $(\tau_{i+1} - \tau_i)_{i \ge 1}$ are i.i.d. (this follows from Kuczek's argument (see [9])).
- Provided M has been fixed sufficiently large for each integer $r \ge 1$, $\mathbb{P}(\tau_{i+1} - \tau_i \ge rM) \le \mathrm{e}^{-rMc/4}$. This follows from the fact that random variable $\tau_{i+1} - \tau_i$ is simply the random variable τ_1 conditioned on an event of probability at least $1/2$ (provided M was fixed large).

Thus by elementary properties of geometric random variables we have

$$\mathbb{P}(\tau_{n/2M} > n) \le \mathbb{P}\left(Y \ge \frac{n}{2M}\right) \le \mathrm{e}^{-cn}$$

for $Y \sim B\left(\frac{n}{M}, \mathrm{e}^{-cM/4}\right)$, $c > 0$, and M large. This completes the proof of the lemma. □

Given that the path of the random walk satisfies the condition of this lemma, we call the (not uniquely defined) points $i_1, i_2, \dots$ *regular* points.

Given this result, we consider the M-skeleton induced by the values $\{X^\kappa(jM), 0 \le j \le n/M\}$, discretized via spatial cubes of length $M/8$ (rather than $2M$ as in the proof of Theorem 1.3). It is to be noted that if $(X^\kappa(t))_{0\le t\le n}$ satisfies the claim for Lemma 4.1 and $y_0 := 0, y_1, y_2, \dots, y_{n/M}$ with $y_k \in \mathbb{Z}^4$, $0 \le k \le n/M$, is its M-skeleton, namely,

$$X^\kappa(kM) \in C_{y_k} := \prod_{j=1}^{4} \left[y_k^{(j)} \frac{M}{8}, (y_k^{(j)} + 1)\frac{M}{8} \right), \qquad 0 \le k \le n/M, \tag{25}$$

where $y_k^{(j)}$ denotes the jth coordinate of y_k (we suppose without loss of generality that M is a multiple of 8), then, by (24) and (25), we must have

$$y_{i_j}^{(1)} - 4k \le y_{i_j+k}^{(1)} + 1.$$

In particular, we must have

$$y_{i_{j'}}^{(1)} - 4(i_j - i_{j'}) \le y_{i_j}^{(1)} + 1 \qquad \forall i_{j'} < i_j. \tag{26}$$

In the following we modify the definition of appropriate skeletons by adding in the requirement that the skeleton must possess at least $n/2M$ indices $i_1, i_2, \dots, i_{n/2M}$ with the corresponding y_{i_j} satisfying (26). We note that the resizing of the cubes makes the notion of acceptability a little more stringent but does not change the essentials.

Remark first that Lemma 2.5 is still valid in our new setting. Lemma 4.1 immediately gives that with this new definition, Lemma 2.4 remains true. Of course since this definition is more restrictive we have

$$|\Xi_A| \le K^{n/M}$$

for K as in Lemma 2.1.

In fact in our program all that remains to do, i.e., in any substantive way different from the proof of Theorem 1.3, is to give a bound on the probability of B_E for appropriate B_E. This is the content of the lemma below (analogous to Lemma 2.2). Given this lemma, we can then proceed exactly as with the proof of Theorem 1.3.

Lemma 4.2. *For any skeleton $(y_k)_{0\le k\le n/M}$ in Ξ_A, the probability that ξ is not good for $(y_k)_{0\le k\le n/M}$, i.e.,*

$$\nexists \frac{n}{4M} \text{ indices } 1 \le j \le \frac{n}{M} : \xi_s(z) = 0 \ \ \forall s \in [jM, jM+1] \text{ for some } z \in C_{y_j},$$

is less than $(4K)^{-n/M}$.

Proof. We note that proving the analogous result for Theorem 1.3, we did not need our skeleton to be in Ξ_A, the proof worked over any skeleton. For us however it is vital that our skeleton satisfies (26).

We consider a skeleton in Ξ_A. Let the first $n/2M$ regular points of our skeleton be $i_1, i_2, \ldots, i_{n/2M}$. For each $1 \le i_j \le n/2M$, we choose R points

$$x_1^{i_j}, \ldots, x_R^{i_j} \in C_{y_{i_j}}$$

so spread out that for random walks $(X(t))_{t \ge 0}$ as in (4) beginning at the points $x_k^{i_j}$, $k = 1, \ldots, R$, the chance that two of them meet is less than $0 < \epsilon \ll 1$.

Now, consider $i_{j'} < i_j$ and the probability that a random walk starting at $(x_k^{i_j}, i_j M)$ meets a random walk starting at $(x_{k'}^{i_{j'}}, i_{j'} M)$ satisfies the following lemma.

Lemma 4.3. *For $i_{j'} < i_j$, there exits $K > 0$ such that*

$$\mathbb{P}\left(X^{x_k^{i_j}, i_j M} \text{ meets } X^{x_{k'}^{i_{j'}}, i_{j'} M} \right) \le \frac{K}{M^2 (i_j - i_{j'})^2}.$$

Proof. The important point is that since our skeleton is in Ξ_A,

$$\left(x_{k'}^{i_{j'}} \right)^{(1)} \le \left(x_k^{i_j} \right)^{(1)} + (i_j - i_{j'}) \frac{M}{2} + \frac{M}{4}, \tag{27}$$

and so we have

$$\begin{aligned}
&\mathbb{P}\left(X^{x_k^{i_j}, i_j M} \text{ meets } X^{x_{k'}^{i_{j'}}, i_{j'} M} \right) \\
&\quad \le \mathbb{P}\Bigg(\left(X^{x_k^{i_j}, i_j M} \right)^{(1)} ((i_j - i_{j'})M) \ge (x_k^{i_j})^{(1)} + (i_j - i_{j'}) \frac{3M}{4}, \\
&\qquad\qquad X^{x_k^{i_j}, i_j M} \text{ meets } X^{x_{k'}^{i_{j'}}, i_{j'} M} \Bigg) \\
&\qquad + \mathbb{P}\left(\left(X^{x_k^{i_j}, i_j M} \right)^{(1)} ((i_j - i_{j'})M) \le (x_k^{i_j})^{(1)} + (i_j - i_{j'}) \frac{3M}{4} \right).
\end{aligned}$$

By standard large deviation bounds,

$$\mathbb{P}\left(\left(X^{x_k^{i_j}, i_j M} \right)^{(1)} ((i_j - i_{j'})M) \le (x_k^{i_j})^{(1)} + (i_j - i_{j'}) \frac{3M}{4} \right) \le \mathrm{e}^{-CM(i_j - i_{j'})},$$

for some universal $C \in (0, \infty)$. For the other term, we have the following lemma.

Lemma 4.4. *For two independent processes* $X = (X(t))_{t\geq 0}$ *and* $Y = (Y(t))_{t\geq 0}$ *with* $X(0) = x \in \mathbb{Z}^4$ *and* $Y(0) = y \in \mathbb{Z}^4$*, the probability that* X *ever meets* Y *is bounded by* $K/\|x-y\|_\infty^2$.

Proof. $X - Y$ is not exactly a simple random walk, but it is a symmetric random walk and so local CLT gives appropriate random walks bounds (see, e.g., [10]). □

From this and inequality (27), we have

$$\mathbb{P}\Bigg(X^{x_k^{i_j},i_jM} \text{ meets } X^{x_{k'}^{i_{j'}},i_{j'}M}\Bigg|\left(X^{x_k^{i_j},i_jM}\right)^{(1)}((i_j - i_{j'})M)$$

$$\geq \left(x_k^{i_j}\right)^{(1)} + (i_j - i_{j'})\frac{3M}{4}\Bigg) \leq \frac{K}{M^2(i_j - i_{j'})^2}.$$

□

Thus, for any R large but fixed, we can choose M so that for all skeletons in Ξ_A and each $i_{j'}$, we have

$$\sum_{i_j<i_{j'}}\sum_{k,k'}\mathbb{P}\left(X^{x_k^{i_j},i_jM} \quad \text{meets} \quad X^{x_{k'}^{i_{j'}},i_{j'}M}\right) \leq \frac{R^2K}{M^2}\sum_{r=1}^{+\infty}\frac{1}{r^2}$$

$$\leq \frac{R^2K'}{M^2} < \epsilon^2$$

with M chosen sufficiently large, which is analogous to (13). From this point on, the rest follows as for the proof of Lemma 2.2. □

References

1. Bramson, M., Cox, J.T., Griffeath D.: Occupation time large deviations of the voter model. Probab. Theory Relat. Fields **77**, 401–413 (1988)
2. Carmona, R.A., Molchanov, S.A.: Parabolic Anderson Problem and Intermittency. AMS Memoir 518. American Mathematical Society, Providence, RI (1994)
3. Clifford, P., Sudbury, A.: A model for spatial conflict. Biometrika **60**, 581–588 (1973)
4. Durrett, R.: Probability: Theory and Examples, 3rd edn. Thomson, Brooks/Cole, Duxbury Advanced Series (2005)
5. Gärtner, J., den Hollander, F.: Intermittency in a catalytic random medium. Ann. Probab. **34**, 2219–2287 (2006)
6. Gärtner, J., den Hollander, F., Maillard, G.: Intermittency on catalysts: Symmetric exclusion. Electron. J. Probab. **12**, 516–573 (2007)
7. Gärtner, J., den Hollander, F., Maillard, G.: Intermittency on catalysts: Voter model. Ann. Probab. **38**, 2066–2102 (2010)
8. Holley, R.A., Liggett, T.M.: Ergodic theorems for weakly interacting infinte systems and the voter model. Ann. Probab. **3**, 643–663 (1975)

9. Kuczek, T.: The central limit theorem for the right edge of supercritical oriented percolation. Ann. Probab. **17**, 1322–1332 (1989)
10. Lawler, G.F., Limic, V.: Random Walk: A Modern Introduction. Cambridge Studies in Advanced Math., vol. 123. Cambridge University Press, Cambridge (2010)
11. Liggett, T.M.: Interacting Particle Systems. Grundlehren der Mathematischen Wissenschaften, vol. 276. Springer, New York (1985)
12. Spitzer, F.: Principles of Random Walk, 2nd edn. Springer, Berlin (1976)

Precise Asymptotics for the Parabolic Anderson Model with a Moving Catalyst or Trap

Adrian Schnitzler and Tilman Wolff

Abstract We consider the solution $u\colon [0,\infty) \times \mathbb{Z}^d \to [0,\infty)$ to the parabolic Anderson model, where the potential is given by $(t,x) \mapsto \gamma\delta_{Y_t}(x)$ with Y a simple symmetric random walk on $\mathbb{Z}^d$. Depending on the parameter $\gamma \in [-\infty,\infty)$, the potential is interpreted as a randomly moving catalyst or trap.

In the trap case, i.e., $\gamma < 0$, we look at the annealed time asymptotics in terms of the first moment of u. Given a localized initial condition, we derive the asymptotic rate of decay to zero in dimensions 1 and 2 up to equivalence and characterize the limit in dimensions 3 and higher in terms of the Green's function of a random walk. For a homogeneous initial condition, we give a characterisation of the limit in dimension 1 and show that the moments remain constant for all time in dimensions 2 and higher.

In the case of a moving catalyst ($\gamma > 0$), we consider the solution u from the perspective of the catalyst, i.e., the expression $u(t, Y_t + x)$. Focusing on the cases where moments grow exponentially fast (that is, γ sufficiently large), we describe the moment asymptotics of the expression above up to equivalence. Here, it is crucial to prove the existence of a principal eigenfunction of the corresponding Hamilton operator. While this is well-established for the first moment, we have found an extension to higher moments.

AMS 2010 Subject Classification. Primary 60K37, 82C44; Secondary 60H25.

A. Schnitzler (✉)
Technische Universität Berlin, Institut für Mathematik, Straße des 17. Juni 136, 10623 Berlin, Germany
e-mail: schnitzler@math.tu-berlin.de

T. Wolff
Weierstrass Institute for Applied Analysis and Stochastics, Mohrenstraße 39, 10117 Berlin, Germany
e-mail: wolff@wias-berlin.de

J.-D. Deuschel et al. (eds.), *Probability in Complex Physical Systems*, Springer Proceedings in Mathematics 11, DOI 10.1007/978-3-642-23811-6_4,

1 Introduction

The parabolic Anderson model (PAM) is the heat equation on the lattice with a random potential, given by

$$\begin{cases} \frac{\partial}{\partial t}u(t,x) = \kappa \Delta u(t,x) + \xi(t,x)u(t,x), & (t,x) \in (0,\infty) \times \mathbb{Z}^d, \\ u(0,x) = u_0(x), & x \in \mathbb{Z}^d, \end{cases} \tag{1}$$

where $\kappa > 0$ denotes a diffusion constant, u_0 a nonnegative function and Δ the discrete Laplacian, defined by

$$\Delta f(x) := \sum_{\substack{y \in \mathbb{Z}^d: \\ |x-y|=1}} [f(y) - f(x)], \qquad x \in \mathbb{Z}^d,\ f: \mathbb{Z}^d \to \mathbb{R}.$$

Furthermore, $\xi: [0,\infty) \times \mathbb{Z}^d \longrightarrow \mathbb{R}$ is a space- and time-dependent random potential. We deal with the special case that the potential is given by

$$\xi(t,x) = \gamma \delta_{Y_t}(x), \qquad (t,x) \in [0,\infty) \times \mathbb{Z}^d,$$

with a simple symmetric random walk Y with generator $\rho\Delta$ that starts in the origin and a parameter $\gamma \in [-\infty,\infty)$ called *coupling constant*. In this paper, we analyse the large time asymptotics after averaging over the potential which is usually referred to as *annealed asymptotics*. We denote expectation with respect to the potential ξ by $\langle \cdot \rangle$.

One possible interpretation of this system arises from chemistry. Here, $u(t,x)$ describes the concentration of reactant particles in a point x at time t in the presence of a randomly moving particle. In the case $\gamma < 0$, the particle acts as a decatalyst (or trap) that kills reactant particles with rate $-\gamma$ at its position. In the case of positive γ, we consider a catalyst particle that causes reactants to multiply with rate γ. In both cases, $\langle u(t,x) \rangle$ is interpreted as the averaged concentration. For further interpretations and an overview over the PAM, see, for instance [2, 13, 18] and [12].

Annealed asymptotics in the case of a positive coupling constant γ have already been investigated in [5]. In the present work, we derive similar results with regard to the expression $\tilde{u}(t,x) := u(t, Y_t + x)$, which can be interpreted as the particle concentration in a neighbourhood of the catalyst. In addition to logarithmic asymptotics in terms of Lyapunov exponents, we derive asymptotics up to equivalence for most of the parameter choices where exponential growth is observed.

The case that γ is negative has to the best of our knowledge not been investigated so far. Its analysis relies on techniques quite different from those in the catalyst case as a functional analytic approach proves unfeasible here. We calculate moment limits dependent on the model parameters and, in the case of moment convergence towards zero, specify the convergence speed up to equivalence.

Whereas the PAM with time-independent potential or white-noise potential is well understood, some other time-dependent potentials have just been examined recently. In [6, 11], and [14], for instance, the authors investigate the case of infinitely many randomly moving catalysts. In [3], the authors deal with the case of finitely many catalysts, whereas the article [4] is dedicated to a model similar to the case of infinitely many moving traps. Further examples of time-dependent potentials can be found in [7, 8, 10, 17], and the recent survey [9]. Within these proceedings, [15, 16] and [19] deal with the parabolic Anderson model with time-independent potential.

In Sect. 2.1, we analyze the PAM with localized initial condition $u_0 = \delta_z$ and $\gamma < 0$. Let

$$M_z(t) := \Big\langle \sum_{x\in\mathbb{Z}^d} u(t,x)\Big\rangle, \qquad (t,z) \in [0,\infty) \times \mathbb{Z}^d,$$

denote the expected total mass of the system at time t if the solution is initially localized in z and the trap starts in the origin. We find

Theorem 1. *For $d = 1, 2$ and every $z \in \mathbb{Z}^d$,*

$$(i) \qquad M_z(t) \sim \tfrac{2}{\sqrt{\pi}} \tfrac{\sqrt{\kappa+\rho}}{-\gamma} t^{-\frac{1}{2}}, \qquad t \to \infty \qquad \text{for} \qquad d = 1;$$

$$(ii) \qquad M_z(t) \sim 4\pi \tfrac{\kappa+\rho}{-\gamma} (\log t)^{-1}, \qquad t \to \infty \qquad \text{for} \qquad d = 2,$$

and

Theorem 2. *For $d \geq 3$ and every $z \in \mathbb{Z}^d$,*

$$\lim_{t\to\infty} M_z(t) = 1 + \frac{\gamma}{\kappa + \rho - \gamma G_1(0)} G_1(z),$$

where G_κ denotes the Green's function of a random walk with generator $\kappa\Delta$.

Remark 1. Theorems 1 and 2 can be generalized to all initial conditions with compact support without much effort.

In Sect. 2.2, we analyze the case of a homogeneous initial condition $u_0 \equiv 1$. We find that in dimensions 2 and higher, the average total mass in each point remains constant for all t. This seems surprising since a symmetric random walk is recurrent in dimensions 1 and 2, but it follows by a rescaling argument and the fact that a Brownian motion is point recurrent only in dimension 1. In dimension 1, we give a representation of the asymptotic mass that depends on

$$a := \kappa/\rho,$$

but not on the strength of the potential γ. Let

$$m_x(t) := \langle u(t,x)\rangle, \qquad (t,x) \in [0,\infty) \times \mathbb{Z}^d,$$

denote the expected mass at time t in the lattice point x. The main results of this section are for $d = 1$,

Theorem 3. *For all $x \in \mathbb{Z}$,*

$$\lim_{t\to\infty} m_x(t) = 1 - \frac{1}{\pi}\int_0^1 \mathrm{d}s\, \frac{\sqrt{(1+a)(1-s)s + \frac{as^2}{1+a}}}{as^2\left(1+\frac{1}{(1+a)^2}\right)+s},$$

and for higher dimensions

Theorem 4. *For $d \geq 2$ and all $x \in \mathbb{Z}^d$,*

$$\lim_{t\to\infty} m_x(t) = 1.$$

Remark 2. Even though the formula in Theorem 3 looks quite clumsy, we find that $\lim_{t\to\infty} m_x(t)$ is decreasing in a. It tends to $1/2$ as a tends to zero and it tends to zero as a tends to infinity.

The third section is dedicated to analysing the leading order asymptotics of moments of the PAM solution from the perspective of the catalyst, i.e., we consider $\gamma > 0$ and the expression $\tilde{u}(t,x) := u(t, Y_t + x)$. For $p \in \mathbb{N}$ and $x = (x_1, \ldots, x_p) \in \mathbb{Z}^{\mathrm{pd}}$, we denote by

$$\tilde{m}_x^p(t) := \left\langle \prod_{i=1}^p \tilde{u}(t, x_i)\right\rangle$$

the p-th mixed moment at x. Moreover, introduce the p-th Hamilton operator on $l^\infty\left(\mathbb{Z}^{\mathrm{pd}}\right)$ by

$$\mathscr{H}^p := \mathscr{A}^p + \gamma V^{\mathrm{p}},$$

where the potential V^{p} is defined as $(V^{\mathrm{p}} f)(x) = \sum_{i=1}^p \delta_0(x_i) f(x)$, and $\mathscr{A}^p$ acts on $l^\infty\left(\mathbb{Z}^{\mathrm{pd}}\right)$ as

$$\mathscr{A}^p f(x) = \kappa \sum_{\substack{e\in\mathbb{Z}^{\mathrm{pd}}\\|e|=1}} (f(x+e) - f(x)) + \rho \sum_{\substack{e\in\mathbb{Z}^{d}\\|e|=1}} \left(f(x_1+e, \ldots, x_p+e) - f(x)\right).$$

Here, the first term represents the random movement of a collection of p independent random walks accounting for particle diffusion, and the second term arises from the shift by the position of the catalyst. By application of the well-established Feynman–Kac formula and calculating the generator of the resulting semigroup, we obtain the operator representation

$$\tilde{m}_x^p(t) = \left(\mathrm{e}^{t\mathscr{H}^p}\mathbb{1}\right)(x), \qquad x \in \mathbb{Z}^{\mathrm{pd}}. \tag{2}$$

This gives the connection between large time moment asymptotics and spectral analysis of the above Hamiltonian. Let us denote by λ_p the supremum of the l^2-spectrum of $\mathscr{H}^p$. Gärtner and Heydenreich [5] have shown that, for all $p \in \mathbb{N}$ and independently of $x \in \mathbb{Z}^d$,

$$\lim_{t\to\infty} (1/t) \log\langle u(t,x)^p \rangle = \lambda_p.$$

This limit is called p-th *Lyapunov exponent.* It can be shown by similar methods that just as well

$$\lim_{t\to\infty} (1/t) \log \tilde{m}_x^p (t) = \lambda_p, \qquad x \in \mathbb{Z}^{\mathrm{pd}}.$$

However, this does not enable us to derive large time asymptotics up to equivalence. Assuming the existence of an eigenfunction (v_p) corresponding to λ_p with certain properties, we could on a heuristic level decompose the right hand side of (2) as

$$\tilde{m}_x^p (t) = \mathrm{e}^{t\lambda_p} (\mathbb{1}, v_p)_{l^2} v_p(x) + o(\mathrm{e}^{t\lambda_p}), \qquad x \in \mathbb{Z}^{\mathrm{pd}}.$$

Our next main result contains criteria under which this is indeed possible.

Theorem 5. *Fix $\kappa > 0$, $\rho > 0$, and let one of the following conditions be satisfied:*

(i) $p = 1$ *or* $p = 2$, γ *large enough to ensure* $\lambda_p > 0$,
(ii) $p \in \mathbb{N}$, $\gamma > 4d\ (\kappa p + \rho)$.

Then, there exists a strictly positive and summable l^2-eigenfunction v_p of $\mathscr{H}^p$ corresponding to $\lambda_p > 0$. Assuming v_p to be normed in $l^2\left(\mathbb{Z}^{\mathrm{pd}}\right)$, the large time asymptotics of the p-th moment are given by

$$\tilde{m}_x^p (t) \sim \mathrm{e}^{\lambda_p t}\, v_p(x)\, \left\|v_p\right\|_1, \qquad t \to \infty, \tag{3}$$

where $\|\cdot\|_1$ denotes the norm in $l^1\left(\mathbb{Z}^{\mathrm{pd}}\right)$.

Remark 3. In the case $p = 1$, λ_p is strictly positive if and only if $1/\gamma < G_{\kappa+\rho}(0)$. In this case, the existence of a suitable eigenfunction has been known for quite a while, see e.g. [2] or [6].

Remark 4. For the cases $p \geq 2$, the condition $1/\gamma < pG_{\kappa+\rho}(0)$ is sufficient to have positive exponential growth (i.e., $\lambda_p > 0$). The condition $\gamma > 4d\ (\kappa p + \rho)$ also implies exponential growth of the p-th moment.

2 Moving Trap

This section is devoted to the case $\gamma < 0$. Our main proof tool is the Feynman–Kac representation of the solution u given by

$$u(t,x) = \mathbb{E}_x^X \exp\left\{\gamma \int_0^t \delta_{Y_{t-s}}(X_s)\, \mathrm{d}s\right\} u_0(X_t), \qquad (t,x) \in [0,\infty) \times \mathbb{Z}^d.$$

$\mathbb{E}_z^X$, $\mathbb{E}_z^Y$ and $\mathbb{E}_z^Z$ denote the expectation of a random walk with generator $\kappa\Delta$, $\rho\Delta$ and $(\kappa+\rho)\Delta$, respectively. The subscript z indicates the starting point and the corresponding probability measures will be denoted by $\mathbb{P}_z^{\cdot}$. By

$$p_t(z) = \mathbb{P}_0^X\left(X_{t/\kappa} = z\right) = \mathbb{P}_z^X\left(X_{t/\kappa} = 0\right),$$

we denote the transition probability of a random walk with generator Δ.

2.1 Localized Initial Condition

In this section, we prove Theorems 1 and 2. With the help of the Feynman–Kac representation and a time reversal we find that, for all $t \geq 0$ and $z \in \mathbb{Z}^d$,

$$M_z(t) = \mathbb{E}_z^X \mathbb{E}_0^Y \exp\left\{\gamma \int_0^t \delta_0(X_s - Y_s)\, \mathrm{d}s\right\} = \mathbb{E}_z^Z \exp\left\{\gamma \int_0^t \delta_0(Z_s)\, \mathrm{d}s\right\}.$$

2.1.1 Dimensions 1 and 2

We start with the dimensions where the random walk is recurrent.

Proof (Theorem 1). Using the semi-group representation of the resolvent $(\lambda - (\kappa+\rho)\Delta)^{-1}$, we find that

$$\begin{aligned} r_\lambda^{\kappa+\rho}(z) &:= \int_0^\infty \mathrm{d}t\, \mathrm{e}^{-\lambda t} \mathbb{E}_z^Z \exp\left\{\gamma \int_0^t \delta_0(Z_s)\, \mathrm{d}s\right\} \\ &= \frac{1}{\lambda} + \gamma\left(\int_0^\infty \mathrm{d}t\, \mathrm{e}^{-\lambda t} p_{(\kappa+\rho)t}(z)\right) r_\lambda^{\kappa+\rho}(0). \end{aligned}$$

This implies, for all $\lambda > 0$,

$$\int_0^\infty \mathrm{d}t\, \mathrm{e}^{-\lambda t} M_0(t) = \left(\lambda\left(1 - \gamma \int_0^\infty \mathrm{d}t\, \mathrm{e}^{-\lambda t} p_{(\kappa+\rho)t}(0)\right)\right)^{-1}.$$

Now, the claim for $z = 0$ follows by a standard Tauberian theorem. The case $z \neq 0$ follows due to the recurrence of Z. □

2.1.2 Dimensions 3 and Higher

A Tauberian theorem is not applicable in transient dimensions because here the expected number of particles does not converge to zero.

Proof (Theorem 2). Let

$$v(z) := \lim_{t\to\infty} M_z(t) = \mathbb{E}_z^Z \exp\left\{\gamma \int_0^\infty \delta_0(Z_s)\,\mathrm{d}s\right\}, \qquad z \in \mathbb{Z}^d.$$

Notice that the Green's function $G_{\kappa+\rho}$ is finite in transient dimensions and admits the following probabilistic representation.

$$G_{\kappa+\rho}(z) = \mathbb{E}_z^Z \int_0^\infty \delta_0(Z_s)\,\mathrm{d}s, \qquad z \in \mathbb{Z}^d.$$

That implies $v(z) \in (0,1)$ for all $z \in \mathbb{Z}^d$. Furthermore, we find that v is the unique solution to the following boundary problem

$$\begin{cases} (\kappa+\rho)\Delta v(z) + \gamma\delta_0(z)v(z) = 0, & z \in \mathbb{Z}^d, \\ \lim_{|z|\to\infty} v(z) = 1. \end{cases}$$

Hence, for all $z \in \mathbb{Z}^d$,

$$v(z) = 1 + \frac{\gamma}{\kappa+\rho-\gamma G_1(0)} G_1(z).$$

□

2.2 *Homogeneous Initial Condition*

In this section, we prove Theorems 3 and 4. For a homogeneous initial condition the Feynman–Kac representation yields, for all $t \geq 0$ and $x \in \mathbb{Z}^d$,

$$m_x(t) = \sum_{y\in\mathbb{Z}^d} \mathbb{E}_x^X \mathbb{E}_y^Y \exp\left\{\gamma \int_0^t \delta_0(X_s - Y_s)\,\mathrm{d}s\right\} \delta_0(Y_t).$$

2.2.1 Dimension 1

Let $\tau := \inf\{t \geq 0 \colon X_t = Y_t\} = \inf\{t \geq 0 \colon Z_t = 0\}$ be the first hitting time of X and Y. The density of τ with respect to $\mathbb{P}_z^Z$, $z \neq 0$, will be denoted by f_τ^z. To prove Theorem 3, we split $m_x(t)$ into two parts

$$\widetilde{m}_x(t) := \sum_{y \in \mathbb{Z}^d} \mathbb{E}_x^X \mathbb{E}_y^Y \mathbb{1}_{\tau > t} \delta_0(Y_t),$$

where X and Y have not met up to time t, and

$$\widehat{m}_x(t) := \sum_{y \in \mathbb{Z}^d} \mathbb{E}_x^X \mathbb{E}_y^Y \mathbb{1}_{\tau \leq t} \exp\left\{\gamma \int_0^t \delta_0\left(X_s - Y_s\right) \mathrm{d}s\right\} \delta_0\left(Y_t\right),$$

where they have already met by time t. The next proposition shows that $\widehat{m}_x$ is asymptotically negligible. Notice that this implies that there is no difference between the hard trap ($\gamma = -\infty$) and the soft trap ($\gamma \in (-\infty, 0)$) case because $\widetilde{m}_x$ does not depend on γ.

Proposition 1. *For all $x \in \mathbb{Z}$,*

$$\lim_{t \to \infty} \widehat{m}_x(t) = 0.$$

Proof. We can assume without loss of generality that $x = 0$ since Z is recurrent. Let $\sigma := \inf\{t \geq 0 \colon Z_t \neq Z_0\}$ be the the first jumping time of Z. Furthermore, for $t \geq 0$ let

$$w(t) := \mathbb{E}_0^Z \exp\left\{\gamma \int_0^t \delta_0\left(Z_s\right) \mathrm{d}s\right\} \delta_0\left(Z_t\right).$$

In a first step, we give an upper bound for the rate of decay of w. Let us abbreviate $\alpha := 2(\kappa + \rho)$. Using the strong Markov property of Z, we find

$$\begin{aligned} w(t) &= \mathbb{E}_0^Z \mathbb{1}_{\sigma > t} \mathrm{e}^{\gamma t} + \mathbb{E}_0^Z \mathbb{1}_{\sigma \leq t} \mathrm{e}^{\gamma \sigma} \left[\mathbb{E}_{Z_\sigma}^Z \exp\left\{\gamma \int_0^{t-s} \delta_0\left(Z_u\right) \mathrm{d}u\right\} \delta_0\left(Z_{t-s}\right)\right]_{s=\sigma} \\ &= \mathrm{e}^{(\gamma-\alpha)t} + \alpha \int_0^t \mathrm{d}s\, \mathrm{e}^{(\gamma-\alpha)s} \mathbb{E}_1^Z \mathbb{1}_{\tau \leq t} \left[\mathbb{E}_0^Z \exp\left\{\gamma \int_0^{t-s-r} \delta_0\left(Z_u\right) \mathrm{d}u\right\} \delta_0\left(Z_{t-s-r}\right)\right]_{r=\tau} \\ &= 1 - E\left(t\right) + \left(\frac{\alpha}{\alpha - \gamma} \left(\psi * f_\tau^1\right) * w\right)(t). \end{aligned}$$

Here, E denotes the distribution function of an exponentially distributed random variable with parameter $\alpha - \gamma$ and ψ denotes the corresponding density. By iteration we find that, for any $k \geq 1$,

$$w = (1-E) * \sum_{n=0}^{k} \left(\frac{\alpha}{\alpha-\gamma}\right)^n \psi^{*n} * f_\tau^{1*n} + \left(\frac{\alpha}{\alpha-\gamma}\right)^{(k+1)} \psi^{*(k+1)} * f_\tau^{1*(k+1)} * w.$$

Since there exists $C_1 > 0$ such that $f_\tau^z(t) \leq C_1 (1+t)^{-3/2}$ for all $z \neq 0$ and $t > 0$, we see that asymptotically

$$\begin{aligned} w(t) &\sim \left((1-E) * \Big(\sum_{n=0}^{\infty} \left(\frac{\alpha}{\alpha-\gamma}\right)^n \psi^{*n} * f_\tau^{1*n} \Big) \right)(t) \\ &\leq (1-E(t)) * \left(C_1 (1+t)^{-3/2} \right) = C_2 (1+t)^{-3/2}, \qquad t \to \infty, \end{aligned}$$

where C_2 is a positive constant. Let $Z^{(1)} := X - Y$ and $Z^{(2)} := X + Y$. Then it follows by Hölder's inequality that

$$\begin{aligned} \widehat{m}_0(t) &= \sum_{z,y\in\mathbb{Z}} \mathbb{E}_{-y}^{Z^{(1)}}, \mathbb{E}_y^{Z^{(2)}} \mathbb{1}_{\tau\leq t} \exp\left\{ -\gamma \int_0^t \delta_0\left(Z_s^{(1)}\right) \mathrm{d}s \right\} \delta_z\left(Z_t^{(1)}\right) \delta_z\left(Z_t^{(2)}\right) \\ &\leq \sum_{z,y\in\mathbb{Z}} \left(\mathbb{E}_{-y}^{Z^{(1)}} \mathbb{1}_{\tau\leq t} \exp\left\{ -\frac{3}{2}\gamma \int_0^t \delta_0\left(Z_s^{(1)}\right) \mathrm{d}s \right\} \delta_z\left(Z_t^{(1)}\right) \right)^{2/3} \\ &\quad \times \left(\mathbb{E}_y^{Z^{(2)}} \delta_z(Z_t^{(2)}) \right)^{1/3}. \end{aligned}$$

For $t \geq 0$ let

$$h(t) := \mathbb{E}_0^{Z^{(1)}} \exp\left\{ -\frac{3}{2}\gamma \int_0^t \delta_0\left(Z_s^{(1)}\right) \mathrm{d}s \right\} \delta_0\left(Z_t^{(1)}\right).$$

Obviously h admits the same asymptotic behaviour as w. Fix $K > 0$. The strong Markov property and the central limit theorem yield

$$\begin{aligned} \widehat{m}_0(t) &\leq \sum_{z,y\in\mathbb{Z}} \left(\mathbb{E}_{-y}^{Z^{(1)}} \mathbb{1}_{\tau\leq t} \left[\mathbb{E}_z^{Z^{(1)}} \mathbb{1}_{\tilde{\tau}\leq t-s} \left[h(t-s-r) \right]_{r=\tilde{\tau}} \right]_{s=\tau} \right)^{2/3} (p_{2(\kappa+\rho)t}(y,z))^{1/3} \\ &\sim \sum_{z,y\leq|K|\sqrt{t}} \left(\mathbb{E}_{-y}^{Z^{(1)}} \mathbb{1}_{\tau\leq t} \left[\mathbb{E}_z^{Z^{(1)}} \mathbb{1}_{\tilde{\tau}\leq t-s} \left[h(t-s-r) \right]_{r=\tilde{\tau}} \right]_{s=\tau} \right)^{2/3} \\ &\quad \times \left(p_{2(\kappa+\rho)t}(y,z) \right)^{1/3} \leq C K^2 t^{-1/6} \overset{t\to\infty}{\longrightarrow} 0. \end{aligned}$$

Here, C is a positive constant. This proves the claim . □

Now we show what $\widetilde{m}_x$ asymptotically looks like. Recall that $a = \kappa/\rho$.

Proposition 2. *For all $x \in \mathbb{Z}$,*

$$\lim_{t\to\infty} \widetilde{m}_x(t) = 1 - \frac{1}{\pi}\int_0^1 \mathrm{d}s \, \frac{\sqrt{(1+a)(1-s)s + \frac{as^2}{1+a}}}{as^2\left(1+\frac{1}{(1+a)^2}\right)+s}.$$

Proof. Because of the strong Markov property of X and Y, we find

$$\begin{aligned}\widetilde{m}_x(t) &= \sum_{y\in\mathbb{Z}} \mathbb{E}_x^X \mathbb{E}_y^Y \delta_0(Y_t) - \sum_{y\in\mathbb{Z}} \mathbb{E}_x^X \mathbb{E}_y^Y \mathbb{1}_{\tau\le t}\delta_0(Y_t)\\ &= 1 - \sum_{y\in\mathbb{Z}} \mathbb{E}_x^X \mathbb{E}_y^Y \mathbb{1}_{\tau\le t}[\mathbb{E}_{Y_\tau}^X \mathbb{E}_{Y_\tau}^Y \delta_0(Y_{t-s})]_{s=\tau}\\ &= 1 - \sum_{y\in\mathbb{Z}} \mathbb{E}_x^X \mathbb{E}_y^Y \mathbb{1}_{\tau\le t}\, p_{\rho(t-\tau)}(Y_\tau).\end{aligned}$$

It follows by Donsker's invariance principle that

$$\lim_{t\to\infty}\sum_{y\in\mathbb{Z}} \mathbb{E}_x^X \mathbb{E}_y^Y \mathbb{1}_{\tau\le t}\, p_{\rho(t-\tau)}(Y_\tau) = \int_{-\infty}^{\infty} \mathrm{d}y\, \mathbb{E}_0^{W^{(1)}}\mathbb{E}_0^{W^{(2)}} \mathbb{1}_{\tau_y^{(W)}\le 1}\, p^{(\mathrm{G})}_{\rho(1-\tau_y^{(W)})}\left(W^{(2)}_{\tau_y^{(W)}}\right).$$

Here $W^{(1)}$ and $W^{(2)}$ denote two independent Brownian motions that start in the origin with variance 2κ and 2ρ, respectively. Their expectations are denoted by $\mathbb{E}_0^{W^{(1)}}$ and $\mathbb{E}_0^{W^{(2)}}$, respectively. Moreover, $\tau_y^{(W)} := \inf\{t > 0 \colon W_t^{(1)} - W_t^{(2)} = y\}$ and $p_s^{(\mathrm{G})}$ denotes a Gaussian density with variance $2s$.

Indeed, the application of Donsker's invariance principle is not trivial because we have to sum over all $x \in \mathbb{Z}$, where it cannot be applied uniformly.

Let $W^{(-)} := W^{(1)} - W^{(2)}$, $W^{(+)} := W^{(1)} + \frac{\kappa}{\rho}W^{(2)}$ and $\tau_y^{(-)} := \inf\{t \ge 0\colon W_t^{(-)} = y\}$. Notice that $W^{(-)}$ and $W^{(+)}$ are independent. It follows that

$$\begin{aligned}&\int_{-\infty}^{\infty} \mathrm{d}y\, \mathbb{E}_0^{W^{(1)}}\mathbb{E}_0^{W^{(2)}} \mathbb{1}_{\tau_y^{(W)}\le 1}\, p^{(\mathrm{G})}_{\rho\left(1-\tau_y^{(W)}\right)}\left(W^{(2)}_{\tau_y^{(W)}}\right)\\ &= \int_{-\infty}^{\infty} \mathrm{d}y\, \mathbb{E}_0^{W^{(-)}}\mathbb{E}_y^{W^{(+)}} \mathbb{1}_{\tau_y^{(-)}\le 1}\, p^{(\mathrm{G})}_{\rho\left(1-\tau_y^{(-)}\right)}\left(-\frac{\rho}{\kappa+\rho}W^{(+)}_{\tau_y^{(-)}}\right)\\ &= \int_{-\infty}^{\infty} \mathrm{d}y\, \mathbb{E}_0^{W^{(-)}} \mathbb{1}_{\tau_y^{(-)}\le 1}\, p^{(\mathrm{G})}_{\rho\left(1-\tau_y^{(-)}\right)+\frac{\kappa\rho^2\tau_y^{(-)}}{(\kappa+\rho)^2}}(y)\end{aligned}$$

$$= \int_{-\infty}^{\infty} \mathrm{d}y \int_{0}^{1} \mathrm{d}s \, \frac{|y| \exp\left\{-\frac{y^2}{2(\kappa+\rho)s}\right\}}{s\sqrt{2\pi(\kappa+\rho)s}} \, \frac{\exp\left\{-\frac{y^2}{2\rho(1-s)+\frac{2\kappa\rho^2 s}{(\kappa+\rho)^2}}\right\}}{\sqrt{2\pi\left[2\rho(1-s)+\frac{2\kappa\rho^2 s}{(\kappa+\rho)^2}\right]}}$$

$$= \frac{1}{\pi} \int_{0}^{1} \mathrm{d}s \, \frac{\sqrt{(\kappa+\rho)\rho(1-s)s + \frac{\kappa\rho^2 s^2}{(\kappa+\rho)}}}{\kappa^2 s^2 + \rho s + \frac{\kappa\rho^2 s^2}{(\kappa+\rho)^2}}.$$

Now the claim follows by substituting a. □

Theorem 3 follows immediately from Propositions 1 and 2.

2.2.2 Dimensions 2 and Higher

In dimensions 2 and higher, we find that asymptotically the expected mass remains constant because a Brownian motion is point recurrent only in dimension 1.

Proof (Theorem 4). Let $\tau_\varepsilon^{(Z)} := \inf\{t \geq 0\colon Z_t \in B_\varepsilon(0)\}$ be the first time that the process Z hits the centered ball $B_\varepsilon(0)$ with radius $\varepsilon > 0$, and let

$$\overline{m}_x(t) := \sum_{y \in \mathbb{Z}^d} \mathbb{E}_x^X \mathbb{E}_y^Y \mathbb{1}_{\tau^{(Z)}_{\varepsilon\sqrt{t}} > t} \delta_0(Y_t).$$

Similarly as in the case d = 1 we find with the help of Donsker's invariance principle that

$$\lim_{t\to\infty} 1 - \overline{m}_0(t) = \lim_{\varepsilon\to 0} \int_{\mathbb{R}^d} \mathrm{d}x \, \mathbb{P}_x^W \left(\tau_\varepsilon^{(W)} \leq \frac{1}{2}\right).$$

However, for d $\geq$ 2 and $x \neq 0$,

$$\lim_{\varepsilon\to 0} \mathbb{P}_x^W \left(\tau_\varepsilon^{(W)} \leq \frac{1}{2}\right) = \mathbb{P}_x^W \left(\bigcap_{\varepsilon>0} \left\{\tau_\varepsilon^{(W)} \leq \frac{1}{2}\right\}\right) = \mathbb{P}_x^W \left(\tau_{0+}^{(W)} \leq \frac{1}{2}\right) = 0.$$

Hence, it follows by monotone convergence that $\lim_{t\to\infty} \overline{m}_0(t) = 1$ which implies that $\lim_{t\to\infty} m_x(t) = 1$ for all $x \in \mathbb{Z}^d$. □

3 Moving Catalyst

In this section, we stick to the homogeneous initial condition $u_0 \equiv 1$ and examine the case of a randomly moving catalyst, i.e., we consider $\gamma > 0$.

3.1 Spectral Properties of Higher-Order Anderson Hamiltonians

Throughout this section, we write $\lambda_p := \sup \sigma(\mathscr{H}^p)$ for all $p \in \mathbb{N}$. Considering the first Hamilton operator $\mathscr{H}^1$ given by

$$\mathscr{H}^1 := (\kappa + \rho)\,\Delta + \gamma\delta_0,$$

the existence of an eigenfunction $v_1 \in l^2(\mathbb{Z}^d)$ corresponding to its largest spectral value, provided that this value is greater than zero, has been widely known for some time. The following theorem extends this to the case $p = 2$ and constitutes the main statement of this section:

Theorem 6. *Assume $\lambda_2 > 0$. Then, λ_2 is isolated in the point spectrum of $\mathscr{H}^2$ with one-dimensional eigenspace. The corresponding eigenfunction may be chosen strictly positive.*

For a start, we restrict the operator to the subspace of component-wise symmetric functions

$$\mathfrak{S}^2 := \left\{ f \in l^2\left(\mathbb{Z}^{2d}\right) \mid f(x,y) = f(y,x) \quad \forall x,y \in \mathbb{Z}^{2d} \right\},$$

which is obviously closed in $l^2\left(\mathbb{Z}^{2d}\right)$. Recall the definition of the operators $\mathscr{A}^2$ and V^2 from Sect. 1 and define $\tilde{\mathscr{A}}^2$ and $\tilde{V}^2$ as their restrictions on the set $\mathfrak{S}^2$ above. In the same manner, we denote by $\tilde{\mathscr{H}}^2$ the restricted second-order Hamilton operator. The reader may easily retrace these operators are endomorphisms on $\mathfrak{S}^2$. In particular, $\tilde{\mathscr{H}}^2$ is a self-adjoint operator on the Hilbert space $\mathfrak{S}^2$, and it is essential that the supremum of its spectrum coincides with λ_2, which can be shown elementarily. Each eigenfunction of $\tilde{\mathscr{H}}^2$ corresponding to λ_2 is an eigenfunction of $\mathscr{H}^2$ as well. Moreover, we expect that an eigenfunction of $\mathscr{H}^2$ is, or at least could be chosen as, an element of $\mathfrak{S}^2$. In view of that, passing over to $\mathfrak{S}^2$ is just a natural approach. In the next step, we write $\tilde{\mathscr{A}}^2 + \gamma\tilde{V}^2$ rather than $\tilde{\mathscr{H}}^2$ in order to emphasize the dependence on the potential parameter γ, and we establish a further translation of the main task:

Lemma 1. *Suppose $\lambda > 0$. Then, the resolvent operator $\tilde{R}_\lambda := (\lambda - \tilde{\mathscr{A}}^2)^{-1}$ exists on $\mathfrak{S}^2$, and for all $\gamma > 0$, we have*

(i)

$$\lambda \in \sigma\left(\tilde{\mathscr{A}}^2 + \gamma\tilde{V}^2\right) \quad \Longleftrightarrow \quad \gamma^{-1} \in \sigma\left(\tilde{R}_\lambda\tilde{V}^2\right),$$

(ii)

$$\lambda = \sup \sigma\left(\tilde{\mathscr{A}}^2 + \gamma\tilde{V}^2\right) \quad \Longrightarrow \quad \gamma^{-1} = \sup \sigma\left(\tilde{R}_\lambda\tilde{V}^2\right).$$

Moreover, for $v \in \mathfrak{S}^2$ and $\gamma > 0$,

(iii)

$$\left(\tilde{\mathscr{A}}^2 + \gamma\tilde{V}^2\right)v = \lambda v \quad \Longleftrightarrow \quad \left(\tilde{R}_\lambda\tilde{V}^2\right)v = \frac{1}{\gamma}v.$$

Proof. A Fourier transform reveals that the spectrum of $\mathscr{A}^2$ is concentrated on the negative half-axis, thus $(\lambda - \mathscr{A}^2)^{-1}$ exists on $l^2\left(\mathbb{Z}^{2d}\right)$ for all $\lambda > 0$. In particular, it exists on $\mathfrak{S}^2$, and then it coincides with $(\lambda - \tilde{\mathscr{A}}^2)^{-1}$ as $\tilde{\mathscr{A}}^2$ is an endomorphism on $\mathfrak{S}^2$. Assertions (i) and (iii) follow by rearranging the equations considered and applying the resolvent operator. The second relation is shown using the Rayleigh-Ritz formula. □

As a next step, we introduce an operator $\tilde{T}_\lambda$ on $l^2\left(\mathbb{Z}^d\right)$ having the same spectrum and the same point spectrum as $\tilde{R}_\lambda \tilde{V}^2$ and that admits the decomposition $\tilde{T}_\lambda = \tilde{T}^{(1)} + \tilde{T}^{(2)}$. Here, $\tilde{T}^{(1)}$ is compact and the supremum of $\sigma(\tilde{T}^{(2)})$ is strictly smaller than the supremum of $\sigma(\tilde{T}_\lambda)$. Then, we use Weyl's theorem to obtain that the largest value in $\sigma(\tilde{T}_\lambda)$ belongs to the point spectrum $\sigma_{\rm p}(\tilde{T}_\lambda)$. The resolvent $R_\lambda := \left(\lambda - \mathscr{A}^2\right)^{-1}$ admits the representation

$$\left(R_\lambda f\right)(x_1, x_2) = \sum_{y_1, y_2 \in \mathbb{Z}^d} r_\lambda^{(2)}(y_1 - x_1, y_2 - x_2) f(y_1, y_2), \quad x_1, x_2 \in \mathbb{Z}^d,$$

where the resolvent kernel $r_\lambda^{(2)} : \mathbb{Z}^{2d} \to (0, \infty)$ is defined as

$$r_\lambda^{(2)}(x_1, x_2) := \int_0^\infty \mathrm{d}t\, \mathrm{e}^{-\lambda t}\, \mathbb{P}_0\left(Z_t = (x_1, x_2)\right), \quad x_1, x_2 \in \mathbb{Z}^d.$$

Here, Z is a random walk on $\mathbb{Z}^{2d}$ with generator $\mathscr{A}^2$. Then, we obtain

$$\left(R_\lambda V^2 f\right)(x_1, x_2) = \sum_{y_1, y_2 \in \mathbb{Z}^d} r_\lambda^{(2)}(y_1 - x_1, y_2 - x_2)\left[\delta_0(y_1) + \delta_0(y_2)\right] f(y_1, y_2). \tag{4}$$

If we assume $f \in \mathfrak{S}^2$, we get

$$\left(\tilde{R}_\lambda \tilde{V}^2 f\right)(x_1, x_2) = \sum_{y \in \mathbb{Z}^d} \left[r_\lambda^{(2)}(y - x_1, -x_2) + r_\lambda^{(2)}(-x_1, y - x_2)\right] f(y, 0),$$

for $x_1, x_2 \in \mathbb{Z}^d$, and in particular

$$\left(\tilde{R}_\lambda \tilde{V}^2 f\right)(x, 0) = \sum_{y \in \mathbb{Z}^d} \left[r_\lambda^{(2)}(y - x, 0) + r_\lambda^{(2)}(-x, y)\right] f(y, 0),$$

for $x \in \mathbb{Z}^d$. Let us therefore introduce the operator $\tilde{T}_\lambda = \tilde{T}^{(1)} + \tilde{T}^{(2)}$ on $l^2\left(\mathbb{Z}^d\right)$ with

$$\tilde{T}^{(1)} \tilde{f}(x) := \sum_{y \in \mathbb{Z}^d} r_\lambda^{(2)}(-x, y) \tilde{f}(y), \quad \tilde{T}^{(2)} \tilde{f}(x) := \sum_{y \in \mathbb{Z}^d} r_\lambda^{(2)}(y - x, 0) \tilde{f}(y), \; x \in \mathbb{Z}^d.$$

Both operators are apparently self-adjoint. The lemma below identifies the spectra and point spectra of $\tilde{T}_\lambda$ and $\tilde{R}_\lambda \tilde{V}^2$:

Lemma 2. *For all $\lambda > 0$,*

$$\sigma\left(\tilde{T}_\lambda\right) = \sigma\left(\tilde{R}_\lambda \tilde{V}^2\right), \quad \sigma_p\left(\tilde{T}_\lambda\right) = \sigma_p\left(\tilde{R}_\lambda \tilde{V}^2\right).$$

Proof. The crucial and least intuitive part is to show that

$$\mu - \tilde{T}_\lambda \text{ surjective} \Rightarrow \mu - \tilde{R}_\lambda \tilde{V}^2 \text{ surjective.} \tag{5}$$

All other implications are rather straightforward and we omit them for the sake of conciseness. Assume $\mu - \tilde{T}_\lambda$ is surjective and choose $g \in \mathfrak{S}^2$. Define $\tilde{g}(x) := g(x,0)$ for $x \in \mathbb{Z}^d$. There exists $\tilde{f} \in l^2\left(\mathbb{Z}^d\right)$ with $\left(\mu - \tilde{T}_\lambda\right)\tilde{f} = \tilde{g}$ by assumption. We define

$$f(x_1,x_2) := \tilde{f}(x_1)\delta_0(x_2) + \tilde{f}(x_2)\delta_0(x_1) - \delta_0(x_1)\delta_0(x_2)\tilde{f}(0), \qquad x_1, x_2 \in \mathbb{Z}^d$$

and then, for $x_1, x_2 \in \mathbb{Z}^d$,

$$F(x_1,x_2) := \begin{cases} f(x_1,x_2), & x_1 = 0 \text{ or } x_2 = 0; \\ \mu^{-1}\left(\tilde{R}_\lambda \tilde{V}^2 f(x_1,x_2) + g(x_1,x_2)\right), & \text{else.} \end{cases}$$

We realize that $F \in \mathfrak{S}^2$ and proceed by showing that F is the desired function satisfying $\left(\mu - \tilde{R}_\lambda \tilde{V}^2\right)F = g$. Note that $\tilde{R}_\lambda \tilde{V}^2 F(x_1,x_2) = \tilde{R}_\lambda \tilde{V}^2 f(x_1,x_2)$ for all $x_1, x_2 \in \mathbb{Z}^d$. In the first place, we have

$$\begin{aligned} \tilde{R}_\lambda \tilde{V}^2 F(x_1,0) &= \sum_{y \in \mathbb{Z}^d} \left[r_\lambda^{(2)}(y - x_1, 0) + r_\lambda^{(2)}(-x_1, y)\right] f(y,0) \\ &= \tilde{T}_\lambda \tilde{f}(x_1) = \mu \tilde{f}(x_1) - \tilde{g}(x_1) \\ &= \mu F(x_1,0) - g(x_1,0), \qquad x_1 \in \mathbb{Z}^d, \end{aligned} \tag{6}$$

and by symmetry $\tilde{R}_\lambda \tilde{V}^2 F(0,x_2) = \mu F(0,x_2) - g(0,x_2)$ for $x_2 \in \mathbb{Z}^d$. Moreover,

$$\begin{aligned} \mu F(x_1,x_2) - \tilde{R}_\lambda \tilde{V}^2 F(x_1,x_2) &= \tilde{R}_\lambda \tilde{V}^2 f(x_1,x_2) + g(x_1,x_2) - \tilde{R}_\lambda \tilde{V}^2 f(x_1,x_2) \\ &= \quad g(x_1,x_2), \qquad x_1, x_2 \in \mathbb{Z}^d, x_1, x_2 \neq 0. \end{aligned} \tag{7}$$

Equations (6) and (7) yield the desired result $(\mu - \tilde{R}_\lambda \tilde{V}^2)F = g$. Thus, we have shown (5). □

In the next step, we are able to calculate the supremum of the spectrum of $\tilde{T}^{(2)}$. Its value is given in terms of the Laplace resolvent kernel r_λ^κ defined by

$$r_\lambda^\kappa(x) := \int_0^\infty \mathrm{d}t\, \mathrm{e}^{-\lambda t} \mathbb{P}_0 (X_t = x), \quad x \in \mathbb{Z}^d, \tag{8}$$

with X a random walk on $\mathbb{Z}^d$ with generator $\kappa\Delta$.

Lemma 3. *We have* $\sup \sigma(\tilde{T}^{(2)}) = \|\tilde{T}^{(2)}\|_2 = r_\lambda^{\kappa+\rho}(0)$.

Proof. It will be sufficient to show that

$$\sup \left\{|\mu| : \mu \in \sigma(\tilde{T}^{(2)})\right\} = r_\lambda^{\kappa+\rho}(0). \tag{9}$$

The proof involves a Fourier transform (which we denote by $\mathscr{F}$) of the operator $\tilde{T}^{(2)}$. For $\hat{f} \in L^2\left([-\pi,\pi)^d\right)$, the transformed operator reads

$$\begin{aligned}
\hat{T}^{(2)} \hat{f}(l) &= (2\pi)^{-d} \sum_{x\in\mathbb{Z}^d} \mathrm{e}^{\mathrm{i}(l,x)} \sum_{y\in\mathbb{Z}^d} r_\lambda^{(2)}(y-x,0) \int_{([-\pi,\pi)^d)} \mathrm{d}k\, \mathrm{e}^{-\mathrm{i}(k,y)} \hat{f}(k) \\
&= (2\pi)^{-d} \sum_{y\in\mathbb{Z}^d} \mathrm{e}^{\mathrm{i}(l,y)} \int_{([-\pi,\pi)^d)} \mathrm{d}k\, \mathrm{e}^{-\mathrm{i}(k,y)} \hat{f}(k) \sum_{y\in\mathbb{Z}^d} \mathrm{e}^{\mathrm{i}(l,x-y)} r_\lambda^{(2)}(x-y,0) \\
&= \left(\mathscr{F}\mathscr{F}^{-1}\hat{f}\right)(l) \left(\sum_{z\in\mathbb{Z}^d} \mathrm{e}^{\mathrm{i}(l,z)} r_\lambda^{(2)}(z,0)\right), \qquad l \in [-\pi,\pi)^{\mathrm{d}}.
\end{aligned}$$

Thus, $\hat{T}^{(2)}$ is a multiplication operator and the multiplier

$$\hat{r}(l) := \sum_{z\in\mathbb{Z}^d} \mathrm{e}^{\mathrm{i}(l,z)} r_\lambda^{(2)}(z,0), \qquad l \in [-\pi,\pi)^{\mathrm{d}}$$

is obviously continuous. Hence, its spectrum is just the closure of the range of that multiplier. As each of the two components of a random walk on $\mathbb{Z}^{2d}$ with generator $\mathscr{A}^2$ is just a random walk on $\mathbb{Z}^d$ with generator $(\kappa+\rho)\Delta$, we have

$$\sup_{l\in[-\pi,\pi)^d} |\hat{r}(l)| = \hat{r}(0) = \sum_{z\in\mathbb{Z}^d} r_\lambda^{(2)}(z,0) = r_\lambda^{\kappa+\rho}(0),$$

and equation (9) follows by taking into account that the Fourier transform is an isometry. □

Lemma 4. *Suppose* $\lambda_2 = \sup\sigma\left(\mathscr{H}^2\right)$. *Then, the operator* $\tilde{T}_{\lambda_2} = \tilde{T}^{(1)} + \tilde{T}^{(2)}$ *has a strictly positive eigenfunction* $\tilde{v}$ *corresponding to its largest spectral value* $1/\gamma$. *This value is isolated in the spectrum.*

Proof. At first, we realize that $\tilde{T}^{(1)}$ belongs to the trace class as

$$\sum_{x\in\mathbb{Z}^d} \left(\tilde{T}^{(1)}\delta_x, \delta_x\right) < \int_0^\infty \mathrm{d}t\, \mathrm{e}^{-\lambda_2 t} < \infty,$$

and therefore $\tilde{T}^{(1)}$ is compact. Then, we explain why $\sup\sigma(\tilde{T}^{(1)} + \tilde{T}^{(2)}) > \sup\sigma(\tilde{T}^{(2)})$, which together with Weyl's theorem (see e.g., [20]) yields the existence of an eigenfunction. In the end, it remains to show that we may choose this eigenfunction strictly positive.

In order to show that $\sup\sigma(\tilde{T}^{(1)} + \tilde{T}^{(2)}) > \sup\sigma(\tilde{T}^{(2)})$, we recall that

$$\sup\sigma\left(\tilde{T}_{\lambda_2}\right) = \sup\sigma\left(\tilde{R}_{\lambda_2}\tilde{V}^2\right) = \frac{1}{\gamma}$$

by Lemmas 1 and 2, and the supremum of $\sigma(\tilde{T}^{(2)})$ is equal to $r_{\lambda_2}^{\kappa+\rho}(0)$ by Lemma 3. Therefore, it suffices to show that

$$\frac{1}{\gamma} > r_{\lambda_2}^{\kappa+\rho}(0). \tag{10}$$

Let $\lambda_1 := \sup\sigma\left(\mathscr{H}^1\right)$. In case $\lambda_1 > 0$, it is well-known that

$$\frac{1}{\gamma} = r_{\lambda_1}^{\kappa+\rho}(0),$$

compare, e.g., Carmona and Molchanov [2]. Moreover, as λ_1 and λ_2 are the exponential growth rates of the first and second moment of $\tilde{u}(t,x)$, Hölder's inequality yields

$$\lambda_1 \leq \frac{1}{2}\lambda_2,$$

thus a fortiori $\lambda_1 < \lambda_2$. As $\lambda \mapsto r_\lambda^{\kappa+\rho}(0)$ is strictly decreasing (see e.g., (8)),

$$\frac{1}{\gamma} = r_{\lambda_1}^{\kappa+\rho}(0) > r_{\lambda_2}^{\kappa+\rho}(0)$$

and we have shown (10) for the case $\lambda_1 > 0$. In case $\lambda_1 = 0$, we have

$$\frac{1}{\gamma} \geq G_{\kappa+\rho}(0),$$

and we arrive at (10) as $G_{\kappa+\rho}(0) > r_\lambda^{\kappa+\rho}(0)$ for all $\lambda > 0$. Weyl's theorem now states that $1/\gamma$ belongs to the discrete spectrum of $\tilde{T}_{\lambda_2}$ since $\tilde{T}^{(1)}$ is compact. Consequently, the value $1/\gamma$ is isolated in the point spectrum. Finally, we show that a corresponding eigenfunction $\tilde{v}$ may be chosen strictly positive. It suffices to show that $\tilde{T}_{\lambda_2}$ is positive in the sense that it maps nonnegative, non-zero functions to positive functions. Choose a nonnegative function f arbitrarily and assume $f(y_1) > 0$ for some $y_1 \in \mathbb{Z}^d$. Then, for all $x \in \mathbb{Z}^d$,

$$\tilde{T}_{\lambda_2} f(x) \geq \left[r^{(2)}_{\lambda_2}(y_1 - x, 0) + r^{(2)}_{\lambda_2}(-x, y_1) \right] f(y_1) > 0.$$

Consequently, we may choose $\tilde{v}$ strictly positive, and the proof is complete. □

Let us now prove the main result of this section:

Proof (Theorem 6). Let $\lambda_2 = \sup \sigma\left(\mathscr{H}^2\right)$. The preceding lemma states that there exists a strictly positive function $\tilde{v} \in l^2\left(\mathbb{Z}^d\right)$ with $T_{\lambda_2}\tilde{v} = (1/\gamma)\tilde{v}$. By Lemma 2, there exists $v_2 \in \mathfrak{S}^2$ with $\tilde{R}_{\lambda_2}\tilde{V}^2 v_2 = (1/\gamma)v_2$, as point spectra of both operators coincide. Naturally, v_2 is also an eigenfunction of $R_{\lambda_2}V^{\mathrm{p}}$ on $l^2\left(\mathbb{Z}^{2d}\right)$. We easily verify that

$$R_{\lambda_2}V^{\mathrm{p}} f > 0$$

for all nonnegative, non-zero $f \in l^2\left(\mathbb{Z}^{2d}\right)$, thus v_2 may be chosen strictly positive. Now Lemma 1 yields that v_2 is an eigenfunction of $\tilde{\mathscr{H}}^2$ and $\mathscr{H}^2$ corresponding to λ_2.

In order to show that its corresponding eigenspace is one-dimensional, let $(w_i)_{i\in I}$ represent an orthonormal basis of this eigenspace. The w_i are principal eigenfunctions of $R_{\lambda_2}V^{\mathrm{p}}$ that maps nonnegative, non-zero functions to positive functions. Hence we may choose all w_i strictly positive. As two strictly positive functions in $l^2\left(\mathbb{Z}^{\mathrm{pd}}\right)$ cannot be orthogonal, it follows that $|I| = 1$, i.e., the eigenspace corresponding to λ_2 is one-dimensional. □

We will additionally need that the largest eigenvalue λ_2 is isolated in the spectrum of $\tilde{\mathscr{H}}^2$ in order to describe the asymptotic moment behaviour:

Lemma 5. *The value λ_2 is isolated in $\sigma(\tilde{\mathscr{H}}^2)$.*

Proof. We know by Lemma 4 that $\gamma^{-1} = \sup \sigma(\tilde{T}_{\lambda_2})$ is an isolated eigenvalue, so there exists $\tilde{\delta} > 0$ with

$$\left[\gamma^{-1} - \tilde{\delta}, \gamma^{-1} + \tilde{\delta}\right] \cap \sigma\left(\tilde{T}_{\lambda_2}\right) = \left\{\gamma^{-1}\right\}.$$

Define now δ_1 small enough to ensure

$$\left\| \tilde{T}_{\lambda_2 - \varepsilon} - \tilde{T}_{\lambda_2} \right\|_2 < \tilde{\delta}/2 \text{ for all } \varepsilon \text{ with } 0 < \varepsilon \leq \delta_1.$$

It is quickly verified that this is always possible (e.g., by the mean value theorem). We can show with a similar argument that $\sup \sigma(\tilde{T}_\lambda)$ depends continuously on λ, making it possible to find δ_2 small enough to satisfy

$$\sup \sigma\left(\tilde{T}_{\lambda_2 - \varepsilon}\right) - \sup \sigma\left(\tilde{T}_{\lambda_2}\right) \leq \tilde{\delta}/2 \text{ for all } \varepsilon \text{ with } 0 < \varepsilon \leq \delta_2.$$

If we choose now $\varepsilon < \delta_1 \wedge \delta_2$, it follows

$$\gamma^{-1} \neq \sup \sigma\left(\tilde{T}_{\lambda_2 - \varepsilon}\right) \in \left[\gamma^{-1} - \tilde{\delta}/2, \gamma^{-1} + \tilde{\delta}/2\right],$$

and by Theorem 8 below, the interval $[\gamma^{-1} - \tilde{\delta}/2, \gamma^{-1} + \tilde{\delta}/2]$ contains exactly one element of the spectrum of $\tilde{T}_{\lambda_2} + (\tilde{T}_{\lambda_2-\varepsilon} - \tilde{T}_{\lambda_2}) = \tilde{T}_{\lambda_2-\varepsilon}$. Therefore, $\gamma^{-1} \in \rho(\tilde{T}_{\lambda_2-\varepsilon})$ and it follows that $\lambda_2 - \varepsilon \in \rho(\tilde{\mathscr{H}}^2)$ in the usual way by Lemmas 1 and 2. Thus, λ_2 is isolated in $\sigma(\tilde{\mathscr{H}}^2)$. □

Let us in the following present a sufficient condition for the existence of an eigenfunction with the desired properties that holds for general $p \in \mathbb{N}$:

Theorem 7. *Suppose $p \in \mathbb{N}$ and $\gamma > 4d\,(\kappa p + \rho)$. Then, $\lambda_p = \sup \sigma\,(\mathscr{H}^p)$ is positive, isolated in the spectrum and belongs to the point spectrum with one-dimensional eigenspace. The corresponding eigenfunction may be chosen strictly positive.*

The proof relies on the following theorem from perturbation theory of bounded operators. It describes the behaviour of an isolated eigenvalue under a bounded perturbation that is sufficiently small in a certain sense, see Birman and Solomjak [1], Chap. 9.4 for a proof.

Theorem 8. *Let T, S denote two self-adjoint operators on a Hilbert space. Suppose $\mu \in \sigma_p\,(T)$ with multiplicity r and*

$$[\mu - \varepsilon, \mu + \varepsilon] \cap \sigma\,(T) = \{\mu\}\,,$$

for some $\varepsilon > 0$. Moreover, assume $\sigma\,(S) \subset [\delta_1, \delta_2]$ for some $\delta_1 < \delta_2 \in \mathbb{R}$ with $\delta_2 - \delta_1 < \varepsilon$. Then, the set

$$[\mu + \delta_1, \mu + \delta_2] \cap \sigma\,(T + S)\,,$$

contains only isolated eigenvalues of $T + S$ whose sum of multiplicities equals r.

Proof (Theorem 7). We have $\mathscr{H}^p = \mathscr{A}^p + \gamma V^p$, and the idea is to understand the generator $\mathscr{A}^p$ as a perturbation of the potential γV^p. With increasing γ, the perturbation $\mathscr{A}^p$ remains relatively small, which allows an application of Lemma 8. As γV^p is a multiplication, its spectrum coincides with the essential range of the multiplier and we easily verify that γp is the largest eigenvalue of γV^p and has one-dimensional eigenspace. Moreover,

$$\sigma\,(\gamma V^p) \cap [\gamma p - \gamma, \gamma p + \gamma] = \{\gamma p\}\,,$$

and we may show by a Fourier transform that

$$\sigma\,(\mathscr{A}^p) \subset [-4d\,(\kappa p + \rho)\,, 0]\,.$$

Theorem 8 now yields that the set

$$\sigma\,(\mathscr{H}^p) \cap [\gamma p - \gamma, \gamma p]$$

contains exactly one element, which is an eigenvalue with multiplicity one. This element must be $\lambda_p = \sup \sigma(\mathscr{H}^p)$ due to the nonpositive definiteness of $\mathscr{A}^p$. It remains to show that the corresponding eigenfunction may be chosen strictly positive. To that purpose, we consider that v_p is also an eigenfunction of $e^{\mathscr{H}^p}$ corresponding to its largest eigenvalue e^{λ_p}. Employing the Feynman–Kac representation of this operator, we see that it maps nonnegative, non-zero functions to strictly positive functions. This means that all principal eigenfunctions are either strictly positive or strictly negative. □

3.2 Application to Annealed Higher Moment Asymptotics

A natural approach to more exact asymptotics of mixed moments, and the main idea proving Theorem 5, is to decompose the semigroup representation

$$\tilde{m}_x^p(t) = \left(e^{t\mathscr{H}^p}\mathbb{1}\right)(x).$$

Certainly we must consider the initial condition $\mathbb{1}$ as an appropriate limit of l^2-functions when attempting a rigorous proof. With $\{E_\mu | \mu \in \mathbb{R}\}$, the family of spectral projectors associated with $\mathscr{H}^p$, the spectral theorem for self-adjoint operators yields

$$\tilde{m}_x^p(t) = e^{\lambda_p t}\left(\mathbb{1}, v_p\right) v_p(x) + \int_{-\infty}^{\lambda_p-\varepsilon} e^{\mu t}\, dE_\mu(\mathbb{1})(x), \tag{11}$$

for some $\varepsilon > 0$ small enough. Here, $\lambda_p = \sup\sigma(\mathscr{H}^p)$ must be a positive eigenvalue with multiplicity one that is isolated in $\sigma(\mathscr{H}^p)$, and v_p is a strictly positive and l^2-normed eigenfunction corresponding to λ_p. Beyond that, we need that $\left(\mathbb{1}, v_p\right) < \infty$. If these requirements are met, we may asymptotically neglect the last term in (11). In order to prove Theorem 5, we need the following two auxiliary lemmas that are of pure technical nature and thus given without a proof. The first one enables us to approximate the homogeneous initial condition with l^2-functions. Let $\mathbf{Q}_t^p := \mathbb{Z}^{pd} \cap [-t,t]^{pd}$ for $t > 0$.

Lemma 6. *For all $x \in \mathbb{Z}^{pd}$,*

$$\left(e^{t\mathscr{H}^p}\mathbb{1}_{\mathbf{Q}_{t^2}^p}\right)(x) \sim \tilde{m}_x^p(t), \qquad t \to \infty.$$

The second auxiliary lemma ensures that the considered principal eigenfunctions are summable:

Lemma 7. *Suppose $\lambda_p = \sup\sigma(\mathscr{H}^p) > 0$ and there exists a corresponding eigenfunction $v_p \in l^2(\mathbb{Z}^{pd})$. Then, $v_p \in l^1(\mathbb{Z}^{pd})$.*

Let us now give a concise proof of the main statement:

Proof (Theorem 5). It suffices to show that, for all $x \in \mathbb{Z}^{\mathrm{pd}}$,

$$e^{-\lambda_p t} \tilde{m}_x^p(t) \longrightarrow \|v_p\|_1 v_p(x) \tag{12}$$

as t approaches infinity. The spectral representation of $e^{t\mathscr{H}^p}$ yields

$$e^{-t\lambda_p + t\mathscr{H}^p} \mathbb{1}_{\mathbf{Q}_{t^2}^p} = \left(\mathbb{1}_{\mathbf{Q}_{t^2}^p}, v_p\right) v_p + \int_{-\infty}^{\lambda_p - \varepsilon} e^{t(\mu - \lambda_p)} \, dE_\mu \left(\mathbb{1}_{\mathbf{Q}_{t^2}^p}\right),$$

for some $\varepsilon > 0$ as λ_p is isolated in the spectrum. For t large enough, the l^2-norm of the integral is roughly estimated from above by

$$\left\| \int_{-\infty}^{\lambda_p - \varepsilon} e^{t\mu - t\lambda_p} \, dE_\mu \left(\mathbb{1}_{\mathbf{Q}_{t^2}^p}\right) \right\|_2 \le e^{-t\varepsilon} \left\| \mathbb{1}_{\mathbf{Q}_{t^2}^p} \right\|_2 \le \left(2t^2\right)^{\frac{pd}{2}} e^{-t\varepsilon},$$

which means we may neglect this term and have

$$e^{-t\lambda_p + t\mathscr{H}^p} \mathbb{1}_{\mathbf{Q}_{t^2}^p} \xrightarrow{l^2} \|v_p\|_1 v_p, \qquad t \to \infty. \tag{13}$$

The limit (12) now follows by equation (13), Lemma 6 and the triangle inequality. This completes the proof. □

Acknowledgement The results of this paper have been derived in two theses under the supervision of Jürgen Gärtner whom we would like to thank for his invaluable support.

References

1. Birman, M.S., Solomjak, M.Z.: Spectral Theory of Selfadjoint Operators in Hilbert Space, Reidel, Dordrecht (1980)
2. Carmona, C., Molchanov, S.A.: Parabolic Anderson problem and intermittency. Mem. Amer. Math. Soc. **108**(3), (1994)
3. Castell, F., Gün, O., Maillard, G.: Parabolic Anderson model with a finite number of moving catalysts. In: Deuschel, J.-D., Gentz, B., König, W., von Renesse, M.-K., Scheutzow, M., Schmock, U. (eds.) Probability in Complex Physical Systems, Springer Proceedings in Mathematics, Vol. 11. Springer, Heidelberg (2012)
4. Drewitz, A., Gärtner, J., Ramírez, A., Sun, R.: Survival probability of a random walk among a Poisson system of moving traps. In: Deuschel, J.-D., Gentz, B., König, W., von Renesse, M.-K., Scheutzow, M., Schmock, U. (eds.) Probability in Complex Physical Systems, Springer Proceedings in Mathematics, Vol. 11. Springer, Heidelberg (2012)
5. Gärtner, J., Heydenreich, M.: Annealed asymptotics for the parabolic Anderson model with a moving catalyst. Stoch. Process. Appl. **116**(11), 1511–1529 (2006)
6. Gärtner, J., den Hollander, F.: Intermittency in a catalytic random medium. Ann. Probab. **34**(6), 2219–2287 (2006)
7. Gärtner, J., den Hollander, F., Maillard, G.: Intermittency on catalysts: Symmetric exclusion. Elec. J. Prob. **12**, 516–573 (2007)

8. Gärtner, J., den Hollander, F., Maillard, G.: Intermittency on catalysts: three-dimensional simple symmetric exclusion. Elec. J. Prob. **14**, 2091–2129 (2009)
9. Gärtner, J., den Hollander F., Maillard, G.: Intermittency on catalysts. In: Blath, J., Mörters, P., Scheutzow, M. (eds.) Trends in Stochastic Analysis, London Mathematical Society Lecture Note Series 353. pp. 235–248. Cambridge University Press, Cambridge (2009)
10. Gärtner, J., den Hollander, F., Maillard, G.: Intermittency on catalysts: voter model. Ann. Probab. **38**(5), 2066–2102 (2010)
11. Gärtner, J., den Hollander, F., Maillard, G.: Quenched Lyapunov exponent for the parabolic Anderson model in a dynamic random environment. In: Deuschel, J.-D., Gentz, B., König, W., von Renesse, M.-K., Scheutzow, M., Schmock, U. (eds.) Probability in Complex Physical Systems, Springer Proceedings in Mathematics, Vol. 11. Springer, Heidelberg (2012)
12. Gärtner, J., König, W.: The parabolic Anderson model. In: Deuschel, J.-D., Greven, A. (eds.) Interacting Stochastic Systems. pp. 153–179. Springer (2005)
13. Gärtner, J., Molchanov, S.A.: Parabolic problems for the Anderson model: I. Intermittency and related topics. Commun. Math. Phys. **132**(3), 613–655 (1990)
14. Kesten, H., Sidoravicius, V.: Branching random walk with catalysts. Elec. J. Prob. **8**, 1–51 (2003)
15. König, W., Schmidt, S.: The parabolic Anderson model with acceleration and deceleration. In: Deuschel, J.-D., Gentz, B., König, W., von Renesse, M.-K., Scheutzow, M., Schmock, U. (eds.) Probability in Complex Physical Systems, Springer Proceedings in Mathematics, Vol. 11. Springer, Heidelberg (2012)
16. Lacoin, H., Mörters, P.: A scaling limit theorem for the parabolic Anderson model with exponential potential. In: Deuschel, J.-D., Gentz, B., König, W., von Renesse, M.-K., Scheutzow, M., Schmock, U. (eds.) Probability in Complex Physical Systems, Springer Proceedings in Mathematics, Vol. 11. Springer, Heidelberg (2012)
17. Maillard, G., Mountford, T., Schöpfer, S.: Parabolic Anderson model with voter catalysts: Dichotomy in the behavior of Lyapunov exponents. In: Deuschel, J.-D., Gentz, B., König, W., von Renesse, M.-K., Scheutzow, M., Schmock, U. (eds.) Probability in Complex Physical Systems, Springer Proceedings in Mathematics, Vol. 11. Springer, Heidelberg (2012)
18. Molchanov, S.A.: Lectures on random media, Lecture Notes in Math. **1581**, 242–411 (1994)
19. Molchanov, S., Zhang, H.: Parabolic Anderson model with the long range basic Hamiltonian and Weibull type random potential. In: Deuschel, J.-D., Gentz, B., König, W., von Renesse, M.-K., Scheutzow, M., Schmock, U. (eds.) Probability in Complex Physical Systems, Springer Proceedings in Mathematics, Vol. 11. Springer, Heidelberg (2012)
20. Reed, M., Simon, B.: Methods of Modern Mathematical Physics. I. Functional Analysis. Academic Press, New York (1972)

Parabolic Anderson Model with a Finite Number of Moving Catalysts

Fabienne Castell, Onur Gün, and Grégory Maillard

Abstract We consider the parabolic Anderson model (PAM) which is given by the equation $\partial u/\partial t = \kappa \Delta u + \xi u$ with $u\colon \mathbb{Z}^d \times [0,\infty) \to \mathbb{R}$, where $\kappa \in [0,\infty)$ is the diffusion constant, Δ is the discrete Laplacian, and $\xi\colon \mathbb{Z}^d \times [0,\infty) \to \mathbb{R}$ is a space–time random environment. The solution of this equation describes the evolution of a "reactant" u under the influence of a "catalyst" ξ.

In the present paper, we focus on the case where ξ is a system of n independent simple random walks each with step rate $2d\rho$ and starting from the origin. We study the *annealed* Lyapunov exponents, i.e., the exponential growth rates of the successive moments of u w.r.t. ξ and show that these exponents, as a function of the diffusion constant κ and the rate constant ρ, behave differently depending on the dimension d. In particular, we give a description of the intermittent behavior of the system in terms of the annealed Lyapunov exponents, depicting how the total mass of u concentrates as $t \to \infty$. Our results are both a generalization and an extension of the work of Gärtner and Heydenreich [3], where only the case $n = 1$ was investigated.

F. Castell (✉)
CMI-LATP, Université de Provence, 39 rue F. Joliot-Curie, 13453 Marseille Cedex 13, France
e-mail: castell@cmi.univ-mrs.fr

O. Gün
CMI-LATP, Université de Provence, 39 rue F. Joliot-Curie, 13453 Marseille Cedex 13, France
e-mail: gun@cmi.univ-mrs.fr

G. Maillard
CMI-LATP, Université de Provence, 39 rue F. Joliot-Curie, 13453 Marseille Cedex 13, France

EURANDOM, P.O. Box 513, 5600 MB Eindhoven, The Netherlands
e-mail: maillard@cmi.univ-mrs.fr

J.-D. Deuschel et al. (eds.), *Probability in Complex Physical Systems*, Springer Proceedings in Mathematics 11, DOI 10.1007/978-3-642-23811-6_5,

1 Introduction

1.1 Model

The parabolic Anderson model (PAM) is the partial differential equation

$$\begin{cases} \frac{\partial}{\partial t} u(x,t) = \kappa \Delta u(x,t) + \xi(x,t) u(x,t), \\ u(x,0) = 1, \end{cases} \qquad x \in \mathbb{Z}^d,\ t \geq 0. \tag{1}$$

Here, the u-field is $\mathbb{R}$-valued, $\kappa \in [0,\infty)$ is the diffusion constant, Δ is the discrete Laplacian acting on u as

$$\Delta u(x,t) = \sum_{\substack{y \in \mathbb{Z}^d \\ y \sim x}} [u(y,t) - u(x,t)],$$

($y \sim x$ meaning that y is nearest neighbor of x), and

$$\xi = (\xi_t)_{t \geq 0} \quad \text{with} \quad \xi_t = \{\xi(x,t)\colon x \in \mathbb{Z}^d\},$$

is an $\mathbb{R}$-valued random field that evolves with time and that drives the equation.

One interpretation of (1) comes from population dynamics by considering a system of two types of particles A and B. A-particles represent "catalysts," B-particles represent "reactants" and the dynamics is subject to the following rules:

- A-particles evolve independently of B-particles according to a prescribed dynamics with $\xi(x,t)$ denoting the number of A-particles at site x at time t;
- B-particles perform independent simple random walks at rate $2d\kappa$ and split into two at a rate that is equal to the number of A-particles present at the same location;
- The initial configuration of B-particles is that there is exactly one particle at each lattice site.

Then, under the above rules, $u(x,t)$ represents the average number of B-particles at site x at time t conditioned on the evolution of the A-particles.

It is possible to add that B-particles die at rate $\delta \in [0,\infty)$. This leads to the trivial transformation $u(x,t) \to u(x,t)e^{\delta t}$. We will hereafter assume that $\delta = 0$. It is also possible to add a coupling constant $\gamma \in (0,\infty)$ in front of the ξ-term in (1), but this can be reduced to $\gamma = 1$ by a scaling argument.

In what follows, we focus on the case where

$$\xi(x,t) = \sum_{k=1}^{n} \delta_x\big(Y_k^\rho(t)\big), \tag{2}$$

with $\{Y_k^\rho : 1 \le k \le n\}$ a family of n independent simple random walks, where for each $k \in \{1,\dots,n\}$, $Y_k^\rho = (Y_k^\rho(t))_{t\ge 0}$ is a simple random walk with step rate $2d\rho$ starting from the origin. We write $\mathbb{P}_0^{\otimes n}$ and $\mathbb{E}_0^{\otimes n}$ to denote respectively the law and the expectation of the family of n independent simple random walks $\{Y_k^\rho : 1 \le k \le n\}$ where initially all of the walkers are located at 0.

The rest of the section is organized as follows. In Sect. 1.2, we define the *annealed Lyapunov exponents* and introduce the *intermittency* phenomenon. In Sect. 1.3, we review some related models from the literature. In Sect. 1.4, we state our main results, and finally, in Sect. 1.5, we give some further comments and add few results and conjectures.

1.2 Lyapunov Exponents and Intermittency

Our focus will be on the *annealed* Lyapunov exponents that describe the exponential growth rate of the successive moments of the solution of (1).

By the Feynman–Kac formula, the solution of (1) reads

$$u(x,t) = \mathrm{E}_x\left(\exp\left[\int_0^t \xi\,(X^\kappa(s), t-s)\,\mathrm{d}s\right]\right), \tag{3}$$

where $X^\kappa = (X^\kappa(t))_{t\ge 0}$ is the simple random walk on $\mathbb{Z}^d$ with step rate $2d\kappa$ and E_x denotes expectation with respect to X^κ given $X^\kappa(0) = x$. The connection between the parabolic Anderson equation (1) with random time-independent potential ξ and the Feyman–Kac functional (3) is well understood (see e.g. Gärtner and Molchanov [10]) and can be easily extended to the time-dependent potential setting. Taking into account our choice of catalytic medium in (2) we define $\Lambda_p(t)$ as

$$\begin{aligned}\Lambda_p(t) &= \frac{1}{t}\log \mathbb{E}_0^{\otimes n}\left([u(x,t)]^p\right)^{1/p} \\ &= \frac{1}{pt}\log\left(\mathbb{E}_0^{\otimes n}\otimes \mathrm{E}_x^{\otimes p}\right)\left(\exp\left[\sum_{j=1}^p\sum_{k=1}^n\int_0^t \delta_0\big(X_j^\kappa(s) - Y_k^\rho(t-s)\big)\,\mathrm{d}s\right]\right),\end{aligned} \tag{4}$$

where $\{X_j^\kappa : 1 \le j \le p\}$ is a family of p independent copies of X^κ and $\mathrm{E}_x^{\otimes p}$ stands for the expectation of this family with $X_j^\kappa(0) = x$ for all j.

If the last quantity admits a limit as $t \to \infty$, we define

$$\lambda_p := \lim_{t\to\infty} \Lambda_p(t), \tag{5}$$

to be the p-th (annealed) Lyapunov exponent of the solution u of the PAM (1).

We will see in Theorem 1.1 that the limit in (5) exists and is independent of x. Hence, we suppress x in the notation. However, λ_p is clearly a function of n, d, κ and ρ. In what follows, our main focus will be to analyze the dependence of λ_p on the parameters n, p, κ and ρ, therefore we will often write $\lambda_p^{(n)}(\kappa, \rho)$.

In particular, our main subject of interest will be to draw the qualitative picture of *intermittency* for these systems. First, note that by the moment inequality we have

$$\lambda_p^{(n)} \geq \lambda_{p-1}^{(n)}, \tag{6}$$

for all $p \in \mathbb{N} \setminus \{1\}$. The system (or the solution of the system) (1) is said to be *p-intermittent* if the above inequality is strict, namely,

$$\lambda_p^{(n)} > \lambda_{p-1}^{(n)}. \tag{7}$$

The system is *fully intermittent* if (7) holds for all $p \in \mathbb{N} \setminus \{1\}$. We will sometimes say that the system is *partially intermittent* if it is p-intermittent for some $p \in \mathbb{N} \setminus \{1\}$.

Also note that, using Hölder's inequality, p-intermittency implies q-intermittency for all $q \geq p$ (see e.g. [3], Lemma 3.1). Thus, for any fixed $n \in \mathbb{N}$, p-intermittency in fact implies that

$$\lambda_q^{(n)} > \lambda_{q-1}^{(n)} \quad \forall q \geq p,$$

and 2-intermittency means full intermittency.

Geometrically, intermittency corresponds to the solution being *asymptotically concentrated* on a thin set, which is expected to consist of "islands" located far from each other (see [9], Sect. 1 and references therein for more details). Here, due to the lack of ergodicity, such a geometric picture of intermittency is not available. Nevertheless, (7) can still be interpreted as the p-th moment of u being generated by some exponentially rare event (see [3], Sect. 1.2 for a more detailed analysis).

1.3 Literature

The behavior of the *annealed* Lyapunov exponents and particularly the problem of intermittency for the PAM in a *space–time* random environment was subject to various studies. Carmona and Molchanov [2] obtained an essentially complete qualitative description of the annealed Lyapunov exponents and intermittency when ξ is white noise, i.e.,

$$\xi(x,t) = \frac{\partial}{\partial t} W(x,t),$$

where $W = (W_t)_{t\geq 0}$ with $W_t = \{W(x,t)\colon x \in \mathbb{Z}^d\}$ is a field of independent Brownian motions. In particular, it was shown that $\lambda_1 = 1/2$ for all $d \geq 1$ and, $\lambda_p > 1/2$ for $p \in \mathbb{N} \setminus \{1\}$ in $d = 1, 2$. It is also proved that for $d \geq 3$, there exist $0 < \kappa_2 \leq \kappa_3 \leq \ldots$ satisfying

$$\lambda_p(\kappa) - \frac{1}{2} \begin{cases} > 0, & \text{for } \kappa \in [0, \kappa_p), \\ = 0, & \text{for } \kappa \in [\kappa_p, \infty), \end{cases} \qquad p \in \mathbb{N} \setminus \{1\}.$$

Further refinements on the behavior of the Lyapunov exponents were obtained in Greven and den Hollander [11]. Upper and lower bounds on κ_p were derived, and the asymptotics of κ_p as $p \to \infty$ was computed. In addition, it was proved that the κ_p's are distinct for d large enough.

More recently various models where ξ is *non-Gaussian* were investigated. Kesten and Sidoravicius [13] and Gärtner and den Hollander [4] considered the case where ξ is given by a Poisson field of independent simple random walks. In [13], the survival versus extinction of the system is studied. In [4], the moment asymptotics were studied and a partial picture of intermittency, depending on the parameters d and κ, was obtained. The case where ξ is a single random walk – corresponding to $n = 1$ case in our setting – was studied by Gärtner and Heydenreich [3]. Analogous results to those contained in Theorems 1.1, 1.2 and Corollary 1.1(i) were obtained.

The investigation of annealed Lyapunov behavior and intermittency was extented to non-Gaussian and *space correlated* potentials in Gärtner, den Hollander and Maillard, in [5] and [7], for the case where ξ is an exclusion process with symmetric random walk transition kernel, starting form a Bernoulli product measure. Later Gärtner, den Hollander and Maillard [8], and Maillard, Mountford and Schöpfer [14], studied the case where ξ is a voter model starting either from Bernoulli product measure or from equilibrium (see Gärtner, den Hollander and Maillard [6], for an overview).

1.4 Main Results

Our first theorem states that the Lyapunov exponents exist and behave nicely as a function of κ and ρ. It will be proved in Sect. 2.

Theorem 1.1 (Existence and first properties). *Let $d \geq 1$ and $n, p \in \mathbb{N}$.*

(i) For all $\kappa, \rho \in [0, \infty)$, the limit in (5) *exists, is finite, and is independent of x if $(\kappa, \rho) \neq (0, 0)$.*
(ii) On $[0, \infty)^2$, $(\kappa, \rho) \mapsto \lambda_p^{(n)}(\kappa, \rho)$ is continuous, convex and non-increasing in both κ and ρ.

Let $G_d(x)$ be the Green function at lattice site x of simple random walk stepping at rate $2d$ and

$$\mu(\kappa) = \sup \mathrm{Sp}(\kappa\Delta + \delta_0), \tag{8}$$

be the supremum of the spectrum of the operator $\kappa\Delta+\delta_0$ in $l^2(\mathbb{Z}^d)$. It is well-known that (see, e.g., [4], Lemma 1.3) $\mathrm{Sp}(\kappa\Delta + \delta_0) = [-4d\kappa, 0] \cup \{\mu(\kappa)\}$ with

$$\mu(\kappa) \begin{cases} = 0, & \text{if } \kappa \geq G_d(0), \\ > 0, & \text{if } \kappa < G_d(0). \end{cases} \tag{9}$$

Furthermore, $\kappa \mapsto \mu(\kappa)$ is continuous, non-increasing and convex on $[0,\infty)$, and strictly decreasing on $[0, G_d(0)]$.

The next theorem gives the limiting behavior of $\lambda_p^{(n)}$ as $\kappa \downarrow 0$ and $\kappa \to \infty$, and describes a region of κ where $\lambda_p^{(n)} = 0$. Note that by symmetry, $\lambda_p^{(n)}(\kappa,\rho) = \frac{n}{p}\lambda_n^{(p)}(\rho,\kappa)$, for all $n, p \in \mathbb{N}$ and $\kappa,\rho \in [0,\infty)$. Therefore, the κ-dependence described below can be transcribed in terms of ρ-dependence.

Theorem 1.2 (**κ- ρ-dependence**). *Let $n, p \in \mathbb{N}$ and $\rho \in [0,\infty)$.*

(i) For all $d \geq 1$, $\lim_{\kappa\downarrow 0} \lambda_p^{(n)}(\kappa,\rho) = \lambda_p^{(n)}(0,\rho) = n\mu(\rho/p)$.
(ii) If $1 \leq d \leq 2$, then $\lambda_p^{(n)}(\kappa,\rho) > 0$ for all $\kappa \in [0,\infty)$. Moreover, $\kappa \mapsto \lambda_p^{(n)}(\kappa,\rho)$ is strictly decreasing with $\lim_{\kappa\to\infty} \lambda_p^{(n)}(\kappa,\rho) = 0$ (see Fig. 1).
(iii) If $d \geq 3$, then $\lambda_p^{(n)}(\kappa,\rho) = 0$ for all $\kappa \in [nG_d(0),\infty)$ (see Fig. 2).

Our next result describes the limiting behavior of $\lambda_p^{(n)}$ as $p \to \infty$ and $n \to \infty$.

Theorem 1.3 (**n- p-dependence**). *Let $d \geq 1$ and $\kappa,\rho \in [0,\infty)$.*

(i) For all $n \in \mathbb{N}$, $\lim_{p\to\infty} \lambda_p^{(n)}(\kappa,\rho) = n\mu(\kappa/n)$ (see Fig. 1–2);
(ii) For all $p > \rho/G_d(0)$, $\lim_{n\to\infty} \lambda_p^{(n)}(\kappa,\rho) = +\infty$;

(iii) For all $p \leq \rho/G_d(0)$ and $n \in \mathbb{N}$, $\lambda_p^{(n)}(\kappa,\rho) = 0$.

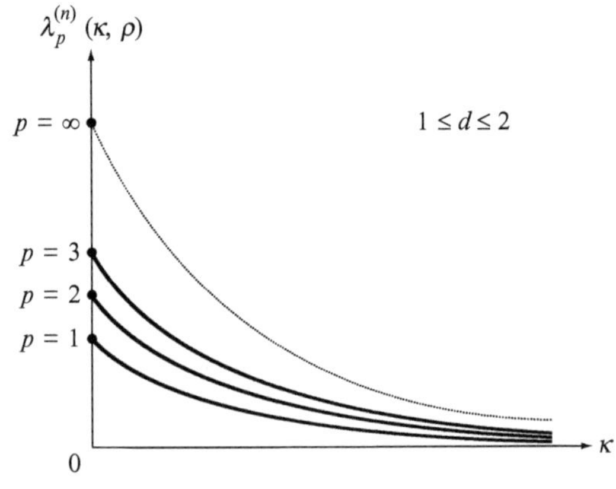

Fig. 1 For $1 \leq d \leq 2$, the system is partially intermittent. Full intermittency is conjectured, and proved for $n = 1, 2$

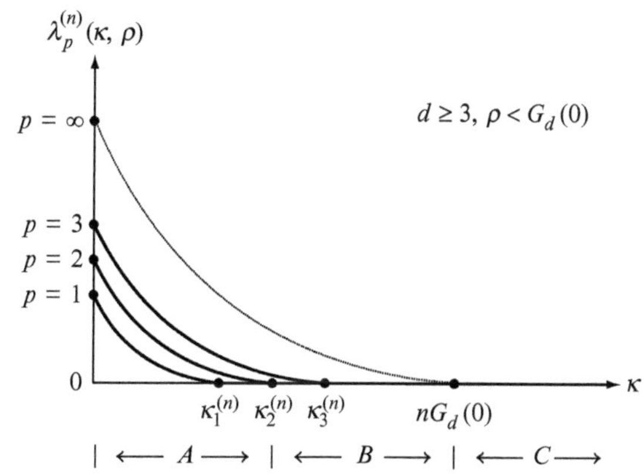

Fig. 2 For $d \geq 3$ and $\rho < G_d(0)$, the system is partially intermittent on $A \cup B$ and not intermittent on C. Full intermittency on A is conjectured, and proved for $n = 1, 2$

By part (ii) of Theorem 1.1, $\lambda_p^{(n)}(\kappa, \rho)$ is non-increasing in κ. Hence, we can define $\{\kappa_p^{(n)}(\rho) \colon p \in \mathbb{N}\}$ as the non-decreasing sequence of critical κ's for which

$$\lambda_p^{(n)}(\kappa,\rho) \begin{cases} > 0, & \text{for } \kappa \in \left[0, \kappa_p^{(n)}(\rho)\right), \\ = 0, & \text{for } \kappa \in \left[\kappa_p^{(n)}(\rho), \infty\right), \end{cases} \qquad p \in \mathbb{N}. \tag{10}$$

As a consequence of Theorems 1.1 and 1.2 we have,

$$\begin{cases} \kappa_p^{(n)}(\rho) = \infty, & \text{if } 1 \leq d \leq 2, \\ 0 < \kappa_p^{(n)}(\rho) < \infty, & \text{if } d \geq 3 \text{ and } p > \rho/G_d(0), \\ \kappa_p^{(n)}(\rho) = 0, & \text{if } d \geq 3 \text{ and } p \leq \rho/G_d(0). \end{cases} \tag{11}$$

Our fourth theorem, which gives bounds on $\kappa_p^{(n)}(\rho)$ for $d \geq 3$, will be proved in Sect. 4. For this theorem, we need to define the inverse of the function $\mu(\kappa)$. Note that by (8) and (9), we have $\mu(0) = 1$ and $\mu(G_d(0)) = 0$. It is easy to see that $\mu(\kappa)$ restricted to the domain $[0, G_d(0)]$ is invertible with an inverse function $\mu^{-1} \colon [0,1] \to [0, G_d(0)]$. We extend μ^{-1} to $[0,\infty)$ by declaring $\mu^{-1}(t) = 0$ for $t > 1$. Denote

$$\alpha_d = \frac{G_d(0)}{2d\,\|G_d\|_2^2} \in [0,\infty)\,, \tag{12}$$

where $\|G_d\|_2$ is the l_2 norm of G_d. Since $\|G_d\|_2 < \infty$ if and only if $d \geq 5$, $\alpha_d = 0$ for $d \in \{3,4\}$.

Theorem 1.4 (Critical κ's). *Let $n, p \in \mathbb{N}$.*

(i) If $d \geq 3$, then $\rho \in [0,\infty) \mapsto \kappa_p^{(n)}(\rho)$ is a continuous, non-increasing and convex function such that

$$\max\left(\frac{n}{4d}\,\mu(\rho/p), n\mu^{-1}(4d\rho/p)\right) \le \kappa_p^{(n)}(\rho) \le nG_d(0)\left(1-\frac{\rho}{pG_d(0)}\right)_+. \tag{13}$$

(ii) If $d \ge 5$, then

$$\kappa_p^{(n)}(\rho) \ge \left(nG_d(0) - \rho\frac{n}{p\alpha_d}\right)_+. \tag{14}$$

(iii) If $d \ge 5$ and $p \in \mathbb{N}\setminus\{1\}$ are such that $\alpha_d > \frac{p-1}{p}$, then

$$\kappa_{p-1}^{(n)}(\rho) < \kappa_p^{(n)}(\rho) \quad \forall \rho \in (0, pG_d(0)). \tag{15}$$

Note that the condition $\alpha_d > \frac{p-1}{p}$ is always true if d is large enough by the following lemma, whose proof is given in the appendix.

Lemma 1.1. *If $d \ge 3$, then $\alpha_d \le 1$ and $\lim_{d\to\infty}\alpha_d = 1$.*

As a consequence of the previous statements, our next result gives some general intermittency properties for all dimensions, and describes several regimes in the intermittent behavior of the system.

Corollary 1.1 (Intermittency). *Let $n \in \mathbb{N}$.*

(i) If $d \ge 1$, then (see Fig. 2)

- *For $\kappa \in [0, nG_d(0))$ there exists $p \ge 2$ such that the system is p-intermittent;*
- *For $\kappa \in [nG_d(0), \infty)$ the system is not intermittent.*

(ii) Fix $p \in \mathbb{N}\setminus\{1\}$. If d is large enough (such that $\alpha_d > (p-1)/p$), then for all $q \in \{2, \ldots, p\}$, $\rho \in [0, qG_d(0))$ and $\kappa \in (\kappa_{q-1}^{(n)}(\rho), \kappa_q^{(n)}(\rho))$, the system is q-intermittent (see Fig. 3).

Note that since $G_d(0) = \infty$ for $d = 1, 2$, Corollary 1.1(i) implies that for dimensions 1 and 2 the system is always p-intermittent for some p. Some other partial results about intermittency are given in sect. 1.4 (see also figures).

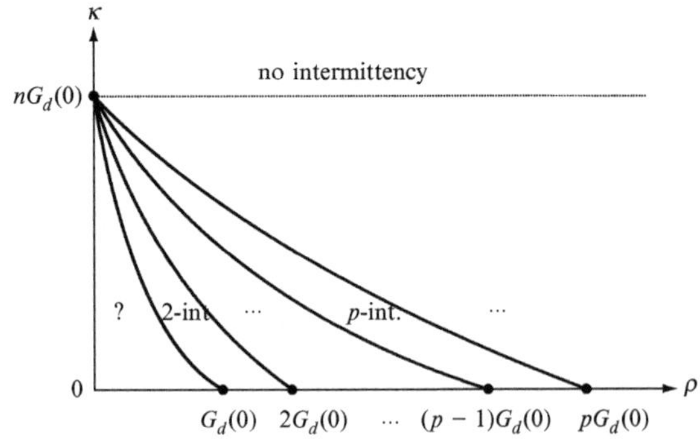

Fig. 3 Phase diagram of intermittency when d is large enough. The bold curves represent $\rho \in [0,\infty) \mapsto \kappa_q^{(n)}(\rho)$, $q = 1, \ldots, p$. In the "?" region, full intermittency is proved in a small neighborhood of 0

1.5 Discussion

Our results can be extended to various different random medium. For example, consider the system of catalysts given by a collection of independent random walks where there is one walker starting from each site of a large box. More precisely, let D_{R} denote the box in $\mathbb{Z}^d$ with side length R. Consider the random medium

$$\xi(x,t) = \sum_{k \in D_{\mathrm{R}}} \delta_x(Y_k^\rho(t)),$$

with $\{Y_k^\rho : k \in D_{\mathrm{R}}\}$ a family of R^d simple random walks, where for each $k \in D_{\mathrm{R}}$, Y_k^ρ is a simple random walk with step rate $2d\rho$ starting from $Y_k^\rho(0) = k$. For a fixed size box, there is a positive probability that all the random walks meet at the origin in finite time. Then, it is easy to see that the Lyapunov exponents are the same as in the case of n independent random walks starting from the origin where $n = R^d$. An interesting set up would be case where the length of the initial box grows with time. A natural question arises as whether the large time limit would be related to the case of Poisson field of simple random walks, considered in [4], or it would have different behaviour depending on how fast the size of the box grows with time.

Let us now discuss some facts about the intermittent picture. First of all, as one can easily guess from (4), $\lambda_p^{(n)}(\kappa,\rho)$ is the top of the spectrum of the operator $\mathscr{L}_p$ where for $f(x_1,\cdots,x_p,y_1,\cdots,y_n)$ in $l_2(\mathbb{Z}^{d(p+n)})$ $\mathscr{L}_p$, is defined by:

$$\mathscr{L}_p(f) = \kappa \sum_{k=1}^{p} \Delta_{x_k} f + \rho \sum_{j=1}^{n} \Delta_{y_j} f + I_p f. \tag{16}$$

Here,

$$I_p f(x_1,\cdots,x_p,y_1,\cdots,y_n) = \sum_{k=1}^{n}\sum_{j=1}^{p} \delta_0(x_j - y_k) f(x_1,\cdots,x_p,y_1,\cdots,y_n).$$

This is the meaning of equation (20) of Sect. 2 from which most of our results are derived. The following proposition links full intermittency and existence of an eigenfunction corresponding to $\lambda_1^{(n)}(\kappa,\rho)$.

Proposition 1.1. *If there exists* $f \in l_2(\mathbb{Z}^{d(1+n)})$ *with* $\|f\|_2 = 1$, *such that* $\mathscr{L}_1(f) = \lambda_1^{(n)}(\kappa,\rho) f$, *then* $\lambda_2^{(n)}(\kappa,\rho) > \lambda_1^{(n)}(\kappa,\rho)$, *and the system is fully intermittent.*

Proposition 1.1 is proved in the appendix. The existence of an eigenfunction corresponding to $\lambda_1^{(n)}(\kappa,\rho)$ (and therefore full intermittency) was proved in the following cases:

- $n = 1,2$ and $\kappa + \rho < nG_d(0)$. This is done in [3] for $n = 1$, and in [15] for $n = 2$.
- $n \geq 3$ and $4d(\rho n + \kappa) < 1$ in [15].

To prove these results, in [3] and [15] $\lambda_1^{(n)}(\kappa,\rho)$ was expressed as the top of the spectrum of the operator $\mathscr{H} = \mathscr{B} + \sum_{j=1}^n \delta_0(z_j)$, where $\mathscr{B}$ is the generator of the Markov process $Z(t) = (X_1^\kappa(t) - Y_1^\rho(t), \cdots, X_1^\kappa(t) - Y_n^\rho(t))$ (see (4)). For $n = 1$, $\mathscr{H}$ is just $(\kappa+\rho)\Delta + \delta_0$, which is a compact perturbation of $(\kappa+\rho)\Delta$. This fact easily implies the existence of an eigenfunction corresponding to $\lambda_1^{(1)}(\kappa,\rho)$. However, this is no more the case as soon as $n \geq 2$. In [15], Schnitzler and Wolff considered $\mathscr{B}$ as a perturbation of $\sum_{j=1}^n \delta_0(z_j)$, leading to the results for $n \geq 2$. Expressing $\lambda_p^{(n)}(\kappa,\rho)$ in terms of the process $(Z(t))_{t\geq 0}$ does not seem very fruitful in cases other than the one treated in [3] and[15]. Therefore, it appeared to us more natural and more tractable to express $\lambda_p^{(n)}(\kappa,\rho)$ in terms of the process $(X_1^\kappa(t), \cdots, X_p^\kappa(t), Y_1^\rho(t), \cdots, Y_n^\rho(t))$. We complete the intermittent picture by the following conjecture:

Conjecture 1 (Intermittency). Fix $n \in \mathbb{N}$. Then (see Fig. 1–2),

(1) For $1 \leq d \leq 2$, the system is full intermittent (proved for $n = 1, 2$);

(2) For $d \geq 3$, the intermittency vanishes as κ increases. More precisely, for $d \geq 3$, there are three different regimes:

A: For $\kappa \in [0, \kappa_2^{(n)})$, the system is full intermittent (proved in a small neighborhood of 0);
B: For $\kappa \in [\kappa_2^{(n)}, nG_d(0))$, there exists $p = p(\kappa) \geq 3$ such that the system is q-intermittent for all $q \geq p$;
C: For $\kappa \in [nG_d(0), \infty)$, the system is not p-intermittent for any $p \geq 2$.

To complete Theorem 1.4, we close with a conjecture about critical κ's, whose analogue for white noise potential was conjectured in Carmona and Molchanov [2] and partially proved in Greven and den Hollander [11]:

Conjecture 2 (Critical κ's). For all fixed $n \geq 1$ and d large enough, the $\kappa_p^{(n)}$'s are distinct (see Fig. 2).

2 Proof of Theorem 1.1

Step 1: We first prove that if the limit in (5) exists for $x = 0$, then it exists for all $x \in \mathbb{Z}^d$ and does not depend on x as soon as $(\kappa,\rho) \neq (0,0)$. To this end, let us introduce some notations. For any $t > 0$, we denote

$$Y_t = (Y_1^\rho(t), \cdots, Y_n^\rho(t)) \in \mathbb{Z}^{\mathrm{dn}}\ ,\ X_t = (X_1^\kappa(t), \cdots, X_p^\kappa(t)) \in \mathbb{Z}^{\mathrm{dp}}.$$

For $(x,y) \in \mathbb{Z}^{\mathrm{dp}} \times \mathbb{Z}^{\mathrm{dn}}$, $\mathbb{E}_{x,y}^{X,Y}$ denotes the expectation under the law of $(X_t, Y_t)_{t\geq 0}$ starting from (x,y). The same notation is used for $x \in \mathbb{Z}^d$ and $y \in \mathbb{Z}^d$. In that case, it means that $X_0 = (x, \cdots, x)$, $Y_0 = (y, \cdots, y)$ and $\mathbb{E}_{x,y}^{X,Y} = \mathbb{E}_y^{\otimes n} \otimes \mathrm{E}_x^{\otimes p}$. Finally, for $x = (x_1, \cdots, x_p) \in \mathbb{Z}^{\mathrm{dp}}$ and $y = (y_1, \cdots, y_n) \in \mathbb{Z}^{\mathrm{dn}}$, set

$$I_p(x,y) = \sum_{j=1}^{p}\sum_{k=1}^{n} \delta_0(x_j - y_k). \tag{17}$$

Then, by time reversal for Y in (4), for all $x \in \mathbb{Z}^d$ and $t > 0$,

$$\mathbb{E}_0^{\otimes n}\left[u(x,t)^p\right] = \sum_{z \in \mathbb{Z}^{\mathrm{dn}}} \mathbb{E}_{x,z}^{X,Y}\left[\exp\left(\int_0^t I_p(X_s,Y_s)\,\mathrm{d}s\right)\delta_0(Y_t)\right]. \tag{18}$$

Using the Markov property at time 1 and the fact that $1 \leq \exp\left(\int_0^1 I_p(X_s,Y_s)\,\mathrm{d}s\right)$, we get for x_1 and x_2 any fixed points in $\mathbb{Z}^d$,

$$\begin{aligned}\mathbb{E}_0^{\otimes n}\left[u(x_1,t)^p\right] &\geq \sum_{z \in \mathbb{Z}^{\mathrm{dn}}} \mathbb{E}_{x_1,z}^{X,Y}\left[\delta_{(x_2,\cdots,x_2)}(X_1)\delta_z(Y_1)\exp\left(\int_1^t I_p(X_s,Y_s)\,\mathrm{d}s\right)\delta_0(Y_t)\right] \\ &= \left(p_1^\kappa\left(x_1,x_2\right)\right)^p \left(p_1^\rho(0,0)\right)^n \mathbb{E}_0^{\otimes n}\left(\left[u\left(x_2,t-1\right)\right]^p\right),\end{aligned}$$

where p_t^ν is the transition kernel of a simple random walk on $\mathbb{Z}^d$ with step rate $2\mathrm{d}\nu$. This proves the independence of λ_p w.r.t. x as soon as $\kappa > 0$, since in this case for all $x_1, x_2 \in \mathbb{Z}^d$, $p_1^\kappa(x_1,x_2) > 0$.

For $\kappa = 0$, since the X-particles do not move, we have

$$\mathbb{E}_0^{\otimes n}\left[u(x_1,t)^p\right] = \mathbb{E}_0\left[\exp\left(p\int_0^t \delta_{x_1}(Y_1^\rho(s))\,\mathrm{d}s\right)\right]^n. \tag{19}$$

The same reasoning leads now to

$$\mathbb{E}_0^{\otimes n}\left[u(x_1,t)^p\right] \geq p_1^\rho(0,x_1-x_2)^n \mathbb{E}_0^{\otimes n}\left(\left[u(x_2,t-1)\right]^p\right).$$

Step 2: Variational representation. From now on, we restrict our attention to the case $x = 0$. The aim of this step is to give a variational representation of $\lambda_p^{(n)}(\kappa,\rho)$. To this end, we introduce further notations. Let $(e_1,\cdots,e_d)$ be the canonical basis of $\mathbb{R}^d$. For $x = (x_1,\cdots,x_p) \in \mathbb{Z}^{\mathrm{dp}}$, and $f : (x,y) \in \mathbb{Z}^{\mathrm{dp}} \times \mathbb{Z}^{\mathrm{dn}} \mapsto \mathbb{R}$, we set

$$\nabla_x f(x,y) = \left(\nabla_{x_1} f(x,y),\cdots,\nabla_{x_p} f(x,y)\right) \in \mathbb{R}^{\mathrm{dp}},$$

where for $j \in \{1,\cdots,p\}$, and $i \in \{1,\cdots,d\}$,

$$\left\langle \nabla_{x_j} f(x,y), e_i\right\rangle = f(x_1,\cdots,x_j+e_i,\cdots,x_p,y) - f(x,y).$$

The same notation is used for the y-coordinates, so that $\nabla_y f(x,y) \in \mathbb{R}^{\mathrm{dn}}$. We also define

$$\begin{aligned}\Delta_x f(x,y) &= \sum_{j=1}^{p} \Delta_{x_j} f(x,y)\\ &= \sum_{j=1}^{p} \sum_{\substack{z_j \in \mathbb{Z}^d\\ z_j \sim x_j}} \left[f(x_1,\cdots,z_j,\cdots,x_p,y) - f(x_1,\cdots,x_j,\cdots,x_p,y)\right].\end{aligned}$$

Proposition 2.1. *Let $d \geq 1$ and $n, p \in \mathbb{N}$. For all $\kappa, \rho \in [0,\infty)$,*

$$\begin{aligned}\lambda_p^{(n)}(\kappa,\rho) &= \lim_{t\to\infty} \frac{1}{pt} \log \mathbb{E}_0^{\otimes n}\left[u(0,t)^p\right]\\ &= \frac{1}{p} \sup_{\substack{f\in l^2(\mathbb{Z}^{\mathrm{dp}}\times\mathbb{Z}^{\mathrm{dn}})\\ \|f\|_2=1}} \left\{-\kappa \left\|\nabla_x f\right\|_2^2 - \rho \left\|\nabla_y f\right\|_2^2 + \sum_{(x,y)} I_p(x,y) f^2(x,y)\right\}.\end{aligned} \tag{20}$$

Proof. **Upper bound.** For a positive integer m, let B_R^m denote the ball in $\mathbb{Z}^{\mathrm{dm}}$ of radius $R = t\log(t)$ centred at the origin. We first prove the following lemma which states we can restrict (18) to X paths being in B_R^p at time t and Y paths starting from B_R^n.

Lemma 2.1. *As $t \to \infty$,*

$$\mathbb{E}_0^{\otimes n}\left[u(x,t)^p\right] = (1+o(1)) \sum_{z\in B_R^n} \mathbb{E}_{0,z}^{X,Y}\left[\exp\left(\int_0^t I_p(X_s,Y_s)\,\mathrm{d}s\right)\delta_0(Y_t)\ \mathbb{1}_{(B_R^p)}(X_t)\right]. \tag{21}$$

Proof. It is enough to prove that

$$r(t) := \frac{\mathbb{E}_0^{\otimes n}\left[u(x,t)^p\right] - \sum_{z\in B_R^n} \mathbb{E}_{0,z}^{X,Y}\left[\exp\left(\int_0^t I_p(X_s,Y_s)\,\mathrm{d}s\right)\delta_0(Y_t)\ \mathbb{1}_{(B_R^p)}(X_t)\right]}{\mathbb{E}_0^{\otimes n}\left[u(x,t)^p\right]}, \tag{22}$$

converges to 0 as $t \to \infty$. Using the trivial bounds

$$1 \leq \exp\left(\int_0^t I_p(X_s,Y_s)\,\mathrm{d}s\right) \leq \exp(tnp), \tag{23}$$

and splitting the sum in (18), we get

$$
\begin{aligned}
r(t) &\le \frac{e^{\mathrm{tnp}}}{\sum\limits_{z\in Z^{\mathrm{dn}}} \mathbb{E}_{0,z}^{X,Y}\,[\delta_0(Y_t)]} \left(\sum_{z\notin B_R^n} \mathbb{E}_{0,z}^{X,Y}\,[\delta_0(Y_t)] + \sum_{z\in B_R^n} \mathbb{E}_{0,z}^{X,Y}\left[\delta_0(Y_t)\,\mathbb{1}_{(B_R^p)^c}(X_t)\right]\right) \\
&\le \frac{e^{\mathrm{tnp}}}{\sum\limits_{z\in \mathbb{Z}^{\mathrm{dn}}} \mathbb{P}_z(Y_t=0)} \left(\sum_{z\notin B_R^n} \mathbb{P}_z(Y_t=0) + \mathbb{P}_0(X_t\notin B_R^p) \sum_{z\in B_R^n} \mathbb{P}_z(Y_t=0)\right) \\
&\le e^{\mathrm{tnp}}(\mathbb{P}_0(Y_t\notin B_R^n) + \mathbb{P}_0(X_t\notin B_R^p)),
\end{aligned}
$$

where for the last two inequalities we used the time reversal of Y. We have for $R = t\log(t)$ and large enough t

$$
\begin{aligned}
\mathbb{P}_0(Y_1^{\rho}(t)\notin B_R^1) &\le \exp[-C(d,\rho)t\log(t)], \\
\mathbb{P}_0(X_1^{\kappa}(t)\notin B_R^1) &\le \exp[-C(d,\kappa)t\log(t)]
\end{aligned}
\tag{24}
$$

for some positive constants $C(d,\rho)$ and $C(d,\kappa)$ (see for instance Lemma 4.3 in [10]). Using this, we get

$$
r(t) \le e^{\mathrm{tnp}}\left(ne^{-C(d,\rho)t\log t} + pe^{-C(d,\kappa)t\log t}\right) \overset{t\to\infty}{\longrightarrow} 0.
$$

This finishes the proof of the lemma. □

Using Lemma 2.1 it is enough to study the existence of

$$
\begin{aligned}
&\lim_{t\to\infty}\frac{1}{t}\log\sum_{z\in B_R^n}\mathbb{E}_{0,z}^{X,Y}\left[\exp\left(\int_0^t I_p(X_s,Y_s)\,\mathrm{d}s\right)\delta_0(Y_t)\,\mathbb{1}_{B_R^p}(X_t)\right] \\
&= \lim_{t\to\infty}\frac{1}{t}\log\left\langle f_1, e^{t\mathscr{L}_p} f_2\right\rangle,
\end{aligned}
$$

where $f_1 : (x,y)\in\mathbb{Z}^{\mathrm{dp}}\times\mathbb{Z}^{\mathrm{dn}} \mapsto \delta_0(x)\,\mathbb{1}_{B_R^n}(y)$, $f_2 : (x,y)\in\mathbb{Z}^{\mathrm{dp}}\times\mathbb{Z}^{\mathrm{dn}} \mapsto \mathbb{1}_{B_R^p}(x)\delta_0(y)$, and $\mathscr{L}_p$ is the bounded self-adjoint operator in $l^2(\mathbb{Z}^{\mathrm{dp}}\times\mathbb{Z}^{\mathrm{dn}})$ defined by

$$
\mathscr{L}_p f(x,y) = \kappa\Delta_x f(x,y) + \rho\Delta_y f(x,y) + I_p(x,y)f(x,y), \quad (x,y)\in\mathbb{Z}^{\mathrm{dp}}\times\mathbb{Z}^{\mathrm{dn}}.
$$

For a linear operator $\mathscr{L}$ on $l^2(\mathbb{Z}^{\mathrm{dp}}\times\mathbb{Z}^{\mathrm{dn}})$, we define

$$
\|\mathscr{L}\|_{2,2} := \sup_{\substack{f\in l^2(\mathbb{Z}^{\mathrm{dp}}\times\mathbb{Z}^{\mathrm{dn}})\\ \|f\|_2=1}} \langle f,\mathscr{L}f\rangle .
$$

Note that we have

$$\left\langle f_1, e^{t\mathscr{L}_p} f_2\right\rangle \leq \|f_1\|_2 \left\|e^{t\mathscr{L}_p}\right\|_{2,2} \|f_2\|_2 = C(d,n,p)R^{d(n+p)/2} \left\|e^{t\mathscr{L}_p}\right\|_{2,2},$$

for some constant $C(d,n,p) > 0$. Thus,

$$\lim_{t\to\infty} \frac{1}{t} \log\left\langle f_1, e^{t\mathscr{L}_p} f_2\right\rangle \leq \left\|\mathscr{L}_p\right\|_{2,2} = \sup_{\substack{f\in l^2(\mathbb{Z}^{dp}\times\mathbb{Z}^{dn})\\ \|f\|_2=1}} \left\langle f, \mathscr{L}_p f\right\rangle,$$

which is the upper bound in (20).

Lower bound. By (18) with $x = 0$, it follows that

$$\begin{aligned}
\mathbb{E}_0^{\otimes n}\left[u(0,t)^p\right] &\geq \mathbb{E}_{0,0}^{X,Y}\left[\exp\left(\int_0^t I_p(X_s,Y_s)\,\mathrm{d}s\right)\delta_0(X_t)\delta_0(Y_t)\right] \\
&= \left\langle \delta_0\otimes\delta_0, e^{t\mathscr{L}_p}(\delta_0\otimes\delta_0)\right\rangle = \left\|e^{\frac{t}{2}\mathscr{L}_p}(\delta_0\otimes\delta_0)\right\|_2^2 \\
&= \sum_{x\in\mathbb{Z}^{dp}}\sum_{y\in\mathbb{Z}^{dn}} \left(e^{\frac{t}{2}\mathscr{L}_p}(\delta_0\otimes\delta_0)(x,y)\right)^2.
\end{aligned}$$

Restricting the sum over $B_R^p \times B_R^n$, and applying Jensen's inequality, we get

$$\begin{aligned}
&\mathbb{E}_0^{\otimes n}\left[u(0,t)^p\right] \\
&\quad\geq \sum_{x\in B_R^p}\sum_{y\in B_R^n} \left(\mathrm{e}^{\frac{t}{2}\mathscr{L}_p}(\delta_0\otimes\delta_0)(x,y)\right)^2 \\
&\quad\geq \frac{1}{|B_R^n|}\frac{1}{|B_R^p|}\left(\sum_{x\in B_R^p}\sum_{y\in B_R^n} \mathrm{e}^{\frac{t}{2}\mathscr{L}_p}(\delta_0\otimes\delta_0)(x,y)\right)^2 \\
&\quad= \frac{C(d,n,p)}{R^{d(n+p)}}\left(\sum_{x\in B_R^p}\sum_{y\in B_R^n} \mathbb{E}_{x,y}^{X,Y}\left[\exp\left(\int_0^{t/2} I_p(X_s,Y_s)\,\mathrm{d}s\right)\delta_0(X_{t/2})\delta_0(Y_{t/2})\right]\right)^2 \\
&\quad= \frac{C(d,n,p)}{R^{d(n+p)}}\left(\mathbb{E}_{0,0}^{X,Y}\left[\exp\left(\int_0^{t/2} I_p(X_s,Y_s)\,\mathrm{d}s\right)\mathbb{1}_{B_R^p}(X_{t/2})\,\mathbb{1}_{B_R^n}(Y_{t/2})\right]\right)^2.
\end{aligned}$$

Taking $R = t\log(t)$, we obtain that

$$\liminf_{t\to\infty}\frac{1}{t}\log\mathbb{E}_0^{\otimes n}\left[u(0,t)^p\right]$$
$$\geq \liminf_{t\to\infty}\frac{2}{t}\log\mathbb{E}_{0,0}^{X,Y}\left[\exp\left(\int_0^{t/2} I_p(X_s,Y_s)\,\mathrm{d}s\right)\mathbb{1}_{B_R^p}(X_{t/2})\,\mathbb{1}_{B_R^n}(Y_{t/2})\right].$$

On the other hand, by (23), (24) and our choice of R, we have

$$\mathbb{E}_{0,0}^{X,Y}\left[\exp\left(\int_0^{t/2} I_p(X_s,Y_s)\,\mathrm{d}s\right)\mathbb{1}_{(B_R^p\times B_R^n)^c}(X_{t/2},Y_{t/2})\right]$$
$$\leq \exp\left(\frac{tnp}{2}\right)\mathbb{P}_0(X_{t/2}\notin B_R^p)\mathbb{P}_0(Y_{t/2}\notin B_R^n)$$
$$\leq np\exp\left[\frac{tnp}{2}-\big(C(d,\rho)+C(d,\kappa)\big)t\log(t)\right],$$

and therefore, with a similary reasoning as in the proof of Lemma 2.1 we get

$$\liminf_{t\to\infty}\frac{1}{t}\log\mathbb{E}_0^{\otimes n}\left[u(0,t)^p\right]\geq \liminf_{t\to\infty}\frac{2}{t}\log\mathbb{E}_{0,0}^{X,Y}\left[\exp\left(\int_0^{t/2} I_p(X_s,Y_s)\,\mathrm{d}s\right)\right].$$

Now, the occupation measure $\frac{1}{t}\int_0^t \delta_{(X_s,Y_s)}\,\mathrm{d}s$ satisfies a weak large deviations principle (LDP) in the space $\mathcal{M}_1(\mathbb{Z}^{\mathrm{dp}}\times\mathbb{Z}^{\mathrm{dn}})$ of probability measures on $\mathbb{Z}^{\mathrm{dp}}\times\mathbb{Z}^{\mathrm{dn}}$, endowed with the weak topology. The speed of this LDP is t, and the rate function is given for all $\nu\in\mathcal{M}_1(\mathbb{Z}^{\mathrm{dp}}\times\mathbb{Z}^{\mathrm{dn}})$ by

$$J(\nu)=\kappa\left\|\nabla_x\sqrt{\nu}\right\|_2^2+\rho\left\|\nabla_y\sqrt{\nu}\right\|_2^2,$$

(see, e.g., den Hollander [12], Sect. IV.4). Since I is bounded, the lower bound in Varadhan's integral lemma (see, e.g., den Hollander [12], Sect. III.3) yields

$$\liminf_{t\to\infty}\frac{1}{t}\log\mathbb{E}_0^{\otimes n}\left[u(0,t)^p\right]\geq \sup_{\nu\in\mathcal{M}_1(\mathbb{Z}^{\mathrm{dp}}\times\mathbb{Z}^{\mathrm{dn}})}\left\{\sum_{(x,y)} I_p(x,y)\nu(x,y)-J(\nu)\right\}.$$

Setting $f(x,y)=\sqrt{\nu}(x,y)$ gives then the lower bound in (20). □

Step 3: Properties of $\lambda_p^{(n)}$. Since $0\leq I_p(x,y)\leq np$, we clearly have $0\leq\lambda_p^{(n)}\leq n$. Using representation (20), we can conclude that the function $(\kappa,\rho)\mapsto\lambda_p^{(n)}(\kappa,\rho)$

is convex and non-increasing in κ and ρ. Moreover, $\lambda_p^{(n)}(\kappa,\rho)$ is lower semi-continuous since it is supremum of functions that are linear in κ and ρ. Finally, since every finite convex function is also upper semi-continuous, $\lambda_p^{(n)}$ is upper semi-continuous. Hence, $\lambda_p^{(n)}(\kappa,\rho)$ is continuous.

3 Proof of Theorems 1.2–1.3

By symmetry, note that for all $n, p \in \mathbb{N}$ and $\kappa, \rho \in [0,\infty)$,

$$\lambda_p^{(n)}(\kappa,\rho) = \frac{n}{p}\lambda_n^{(p)}(\rho,\kappa). \tag{25}$$

3.1 Proof of Theorem 1.2

Proof of (i) By continuity, $\lim_{\kappa\to 0}\lambda_p^{(n)}(\kappa,\rho) = \lambda_p^{(n)}(0,\rho)$. Now for $\kappa = 0$, the X particles do not move so that $\mathbb{E}_0^{\otimes n}\left[u(0,t)^p\right] = \mathbb{E}_0\left(\exp\left(pL_t^Y(0)\right)\right)^n$ (see (19)), where $L_t^Y(0)$ is the local time at 0 of a simple random walk in $\mathbb{Z}^d$ with rate $2d\rho$. Using the LDP for L_t^Y, we obtain

$$\lambda_p^{(n)}(0,\rho) = \frac{n}{p}\sup_{\substack{f\in l^2(\mathbb{Z}^d)\\ \|f\|_2=1}}\langle f, (\rho\Delta + p\delta_0)f\rangle = n\mu(\rho/p).$$

Proof of (ii) For all $n, p \in \mathbb{N}$ and $\kappa, \rho \in [0,\infty)$, we have

$$\lambda_p^{(n)}(\kappa,\rho) \geq \lambda_1^{(n)}(\kappa,\rho) = n\lambda_n^{(1)}(\rho,\kappa) \geq n\lambda_1^{(1)}(\rho,\kappa) = n\mu(\kappa+\rho),$$

where the last equality is proved in [3] and comes from the fact that $X_t^1 - Y_t^1$ is a simple random walk in $\mathbb{Z}^d$ with jump rate $2d(\kappa+\rho)$. Since $G_d(0) = \infty$ for $d = 1, 2$, it follows from (9) that $\lambda_p^{(n)}(\kappa,\rho) > 0$ for $d = 1, 2$.

Let us prove that $\lim_{\kappa\to\infty}\lambda_p^{(n)}(\kappa,\rho) = 0$. By monotonicity in ρ,

$$\lambda_p^{(n)}(\kappa,\rho) \leq \lambda_p^{(n)}(\kappa,0) = n\mu(\kappa/n). \tag{26}$$

Hence, the only thing to prove is that $\lim_{\kappa\to\infty}\mu(\kappa) = 0$. To this end, one can use the discrete Gagliardo-Nirenberg inequality: there exists a constant C such that for all $f : \mathbb{Z}^d \mapsto \mathbb{R}$,

$$\text{for } d = 1, \quad \|f\|_\infty^2 \leq C\,\|f\|_2\,\|\nabla f\|_2\,; \tag{27}$$

$$\text{for } d = 2, \quad \|f\|_4^2 \leq C\,\|f\|_2\,\|\nabla f\|_2\,. \tag{28}$$

The proof of these inequalities follows the same lines as the proof of the usual Gagliardo-Nirenberg inequality (see, e.g., Brezis [1]). For completeness, a short proof is given in the appendix. From (27) and (28), we get for all $f \in l_2(\mathbb{Z}^d)$ with $\|f\|_2 = 1$,

$$-\kappa \|\nabla f\|_2^2 + f(0)^2 \le \begin{cases} -\kappa \|\nabla f\|_2^2 + \|f\|_\infty^2 & \text{for } d = 1 \\ -\kappa \|\nabla f\|_2^2 + \|f\|_4^2 & \text{for } d = 2 \end{cases}$$
$$\le -\kappa \|\nabla f\|_2^2 + C \|\nabla f\|_2 .$$

Taking the supremum over f yields

$$\mu(\kappa) \le \sup_{x \ge 0} \left(-\kappa x^2 + Cx\right) = \frac{C^2}{4\kappa}.$$

The strict monotonicity is now an easy consequence of the fact that $\kappa \mapsto \lambda_p^{(n)}(\kappa, \rho)$ is convex, positive, non-increasing, and tends to 0 as $\kappa \to \infty$.

Proof of (iii). By (25) and (26), we get

$$\lambda_p^{(n)}(\kappa, \rho) \le n \min\left(\mu(\kappa/n), \mu(\rho/p)\right). \tag{29}$$

Then the claim follows by (9).

3.2 *Proof of Theorem 1.3*

Proof of (i). Fix $\epsilon > 0$. Let f approaching the supremum in the variational representation (20) of $\lambda_p^{(n)}(\kappa, 0)$, so that

$$p\lambda_p^{(n)}(\kappa, 0) - \epsilon \le -\kappa \|\nabla_x f\|_2^2 + \sum_{x \in \mathbb{Z}^{\mathrm{dp}}} \sum_{y \in \mathbb{Z}^{\mathrm{dn}}} I_p(x, y) f^2(x, y)$$
$$\le p\lambda_p^{(n)}(\kappa, \rho) + \rho \sup_{\substack{f \in l^2(\mathbb{Z}^{\mathrm{dp}} \times \mathbb{Z}^{\mathrm{dn}}) \\ \|f\|_2 = 1}} \left\|\nabla_y f\right\|_2^2 .$$

For $x \in \mathbb{Z}^{\mathrm{dp}}$, set $f_x : y \in \mathbb{Z}^{\mathrm{dn}} \mapsto f(x, y)$. Since the bottom of the spectrum of Δ in $l^2(\mathbb{Z}^{\mathrm{dn}})$ is $-4dn$,

$$\sum_{y \in \mathbb{Z}^{\mathrm{dn}}} \left\|\nabla_y f_x(y)\right\|_2^2 \le 4dn \sum_{y \in \mathbb{Z}^{\mathrm{dn}}} f_x^2(y) ,$$

for all $x \in \mathbb{Z}^{\mathrm{dp}}$. Hence,

$$\sum_{x\in\mathbb{Z}^{\mathrm{dp}}}\sum_{y\in\mathbb{Z}^{\mathrm{dn}}} \left\|\nabla_y f_x(y)\right\|_2^2 \le 4dn \sum_{x\in\mathbb{Z}^{\mathrm{dp}}}\sum_{y\in\mathbb{Z}^{\mathrm{dn}}} f_x^2(y) = 4\mathrm{d}n\,.$$

Therefore, for all $\epsilon > 0$,

$$p\lambda_p^{(n)}(\kappa,0) - \epsilon \le p\lambda_p^{(n)}(\kappa,\rho) + 4\mathrm{d}n\rho.$$

Letting $\epsilon \to 0$ yields,

$$\lambda_p^{(n)}(\kappa,0) - \frac{4\mathrm{d}n\rho}{p} \le \lambda_p^{(n)}(\kappa,\rho) \le \lambda_p^{(n)}(\kappa,0)\,, \tag{30}$$

which, after letting $p \to \infty$, gives the claim.

Proof of (ii). By (25), $\lim_{n\to\infty}\lambda_p^{(n)}(\kappa,\rho) = \lim_{n\to\infty}\frac{n}{p}\lambda_n^{(p)}(\rho,\kappa)$ and by (i),

$$\lim_{n\to\infty}\lambda_n^{(p)}(\rho,\kappa) \ge \lambda_n^{(p)}(\rho,0) = p\mu(\rho/p) > 0\,,\ \text{for } p > \rho/G_d(0).$$

Hence, for $p > \rho/G_d(0)$, $\lim_{n\to\infty}\lambda_p^{(n)}(\kappa,\rho) = +\infty$.

Proof of (iii). This is a direct consequence of Theorem 1.2(iii).

4 Proof of Theorem 1.4

Proof of (i). We first prove that

$$\kappa_p^{(n)}(\rho) = \sup_{\substack{f\in l_2(\mathbb{Z}^{\mathrm{dp}}\times\mathbb{Z}^{\mathrm{dn}})\\ \|f\|_2=1}} \frac{\sum_{x,y} I_p(x,y)f^2(x,y) - \rho\left\|\nabla_y f\right\|_2^2}{\|\nabla_x f\|_2^2}\,, \tag{31}$$

with I defined as in (17). Indeed, let us denote by S the supremum in the right-hand side of (31).

If $\kappa \ge \kappa_p^{(n)}(\rho)$, then $\lambda_p^{(n)}(\kappa,\rho) = 0$. Therefore, using (20), for all $f \in l_2(\mathbb{Z}^{\mathrm{dp}} \times \mathbb{Z}^{\mathrm{dn}})$ such that $\|f\|_2 = 1$,

$$\sum_{x\in\mathbb{Z}^{\mathrm{dp}}}\sum_{y\in\mathbb{Z}^{\mathrm{dn}}} I_p(x,y)f^2(x,y) - \rho\left\|\nabla_y f\right\|_2^2 \le \kappa\left\|\nabla_x f\right\|_2^2,$$

so that $\kappa \ge S$. Hence, $\kappa_p^{(n)}(\rho) \ge S$. On the opposite direction, we can assume that $S < \infty$. Then, by definition of S, for all $f \in l_2(\mathbb{Z}^{\mathrm{dp}} \times \mathbb{Z}^{\mathrm{dn}})$ such that $\|f\|_2 = 1$,

$$\sum_{x \in \mathbb{Z}^{\mathrm{dp}}} \sum_{y \in \mathbb{Z}^{\mathrm{dn}}} I_p(x,y) f^2(x,y) - \rho \left\| \nabla_y f \right\|_2^2 \leq S \left\| \nabla_x f \right\|_2^2.$$

Thus, for all $f \in l_2(\mathbb{Z}^{\mathrm{dp}} \times \mathbb{Z}^{\mathrm{dn}})$ such that $\|f\|_2 = 1$, and all $\kappa \geq S$,

$$\sum_{x \in \mathbb{Z}^{\mathrm{dp}}} \sum_{y \in \mathbb{Z}^{\mathrm{dn}}} I_p(x,y) f^2(x,y) - \rho \left\| \nabla_y f \right\|_2^2 - \kappa \left\| \nabla_x f \right\|_2^2 \leq (S - \kappa) \left\| \nabla_x f \right\|_2^2 \leq 0.$$

Hence, for all $\kappa \geq S$, $\lambda_p^{(n)}(\kappa,\rho) = 0$, i.e., $\kappa \geq \kappa_p^{(n)}(\rho)$. Hence, $S \geq \kappa_p^{(n)}(\rho)$. This proves (31).

Since $\rho \mapsto \kappa_p^{(n)}(\rho)$ is a supremum of linear functions, it is lower semi-continuous and convex. It is also obvious that $\rho \mapsto \kappa_p^{(n)}(\rho)$ is non-increasing. The continuity follows then from the finiteness of $\kappa_p^{(n)}(\rho)$.

The lower bound in (13) is a direct consequence of (30). Indeed, since $\lambda_p^{(n)}(\kappa, 0) = n\mu(\kappa/n)$, it follows from (30) that if $\mu(\kappa/n) > 4\mathrm{d}\rho/p$, then $\kappa < \kappa_p^{(n)}(\rho)$. This yields the bound:

$$\kappa_p^{(n)}(\rho) \geq n\mu^{-1}(4\mathrm{d}\rho/p).$$

Using the symmetry relation (25), we also get from (30) that

$$\lambda_p^{(n)}(\kappa,\rho) \geq n\mu(\rho/p) - 4d\kappa.$$

This leads to $\kappa_p^{(n)}(\rho) \geq \frac{n}{4d}\mu(\rho/p)$. Hence, if $\rho/p < G_d(0)$, $\kappa_p^{(n)}(\rho) > 0$. We have already seen that $\kappa_p^{(n)}(\rho) = 0$ if $\rho/p \geq G_d(0)$. Since $\lambda_p^{(n)}(\kappa, 0) = n\mu(\kappa/n)$, it follows that $\kappa_p^{(n)}(0) = nG_d(0)$. Using convexity, we have, for all $\rho \in [0, pG_d(0)]$,

$$\kappa_p^{(n)}(\rho) \leq \frac{\kappa_p^{(n)}(pG_d(0)) - \kappa_p^{(n)}(0)}{pG_d(0)}\rho + \kappa_p^{(n)}(0) = n\left(G_d(0) - \rho/p\right).$$

Since $\kappa_p^{(n)}(\rho) = 0$ if $\rho/p \geq G_d(0)$, then the upper bound in (13) is proved.

Proof of (ii). To prove (14), let f_0 be the function

$$f_0(x,y) = \prod_{i=1}^{p} \frac{G_d(x_i)}{\|G_d\|_2} \prod_{j=1}^{n} \delta_0(y_j).$$

Note that for $d \geq 5$, $\|G_d\|_2 < \infty$, so that f_0 is well-defined, and has l_2-norm equal to 1. From (31), we get

$$\kappa_p^{(n)}(\rho) \geq \frac{\sum_{x,y} I_p(x,y) f_0^2(x,y) - \rho \left\| \nabla_y f_0 \right\|_2^2}{\|\nabla_x f_0\|_2^2}.$$

An easy computation then gives

$$\sum_{x,y} I_p(x,y) f_0^2(x,y) = np\frac{G_d^2(0)}{\|G_d\|_2^2},$$

$$\left\|\nabla_y f_0\right\|_2^2 = n \left\|\nabla_{y_1}\delta_0\right\|_2^2 = 2\mathrm{d}n\,,$$

and

$$\|\nabla_x f_0\|_2^2 = p\frac{\|\nabla_{x_1} G_d\|_2^2}{\|G_d\|_2^2} = p\frac{G_d(0)}{\|G_d\|_2^2},$$

since $\|\nabla_{x_1} G_d\|_2^2 = \langle G_d, -\Delta G_d\rangle = \langle G_d, \delta_0\rangle = G_d(0)$. This gives (14).

Proof of (iii). The inequality (15) is clear if $\rho \in [(p-1)G_d(0), pG_d(0))$, since in this case, $\kappa_{p-1}^{(n)}(\rho) = 0 < \kappa_p^{(n)}(\rho)$. We assume therefore that $\rho \in (0, (p-1)G_d(0))$. From (13), we have $\kappa_{p-1}^{(n)}(\rho) \leq nG_d(0) - \rho n/(p-1)$, whereas, from (14), $\kappa_p^{(n)}(\rho) \geq nG_d(0) - \rho n/(p\alpha_d)$. Hence, $\kappa_{p-1}^{(n)}(\rho) < \kappa_p^{(n)}(\rho)$ as soon as $\alpha_d > \frac{p-1}{p}$. This gives the claim.

5 Proof of Corollary 1.1

Proof of (i). The function $p \mapsto \lambda_p^{(n)}(\kappa,\rho)$ increases from $\lambda_1^{(n)}(\kappa,\rho)$ to $n\mu(\kappa/n)$. Hence, there exists p such that $\lambda_p^{(n)}(\kappa,\rho) < \lambda_{p+1}^{(n)}(\kappa,\rho)$ as soon as $\lambda_1^{(n)}(\kappa,\rho) < n\mu(\kappa/n)$. But $n\mu(\kappa/n) = \lambda_1^{(n)}(\kappa,0)$. Hence, if $\lambda_1^{(n)}(\kappa,\rho) = n\mu(\kappa/n)$, the convex decreasing function $\rho \mapsto \lambda_1^{(n)}(\kappa,\rho)$ is constant. Being equal to 0 for $\rho \geq G_d(0)$, we get that $n\mu(\kappa/n) = 0$, which can not be the case if $\kappa < nG_d(0)$. This ends the proof of the first part.

If $\kappa \geq nG_d(0)$, then $\lambda_p^{(n)}(\kappa,\rho) = 0$, for all $p \geq 1$, and the system is not intermittent. This proves the second part.

Proof of (ii). For all $p \in \mathbb{N}\setminus\{1\}$ by Lemma 1.1 for d large enough, we have $\alpha_d > \frac{p-1}{p}$. This implies that $\alpha_d > \frac{q-1}{q}$ for all $q \in \mathbb{N}\setminus\{1\}$ and $q \leq p$. Hence, by Theorem 1.4(iii), for all $q \in \mathbb{N}\setminus\{1\}$ with $q \leq p$ we have $\kappa_{q-1}^{(n)}(\rho) < \kappa_q^{(n)}(\rho)$, for all $\rho \in (0, pG_d(0))$. Hence, in the domain

$$\left\{(\kappa,\rho)\colon \rho \in (0, qG_d(0))\,,\ \kappa_{q-1}^{(n)}(\rho) \leq \kappa < \kappa_q^{(n)}(\rho)\right\},$$

one has

$$\lambda_1^{(n)}(\kappa,\rho) = \cdots = \lambda_{q-1}^{(n)}(\kappa,\rho) = 0 < \lambda_q^{(n)}(\kappa,\rho)\,,$$

which proves the desired result.

Acknowledgement The research in this paper was supported by the ANR-project MEMEMO.

Appendix: Proof of Lemma 1.1

For a function $f : \mathbb{Z}^d \mapsto \mathbb{R}$, let $\hat{f}$ denote the Fourier transform of f:

$$\hat{f}(\theta) = \sum_{x \in \mathbb{Z}^d} \mathrm{e}^{i\langle \theta, x\rangle} f(x) \quad \forall \theta \in [0, 2\pi]^d.$$

Then, the inverse Fourier transform is given by

$$f(x) = \frac{1}{(2\pi)^d} \int_{[0,2\pi]^d} \mathrm{e}^{-i\langle \theta, x\rangle} \hat{f}(\theta)\, \mathrm{d}\theta\,,$$

and the Plancherel's formula reads

$$\sum_{x \in \mathbb{Z}^d} f^2(x) = \frac{1}{(2\pi)^d} \int_{[0,2\pi]^d} |\hat{f}(\theta)|^2\, \mathrm{d}\theta.$$

Using the equation $\Delta G_d = -\delta_0$, we get that

$$\hat{G}_d(\theta) = \frac{1}{2\sum_{i=1}^d (1-\cos(\theta_i))}.$$

Hence,

$$\begin{aligned} G_d(0) &= \frac{1}{(2\pi)^d} \int_{[0,2\pi]^d} \frac{\mathrm{d}\theta}{2\sum_{i=1}^d (1-\cos(\theta_i))} \\ &= \frac{1}{\pi^d} \int_{[0,\pi]^d} \frac{\mathrm{d}\theta}{2\sum_{i=1}^d (1-\cos(\theta_i))} \\ &= \mathbb{E}\left[\frac{1}{2\sum_{i=1}^d (1-\cos(\Theta_i))}\right] \end{aligned}$$

where the random variables (Θ_i) are i.i.d. with uniform distribution on $[0, \pi]$. Moreover, by Plancherel's formula we have

$$\|G_d\|_2^2 = \frac{1}{(2\pi)^d} \int_{[0,2\pi]^d} \frac{d\theta}{\left(2\sum_{i=1}^d (1-\cos(\theta_i))\right)^2} = \mathbb{E}\left[\frac{1}{\left(2\sum_{i=1}^d (1-\cos(\Theta_i))\right)^2}\right].$$

Thus,

$$\alpha_d = \frac{G_d(0)}{2d\, \|G_d\|_2^2} = \frac{\mathbb{E}\left[\frac{1}{\bar{S}_d}\right]}{\mathbb{E}\left[\frac{1}{\bar{S}_d^2}\right]},$$

where $\bar{S}_d = \frac{1}{d}\sum_{i=1}^d (1-\cos(\Theta_i))$. Applying Hölder's and Jensen's inequality, we get that

$$\alpha_d \le \frac{1}{\sqrt{\mathbb{E}\left[\frac{1}{\bar{S}_d^2}\right]}} \le \mathbb{E}(\bar{S}_d) = 1.$$

By the law of large numbers, $\bar{S}_d$ converges almost surely to $\mathbb{E}\left[1-\cos(\Theta)\right] = 1$ as d tends to infinity. We are now going to prove that $\bar{S}_d^{-2}$ is uniformly integrable by showing that for all $p > 2$,

$$\sup_{d>2p} \mathbb{E}\left[\bar{S}_d^{-p}\right] < \infty. \tag{32}$$

Indeed, let $\epsilon \in (0,\pi)$ be a small positive number to be fixed later. Let

$$\mathscr{I} = \{i \in \{1,\cdots,d\} : 0 \le \Theta_i \le \epsilon\}.$$

$$\bar{S}_d \ge \frac{1}{d}\sum_{i\notin\mathscr{I}}(1-\cos(\epsilon)) + \frac{c_\epsilon}{d}\sum_{i\in\mathscr{I}}\Theta_i^2,$$

where $c_\epsilon = \inf_{0\le\theta\le\epsilon}\frac{1-\cos(\theta)}{\theta^2} \to 1/2$ when $\epsilon \to 0$. Therefore,

$$\mathbb{E}\left[\bar{S}_d^{-p}\right] \le d^p \sum_{k=0}^{d} \sum_{\substack{I\subset\{1,\cdots,d\}\\ |I|=k}} \mathbb{E}\left[\frac{\mathbb{1}_{\mathscr{I}=I}}{\left((1-\cos(\epsilon))(d-k) + c_\epsilon\sum_{i\in I}\Theta_i^2\right)^p}\right].$$

Since the last expectation only depends on $|I|$, we get

$$\mathbb{E}\left[\bar{S}_d^{-p}\right] \le d^p \sum_{k=0}^{d} \binom{d}{k} a(k,\epsilon,d),$$

with

$$a(k,\epsilon,d) := \frac{1}{\pi^d}\int_{\substack{0\le\theta_1,\cdots,\theta_k\le\epsilon\\ \epsilon\le\theta_{k+1},\cdots,\theta_d\le\pi}} \frac{\mathrm{d}\theta_1\cdots\mathrm{d}\theta_d}{\left((1-\cos(\epsilon))(d-k) + c_\epsilon(\theta_1^2+\cdots+\theta_k^2)\right)^p}.$$

Let ω_d denote the volume of the d-dimensional unit ball. For $k = d$,

$$a(d,\epsilon,d) = \frac{1}{\pi^d}\int_{0\le\theta_1,\cdots,\theta_d\le\epsilon}\frac{\mathrm{d}\theta_1\cdots\mathrm{d}\theta_d}{c_\epsilon^p\,\|\theta\|^{2p}}$$
$$\le \frac{1}{c_\epsilon^p\pi^d}\omega_d\int_0^{\sqrt{d}\epsilon} r^{d-2p-1}\mathrm{d}r$$
$$= \left(\frac{\epsilon}{\pi}\right)^d\frac{1}{(c_\epsilon\epsilon^2)^p}d^{\frac{d}{2}-p}\frac{\omega_d}{d-2p},$$

for $d > 2p$.

Note that for large d, $\omega_d \simeq \frac{(2e\pi)^{d/2}}{\sqrt{\pi d}d^{d/2}}$. Therefore, as $d \to \infty$

$$d^p\binom{d}{d}a(d,\epsilon,d) = O\left(d^{-3/2}(\epsilon^2 2e/\pi)^{d/2}\right).$$

If ϵ is chosen so that $\epsilon^2 \le \pi/(2e)$, we obtain that $\lim_{d\to\infty} d^p\binom{d}{d}a(d,\epsilon,d) = 0$.

For $k \le d-1$,

$$a(k,\epsilon,d) \le \frac{1}{(1-\cos(\epsilon))^p}\frac{1}{(d-k)^p}\left(\frac{\epsilon}{\pi}\right)^k\left(1-\frac{\epsilon}{\pi}\right)^{d-k},$$

and $d^p\binom{d}{k}a(k,\epsilon,d) \le \frac{1}{(1-\cos(\epsilon))^p}\mathbb{E}\left[\,\mathbb{1}_{N=k}(1-N/d)^{-p}\right]$, where N is a Binomial random variable with parameters d and ϵ/π. Hence, for $\epsilon < \min(\pi, \sqrt{\pi/(2e)})$,

$$\mathbb{E}\left[\frac{1}{\bar{S}_d^p}\right] \le \frac{1}{(1-\cos(\epsilon))^p}\mathbb{E}\left[\,\mathbb{1}_{N\le d-1}(1-N/d)^{-p}\right] + O\left(d^{-3/2}\right)$$
$$\le \frac{d^p}{(1-\cos(\epsilon))^p}\mathbb{P}\left[d\frac{2\epsilon}{\pi}\le N\le d-1\right]$$
$$+\frac{1}{(1-\cos(\epsilon))^p(1-\frac{2\epsilon}{\pi})^p} + O\left(d^{-3/2}\right).$$

Now, by the large deviations principle satisfied by N/d, there is an $i(\epsilon) > 0$ such that $\mathbb{P}\left[N \ge \mathrm{d}2\epsilon/\pi\right] \le \exp(-\mathrm{d}i(\epsilon))$. This ends the proof of (32).

Using the uniform integrability (32), and the fact that $\bar{S}_d$ converges a.s. to 1, we obtain that $\mathbb{E}\left[\frac{1}{\bar{S}_d}\right]$ and $\mathbb{E}\left[\frac{1}{\bar{S}_d^2}\right]$ both converge to 1, when d goes to infinity.

Appendix: Proof of Proposition 1.1.

Let $f \in l_2(\mathbb{Z}^{d(1+n)})$ with $\|f\|_2 = 1$, such that $\mathscr{L}_1 f = \lambda_1^{(n)}(\kappa,\rho) f$. Define

$$\tilde{f}(x_1,x_2,y) = f(x_1,y) f(x_2,y), \quad x_1, y_1 \in \mathbb{Z}^d,\ y \in \mathbb{Z}^{\mathrm{dn}}.$$

Since

$$\sum_{x_1,x_2,y} \tilde{f}^2(x_1,x_2,y) = \sum_y \left(\sum_x f^2(x,y)\right)^2 \le \left(\sup_y \sum_x f^2(x,y)\right) \|f\|_2^2 \le \|f\|_2^4,$$

it follows that $\tilde{f}$ is in $l_2(\mathbb{Z}^{d(2+n)})$. A simple computation yields

$$\Delta_{x_1} \tilde{f}(x_1,x_2,y) = f(x_2,y)\Delta_x f(x_1,y), \;\; \Delta_{x_2} \tilde{f}(x_1,x_2,y) = f(x_1,y)\Delta_x f(x_2,y),$$

and

$$\begin{aligned}\Delta_y \tilde{f}(x_1,x_2,y) = {} & f(x_2,y)\Delta_y f(x_1,y) + f(x_1,y)\Delta_y f(x_2,y) \\ & + \sum_{z\sim y} (f(x_1,z) - f(x_1,y))(f(x_2,z) - f(x_2,y)).\end{aligned}$$

Since

$$I_2 \tilde{f}(x_1,x_2,y) = f(x_2,y) I_1 f(x_1,y) + f(x_1,y) I_1 f(x_2,y),$$

(recalling (17)), this leads to

$$\begin{aligned}\mathscr{L}_2 \tilde{f}(x_1,x_2,y) = {} & 2\lambda_1^{(n)}(\kappa,\rho)\tilde{f}(x_1,x_2,y) \\ & + \rho \sum_{z\sim y} \big(f(x_1,z) - f(x_1,y)\big)\big(f(x_2,z) - f(x_2,y)\big)\end{aligned}$$

(recalling (16). Therefore,

$$\begin{aligned}\lambda_2^{(n)}(\kappa,\rho) \left\|\tilde{f}\right\|_2^2 & \ge \frac{1}{2}\left\langle \tilde{f}, \mathscr{L}_2 \tilde{f}\right\rangle \\ & = \lambda_1^{(n)}(\kappa,\rho)\left\|\tilde{f}\right\|_2^2 + \frac{\rho}{2}\sum_{y,z\sim y}\left(\sum_x f(x,y)(f(x,z) - f(x,y))\right)^2.\end{aligned}$$

Note that

$$\sum_{y,z\sim y}\left(\sum_x f(x,y)(f(x,z) - f(x,y))\right)^2 \ge 0,$$

with equality to 0 if and only if for all y and $z \sim y$, $\sum_x f(x,y)(f(x,z) - f(x,y)) = 0$. Interchanging the role of z and y yields $\sum_x (f(x,z) - f(x,y))^2 = 0$, so that for all x, y and $z \sim y$, $f(x,z) = f(x,y)$. Hence, for all x, y, $f(x,y) = f(x,0)$. This is impossible since $\|f\|_2 = 1$. Thus, $\lambda_2^{(n)}(\kappa,\rho) > \lambda_1^{(n)}(\kappa,\rho)$. □

Appendix: Proof of the Discrete Gagliardo–Nirenberg Inequality

Proof for $d = 1$. One can assume that $\|f\|_2 < \infty$, otherwise there is nothing to prove. Hence $\lim_{|x|\to\infty} |f(x)| = 0$, and by the Cauchy–Schwarz inequality, we have for all $x \in \mathbb{Z}$,

$$
\begin{aligned}
f^2(x) &= \sum_{j=-\infty}^{x} f^2(j) - f^2(j-1) \\
&\le \sum_{j=-\infty}^{+\infty} |f(j) - f(j-1)| \, (|f(j)| + |f(j-1)|) \\
&\le 2 \sqrt{\sum_j |f(j) - f(j-1)|^2} \sqrt{\sum_j f^2(j)} \\
&= 2 \, \|f\|_2 \, \|\nabla f\|_2 \,,
\end{aligned}
$$

which proves (27) with $C = 2$.

Proof for $d = 2$. Here again, one can assume that $\|f\|_2 < \infty$, and consequently $\lim_{|x_1|\to\infty} |f(x_1,x_2)| = 0$. Then, by the Cauchy–Schwarz inequality, we have for all $x_1, x_2 \in \mathbb{Z}$,

$$
\begin{aligned}
f^2(x_1,x_2) &= \sum_{j_1=-\infty}^{x_1} f^2(j_1,x_2) - f^2(j_1-1,x_2) \\
&\le \sum_{j_1=-\infty}^{+\infty} |f(j_1,x_2) - f(j_1-1,x_2)| \, (|f(j_1,x_2)| + |f(j_1-1,x_2)|) \\
&\le 2 \sqrt{\sum_{j_1} |f(j_1,x_2) - f(j_1-1,x_2)|^2} \sqrt{\sum_{j_1} f^2(j_1,x_2)} \\
&= 2 \, \|f(\cdot,x_2)\|_2 \sqrt{\sum_{j_1} |\nabla_{x_1} f(j_1,x_2)|^2} := 2\tilde{f}_1(x_2).
\end{aligned}
$$

Similarly, we have

$$f^2(x_1,x_2) \le 2\,\|f(x_1,\cdot)\|_2 \sqrt{\sum_{j_2} |\nabla_{x_2} f(x_1,j_2)|^2} := 2\tilde{f}_2(x_1).$$

Thus

$$\sum_{x_1,x_2} f^4(x_1,x_2) \le 4\left(\sum_{x_2} \tilde{f}_1(x_2)\right)\left(\sum_{x_1} \tilde{f}_2(x_1)\right).$$

Since

$$\begin{aligned}\sum_{x_2} \tilde{f}_1(x_2) &= \sum_{x_2} \|f(\cdot,x_2)\|_2 \sqrt{\sum_{j_1} |\nabla_{x_1} f(j_1,x_2)|^2}\\ &\le \sqrt{\sum_{x_2} \|f(\cdot,x_2)\|_2^2}\sqrt{\sum_{x_2}\sum_{j_1} |\nabla_{x_1} f(j_1,x_2)|^2}\\ &\le \|f\|_2\,\|\nabla f\|_2\,,\end{aligned}$$

and the same being true for $\sum_{x_1} \tilde{f}_2(x_1)$, it follows that

$$\|f\|_4^4 \le 4\,\|f\|_2^2\,\|\nabla f\|_2^2\,,$$

which proves (28) with $C = 2$.

References

1. Brezis, H.: Analyse fonctionnelle: Théorie et applications. Collection Mathématiques Appliquées pour la Maîtrise, Masson, Paris (1983)
2. Carmona R.A., Molchanov, S.A.: Parabolic Anderson Problem and Intermittency. AMS Memoir 518, American Mathematical Society, Providence RI (1994)
3. Gärtner, J., Heydenreich, M.: Annealed asymptotics for the parabolic Anderson model with a moving catalyst. Stoch. Proc. Appl. **116**, 1511–1529 (2006)
4. Gärtner, J., den Hollander, F.: Intermittency in a catalytic random medium. Ann. Probab. **34**, 2219–2287 (2006)
5. Gärtner J., den Hollander, F., Maillard, G.: Intermittency on catalysts: symmetric exclusion. Electronic J. Probab. **12**, 516–573 (2007)
6. Gärtner, J., den Hollander, F., Maillard, G.: Intermittency on catalysts. In: Blath, J., Mörters P., Scheutzow, M. (eds.) Trends in Stochastic Analysis, London Mathematical Society Lecture Note Series 353, pp. 235-248. Cambridge University Press, Cambridge (2009)
7. Gärtner, J., den Hollander, F., Maillard, G.: Intermittency on catalysts: three-dimensional simple symmetric exclusion. Electronic J. Probab. **72**, 2091–2129 (2009)
8. Gärtner, J., den Hollander, F., Maillard, G.: Intermittency on catalysts: voter model. Ann. Probab. **38**, 2066–2102 (2010)

9. Gärtner, J., König, W.: The parabolic Anderson model. In: Deuschel, J.-D., Greven, A. (eds.) Interacting Stochastic Systems, pp. 153–179. Springer, Berlin (2005)
10. Gärtner, J., Molchanov, S.A.: Parabolic problems for the Anderson model. Commun. Math. Phys. **132**, 613–655 (1990)
11. Greven, A., den Hollander, F.: Phase transition for the long-time behavior of interacting diffusions. Ann. Probab. **35**, 1250–1306 (2007)
12. den Hollander, F.: Large Deviations. Fields Institute Monographs 14, American Mathematical Society, Providence, RI (2000)
13. Kesten, H., Sidoravicius, V.: Branching random walk with catalysts. Electr. J. Prob. **8**, 1–51 (2003)
14. Maillard, G., Mountford, T., Schöpfer, S.: Parabolic Anderson model with voter catalysts: Dichotomy in the behavior of Lyapunov exponents. In: Deuschel, J.-D., Gentz, B., König, W., von Renesse, M.-K., Scheutzow, M., Schmock, U., (eds.), Probability in Complex Physical Systems, Vol. 11, pp. 33–68. Springer, Heidelberg (2012)
15. Schnitzler, A., Wolff, T.: Precise asymptotics for the parabolic Anderson model with a moving catalyst or trap. In: Deuschel, J.-D., Gentz, B., König, W., von Renesse, M.-K., Scheutzow, M., Schmock, U., (eds.), Probability in Complex physical systems, Vol. 11, pp. 69–89. Springer, Heldelberg (2012)

Survival Probability of a Random Walk Among a Poisson System of Moving Traps

Alexander Drewitz, Jürgen Gärtner, Alejandro F. Ramírez, and Rongfeng Sun

Abstract We review some old and prove some new results on the survival probability of a random walk among a Poisson system of moving traps on $\mathbb{Z}^d$, which can also be interpreted as the solution of a parabolic Anderson model with a random time-dependent potential. We show that the annealed survival probability decays asymptotically as $e^{-\lambda_1\sqrt{t}}$ for $d = 1$, as $e^{-\lambda_2 t/\log t}$ for $d = 2$, and as $e^{-\lambda_d t}$ for $d \geq 3$, where λ_1 and λ_2 can be identified explicitly. In addition, we show that the quenched survival probability decays asymptotically as $e^{-\tilde{\lambda}_d t}$, with $\tilde{\lambda}_d > 0$ for all $d \geq 1$. A key ingredient in bounding the annealed survival probability is what is known in the physics literature as the Pascal principle, which asserts that the annealed survival probability is maximized if the random walk stays at a fixed position. A corollary of independent interest is that the expected cardinality of the range of a continuous time symmetric random walk increases under perturbation by a deterministic path.

AMS 2010 subject classification: 60K37, 60K35, 82C22.

A. Drewitz
Departement Mathematik, Eidgenössische Technische Hochschule Zürich, Rämistrasse 101, 8092 Zürich, Switzerland
e-mail: alexander.drewitz@math.ethz.ch

J. Gärtner
Institut für Mathematik, Technische Universität Berlin, Sekr. MA 7-5, Str. des 17. Juni 136, 10623 Berlin, Germany
e-mail: jg@mail.math.tu-berlin.de

A.F. Ramírez
Facultad de Matemáticas, Pontificia Universidad Católica de Chile, Vicuña Mackenna 4860, Macul, Santiago, Chile
e-mail: aramirez@mat.puc.cl

R. Sun (✉)
Department of Mathematics, National University of Singapore, 10 Lower Kent Ridge Road, 119076 Singapore
e-mail: matsr@nus.edu.sg

J.-D. Deuschel et al. (eds.), *Probability in Complex Physical Systems*, Springer Proceedings in Mathematics 11, DOI 10.1007/978-3-642-23811-6_6,

1 Introduction

1.1 Model and Results

Let $X := (X(t))_{t\geq 0}$ be a simple symmetric random walk on $\mathbb{Z}^d$ with jump rate $\kappa \geq 0$, and let $(Y_j^y)_{1\leq j\leq N_y, y\in\mathbb{Z}^d}$ be a collection of independent simple symmetric random walks on $\mathbb{Z}^d$ with jump rate $\rho > 0$, where N_y is the number of walks that start at each $y \in \mathbb{Z}^d$ at time 0, $(N_y)_{y\in\mathbb{Z}^d}$ are i.i.d. Poisson distributed with mean $\nu > 0$, and $Y_j^y := (Y_j^y(t))_{t\geq 0}$ denotes the jth walk starting at y at time 0. Let us denote the number of walks Y at position $x \in \mathbb{Z}^d$ at time $t \geq 0$ by

$$\xi(t,x) := \sum_{y\in\mathbb{Z}^d, 1\leq j\leq N_y} \delta_x(Y_j^y(t)). \tag{1}$$

It is easy to see that for each $t \geq 0$, $(\xi(t,x))_{x\in\mathbb{Z}^d}$ are i.i.d. Poisson distributed with mean ν, so that $(\xi(t,\cdot))_{t\geq 0}$ is a stationary process, and furthermore it is reversible in the sense that $(\xi(t,\cdot))_{0\leq t\leq T}$ is equally distributed with $(\xi(T-t,\cdot))_{0\leq t\leq T}$. We will interpret the collection of walks Y as traps, and at each time t, the walk X is killed with rate $\gamma\xi(t,X(t))$ for some parameter $\gamma > 0$. Conditional on the realization of the field of traps ξ, the probability that the walk X survives by time t is given by

$$Z_{t,\xi}^{\gamma} := \mathbb{E}_0^X \left[\exp\left\{ -\gamma \int_0^t \xi(s, X(s))\, \mathrm{d}s \right\} \right], \tag{2}$$

where $\mathbb{E}_0^X$ denotes expectation with respect to X with $X(0) = 0$. We call this the *quenched survival probability*, which depends on the random medium ξ. When we furthermore average over ξ, which we denote by $\mathbb{E}^\xi$, we obtain the *annealed survival probability*

$$\mathbb{E}^\xi \left[Z_{t,\xi}^{\gamma} \right] = \mathbb{E}^\xi \mathbb{E}_0^X \left[\exp\left\{ -\gamma \int_0^t \xi(s, X(s))\, \mathrm{d}s \right\} \right]. \tag{3}$$

We will study the long time behavior of the annealed and quenched survival probabilities and, in particular, identify their rate of decay and their dependence on the spatial dimension d and the parameters κ, ρ, ν, and γ.

Here are our main results on the decay rate of the annealed and quenched survival probabilities.

Theorem 1.1 [Annealed survival probability]. *Assume that* $\gamma \in (0, \infty]$, $\kappa \geq 0$, $\rho > 0$, *and* $\nu > 0$, *then*

$$\mathbb{E}^\xi[Z^\gamma_{t,\xi}] = \begin{cases} \exp\Big\{-\nu\sqrt{\frac{8\rho t}{\pi}}(1+o(1))\Big\}, & d=1, \\ \exp\Big\{-\nu\pi\rho\frac{t}{\log t}(1+o(1))\Big\}, & d=2, \\ \exp\Big\{-\lambda_{d,\gamma,\kappa,\rho,\nu}\, t(1+o(1))\Big\}, & d\geq 3, \end{cases} \tag{4}$$

where $\lambda_{d,\gamma,\kappa,\rho,\nu}$ depends on d, γ, κ, ρ, ν, and is called the annealed Lyapunov exponent. Furthermore, $\lambda_{d,\gamma,\kappa,\rho,\nu} \geq \lambda_{d,\gamma,0,\rho,\nu} = \nu\gamma/(1+\frac{\gamma G_d(0)}{\rho})$, where $G_d(0) := \int_0^\infty p_t(0)\,\mathrm{d}t$ is the Green function of a simple symmetric random walk on $\mathbb{Z}^d$ with jump rate 1 and transition kernel $p_t(\cdot)$.

Note that in dimensions 1 and 2, the annealed survival probability decays sub-exponentially, and the prefactor in front of the decay rate is surprisingly independent of $\gamma \in (0,\infty]$ and $\kappa \geq 0$. The key ingredient in the proof is what is known in the physics literature as the *Pascal principle*, which asserts that in (3), if we condition on the random walk trajectory X, then the annealed survival probability is maximized when $X \equiv 0$. The discrete time version of the Pascal principle was proved by Moreau, Oshanin, Bénichou, and Coppey in [19, 20]. We will include the proof for the reader's convenience. As a corollary of the Pascal principle, we will show in Corollary 2.1 that the expected cardinality of the range of a continuous time symmetric random walk increases under perturbation by a deterministic path.

In contrast to the annealed case, the quenched survival probability always decays exponentially.

Theorem 1.2 [Quenched survival probability]. *Assume that $d \geq 1$, $\gamma > 0$, $\kappa \geq 0$, $\rho > 0$ and $\nu > 0$. Then there exists deterministic $\tilde\lambda_{d,\gamma,\kappa,\rho,\nu}$ depending on d,γ,κ,ρ,ν, called the quenched Lyapunov exponent, such that $\mathbb{P}^\xi$-a.s.,*

$$Z^\gamma_{t,\xi} = \exp\big\{-\tilde\lambda_{d,\gamma,\kappa,\rho,\nu}\, t(1+o(1))\big\} \quad \text{as } t\to\infty. \tag{5}$$

Furthermore, $0 < \tilde\lambda_{d,\gamma,\kappa,\rho,\nu} \leq \gamma\nu + \kappa$ for all $d \geq 1, \gamma > 0, \kappa \geq 0, \rho > 0$ and $\nu > 0$.

Remark. When $\gamma < 0$, $Z^\gamma_{t,\xi}$ can be interpreted as the expected number of branching random walks in the catalytic medium ξ. See Sect. 1.3 for more discussion on this model. As will be outlined at the end of Sect. 4.1, (5) also holds in this case, and lies in the interval $[-\gamma\nu-\kappa,\infty)$.

In Proposition 3.2 below, we will also give an upper bound of the same order as in Theorem 1.1 for the survival probability $\mathbb{E}^\xi[Z^\gamma_{t,\xi}]$, where $(\xi(0,x))_{x\in\mathbb{Z}^d}$ is deterministic and satisfies some constraints. These constraints hold asymptotically a.s. for i.i.d. Poisson distributed $(\xi(0,x))_{x\in\mathbb{Z}^d}$. Therefore, we call this a *semi-annealed* bound, which we will use in Sect. 3 to obtain sub-exponential bounds on the quenched survival probability in dimensions 1 and 2.

1.2 Relation to the Parabolic Anderson Model

The annealed and quenched survival probabilities $Z^{\gamma}_{t,\xi}$ and $\mathbb{E}^{\xi}[Z^{\gamma}_{t,\xi}]$ are closely related to the solution of the parabolic Anderson model (PAM), namely, the solution of the following parabolic equation with random potential ξ:

$$\begin{aligned} \frac{\partial}{\partial t}u(t,x) &= \kappa\Delta u(t,x) - \gamma\,\xi(t,x)\,u(t,x), \\ u(0,x) &= 1, \end{aligned} \qquad x \in \mathbb{Z}^d,\ t \geq 0, \tag{6}$$

where γ, κ, and, ξ are as before, and $\Delta f(x) = \frac{1}{2d}\sum_{\|y-x\|=1}(f(y) - f(x))$ is the discrete Laplacian on $\mathbb{Z}^d$, which is also the generator of a simple symmetric random walk on $\mathbb{Z}^d$ with jump rate 1.

By the Feynman–Kac formula, the solution u admits the representation

$$u(t,0) = \mathbb{E}^X_0\left[\exp\left\{-\gamma\int_0^t \xi(t-s, X(s))\,\mathrm{d}s\right\}\right], \tag{7}$$

which differs from $Z^{\gamma}_{t,\xi}$ in (2) by a time reversal in ξ. When we average $u(t,0)$ over the random field ξ, by the reversibility of $(\xi(t,\cdot))_{0\leq s\leq t}$, we have

$$\begin{aligned} \mathbb{E}^{\xi}[u(t,0)] &= \mathbb{E}^{\xi}\mathbb{E}^X_0\left[\exp\left\{-\gamma\int_0^t \xi(t-s, X(s))\,\mathrm{d}s\right\}\right] \\ &= \mathbb{E}^X_0\,\mathbb{E}^{\xi}\left[\exp\left\{-\gamma\int_0^t \xi(s, X(s))\,\mathrm{d}s\right\}\right] = \mathbb{E}^{\xi}[Z^{\gamma}_{t,\xi}]. \end{aligned} \tag{8}$$

Therefore, Theorem 1.1 also applies to the annealed solution $\mathbb{E}^{\xi}[u(t,0)]$. Despite the difference between $Z^{\gamma}_{t,\xi}$ and $u(t,0)$ due to time reversal, Theorem 1.2 also holds with $u(t,0)$ in place of $Z^{\gamma}_{t,\xi}$.

Theorem 1.3 [Quenched solution of PAM]. *Let $d \geq 1$, $\gamma > 0$, $\kappa \geq 0$, $\rho > 0$, $\nu > 0$, and $\tilde{\lambda}_{d,\gamma,\kappa,\rho,\nu} > 0$ be the same as in Theorem 1.2. Then $\mathbb{P}^{\xi}$-a.s.,*

$$u(t,0) = \exp\left\{-\tilde{\lambda}_{d,\gamma,\kappa,\rho,\nu}\,t(1+o(1))\right\} \quad \text{as } t \to \infty. \tag{9}$$

Remark. By Theorem 1.2 and the remark following it, for any $\gamma \in \mathbb{R}$, $t^{-1}\log u(t,0)$ converges in probability to $-\tilde{\lambda}_{d,\gamma,\kappa,\rho,\nu}$ because $u(t,0)$ is equally distributed with $Z^{\gamma}_{t,\xi}$. However, we were only able to strengthen this to almost sure convergence for the $\gamma > 0$ case, but not for $\gamma < 0$. For a broader investigation of the case $\gamma < 0$, see Gärtner et al. [14], which is also contained in the present volume.

1.3 Review of Related Results

The study of trapping problems has a long history in the mathematics and physics literature. We review some models and results that are most relevant to our problem.

1.3.1 Immobile Traps

Extensive studies have been carried out for the case of immobile traps, i.e., $\rho = 0$ and $\xi(t,\cdot) \equiv \xi(0,\cdot)$ for all $t \geq 0$. A continuum version is Brownian motion among Poissonian obstacles, where a ball of size 1 is placed and centered at each point of a mean density 1 homogeneous Poisson point process in $\mathbb{R}^d$, acting as traps or obstacles, and an independent Brownian motion starts at the origin and is killed at rate γ times the number of obstacles it is contained in. Using a large deviation principle for the Brownian motion occupation time measure, Donsker and Varadhan [7] showed that the annealed survival probability decays asymptotically as $\exp\{-C_{d,\gamma} t^{\frac{d}{d+2}}(1 + o(1))\}$. Using spectral techniques, Sznitman [24] later developed a coarse graining method, known as the *method of enlargement of obstacles*, to show that the quenched survival probability decays asymptotically as $\exp\{-\bar{C}_{d,\gamma}\frac{t}{(\log t)^{2/d}}(1 + o(1))\}$. Similar results have also been obtained for random walks among immobile Bernoulli traps (i.e., $\xi(0,x) \in \{0,1\}$), see e.g., [1, 2, 4, 8]. Traps with a more general form of the trapping potential ξ have also been studied in the context of the parabolic Anderson model (see e.g. Biskup and König [3]), where alternative techniques to the method of enlargement of obstacles were developed and the order of sub-exponential decay of the survival probabilities may vary depending on the distribution of ξ. Compared to our results in Theorems 1.1 and 1.2 we note that when the traps are moving, both the annealed and quenched survival probabilities decay faster than when the traps are immobile. The heuristic reason is that the walk survives by finding large space–time regions void of traps, which are easily destroyed if the traps are moving. Another example is a Brownian motion among Poissonian obstacles where the obstacles move with a deterministic drift. It has been shown that the annealed and quenched survival probabilities decay exponentially if the drift is sufficiently large, see e.g., [24, Thms. 5.4.7 and 5.4.9].

1.3.2 Mobile Traps

The model we consider here has in fact been studied earlier by Redig in [22], where he considered a trapping potential ξ generated by a reversible Markov process, such as a Poisson system of random walks, or the symmetric exclusion process in equilibrium. Using spectral techniques applied to the process of moving traps viewed from the random walk, he established an exponentially decaying upper bound for the annealed survival probability when the empirical distribution of the trapping potential, $\frac{1}{t}\int_0^t \xi(s,0)\,\mathrm{d}s$, satisfies a large deviation principle with scale t. This applies, for instance, to ξ generated from either a Poisson system of independent random walks or the symmetric exclusion process in equilibrium, in dimensions $d \geq 3$.

1.3.3 Annihilating Two-Type Random Walks

In [5], Bramson and Lebowitz studied a model from chemical physics, where there are two types of particles, As and Bs, both starting initially with an i.i.d. Poisson distribution on $\mathbb{Z}^d$ with density $\rho_A(0)$ and $\rho_B(0)$, respectively. All particles perform independent simple symmetric random walk with jump rate 1, particles of the same type do not interact, and when two particles of opposite types meet, they annihilate each other. This system models a chemical reaction $A + B \to$ inert. It was shown in [5] that when $\rho_A(0) = \rho_B(0) > 0$, then $\rho_A(t)$ and $\rho_B(t)$ (the densities of the A and B particles at time t) decay with the order $t^{-d/4}$ in dimensions $1 \leq d \leq 4$, and decay with the order t^{-1} in $d \geq 4$. When $\rho_A(0) > \rho_B(0) > 0$, it was shown that $\rho_A(t) \to \rho_A(0) - \rho_B(0)$ as $t \to \infty$, and $-\log \rho_B(t)$ increases with the order $\sqrt{t}$ in $d = 1$, $t/\log t$ in $d = 2$, and t in $d \geq 3$, which is the same as in Theorem 1.1. Heuristically, as $\rho_B(t) \to 0$ and $\rho_A(t) \to \rho_A(0) - \rho_B(0) > 0$, we can effectively model the B particles as uncorrelated single random walks among a Poisson field of moving traps with density $\rho_A(0) - \rho_B(0)$. In light of Theorem 1.1, it is natural to conjecture that $\rho_B(t)$ decays exactly as prescribed in Theorem 1.1 with $\nu = \rho_A(0) - \rho_B(0)$ and $\gamma = \infty$, whence we obtain not only the logarithmic order of decay as in [5], but also the constant prefactor. However, we will not address this issue here.

1.3.4 Random Walk Among Moving Catalysts

Instead of considering ξ as a field of moving traps, we may consider it as a field of moving catalysts for a system of branching random walks which we call reactants. At time 0, a single reactant starts at the origin which undergoes branching. Independently, each reactant performs simple symmetric random walk on $\mathbb{Z}^d$ with jump rate κ, and undergoes binary branching with rate $|\gamma|\xi(t,x)$ when the reactant is at position x at time t. This model was studied by Kesten and Sidoravicius in [15], and in the setting of the parabolic Anderson model, studied by Gärtner and den Hollander in [11]. For the catalytic model, γ is negative in (2), (3), (7), and (8), and $\mathbb{Z}^{\gamma}_{t,\xi}$ and $\mathbb{E}^{\xi}[Z^{\gamma}_{t,\xi}]$ now represent the quenched, resp. annealed, expected number of reactants at time t. It was shown in [11] that $\mathbb{E}^{\xi}[Z^{\gamma}_{t,\xi}]$ grows double exponentially fast (i.e., $t^{-1}\log\log \mathbb{E}^{\xi}[Z^{\gamma}_{t,\xi}]$ tends to a positive limit as $t \to \infty$) for all $\gamma < 0$ in dimensions $d = 1$ and 2. In $d \geq 3$, there exists a critical $\gamma_{c,d} < 0$ such that $\mathbb{E}^{\xi}[Z^{\gamma}_{t,\xi}]$ grows double exponentially for $\gamma < \gamma_{c,d}$, and grows exponentially (i.e., $t^{-1}\log \mathbb{E}^{\xi}[Z^{\gamma}_{t,\xi}]$ tends to a positive limit as $t \to \infty$) for all $\gamma \in (\gamma_{c,d}, 0)$. In the quenched case, however, it was shown in [15] that $Z^{\gamma}_{t,\xi}$ only exhibits exponential growth (with $\log Z^{\gamma}_{t,\xi}$ shown to be of order t) regardless of the dimension $d \geq 1$ and the strength of interaction $\gamma < 0$. Such dimension dependence bears similarities with our results for the trap model in Theorems 1.1 and 1.2.

1.3.5 Directed Polymer in a Random Medium

We used $Z^{\gamma}_{t,\xi}$ to denote the survival probability, because $Z^{\gamma}_{t,\xi}$ and $\mathbb{E}^{\xi}[Z^{\gamma}_{t,\xi}]$ are in fact the quenched and the annealed partition functions, respectively, of a directed polymer model in a random time-dependent potential ξ at inverse temperature γ. The directed polymer is modeled by $(X(s))_{0\leq s\leq t}$. In the polymer measure, a trajectory $(X(s))_{0\leq s\leq t}$ is reweighted by the survival probability of a random walk following that trajectory in the environment ξ. Namely, we define a change of measure on $(X(s))_{0\leq s\leq t}$ with density $\mathrm{e}^{-\gamma\int_0^t \xi(s,X(s))\,\mathrm{d}s}/Z^{\gamma}_{t,\xi}$ in the quenched model, and with density $\mathbb{E}^{\xi}[\mathrm{e}^{-\gamma\int_0^t \xi(s,X(s))\,\mathrm{d}s}]/\mathbb{E}^{\xi}[Z^{\gamma}_{t,\xi}]$ in the annealed model. Qualitatively, the polymer measure favors trajectories which seek out space–time regions void of traps. However, a more quantitative geometric characterization as was carried out for the case of immobile traps (see e.g. [24]) is still lacking.

For readers interested in more background on the problem of a Brownian motion (or random walk) in time-independent potential, we refer to the book by Sznitman [24] on Brownian motion among Poissonian obstacles, and the survey by Gärtner and König [12] on the parabolic Anderson model. For readers interested in more recent studies of a random walk in time-dependent catalytic environments, we refer to the survey by Gärtner, den Hollander, and Maillard [13]. For readers interested in more recent studies of the trapping problem in the physics literature, we refer to the papers of Moreau, Oshanin, Bénichou, and Coppey [19, 20] and the references therein.

After the completion of this paper, we learnt that the continuum analogue of our model, i.e., the study of the survival probability of a Brownian motion among a Poisson field of moving obstacles, has recently been carried out by Peres, Sinclair, Sousi, and Stauffer in [21]. See Theorems 1.1 and 3.5 therein.

1.4 Outline

The rest of this paper is organized as follows. Section 2 is devoted to the proof of Theorem 1.1 on the annealed survival probability, where the so-called Pascal principle will be introduced. In Sect. 3, we give a preliminary upper bound on the quenched survival probability in dimensions 1 and 2, as well as an upper bound for a semi-annealed system. Finally, in Sect. 4, we prove the existence of the quenched Lyapunov exponent in Theorems 1.2 and 1.3 via a shape theorem, and we show that the quenched Lyapunov exponent is always positive.

2 Annealed Survival Probability

In this section, we prove Theorem 1.1. We start with a proof in Sect. 2.1 of the existence of the annealed Lyapunov exponent $\lambda_{d,\gamma,\kappa,\rho,\nu}$. Our proof follows the same argument as for the catalytic model with $\gamma < 0$ in Gärtner and den Hollander [11], which is based on a special representation of $\mathbb{E}^{\xi}[Z^{\gamma}_{t,\xi}]$ after integrating out the Poisson random field ξ, which then allows us to apply the subadditivity lemma. In Sect. 2.2, we prove Theorem 1.1 for the special case $\kappa = 0$, i.e., $X \equiv 0$, relying on exact calculations. Sections 2.3 and 2.4 prove, respectively, the lower and upper bound on $\mathbb{E}^{\xi}[Z^{\gamma}_{t,\xi}]$ in Theorem 1.1, for $d = 1, 2$ and general $\kappa > 0$. The lower bound is obtained by creating a space–time box void of traps and forcing X to stay inside the box, while the upper bound is based on the so-called Pascal principle, first introduced in the physics literature by Moreau et al. [19, 20]. In Sect. 2.4, we will also prove the aforementioned Corollary 2.1 on the range of a symmetric random walk.

2.1 Existence of the Annealed Lyapunov Exponent

In this section, we prove the existence of the annealed Lyapunov exponent

$$\lambda = \lambda_{d,\gamma,\kappa,\rho,\nu} := -\lim_{t\to\infty} \frac{1}{t} \log \mathbb{E}^{\xi}[Z^{\gamma}_{t,\xi}]. \tag{10}$$

Remark. Clearly, $\lambda \geq 0$, and Theorem 1.1 will imply that λ always equals 0 in dimensions $d = 1, 2$. For $d \geq 3$, the lower bound for the quenched survival probability in Theorem 1.2 will imply that $\lambda < \gamma\nu + \kappa < \infty$, while an exact calculation of λ for the case $\kappa = 0$ in Sect. 2.2 and the Pascal principle in Sect. 2.4 will imply that $\lambda > 0$ for all $\gamma, \nu, \rho > 0$, and $\kappa \geq 0$.

Proof of (10). The proof is similar to that for the catalytic model with $\gamma < 0$ in [11]. As in [11], we can integrate out the Poisson system ξ to obtain

$$\begin{aligned}\mathbb{E}^{\xi}[Z^{\gamma}_{t,\xi}] = \mathbb{E}^{\xi}[u(t,0)] &= \mathbb{E}^X_0 \mathbb{E}^{\xi}\left[\exp\left\{-\gamma \int_0^t \xi(t-s, X(s))\,\mathrm{d}s\right\}\right] \\ &= \mathbb{E}^X_0\left[\exp\left\{\nu \sum_{y\in\mathbb{Z}^d} (v_X(t,y) - 1)\right\}\right],\end{aligned} \tag{11}$$

where conditional on X,

$$v_X(t,y) = \mathbb{E}^Y_y\left[\exp\left\{-\gamma \int_0^t \delta_0(Y(s) - X(t-s))\,\mathrm{d}s\right\}\right] \tag{12}$$

with $\mathbb{E}_y^Y[\cdot]$ denoting expectation with respect to a simple symmetric random walk Y with jump rate ρ and $Y(0) = y$. By the Feynman–Kac formula, $(v_X(t, y))_{t\geq 0, y\in\mathbb{Z}^d}$ solves the equation

$$\begin{aligned} \frac{\partial}{\partial t} v_X(t, y) &= \rho \Delta v_X(t, y) - \gamma \delta_{X(t)}(y)\, v_X(t, y), \qquad y \in \mathbb{Z}^d,\ t \geq 0, \\ v_X(0, \cdot) &\equiv 1, \end{aligned} \tag{13}$$

which implies that $\Sigma_X(t) := \sum_{y\in\mathbb{Z}^d}(v_X(t, y) - 1)$ is the solution of the equation

$$\begin{aligned} \frac{\mathrm{d}}{\mathrm{d}t}\Sigma_X(t) &= -\gamma v_X(t, X(t)), \\ \Sigma_X(0) &= 0. \end{aligned} \tag{14}$$

Hence, $\Sigma_X(t) = -\gamma \int_0^t v_X(s, X(s))\,\mathrm{d}s$, and the representation (11) becomes

$$\mathbb{E}^\xi[Z_{t,\xi}^\gamma] = \mathbb{E}_0^X\left[\exp\left\{-\nu\gamma \int_0^t v_X(s, X(s))\,\mathrm{d}s\right\}\right]. \tag{15}$$

We now observe that for $t_1, t_2 > 0$,

$$\begin{aligned} \mathbb{E}^\xi[Z_{t_1+t_2,\xi}^\gamma] &= \mathbb{E}_0^X\left[\exp\left\{-\nu\gamma \int_0^{t_1} v_X(s, X(s))\,\mathrm{d}s\right\}\exp\left\{-\nu\gamma \int_{t_1}^{t_1+t_2} v_X(s, X(s))\,\mathrm{d}s\right\}\right] \\ &\geq \mathbb{E}_0^X\left[\exp\left\{-\nu\gamma \int_0^{t_1} v_X(s, X(s))\,\mathrm{d}s\right\}\exp\left\{-\nu\gamma \int_0^{t_2} v_{\theta_{t_1}X}(s, (\theta_{t_1}X)(s))\,\mathrm{d}s\right\}\right] \\ &= \mathbb{E}^\xi[Z_{t_1,\xi}^\gamma]\mathbb{E}^\xi[Z_{t_2,\xi}^\gamma], \end{aligned} \tag{16}$$

where $\theta_{t_1}X := ((\theta_{t_1}X)(s))_{s\geq 0} = (X(t_1+s) - X(t_1))_{s\geq 0}$, we used the independence of $(X(s))_{0\leq s\leq t_1}$ and $((\theta_{t_1}X)(s))_{0\leq s\leq t_2}$, and the fact that for $s > t_1$,

$$\begin{aligned} v_X(s, X(s)) &= \mathbb{E}_{X(s)}^Y\left[\exp\left\{-\gamma \int_0^s \delta_0(Y(r) - X(s-r))\,\mathrm{d}r\right\}\right] \\ &\leq \mathbb{E}_{X(s)}^Y\left[\exp\left\{-\gamma \int_0^{s-t_1} \delta_0(Y(r) - X(s-r))\,\mathrm{d}r\right\}\right] \\ &= v_{\theta_{t_1}X}(s - t_1, (\theta_{t_1}X)(s - t_1)). \end{aligned}$$

From (16), we deduce that $-\log \mathbb{E}^{\xi}[Z^{\gamma}_{t,\xi}]$ is subadditive in t, and hence the limit in (10) exists and

$$\lambda_{d,\gamma,\kappa,\rho,\nu} = -\sup_{t>0} \frac{1}{t} \log \mathbb{E}^{\xi}[Z^{\gamma}_{t,\xi}]. \tag{17}$$

□

2.2 Special Case $\kappa = 0$

In this section, we prove Theorem 1.1 for the case $\kappa = 0$, which will be useful for lower bounding $\mathbb{E}^{\xi}[Z^{\gamma}_{t,\xi}]$ for general $\kappa > 0$, as well as for providing an upper bound on $\mathbb{E}^{\xi}[Z^{\gamma}_{t,\xi}]$ by the Pascal principle.

Proof Theorem 1.1 for $\kappa = 0$. We first treat the case $\gamma \in (0, \infty)$. When $\kappa = 0$, (15) becomes

$$\mathbb{E}^{\xi}[Z^{\gamma}_{t,\xi}] = \exp\left\{-\nu\gamma \int_0^t v_0(s,0)\,\mathrm{d}s\right\}, \tag{18}$$

where v_0 is the solution of (13) with $X \equiv 0$. It then suffices to analyze the asymptotics of $v_0(t,0)$ as $t \to \infty$. Note that the representation (12) for $v_0(t,0)$ becomes

$$v_0(t,0) = \mathbb{E}^Y_0\left[\mathrm{e}^{-\gamma \int_0^t \delta_0(Y(s))\,\mathrm{d}s}\right], \tag{19}$$

which is the Laplace transform of the local time of Y at the origin. For $d = 1, 2$, $v_0(t,0) \downarrow 0$ as $t \uparrow \infty$ by the recurrence of simple random walks, while for $d \geq 3$, $v_0(t,0) \downarrow C_d$ for some $C_d > 0$ by transience.

By Duhamel's principle (see e.g., [9, pp. 49] for a continuous-space version), we have the following integral representation for the solution v_X of (13),

$$v_X(t,y) = 1 - \gamma \int_0^t p_{\rho s}\big(y - X(t-s)\big)\, v_X\big(t-s, X(t-s)\big)\,\mathrm{d}s, \tag{20}$$

where $p_s(\cdot)$ is the transition probability kernel of a rate 1 simple symmetric random walk on $\mathbb{Z}^d$. When $X \equiv 0$, we obtain

$$v_0(t,0) = 1 - \gamma \int_0^t p_{\rho s}(0) v_0(t-s,0)\,\mathrm{d}s. \tag{21}$$

Denote the Laplace transforms (in t) of $v_0(t,0)$ and $p_t(0)$ by

$$\hat{v}_0(\lambda) = \int_0^\infty \mathrm{e}^{-\lambda t} v_0(t,0)\,\mathrm{d}t, \qquad \hat{p}(\lambda) = \int_0^\infty \mathrm{e}^{-\lambda t} p_t(0)\,\mathrm{d}t. \tag{22}$$

Taking Laplace transform in (21) and solving for $\hat{v}_0(\lambda)$ then gives

$$\hat{v}_0(\lambda) = \frac{1}{\lambda} \cdot \frac{\rho}{\rho + \gamma\, \hat{p}(\lambda/\rho)}. \tag{23}$$

We can apply the local central limit theorem for continuous time simple random walks in $d = 1$ and 2 (i.e., $p_t(0) = \left(\frac{d}{2\pi t}\right)^{d/2}(1 + o(1))$ as $t \to \infty$) to obtain the following asymptotics for $\hat{p}(\lambda)$ as $\lambda \downarrow 0$,

$$\hat{p}(\lambda) = \begin{cases} \dfrac{1}{\sqrt{2\lambda}}(1 + o(1)), & d = 1, \\[2mm] \dfrac{\ln\left(\frac{1}{\lambda}\right)}{\pi}(1 + o(1)), & d = 2, \\[2mm] G_d(0)(1 + o(1)), & d \geq 3, \end{cases} \tag{24}$$

with $G_d(0) = \int_0^\infty p_t(0)\,\mathrm{d}t$, which translates into the following asymptotics for $\hat{v}_0(\lambda)$ as $\lambda \downarrow 0$:

$$\hat{v}_0(\lambda) = \begin{cases} \dfrac{\sqrt{2\rho}}{\gamma} \cdot \dfrac{1}{\sqrt{\lambda}}(1 + o(1)), & d = 1, \\[2mm] \dfrac{\pi\rho}{\gamma} \cdot \dfrac{1}{\lambda \ln\left(\frac{1}{\lambda}\right)}(1 + o(1)), & d = 2, \\[2mm] \dfrac{\rho}{\rho + \gamma G_d(0)} \cdot \dfrac{1}{\lambda}(1 + o(1)), & d \geq 3. \end{cases} \tag{25}$$

Since $v_0(t, 0)$ is monotonically decreasing in t by (19), by Karamata's Tauberian theorem (see e.g. [10, Chap. XIII.5, Thm. 4]), we have the following asymptotics for $v_0(t, 0)$ as $t \to \infty$,

$$v_0(t, 0) = \begin{cases} \dfrac{1}{\gamma}\sqrt{\dfrac{2\rho}{\pi}} \cdot \dfrac{1}{\sqrt{t}}(1 + o(1)), & d = 1, \\[2mm] \dfrac{\pi\rho}{\gamma} \cdot \dfrac{1}{\ln t}(1 + o(1)), & d = 2, \\[2mm] \dfrac{\rho}{\rho + \gamma G_d(0)}(1 + o(1)), & d \geq 3, \end{cases} \tag{26}$$

which by (18) implies Theorem 1.1 for $\kappa = 0$ and $\gamma \in (0, \infty)$.

When $\kappa = 0$ and $\gamma = \infty$, we have

$$\mathbb{E}^{\xi}[Z^{\gamma}_{t,\xi}] = \mathbb{P}\Big(\xi(s, 0) = 0\ \forall\, s \in [0, t]\Big) = \exp\Big\{ -\nu \sum_{y \in \mathbb{Z}^d} \psi(t, y)\Big\},$$

where $\psi(t,y) = \mathbb{P}_y^Y(\exists s \in [0,t] : Y(s) = 0)$ for a jump rate ρ simple symmetric random walk Y starting from y. Note further that $\psi(t,y)$ solves the parabolic equation

$$\frac{\partial}{\partial t}\psi(t,y) = \rho\Delta\psi(t,y), \qquad y \neq 0, t \geq 0, \tag{27}$$

with boundary conditions $\psi(\cdot,0) \equiv 1$ and $\psi(0,\cdot) \equiv 0$. Therefore, $\sum_{y\in\mathbb{Z}^d}\psi(t,y)$ solves the equation

$$\frac{\mathrm{d}}{\mathrm{d}t}\sum_{y\in\mathbb{Z}^d}\psi(t,y) = -\rho\Delta\psi(t,0) = \rho(1-\psi(t,e_1)) = \rho\phi(t,e_1), \tag{28}$$

where $e_1 = (1,0,\ldots,0)$, $\phi(t,e_1) := 1-\psi(t,e_1)$, and we have used the fact that $\sum_{x\in\mathbb{Z}^d}\Delta\psi(t,x) = 0$ and the symmetry of the simple symmetric random walk. Therefore,

$$\mathbb{E}^\xi[Z_{t,\xi}^\gamma] = \exp\left\{-\nu\rho\int_0^t \phi(s,e_1)\,\mathrm{d}s\right\}. \tag{29}$$

By generating function calculations and Tauberian theorems (see e.g. [17, Sect. 2.4] or [23, Sect. 32, P3]), it is known that $\phi(t,e_1)$, which is the probability that a rate 1 simple random walk starting from e_1 does not hit 0 before time ρt, has the asymptotics $\phi(t,e_1) = \sqrt{\frac{2}{\pi\rho t}}(1+o(1))$ for $d=1$, $\phi(t,e_1) = \frac{\pi}{\ln t}(1+o(1))$ for $d=2$, and $\phi(t,e_1) = G_d(0)^{-1}(1+o(1))$ for $d\geq 3$. Therefore, as $t\to\infty$,

$$\log\mathbb{E}^\xi[Z_{t,\xi}^\gamma] = \begin{cases} -\nu\sqrt{\dfrac{8\rho t}{\pi}}(1+o(1)), & d=1,\\[2mm] -\nu\pi\dfrac{\rho t}{\ln t}(1+o(1)), & d=2,\\[2mm] -\nu\dfrac{\rho t}{G_d(0)}(1+o(1)), & d\geq 3,\end{cases} \tag{30}$$

which proves Theorem 1.1 for $\kappa = 0$ and $\gamma = \infty$. □

Remark. When $\kappa = 0$ so that $X \equiv 0$, the representation (18) allows us to easily compute the Laplace transform of $D_t := \frac{1}{t}\int_0^t \xi(s,0)\,\mathrm{d}s$, since $\mathbb{E}^\xi[Z_{t,\xi}^\gamma] = \mathbb{E}^\xi[\exp\{-\gamma t D_t\}]$. By replacing γt with a suitable scale $\lambda t/a_t$, where $\lambda \in \mathbb{R}$, $a_t = \sqrt{t}$ for $d=1$, $a_t = \log t$ for $d=2$, and $a_t = 1$ for $d\geq 3$, we can identify

$$\Psi(-\lambda) := \lim_{t\to\infty}\frac{a_t}{t}\mathbb{E}^\xi\left[\exp\left\{-\frac{\lambda t}{a_t}D_t\right\}\right]$$

using the asymptotics in (26). As shown in Cox and Griffeath [6], applying the Gärtner–Ellis theorem then leads to a large deviation principle for D_t with scale

t/a_t, except that in [6], the derivation of $\Psi(-\lambda)$ was by Taylor expansion in λ, which can be greatly simplified if we use the representation from (18) instead.

2.3 Lower Bound on the Annealed Survival Probability

In this section, we prove the lower bound on $\mathbb{E}^\xi[Z^\gamma_{t,\xi}]$ in Theorem 1.1 for dimensions $d = 1$ and 2, i.e.,

Lemma 2.1. *For all $\gamma \in (0, \infty]$, $\kappa \geq 0$, $\rho > 0$, and $\nu > 0$, we have*

$$\begin{aligned} \liminf_{t\to\infty} \frac{1}{\sqrt{t}} \log \mathbb{E}^\xi[Z^\gamma_{t,\xi}] &\geq -\nu\sqrt{\frac{8\rho}{\pi}}, \qquad & d = 1, \\ \liminf_{t\to\infty} \frac{\ln t}{t} \log \mathbb{E}^\xi[Z^\gamma_{t,\xi}] &\geq -\nu\pi\rho, \qquad & d = 2. \end{aligned} \tag{31}$$

Proof. The basic strategy is the same as for the case of immobile traps, namely, we force the environment ξ to create a ball B_{R_t} of radius R_t around the origin, which remains void of traps up to time t, and we force the random walk X to stay inside B_{R_t} up to time t. This leads to a lower bound on the survival probability that is independent of $\gamma \in (0, \infty]$ and $\kappa \geq 0$. Surprisingly, in dimensions $d = 1$ and 2, this lower bound turns out to be sharp, which can be attributed to the larger fluctuation of the random field ξ in $d = 1$ and 2, which makes it easier to create space–time regions void of traps. Note that it is clearly more costly to maintain the same space–time region void of traps than in the case when the traps are immobile.

Recall that ξ is the counting field of a family of independent random walks $\{Y^y_j\}_{y\in\mathbb{Z}^d, 1\leq j\leq N_y}$, where $\{N_y\}_{y\in\mathbb{Z}^d}$ are i.i.d. Poisson random variables with mean ν. Let B_r denote the ball of radius r, i.e., $B_r = \{x \in \mathbb{Z}^d : \|x\|_\infty \leq r\}$. For a scale function $1 << R_t << \sqrt{t}$ to be chosen later, let E_t denote the event that $N_y = 0$ for all $y \in B_{R_t}$. Let F_t denote the event that $Y^y_j(s) \notin B_{R_t}$ for all $y \notin B_{R_t}$, $1 \leq j \leq N_y$, and $s \in [0, t]$; furthermore, let G_t denote the event that X with $X(0) = 0$ does not leave B_{R_t} before time t. Then by (3),

$$\mathbb{E}^\xi[Z^\gamma_{t,\xi}] \geq \mathbb{P}(E_t \cap F_t \cap G_t) = \mathbb{P}(E_t)\mathbb{P}(F_t)\mathbb{P}(G_t). \tag{32}$$

Note that $\mathbb{P}(E_t) = \mathrm{e}^{-\nu(2R_t+1)^d}$. To estimate $\mathbb{P}(G_t)$, note that by Donsker's invariance principle if $1 << R_t << \sqrt{t}$ as $t \to \infty$, then there exists $\alpha > 0$ such that for all t sufficiently large,

$$\inf_{x\in B_{\sqrt{t}/2}} \mathbb{P}\Big(X(s) \in B_{\sqrt{t}}\ \forall\, s \in [0, t]\,,\ X(t) \in B_{\sqrt{t}/2}\Big|X(0) = x\Big) \geq \alpha. \tag{33}$$

By partitioning $[0, t]$ into intervals of length R_t^2 and applying the Markov property, we obtain

$$\begin{aligned}\mathbb{P}(G_t) &\geq \mathbb{P}\Big(X(s) \in B_{R_t}\ \forall\, s \in [(i-1)R_t^2, iR_t^2], \\ &\qquad \text{and } X(iR_t^2) \in B_{R_t/2},\ i = 1, 2, \cdots, \lceil t/R_t^2 \rceil\Big) \\ &\geq \alpha^{t/R_t^2} = \mathrm{e}^{t \ln \alpha / R_t^2}. \end{aligned} \tag{34}$$

To estimate $\mathbb{P}(F_t)$, let $\tilde{F}_t$ denote the event that $Y_j^y(s) \neq 0$ for all $y \in \mathbb{Z}^d$, $1 \leq j \leq N_y$, and $s \in [0, t]$. Note that $\mathbb{P}(\tilde{F}_t)$ is precisely the annealed survival probability $\mathbb{E}^\xi[Z_{t,\xi}^\gamma]$ when $\kappa = 0$ and $\gamma = \infty$, which satisfies the asymptotics in Theorem 1.1 by our calculations in Sect. 2.2. We next compare $\mathbb{P}(F_t)$ with $\mathbb{P}(\tilde{F}_t)$.

For a jump rate ρ simple random walk Y starting from $y \in \mathbb{Z}^d$, let $\tau_{B_{R_t}}$ denote the stopping time when Y first enters B_{R_t}, and τ_0 the stopping time when Y first visits 0. Then standard computations yield

$$\ln \mathbb{P}(F_t) = -\nu \sum_{y \in \mathbb{Z}^d \backslash B_{R_t}} \mathbb{P}_y^Y(\tau_{B_{R_t}} \leq t), \tag{35}$$

and a similar identity holds for $\ln \mathbb{P}(\tilde{F}_t)$ with B_{R_t} replaced by B_0. Note that

$$\sum_{y \in \mathbb{Z}^d \backslash B_{R_t}} \mathbb{P}_y^Y(\tau_{B_{R_t}} \leq t) \geq \sum_{y \in \mathbb{Z}^d \backslash B_{R_t}} \mathbb{P}_y^Y(\tau_0 \leq t) = \sum_{y \in \mathbb{Z}^d} \mathbb{P}_y^Y(\tau_0 \leq t) - \sum_{y \in B_{R_t}} \mathbb{P}_y^Y(\tau_0 \leq t).$$

Hence,

$$\ln \mathbb{P}(F_t) \leq \ln \mathbb{P}(\tilde{F}_t) + \nu \sum_{y \in B_{R_t}} \mathbb{P}_y^Y(\tau_0 \leq t) \leq \ln \mathbb{P}(\tilde{F}_t) + \nu(2R_t + 1)^d. \tag{36}$$

On the other hand, for $\epsilon > 0$, we have

$$\begin{aligned}\sum_{y \in \mathbb{Z}^d} \mathbb{P}_y^Y(\tau_0 \leq t + \epsilon t) &\geq \sum_{y \in \mathbb{Z}^d \backslash B_{R_t}} \mathbb{P}_y^Y\big(\tau_{B_{R_t}} \leq t, \tau_0 \leq t + \epsilon t\big) \\ &\geq \inf_{z \in \partial B_{R_t}} \mathbb{P}_z^Y(\tau_0 \leq \epsilon t) \sum_{y \in \mathbb{Z}^d \backslash B_{R_t}} \mathbb{P}_y^Y\big(\tau_{B_{R_t}} \leq t\big),\end{aligned}$$

where we used the strong Markov property. Therefore,

$$\sum_{y \in \mathbb{Z}^d \backslash B_{R_t}} \mathbb{P}_y^Y(\tau_{B_{R_t}} \leq t) \leq \frac{\sum_{y \in \mathbb{Z}^d} \mathbb{P}_y^Y(\tau_0 \leq t + \epsilon t)}{\inf_{z \in \partial B_{R_t}} \mathbb{P}_z^Y(\tau_0 \leq \epsilon t)},$$

and hence by (35),

$$\ln \mathbb{P}(F_t) \geq \frac{\ln \mathbb{P}(\tilde{F}_{t+\epsilon t})}{\inf_{z \in \partial B_{R_t}} \mathbb{P}_z^Y(\tau_0 \leq \epsilon t)}. \tag{37}$$

We now choose R_t for $d = 1$ and 2. For $d = 1$, let $R_t = \sqrt{t/\ln t}$, which is by no means the unique scale appropriate. Clearly, $\inf_{z\in\partial B_{\sqrt{t/\ln t}}} \mathbb{P}_z^Y(\tau_0 \le \epsilon t) \to 1$ as $t \to \infty$. By (36) and (37), the fact that $\mathbb{P}(\tilde{F}_t)$ satisfies the asymptotics in Theorem 1.1 for $\kappa = 0$ and $\gamma = \infty$, and that $\epsilon > 0$ can be made arbitrarily small, we obtain

$$\ln \mathbb{P}(F_t) = -\nu\sqrt{\frac{8\rho t}{\pi}}(1 + o(1)) = \ln \mathbb{P}(\tilde{F}_t).$$

Furthermore, for $R_t = \sqrt{t/\ln t}$ we have

$$\ln \mathbb{P}(E_t) = -\nu(2\sqrt{t/\ln t} + 1) \qquad \text{and} \qquad \ln \mathbb{P}(G_t) \ge \ln\alpha \ln t,$$

whence substituting these asymptotics into (32) gives (31) for $d = 1$.

For $d = 2$, let $R_t = \ln t$. Then we have $\inf_{z\in\partial B_{\ln t}} \mathbb{P}_z^Y(\tau_0 \le \epsilon t) \to 1$ as $t \to \infty$, which is an easy consequence of [17, Exercise 1.6.8]. By the same argument as for $d = 1$, we have

$$\ln \mathbb{P}(F_t) = -\nu\pi\rho\frac{t}{\ln t}(1 + o(1)) = \ln \mathbb{P}(\tilde{F}_t).$$

Together with the asymptotics

$$\ln \mathbb{P}(E_t) = -\nu(2\ln t + 1)^2 \qquad \text{and} \qquad \ln \mathbb{P}(G_t) \ge \frac{t\ln\alpha}{\ln^2 t},$$

we deduce from (32) the desired bound in (31) for $d = 2$. □

2.4 *Upper Bound on the Annealed Survival Probability: The Pascal Principle*

In this section, we present an upper bound on the annealed survival probability, called the Pascal principle.

Proposition 2.1 [Pascal principle]. *Let ξ be the random field generated by a collection of irreducible symmetric random walks $\{Y_j^y\}_{y\in\mathbb{Z}^d, 1\le j\le N_y}$ on $\mathbb{Z}^d$ with jump rate $\rho > 0$. Then for all piecewise constant $X : [0,t] \to \mathbb{Z}^d$ with a finite number of discontinuities, we have*

$$\mathbb{E}^\xi\left[\exp\left\{-\gamma\int_0^t \xi(s, X(s))\,ds\right\}\right] \le \mathbb{E}^\xi\left[\exp\left\{-\gamma\int_0^t \xi(s,0)\,ds\right\}\right]. \tag{38}$$

In words, conditional on the random walk X, the annealed survival probability is maximized when $X \equiv 0$. The discrete time version of this result was first proved by Moreau et al. in [19, 20], where they named it the *Pascal principle*, because Pascal once asserted that *all misfortune of men comes from the fact that he does not stay peacefully in his room.* The Pascal principle together with the proof of Theorem 1.1 for $\kappa = 0$ in Sect. 2.2 implies the desired upper bound on the annealed survival probability in Theorem 1.1 for dimensions $d = 1, 2$, and it also shows that for $d \geq 3$, the annealed Lyapunov exponent $\lambda_{d,\gamma,\kappa,\rho,\nu}$ is always bounded from below by $\lambda_{d,\gamma,0,\rho,\nu} = \nu\gamma/(1 + \frac{\gamma G_d(0)}{\rho})$.

We present below the proof of the discrete time version of the Pascal principle from [20], which being written as a physics paper, can be hard for the reader to separate the rigorous arguments from the non-rigorous ones. We then deduce the continuous time version, Proposition 2.1, by discrete approximation. As a byproduct, we will show in Corollary 2.1 that the expected cardinality of the range of a continuous time symmetric random walk increases under perturbation by a deterministic path.

Moreau et al. considered in [20] a discrete time random walk among a Poisson field of moving traps, defined as follows. Let $\bar{X}$ be a discrete time mean zero random walk on $\mathbb{Z}^d$ with $\bar{X}_0 = 0$. Let $\{N_y\}_{y\in\mathbb{Z}^d}$ be i.i.d. Poisson random variables with mean ν, and let $\{\bar{Y}_j^y\}_{y\in\mathbb{Z}^d, 1\leq j\leq N_y}$ be a family of independent *symmetric* random walks on $\mathbb{Z}^d$, where $\bar{Y}_j^y$ denotes the jth random walk starting from y at time 0. Let

$$\bar{\xi}(n,x) := \sum_{y\in\mathbb{Z}^d, 1\leq j\leq N_y} \delta_x(\bar{Y}_j^y(n)). \tag{39}$$

Fix $0 \leq q \leq 1$, which will be the trapping probability. The dynamics of $\bar{X}$ is such that $\bar{X}$ moves independently of the traps $\{\bar{Y}_j^y\}_{y\in\mathbb{Z}^d, 1\leq j\leq N_y}$, and at each time $n \geq 0$, $\bar{X}$ is killed with probability $1-(1-q)^{\bar{\xi}(n,\bar{X}(n))}$. Namely, each trap at the time–space lattice site $(n, \bar{X}(n))$ tries independently to capture $\bar{X}$ with probability q. Given a realization of $\bar{X}$, let $\bar{\sigma}^{\bar{X}}(n)$ denote the probability that $\bar{X}$ has survived till time n. Then analogous to (11), we have

$$\bar{\sigma}^{\bar{X}}(n) = \mathbb{E}^{\bar{\xi}}\Big[(1-q)^{\sum_{i=0}^{n}\bar{\xi}(i,\bar{X}(i))}\Big] = \exp\Big\{-\nu\sum_{y\in\mathbb{Z}^d}\bar{w}^{q,\bar{X}}(n,y)\Big\}, \tag{40}$$

where if we let $\bar{Y}$ denote a random walk with the same jump kernel as $\bar{Y}_j^y$, then

$$\bar{w}^{q,\bar{X}}(n,y) := 1 - \mathbb{E}_y^{\bar{Y}}\Big[(1-q)^{\sum_{i=0}^{n} 1_{\{\bar{Y}(i)=\bar{X}(i)\}}}\Big]. \tag{41}$$

The main result we need from Moreau et al. [20] is the following discrete time Pascal principle.

Lemma 2.2. [Pascal principle in discrete time [20]]. *Let $\bar{Y}$ be an irreducible symmetric random walk on $\mathbb{Z}^d$ with $\mathbb{P}_0^{\bar{Y}}(\bar{Y}(1) = 0) \geq 1/2$. Then for all $q \in [0, 1]$, $n \in \mathbb{N}_0$ and $\bar{X} : \mathbb{N}_0 \to \mathbb{Z}^d$, we have*

$$\sum_{y \in \mathbb{Z}^d} \bar{w}^{q,\bar{X}}(n, y) \geq \sum_{y \in \mathbb{Z}^d} \bar{w}^{q,0}(n, y), \tag{42}$$

and hence $\bar{\sigma}^{\bar{X}}(n) \leq \bar{\sigma}^0(n)$, where $\bar{w}^{q,0}$ and $\bar{\sigma}^0$ denote $\bar{w}^{q,\bar{X}}$ and $\bar{\sigma}^{\bar{X}}$ with $\bar{X} \equiv 0$.

Proof. The argument we present here is extracted from [20]. First note that the assumption $\bar{Y}$ is symmetric implies that the Fourier transform $f(k) := \mathbb{E}_0^{\bar{Y}}[e^{i\langle k, \bar{Y}(1)\rangle}]$ is real for all $k \in [-\pi, \pi]^d$. The assumption $\mathbb{P}_0^{\bar{Y}}(\bar{Y}(1) = 0) \geq 1/2$ guarantees that $f(k) \in [0, 1]$. If we let $p_n^{\bar{Y}}(y)$ denote the n-step transition probability kernel of $\bar{Y}$, then by Fourier inversion, we have

$$\begin{aligned} p_n^{\bar{Y}}(0) &\geq p_n^{\bar{Y}}(y), \\ p_n^{\bar{Y}}(0) &\geq p_{n+1}^{\bar{Y}}(0) \end{aligned} \qquad \text{for all } n \geq 0,\ y \in \mathbb{Z}^d. \tag{43}$$

If we now regard $\bar{X}$ as a trap, then $\bar{w}^{q,\bar{X}}(n, y)$ can be interpreted as the probability that a random walk $\bar{Y}$ starting from y gets trapped by $\bar{X}$ by time n, where each time $\bar{Y}$ and $\bar{X}$ coincide, $\bar{Y}$ is trapped by $\bar{X}$ with probability q. More precisely, let Z_i, $i \in \mathbb{N}_0$, be i.i.d. Bernoulli random variables with mean q, where $Z_i = 1$ means that the trap at $(i, \bar{X}(i))$ is open. Then $\bar{X}$ is killed at the stopping time

$$\tau_{\bar{X}}(\bar{Y}) := \min\{i \geq 0 : \bar{Y}(i) = \bar{X}(i), Z_i = 1\}, \tag{44}$$

and $\bar{w}^{q,\bar{X}}(n, y) = \mathbb{P}_y^{\bar{Y}}(\tau_{\bar{X}} \leq n)$.

We examine the following auxiliary quantity, where by decomposition with respect to $\tau_{\bar{X}}$, we have

$$\begin{aligned} q &= \sum_{y \in \mathbb{Z}^d} \mathbb{P}_y^{\bar{Y}}\Big(\bar{Y}(n) = \bar{X}(n), Z_n = 1\Big) \\ &= \sum_{k=0}^{n-1} \sum_{y \in \mathbb{Z}^d} \mathbb{P}_y^{\bar{Y}}(\tau_{\bar{X}} = k) p_{n-k}^{\bar{Y}}(\bar{X}(n) - \bar{X}(k))\, q + \sum_{y \in \mathbb{Z}^d} \mathbb{P}_y^{\bar{Y}}(\tau_{\bar{X}} = n) \\ &\leq q \sum_{k=0}^{n-1} \sum_{y \in \mathbb{Z}^d} \mathbb{P}_y^{\bar{Y}}(\tau_{\bar{X}} = k) p_{n-k}^{\bar{Y}}(0) + \sum_{y \in \mathbb{Z}^d} \mathbb{P}_y^{\bar{Y}}(\tau_{\bar{X}} = n), \end{aligned} \tag{45}$$

where in the inequality we used (43). Similarly, when $\bar{X}$ is replaced by $\bar{X} \equiv 0$, we have

$$q = \sum_{y \in \mathbb{Z}^d} \mathbb{P}_y^{\bar{Y}} \Big(\bar{Y}(n) = 0, Z_n = 1 \Big)$$

$$= q \sum_{k=0}^{n-1} \sum_{y \in \mathbb{Z}^d} \mathbb{P}_y^{\bar{Y}} (\tau_0 = k) p_{n-k}^{\bar{Y}}(0) + \sum_{y \in \mathbb{Z}^d} \mathbb{P}_y^{\bar{Y}} (\tau_0 = n). \tag{46}$$

Denote

$$\begin{aligned} S_n^{\bar{X}} &:= \sum_{y \in \mathbb{Z}^d} \bar{w}^{q,\bar{X}}(n,y) = \sum_{y \in \mathbb{Z}^d} \mathbb{P}_y^{\bar{Y}} (\tau_{\bar{X}} \le n), \\ S_n^0 &:= \sum_{y \in \mathbb{Z}^d} \bar{w}^{q,0}(n,y) = \sum_{y \in \mathbb{Z}^d} \mathbb{P}_y^{\bar{Y}} (\tau_0 \le n). \end{aligned} \tag{47}$$

Note that $S_0^{\bar{X}} = S_0^0 = q$, and $\sum_{y \in \mathbb{Z}^d} \mathbb{P}_y^{\bar{Y}} (\tau_{\bar{X}} = k) = S_k^{\bar{X}} - S_{k-1}^{\bar{X}}$, $\sum_{y \in \mathbb{Z}^d} \mathbb{P}_y^{\bar{Y}} (\tau_0 = k) = S_k^0 - S_{k-1}^0$, where we set $S_{-1}^{\bar{X}} = S_{-1}^0 = 0$. Together with (45) and (46), this gives

$$q \sum_{k=0}^{n-1} p_{n-k}^{\bar{Y}}(0)(S_k^0 - S_{k-1}^0) + S_n^0 - S_{n-1}^0 \le q \sum_{k=0}^{n-1} p_{n-k}^{\bar{Y}}(0)(S_k^{\bar{X}} - S_{k-1}^{\bar{X}}) + S_n^{\bar{X}} - S_{n-1}^{\bar{X}}.$$

Rearranging terms, we obtain

$$S_n^{\bar{X}} - S_n^0 \ge \big(1 - q p_1^{\bar{Y}}(0)\big)(S_{n-1}^{\bar{X}} - S_{n-1}^0) + q \sum_{k=0}^{n-2} \Big(p_{n-k-1}^{\bar{Y}}(0) - p_{n-k}^{\bar{Y}}(0) \Big)(S_k^{\bar{X}} - S_k^0). \tag{48}$$

This sets up an induction bound for $S_n^{\bar{X}} - S_n^0$. Since $S_0^{\bar{X}} - S_0^0 = 0$, $1 - q p_1^{\bar{Y}}(0) \ge 0$, and $p_k^{\bar{Y}}(0)$ is decreasing in k by (43), it follows that $S_n^{\bar{X}} \ge S_n^0$ for all $n \in \mathbb{N}_0$, which is precisely (42). □

Proof of Proposition 2.1. Integrating out ξ on both sides of (38) as in (11) shows that (38) is equivalent to

$$\sum_{y \in \mathbb{Z}^d} w^{\gamma,X}(t,y) \ge \sum_{y \in \mathbb{Z}^d} w^{\gamma,0}(t,y), \tag{49}$$

where

$$w^{\gamma,X}(t,y) := 1 - \mathbb{E}_y^Y \left[\exp\Big\{ -\gamma \int_0^t \delta_0(Y(s) - X(s))\, \mathrm{d}s \Big\} \right]. \tag{50}$$

For $n \in \mathbb{N}$, let $Y^{(n)}(k) = Y(\frac{kt}{n})$ and $X^{(n)}(k) = X(\frac{t}{n})$ for $k \in \mathbb{N}_0$. Clearly, $Y^{(n)}$ is symmetric, and for n sufficiently large, $\mathbb{P}_0^{Y^{(n)}}(Y^{(n)}(1) = 0) \ge 1/2$. Therefore, we can apply Lemma 2.2 with $\bar{Y} = Y^{(n)}$, $\bar{X} = X^{(n)}$, and $q = q^{(n)} = \gamma t / n$ to obtain

$$\sum_{y\in\mathbb{Z}^d} \bar{w}^{\gamma t/n, X^{(n)}}(n,y) \geq \sum_{y\in\mathbb{Z}^d} \bar{w}^{\gamma t/n,0}(n,y). \tag{51}$$

By (41) and the definition of $Y^{(n)}$ and $X^{(n)}$, we have

$$\begin{aligned}\bar{w}^{\gamma t/n, X^{(n)}}(n,y) &= 1 - \mathbb{E}_y^{Y^{(n)}}\left[\left(1-\frac{\gamma t}{n}\right)^{\sum_{k=0}^n 1_{\{Y^{(n)}(k)=X^{(n)}(k)\}}}\right] \\ &= 1 - \mathbb{E}_y^{Y}\left[\left(1-\frac{\gamma t}{n}\right)^{\sum_{k=0}^n 1_{\{Y(kt/n)=X(kt/n)\}}}\right].\end{aligned}$$

By the assumption that X is a random walk path which is necessarily piecewise constant with a finite number of discontinuities, for a.s. all realization of Y, we have

$$\lim_{n\to\infty}\left(1-\frac{\gamma t}{n}\right)^{\sum_{k=0}^n 1_{\{Y(kt/n)=X(kt/n)\}}} = \exp\left\{-\gamma\int_0^t \delta_0(Y(s)-X(s))\,\mathrm{d}s\right\}.$$

Therefore by the bounded convergence theorem, $\lim_{n\to\infty}\bar{w}^{\gamma t/n,X^{(n)}}(n,y)=w^{\gamma,X}(t,y)$. By the same argument, $\lim_{n\to\infty}\bar{w}^{\gamma t/n,0}(n,y) = w^{\gamma,0}(t,y)$. Next we note that $w^{\gamma t/n,X^{(n)}}(n,y)$ is the probability that $Y^{(n)}$ is trapped by $X^{(n)}$ before time n. Since $Y^{(n)}$ and $X^{(n)}$ are embedded in Y and X, we have $w^{\gamma t/n,X^{(n)}}(n,y) \leq \mathbb{P}_y^Y(\tau_X \leq t)$ uniformly in n, where $\tau_X = \inf\{s\geq 0 : Y(s) = X(s)\}$. Clearly, $\sum_{y\in\mathbb{Z}^d}\mathbb{P}_y^Y(\tau_X \leq t) < \infty$. Similarly $w^{\gamma t/n,0}(n,y) \leq \mathbb{P}_y^Y(\tau_0 \leq t)$ uniformly in n and $\sum_{y\in\mathbb{Z}^d}\mathbb{P}_y^Y(\tau_0\leq t) < \infty$. Therefore, we can send $n\to\infty$ and apply the dominated convergence theorem in (51), from which (49) then follows. □

The Pascal principle in Lemma 2.2 and Proposition 2.1 have the following interesting consequence for the range of a symmetric random walk, which we denote by $R_t(X) = \{y\in\mathbb{Z}^d : X(s) = y \text{ for some } 0\leq s\leq t\}$.

Corollary 2.1 [Increase of expected cardinality of range under perturbation]. *Let $\bar{Y}$ and $\bar{X}$ be discrete time random walks as in Lemma 2.2. Let Y be a continuous time irreducible symmetric random walk on $\mathbb{Z}^d$ with jump rate $\rho > 0$, and let $X : [0,t]\to\mathbb{Z}^d$ be piecewise constant with a finite number of discontinuities. Then for all $n\in\mathbb{N}_0$, respectively $t\geq 0$, we have*

$$\begin{aligned}\mathbb{E}_0^{\bar{Y}}\big[|R_n(\bar{Y}-\bar{X})|\big] &\geq \mathbb{E}_0^{\bar{Y}}\big[|R_n(\bar{Y})|\big],\\ \mathbb{E}_0^{Y}\big[|R_t(Y-X)|\big] &\geq \mathbb{E}_0^{Y}\big[|R_t(Y)|\big],\end{aligned} \tag{52}$$

where $|\cdot|$ denotes the cardinality of the set.

Proof. The first inequality in (2.1) for discrete time random walks follows from the observation that

$$
\begin{aligned}
\sum_{y\in\mathbb{Z}^d} \mathbb{P}_y^{\bar{Y}}(\tau_{\bar{X}} \le n) &= \sum_{y\in\mathbb{Z}^d} \mathbb{P}_0^{\bar{Y}}\big(\bar{Y}(i) - \bar{X}(i) = y \text{ for some } 0 \le i \le n\big) \\
&= \mathbb{E}_0^{\bar{Y}}\big[|R_n(\bar{Y} - \bar{X})|\big], \\
\sum_{y\in\mathbb{Z}^d} \mathbb{P}_y^{\bar{Y}}(\tau_0 \le n) &= \sum_{y\in\mathbb{Z}^d} \mathbb{P}_0^{\bar{Y}}\big(\bar{Y}(i) = y \text{ for some } 0 \le i \le n\big) = \mathbb{E}_0^{\bar{Y}}\big[|R_n(\bar{Y})|\big],
\end{aligned}
\tag{53}
$$

where $\tau_{\bar{X}} = \min\{i \ge 0 : \bar{Y}_i = \bar{X}_i\}$ and $\tau_0 = \min\{i \ge 0 : \bar{Y}_i = 0\}$, which combined with Lemma 2.2 for $q = 1$ gives precisely

$$
\sum_{y\in\mathbb{Z}^d} \mathbb{P}_y^{\bar{Y}}(\tau_{\bar{X}} \le n) \ge \sum_{y\in\mathbb{Z}^d} \mathbb{P}_y^{\bar{Y}}(\tau_0 \le n). \tag{54}
$$

The continuous time case follows by similar considerations, where we apply Proposition 2.1 with $\gamma = \infty$, or rather $\gamma > 0$ with $\gamma \uparrow \infty$. □

3 Quenched and Semi-Annealed Upper Bounds

In this section, we prove sub-exponential upper bounds on the quenched survival probability in dimensions 1 and 2 (the exponential upper bound in dimensions 3 and higher follows trivially from the annealed upper bound by Jensen's inequality and Borel–Cantelli). Although they will be superseded later by a proof of exponential decay using sophisticated results of Kesten and Sidoravicius [16], the proof we present here is relatively simple and self-contained. Along the way, we will also prove an upper bound (Proposition 3.2) on the annealed survival probability of a random walk in a random field of traps ξ with deterministic initial condition, which we call a semi-annealed bound.

Proposition 3.1 [Sub-exponential upper bound on $Z^\gamma_{t,\xi}$]. *There exist constants $C_1, C_2 > 0$ depending on $\gamma, \kappa, \rho, \nu > 0$ such that a.s. with respect to ξ, we have*

$$
\begin{aligned}
\limsup_{t\to\infty} \frac{\log t}{t} \log Z^\gamma_{t,\xi} &\le -C_1, & d &= 1, \\
\limsup_{t\to\infty} \frac{\log\log t}{t} \log Z^\gamma_{t,\xi} &\le -C_2, & d &= 2.
\end{aligned}
\tag{55}
$$

The same bounds hold if we replace $Z^\gamma_{t,\xi}$ by $u(t,0)$ as in Theorem 1.3.

Proof. The proof is based on coarse graining combined with the annealed bound in Theorem 1.1. Let us focus on dimension $d = 1$ first. Let X be a random walk as in (2), and let $M(t) := \sup_{0\le s\le t} |X(s)|_\infty$. The first step is to note that by basic large deviation estimates for X,

$$\mathbb{E}_0^X\Big[\exp\Big\{-\gamma\int_0^t \xi(s,X(s))\,\mathrm{d}s\Big\}1_{\{M_t\geq t\}}\Big]\leq \mathbb{P}_0^X(M_t\geq t)\leq \mathrm{e}^{-Ct}$$

for some $C>0$ depending only on κ. Therefore to show (55), it suffices to prove that

$$\mathbb{E}_0^X\Big[\exp\Big\{-\gamma\int_0^t \xi(s,X(s))\,\mathrm{d}s\Big\}1_{\{M_t<t\}}\Big]\leq \mathrm{e}^{-\frac{Ct}{\log t}} \tag{56}$$

for some $C>0$ for all t sufficiently large. Since the integrand in the definition of $Z^{\gamma}_{t,\xi}$ is monotone in t, we may even restrict our attention to $t\in\mathbb{N}$.

The second step is to introduce a coarse graining scale $L_t:=A\log t$ for some $A>0$, and partition the space–time region $[-2t,2t]\times[0,t]$ into blocks of the form $\Lambda_{i,k}:=[(i-1)L_t,iL_t)\times[(k-1)L_t^2,kL_t^2)$ for $i,k\in\mathbb{Z}$ with $-\frac{2t}{L_t}+1\leq i\leq\frac{2t}{L_t}$ and $1\leq k\leq\frac{t}{L_t^2}$. We say a block

$$\Lambda_{i,k}\text{ is good if}\quad \sum_{(i-1)L_t\leq x<iL_t}\xi((k-1)L_t^2,x)\geq\frac{\nu L_t}{2}.$$

Since for each $s\geq 0$, $(\xi(s,x))_{x\in\mathbb{Z}}$ are i.i.d. Poisson distributed with mean ν, by basic large deviation estimates for Poisson random variables, there exists $C>0$ such that for all $t>1$,

$$\mathbb{P}(\Lambda_{i,k}\text{ is bad})\leq \mathrm{e}^{-C\nu L_t}.$$

Let $G_t(\xi)$ be the event that all the blocks $\Lambda_{i,k}$ in $[-2t,2t]\times[0,t]$ are good. Then

$$\mathbb{P}(G_t^c(\xi))\leq\frac{4t^2}{L_t^3}\mathrm{e}^{-C\nu L_t}=\frac{4}{A^3(\log t)^3t^{C\nu A-2}},$$

which is summable in $t\geq 2$, $t\in\mathbb{N}$ if A is chosen sufficiently large. Therefore by Borel–Cantelli, a.s. with respect to ξ, for all $t\in\mathbb{N}$ sufficiently large, the event $G_t(\xi)$ occurs. To prove (55), it then suffices to prove

$$1_{G_t(\xi)}\mathbb{E}_0^X\Big[\exp\Big\{-\gamma\int_0^t \xi(s,X(s))\,\mathrm{d}s\Big\}1_{\{M_t<t\}}\Big]\leq \mathrm{e}^{-\frac{Ct}{\log t}} \tag{57}$$

almost surely for all $t\in\mathbb{N}$ sufficiently large.

The third step is applying an annealing bound. More precisely, to show (57), it suffices to average over ξ and show that

$$\mathbb{E}^{\xi}\mathbb{E}_0^X\Big[\exp\Big\{-\gamma\int_0^t \xi(s,X(s))\,\mathrm{d}s\Big\}1_{\{M_t<t\}}1_{G_t(\xi)}\Big]\leq \mathrm{e}^{-\frac{2Ct}{\log t}} \tag{58}$$

for some $C > 0$ for all $t \in \mathbb{N}$ sufficiently large. Indeed, (58) implies that

$$\mathbb{P}^\xi\Big(1_{G_t(\xi)}\mathbb{E}_0^X\Big[\exp\Big\{-\gamma\int_0^t \xi(s,X(s))\,ds\Big\}1_{\{M_t<t\}}\Big] > e^{-\frac{Ct}{\log t}}\Big) \le e^{-\frac{Ct}{\log t}},$$

from which (57) then follows by Borel–Cantelli.

To prove (58), let us denote $Z_k := \exp\{-\gamma\int_{(k-1)L_t^2}^{kL_t^2}\xi(s,X(s))\,ds\}$, and let $\mathcal{F}_k$ be the σ-field generated by $(X_s,\xi(s,\cdot))_{0\le s\le kL_t^2}$. Replacing L_t^2 by $t/\lfloor t/L_t^2\rfloor$ if necessary, we may assume without loss of generality that $t/L_t^2 = t/(A\log t)^2 \in \mathbb{N}$. Then

$$\begin{aligned}\mathbb{E}^\xi\mathbb{E}_0^X\Big[\exp\Big\{-\gamma\int_0^t \xi(s,X(s))\,ds\Big\}1_{\{M_t<t\}}1_{G_t(\xi)}\Big] &= \mathbb{E}^\xi\mathbb{E}_0^X\Big[1_{\{M_t<t\}}1_{G_t(\xi)}\prod_{k=1}^{t/L_t^2} Z_k\Big]\\ &= \mathbb{E}^\xi\mathbb{E}_0^X\Big[1_{\{M_t<t\}}1_{G_t(\xi)}\prod_{k=1}^{t/L_t^2}\frac{Z_k}{\mathbb{E}^\xi\mathbb{E}_0^X[Z_k|\mathcal{F}_{k-1}]}\prod_{k=1}^{t/L_t^2}\mathbb{E}^\xi\mathbb{E}_0^X[Z_k|\mathcal{F}_{k-1}]\Big].\end{aligned} \tag{59}$$

By Proposition 3.2 below, on the event $|X((k-1)L_t^2)|_\infty < t$ and $\Lambda_{i,k}$ is good for all $-\frac{2t}{L_t}+1 \le i \le \frac{2t}{L_t}$, which is an event in $\mathcal{F}_{k-1}$, we have

$$\mathbb{E}^\xi\mathbb{E}_0^X[Z_k|\mathcal{F}_{k-1}] = \mathbb{E}^\xi\mathbb{E}_0^X\Big[\exp\Big\{-\gamma\int_{(k-1)L_t^2}^{kL_t^2}\xi(s,X(s))\,ds\Big\}\Big|\mathcal{F}_{k-1}\Big] \le e^{-CL_t} \tag{60}$$

for some $C > 0$ depending on γ,κ,ρ,ν. Substituting this bound into (59) for $1 \le k \le t/L_t^2$ and using the fact that $\prod_{k=1}^{t/L_t^2}\frac{Z_k}{\mathbb{E}^\xi\mathbb{E}_0^X[Z_k|\mathcal{F}_{k-1}]}$ is a martingale then gives the desired bound $e^{-Ct/L_t} = e^{-Ct/(A\log t)}$ for (58).

For dimension $d = 2$, the proof is similar. We choose $L_t = A\log t$ with A sufficiently large. We partition the space–time region $[-2t,2t]^2\times[0,t]$ into blocks of the form $\Lambda_{i,j,k} := [(i-1)L_t, iL_t)\times[(j-1)L_t, jL_t)\times[(k-1)L_t^2, kL_t^2)$, and we define good blocks and bad blocks as before. Applying Proposition 3.2 below then gives an upper bound of $\exp\{-C\frac{t}{L_t^2}\frac{L_t^2}{\log L_t}\} = \exp\{-\frac{Ct}{\log A+\log\log t}\}$, analogous to (58).

Finally, we note that the arguments also apply to the solution of the parabolic Anderson model

$$u(t,0) = \mathbb{E}_0^X\Big[\exp\Big\{-\gamma\int_0^t \xi(t-s,X(s))\,ds\Big\}\Big].$$

The only difference lies in passing the result (55) from $t \in \mathbb{N}$ to $t \in \mathbb{R}$, due to the lack of monotonicity of $u(t,0)$ in t. This can be easily overcome by the observation that for $n-1 < t < n$ with $n \in \mathbb{N}$,

$$u(n,0) \geq \mathrm{e}^{-\kappa(n-t)} \mathrm{e}^{-\gamma \int_t^n \xi(r,0)\,\mathrm{d}r} u(t,0),$$

and the fact that almost surely $\int_i^{i+1} \xi(r,0)\,\mathrm{d}r \leq \sqrt{i}$ for all i large by Borel–Cantelli because $\int_0^1 \xi(r,0)\,\mathrm{d}r$ has finite exponential moments. □

The following is a partial analogue of Theorem 1.1 for ξ with deterministic initial conditions.

Proposition 3.2 [Semi-annealed upper bound]. *Let ξ be defined as in (1) with deterministic initial condition $(\xi(0,x))_{x\in\mathbb{Z}^d}$. For $L > 0$ and $\vec{i} = (i_1,\ldots,i_d) \in \mathbb{Z}^d$, let $B_{L,\vec{i}} := [(i_1-1)L, i_1L) \times \cdots \times [(i_d-1)L, i_dL)$. Assume that there exist $a > 2$ and $\nu > 0$ such that for all $\vec{i} \in [-3L^a, 3L^a]^d$, $\sum_{x\in B_{L,\vec{i}}} \xi(0,x) \geq \nu L^d$. Then there exist constants $C_d > 0$, $d \geq 1$, such that for all L sufficiently large and for all $x \in \mathbb{Z}^d$ with $|x|_\infty \leq L$, we have*

$$\mathbb{E}^\xi \mathbb{E}_x^X\Big[\exp\Big\{-\gamma \int_0^{L^2} \xi(s, X(s))\,\mathrm{d}s\Big\}\Big] \leq \begin{cases} \mathrm{e}^{-C_1 L}, & d=1, \\ \mathrm{e}^{-C_2 \frac{L^2}{\log L}}, & d=2, \\ \mathrm{e}^{-C_d L^2}, & d\geq 3. \end{cases} \tag{61}$$

The same is true if we replace $\int_0^{L^2} \xi(s, X(s))\,\mathrm{d}s$ by $\int_0^{L^2} \xi(s, X(L^2-s))\,\mathrm{d}s$.

Proof. The basic strategy is to dominate $(\xi(L^2/2, x))_{|x|_\infty < 2L^a}$ from below by i.i.d. Poisson random variables, which then allows us to apply Theorem 1.1. We proceed as follows.

Let ξ be generated by independent random walks $(Y_j^y)_{y\in\mathbb{Z}^d, 1\leq j\leq \xi(0,y)}$ as in (1), and let $\bar{\xi}$ be generated by a separate system of independent random walks $(\bar{Y}_j^y)_{y\in\mathbb{Z}^d, 1\leq j\leq \bar{\xi}(0,y)}$, where $(\bar{\xi}(0,y))_{y\in\mathbb{Z}^d}$ are i.i.d. Poisson distributed with mean $\bar{\nu}$. Choose any $\bar{\nu} \in (0,\nu)$. Then by large deviation estimates for Poisson random variables,

$$\begin{aligned} \mathbb{P}^{\bar{\xi}}(G_L^c) &:= \mathbb{P}^{\bar{\xi}}\Big(\sum_{x\in B_{L,\vec{i}}} \bar{\xi}(0,x) \geq \sum_{x\in B_{L,\vec{i}}} \xi(0,x) \text{ for some } \vec{i} \in [-3L^a, 3L^a]^d\Big) \\ &\leq \sum_{\vec{i}\in[-3L^a,3L^a]^d} \mathbb{P}^{\bar{\xi}}\Big(\sum_{x\in B_{L,\vec{i}}} \bar{\xi}(0,x) \geq \nu L^d\Big) \leq 6^d L^{ad} \mathrm{e}^{-C_{\nu,\bar{\nu}} L^d}. \end{aligned} \tag{62}$$

On the event G_L, we will construct a coupling between $(Y_j^y)_{y\in\mathbb{Z}^d, 1\leq j\leq \xi(0,y)}$ and $(\bar{Y}_j^y)_{y\in\mathbb{Z}^d, 1\leq j\leq \bar{\xi}(0,y)}$ as follows. For each walk $\bar{Y}_j^y$ with $1 \leq j \leq \bar{\xi}(0,y)$ and

$y \in B_{L,\vec{i}}$ for some $\vec{i} \in [-3L^a, 3L^a]^d$, we can match $\bar{Y}_j^y$ with a distinct walk Y_k^z for some $z \in B_{L,\vec{i}}$ and $1 \le k \le \xi(0,z)$, which is possible on the event G_L.

Independently for each pair of walks $(\bar{Y}_j^y, Y_k^z)$, we will couple their coordinates as follows: For $1 \le i \le d$, the ith coordinates of the two walks evolve independently until the first time that their difference is of even parity. Note that this is the case either at time 0 already or at the first time when one of the coordinates changes. From then on, the ith coordinates are coupled in such a way that they always jump at the same time and their jumps are always opposites until the first time when the two coordinates coincide. From that time onward, the two coordinates always perform the same jumps at the same time. For walks in the ξ and $\bar{\xi}$ system which have not been paired up, we let them evolve independently. Note that such a coupling preserves the law of ξ (resp. $\bar{\xi}$), and each coupled pair $(\bar{Y}_j^y, Y_k^z)$ is successfully coupled in the sense that $\bar{Y}_j^y(L^2/2) = Y_k^z(L^2/2)$ if the trajectory of $\bar{Y}_j^y$ is in the event

$$E_j^y := \Big\{ \sup_{0\le t\le L^2/2} (\bar{Y}_j^y(t) - \bar{Y}_j^y(0))_i \in [L/2, L] \quad \text{and}$$
$$\inf_{0\le t\le L^2/2} (\bar{Y}_j^y(t) - \bar{Y}_j^y(0))_i \in [-L, -L/2] \quad \forall\, 1 \le i \le d \Big\},$$

because $|y - z|_\infty \le L$ by our choice of pairing of $\bar{Y}_j^y$ and Y_k^z. Then by our coupling of $\bar{\xi}$ and ξ, on the event G_L, we have

$$\xi(L^2/2, x) \ \ge\ \zeta(x) := \sum_{\substack{y\in\mathbb{Z}^d,\\ 1\le j\le \bar{\xi}(0,y)}} 1_{E_j^y} 1_{\{\bar{Y}_j^y(L^2/2)=x\}} \qquad \text{for all } |x|_\infty \le 2L^a. \tag{63}$$

Now observe that because $(\bar{\xi}(0,x))_{x\in\mathbb{Z}^d}$ are i.i.d. Poisson with mean $\bar{\nu}$, and $(\bar{Y}_j^y)_{y\in\mathbb{Z}^d, 1\le j\le\bar{\xi}(0,y)}$ are independent, $(\zeta(x))_{x\in\mathbb{Z}^d}$ are also i.i.d. Poisson distributed with mean $\alpha := \bar{\nu}\mathbb{P}^{\bar{\xi}}(E_j^y) = \bar{\nu}\mathbb{P}^{\bar{\xi}}(E_1^0)$, which is bounded away from 0 uniformly in L by the properties of simple symmetric random walks. This achieves the desired stochastic domination of ξ at time $L^2/2$. Let $\zeta_L(t,\cdot)$ denote the counting field of independent random walks as in (1) with initial condition $\zeta_L(0,y) = \zeta(y)1_{\{|y|_\infty\le 2L^a\}}$. Then using (63), uniformly in $x \in \mathbb{Z}^d$ with $|x|_\infty \le L$, we have

$$\mathbb{E}^{\xi}\mathbb{E}_x^X\Big[\exp\Big\{-\gamma\int_0^{L^2}\xi(s,X(s))\,\mathrm{d}s\Big\}\Big] = \mathbb{E}^{\xi,\bar{\xi}}\mathbb{E}_x^X\Big[\exp\Big\{-\gamma\int_0^{L^2}\xi(s,X(s))\,\mathrm{d}s\Big\}\Big]$$
$$\le \mathbb{P}^{\bar{\xi}}(G_L^c) + \mathbb{P}_x^X(|X(L^2/2)|_\infty > L^2)$$
$$+ \sup_{|x|_\infty\le L^2} \mathbb{E}^{\zeta_L}\mathbb{E}_x^X\Big[\exp\Big\{-\gamma\int_0^{L^2/2}\zeta_L(s,X(s))\,\mathrm{d}s\Big\}\Big]$$

$$\leq 6^d L^{ad} \mathrm{e}^{-C_{\nu,\bar{\nu}} L^d} + \mathrm{e}^{-CL^2}$$
$$+ \sup_{|x|_\infty \leq L^2} \mathbb{E}^{\zeta_L} \mathbb{E}_x^X \left[\exp\left\{ -\gamma \int_0^{L^2/2} \zeta_L(s, X(s))\, \mathrm{d}s \right\} \right]. \tag{64}$$

By the same argument as for (11), we have

$$\mathbb{E}^{\zeta_L} \mathbb{E}_x^X \left[\exp\left\{ -\gamma \int_0^{L^2/2} \zeta_L(s, X(s))\, \mathrm{d}s \right\} \right] = \mathbb{E}_x^X \left[\exp\left\{ -\alpha \sum_{|y|_\infty \leq 2L^a} (1 - v_X(L^2/2, y)) \right\} \right], \tag{65}$$

where

$$v_X(L^2/2, y) = \mathbb{E}_y^Y \left[\exp\left\{ -\gamma \int_0^{L^2/2} \delta_0(Y(s) - X(s))\, \mathrm{d}s \right\} \right].$$

To bound (65), note that by a union bound in combination with Azuma's inequality we obtain,

$$\sup_{|x|_\infty \leq L^2} \mathbb{P}_x^X \Big(\sup_{0 \leq s \leq L^2/2} |X(s)|_\infty > 2L^2 \Big) \leq \mathrm{e}^{-CL^2}. \tag{66}$$

On the complementary event $\{\sup_{0 \leq s \leq L^2/2} |X(s)|_\infty \leq 2L^2\}$, we have

$$1 - v_X(L^2/2, y) \leq \mathbb{P}_y^Y(\tau_{2L^2} \leq L^2/2) \leq \mathbb{P}(\mathcal{P}_{L^2/2} \geq |y|_\infty - 2L^2),$$

where $\tau_{2L^2} := \inf\{s \geq 0 : |Y(s)|_\infty \leq 2L^2\}$, and $\mathcal{P}_{L^2/2}$ is a Poisson random variable with mean $\rho L^2/2$, which counts the number of jumps of Y before time $L^2/2$. Therefore for L sufficiently large,

$$\sum_{|y|_\infty > 2L^a} (1 - v_X(L^2/2, y)) \leq \sum_{|y|_\infty > 2L^a} \mathbb{P}(\mathcal{P}_{L^2/2} \geq |y|_\infty - 2L^2)$$
$$\leq C \sum_{r=2L^a}^{\infty} \mathbb{P}(\mathcal{P}_{L^2/2} \geq r/2)\, r^{d-1} \leq C \mathbb{E}[\mathcal{P}_{L^2/2}^k] \sum_{r=2L^a}^{\infty} r^{d-k-1}$$
$$\leq C(\rho L^2/2)^k (2L^a)^{d-k} \leq C L^{-(a-2)k + ad} \leq 1, \tag{67}$$

where we have changed the values of the constant C (independent of L) from line to line, and the last inequality holds for all L large if we choose k large enough. Substituting the bounds (66) and (67) into (65) then gives the following bound uniformly for $x \in \mathbb{Z}^d$ with $|x|_\infty \leq L^2$:

$$\mathbb{E}^{\zeta_L}\mathbb{E}_x^X\Bigg[\exp\Bigg\{-\gamma\int_0^{L^2/2}\zeta_L(s,X(s))\,\mathrm{d}s\Bigg\}\Bigg]$$
$$\le \mathbb{P}_x^X\big(\sup_{0\le s\le L^2/2}|X(s)|_\infty > 2L^2\big) + \mathbb{E}_x^X\Big[\exp\Big\{-\alpha\sum_{y\in\mathbb{Z}^d}(1-\nu_X(L^2/2,y))+1\Big\}\Big]$$
$$\le \mathrm{e}^{-CL^2} + e\mathbb{E}_x^X\Big[\exp\Big\{-\alpha\sum_{y\in\mathbb{Z}^d}(1-\nu_X(L^2/2,y))\Big\}\Big],$$

where by the representation (11), the expectation is precisely the annealed survival probability of a random walk among a Poisson field of traps with density α, for which the bounds in (4) apply with ν replaced by α and t by $L^2/2$. Substituting this bound back into (64) then gives (61). The same proof applies when we reverse the time direction of X in (61). □

4 Existence and Positivity of the Quenched Lyapunov Exponent

In this section, we prove Theorems 1.2 and 1.3. In Sect. 4.1, we state a shape theorem which implies the existence of the quenched Lyapunov exponent for the quenched survival probability $Z^\gamma_{t,\xi}$. In Sect. 4.2, we prove the stated shape theorem for bounded ergodic random fields. In Sect. 4.3, we show how to deduce the existence of the quenched Lyapunov exponent for the solution of the parabolic Anderson model from what we already know for the quenched survival probability. Finally in Sect. 4.4, we prove the positivity of the quenched Lyapunov exponent, which concludes the proof of Theorems 1.2 and 1.3.

4.1 Shape Theorem and the Quenched Lyapunov Exponent

In this section, we focus exclusively on the quenched survival probability $Z^\gamma_{t,\xi}$. The approach we adopt in proving the existence of the quenched Lyapunov exponent for $Z^\gamma_{t,\xi}$ uses the subadditive ergodic theorem and follows ideas used by Varadhan in [25] to prove the quenched large deviation principle for random walks in random environments.

For $s \ge 0$ and $x \in \mathbb{Z}^d$, let $\mathbb{P}^X_{x,s}$ and $\mathbb{E}^X_{x,s}$ denote, respectively, probability and expectation for a jump rate κ simple symmetric random walk X, starting from x at time s. For each $0 \le s < t$ and $x, y \in \mathbb{Z}^d$, define

$$\begin{aligned} e(s,t,x,y,\xi) &:= \mathbb{E}^X_{x,s}\Bigg[\exp\Bigg\{-\gamma\int_s^t \xi(u,X(u))\,\mathrm{d}u\Bigg\}1_{\{X(t)=y\}}\Bigg], \\ a(s,t,x,y,\xi) &:= -\log e(s,t,x,y,\xi). \end{aligned} \tag{68}$$

We call $a(s,t,x,y,\xi)$ the *point-to-point* passage function from x to y between times s and t. We will prove the following shape theorem for $a(0,t,0,y,\xi)$.

Theorem 4.1 [Shape theorem]. *There exists a deterministic convex function $\alpha : \mathbb{R}^d \to \mathbb{R}$, which we call the shape function, such that $\mathbb{P}^{\xi}$-a.s., for any compact $K \subset \mathbb{R}^d$,*

$$\lim_{t\to\infty} \sup_{y\in tK\cap\mathbb{Z}^d} |t^{-1}a(0,t,0,y,\xi) - \alpha(y/t)| = 0. \tag{69}$$

Furthermore, for any $M > 0$, we can find a compact $K \subset \mathbb{R}^d$ such that

$$\limsup_{t\to\infty} \frac{1}{t} \log \mathbb{E}_0^X \left[\exp\left\{ -\gamma \int_0^t \xi(s, X(s))\, \mathrm{d}s \right\} 1_{\{X(t)\notin tK\}} \right] \le -M. \tag{70}$$

Remark. Note that (5) in Theorem 1.2 follows easily from Theorem 4.1, which we leave to the reader as an exercise. In particular, the quenched Lyapunov exponent satisfies

$$\tilde{\lambda}_{d,\gamma,\kappa,\rho,\nu} = \inf_{y\in\mathbb{R}^d} \alpha(y) = \alpha(0) = \lim_{t\to\infty} t^{-1}a(0,t,0,0,\xi), \tag{71}$$

where $\inf_{y\in\mathbb{R}^d} \alpha(y) = \alpha(0)$ follows from the convexity and symmetry of α since ξ is symmetric.

The unboundedness of the random field ξ creates complications for the proof of Theorem 4.1. Therefore, we first replace ξ by $\xi_N := (\max\{\xi(s,x), N\})_{s\ge 0, x\in\mathbb{Z}^d}$ for some large $N > 0$ and prove a corresponding shape theorem, then use almost sure properties of ξ established by Kesten and Sidoravicius in [15] to control the error caused by the truncation.

Theorem 4.2 [Shape theorem for bounded ergodic potentials]. *Let $\zeta := (\zeta(s,x))_{s\ge 0, x\in\mathbb{Z}^d}$ be a real-valued random field, which is ergodic with respect to the shift map $\theta_{r,z}\zeta := (\zeta(s+r, x+z))_{s\ge 0, x\in\mathbb{Z}^d}$, for all $r \ge 0$ and $z \in \mathbb{Z}^d$. Assume further that $|\zeta(0,0)| \le A$ a.s. for some $A > 0$. Then the conclusions of Theorem 4.1 hold with ξ replaced by ζ.*

Remark. Note that Theorem 4.2 can be applied to the occupation field of the exclusion process or the voter model in an ergodic equilibrium, which in particular implies the existence of the corresponding quenched Lyapunov exponents.

Before we prove Theorem 4.2 in the next section, let us first show how to deduce Theorem 4.1 from Theorem 4.2, using almost sure bounds on ξ from [15].

Proof of Theorem 4.1. Note that since ξ is non-negative, (70) follows from elementary large deviation estimates for the random walk X, if we take K to be a large enough closed ball centered at the origin, which we fix for the rest of the proof.

By applying Theorem 4.2 to the truncated random field ξ_N, we have that for each $N > 0$, there exists a convex shape function $\alpha_N : \mathbb{R}^d \to \mathbb{R}$ such that (69) holds with ξ replaced by ξ_N and α replaced by α_N. Note that α_N is monotonically increasing in N, and its limit α is necessarily convex. To prove (69), it then suffices to show that, for any $\epsilon > 0$, we can choose N sufficiently large such that $\mathbb{P}^\xi$-a.s.,

$$\frac{1}{t} \sup_{y \in tK \cap \mathbb{Z}^d} |a(0,t,0,y,\xi) - a(0,t,0,y,\xi_N)| \leq \epsilon \qquad \text{for all } t \text{ sufficiently large.} \tag{72}$$

To prove (72), we will need Lemma 15 from [15], which by Borel–Cantelli implies that there exist positive constants C_0, C_1, C_2, C_3, C_4 with $C_0 > 1$, such that if Ξ_l denotes the space of all possible random walk trajectories $\pi : [0,t] \to \mathbb{Z}^d$, which contain exactly l jumps and are contained in the rectangle $[-C_1 t \log t, C_1 t \log t]^d$, then $\mathbb{P}^\xi$-a.s., for all $t \in \mathbb{N}$ sufficiently large, we have

$$\sup_{\pi \in \Xi_l} \int_0^t \xi(s,\pi(s)) 1_{\{\xi(s,\pi(s)) \geq C_2 \nu C_0^{dm}\}} \, \mathrm{d}s \leq (t+l) \sum_{r=m}^{\infty} C_3 C_0^{r(d+6)+d} \mathrm{e}^{-C_4 C_0^{r/4}} \qquad \forall\, m \in \mathbb{N}, l \geq 0, \tag{73}$$

where $A_m := \sum_{r=m}^{\infty} C_3 C_0^{r(d+6)+d} \mathrm{e}^{-C_4 C_0^{r/4}} \to 0$ as $m \to \infty$.

One important consequence of (73) is that

$$0 < \sup_{y \in K} \alpha(y) < \infty. \tag{74}$$

Indeed, if $l_t(X)$ denotes the number of jumps of X on the time interval $[0,t]$, then

$$\begin{aligned}
\sup_{y \in K} \alpha(y) &\leq \lim_{t \to \infty} -t^{-1} \log \inf_{y \in tK \cap \mathbb{Z}^d} \mathbb{E}_0^X \left[\exp\left\{ -\gamma \int_0^t \xi(s, X(s)) \, \mathrm{d}s \right\} 1_{\{X(t)=y\}} \right] \\
&\leq \lim_{t \to \infty} -t^{-1} \log \inf_{y \in tK \cap \mathbb{Z}^d} \mathbb{E}_0^X \left[\exp\left\{ -\gamma \int_0^t \xi(s, X(s)) \, \mathrm{d}s \right\} \right. \\
&\qquad \left. 1_{\{X(t)=y, l_t(X) \leq 2D(K)t, X \in \Xi_{l_t(X)}\}} \right],
\end{aligned}$$

where $D(K) := \sup_{y \in K} |y|_1$. We can then apply (73) and large deviation estimates for random walks to the above bound to deduce $\sup_{y \in K} \alpha(y) < \infty$. The fact that $\sup_{y \in K} \alpha(y) > 0$ for a large ball K again follows from basic large deviation estimates.

By large deviation estimates, we can find B large enough such that

$$\mathbb{P}_0^X(l_t(X) \geq Bt \text{ or } X \notin \Xi_{l_t(X)}) \leq \mathrm{e}^{-2\sup_{y\in K}\alpha(y)t} \qquad \text{for all } t \text{ sufficiently large.} \tag{75}$$

Let $N = C_2 \nu C_0^{dm}$. Then by (73), $\mathbb{P}^\xi$-a.s., uniformly in $y \in \mathbb{Z}^d$ and for all t large, we have

$$\begin{aligned} e(0,t,0,y,\xi) &\geq \mathrm{e}^{-(1+B)A_m\gamma t}\mathbb{E}_0^X\Bigg[\exp\Bigg\{-\gamma\int_0^t \xi_N(s,X(s))\,\mathrm{d}s\Bigg\} \\ &\qquad\qquad \times 1_{\{X(t)=y, l_t(X)\leq Bt, X\in\Xi_{l_t(X)}\}}\Bigg] \\ &\geq \mathrm{e}^{-(1+B)A_m\gamma t}(e(0,t,0,y,\xi_N) - \mathrm{e}^{-2\sup_{y\in K}\alpha(y)t}), \end{aligned} \tag{76}$$

where in the last inequality we applied (75). Since $-t^{-1}\log e(t,0,y,\xi_N) \to \alpha_N(y/t)$ uniformly for $y \in tK \cap \mathbb{Z}^d$ by Theorem 4.2, and $\sup_{y\in K}\alpha_N(y) \leq \sup_{y\in K}\alpha(y)$, (76) implies that $\mathbb{P}^\xi$-a.s., uniformly in $y \in tK \cap \mathbb{Z}^d$ and for all t large, we have

$$t^{-1}a(0,t,0,y,\xi) \leq t^{-1}a(0,t,0,y,\xi_N) + (1+B)A_m\gamma + o(1).$$

Since $a(0,t,0,y,\xi) \geq a(0,t,0,y,\xi_N)$, and A_m can be made arbitrarily small by choosing m sufficiently large, (72) then follows. □

Remark. Theorem 4.1 in fact holds for the catalytic case as well, where we take $\gamma < 0$ in (68) and (70). This implies the existence of the quenched Lyapunov exponent in Theorem 1.2 for the catalytic case, where we set $\gamma < 0$ in the definition of $Z_{t,\xi}^\gamma$. Indeed, Theorem 4.2 still applies to the truncated field ξ_N. To control the error caused by the truncation, the following modifications are needed in the proof of Theorem 4.1. To prove (70), we need to apply (73). More precisely, we need to first consider trajectories $(X_x)_{0\leq s\leq t}$, which are not contained in $[-C_1 t\log t, C_1 t\log t]^d$. The contribution from these trajectories can be shown to decay super-exponentially in t by large deviation estimates and a bound on ξ given in (2.37) of [15, Lemma 4]. For X which lies inside $[-C_1 t\log t, C_1 t\log t]^d$, we can then use (73) and large deviations to deduce (70). In contrast to (76), we need to upper bound $e(0,t,0,y,\xi)$ in terms of $e(0,t,0,y,\xi_N)$. The proof is essentially the same, except that in place of (75), we need to show that we can choose B large enough, such that $\mathbb{P}^\xi$-a.s.,

$$\begin{aligned} &\sup_{y\in tK\cap\mathbb{Z}^d}\mathbb{E}_0^X\Bigg[\exp\Bigg\{|\gamma|\int_0^t \xi(s,X(s))\,\mathrm{d}s\Bigg\} 1_{\{X(t)=y\}}1_{\{l_t(X)\geq Bt \text{ or } X\notin\Xi_{l_t(X)}\}}\Bigg] \\ &\leq \inf_{y\in tK\cap\mathbb{Z}^d}\mathbb{P}_0^X(X_t = y). \end{aligned} \tag{77}$$

This can be proved by appealing to (70) and applying (73) and large deviation estimates.

4.2 Proof of Shape Theorem for Bounded Ergodic Potentials

In this section, we prove Theorem 4.2. From now on, let $\mathbb{Q}_+$ denote the set of positive rationals, and let $\mathbb{Q}^d$ denote the set of points in $\mathbb{R}^d$ with rational coordinates. We start with the following auxiliary result.

Lemma 4.1. *There exists a deterministic function* $\alpha : \mathbb{Q}^d \to [-\gamma A, \infty)$ *such that for every* $y \in \mathbb{Q}^d$,

$$\lim_{\substack{t\to\infty \\ ty\in\mathbb{Z}^d}} t^{-1} a(0,t,0,ty,\zeta) = \alpha(y) \qquad \mathbb{P}^\zeta - a.s. \tag{78}$$

Proof. Since we assume $y \in \mathbb{Q}^d$ and $ty \in \mathbb{Z}^d$ in (78), without loss of generality, it suffices to consider $y \in \mathbb{Z}^d$ and $t \in \mathbb{N}$. Note that by the definition of the passage function a in (68), $\mathbb{P}^\zeta$-a.s.,

$$a(t_1,t_3,x_1,x_3,\zeta) \le a(t_1,t_2,x_1,x_2,\zeta) + a(t_2,t_3,x_2,x_3,\zeta) \qquad \forall\, t_1 < t_2 < t_3,\ x_1,x_2,x_3 \in \mathbb{Z}^d. \tag{79}$$

Together with our assumption on the ergodicity of ζ, this implies that the two-parameter family $a(s,t,sy,ty,\zeta)$, $0 \le s \le t$ with $s,t \in \mathbb{Z}$, satisfies the conditions of Kingman's subadditive ergodic theorem (see e.g., [18]). Therefore, there exists a deterministic constant $\alpha(y)$ such that (78) holds. The fact that $\alpha(y) \ge -\gamma A$ follows by bounding ζ from above by the uniform bound A, and $\alpha(y) < \infty$ follows from large deviation bounds for the random walk X. □

To extend the definition of $\alpha(y)$ in Lemma 4.1 to $y \notin \mathbb{Q}^d$ and to prove the uniform convergence in (69), we need to establish equicontinuity of $t^{-1}a(0,t,0,ty,\zeta)$ in y, as $t \to \infty$. For that, we first need a large deviation estimate for the random walk X.

Lemma 4.2. *Let X be a jump rate κ simple symmetric random walk on $\mathbb{Z}^d$ with $X(0) = 0$. Then for every $t > 0$ and $x \in \mathbb{Z}^d$, we have*

$$\mathbb{P}_0^X(X(t) = x) = \frac{e^{-J(\frac{x}{t})t}}{(2\pi t)^{\frac{d}{2}} \Pi_{i=1}^d \left(\frac{x_i^2}{t^2} + \frac{\kappa^2}{d^2}\right)^{1/4}} (1 + o(1)), \tag{80}$$

where

$$J(x) := \sum_{i=1}^d \frac{\kappa}{d} j\left(\frac{dx_i}{\kappa}\right) \qquad \text{with} \quad j(y) := y \sinh^{-1} y - \sqrt{y^2+1} + 1,$$

and the error term $o(1)$ tends to zero as $t \to \infty$ uniformly in $x \in tK \cap \mathbb{Z}^d$, for any compact $K \subset \mathbb{R}^d$.

Proof. Since the coordinates of X are independent, it suffices to consider the case X is a rate κ/d simple symmetric random walk on $\mathbb{Z}$. Let $\sigma := t/\lceil t \rceil$. Let $Z_1^\lambda, \ldots, Z_{\lceil t \rceil}^\lambda$ be i.i.d. with

$$\mathbb{P}(Z_1^\lambda = y) = \mathbb{P}(X(\sigma) = y) \mathrm{e}^{\lambda y - \Phi(\lambda)}, \qquad y \in \mathbb{Z},$$

where

$$\Phi(\lambda) = \log \mathbb{E}\left[\mathrm{e}^{\lambda X(\sigma)}\right] = \frac{\sigma\kappa}{d}(\cosh\lambda - 1).$$

Note that

$$\mathbb{E}[Z_1^\lambda] = \frac{\mathrm{d}\Phi}{\mathrm{d}\lambda}(\lambda) = \frac{\sigma\kappa}{d}\sinh\lambda \qquad \text{and} \qquad \mathrm{Var}(Z_1^\lambda) = \frac{\mathrm{d}^2\Phi}{\mathrm{d}^2\lambda}(\lambda) = \frac{\sigma\kappa}{d}\cosh\lambda.$$

We shall set $\lambda = \sinh^{-1}(\frac{dx}{\kappa t})$ so that $\mathbb{E}[Z_1^\lambda] = x/\lceil t \rceil$. If we let $S_{\lceil t \rceil} := \sum_{i=1}^{\lceil t \rceil} Z_i^\lambda$, then observe that

$$\mathbb{P}_0^X(X(t) = x) = \mathbb{P}(S_{\lceil t \rceil} = x)\mathrm{e}^{-\lambda x + \lceil t \rceil \Phi(\lambda)} = \mathbb{P}(S_{\lceil t \rceil} = x)\mathrm{e}^{-\frac{\kappa}{d} j(\frac{dx}{\kappa t})t}.$$

Note that $S_{\lceil t \rceil} - x$ has mean 0, variance $t\sqrt{\frac{x^2}{t^2} + \frac{\kappa^2}{d^2}}$, and characteristic function

$$\mathrm{e}^{\lceil t \rceil \left(\Phi(ik+\lambda) - \Phi(\lambda)\right) - ikx} = \mathrm{e}^{ix(\sin k - k) - t\sqrt{\frac{x^2}{t^2} + \frac{\kappa^2}{d^2}}(1-\cos k)}.$$

Applying Fourier inversion then gives (80). □

With the help of Lemma 4.2, we can control the modulus of continuity of $t^{-1}a(0, t, 0, ty, \zeta)$.

Lemma 4.3. *Let K be any compact subset of $\mathbb{R}^d$. There exists $\phi_K : (0, \infty) \to (0, \infty)$ with $\lim_{r \downarrow 0} \phi_K(r) = 0$, such that for any $\epsilon > 0$, $\mathbb{P}^\zeta$-a.s., we have*

$$\limsup_{t\to\infty} \sup_{\substack{x,y \in tK \cap \mathbb{Z}^d \\ \|x-y\| \le \epsilon t}} t^{-1}|a(0,t,0,x,\zeta) - a(0,t,0,y,\zeta)| \le \phi_K(\epsilon). \tag{81}$$

Proof. Let $K \subset \mathbb{R}^d$ be compact. It suffices to consider $\epsilon \in (0, 1/2)$, which we also fix from now on. First note that, by Lemma 4.2, $\mathbb{P}^\zeta$-a.s.,

$$\inf_{z \in tK \cap \mathbb{Z}^d} e(0,t,0,z,\zeta) \ge \mathrm{e}^{-At} \inf_{z \in tK \cap \mathbb{Z}^d} \mathbb{P}_0^X(X_t = z) \ge \mathrm{e}^{-(A+1)t - \sup_{u\in K} J(u)\, t} \tag{82}$$

for all t sufficiently large.

Also note that for all $z \in \mathbb{Z}^d$ and $t > 0$,

$$e(0,t,0,z,\zeta) = \sum_{w\in\mathbb{Z}^d} e(0,(1-\epsilon)t,0,w,\zeta)\, e((1-\epsilon)t,t,w,z,\zeta). \tag{83}$$

By large deviation estimates, we can choose a ball B_R centered at the origin with radius R large enough and independent of ϵ, such that $K \subset B_R$ and $\mathbb{P}^\zeta$-a.s.,

$$\begin{aligned}\sup_{z\in\mathbb{Z}^d} &\sum_{\substack{w\in\mathbb{Z}^d\\ w\notin tB_R}} e(0,(1-\epsilon)t,0,w,\zeta)\, e((1-\epsilon)t,t,w,z,\zeta)\\ &\le \mathbb{P}_0^X\big(X((1-\epsilon)t) \notin tB_R\big)\mathrm{e}^{At} \le \mathrm{e}^{-(A+2)t-\sup_{u\in K} J(u)\,t}\end{aligned}$$

for all t sufficiently large. In view of (82), the dominant contribution in (83) comes from $w \in tB_R \cap \mathbb{Z}^d$. Therefore to prove (81), it suffices to verify

$$\begin{aligned}\limsup_{t\to\infty} \sup_{\substack{x,y\in tK\cap\mathbb{Z}^d\\ \|x-y\|\le\epsilon t}} t^{-1}\left|\log\frac{\sum_{w\in tB_R\cap\mathbb{Z}^d} e(0,(1-\epsilon)t,0,w,\zeta)\, e((1-\epsilon)t,t,w,y,\zeta)}{\sum_{w\in tB_R\cap\mathbb{Z}^d} e(0,(1-\epsilon)t,0,w,\zeta)\, e((1-\epsilon)t,t,w,x,\zeta)}\right|\\ \le \phi_K(\epsilon).\end{aligned} \tag{84}$$

Note that $\mathbb{P}^\zeta$-a.s., and uniformly in $x,y \in tK \cap \mathbb{Z}^d$ with $\|x-y\| \le \epsilon t$,

$$\begin{aligned}&\frac{\sum_{w\in tB_R\cap\mathbb{Z}^d} e(0,(1-\epsilon)t,0,w,\zeta)\, e((1-\epsilon)t,t,w,y,\zeta)}{\sum_{w\in tB_R\cap\mathbb{Z}^d} e(0,(1-\epsilon)t,0,w,\zeta)\, e((1-\epsilon)t,t,w,x,\zeta)}\\ &\le \sup_{w\in tB_R\cap\mathbb{Z}^d} \frac{e((1-\epsilon)t,t,w,y,\zeta)}{e((1-\epsilon)t,t,w,x,\zeta)} \le \sup_{w\in tB_R\cap\mathbb{Z}^d} \frac{\mathbb{P}_0^X(X(\epsilon t)=y-w)}{\mathbb{P}_0^X(X(\epsilon t)=x-w)}\,\mathrm{e}^{2A\epsilon t}\\ &\le \exp\left\{\epsilon t \sup_{w\in tB_R\cap\mathbb{Z}^d}\left(J\Big(\frac{x-w}{\epsilon t}\Big) - J\Big(\frac{y-w}{\epsilon t}\Big)\right) + 3A\epsilon t\right\}\\ &\le \exp\left\{\epsilon t \sup_{u,v\in B_{2R/\epsilon},\|u-v\|\le 1} |J(u)-J(v)| + 3A\epsilon t\right\}\end{aligned}$$

for all t sufficiently large, where we applied Lemma 4.2, and $B_{2R/\epsilon}$ denotes the ball of radius $2R/\epsilon$, centered at the origin. Therefore, (84) holds with

$$\phi_K(\epsilon) = 3A\epsilon + \epsilon \sup_{u,v\in B_{2R/\epsilon},\|u-v\|\le 1} |J(u)-J(v)|.$$

It only remains to verify that $\phi_K(\epsilon) \downarrow 0$ as $\epsilon \downarrow 0$, which is easy to check from the definition of J. □

Proof of Theorem 4.2. Because ζ is uniformly bounded, (70) follows by large deviation estimates for the number of jumps of X up to time t. Lemma 4.3 implies that for each compact $K \subset \mathbb{R}^d$, the function α in Lemma 4.1 satisfies

$$\sup_{\substack{u,v \in K \cap \mathbb{Q}^d \\ \|u-v\| \leq \epsilon}} |\alpha(u) - \alpha(v)| \leq \phi_K(\epsilon) \qquad \text{for all } \epsilon > 0. \tag{85}$$

This allows us to extend α to a continuous function on $\mathbb{R}^d$.

To prove (69), it suffices to show that for each $\delta > 0$,

$$\limsup_{t \to \infty} \sup_{y \in tK \cap \mathbb{Z}^d} |t^{-1} a(0,t,0,y,\xi) - \alpha(y/t)| \leq \delta. \tag{86}$$

We can choose an ϵ such that $\phi_K(\epsilon) < \delta/3$. We can then find a finite number of points $x_1, \ldots, x_m \in \mathbb{Q}^d$ which form an ϵ-net in K, and along a subsequence of times of the form $t_n = n\sigma$ with $\sigma x_i \in \mathbb{Z}^d$ for all x_i, we have $t_n^{-1} a(0, t_n, 0, t_n x_i) \to \alpha(x_i)$ a.s. The uniform control of modulus of continuity provided by Lemma 4.3 and (85) then implies (86) along t_n. This can be transferred to $t \to \infty$ along $\mathbb{R}$ using

$$\begin{aligned} e(0,t,0,y,\zeta) &\geq e(0,s,0,y,\zeta)\, e(s,t,y,y,\zeta) \\ &\geq e(0,s,0,y,\zeta) \mathrm{e}^{-(\kappa+\gamma A)(t-s)} \quad \text{for } s < t. \end{aligned}$$

Finally, to prove the convexity of α, let $x, y \in \mathbb{R}^d$ and $\beta \in (0,1)$. Then $\mathbb{P}^{\zeta}$-a.s., we have

$$\begin{aligned} a(0,t_n,0,\beta y_n + (1-\beta)x_n, \zeta) &\leq a(0,\beta t_n, 0, \beta y_n, \zeta) \\ &+ a(\beta t_n, t_n, \beta y_n, \beta y_n + (1-\beta)x_n, \zeta), \end{aligned}$$

where we take sequences t_n, x_n, y_n with $t_n \to \infty$, $x_n/t_n \to x$, $y_n/t_n \to y$, and $\beta y_n, (1-\beta)x_n \in \mathbb{Z}^d$. By Lemma 4.1, the first term divided by t_n converges a.s. to $\alpha(\beta y + (1-\beta)x)$, the second term divided by βt_n converges a.s. to $\alpha(y)$, while the last term divided by $(1-\beta)t_n$ converges in probability to $\alpha(x)$ by translation invariance. The convexity of α then follows. □

4.3 Existence of the Quenched Lyapunov Exponent for the PAM

Proof of (9) in Theorem 1.3. Since $Z^{\gamma}_{t,\xi}$ is equally distributed with $u(t,0)$ for each $t \geq 0$, $-t^{-1} \log u(t,0)$ converges in probability to the quenched Lyapunov exponent $\tilde{\lambda}_{d,\gamma,\kappa,\rho,\nu}$. It only remains to verify the almost sure convergence. We will bound the variance of $\log u(t,0)$, which is the same as that of $\log Z^{\gamma}_{t,\xi}$, and then apply Borel–Cantelli.

Assume that $t \in \mathbb{N}$. Note that we can write ξ as a sum of i.i.d. random fields $(\xi_i(s,x))_{s \geq 0, x \in \mathbb{Z}^d}$, each of which is defined from a Poisson system of independent random walks with density ν/t, in the same way as ξ. Then we can perform a martingale decomposition and write

$$
\begin{aligned}
\log Z^{\gamma}_{t,\xi} - \mathbb{E}^{\xi}[\log Z^{\gamma}_{t,\xi}] &= \sum_{i=1}^{t} V_i \\
&:= \sum_{i=1}^{t} \Big(\mathbb{E}^{\xi}[\log Z^{\gamma}_{t,\xi} | \xi_1, \dots, \xi_i] - \mathbb{E}^{\xi}[\log Z^{\gamma}_{t,\xi} | \xi_1, \dots, \xi_{i-1}] \Big),
\end{aligned}
$$

and hence $\mathrm{Var}(\log Z^{\gamma}_{t,\xi}) = \sum_{i=1}^{t} \mathbb{E}^{\xi}[V_i^2]$.

For each $1 \le i \le t$, we have

$$
\begin{aligned}
V_i &= \mathbb{E}^{\xi_{i+1},\dots,\xi_t}\big[\log Z^{\gamma}_{t,\xi} - \mathbb{E}^{\xi_i}[\log Z^{\gamma}_{t,\xi}]\big] \\
&= \mathbb{E}^{\xi_{i+1},\dots,\xi_t}\mathbb{E}^{\xi_i'} \left[\log \frac{\mathbb{E}_0^X \Big[\mathrm{e}^{-\gamma \int_0^t \left(\sum_{1\le j\le t, j\neq i} \xi_j(s,X(s)) + \xi_i(s,X(s)) \right) \mathrm{d}s} \Big]}{\mathbb{E}_0^X \Big[\mathrm{e}^{-\gamma \int_0^t \left(\sum_{1\le j\le t, j\neq i} \xi_j(s,X(s)) + \xi_i'(s,X(s)) \right) \mathrm{d}s} \Big]} \right] \\
&= \mathbb{E}^{\xi_{i+1},\dots,\xi_t}\mathbb{E}^{\xi_i'} \left[\log \mathbb{E}^{X,i}\Big[\mathrm{e}^{-\gamma \int_0^t \xi_i(s,X(s))\,\mathrm{d}s}\Big] - \log \mathbb{E}^{X,i}\Big[\mathrm{e}^{-\gamma \int_0^t \xi_i'(s,X(s))\,\mathrm{d}s}\Big] \right],
\end{aligned}
$$

where ξ_i' denotes an independent copy of ξ_i, and $\mathbb{E}^{X,i}$ denotes expectation with respect to the Gibbs transform of the random walk path measure $\mathbb{P}_0^X$, with Gibbs weight $\mathrm{e}^{-\gamma \int_0^t \sum_{1\le j\le t, j\neq i} \xi_j(s,X(s))\,\mathrm{d}s}$. Then by Jensen's inequality,

$$
\begin{aligned}
\mathbb{E}^{\xi}[V_i^2] &\le \mathbb{E}^{\xi,\xi_i'}\Big[\Big(\log \mathbb{E}^{X,i}\Big[\mathrm{e}^{-\gamma \int_0^t \xi_i(s,X(s))\,\mathrm{d}s}\Big] - \log \mathbb{E}^{X,i}\Big[\mathrm{e}^{-\gamma \int_0^t \xi_i'(s,X(s))\,\mathrm{d}s}\Big]\Big)^2\Big] \\
&\le 2\mathbb{E}^{\xi,\xi_i'}\Big[\Big(\log \mathbb{E}^{X,i}\Big[\mathrm{e}^{-\gamma \int_0^t \xi_i(s,X(s))\,\mathrm{d}s}\Big]\Big)^2\Big] \\
&\quad + 2\mathbb{E}^{\xi,\xi_i'}\Big[\Big(\log \mathbb{E}^{X,i}\Big[\mathrm{e}^{-\gamma \int_0^t \xi_i'(s,X(s))\,\mathrm{d}s}\Big]\Big)^2\Big] \\
&= 4\mathbb{E}^{\xi}\Big[\Big(\log \mathbb{E}^{X,i}\Big[\mathrm{e}^{-\gamma \int_0^t \xi_i(s,X(s))\,\mathrm{d}s}\Big]\Big)^2\Big] \\
&\le 4\mathbb{E}^{\xi}\Big[\Big(\mathbb{E}^{X,i}\Big[\gamma \int_0^t \xi_i(s,X(s))\,\mathrm{d}s\Big]\Big)^2\Big] \\
&\le 4\gamma^2 \mathbb{E}^{\xi}\mathbb{E}^{X,i}\Big[\Big(\int_0^t \xi_i(s,X(s))\,\mathrm{d}s\Big)^2\Big]
\end{aligned}
$$

$$= 4\gamma^2 \mathbb{E}^{X,i} \mathbb{E}^{\xi_i} \Big[\Big(\int_0^t \xi_i(s, X(s))\, \mathrm{d}s \Big)^2 \Big],$$

where in the third line we used the exchangeability of $\{\xi_i, \xi_i'\}$, and in the fourth line we applied Jensen's inequality[1] to the non-negative convex function $-\log x$ on the interval $(0, 1]$.

Note that for any realization of $((X(s))_{0 \le s \le t}$, we have

$$\mathbb{E}^{\xi_i} \Big[\Big(\int_0^t \xi_i(s, X(s))\, \mathrm{d}s \Big)^2 \Big]$$

$$= 2 \iint_{0<u<v<t} \mathbb{E}^{\xi_i} [\xi_i(u, X(u)) \xi_i(v, X(v))]\, \mathrm{d}u\, \mathrm{d}v$$

$$= 2 \iint_{0<u<v<t} \Big(\frac{\nu^2}{t^2} \sum_{\substack{y \in \mathbb{Z}^d \\ y \neq X(u)}} \mathbb{P}^Y_{y,u}(Y(v) = X(v))$$

$$+ \Big(\frac{\nu^2}{t^2} + \frac{\nu}{t} \Big) \mathbb{P}^Y_{X(u),u}(Y(v) = X(v)) \Big)\, \mathrm{d}u\, \mathrm{d}v$$

$$\le 2\nu^2 + 2\nu \int_0^t \mathbb{P}^Y_{0,0}(Y(s) = 0)\, \mathrm{d}s,$$

where $\mathbb{P}^Y_{y,s}$ denotes probability for a simple symmetric random walk on $\mathbb{Z}^d$ with jump rate ρ, starting from y at time s, and in the last line we used that $\mathbb{P}^Y_{0,0}(Y(s) = y)$ is maximized at $y = 0$ for all $s \ge 0$. Combined with the previous bounds, we obtain

$$\mathrm{Var}(\log u(t, 0)) = \mathrm{Var}(\log Z^\gamma_{t,\xi}) = \sum_{i=1}^t \mathbb{E}^\xi [V_i^2]$$

$$\le 8\gamma^2 \nu^2 t + 8\gamma^2 \nu t \int_0^t \mathbb{P}^Y_{0,0}(Y(s) = 0)\, \mathrm{d}s \le C t^{\frac{3}{2}}$$

for some $C > 0$, since $\int_0^t \mathbb{P}^Y_{0,0}(Y(s) = 0)\, \mathrm{d}s$ is of order $\sqrt{t}$ in dimension $d = 1$, of order $\log t$ in $d = 2$, and converges in $d \ge 3$. Therefore for any $\epsilon > 0$,

$$\mathbb{P}^\xi \Big(\big| \log u(t, 0) - \mathbb{E}^\xi [\log u(t, 0)] \big| \ge \epsilon t \Big) \le \frac{C}{\epsilon^2 \sqrt{t}},$$

[1] Note that this is where the proof fails for the $\gamma < 0$ case.

which by Borel–Cantelli implies that along the sequence $t_n = n^3$, $n \in \mathbb{N}$, we have almost sure convergence of $-t^{-1} \log u(t,0)$ to the quenched Lyapunov exponent $\tilde{\lambda}_{d,\gamma,\kappa,\rho,\nu}$.

To extend the almost sure convergence to $t \to \infty$ along $\mathbb{R}$, consider $t \in [t_n, t_{n+1})$ for some $n \in \mathbb{N}$. As at the end of the proof of Proposition 3.1, we have

$$u(t,0) \geq \mathrm{e}^{-\kappa(t-t_n)} \mathrm{e}^{-\gamma \int_{t_n}^{t} \xi(s,0)\,\mathrm{d}s} u(t_n,0),$$

$$u(t,0) \leq \mathrm{e}^{\kappa(t_{n+1}-t)} \mathrm{e}^{\gamma \int_{t}^{t_{n+1}} \xi(s,0)\,\mathrm{d}s} u(t_{n+1},0).$$

Note that $(t_{n+1}-t_n)/t_n \to 0$ as $n \to \infty$, and we claim that also $t_n^{-1} \int_{t_n}^{t_{n+1}} \xi(s,0)\,\mathrm{d}s \to 0$ a.s. as $n \to \infty$, which then implies the desired almost sure convergence of $t^{-1} \log u(t,0)$ as $t \to \infty$ along $\mathbb{R}$. Indeed, since $\int_0^1 \xi(s,0)\,\mathrm{d}s$ has finite exponential moments, as can be seen from (15) applied to the case $\gamma < 0$ and $X \equiv 0$, we have exponential tail bounds on $\int_0^1 \xi(s,0)\,\mathrm{d}s$, which by Borel–Cantelli implies that a.s. $\sup_{0 \leq i < m} \int_i^{i+1} \xi(s,0)\,\mathrm{d}s \leq \log m$ for all $m \in \mathbb{N}$ sufficiently large. The above claim then follows. □

4.4 Positivity of the Quenched Lyapunov Exponent

In this section, we conclude the proof of Theorems 1.2 and 1.3 by showing that the quenched Lyapunov exponent $\tilde{\lambda}_{d,\gamma,\kappa,\rho,\nu}$ is positive in all dimensions. The strategy is as follows: Employing a result of Kesten and Sidoravicius [16, Prop. 8], we deduce that $\mathbb{P}^{\xi}$-a.s. for eventually all integer time points t, sufficiently many X paths encounter a ξ-particle close-by for of order t many integer time points. Using the Markov property, we then show that with positive $\mathbb{P}_0^X$ probability, X moves to a close-by ξ-particle (which itself stays at its site for some time) within a very short time interval and collects some local time with this ξ-particle. This then implies the desired exponential decay.

Proof of Theorems 1.2 and 1.3. Since we have shown the quenched Lyapunov exponent $\tilde{\lambda}_{d,\gamma,\kappa,\rho,\nu}$ in Theorems 1.2 and 1.3 to be the same, it suffices to consider only Theorem 1.2. Note that the upper bound on $\tilde{\lambda}_{d,\gamma,\kappa,\rho,\nu}$ in Theorem 1.2 follows trivially by requiring the walk X to stay at the origin. To show $\tilde{\lambda}_{d,\gamma,\kappa,\rho,\nu} > 0$, we will make the strategy outlined above precise. In compliance with [16], we let C_0 and $r > 0$ be large integers and for $\vec{i} \in \mathbb{Z}^d$ define the cubes

$$\mathcal{Q}_r(\vec{i}) := \prod_{j=1}^{d} [i_j, i_j + C_0^r).$$

In a slight abuse of common notation, let $D([0,\infty),\mathbb{Z}^d)$ denote the Skorohod space restricted to those functions that start in 0 at time 0 and have nearest neighbor jumps only. Then set

$$J_k := \big\{\Phi \in D([0,\infty),\mathbb{Z}^d) : \Phi \text{ jumps at most } dC_0^r(\kappa \vee 1)k \text{ times up to time } k\big\}.$$

For integer times $t > 0$ define

$$\Xi(t) := \bigcap_{k=\lfloor t/4\rfloor}^{t} J_k.$$

Then standard large deviation bounds yield

$$\mathbb{P}_0^X\big(X \in \Xi(t)^c\big) \le \mathrm{e}^{-c(t+o(t))}, \tag{87}$$

for some $c > 0$. In addition, define the cube

$$\mathcal{C}_t := [-dC_0^r(\kappa \vee 1)t, dC_0^r(\kappa \vee 1)t]^d \cap \mathbb{Z}^d,$$

as well as for arbitrary $t \in \mathbb{N}$, $k \in \{0,\dots,t\}$, $\Phi \in \Xi(t)$ and $\epsilon \ge 0$ the events

$$\begin{aligned} A(t,\Phi,k,\epsilon) := \big\{\exists \vec{i} \in \mathcal{C}_t \,:\, \Phi(k) \in \mathcal{Q}_r(\vec{i}) \text{ and } \exists\, y \in \mathcal{Q}_r(\vec{i}) \,:\, \xi(s,y) \\ \ge 1\,\forall\, s \in [k, k+\epsilon/\rho]\big\} \end{aligned}$$

and

$$G(t) := \bigcap_{\Phi\in\Xi(t)} \Big\{ \sum_{k\in\{\lfloor t/4\rfloor,\cdots,t-1\}} 1_{A(t,\Phi,k,\epsilon)} \ge \epsilon t\Big\},$$

which both depend on ξ.

For ϵ small enough, using Borel–Cantelli, it is a consequence of [16, Prop. 8] that $\mathbb{P}^\xi$-a.s., $G(t)$ occurs for eventually all $t \in \mathbb{N}$. Indeed, denoting by $\Xi(t)|_{\{\lfloor t/4\rfloor,\dots,t\}}$ the subset of $(\mathbb{Z}^d)^{\{\lfloor t/4\rfloor,\dots,t\}}$ obtained by restricting each element of $\Xi(t)$ to the domain $\{\lfloor t/4\rfloor,\dots,t\}$, we estimate

$$\begin{aligned} \mathbb{P}^\xi\big(G(t)^c\big) &\le \mathbb{P}^\xi\Bigg(\bigcup_{\Phi\in\Xi(t)}\Bigg\{\sum_{k\in\{\lfloor t/4\rfloor,\dots,t-1\}} 1_{A(t,\Phi,k,\epsilon)} \le \epsilon t\Bigg\}\Bigg) \\ &\le \mathbb{P}^\xi\Bigg(\bigcup_{\Phi\in\Xi(t)}\Bigg\{\sum_{k\in\{\lfloor t/4\rfloor,\dots,t-1\}} 1_{A(t,\Phi,k,0)} \le \frac{t}{2}\Bigg\}\Bigg) \\ &\quad + \big|\Xi(t)|_{\{\lfloor t/4\rfloor,\dots,t\}}\big| \times \max_{\Phi\in\Xi(t)} \mathbb{P}^\xi\Bigg(\sum_{k\in\{\lfloor t/4\rfloor,\dots,t-1\}} 1_{A(t,\Phi,k,0)} \ge \frac{t}{2}, \end{aligned}$$

$$\sum_{k\in\{\lfloor t/4\rfloor,\dots,t-1\}} 1_{A(t,\Phi,k,\epsilon)} \le \epsilon t \Bigg)$$

$$\le \mathbb{P}^\xi\Bigg(\bigcup_{\Phi\in\Xi(t)}\Bigg\{\sum_{k\in\{\lfloor t/4\rfloor,\dots,t-1\}} 1_{A(t,\Phi,k,0)} \le \frac{t}{2}\Bigg\}\Bigg)$$

$$+\left|\Xi(t)|_{\{\lfloor t/4\rfloor,\dots,t\}}\right| \times \mathbb{P}\Bigg(\sum_{i=1}^{t/2} p_{i,\epsilon} \le \epsilon t\Bigg), \tag{88}$$

where in the last step we observed that, given $\Phi \in \Xi(t)$, by the strong Markov property of ξ applied successively to the stopping times $\tau_i := \inf\{j \ge \lfloor t/4\rfloor : \sum_{k=\lfloor t/4\rfloor}^{j} 1_{A(t,\Phi,k,0)} = i\}$, we can couple ξ with a sequence of i.i.d. Bernoulli random variables $(p_{i,\epsilon})_{i\in\mathbb{N}}$ with

$$\mathbb{P}(p_{1,\epsilon} = 1) = \mathbb{P}^\xi\big(Y_1^0(s) = 0\,\forall s \in [0,\epsilon/\rho]\,\big|\,\xi(0,0) \ge 1\big),$$

such that $1_{A(t,\Phi,\tau_i,\epsilon)} \ge p_{i,\epsilon}$ a.s. for all $i \in \mathbb{N}$, and hence $\sum_{k\in\{\lfloor t/4\rfloor,\dots,t-1\}} 1_{A(t,\Phi,k,\epsilon)} \ge \sum_{i=1}^{t/2} p_{i,\epsilon}$ on the event $\sum_{k\in\{\lfloor t/4\rfloor,\dots,t-1\}} 1_{A(t,\Phi,k,0)} \ge \frac{t}{2}$. Here, $p_{i,\epsilon}$ corresponds to the event that given $A(t,\Phi,\tau_i,0)$, a chosen Y-particle, which is close to Φ at time τ_i, does not jump on the time interval $[\tau_i, \tau_i + \epsilon/\rho]$.

By [16, Prop. 8], the first term in (88) is bounded from above by $1/t^2$ for t large enough. For the second term we have $\left|\Xi(t)|_{\{\lfloor t/4\rfloor,\dots,t\}}\right| \le e^{Ct}$ for some $C > 0$ and all t, while large deviations yield that we can find $\epsilon > 0$ such that

$$\mathbb{P}\Big(\sum_{k=1}^{t/2} p_{k,\epsilon} \le \epsilon t\Big) \le e^{-2Ct}$$

for t large enough. From now on we fix such an ϵ. Borel–Cantelli then yields that $\mathbb{P}^\xi$-a.s., $G(t)$ holds for all $t \in \mathbb{N}$ large enough.

Next observe that by the strong Markov property of X, we can construct a coupling such that on the event $\sum_{k\in\{\lfloor t/4\rfloor,\dots,t-1\}} 1_{A(t,X,k,\epsilon)} \ge \epsilon t$, the random variable $\int_0^t \xi(s, X(s))\,ds$ almost surely dominates the sum of i.i.d. random variables $(q_{i,\epsilon})_{1\le i\le \epsilon t}$ with

$$\begin{aligned}\mathbb{P}(q_{1,\epsilon} = \epsilon/(2\rho)) \quad &= \alpha := \inf_{y,z\in\mathcal{Q}_r(0)} \mathbb{P}_y^X(X(s) = z\,\forall s \in [\epsilon/(2\rho), \epsilon/\rho)]) > 0,\\ \mathbb{P}(q_{1,\epsilon} = 0) \quad &= 1-\alpha;\end{aligned}$$

$q_{i,\epsilon}$ corresponds to the event that given $\tau_i := \inf\{j \ge \lfloor t/4\rfloor : \sum_{k=\lfloor t/4\rfloor}^{j} 1_{A(t,X,k,\epsilon)} = i\}$, X finds a Y-particle in the ξ field which guarantees the event $A(t,X,\tau_i,\epsilon)$, and then occupies the same position as that Y-particle on the time interval $[\tau_i + \epsilon/(2\rho), \tau_i + \epsilon/\rho]$. Since $\mathbb{P}^\xi$-a.s., $G(t)$ holds for all $t \in \mathbb{N}$ large enough, for such t, we have

$$\mathbb{E}_0^X\Big[\exp\Big\{-\gamma\int_0^t \xi(s,X(s))\,\mathrm{d}s\Big\},\,\Xi(t)\Big]\le\mathbb{E}\Big[\mathrm{e}^{-\gamma\sum_{i=1}^{\epsilon t}q_{i,\epsilon}}\Big]=\big(\alpha\mathrm{e}^{-\gamma\epsilon/(2\rho)}+1-\alpha\big)^{\epsilon t}. \tag{89}$$

Thus, with (87) and (89) we obtain that $\mathbb{P}^\xi$-a.s., for all $t\in\mathbb{N}$ large,

$$\mathbb{E}_0^X\Big[\exp\Big\{-\gamma\int_0^t \xi(s,X(s))\,\mathrm{d}s\Big\}\Big]\le\mathbb{E}_0^X\Big[\exp\Big\{-\gamma\int_0^t \xi(s,X(s))\,\mathrm{d}s\Big\},\,\Xi(t)\Big]$$
$$+\mathbb{P}_0^X\big(X\in\Xi(t)^c\big)\le\mathrm{e}^{-\delta(t+o(t))}$$

for some $\delta>0$. This establishes the desired result along integer t. Since $Z_{t,\xi}^\gamma$ is monotone in t, we deduce that the result holds as stated. □

Acknowledgements We thank Frank den Hollander for bringing [22] to our attention, Alain-Sol Sznitman for suggesting that we prove a shape theorem for the quenched survival probability, and Vladas Sidoravicius for explaining to us [16, Prop. 8], which we use to prove the positivity of the quenched Lyapunov exponent. A.F. Ramírez was partially supported by Fondo Nacional de Desarrollo Científico y Tecnológico grant 1100298. J. Gärtner, R. Sun, and partially A.F. Ramírez were supported by the DFG Forschergruppe 718 Analysis and Stochastics in Complex Physical Systems.

References

1. Antal, P.: Trapping problem for the simple random walk. Dissertation ETH, No 10759 (1994)
2. Antal, P.: Enlargement of obstacles for the simple random walk. Ann. Probab. **23**, 1061–1101 (1995)
3. Biskup, M., König, W.: Long-time tails in the parabolic Anderson model with bounded potential. Ann. Probab. **29**, 636–682 (2001)
4. Bolthausen, E.: Localization of a two-dimensional random walk with an attractive path interaction. Ann. Probab. **22**, 875–918 (1994)
5. Bramson, M., Lebowitz, J.: Asymptotic behavior of densities for two-particle annihilating random walks. J. Statist. Phys. **62**, 297–372 (1991)
6. Cox, T., Griffeath, D.: Large deviations for Poisson systems of independent random walks. Z. Wahrsch. Verw. Gebiete **66**, 543–558 (1984)
7. Donsker, M., Varadhan, S.R.S.: Asymptotics for the Wiener sausage. Comm. Pure Appl. Math. **28**, 525–565 (1975)
8. Donsker, M., Varadhan, S.R.S.: On the number of distinct sites visited by a random walk. Comm. Pure Appl. Math. **32**, 721–747 (1979)
9. Evans, L.C.: Partial differential equations, 2nd edn. Graduate Studies in Mathematicsm, vol. 19. American Mathematical Society, Providence, RI (2010)
10. Feller, W.: An introduction to probability theory and its applications, vol. II. Wiley, New York (1966)
11. Gärtner, J., den Hollander, F.: Intermittency in a catalytic random medium. Ann. Probab. **34**, 2219–2287 (2006)
12. Gärtner, J., König, W.: The parabolic Anderson model. Interacting Stochastic Systems, pp. 153–179. Springer, Berlin (2005)

13. Gärtner, J., den Hollander, F., Maillard, G.: Intermittency on catalysts. Trends in Stochastic Analysis, pp. 235–248, London Math. Soc. Lecture Note Ser., vol. 353. Cambridge University Press, Cambridge (2009)
14. Gärtner, J., den Hollander, F., Maillard, G.: Quenched Lyapunov exponent for the parabolic Anderson model in a dynamic random environment. In: Probability in Complex Physical Systems. Springer, Heidelberg, pp. 159–193 (2012)
15. Kesten, H., Sidoravicius, V.: Branching random walks with catalysts. Electron. J. Probab. **8**, 1–51 (2003)
16. Kesten, H., Sidoravicius, V.: The spread of a rumor or infection in a moving population. Ann. Probab. **33**, 2402–2462 (2005)
17. Lawler, G.F.: Intersections of Random Walks. Birkhäuser Boston (1996)
18. Liggett, T.: An improved subadditive ergodic theorem. Ann. Probab. **13**, 1279–1285 (1985)
19. Moreau, M., Oshanin, G., Bénichou, O., Coppey, M.: Pascal principle for diffusion-controlled trapping reactions. Phys. Rev. E **67**, 045104(R) (2003)
20. Moreau, M., Oshanin, G., Bénichou, O., Coppey, M.: Lattice theory of trapping reactions with mobile species. Phys. Rev. E **69**, 046101 (2004)
21. Peres, Y., Sinclair, A., Sousi, P., Stauffer, A.: Mobile geometric graphs: Detection, coverage and percolation. Proceedings of the 22nd ACM-SIAM Symposium on Discrete Algorithms (50DA), 412–428 (2011)
22. Redig, F.: An exponential upper bound for the survival probability in a dynamic random trap model. J. Stat. Phys. **74**, 815–827 (1994)
23. Spitzer, F.: Principles of Random Walk, 2nd edn. Springer, Berlin (1976)
24. Sznitman, A.S.: Brownian Motion, Obstacles and Random Media. Springer, Berlin (1998)
25. Varadhan, S.R.S.: Large deviations for random walks in a random environment. Comm. Pure Appl. Math. **56**, 1222–1245 (2003)

Quenched Lyapunov Exponent for the Parabolic Anderson Model in a Dynamic Random Environment

Jürgen Gärtner, Frank den Hollander, and Grégory Maillard

Abstract We continue our study of the parabolic Anderson equation $\partial u/\partial t = \kappa\Delta u + \gamma\xi u$ for the space–time field $u\colon \mathbb{Z}^d \times [0,\infty) \to \mathbb{R}$, where $\kappa \in [0,\infty)$ is the diffusion constant, Δ is the discrete Laplacian, $\gamma \in (0,\infty)$ is the coupling constant, and $\xi\colon \mathbb{Z}^d \times [0,\infty) \to \mathbb{R}$ is a space–time random environment that drives the equation. The solution of this equation describes the evolution of a "reactant" u under the influence of a "catalyst" ξ, both living on $\mathbb{Z}^d$.

In earlier work we considered three choices for ξ: independent simple random walks, the symmetric exclusion process, and the symmetric voter model, all in equilibrium at a given density. We analyzed the *annealed* Lyapunov exponents, i.e., the exponential growth rates of the successive moments of u w.r.t. ξ, and showed that these exponents display an interesting dependence on the diffusion constant κ, with qualitatively different behavior in different dimensions d. In the present paper we focus on the *quenched* Lyapunov exponent, i.e., the exponential growth rate of u conditional on ξ.

We first prove existence and derive qualitative properties of the quenched Lyapunov exponent for a general ξ that is stationary and ergodic under translations in space and time and satisfies certain noisiness conditions. After that we focus

J. Gärtner
Institut für Mathematik, Technische Universität Berlin, Strasse des 17. Juni 136, D-10623 Berlin, Germany
e-mail: jg@math.tu-berlin.de

F. den Hollander
Mathematical Institute, Leiden University, P.O. Box 9512, 2300 RA Leiden, The Netherlands
EURANDOM, P.O. Box 513, 5600 MB Eindhoven, The Netherlands
e-mail: denholla@math.leidenuniv.nl

G. Maillard (✉)
CMI-LATP, Université de Provence, 39 rue F. Joliot-Curie, 13453 Marseille Cedex 13, France
EURANDOM, P.O. Box 513, 5600 MB Eindhoven, The Netherlands
e-mail: maillard@cmi.univ-mrs.fr

J.-D. Deuschel et al. (eds.), *Probability in Complex Physical Systems*, Springer Proceedings in Mathematics 11, DOI 10.1007/978-3-642-23811-6_7,

on the three particular choices for ξ mentioned above and derive some further properties. We close by formulating open problems.

1 Introduction

Section 1.1 defines the parabolic Anderson model, Sect. 1.2 introduces the quenched Lyapunov exponent, Sect. 1.3 summarizes what is known in the literature, Sect. 1.4 contains our main results, while Sect. 1.5 provides a discussion of these results and lists open problems.

1.1 Parabolic Anderson Model

The parabolic Anderson model (PAM) is the partial differential equation

$$\frac{\partial}{\partial t}u(x,t) = \kappa\Delta u(x,t) + [\gamma\xi(x,t) - \delta]u(x,t), \qquad x \in \mathbb{Z}^d,\ t \geq 0. \tag{1}$$

Here, the u-field is $\mathbb{R}$-valued, $\kappa \in [0,\infty)$ is the diffusion constant, Δ is the discrete Laplacian acting on u as

$$\Delta u(x,t) = \sum_{\substack{y\in\mathbb{Z}^d\\ \|y-x\|=1}} [u(y,t) - u(x,t)], \tag{2}$$

($\|\cdot\|$ is the Euclidian norm), $\gamma \in [0,\infty)$ is the coupling constant, $\delta \in [0,\infty)$ is the killing constant, while

$$\xi = (\xi_t)_{t\geq 0} \text{ with } \xi_t = \{\xi(x,t)\colon x \in \mathbb{Z}^d\}, \tag{3}$$

is an $\mathbb{R}$-valued random field that evolves with time and that drives the equation. The ξ-field provides a dynamic random environment defined on a probability space $(\Omega, \mathcal{F}, \mathbb{P})$. As initial condition for (1) we take

$$u(x,0) = \delta_0(x), \qquad x \in \mathbb{Z}^d. \tag{4}$$

One interpretation of (1) and (4) comes from population dynamics. Consider a system of two types of particles, A (catalyst) and B (reactant), subject to:

- A-particles evolve autonomously according to a prescribed dynamics with $\xi(x,t)$ denoting the number of A-particles at site x at time t.

- B-particles perform independent random walks at rate $2d\kappa$ and split into two at a rate that is equal to γ times the number of A-particles present at the same location.
- B-particles die at rate δ.
- The initial configuration of B-particles is one particle at site 0 and no particle elsewhere.

Then

$$u(x,t) = \text{the average number of } B\text{-particles at site } x \text{ at time } t \text{ conditioned on the evolution of the } A\text{-particles.} \tag{5}$$

It is possible to remove δ via the trivial transformation $u(x,t) \to u(x,t)\mathrm{e}^{-\delta t}$. In what follows we will therefore put $\delta = 0$.

Throughout the paper, $\mathbb{P}$ denotes the law of ξ and we assume that

$$\bullet \quad \xi \text{ is stationary and ergodic under translations in space and time, } \xi \text{ is not constant, and } \rho = \mathbb{E}(\xi(0,0)) \in \mathbb{R} \tag{6}$$

$$\bullet \quad \forall\, \kappa,\gamma \in [0,\infty)\ \exists\, c = c(\kappa,\gamma) < \infty\colon\ \mathbb{E}(\log u(0,t)) \leq ct\ \ \forall\, t \geq 0 \tag{7}$$

Three choices of ξ will receive special attention:

(1) *Independent Simple Random Walks* (ISRW) [Kipnis and Landim [22], Chap. 1]. Here, $\xi_t \in \Omega = (\mathbb{N} \cup \{0\})^{\mathbb{Z}^d}$ and $\xi(x,t)$ represents the number of particles at site x at time t. Under the ISRW-dynamics particles move around independently as simple random walks stepping at rate 1. We draw ξ_0 according to the equilibrium ν_ρ with density $\rho \in (0,\infty)$, which is a Poisson product measure.

(2) *Symmetric Exclusion Process* (SEP) [Liggett [23], Chap. VIII]. Here, $\xi_t \in \Omega = \{0,1\}^{\mathbb{Z}^d}$ and $\xi(x,t)$ represents the presence ($\xi(x,t) = 1$) or absence ($\xi(x,t) = 0$) of a particle at site x at time t. Under the SEP-dynamics particles move around independently according to an irreducible symmetric random walk transition kernel at rate 1, but subject to the restriction that no two particles can occupy the same site. We draw ξ_0 according to the equilibrium ν_ρ with density $\rho \in (0,1)$, which is a Bernoulli product measure.

(3) *Symmetric Voter Model* (SVM) [Liggett [23], Chap. V]. Here, $\xi_t \in \Omega = \{0,1\}^{\mathbb{Z}^d}$ and $\xi(x,t)$ represents the opinion of a voter at site x at time t. Under the SVM-dynamics each voter imposes its opinion on another voter according to an irreducible symmetric random walk transition kernel at rate 1. We draw ξ_0 according to the equilibrium distribution ν_ρ with density $\rho \in (0,1)$, which is not a product measure.

Note: While ISRW and SEP are conservative and reversible in time, SVM is not. The equilibrium properties of SVM are qualitatively different for recurrent and transient random walk. For *recurrent* random walk all equilibria with $\rho \in (0,1)$

are non-ergodic, namely, $\nu_\rho = (1-\rho)\delta_{\{\eta\equiv 0\}} + \rho\delta_{\{\eta\equiv 1\}}$, and therefore are precluded by (6). For *transient* random walk, on the other hand, there are ergodic equilibria.

1.2 Lyapunov Exponents

Our focus will be on the *quenched* Lyapunov exponent, defined by

$$\lambda_0 = \lim_{t\to\infty} \frac{1}{t} \log u(0,t) \qquad \xi\text{-a.s.} \tag{8}$$

We will be interested in comparing λ_0 with the *annealed* Lyapunov exponents, defined by

$$\lambda_p = \lim_{t\to\infty} \frac{1}{t} \log \mathbb{E}\big([u(0,t)]^p\big)^{1/p}, \qquad p \in \mathbb{N}, \tag{9}$$

which were analyzed in detail in our earlier work (see Sect. 1.3). In (8–9) we pick $x = 0$ as the reference site to monitor the growth of u. However, it is easy to show that the Lyapunov exponents are the same at other sites.

By the Feynman–Kac formula, the solution of (1) reads

$$u(x,t) = \mathrm{E}_x\left(\exp\left[\gamma \int_0^t \xi\,(X^\kappa(s), t-s)\,\mathrm{d}s\right] u\big(X^\kappa(t),0\big)\right), \tag{10}$$

where $X^\kappa = (X^\kappa(t))_{t\geq 0}$ is simple random walk on $\mathbb{Z}^d$ stepping at rate $2d\kappa$ and E_x denotes expectation with respect to X^κ given $X^\kappa(0) = x$. In particular, for our choice in (4), for any $t > 0$ we have

$$\begin{aligned} u(0,t) &= \mathrm{E}_0\Big(\exp\Big[\gamma \int_0^t \xi\big(X^\kappa(s), t-s\big)\,\mathrm{d}s\Big]\delta_0\big(X^\kappa(t)\big)\Big) \\ &= \mathrm{E}_0\Big(\exp\Big[\gamma \int_0^t \xi\big(X^\kappa(s), s\big)\,\mathrm{d}s\Big]\delta_0\big(X^\kappa(t)\big)\Big), \end{aligned} \tag{11}$$

where in the last line we reverse time and use that X^κ is a reversible dynamics. Therefore, we can define

$$\Lambda_0(t) = \frac{1}{t}\log u(0,t) = \frac{1}{t}\log \mathrm{E}_0\Big(\exp\Big[\gamma \int_0^t \xi\big(X^\kappa(s), s\big)\,\mathrm{d}s\Big]u\big(X^\kappa(t),0\big)\Big). \tag{12}$$

If the last quantity ξ-a.s. admits a limit as $t \to \infty$, then

$$\lambda_0 = \lim_{t\to\infty} \Lambda_0(t) \qquad \xi\text{-a.s.}, \tag{13}$$

where the limit is expected to be ξ-a.s. constant.

Clearly, λ_0 is a function of d, κ, γ and the parameters controlling ξ. In what follows, our main focus will be on the dependence on κ, and therefore we will often write $\lambda_0(\kappa)$. Note that $p \mapsto \lambda_p(\kappa)$ is nondecreasing for $p \in \mathbb{N} \cup \{0\}$.

Note: Conditions (6–7) imply that the expectations in (10–12) are strictly positive and finite for all $x \in \mathbb{Z}^d$ and $t \geq 0$, and that $\lambda_0 < \infty$. Moreover, by Jensen's inequality applied to (12) with $u(\cdot, 0)$ given by (4), we have $\mathbb{E}(\Lambda_0(t)) \geq \rho\gamma + \frac{1}{t} \log \mathrm{P}_0(X^\kappa(t) = 0)$ and, since the last term tends to zero as $t \to \infty$, we find that $\lambda_0 \geq \rho\gamma > -\infty$.

1.3 Literature

The behavior of the Lyapunov exponents for the PAM in a *time-dependent* random environment has been the subject of several papers.

1.3.1 White Noise

Carmona and Molchanov [6] obtained a qualitative description of both the *quenched* and the *annealed* Lyapunov exponents when ξ is white noise, i.e.,

$$\xi(x,t) = \frac{\partial}{\partial t} W(x,t), \tag{14}$$

where $W = (W_t)_{t \geq 0}$ with $W_t = \{W(x,t) \colon x \in \mathbb{Z}^d\}$ is a space–time field of independent Brownian motions. This choice is special because the increments of ξ are *independent in space and time*. They showed that if $u(\cdot, 0)$ has compact support (e.g., $u(\cdot, 0) = \delta_0(\cdot)$ as in (4)), then the quenched Lyapunov exponent $\lambda_0(\kappa)$ defined in (8) exists and is constant ξ-a.s., and is independent of $u(\cdot, 0)$. Moreover, they found that the asymptotics of $\lambda_0(\kappa)$ as $\kappa \downarrow 0$ is singular, namely, there are constants $C_1, C_2 \in (0, \infty)$, and $\kappa_0 \in (0, \infty)$ such that

$$C_1 \frac{1}{\log(1/\kappa)} \leq \lambda_0(\kappa) \leq C_2 \frac{\log\log(1/\kappa)}{\log(1/\kappa)} \qquad \forall\, 0 < \kappa \leq \kappa_0. \tag{15}$$

Subsequently, Carmona, Molchanov and Viens [7], Carmona, Koralov and Molchanov [5], and Cranston, Mountford and Shiga [9] proved the existence of λ_0 when $u(\cdot, 0)$ has noncompact support (e.g., $u(\cdot, 0) \equiv 1$), showed that there is a constant $C \in (0, \infty)$ such that

$$\lim_{\kappa \downarrow 0} \log(1/\kappa)\, \lambda_0(\kappa) = C, \tag{16}$$

and proved that

$$\lim_{p \downarrow 0} \lambda_p(\kappa) = \lambda_0(\kappa) \quad \forall\, \kappa \in [0, \infty). \tag{17}$$

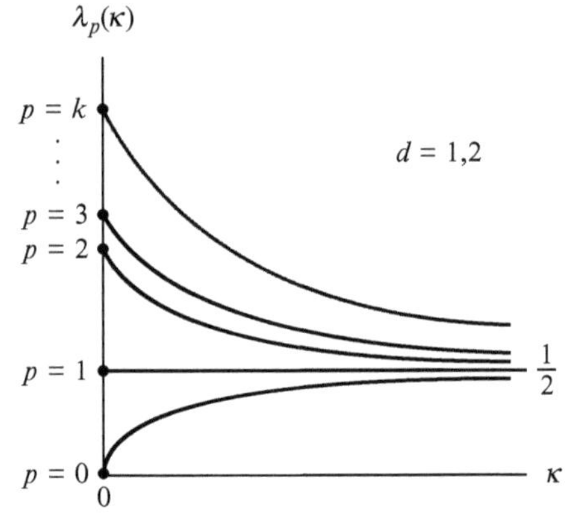

Fig. 1 Quenched and annealed Lyapunov exponents when $d = 1, 2$ for white noise

(These results were later extended to Lévy white noise by Cranston, Mountford and Shiga [10], and to colored noise by Kim, Viens and Vizcarra [20].) Further refinements on the behavior of the Lyapunov exponents were conjectured in Carmona and Molchanov [6] and proved in Greven and den Hollander [18]. In particular, it was shown that $\lambda_1(\kappa) = \frac{1}{2}$ for all $\kappa \in [0, \infty)$, while for the other Lyapunov exponents the following dichotomy holds (see Figs. 1–2):

- $d = 1, 2$: $\lambda_0(\kappa) < \frac{1}{2}$, $\lambda_p(\kappa) > \frac{1}{2}$ for $p \in \mathbb{N}\backslash\{1\}$, for $\kappa \in [0, \infty)$
- $d \geq 3$: there exist $0 < \kappa_0 \leq \kappa_2 \leq \kappa_3 \leq \ldots < \infty$ such that

$$\lambda_0(\kappa) - \tfrac{1}{2} \begin{cases} < 0, & \text{for } \kappa \in [0, \kappa_0), \\ = 0, & \text{for } \kappa \in [\kappa_0, \infty), \end{cases} \tag{18}$$

and

$$\lambda_p(\kappa) - \tfrac{1}{2} \begin{cases} > 0, & \text{for } \kappa \in [0, \kappa_p), \\ = 0, & \text{for } \kappa \in [\kappa_p, \infty), \end{cases} \qquad p \in \mathbb{N}\backslash\{1\}. \tag{19}$$

Moreover, variational formulas for κ_p were derived, which in turn led to upper and lower bounds on κ_p and to the identification of the asymptotics of κ_p for $p \to \infty$ (κ_p grows linearly with p). In addition, it was shown that for every $p \in \mathbb{N}\backslash\{1\}$ there exists a $d(p) < \infty$ such that $\kappa_p < \kappa_{p+1}$ for $d \geq d(p)$. Moreover, it was shown that $\kappa_0 < \kappa_2$ in Birkner, Greven and den Hollander [2] ($d \geq 5$), Birkner and Sun [3] ($d = 4$), Berger and Toninelli [1], Birkner and Sun [4] ($d = 3$). Note that, by Hölder's inequality, all curves in Figs. 1–2 are distinct whenever they are different from $\frac{1}{2}$.

1.3.2 Interacting Particle Systems

Various models where ξ is *dependent in space and time* were looked at subsequently. Kesten and Sidoravicius [19] and Gärtner and den Hollander [13] considered the

Fig. 2 Quenched and annealed Lyapunov exponents when $d \geq 3$ for white noise

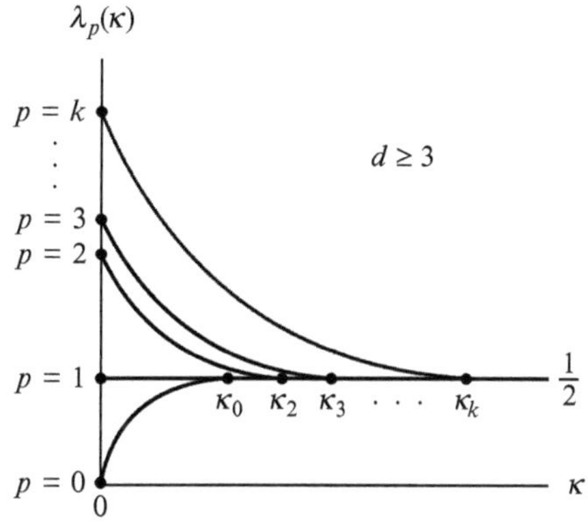

case where ξ is a field of independent simple random walks in Poisson equilibrium (ISRW). The survival versus extinction pattern [19] and the annealed Lyapunov exponents [13] were analyzed in particular, their dependence on d, κ, γ, and ρ. The case where ξ is a single random walk was studied by Gärtner and Heydenreich [12]. Gärtner, den Hollander and Maillard [14], [16], [17] subsequently considered the cases where ξ is an exclusion process with an irreducible symmetric random walk transition kernel starting from a Bernoulli product measure (SEP), respectively, a voter model with an irreducible symmetric *transient* random walk transition kernel starting either from a Bernoulli product measure or from equilibrium SVM. In each of these cases, a fairly complete picture of the behavior of the annealed Lyapunov exponents was obtained, including the presence or absence of *intermittency*, i.e., $\lambda_p(\kappa) > \lambda_{p-1}(\kappa)$ for some or all values of $p \in \mathbb{N}\backslash\{1\}$ and $\kappa \in [0,\infty)$. Several conjectures were formulated as well. In what follows we describe these results in some more detail. We refer the reader to Gärtner, den Hollander and Maillard [15] for an overview.

It was shown in Gärtner and den Hollander [13] and Gärtner, den Hollander and Maillard [14], [16], [17] that for ISRW, SEP, and SVM in equilibrium the function $\kappa \mapsto \lambda_p(\kappa)$ satisfies:

- If $d \geq 1$ and $p \in \mathbb{N}$, then the limit in (9) exists for all $\kappa \in [0,\infty)$. Moreover, if $\lambda_p(0) < \infty$, then $\kappa \mapsto \lambda_p(\kappa)$ is finite, continuous, strictly decreasing, and convex on $[0,\infty)$.
- There are two regimes (we summarize results only for the case where the random walk transition kernel has finite second moment):
  - *Strongly catalytic regime* (see Fig. 3):
    - ISRW: $d = 1,2$, $p \in \mathbb{N}$ or $d \geq 3$, $p \geq 1/\gamma G_d$: $\lambda_p \equiv \infty$ on $[0,\infty)$ (G_d is the Green function at the origin of simple random walk)
    - SEP: $d = 1,2$, $p \in \mathbb{N}$: $\lambda_p \equiv \gamma$ on $[0,\infty)$
    - SVM: $d = 1,2,3,4$, $p \in \mathbb{N}$: $\lambda_p \equiv \gamma$ on $[0,\infty)$

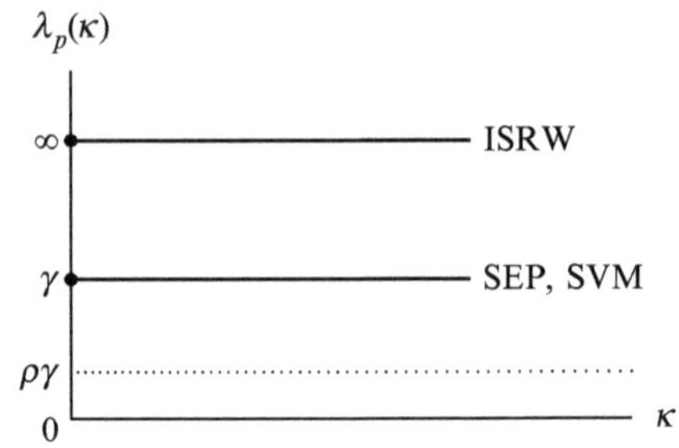

Fig. 3 Triviality of the annealed Lyapunov exponents for ISRW, SEP, and SVM in the strongly catalytic regime

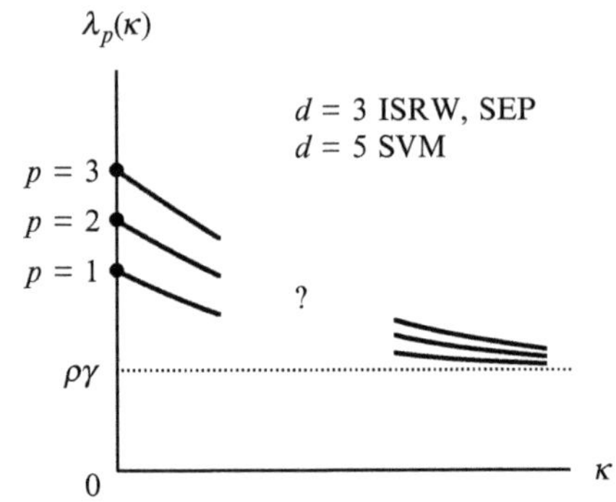

Fig. 4 Nontriviality of the annealed Lyapunov exponents for ISRW, SEP, and SVM in the weakly catalytic regime at the critical dimension

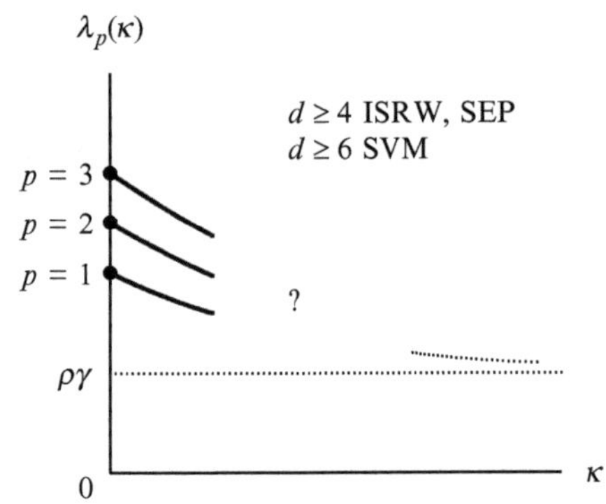

Fig. 5 Nontriviality of the annealed Lyapunov exponents for ISRW, SEP, and SVM in the weakly catalytic regime above the critical dimension

- *Weakly catalytic regime* (see Fig. 4–5):
  - ISRW: $d \geq 3$, $p < 1/\gamma G_d$: $\rho\gamma < \lambda_p < \infty$ on $[0, \infty)$
  - SEP: $d \geq 3$, $p \in \mathbb{N}$: $\rho\gamma < \lambda_p < \gamma$ on $[0, \infty)$
  - SVM: $d \geq 5$, $p \in \mathbb{N}$: $\rho\gamma < \lambda_p < \gamma$ on $[0, \infty)$

- For all three dynamics, in the weakly catalytic regime $\lim_{\kappa\to\infty} \kappa[\lambda_p(\kappa) - \rho\gamma] = C_1 + C_2 p^2 1_{\{d=d_c\}}$ with $C_1, C_2 \in (0, \infty)$ and d_c a critical dimension: $d_c = 3$ for ISRW, SEP and $d_c = 5$ for SVM.
- Intermittent behavior:
  - In the strongly catalytic regime, there is no intermittency for all three dynamics.

– In the weakly catalytic regime, there is full intermittency for:

  · All three dynamics when $0 \leq \kappa \ll 1$
  · ISRW and SEP in $d = 3$ when $\kappa \gg 1$
  · SVM in $d = 5$ when $\kappa \gg 1$

Note: For SVM the convexity of $\kappa \mapsto \lambda_p(\kappa)$ and its scaling behavior for $\kappa \to \infty$ have not actually been proved, but have been argued on heuristic grounds.

Recently, there has been further progress for the case where ξ consists of 1 random walk (Schnitzler and Wolff [25]) or n independent random walks (Castell, Gün and Maillard [8]), ξ is the SVM (Maillard, Mountford and Schöpfer [24]), and for the trapping version of the PAM with $\gamma \in (-\infty, 0)$ (Drewitz, Gärtner, Ramírez and Sun [11]). All these papers appear elsewhere in the present volume.

1.4 Main Results

We have six theorems, all relating to the *quenched* Lyapunov exponent and extending the results on the annealed Lyapunov exponents listed in Sect. 1.3.

Let e be any nearest-neighbor site of 0, and abbreviate

$$I^{\xi}(x,t) = \int_0^t [\xi(x,s) - \rho]\, \mathrm{d}s, \qquad x \in \mathbb{Z}^d,\ t \geq 0. \tag{20}$$

Our first three theorems deal with general ξ and employ four successively stronger *noisiness conditions*:

$$\lim_{t\to\infty} \frac{1}{\log t}\, \mathbb{E}\big(|I^{\xi}(0,t) - I^{\xi}(e,t)|\big) = \infty, \tag{21}$$

$$\liminf_{t\to\infty} \frac{1}{t}\, \mathbb{E}\big(|I^{\xi}(0,t) - I^{\xi}(e,t)|^2\big) > 0, \quad \limsup_{t\to\infty} \frac{1}{t^2}\, \mathbb{E}\big(|I^{\xi}(0,t)|^4\big) < \infty, \tag{22}$$

$$\limsup_{t\to\infty} \frac{1}{t^{2/3}} \log\Big[\sup_{\eta\in\Omega} \mathbb{P}_{\eta}\big(I^{\xi}(0,t) > t^{5/6}\big)\Big] < 0, \tag{23}$$

$$\exists\, c < \infty\colon\ \sup_{\eta\in\Omega} \mathbb{E}_{\eta}\big(\exp\big[\mu I^{\xi}(0,t)\big]\big) \leq \exp[c\mu^2 t] \qquad \forall\, \mu, t > 0, \tag{24}$$

where $\mathbb{P}_{\eta}$ denotes the law of ξ starting from $\xi_0 = \eta$.

Theorem 1.1. *Fix $d \geq 1$, $\kappa \in [0,\infty)$ and $\gamma \in (0,\infty)$. The limit in* (8) *exists $\mathbb{P}$-a.s. and in $\mathbb{P}$-mean, and is finite.*

Theorem 1.2. *Fix $d \geq 1$ and $\gamma \in (0,\infty)$.*
(i) $\lambda_0(0) = \rho\gamma$ and $\rho\gamma < \lambda_0(\kappa) < \infty$ for all $\kappa \in (0,\infty)$ with $\rho = \mathbb{E}(\xi(0,0)) \in \mathbb{R}$.
(ii) $\kappa \mapsto \lambda_0(\kappa)$ is globally Lipschitz outside any neighborhood of 0. Moreover, if ξ

is bounded from above, then the Lipschitz constant at κ *tends to zero as* $\kappa \to \infty$.
(iii) If ξ *satisfies condition* (21) *and is bounded from below, then* $\kappa \mapsto \lambda_0(\kappa)$ *is not Lipschitz at* 0.

Theorem 1.3. *(i) If* ξ *satisfies condition* (22) *and is bounded from below, then*

$$\liminf_{\kappa \downarrow 0} \log(1/\kappa)\,[\lambda_0(\kappa) - \rho\gamma] > 0. \tag{25}$$

(ii) If ξ *is a Markov process that satisfies condition* (23) *and is bounded from above, then*

$$\limsup_{\kappa \downarrow 0} [\log(1/\kappa)]^{1/6}\,[\lambda_0(\kappa) - \rho\gamma] < \infty. \tag{26}$$

(iii) If ξ *is a Markov process that satisfies condition* (24) *and is bounded from above, then*

$$\limsup_{\kappa \downarrow 0} \frac{\log(1/\kappa)}{\log\log(1/\kappa)}\,[\lambda_0(\kappa) - \rho\gamma] < \infty. \tag{27}$$

Our last three theorems deal with ISRW, SEP and SVM.

Theorem 1.4. *For ISRW, SEP, and SVM in the weakly catalytic regime,* $\lim_{\kappa\to\infty} \lambda_0(\kappa) = \rho\gamma$.

Theorem 1.5. *ISRW and SEP satisfy conditions* (21) *and* (22).

Theorem 1.6. *For ISRW in the strongly catalytic regime,* $\lambda_0(\kappa) < \lambda_1(\kappa)$ *for all* $\kappa \in [0, \infty)$.

Theorems 1.1–1.3 wil be proved in Sect. 2, Theorems 1.4–1.6 in Sect. 3.
Note: Theorem 1.4 extends to voter models that are nonnecessarily symmetric (see Sect. 3.1).

1.5 Discussion and Open Problems

1. Figure 6 graphically summarizes the results in Theorems 1.1–1.3 and what we expect to be provable with a little more effort. The main message of this figure is that the *qualitative* behavior of $\kappa \mapsto \lambda_0(\kappa)$ is well understood, including the logarithmic singularity at $\kappa = 0$. Note that Theorems 1.2 and 1.3(i) do not imply continuity at $\kappa = 0$, while Theorems 1.3(ii–iii) do.

2. Figures 7–9 summarize how we expect $\kappa \mapsto \lambda_0(\kappa)$ to compare with $\kappa \mapsto \lambda_1(\kappa)$ for the three dynamics.

3. Conditions (6–7) are trivially satisfied for SEP and SVM, because ξ is bounded. For ISRW they follow from Kesten and Sidoravicius [19], Theorem 2.

4. Conditions (21–22) are *weak* while conditions (23–24) are *strong*. Theorem 1.5 states that conditions (21–22) are satisfied for ISRW and SEP. We will see in

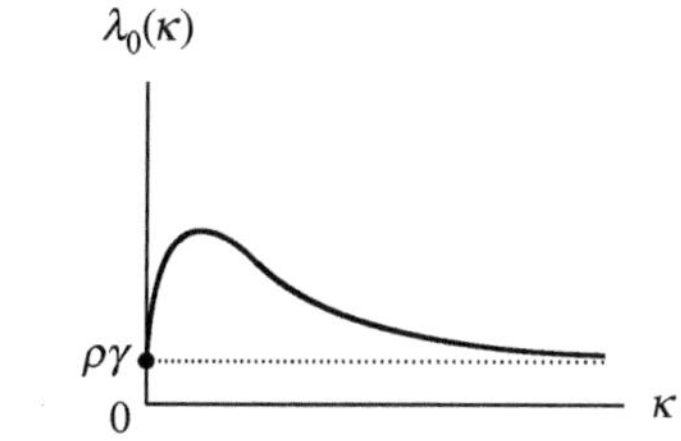

Fig. 6 Conjectured behavior of the quenched Lyapunov exponent

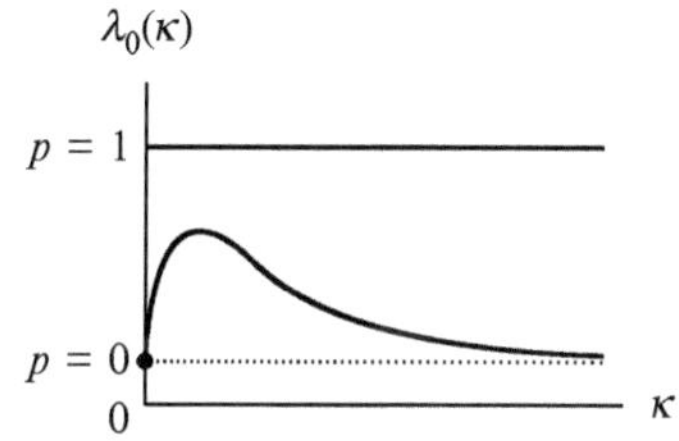

Fig. 7 Conjectured behavior for ISRW, SEP, and SVM below the critical dimension

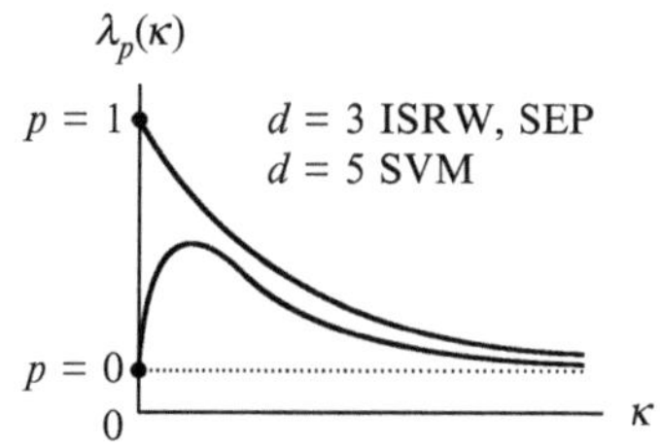

Fig. 8 Conjectured behavior for ISRW, SEP, and SVM at the critical dimension

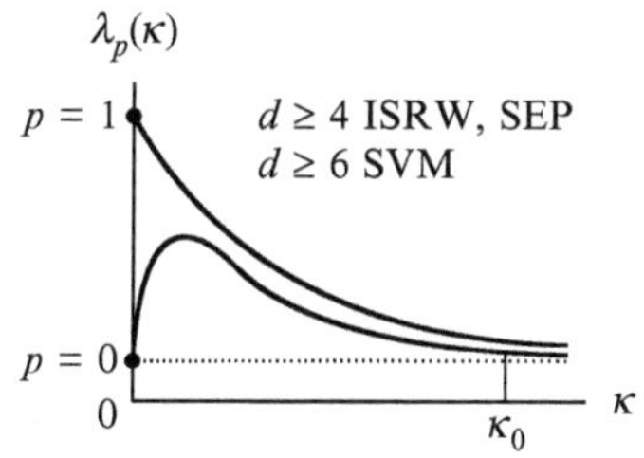

Fig. 9 Conjectured behavior for ISRW, SEP, and SVM above the critical dimension

Sect. 3.2 that, most likely, they are satisfied for SVM as well. Conditions (23–24) fail for the three dynamics, but are satisfied, e.g., for spin-flip dynamics in the so-called "$M < \epsilon$ regime" (see Liggett [23], Sect. I.3). (The verification of this statement is left to the reader.)

5. The following problems remain open:

- Extend Theorem 1.1 to the initial condition $u(\cdot, 0) \equiv 1$, and show that λ_0 is the same as for the initial condition $u(\cdot, 0) = \delta_0(\cdot)$ assumed in (4). (The proof of Theorem 1.1 in Sect. 2.1 shows that it is straightforward to do this extension for $u(\cdot, 0)$ symmetric with bounded support. Recall the remark made prior to (12).)

- Prove that $\lim_{\kappa \downarrow 0} \lambda_0(\kappa) = \rho\gamma$ and $\lim_{\kappa \to \infty} \lambda_0(\kappa) = \rho\gamma$ under conditions (6–7) alone. (These limits correspond to time ergodicity, respectively, space ergodicity of ξ, but are nontrivial because they require some control on the fluctuations of ξ.)
- Prove Theorems 1.2(ii–iii) without the boundedness assumptions on ξ. Prove Theorem 1.3(i) under condition (21) alone. (The proof of Theorem 1.2(iii) in Sect. 2.4 shows that $\lambda_0(\kappa) - \rho\gamma$ stays above any positive power of κ as $\kappa \downarrow 0$.) Improve Theorems 1.3(ii–iii) by establishing under what conditions the upper bounds in (26–27) can be made to match the lower bound in (25).
- Extend Theorems 1.4–1.6 by proving the qualitative behavior for the three dynamics conjectured in Figs. 7–9. (For white noise dynamics the curves successively merge for all $d \geq 3$ (see Figs. 1–2).)
- For the three dynamics in the weakly catalytic regime, find the asymptotics of $\lambda_0(\kappa)$ as $\kappa \to \infty$ and compare this with the asymptotics of $\lambda_p(\kappa)$, $p \in \mathbb{N}$, as $\kappa \to \infty$ (see Figs. 4–5).
- Extend the existence of λ_p to all (noninteger) $p > 0$ and prove that $\lambda_p \downarrow \lambda_0$ as $p \downarrow 0$. (For white noise dynamics this extension is achieved in (17).)

2 Proof of Theorems 1.1–1.3

The proofs of Theorems 1.1–1.3 are given in Sects. 2.1, 2.2–2.4, and 2.5–2.7, respectively. Without loss of generality, we may assume that $\rho = \mathbb{E}(\xi(0,0)) = 0$, by the remark made prior to conditions (6–7).

2.1 Proof of Theorem 1.1

Proof. Recall (4) and (12–13), abbreviate

$$\chi(s,t) = \mathrm{E}_0\left(\exp\left[\gamma \int_0^{t-s} \xi\big(X^\kappa(v), s+v\big)\, dv \right] \delta_0(X^\kappa(t-s)) \right), \qquad 0 \leq s \leq t < \infty, \tag{28}$$

and note that $\chi(0,t) \overset{\mathbb{P}}{=} u(0,t)$. Picking $u \in [s,t]$, inserting $\delta_0(X^\kappa(u-s))$ under the expectation and using the Markov property of X^κ at time $u-s$, we obtain

$$\chi(s,t) \geq \chi(s,u)\, \chi(u,t), \qquad 0 \leq s \leq u \leq t < \infty. \tag{29}$$

Thus, $(s,t) \mapsto \log \chi(s,t)$ is superadditive. By condition (6), the law of $\{\chi(u+s, u+t) : 0 \leq s \leq t < \infty\}$ is the same for all $u \geq 0$. Therefore the superadditive ergodic theorem (see Kingman [21]) implies that

$$\lambda_0 = \lim_{t\to\infty} \frac{1}{t} \log \chi(0,t) \text{ exists } \mathbb{P}\text{-a.s. and in } \mathbb{P}\text{-mean.} \tag{30}$$

We saw at the end of Sect. 1.2 that $\lambda_0 \in [0,\infty)$ (because $\rho = 0$). □

2.2 *Proof of Theorem 1.2(i)*

Proof. The fact that $\lambda_0(0) = 0$ is immediate from (12–13) because $\mathrm{P}_0(X^0(t) = 0) = 1$ for all $t \geq 0$ and $\int_0^t \xi(0,s)\,\mathrm{d}s = o(t)$ ξ-a.s. as $t \to \infty$ by the ergodic theorem (recall condition (6)). We already know that $\lambda_0(\kappa) \in [0,\infty)$ for all $\kappa \in [0,\infty)$. The proof of the strict lower bound $\lambda_0(\kappa) > 0$ for $\kappa \in (0,\infty)$ comes in two steps.

1. Fix $T > 0$ and consider the expression

$$\lambda_0 = \lim_{n\to\infty} \frac{1}{nT} \mathbb{E}\big(\log u(0,nT)\big) \qquad \xi\text{-a.s.} \tag{31}$$

Partition the time interval $[0,nT)$ into n pieces of length T,

$$\mathscr{I}_j = [(j-1)T, jT), \qquad j = 1,\dots,n. \tag{32}$$

Use the Markov property of X^κ at the beginning of each piece, to obtain

$$\begin{aligned} u(0,nT) &= \mathrm{E}_0\Bigg(\exp\Bigg[\gamma \sum_{j=1}^n \int_{\mathscr{I}_j} \xi(X^\kappa(s),s)\,\mathrm{d}s\Bigg]\delta_0(X^\kappa(nT))\Bigg) \\ &= \sum_{x_1,\dots,x_{n-1}\in\mathbb{Z}^d} \prod_{j=1}^n \mathrm{E}_{x_{j-1}}\Bigg(\exp\Bigg[\gamma \int_0^T \xi\big(X^\kappa(s),(j-1)T+s\big)\,\mathrm{d}s\Bigg]\delta_{x_j}(X^\kappa(T))\Bigg) \end{aligned} \tag{33}$$

with $x_0 = x_n = 0$. Next, for $x,y \in \mathbb{Z}^d$, let $\mathrm{E}_{x,y}^{(T)}$ denote the conditional expectation over X^κ given that $X^\kappa(0) = x$ and $X^\kappa(T) = y$, and abbreviate, for $1 \leq j \leq n$,

$$\mathrm{E}_{x,y}^{(T)}(j) = \mathrm{E}_{x,y}^{(T)}\Bigg(\exp\Bigg[\gamma \int_0^T \xi\big(X^\kappa(s),(j-1)T+s\big)\,\mathrm{d}s\Bigg]\Bigg). \tag{34}$$

Then we can write

$$\begin{aligned} &\mathrm{E}_{x_{j-1}}\Bigg(\exp\Bigg[\gamma \int_0^T \xi\big(X^\kappa(s),(j-1)T+s\big)\,\mathrm{d}s\Bigg]\delta_{x_j}(X^\kappa(T))\Bigg) \\ &\quad = p_T^\kappa(x_j - x_{j-1})\,\mathrm{E}_{x_{j-1},x_j}^{(T)}(j), \end{aligned} \tag{35}$$

where we abbreviate $p_T^\kappa(x) = \mathbb{P}_0(X^\kappa(T) = x)$, $x \in \mathbb{Z}^d$. Combined with (33), this gives

$$\begin{aligned} u(0,nT) &= \sum_{x_1,\cdots,x_{n-1}\in\mathbb{Z}^d} \left(\prod_{j=1}^n p_T^\kappa(x_j - x_{j-1})\right)\left(\prod_{j=1}^n \mathrm{E}_{x_{j-1},x_j}^{(T)}(j)\right) \\ &= p_{nT}^\kappa(0)\,\mathrm{E}_{0,0}^{(nT)}\left(\prod_{j=1}^n \mathrm{E}_{X^\kappa((j-1)T),X^\kappa(jT)}^{(T)}(j)\right). \end{aligned} \tag{36}$$

2. To estimate the last expectation in (36), abbreviate $\xi_I = (\xi_t)_{t\in I}$, $I \subset [0,\infty)$ and apply Jensen's inequality to (34), to obtain

$$\mathrm{E}_{x,y}^{(T)}(j) = \exp\left[\gamma \int_0^T \mathrm{E}_{x,y}^{(T)}\Big(\xi\big(X^\kappa(s),(j-1)T+s\big)\Big)\,\mathrm{d}s + C_{x,y}\big(\xi_{[(j-1)T,jT]},T\big)\right] \tag{37}$$

for some $C_{x,y}(\xi_{[(j-1)T,jT]},T)$ that satisfies

$$C_{x,y}\big(\xi_{[(j-1)T,jT]},T\big) > 0 \quad \xi\text{-a.s.} \qquad \forall\, x,y \in \mathbb{Z}^d,\ 1 \le j \le n. \tag{38}$$

Here, the strict positivity is an immediate consequence of the fact that ξ is not constant (recall condition (6)) and $u \mapsto e^u$ is strictly convex. Combining (36–37) and using Jensen's inequality again, this time w.r.t. $\mathrm{E}_{0,0}^{(nT)}$, we obtain

$$\begin{aligned} \mathbb{E}\Big(\log u(0,nT)\Big) &\ge \log p_{nT}^\kappa(0) + \mathbb{E}\left(\mathrm{E}_{0,0}^{(nT)}\left(\sum_{j=1}^n \mathrm{E}_{X^\kappa((j-1)T),X^\kappa(jT)}^{(T)}\right.\right. \\ &\quad \times\left(\gamma\int_0^T \xi\big(X^\kappa(s),(j-1)T+s\big)\,\mathrm{d}s\right. \\ &\quad \left.\left.\left.+ C_{X^\kappa((j-1)T),X^\kappa(jT)}\big(\xi_{[(j-1)T,jT]},T\big)\right)\right)\right) \\ &= \log p_{nT}^\kappa(0) + \mathbb{E}\left(\mathrm{E}_{0,0}^{(nT)}\left(\sum_{j=1}^n \mathrm{E}_{X^\kappa((j-1)T),X^\kappa(jT)}^{(T)}\right.\right. \\ &\quad \left.\left.\times\Big(C_{X^\kappa((j-1)T),X^\kappa(jT)}\big(\xi_{[(j-1)T,jT]},T\big)\Big)\right)\right), \end{aligned} \tag{39}$$

where the middle term in the second line vanishes because of condition (6) and our assumption that $\mathbb{E}(\xi(0,0)) = 0$. After inserting the indicator of the event $\{X^\kappa((j-1)T) = X^\kappa(jT)\}$ for $1 \le j \le n$ in the last expectation in (39), we get

$$
\begin{aligned}
&\mathbb{E}\left(\mathrm{E}_{0,0}^{(nT)}\left(\sum_{j=1}^{n}\mathrm{E}^{(T)}_{X^\kappa((j-1)T),X^\kappa(jT)}\Big(C_{X^\kappa((j-1)T),X^\kappa(jT)}\big(\xi_{[(j-1)T,jT]},T\big)\Big)\right)\right)\\
&\geq \sum_{j=1}^{n}\sum_{z\in\mathbb{Z}^d}\frac{p^\kappa_{(j-1)T}(z)\,p^\kappa_T(0)\,p^\kappa_{(n-j)T}(z)}{p^\kappa_{nT}(0)}\,\mathbb{E}\Big(C_{z,z}\big(\xi_{[(j-1)T,jT]},T\big)\Big)\\
&\geq n\,C_T\,p^\kappa_T(0),
\end{aligned}
\tag{40}
$$

where we abbreviate

$$
C_T=\mathbb{E}\Big(C_{z,z}\big(\xi_{[(j-1)T,jT]},T\big)\Big)>0,
\tag{41}
$$

note that the latter does not depend on j or z, and use that $\sum_{z\in\mathbb{Z}^d}p^\kappa_{(j-1)T}(z)p^\kappa_{(n-j)T}(z)=p^\kappa_{(n-1)T}(0)\geq p^\kappa_{nT}(0)$. Therefore, combining (31) and (39–41), and using that

$$
\lim_{n\to\infty}\frac{1}{nT}\log p^\kappa_{nT}(0)=0,
\tag{42}
$$

we arrive at $\lambda_0\geq(C_T/T)p^\kappa_T(0)>0$. □

2.3 Proof of Theorem 1.2(ii)

Proof. In Step 1 we prove the Lischitz continuity outside any neighborhood of 0 under the restriction that $\xi\leq 1$. This proof is essentially a copy of the proof in Gärtner, den Hollander and Maillard [17] of the Lipschitz continuity of the annealed Lyapunov exponents when ξ is SVM. In Step 2 we explain how to remove the restriction $\xi\leq 1$. In Step 3 we show that the Lipschitz constant tends to zero as $\kappa\to\infty$ when $\xi\leq 1$.

1. Pick $\kappa_1,\kappa_2\in(0,\infty)$ with $\kappa_1<\kappa_2$ arbitrarily. By Girsanov's formula,

$$
\begin{aligned}
&\mathrm{E}_0\left(\exp\left[\gamma\int_0^t\xi(X^{\kappa_2}(s),s)\,\mathrm{d}s\right]\delta_0(X^{\kappa_2}(t))\right)\\
&=\mathrm{E}_0\left(\exp\left[\gamma\int_0^t\xi(X^{\kappa_1}(s),s)\,\mathrm{d}s\right]\delta_0(X^{\kappa_1}(t))\right.\\
&\qquad\left.\times\exp\Big[J(X^{\kappa_1};t)\log(\kappa_2/\kappa_1)-2\mathrm{d}(\kappa_2-\kappa_1)t\Big]\right)\\
&=I+II,
\end{aligned}
\tag{43}
$$

where $J(X^{\kappa_1};t)$ is the number of jumps of X^{κ_1} up to time t, I and II are the contributions coming from the events $\{J(X^{\kappa_1};t)\leq M2\mathrm{d}\kappa_2 t\}$, respectively,

$\{J(X^{\kappa_1};t) > M2d\kappa_2 t\}$, with $M > 1$ to be chosen. Clearly,

$$I \le \exp\Big[\Big(M2d\kappa_2 \log(\kappa_2/\kappa_1) - 2d(\kappa_2-\kappa_1)\Big)t\Big] \times \mathrm{E}_0\Big(\exp\Big[\gamma\int_0^t \xi(X^{\kappa_1}(s),s)\,\mathrm{d}s\Big]\delta_0(X^{\kappa_1}(t))\Big), \tag{44}$$

while

$$II \le \mathrm{e}^{\gamma t}\,\mathrm{P}_0\Big(J(X^{\kappa_2};t) > M2d\kappa_2 t\Big), \tag{45}$$

because we may estimate

$$\int_0^t \xi(X^{\kappa_1}(s),s)\,\mathrm{d}s \le t, \tag{46}$$

and afterwards use Girsanov's formula in the reverse direction. Since $J(X^{\kappa_2};t) = J^*(2d\kappa_2 t)$ with $(J^*(t))_{t\ge 0}$ a rate-1 Poisson process, we have

$$\lim_{t\to\infty}\frac{1}{t}\log \mathrm{P}_0\Big(J(X^{\kappa_2};t) > M2d\kappa_2 t\Big) = -2d\kappa_2\mathscr{I}(M) \tag{47}$$

with

$$\mathscr{I}(M) = \sup_{u\in\mathbb{R}}\big[Mu - (e^u-1)\big] = M\log M - M + 1. \tag{48}$$

Recalling (12–13), we get from (43–47) the upper bound

$$\lambda_0(\kappa_2) \le \big[M2d\kappa_2\log(\kappa_2/\kappa_1) - 2d(\kappa_2-\kappa_1) + \lambda_0(\kappa_1)\big] \vee \big[\gamma - 2d\kappa_2\mathscr{I}(M)\big]. \tag{49}$$

On the other hand, estimating $J(X^{\kappa_1};t) \ge 0$ in (43), we have

$$\mathrm{E}_0\Big(\exp\Big[\gamma\int_0^t \xi(X^{\kappa_2}(s),s)\,\mathrm{d}s\Big]\delta_0(X^{\kappa_2}(t))\Big) \ge \exp[-2d(\kappa_2-\kappa_1)t]\,\mathrm{E}_0\Big(\exp\Big[\gamma\int_0^t \xi(X^{\kappa_1}(s),s)\,\mathrm{d}s\Big]\delta_0(X^{\kappa_1}(t))\Big), \tag{50}$$

which gives the lower bound

$$\lambda_0(\kappa_2) \ge -2d(\kappa_2-\kappa_1) + \lambda_0(\kappa_1). \tag{51}$$

Next, for $\kappa \in (0,\infty)$, define

$$D^+\lambda_0(\kappa) = \limsup_{\epsilon\to 0}\epsilon^{-1}[\lambda_0(\kappa+\epsilon) - \lambda_0(\kappa)],$$
$$D^-\lambda_0(\kappa) = \liminf_{\epsilon\to 0}\epsilon^{-1}[\lambda_0(\kappa+\epsilon) - \lambda_0(\kappa)], \tag{52}$$

where $\epsilon \to 0$ from both sides. Then, in (49) and (51), picking $\kappa_1 = \kappa$ and $\kappa_2 = \kappa+\delta$, respectively, $\kappa_1 = \kappa - \delta$ and $\kappa_2 = \kappa$ with $\delta > 0$ and letting $\delta \downarrow 0$, we get

$$\begin{aligned} D^+\lambda_0(\kappa) &\leq (M-1)2\mathrm{d} \qquad \forall\, M > 1\colon 2\mathrm{d}\kappa\mathscr{I}(M) - \gamma \geq 0, \\ D^-\lambda_0(\kappa) &\geq -2\mathrm{d}. \end{aligned} \tag{53}$$

(The condition in the first line of (53) guarantees that the first term in the right-hand side of (49) is the maximum because $\lambda_0(\kappa) \geq 0$.) Since $\lim_{M\to\infty} \mathscr{I}(M) = \infty$, it follows from (53) that $D^+\lambda_0$ and $D^-\lambda_0$ are bounded outside any neighborhood of $\kappa = 0$.

2. It remains to explain how to remove the restriction $\xi \leq 1$. Without this restriction (46) is no longer true, but by the Cauchy–Schwarz inequality we have

$$\mathrm{II} \leq \mathrm{III} \times \mathrm{IV}, \tag{54}$$

with

$$\mathrm{III} = \left\{ \mathrm{E}_0\left(\exp\left[2\gamma \int_0^t \xi(X^{\kappa_1}(s), s)\, \mathrm{d}s \right] \delta_0(X^{\kappa_1}(t)) \right) \right\}^{1/2} \tag{55}$$

and

$$\begin{aligned} IV &= \left\{ \mathrm{E}_0\left(\exp\left[2J(X^{\kappa_1};t) \log(\kappa_2/\kappa_1) - 4d(\kappa_2-\kappa_1)t \right] \right.\right. \\ &\qquad \left.\left. \times 1\!\!1\{J(X^{\kappa_1};t) > M2\mathrm{d}\kappa_2 t\} \right) \right\}^{1/2} \\ &= \exp\left[\left(d\kappa_1 - 2\mathrm{d}\kappa_2 + d(\kappa_2^2/\kappa_1) \right) t \right] \\ &\qquad \times \left\{ \mathrm{E}_0\left(\exp\left[J(X^{\kappa_1};t) \log\left(\frac{\kappa_2^2/\kappa_1}{\kappa_1} \right) - 2\mathrm{d}\left(\kappa_2^2/\kappa_1 - \kappa_1\right)t \right] \right.\right. \\ &\qquad \left.\left. \times 1\!\!1\{J(X^{\kappa_1};t) > M2\mathrm{d}\kappa_2 t\} \right) \right\}^{1/2} \\ &= \exp\left[\left(d\kappa_1 - 2\mathrm{d}\kappa_2 + d(\kappa_2^2/\kappa_1) \right) t \right] \left\{ \mathrm{P}_0\left(J\left(X^{\kappa_2^2/\kappa_1}; t \right) > M2\mathrm{d}\kappa_2 t \right) \right\}^{1/2}, \end{aligned} \tag{56}$$

where in the last line we use Girsanov's formula in the reverse direction (without ξ). By (12–13) and condition (7), we have $III \leq e^{c_0 t}$ ξ-a.s. for $t \geq 0$ and some $c_0 < \infty$. Therefore, combining (54–56), we get

$$II \le \exp\Big[\Big(c_0 + d\kappa_1 - 2\mathrm{d}\kappa_2 + d(\kappa_2^2/\kappa_1)\Big)t\Big] \Big\{\mathrm{P}_0\Big(J\Big(X^{\kappa_2^2/\kappa_1};t\Big) > M2\mathrm{d}\kappa_2 t\Big)\Big\}^{1/2} \tag{57}$$

instead of (45). The rest of the proof goes along the same lines as in (47–53).

3. Since $\mathscr{I}(M) > 0$ for all $M > 1$, it follows from (53) that $\limsup_{\kappa\to\infty} D^+\lambda_0(\kappa) \le 0$. To prove that $\liminf_{\kappa\to\infty} D^-\lambda_0(\kappa) \ge 0$, we argue as follows. From (43) with $\kappa_1 = \kappa - \delta$ and $\kappa_2 = \kappa$, we get

$$\begin{aligned}
&\mathrm{E}_0\Big(\exp\Big[\gamma\int_0^t \xi(X^\kappa(s),s)\,\mathrm{d}s\Big]\delta_0(X^\kappa(t))\Big)\\
&= \mathrm{E}_0\Big(\exp\Big[\gamma\int_0^t \xi(X^{\kappa-\delta}(s),s)\,\mathrm{d}s\Big]\delta_0(X^{\kappa-\delta}(t))\\
&\quad \times \exp\Big[J(X^{\kappa-\delta};t)\log\big(\tfrac{\kappa}{\kappa-\delta}\big) - 2\mathrm{d}\delta t\Big]\Big)\\
&\ge \mathrm{e}^{-2\mathrm{d}\delta t}\Big[\mathrm{E}_0\Big(\exp\Big[p\gamma\int_0^t \xi(X^{\kappa-\delta}(s),s)\,\mathrm{d}s\Big]\delta_0(X^{\kappa-\delta}(t))\Big)\Big]^{1/p}\\
&\quad \times \Big[\mathrm{E}_0\Big(\exp\Big[qJ(X^{\kappa-\delta};t)\,\log\big(\tfrac{\kappa}{\kappa-\delta}\big)\Big]\Big)\Big]^{1/q}\\
&= \mathrm{e}^{-2\mathrm{d}\delta t}\times \mathrm{I}\times \mathrm{II},
\end{aligned} \tag{58}$$

where we use the reverse Hölder inequality with $(1/p) + (1/q) = 1$ and $-\infty < q < 0 < p < 1$. By direct computation, we have

$$\mathrm{E}_0\Big(\exp\Big[qJ(X^{\kappa-\delta};t)\,\log\big(\tfrac{\kappa}{\kappa-\delta}\big)\Big]\Big) = \exp\Big[-2\mathrm{d}(\kappa-\delta)[1-(\tfrac{\kappa}{\kappa-\delta})^q]t\Big] \tag{59}$$

and hence

$$\frac{1}{\delta t}\log\Big(\mathrm{e}^{-2\mathrm{d}\delta t}\times II\Big) = -2\mathrm{d} - \frac{2\mathrm{d}}{\delta q}(\kappa-\delta)\,[1-(\tfrac{\kappa}{\kappa-\delta})^q]. \tag{60}$$

Moreover, with the help of the additional assumption that $\xi \le 1$, we can estimate

$$I \ge \exp[-(\tfrac{1-p}{p})\gamma t]\Big[\mathrm{E}_0\Big(\exp\Big[\gamma\int_0^t \xi(X^{\kappa-\delta}(s),s)\,\mathrm{d}s\Big]\delta_0(X^{\kappa-\delta}(t))\Big)\Big]^{1/p}. \tag{61}$$

Combining (58) and (60–61), we arrive at (insert $(1-p)/p = -1/q$)

$$\begin{aligned}
&\frac{1}{\delta t}\Bigg[\log \mathrm{E}_0\Bigg(\exp\Bigg[\gamma\int_0^t \xi(X^\kappa(s),s)\,\mathrm{d}s\Bigg]\delta_0(X^\kappa(t))\Bigg)\\
&\quad-\log \mathrm{E}_0\Bigg(\exp\Bigg[\gamma\int_0^t \xi(X^{\kappa-\delta}(s),s)\,\mathrm{d}s\Bigg]\delta_0(X^{\kappa-\delta}(t))\Bigg)\Bigg]\\
&\quad\geq -2\mathrm{d}-\frac{2\mathrm{d}}{\delta q}(\kappa-\delta)\left[1-\left(\tfrac{\kappa}{\kappa-\delta}\right)^q\right]+\frac{\gamma}{\delta q}\\
&\quad-\frac{1}{\delta q t}\log \mathrm{E}_0\Bigg(\exp\Bigg[\gamma\int_0^t \xi(X^{\kappa-\delta}(s),s)\,\mathrm{d}s\Bigg]\delta_0(X^{\kappa-\delta}(t))\Bigg). \qquad (62)
\end{aligned}$$

Let $t\to\infty$ to obtain

$$\frac{1}{\delta}[\lambda_0(\kappa)-\lambda_0(\kappa-\delta)]\geq -2\mathrm{d}-\frac{2\mathrm{d}}{\delta q}(\kappa-\delta)\left[1-\left(\tfrac{\kappa}{\kappa-\delta}\right)^q\right]+\frac{1}{\delta q}\,[\gamma-\lambda_0(\kappa-\delta)]. \qquad (63)$$

Pick $q=-C/\delta$ with $C\in(0,\infty)$ and let $\delta\downarrow 0$ to obtain

$$D^-\lambda_0(\kappa)\geq -2\mathrm{d}+\frac{2\mathrm{d}\kappa}{C}\left(1-e^{-C/\kappa}\right)-\frac{1}{C}\,[\gamma-\lambda_0(\kappa)]. \qquad (64)$$

Let $\kappa\to\infty$ and use that $\lambda_0(\kappa)\geq 0$ to obtain

$$\liminf_{\kappa\to\infty} D^-\lambda_0(\kappa)\geq -\frac{\gamma}{C}. \qquad (65)$$

Finally, let $C\to\infty$ to arrive at the claim. □

2.4 *Proof of Theorem 1.2(iii)*

Proof. Since ξ is assumed to be bounded from below, we may take $\xi\geq -1$ w.l.o.g., because we can scale γ. The proof of Theorem 1.2(iii) is based on the following lemma providing a lower bound for $\lambda_0(\kappa)$ when κ is small enough. Recall (20), and abbreviate

$$E_1(T)=\mathbb{E}\big(|I^\xi(0,T)-I^\xi(e,T)|\big), \qquad T>0. \qquad (66)$$

Lemma 2.1. *For $T\geq 1$ and $\kappa\downarrow 0$,*

$$\lambda_0(\kappa)\geq -\gamma\,\tfrac{1}{T}-2\mathrm{d}\kappa\,\tfrac{T-1}{T}+[1+o_\kappa(1)]\,\tfrac{1}{T}\left[\tfrac{\gamma}{2}E_1(T-1)-\log(1/\kappa)\right]. \qquad (67)$$

Proof. The proof comes in two steps. Recall (4) and (12–13), and write

$$\lambda_0(\kappa)=\lim_{n\to\infty}\frac{1}{nT+1}\log \mathrm{E}_0\Bigg(\exp\Bigg[\gamma\int_0^{nT+1}\xi(X^\kappa(s),s)\,\mathrm{d}s\Bigg]\delta_0\big(X^\kappa(nT+1)\big)\Bigg). \qquad (68)$$

1. Partition the time interval $[0, nT+1)$ as $[\cup_{j=1}^{n+1} \mathscr{B}_j] \cup [\cup_{j=1}^{n} \mathscr{C}_j]$ with

$$\begin{aligned} \mathscr{B}_j &= \big[(j-1)T, (j-1)T+1\big), \quad 1 \le j \le n+1, \\ \mathscr{C}_j &= \big[(j-1)T+1, jT\big), \qquad 1 \le j \le n. \end{aligned} \tag{69}$$

Let

$$I_j^\xi(x) = \int_{\mathscr{C}_j} \xi(x,s)\,\mathrm{d}s \tag{70}$$

and

$$Z_j^\xi = \operatorname{argmax}_{x \in \{0,e\}} I_j^\xi(x), \tag{71}$$

and define the event

$$A^\xi = \left[\bigcap_{j=1}^{n} \left\{X^\kappa(t) = Z_j^\xi \ \forall\, t \in \mathscr{C}_j\right\}\right] \cap \{X^\kappa(nT+1) = 0\}. \tag{72}$$

We may estimate

$$\begin{aligned} &\mathrm{E}_0\left(\exp\left[\gamma \int_0^{nT+1} \xi(X^\kappa(s),s)\,\mathrm{d}s\right] \delta_0\big(X^\kappa(nT+1)\big)\right) \\ &\ge \mathrm{E}_0\left(\exp\left[\gamma \int_0^{nT+1} \xi(X^\kappa(s),s)\,\mathrm{d}s\right] 1\!\!1_{A^\xi}\right) \\ &\ge \mathrm{e}^{-\gamma(n+1)} \exp\left(\gamma \sum_{j=1}^{n} \max\left\{I_j^\xi(0), I_j^\xi(e)\right\}\right) \mathrm{P}_0\big(A^\xi\big). \end{aligned} \tag{73}$$

By the ergodic theorem (recall condition (6)), we have

$$\sum_{j=1}^{n} \max\left\{I_j^\xi(0), I_j^\xi(e)\right\} = [1+o_n(1)]\, n\, \mathbb{E}\big(\max\left\{I_1^\xi(0), I_1^\xi(e)\right\}\big) \quad \xi\text{-a.s. as } n \to \infty. \tag{74}$$

Moreover, we have

$$\mathrm{P}_0\big(A^\xi\big) \ge \Big(\min\left\{p_1^\kappa(0), p_1^\kappa(e)\right\}\Big)^{n+1} \mathrm{e}^{-2d\kappa n(T-1)} = \big(p_1^\kappa(e)\big)^{n+1} \mathrm{e}^{-2d\kappa n(T-1)}, \tag{75}$$

where in the right-hand side the first term is a lower bound for the probability that X^κ moves from 0 to e or vice versa in time 1 in each of the time intervals $\mathscr{B}_j$, while the second term is the probability that X^κ makes no jumps in each of the time intervals $\mathscr{C}_j$.

2. Combining (68) and (73–75), and using that $p_1^\kappa(e) = \kappa[1 + o_\kappa(1)]$ as $\kappa \downarrow 0$, we obtain that

$$\lambda_0(\kappa) \geq -\gamma\tfrac{1}{T} - 2d\kappa\tfrac{T-1}{T} + [1 + o_\kappa(1)]\frac{1}{T}\Big[\gamma\mathbb{E}\big(\max\{I_1^\xi(0), I_1^\xi(e)\}\big) - \log(1/\kappa)\Big]. \tag{76}$$

Because $I_1^\xi(0)$ and $I_1^\xi(e)$ have zero mean, we have

$$\mathbb{E}\big(\max\{I_1^\xi(0), I_1^\xi(e)\}\big) = \tfrac{1}{2}\,\mathbb{E}\big(\big|I_1^\xi(0) - I_1^\xi(e)\big|\big). \tag{77}$$

The expectation in the right-hand side equals $E_1(T-1)$ because $|\mathscr{C}_1| = T - 1$ (recall (66)), and so we get the claim. □

Using Lemma 2.1, we can now complete the proof of Theorem 1.2(iii). By condition (21), for every $c \in (0,\infty)$ we have $E_1(T) \geq c \log T$ for T large enough (depending on c). Pick $\chi \in (0,1)$ and $T = T(\kappa) = \kappa^{-\chi}$ in (67) and let $\kappa \downarrow 0$. Then we obtain

$$\lambda_0(\kappa) \geq [1 + o_\kappa(1)]\left\{-\gamma\kappa^\chi - 2d\kappa + [\tfrac{1}{2}c\gamma\chi - 1]\,\kappa^\chi \log(1/\kappa)\right\}. \tag{78}$$

Finally, pick c large enough so that $\frac{1}{2}c\gamma\chi > 1$. Then, because $\lambda_0(0) = 0$, (78) implies that, for $\kappa \downarrow 0$,

$$\lambda_0(\kappa) - \lambda_0(0) \geq [1 + o_\kappa(1)]\,[\tfrac{1}{2}c\gamma\chi - 1]\,\kappa^\chi \log(1/\kappa), \tag{79}$$

which shows that $\kappa \mapsto \lambda_0(\kappa)$ is not Lipschitz at 0. □

2.5 *Proof of Theorem 1.3(i)*

Proof. Recall (20) and define

$$\begin{aligned} E_k(T) &= \mathbb{E}\big(|I^\xi(0,T) - I^\xi(e,T)|^k\big), \\ \bar{E}_k(T) &= \mathbb{E}\big(|I^\xi(0,T)|^k\big), \end{aligned} \qquad T > 0,\ k \in \mathbb{N}. \tag{80}$$

Estimate, for $N > 0$,

$$\begin{aligned} E_1(T) &= \mathbb{E}\big(|I^\xi(0,T) - I^\xi(e,T)|\big) \\ &\geq \tfrac{1}{2N}\,\mathbb{E}\left(|I^\xi(0,T) - I^\xi(e,T)|^2\, 1\!\!1_{\{|I^\xi(0,T)| \leq N \text{ and } |I^\xi(e,T)| \leq N\}}\right) \\ &= \tfrac{1}{2N}\left[E_2(T) - \mathbb{E}\left(|I^\xi(0,T) - I^\xi(e,T)|^2\, 1\!\!1_{\{|I^\xi(0,T)| > N \text{ or } |I^\xi(e,T)| > N\}}\right)\right]. \end{aligned} \tag{81}$$

By Cauchy–Schwarz,

$$\mathbb{E}\left(|I^{\xi}(0,T)-I^{\xi}(e,T)|^2\,1\!\!1_{\{|I^{\xi}(0,T)|>N \text{ or } |I^{\xi}(e,T)|>N\}}\right)$$
$$\leq [E_4(T)]^{1/2}\left[\mathbb{P}\left(|I^{\xi}(0,T)|>N \text{ or } |I^{\xi}(e,T)|>N\right)\right]^{1/2}. \tag{82}$$

Moreover, by condition (6), $E_4(T) \leq 16\bar{E}_4(T)$ and

$$\mathbb{P}\left(|I^{\xi}(0,T)|>N \text{ or } |I^{\xi}(e,T)|>N\right) \leq \tfrac{2}{N^2}\,\bar{E}_2(T) \leq \tfrac{2}{N^2}\,[\bar{E}_4(T)]^{1/2}. \tag{83}$$

By condition (22), there exist an $a>0$ such that $E_2(T) \geq aT$ and a $b<\infty$ such that $\bar{E}_4(T) \leq bT^2$ for T large enough. Therefore, combining (81–83) and picking $N = cT^{1/2}$ with $c>0$, we obtain

$$E_1(T) \geq A\,T^{1/2} \text{ with } A = \tfrac{1}{2c}\left(a - 2^{5/2}b^{3/4}\tfrac{1}{c}\right), \tag{84}$$

where we note that $A>0$ for c large enough. Inserting this bound into Lemma 2.1 and picking $T = T(\kappa) = B[\log(1/\kappa)]^2$ with $B>0$, we find that, for $\kappa \downarrow 0$,

$$\lambda_0(\kappa) \geq C\,[\log(1/\kappa)]^{-1}\,[1+o_\kappa(1)] \text{ with } C = \tfrac{1}{B}\left(\tfrac{1}{2}\gamma A B^{1/2} - 1\right). \tag{85}$$

Since $C>0$ for $A>0$ and B large enough, this proves the claim in (25). □

2.6 *Proof of Theorem 1.3(iii)*

The proof borrows from Carmona and Molchanov [6], Sect. IV.3.

Proof. Recall (4) and (12–13), estimate

$$\lambda_0(\kappa) \leq \limsup_{n\to\infty} \frac{1}{nT} \log \mathrm{E}_0\left(\exp\left[\gamma \int_0^{nT} \xi(X^\kappa(s),s)\,\mathrm{d}s\right]\right), \tag{86}$$

and pick

$$T = T(\kappa) = K\log(1/\kappa), \qquad K \in (0,\infty), \tag{87}$$

where K is to be chosen later. Partition the time interval $[0,nT)$ into n disjoint time intervals $\mathcal{I}_j$, $1 \leq j \leq n$, defined in (32). Let N_j, $1 \leq j \leq n$, be the number of jumps of X^κ in the time interval $\mathcal{I}_j$, and call $\mathcal{I}_j$ *black* when $N_j>0$ and *white* when $N_j=0$. Using Cauchy–Schwarz, we can split $\lambda_0(\kappa)$ into a black part and a white part, and estimate

$$\lambda_0(\kappa) \leq \mu_0^{(b)}(\kappa) + \mu_0^{(w)}(\kappa), \tag{88}$$

where

$$\mu_0^{(b)}(\kappa) = \limsup_{n\to\infty} \frac{1}{2nT} \log \mathrm{E}_0\left(\exp\left[2\gamma \sum_{\substack{j=1\\N_j>0}}^{n} \int_{\mathscr{I}_j} \xi(X^\kappa(s), s)\, \mathrm{d}s \right]\right), \tag{89}$$

$$\mu_0^{(w)}(\kappa) = \limsup_{n\to\infty} \frac{1}{2nT} \log \mathrm{E}_0\left(\exp\left[2\gamma \sum_{\substack{j=1\\N_j=0}}^{n} \int_{\mathscr{I}_j} \xi(X^\kappa(s), s)\, \mathrm{d}s \right]\right). \tag{90}$$

Lemma 2.2. *If ξ is bounded from above, then there exists a $\delta > 0$ such that*

$$\limsup_{\kappa\downarrow 0}\, (1/\kappa)^\delta\, \mu_0^{(b)}(\kappa) \le 0. \tag{91}$$

Lemma 2.3. *If ξ satisfies condition* (24), *then*

$$\limsup_{\kappa\downarrow 0} \frac{\log(1/\kappa)}{\log\log(1/\kappa)}\, \mu_0^{(w)}(\kappa) < \infty. \tag{92}$$

Theorem 1.3(ii) follows from (88) and Lemmas 2.2 and 2.3. □

We first give the proof of Lemma 2.2.

Proof. Let $N^{(b)} = |\{1 \le j \le n\colon N_j > 0\}|$ be the number of black time intervals. Since ξ is bounded from above (w.l.o.g. we may take $\xi \le 1$, because we can scale γ), we have

$$\begin{aligned}
&\frac{1}{2nT} \log \mathrm{E}_0\left(\exp\left[2\gamma \sum_{\substack{j=1\\N_j>0}}^{n} \int_{\mathscr{I}_j} \xi(X^\kappa(s), s)\, \mathrm{d}s \right]\right)\\
&\le \frac{1}{2nT} \log \mathrm{E}_0\Big(\exp\big[2\gamma T N^{(b)}\big]\Big)\\
&= \frac{1}{2T} \log\left[\left(1-\mathrm{e}^{-2\mathrm{d}\kappa T}\right)e^{2\gamma T} + \mathrm{e}^{-2\mathrm{d}\kappa T}\right]\\
&\le \frac{1}{2T} \log\left[2\mathrm{d}\kappa T \mathrm{e}^{2\gamma T} + 1\right]\\
&\le \frac{1}{2T}\, 2\mathrm{d}\kappa T \mathrm{e}^{2\gamma T}\\
&= \mathrm{d}\kappa^{1-2\gamma K},
\end{aligned} \tag{93}$$

where the first equality uses that the distribution of $N^{(b)}$ is $\mathrm{BIN}(n, 1-\mathrm{e}^{-2\mathrm{d}\kappa T})$, and the second equality uses (87). It follows from (89) and (93) that $\mu_0^{(b)}(\kappa) \le \mathrm{d}\kappa^{1-2\gamma K}$. The claim in (91) therefore follows by picking $0 < K < 1/2\gamma$ and letting $\kappa \downarrow 0$. □

We next give the proof of Lemma 2.3.

Proof. The proof comes in five steps.

1. We begin with some definitions. Define $\Gamma = (\Gamma_1, \dots, \Gamma_n)$ with

$$\Gamma_j = \begin{cases} \{\Delta_1, \dots, \Delta_{N_j}\} & \text{if } \mathscr{I}_j \text{ is black,} \\ \emptyset & \text{if } \mathscr{I}_j \text{ is white,} \end{cases} \tag{94}$$

where $\Delta_1, \dots, \Delta_{N_j}$ are the jumps of X^κ in the time interval $\mathscr{I}_j$ (which takes values in the unit vectors of $\mathbb{Z}^d$). Next, let

$$\Psi = \{\chi \colon \Gamma = \chi\}, \qquad \chi = (\chi_1, \dots, \chi_n), \tag{95}$$

denote the set of possible outcomes of Γ. Since X^κ is stepping at rate $2d\kappa$, the random variable Γ has distribution

$$\mathrm{P}_0(\Gamma = \chi) = e^{-2d\kappa n T} \prod_{j=1}^{n} \frac{(2d\kappa T)^{n_j(\chi)}}{n_j(\chi)!}, \qquad \chi \in \Psi, \tag{96}$$

with $n_j(\chi) = |\chi_j| = |\{\chi_{j,1}, \dots, \chi_{j,n_j(\chi)}\}|$ the number of jumps in χ_j. For $\chi \in \Psi$, we define the event

$$A^{(n)}(\chi; \lambda) = \left\{ \sum_{\substack{j=1 \\ n_j(\chi)=0}}^{n} \int_{\mathscr{I}_j} \xi(x_j(\chi), s)\, ds \geq \lambda \right\}, \qquad \chi \in \Psi,\ \lambda > 0, \tag{97}$$

where $x_j(\chi) = \sum_{i=1}^{j-1} \sum_{k=1}^{n_i(\chi)} \chi_{i,k}$ is the location of χ at the start of χ_j, and λ is to be chosen later. We further define

$$k_l(\chi) = |\{1 \leq j \leq n \colon n_j(\chi) = l\}|, \qquad l \geq 0, \tag{98}$$

which counts the time intervals in which χ makes l jumps, and we note that

$$\sum_{l=0}^{\infty} k_l(\chi) = n. \tag{99}$$

2. With the above definitions, we can now start our proof. Fix $\chi \in \Psi$. By (97) and the exponential Chebychev inequality, we have

$$
\begin{aligned}
\mathbb{P}\big(A^{(n)}(\chi;\lambda)\big) &= \mathbb{P}\left(\sum_{\substack{j=1\\ n_j(\chi)=0}}^{n} \int_{\mathscr{I}_j} \xi(x_j(\chi),s)\,\mathrm{d}s \geq \lambda\right)\\
&\leq \inf_{\mu>0} \mathrm{e}^{-\mu\lambda}\,\mathbb{E}\left(\prod_{\substack{j=1\\ n_j(\chi)=0}}^{n} \exp\left[\mu\int_{\mathscr{I}_j} \xi(x_j(\chi),s)\,\mathrm{d}s\right]\right)\\
&\leq \inf_{\mu>0} \mathrm{e}^{-\mu\lambda}\left[\sup_{\eta\in\Omega} \mathbb{E}_\eta\left(\mathrm{e}^{\mu I^{\xi}(0,T)}\right)\right]^{k_0(\chi)},
\end{aligned}
\tag{100}
$$

where in the second inequality we use the Markov property of ξ at the beginning of the white time intervals, and take the supremum over the starting configuration at the beginning of each of these intervals in order to remove the dependence on $\xi_{(j-1)T}$, $1 \leq j \leq n$ with $n_j(\chi) = 0$, after which we can use (6) and (20). Next, using condition (24) and choosing $\mu = b\lambda/k_0(\chi)T$, we obtain from (100) that

$$
\mathbb{P}\big(A^{(n)}(\chi;\lambda)\big) \leq \exp\left[-\frac{b\lambda^2}{k_0(\chi)T}\right]\left\{\exp\left[c\left(\frac{b\lambda}{k_0(\chi)T}\right)^2 T\right]\right\}^{k_0(\chi)}, \tag{101}
$$

where c is the constant in condition (24). (Note that $A^{(n)}(\chi;\lambda) = \emptyset$ when $k_0(\chi)= 0$, which is a trivial case that can be taken care of afterwards.) By picking $b = 1/2c$, we obtain

$$
\mathbb{P}\big(A^{(n)}(\chi;\lambda)\big) \leq \exp\left[-\frac{1}{4c}\frac{\lambda^2}{k_0(\chi)T}\right]. \tag{102}
$$

3. Our next step is to choose λ, namely,

$$
\lambda = \lambda(\chi) = \sum_{l=0}^{\infty} a_l k_l(\chi) \tag{103}
$$

with

$$
\begin{aligned}
a_0 &= K' \log\log(1/\kappa), & K' &\in (0,\infty),\\
a_l &= lK\log(1/\kappa), & l &\geq 1,
\end{aligned}
\tag{104}
$$

where K is the constant in (87). It follows from (102) after substitution of (103–104) that (recall (87))

$$
\mathbb{P}\Big(A^{(n)}(\chi;\lambda)\Big) \leq \prod_{l=0}^{\infty} \mathrm{e}^{u_l k_l(\chi)} \qquad \forall\, \chi \in \Psi, \tag{105}
$$

where we abbreviate

$$u_0 = -\frac{1}{4cT}a_0^2, \qquad u_l = -\frac{1}{2cT}a_0 a_l = -\frac{1}{2c}a_0 l, \quad l \geq 1. \tag{106}$$

Summing over χ, we obtain

$$\begin{aligned}
\sum_{\chi\in\Psi} \mathbb{P}\left(A^{(n)}(\chi;\lambda)\right) &\leq \sum_{\chi\in\Psi}\left(\prod_{l=0}^{\infty} \mathrm{e}^{u_l k_l(\chi)}\right) \\
&= \sum_{\substack{(k_l)_{l=0}^{\infty} \\ \sum_{l=0}^{\infty} k_l = n}} \left(\frac{n!}{\prod_{l=0}^{\infty} k_l!}\right)\left(\prod_{l=0}^{\infty}(2\mathrm{d})^{lk_l}\right)\left(\prod_{l=0}^{\infty}\mathrm{e}^{u_l k_l}\right) \\
&= \Big(\sum_{l=0}^{\infty}(2\mathrm{d})^l \mathrm{e}^{u_l}\Big)^n,
\end{aligned} \tag{107}$$

where we use that for any sequence $(k_l)_{l=0}^{\infty}$ such that $\sum_{l=0}^{\infty} k_l = n$ (recall (99)) the number of $\chi \in \Psi$ such that $k_l(\chi) = k_l$, $l \geq 0$, equals $(n!/\prod_{l=0}^{\infty} k_l!)\prod_{l=0}^{\infty}(2\mathrm{d})^{lk_l}$ (note that there are $(2\mathrm{d})^l$ different χ_j with $|\chi_j| = l$ for each $1 \leq j \leq n$).

4. By (87) and (104), $T \to \infty$, $a_0 \to \infty$ and $a_0^2/T \downarrow 0$ as $\kappa \downarrow 0$. Hence, for $\kappa \downarrow 0$,

$$\begin{aligned}
\sum_{l=0}^{\infty}(2\mathrm{d})^l e^{u_l} &= \exp\left[-\frac{1}{4cT}a_0^2\right] + \frac{2\mathrm{d}\exp\left[-\frac{1}{2c}a_0\right]}{1-2\mathrm{d}\exp\left[-\frac{1}{2c}a_0\right]} \\
&= 1 - [1+o_\kappa(1)]\,\frac{[K'\log\log(1/\kappa)]^2}{8cK\log(1/\kappa)} + [1+o_\kappa(1)]\,2\mathrm{d}[\log(1/\kappa)]^{-K'/2c} \\
&= 1 - [1+o_\kappa(1)]\,\frac{[K'\log\log(1/\kappa)]^2}{8cK\log(1/\kappa)} < 1,
\end{aligned} \tag{108}$$

where the last equality holds provided we pick $K' > 2c$. It follows from (107–108) that, for $\kappa \downarrow 0$,

$$\sum_{n=1}^{\infty} \mathbb{P}\left(\bigcup_{\chi\in\Psi} A^{(n)}(\chi;\lambda)\right) < \infty. \tag{109}$$

Hence, recalling (97), we conclude that, by the Borel–Cantelli lemma, ξ-a.s. there exists an $n_0(\xi) \in \mathbb{N}$ such that, for all $n \geq n_0(\xi)$,

$$\sum_{\substack{j=1 \\ n_j=0}}^{n} \int_{\mathscr{I}_j} \xi(x_j(\chi), s)\,\mathrm{d}s \leq \lambda = \sum_{l=0}^{\infty} a_l k_l(\chi) \qquad \forall\, \chi \in \Psi. \tag{110}$$

5. The estimate in (110) allows us to proceed as follows. Combining (96), (98), and (110), we obtain, for $n \geq n_0(\xi, \delta, \kappa)$,

$$
\begin{aligned}
&\mathrm{E}_0\left(\exp\left[2\gamma\sum_{\substack{j=1\\N_j=0}}^{n}\int_{\mathscr{I}_j}\xi(X^\kappa(s),s)\,\mathrm{d}s\right]\right)\\
&\quad\leq \mathrm{e}^{-2\mathrm{d}\kappa nT}\sum_{\chi\in\Psi}\left(\prod_{l=0}^{\infty}\left(\frac{(2\mathrm{d}\kappa T)^l}{l!}\right)^{k_l(\chi)}\right)\left(\prod_{l=0}^{\infty}e^{2\gamma a_l k_l(\chi)}\right). \qquad (111)
\end{aligned}
$$

Via the same type of computation as in (107), this leads to

$$
\begin{aligned}
&\mathrm{E}_0\left(\exp\left[2\gamma\sum_{\substack{j=1\\N_j=0}}^{n}\int_{\mathscr{I}_j}\xi(X^\kappa(s),s)\,\mathrm{d}s\right]\right)\\
&\leq \mathrm{e}^{-2\mathrm{d}\kappa nT}\sum_{\substack{(k_l)_{l=0}^{\infty}\\ \sum_{l=0}^{\infty}k_l=n}}\left(\frac{n!}{\prod_{l=0}^{\infty}k_l!}\right)\left(\prod_{l=0}^{\infty}(2\mathrm{d})^{lk_l}\right)\left(\prod_{l=0}^{\infty}\left(\frac{(2\mathrm{d}\kappa T)^l}{l!}\,\mathrm{e}^{2\gamma a_l}\right)^{k_l}\right)\\
&= \mathrm{e}^{-2\mathrm{d}\kappa nT}\left(\sum_{l=0}^{\infty}\frac{((2\mathrm{d})^2\kappa T)^l}{l!}\,\mathrm{e}^{2\gamma a_l}\right)^n. \qquad (112)
\end{aligned}
$$

Hence

$$
\begin{aligned}
&\frac{1}{2nT}\log\mathrm{E}_0\left(\exp\left[2\gamma\sum_{\substack{j=1\\N_j=0}}^{n}\int_{\mathscr{I}_j}\xi(X^\kappa(s),s)\,\mathrm{d}s\right]\right)\\
&\quad\leq -\mathrm{d}\kappa+\frac{1}{2T}\log\left(\sum_{l=0}^{\infty}\frac{((2\mathrm{d})^2\kappa T)^l}{l!}\,\mathrm{e}^{2\gamma a_l}\right). \qquad (113)
\end{aligned}
$$

Note that the right-hand side of (113) does not depend on n. Therefore, letting $n \to \infty$ and recalling (90), we get

$$
\mu_0^{(\mathrm{w})}(\kappa)\leq -\mathrm{d}\kappa+\frac{1}{2T}\log\left(\sum_{l=0}^{\infty}\frac{((2\mathrm{d})^2\kappa T)^l}{l!}\,\mathrm{e}^{2\gamma a_l}\right). \qquad (114)
$$

Finally, by (87) and (104),

$$
\begin{aligned}
\sum_{l=0}^{\infty}\frac{((2\mathrm{d})^2\kappa T)^l}{l!}\mathrm{e}^{2\gamma a_l} &= [\log(1/\kappa)]^{2\gamma K'}+\sum_{l=1}^{\infty}\frac{\left((2\mathrm{d})^2\kappa^{1-2\gamma K}K\log(1/\kappa)\right)^l}{l!}\\
&= [\log(1/\kappa)]^{2\gamma K'}+o_\kappa(1), \qquad \kappa\downarrow 0, \qquad (115)
\end{aligned}
$$

where we recall from the proof of Lemma 2.2 that $0 < K < 1/2\gamma$. Hence

$$\mu_0^{(w)}(\kappa) \leq [1 + o_\kappa(1)] \frac{\gamma K' \log\log(1/\kappa)}{K \log(1/\kappa)}, \qquad \kappa \downarrow 0, \tag{116}$$

which proves the claim in (92). □

2.7 Proof of Theorem 1.3(ii)

Theorem 1.3(ii) follows from (88), Lemma 2.2, and the following modification of Lemma 2.3.

Lemma 2.4. *If ξ satisfies condition (23) and is bounded from above, then*

$$\limsup_{\kappa\downarrow 0}\, [\log(1/\kappa)]^{1/6}\, \mu^{(w)}(\kappa) < \infty. \tag{117}$$

Proof. Most of Steps 1–5 in the proof of Lemma 2.3 can be retained.

1. Recall (94–99). Let

$$f_\epsilon(T) = \sup_{\eta\in\Omega} \mathbb{P}_\eta\left(\int_0^T \xi(0,s)\,\mathrm{d}s > \epsilon T\right), \qquad T > 0, \epsilon = T^{-1/6}. \tag{118}$$

Since $\xi \leq 1$ w.l.o.g., we may estimate

$$\sum_{\substack{j=1 \\ n_j(\chi)=0}}^{n} \int_{\mathcal{I}_j} \xi(x_j(\chi), s)\,\mathrm{d}s \preceq ZT + (k_0(\chi) - Z)\epsilon T, \tag{119}$$

where $\preceq$ means "stochastically dominated by," and Z is the random variable with distribution $\mathrm{P}^* = \mathrm{BIN}(k_0(\chi), f_\epsilon(T))$. With the help of (119), the estimate in (100) can be replaced by

$$\begin{aligned}
\mathbb{P}\big(A^{(n)}(\chi;\lambda)\big) &\leq \mathrm{P}^*\big(ZT + (k_0(\chi) - Z)\epsilon T \geq \lambda\big) \\
&\leq \inf_{\mu>0} \mathrm{e}^{-\mu\lambda}\, \mathrm{E}^*\left(\mathrm{e}^{\mu[ZT+(k_0(\chi)-Z)\epsilon T]}\right) \\
&= \inf_{\mu>0} \mathrm{e}^{-\mu\lambda} \left\{ f_\epsilon(T)\, \mathrm{e}^{\mu T} + [1 - f_\epsilon(T)]\, \mathrm{e}^{\mu\epsilon T} \right\}^{k_0(\chi)}.
\end{aligned} \tag{120}$$

Using condition (24), which implies that there exists a $C \in (0,\infty)$ such that $f_\epsilon(T) \leq \mathrm{e}^{-C\epsilon^2 T}$ for T large enough, and choosing $\mu = C\epsilon^2\lambda/2k_0(\chi)T$, we obtain from (120) that, for T large enough,

$$\mathbb{P}\big(A^{(n)}(\chi;\lambda)\big) \leq \exp\left[-\frac{C\epsilon^2\lambda^2}{2k_0(\chi)T}\right]\left\{\exp\left[\frac{C\epsilon^2\lambda}{2k_0(\chi)} - C\epsilon^2 T\right] + \exp\left[\frac{C\epsilon^3\lambda}{2k_0(\chi)}\right]\right\}^{k_0(\chi)}. \tag{121}$$

2. We choose λ as

$$\lambda = \lambda(\chi) = \sum_{l=0}^{\infty} b_l k_l(\chi) \tag{122}$$

with

$$\begin{aligned} b_0 &= 2\epsilon K \log(1/\kappa) = 2\epsilon T, \\ b_l &= lK \log(1/\kappa) = lT, \quad l \geq 1. \end{aligned} \tag{123}$$

Note that this differs from (104) only for $l = 0$, and that (99) implies, for T large enough,

$$\lambda \geq n2\epsilon T. \tag{124}$$

3. Abbreviate the two exponentials between the braces in the right-hand side of (101) by I and II. Fix $A \in (1,2)$. In what follows we distinguish between two cases: $\lambda > Ak_0(\chi)T$ and $\lambda \leq Ak_0(\chi)T$.

$\lambda > Ak_0(\chi)T$: Abbreviate $\alpha_1 = \frac{1}{4}A > 0$. Neglect the term $-C\epsilon^2 T$ in I, to estimate, for T large enough,

$$\begin{aligned} \mathrm{I} + \mathrm{II} &\leq \exp\left[\frac{C\epsilon^2\lambda}{2k_0(\chi)}\right]\left(1 + \exp\left[-\frac{C\epsilon^2\lambda}{4k_0(\chi)}\right]\right) \\ &\leq \exp\left[\frac{C\epsilon^2\lambda}{2k_0(\chi)}\right]\left(1 + \mathrm{e}^{-\alpha_1 C\epsilon^2 T}\right). \end{aligned} \tag{125}$$

This yields

$$\mathbb{P}\big(A^{(n)}(\chi;\lambda)\big) \leq \exp\left[-\frac{C\epsilon^2\lambda^2}{2k_0(\chi)T} + \frac{C\epsilon^2\lambda}{2}\right]\exp\left[k_0(\chi)\mathrm{e}^{-\alpha_1 C\epsilon^2 T}\right]. \tag{126}$$

$\lambda \leq Ak_0(\chi)T$: Abbreviate $\alpha_2 = 1 - \frac{1}{2}A > 0$. Note that $I \leq \exp[-\alpha_2 C\epsilon^2 T]$ and $II \geq 1$, to estimate

$$\mathrm{I} + \mathrm{II} \leq II\left(1 + \mathrm{e}^{-\alpha_2 C\epsilon^2 T}\right). \tag{127}$$

This yields

$$\mathbb{P}\big(A^{(n)}(\chi;\lambda)\big) \leq \exp\left[-\frac{C\epsilon^2\lambda^2}{2k_0(\chi)T} + \frac{C\epsilon^3\lambda}{2}\right]\exp\left[k_0(\chi)\mathrm{e}^{-\alpha_2 C\epsilon^2 T}\right]. \tag{128}$$

We can combine (126) and (128) into the single estimate

$$\mathbb{P}\big(A^{(n)}(\chi;\lambda)\big) \leq \exp\left[-\frac{C'\epsilon^4\lambda^2}{2k_0(\chi)T}\right]\exp\left[k_0(\chi)\mathrm{e}^{-\alpha CT}\right] \tag{129}$$

for some $C' = C' \in (0,\infty)$ with $\alpha = \alpha_1 \wedge \alpha_2 > 0$. To see why, put $x = \lambda/k_0(\chi)T$, and rewrite the exponent of the first exponential in the right-hand side of (126) and (128) as

$$\tfrac{1}{2}C\epsilon^2 k_0(\chi)T\,(-x^2+x), \quad \text{respectively,} \quad \tfrac{1}{2}C\epsilon^2 k_0(\chi)T\,(-x^2+\epsilon x). \tag{130}$$

In the first case, since $A > 1$, there exists a $B > 0$ such that $-x^2+x \leq -Bx^2$ for all $x \geq A$. In the second case, there exists a $B > 0$ such that $-x^2+\epsilon x \leq -B\epsilon^2 x^2$ for all $x \geq 2\epsilon$. But (124) ensures that $x \geq 2\epsilon n/k_0(\chi) \geq 2\epsilon$. Thus, we indeed get (129) with $C' = CB$.

4. The same estimates as in (105–107) leads us to

$$\sum_{\chi\in\Psi}\mathbb{P}\Big(A^{(n)}(\chi;\lambda)\Big) \leq \Big(\sum_{l=0}^{\infty}(2\mathrm{d})^l e^{\nu_l}\Big)^n, \tag{131}$$

with

$$\nu_0 = -\frac{C'\epsilon^4}{2T}b_0^2 + \mathrm{e}^{-\alpha C\epsilon^2 T}, \qquad \nu_l = -\frac{C'\epsilon^4}{T}b_0 b_l = -C'\epsilon^4 b_0 l, \quad l \geq 1. \tag{132}$$

By (87) and (123), we have

$$\begin{aligned}\sum_{l=0}^{\infty}(2\mathrm{d})^l e^{\nu_l} &= \exp\left[-\frac{C'\epsilon^4}{2T}b_0^2 + \mathrm{e}^{-\alpha C\epsilon^2 T}\right] + \frac{2\mathrm{d}\exp\left[-C'\epsilon^4 b_0\right]}{1-2\mathrm{d}\exp\left[-C'\epsilon^4 b_0\right]} \\ &= \exp\left[-2C' + \mathrm{e}^{-\alpha C T^{2/3}}\right] + \frac{2\mathrm{d}\exp[-2C'T^{1/6}]}{1-2\mathrm{d}\exp[-2C'T^{1/6}]} \\ &< 1,\end{aligned} \tag{133}$$

for T large enough, i.e., κ small enough. This replaces (108). Therefore the analogues of (109–110) hold, i.e., ξ-a.s. there exists an $n_0(\xi) \in \mathbb{N}$ such that, for all $n \geq n_0(\xi)$,

$$\sum_{\substack{j=1\\ n_j=0}}^{n}\int_{\mathscr{I}_j}\xi(x_j(\chi),s)\,\mathrm{d}s \leq \lambda = \sum_{l=0}^{\infty}b_l k_l(\chi) \qquad \forall\,\chi\in\Psi. \tag{134}$$

5. The same estimate as in (111–114) now lead us to

$$\mu_0^{(\mathrm{w})}(\kappa) \le -\mathrm{d}\kappa + \frac{1}{2T}\log\left(\sum_{l=0}^{\infty}\frac{((2\mathrm{d})^2\kappa T)^l}{l!}\,e^{2\gamma b_l}\right). \tag{135}$$

Finally, by (87) and (123),

$$\begin{aligned}\sum_{l=0}^{\infty}\frac{((2\mathrm{d})^2\kappa T)^l}{l!}\mathrm{e}^{2\gamma b_l} &= \mathrm{e}^{4\gamma\epsilon T} + \sum_{l=1}^{\infty}\frac{\left((2\mathrm{d})^2\kappa^{1-2\gamma K}K\log(1/\kappa)\right)^l}{l!} \\ &= \mathrm{e}^{4\gamma\epsilon T} + o_\kappa(1), \qquad \kappa\downarrow 0,\end{aligned} \tag{136}$$

which replaces (115). Hence

$$\mu_0^{(\mathrm{w})}(\kappa) \le [1+o_\kappa(1)]\,2\gamma\epsilon, \qquad \kappa\downarrow 0, \tag{137}$$

which proves the claim in (117). □

3 Proof of Theorems 1.4–1.6

The proofs of Theorems 1.4–1.6 are given in Sects. 3.1–3.3, respectively.

3.1 Proof of Theorem 1.4

Proof. For ISRW, SEP, and SVM in the weakly catalytic regime, it is known that $\lim_{\kappa\to\infty}\lambda_1(\kappa) = \rho\gamma$ (recall Sect. 1.3.2). The claim therefore follows from the fact that $\rho\gamma \le \lambda_0(\kappa) \le \lambda_1(\kappa)$ for all $\kappa \in [0,\infty)$.
Note: The claim extends to nonsymmetric voter models (see [17], Theorems 1.4–1.5). □

3.2 Proof of Theorem 1.5

Proof. It suffices to prove condition (22), because we saw in Sect. 2.5 that condition (22) implies (84), which is stronger than condition (21). Step 1 deals with $E_2(T)$, step 2 with $\bar{E}_4(T)$.

1. Let

$$C(x,t) = \mathbb{E}\big([\xi(0,0)-\rho][\xi(x,t)-\rho]\big), \qquad x\in\mathbb{Z}^d, t\ge 0, \tag{138}$$

denote the two-point correlation function of ξ. By condition (6), we have

$$\begin{aligned} E_2(T) &= \int_0^T \mathrm{d}s \int_0^T \mathrm{d}t \ \mathbb{E}\big([\xi(0,s)-\xi(e,s)][\xi(0,t)-\xi(e,t)]\big) \\ &= 4\int_0^T \mathrm{d}s \int_0^{T-s} \mathrm{d}t \ [C(0,t)-C(e,t)]. \end{aligned} \tag{139}$$

In what follows $G_d(x) = \int_0^\infty p_t(x)\mathrm{d}t$, $x \in \mathbb{Z}^d$, denotes the Green function of simple random walk on $\mathbb{Z}^d$ stepping at rate 1 starting from 0, which is finite if and only if $d \geq 3$. (Recall from Sect. 1.3.2 that SVM with a simple random walk transition kernel is of no interest in $d = 1, 2$.)

Lemma 3.1. *For $x \in \mathbb{Z}^d$ and $t \geq 0$,*

$$C(x,t) = \begin{cases} \rho p_t(x), & \text{ISRW}, \\ \rho(1-\rho)p_t(x), & \text{SEP}, \\ [\rho(1-\rho)/G_d(0)] \int_0^\infty p_{t+u}(x)\,\mathrm{d}u, & \text{SVM}. \end{cases} \tag{140}$$

Proof. For ISRW, we have

$$\xi(x,t) = \sum_{y\in\mathbb{Z}^d} \sum_{j=1}^{N^y} \delta_x\big(Y_j^y(t)\big), \qquad x \in \mathbb{Z}^d, t \geq 0, \tag{141}$$

where $\{N^y\colon y \in \mathbb{Z}^d\}$ are i.i.d. Poisson random variables with mean $\rho \in (0,\infty)$, and $\{Y_j^y\colon y \in \mathbb{Z}^d, 1 \leq j \leq N^y\}$ with $Y_j^y = (Y_j^y(t))_{t\geq 0}$ is a collection of independent simple random walks with jump rate 1 (Y_j^y is the j-th random walk starting from $y \in \mathbb{Z}^d$). Inserting (141) into (138), we get the first line in (140). For SEP and SVM, the claim follows via the graphical representation (see Gärtner, den Hollander and Maillard [14], (1.5.5), and [17], Lemma A.1, respectively). Recall from the remark made at the end of Sect. 1.1 that SVM requires the random walk transition kernel to be *transient*. □

For ISRW, (139–140) yield

$$\frac{1}{T}\, E_2(T) = 4\rho \int_0^T \mathrm{d}t \,\frac{T-t}{T}\, [p_t(0)-p_t(e)], \tag{142}$$

where we note that $p_t(0)-p_t(e) \geq 0$ by the symmetry of the random walk transition kernel. Hence, by monotone convergence,

$$\lim_{T\to\infty} \frac{1}{T} E_2(T) = 4\rho \int_0^\infty \mathrm{d}t\, [p_t(0)-p_t(e)], \tag{143}$$

which is a number in $(0,\infty)$ (see Spitzer [26], Sects. 24 and 29). For SEP, the same computation applies with ρ replaced by $\rho(1-\rho)$. For SVM, (139–140) yield

$$\frac{1}{T}\,E_2(T)=4\frac{\rho(1-\rho)}{G_d(0)}\int_0^\infty \mathrm{d}u\left[\frac{u(2T-u)}{2T}\,1\!\!1_{\{u\leq T\}}+\tfrac{1}{2}T\,1\!\!1_{\{u\geq T\}}\right][p_u(0)-p_u(e)]. \tag{144}$$

Hence, by monotone convergence (estimate $\frac{1}{2}T\leq\frac{1}{2}u$ in the second term of the integrand),

$$\lim_{T\to\infty}\frac{1}{T}E_2(T)=4\frac{\rho(1-\rho)}{G_d(0)}\int_0^\infty \mathrm{d}u\;u\,[p_u(0)-p_u(e)], \tag{145}$$

which again is a number in $(0,\infty)$ (see Spitzer [26], Sect. 24).

2. Let

$$C(x,t;y,u;z,v)=\mathbb{E}\big([\xi(0,0)-\rho][\xi(x,t)-\rho][\xi(y,u)-\rho][\xi(z,v)-\rho]\big),$$
$$x,y,z\in\mathbb{Z}^d,\ 0\leq t\leq u\leq v, \tag{146}$$

denote the four-point correlation function of ξ. Then, by condition (6),

$$\bar{E}_4(T)=4!\int_0^T \mathrm{d}s\int_0^{T-s}\mathrm{d}t\int_t^{T-s}\mathrm{d}u\int_u^{T-s}\mathrm{d}v\;\;C(0,t;0,u;0,v). \tag{147}$$

To prove the second part of (22), we must estimate $C(0,t;0,u;0,v)$. For ISRW, this can be done by using (141), for SEP by using the Markov property and the graphical representation. In both cases the computations are long but straightforward, with leading terms of the form

$$Mp_a(0,0)\,p_b(0,0) \tag{148}$$

with a,b linear in t, u or v, and $M<\infty$. Each of these leading terms, after being integrated as in (147), can be bounded from above by a term of order T^2, and hence $\limsup_{T\to\infty}\bar{E}_4(T)/T^2<\infty$. The details are left to the reader. □

Note: We expect the second part of condition (22) to hold also for SVM. However, the graphical representation, which is based on coalescing random walks, seems too cumbersome to carry through the computations.

3.3 *Proof of Theorem 1.6*

Proof. For ISRW in the strongly catalytic regime, we know that $\lambda_1(\kappa)=\infty$ for all $\kappa\in[0,\infty)$ (recall Fig. 3), while $\lambda_0(\kappa)<\infty$ for all $\kappa\in[0,\infty)$ (by Kesten and Sidoravicius [19], Theorem 2). □

Acknowledgements GM is grateful to CNRS for financial support and to EURANDOM for hospitality. We thank Dirk Erhard for his comments on an earlier draft of the manuscript.

References

1. Berger, Q., Toninelli, F.L.: On the critical point of the random walk pinning model in dimension $d = 3$. Electronic J. Probab. **15**, 654–683 (2010)
2. Birkner, M., Greven, A., den Hollander, F.: Collision local time of transient random walks and intermediate phases in interacting stochastic systems. Electronic J. Probab. **16**, 552–586 (2011)
3. Birkner, M., Sun, R.: Annealed vs quenched critical points for a random walk pinning model. Ann. Inst. H. Poincaré Probab. Statist. **46**, 414–441 (2010)
4. Birkner, M., Sun, R.: Disorder relevance for the random walk pinning model in dimension 3. Ann. Inst. H. Poincaré Probab. Statist. **47**, 259–293 (2011)
5. Carmona, R.A., Koralov, L., Molchanov, S.A.: Asymptotics for the almost-sure Lyapunov exponent for the solution of the parabolic Anderson problem. Random Oper. Stochastic Equat. **9**, 77–86 (2001)
6. Carmona, R.A., Molchanov, S.A.: Parabolic Anderson Problem and Intermittency. AMS Memoir **518**, American Mathematical Society, Providence RI (1994).
7. Carmona, R.A., Molchanov, S.A., Viens, F.: Sharp upper bound on the almost-sure exponential behavior of a stochastic partial differential equation. Random Oper. Stochastic Equat. **4**, 43–49 (1996)
8. Castell, F., Gün, O., Maillard, G.: Parabolic Anderson model with a finite number of moving catalysts. In: Deuschel, J.-D., Gentz, B., König, W., von Renesse, M.-K., Scheutzow, M., Schmock, U., (eds.), Probability in Complex Physical Systems, Vol. 11, pp. 91–117. Springer, Heidelberg (2012)
9. Cranston, M., Mountford, T.S., Shiga, T.: Lyapunov exponents for the parabolic Anderson model. Acta Math. Univ. Comeniane **71**, 163–188 (2002)
10. Cranston, M., Mountford, T.S., Shiga, T.: Lyapunov exponents for the parabolic Anderson model with Lévy noise. Probab. Theory Relat. Fields **132**, 321–355 (2005)
11. Drewitz, A., Gärtner, J., Ramírez, A., Sun, R.: Survival probability of a random walk among a Poisson system of moving traps. In: Deuschel, J.-D., Gentz, B., König, W., von Renesse, M.-K., Scheutzow, M., Schmock, U., (eds.), Probability in Complex physical systems, Vol. 11, pp. 119–158. Springer, Heldelberg (2012)
12. Gärtner, J., Heydenreich, M.: Annealed asymptotics for the parabolic Anderson model with a moving catalyst. Stoch. Proc. Appl. **116**, 1511–1529 (2006)
13. Gärtner, J., den Hollander, F.: Intermittency in a catalytic random medium. Ann. Probab. **34**, 2219–2287 (2006)
14. Gärtner, J., den Hollander, F., Maillard, G.: Intermittency on catalysts: symmetric exclusion. Electronic J. Probab. **12**, 516–573 (2007)
15. Gärtner, J., den Hollander, F., Maillard, G.: Intermittency on catalysts. In: Blath, J., Mörters, P., Scheutzow, M. (eds.) Trends in Stochastic Analysis. London Mathematical Society Lecture Note Series **353**, pp. 235–248. Cambridge University Press, Cambridge (2009)
16. Gärtner, J., den Hollander, F., Maillard, G.: Intermittency on catalysts: Three-dimensional simple symmetric exclusion. Electr. J. Prob. **72**, 2091–2129 (2009)
17. Gärtner, J., den Hollander, F., Maillard, G.: Intermittency on catalysts: Voter model. Ann. Probab. **38**, 2066–2102 (2010)
18. Greven, A., den Hollander, F.: Phase transition for the long-time behavior of interacting diffusions. Ann. Probab. **35**, 1250–1306 (2007)
19. Kesten, H., Sidoravicius, V.: Branching random walk with catalysts. Electr. J. Prob. **8**, 1–51 (2003)

20. Kim, H.-Y., Viens, F., Vizcarra, A.: Lyapunov exponents for stochastic Anderson models with non-gaussian noise. Stoch. Dyn. **8**, 451–473 (2008)
21. Kingman, J.F.C.: Subadditive processes. In: Lecture Notes in Mathematics **539**, pp. 167–223. Springer, Berlin (1976)
22. Kipnis, C., Landim, C.: Scaling Limits of Interacting Particle Systems. Grundlehren der Mathematischen Wissenschaften **320**, Springer, Berlin (1999)
23. Liggett, T.M.: Interacting Particle Systems. Grundlehren der Mathematischen Wissenschaften **276**, Springer, New York (1985)
24. Maillard, G., Mountford, T., Schöpfer, S.: Parabolic Anderson model with voter catalysts: dichotomy in the behavior of Lyapunov exponents. In: Deuschel, J.-D., Gentz, B., König, W., von Renesse, M.-K., Scheutzow, M., Schmock, U., (eds.), Probability in Complex physical systems, Vol. 11, pp. 33–68. Springer, Heidelberg (2012)
25. Schnitzler, A., Wolff, T.: Precise asymptotics for the parabolic Anderson model with a moving catalyst or trap. In: Deuschel, J.-D., Gentz, B., König, W., von Renesse, M.-K., Scheutzow, M., Schmock, U., (eds.), Probability in Complex physical systems, Vol. 11, pp. 69–89. Springer, Heidelberg (2012)
26. Spitzer, F.: Principles of Random Walk. Van Nostrand, Princeton (1964)

Asymptotic Shape and Propagation of Fronts for Growth Models in Dynamic Random Environment

Harry Kesten, Alejandro F. Ramírez, and Vladas Sidoravicius

Abstract We survey recent rigorous results and open problems related to models of Interacting Particle Systems which describe the autocatalytic type reaction $A + B \to 2B$, with diffusion constants of particles being respectively $D_A \geq 0$ and $D_B \geq 0$. Depending on the choice of the values of D_A and D_B, we cover three distinct cases: the so called "rumor or infection spread" model ($D_A > 0$, $D_B > 0$); the Stochastic Combustion process ($D_A = 0$ and $D_B > 0$); and finally the "modified" Diffusion Limited Aggregation, which corresponds to the case $D_A > 0$, $D_B = 0$ with modified transition rule: $A + B \to 2B$ occurs when an A- and a B-particles become nearest neighbors and the A-particle attempts to jump on a vertex where the B-particle is located. Then such jump is suppressed, and A-particle becomes B-particle.

1 Introduction

As the basic model we consider the following interacting particle system: There is a "gas" of particles, each of which performs a continuous time simple random walk on $\mathbb{Z}^d$, with jump rate $D_A \geq 0$. These particles are called A-particles and move independently of each other. In the case $D_A \geq 0$ they are regarded as

---

H. Kesten (✉)
Department of Mathematics, 310 Malott Hall, Cornell University, Ithaca, NY 14853-4201, USA
e-mail: hk21@cornell.edu

A.F. Ramírez
Facultad de Matemáticas, Pontificia Universidad Católica de Chile, Vicuña Mackenna 4860, Macul Santiago, Chile
e-mail: aramirez@mat.puc.cl

V. Sidoravicius
IMPA, Estr. Dona Castorina 110, Jardim Botanico, Rio de Janeiro 22460-320, Brazil
e-mail: vladas@impa.br

J.-D. Deuschel et al. (eds.), *Probability in Complex Physical Systems*,
Springer Proceedings in Mathematics 11, DOI 10.1007/978-3-642-23811-6_8,

individuals who are ignorant of a rumor or are healthy. We assume that we start the system with $N_{\mathrm{A}}(x,0-)$ A-particles at x, and that the $N_{\mathrm{A}}(x,0-)$, $x \in \mathbb{Z}^d$, are i.i.d., mean μ_A Poisson random variables. In addition, there are B-particles which perform continuous time simple random walks with jump rate $D_B \geq 0$. We start with a finite number of B-particles in the system at time 0. In the case $D_B > 0$, B-particles are interpreted as individuals who have heard a certain rumor or who are infected. The B-particles move independently of each other. In the case when $D_B > 0$ the only interaction is that when a B-particle and an A-particle coincide, the latter instantaneously turns into a B-particle. It models the irreversible autocatalytic reaction $A+B \to 2B$, for which corresponding diffusion constants D_A and D_B may differ. In addition to the possible interpretations of such systems mentioned above, in the case of $D_A = 0$ the system can be interpreted as a model for the burning of propellant material, where the B-particles have been interpreted as "packets of energy" which together with A-particles produce more energy, according to the reaction $A + B \to 2B$ (see [27]), and the process is called in the Physics literature the Stochastic Combustion process. It is also known as the "frog" model (see [1] and [27]). Finally, in the case $D_A > 0$ but $D_B = 0$, in order to create spatial growth we modify the transition rule: the transition from A- to B-state happens when an A-particle is at a neighboring vertex to a B-particle and performs an attempt to jump to the vertex where the B-particle is located. Then such jump is suppressed, and a A-particle becomes a B-particle and stops moving. In this case we assume that all A-particles which are at the same neighboring site also become B-particles.

The first basic and natural question to ask is how fast B-particles spread. Specifically, if

$$\widetilde{B}(t) = \{x \in \mathbb{Z}^d \,:\, \text{a } B\text{-particle visits } x \text{ during } [0,t]\},$$

$$B(t) = \widetilde{B}(t) + \left[-\frac{1}{2}, \frac{1}{2}\right]^d,$$

then we are interested in the asymptotic behavior of $B(t)$. The first major goal is to prove a full "shape theorem," saying that $t^{-1}B(t)$ converges almost surely to a non-random set B_0, with the origin as an interior point, so that the true growth rate for $B(t)$ is linear in t (Fig. 1).

The study of these systems was suggested by Frank Spitzer to the first author around 1980. At that time only the case when the A- and B-particles perform the same random walks (i.e., $D_A = D_B > 0$) seems to have been considered.

Shape theorems have a fairly long history and have become the first goal of many investigations of stochastic growth models. To the best of our knowledge Eden (see [9]) was the first one to ask for a shape theorem for his celebrated Eden model. The problem turned out to be a stubborn one. The first real progress was due to Richardson, who proved in [28] a shape theorem not only for the Eden model, but also for a more general class of models, now called Richardson models. In these models one typically thinks of the sites of $\mathbb{Z}^d$ as cells which can be of two types (for instance B and A or infected and susceptible). Cells can change their type to

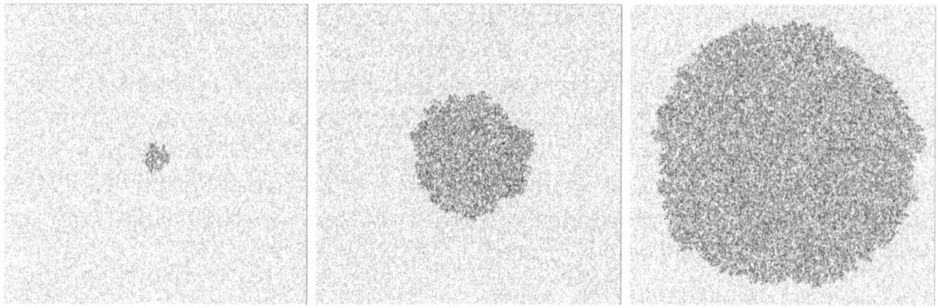

Fig. 1 Three instants of the growth of the cluster of B-particles (*dark collor*) in the case $D_A = D_B > 0$

the type of one of their neighbors according to appropriate rules. One starts with all cells off the origin of type A and cell of type B at the origin and tries to prove a shape theorem for the set of cells of type B at a large time. An important example of such a model is first-passage percolation, which was introduced in [14] (this includes the Eden model, up to a time change). A quite good shape theorem for first-passage percolation is known (see [24], [8], [18]). In more recent first-passage percolation papers even sharper information has been obtained which gives estimates on the rate at which $(1/t)B(t)$ converges to its limit B_0 (see [15] for a survey of such results).

Shape theorems for quite a few variations of Richardson's model and first-passage percolation have been proven (see for instance [4] and [11]), but as far as we know these are all for models in which the cells do not move over time, with one exception. This exception is the Stochastic Combustion Process (or "frog model") which follows the rules given above, but which has $D_A = 0$, i.e., the susceptible or type A cells stand still (see [1] and [27] for this model). The papers [19, 20] may be the first ones which allow both types of particles to move.

In nearly all cases shape theorems are proven by means of Kingman's subadditive ergodic theorem (see [24]). This is also what is used for the Stochastic Combustion process. For this model one can show that the family of random variables $\{T_{x,y}\}$ is subadditive, were $T_{x,y}$ is a version of the first time a particle at y is infected, if one starts with one infected particle at x and one susceptible at each other site. More precisely, the $T_{x,y}$ can all be defined on one probability space such that

$$T_{x,z} \leq T_{x,y} + T_{y,z}, \quad \text{a.s.} \tag{1}$$

for all $x, y, z \in Z^d$ and such that their joint distribution is invariant under translations. Unfortunately this subadditivity property is no longer valid if one allows both types of particles to move. There is no obvious family of random variables with properties like those of the $T_{x,y}$. Nevertheless, subadditivity methods are still heavily used in the proof of the full Shape Theorem in the case $D_A = D_B > 0$. However, now subadditivity is used only for certain "half-space" processes which approximate the true process (see for more details Sect. 2). Moreover, these half-space processes have only approximate superconvolutive properties (in the terminology of [13]), which replaces the "almost sure" subadditivity relation in (1) by a relation between distribution functions. The tool of superconvolutivity in other models with no obvious subadditivity in the strict sense goes back to [28], and was used in [4] and [32], and could be stated in the following way [13]:

Lemma 1. *(H. Kesten, 1974, see [13]) Let $X_s, s \geq 1$, and $Y_{s,t}$, $1 \leq s < t$, be random variables (s, t integers) satisfying the following conditions:*

$$P\{X_{s+t} \leq x\} \geq P\{X_s + X_t' + Y_{s,t} \leq x\}, \tag{2}$$

for all real x and $s, t \geq 0$, where X_t' has the same distribution as X_t but is independent of all X_s,

$$EY_{s,t}^2 \leq C, \quad E|X_1|^2 < \infty, \quad E(X_s^-)^2 < C$$

for all $s, t \geq 1$ and some C (independent of s and t). Then, there exists a $0 \leq \gamma < \infty$ such that

$$\sum_{k=0}^{\infty} P\left\{ |\frac{X_{m2^k}}{m2^k} - \gamma| > \epsilon \right\} < \infty$$

for all $m \geq 1$, $\epsilon > 0$.

However it is not so easy to use superconvolutive property and find right quantities which satisfy (2), in order to prove shape theorems.

In Sect. 2 we focus on the the full Shape Theorem for the case $D_A = D_B > 0$. Essentially we were not able to go beyond the law of large numbers for this model and show fluctuations and Large Deviation Principle for the front. One of the aspects which makes these type of problems difficult is that the process as seen from the front does not converge exponentially fast to its equilibrium. Within a certain class of one-dimensional nonlinear diffusion equations having uniformly traveling wave solutions describing the passage from an unstable to a stable state, it has been observed that for certain initial conditions the velocity of the front at a given time has a rate of relaxation towards its asymptotic value which is algebraic. These are the so called pulled fronts, whose speed is determined by a region of the profile linearized about the unstable solution. For the F-KPP equation, Bramson [3] proved that the speed of the front at a given time is below its asymptotic value and that the convergence is algebraic. In general, the slow relaxation is due to a gapless property of a linear operator governing the convergence of the centered front profile towards the stationary state. However for the Stochastic Combustion model in dimension 1 progress was made in [6] and [2], and it is the content of the Sect. 3. Finally in Sect. 4 we focuss on the modified DLA model.

2 Spread of an Infection in a Moving Population ($D_A > 0,\ D_B > 0$)

In this section we mainly deal with the case $D_A = D_B > 0$. The upper bound for $B(t)$ (see Theorem 1 below) is relatively easy and is proven even for $D_A \neq D_B$. On the other hand the lower bound is much harder. However it turns out that the lower bound for $B(t)$ in Theorem 2, in the case $D_A = D_B$, can be obtained by the methods of [17], which we explain below. It is still an open problem whether $B(t)$ grows linearly with t when $D_A > 0$, but $D_A \neq D_B$. In this case we can only prove that $B(t) \supset \mathscr{C}\big(K_1 t/(\log t)^p\big)$ eventually, for some constants $K_1, p > 0$.

2.1 Shape Theorem

Throughout we shall use $N_{\rm A}(x,t)$ ($N_{\rm B}(x,t)$) to denote the number of A-particles (respectively, B-particles) at position x at time t. $N_{\rm B}$ denotes the total number of B-particles at time 0. We always take $0<N_{\rm B}<\infty$ and consider $N_{\rm B}$, as well as the positions of the initial B-particles, as non-random. At a site x with a B-particle at time 0 all particles immediately turn to B-particles. We write $N_{\rm A}(x,0-)$ for the number of A-particles "just before" the B-particles are added to the system. In accordance with these rules we take $N_{\rm A}(x,0) = 0,\ N_{\rm B}(x,0) = N_{\rm A}(x,0-) + N_{\rm B}(x,0)$ at a site x to which a B-particle is added at time 0. If x does not have a B-particle at time 0, then $N_{\rm A}(x,0) = N_{\rm A}(x,0-)$ and $N_{\rm B}(x,0) = 0$.

Our first theorem states that the rumor/infection cannot spread from the origin faster than linearly in time.

Theorem 1. *For some constant $C_1 < \infty$, and all sufficiently large t,*

$$E\{\text{number of } B\text{-particles with a position outside } \mathscr{C}(C_1 t) \text{ at time } t\} \le 2N_{\rm B}e^{-t}. \tag{3}$$

Consequently it is a.s. the case that

$$B(t) \subset \mathscr{C}(2C_1 t)\text{eventually}. \tag{4}$$

This result holds for any $D_A, D_B \ge 0$ and probably is even valid if one allows the A- and B-particles to perform any random walk with bounded jumps of mean zero.

The proof of Theorem 1 is basically a Peierls argument. It associates to each B-particle, ρ say, present at time t, a so called genealogical path which describes the sequence of B-particles which transmitted the rumor/infection from the initial B-particles to ρ at time t, and also describes the relevant pieces of the paths of these intermediate particles. One proves (3) by taking the expectation of the number of genealogical paths which lead to a B-particle outside $\mathscr{C}(C_1 t)$ at time t.

The next theorem, which is the first important ingredient in proving the Shape Theorem, shows that the rumor/infection spreads at least linearly in time, but only if both the A- and B-particles perform simple random walks with the same jump rate.

Theorem 2. *If $D_A = D_B > 0$, then there exists a constant $C_2 > 0$ such that for each constant $K > 0$*

$$P\{\mathscr{C}(C_2 t) \not\subset B(t)\} \leq \frac{1}{t^K} \text{for all large } t. \tag{5}$$

Consequently, a.s.

$$\mathscr{C}(C_2 t) \subset B(t) \text{ eventually.} \tag{6}$$

To prove the Shape Theorem we will need a form of Theorem 2 which also gives some information about the possible occurrence of A-particles amid the spreading B-particles. More specifically, the same proof as for Theorem 2 can be used to prove the next proposition.

Proposition 1. *If $D_A = D_B$, then for all K there exists a constant $C_3 = C_3(K)$ such that*

$$P\{\text{there is a vertex } z \text{ and an } A\text{-particle at the space–time point } (z,t) \text{ while} \tag{7}$$

$$\text{there also was a } B\text{-particle at } z \text{ at some time } \leq t - C_3[t \log t]^{1/2}\} \tag{8}$$

$$\leq \frac{1}{t^K} \text{for all sufficiently large } t. \tag{9}$$

Consequently, for large t,

$$P\{\text{at time } t \text{ there is a site in } \mathscr{C}(C_2 t/2) \text{ which} \tag{10}$$

$$\text{is occupied by an } A\text{-particle}\} \leq \frac{2}{t^K}. \tag{11}$$

Remark. It can be checked that the constants C_1, C_2 do not depend on the number or positions of the initial B-particles. However, the lower bounds for the times for which (3–6) are valid do depend on these initial data.

The proof of the Theorem 2 is rather involved. To help the intuition, it is best to think of the one-dimensional case, starting with one B-particle at the origin and no other B-particle. All the major difficulties appear already in this special case. Until the last paragraph of these heuristic remarks we therefore take $d = 1$.

In this one-dimensional case, there is for each t a rightmost B-particle, at position $\mathscr{R}(t)$ say, and a leftmost B-particle at position $-\mathscr{L}(t)$. At time t all particles in $[-\mathscr{L}(t), \mathscr{R}(t)]$ are B-particles and all particles outside $[-\mathscr{L}(t), \mathscr{R}(t)]$ are A-particles. Basically we want to show that $\liminf_{t\to\infty} \mathscr{R}(t)/t > 0$ and similarly for $\mathscr{L}(t)$. If there is exactly one particle at $\mathscr{R}(t)$ at time t, then $\mathscr{R}(.)$ behaves like a simple random walk, that is, $P\{\mathscr{R}(t + \mathrm{d}t) = \mathscr{R}(t) \pm 1\} = Dt/2 + O(\mathrm{d}t2)$, with

D standing for the common value of D_A and D_B. However, if there is more than one particle at $\mathscr{R}(t)$ at time t, then the rightmost particle moves one step to the right as soon as one of the particles at $\mathscr{R}(t)$ makes a jump to the right, whereas the rightmost position moves a step to the left only when all particles at $\mathscr{R}(t)$ move to the left. Thus, the rightmost B-particle has a drift to the right at all times when there is more than one particle at $\mathscr{R}(t)$. When there is at least one other particle (of either type) "close to" the rightmost B-particle, then there is a positive probability that in the next time unit another particle will coincide with the rightmost B-particle. This will still provide $\mathscr{R}(.)$ with an upwards drift. By using large deviation estimates for martingales one can see that the only way for $\mathscr{R}(t)/t$ to become small (with a non-negligible probability) is if the particle at $\mathscr{R}(s)$ has for most $s \in [0,t]$ no particle (of any type) nearby. We therefore want to show that the probability of this event goes to 0. One is tempted to try and prove this by studying the environment as seen from the position $\mathscr{R}(t)$. However, this approach seems difficult because the dependence between $\mathscr{R}(t)$ and the particles near $\mathscr{R}(t)$ is very complicated. We have been unable to use this approach. Instead, it turns out to be easier to prove a much stronger property, which uses almost no property of the path $s \mapsto \mathscr{R}(s)$. Roughly speaking we prove that every space–time path $s \mapsto \widehat{\pi}(s)$ with not too many jumps during $[0,t]$ has some particle "near $\widehat{\pi}(s)$ most of the time."

To make this more specific, we introduce some notation. A path $\pi = (x_0, ..., x_m)$ is a sequence of integers with $x_{j+1} - x_j = \pm 1$, $1 \le i \le m$. We regard the x_j as the successive positions of a space–time path $\widehat{\pi}$. There are many space–time paths which traverse the same positions in the same order. A space–time path $\widehat{\pi}$ is specified by giving its successive positions x_i and jump times s_i. For $s_1 < s_2 < \cdots$ we shall sometimes denote the path which jumps to x_i at time s_i by $\widehat{\pi}(\{s_i, x_i\})$. We make the convention that $s_0 = 0$, and unless stated otherwise, $x_0 = 0$. In addition we are here only discussing space–time paths over the time interval $[0,t]$, so we tacitly take $s_m \le t$. $\widehat{\pi}(\{s_i, x_i\})$ is then the path which is at position x_i during $[s_i, s_{i+1})$ for $0 \le i < m$, and at position x_m during $[s_m, t]$. If it is important that the path has exactly m jump times, then we shall write $\widehat{\pi}(\{s_i, x_i\}_{i \le m})$. Throughout this proof we shall only consider paths which are contained in $\mathscr{C}(t \log t) = [-t \log t, t \log t]$. Of particular interest for us is the following class of paths with exactly ℓ jumps:

$$\varXi(\ell, t) = \{\widehat{\pi}(\{s_i, x_i\}_{0 \le i \le \ell}) \text{ with } 0 = s_0 < s_1 < \cdots < s_\ell < t \text{ and } x_i \in \mathscr{C}(t \log t)\}.$$

Instead of using the path followed by $\mathscr{R}(.)$, we shall construct special paths $\widehat{\pi}$ with the property that there is a B-particle at $(\widehat{\pi}(s), s)$ for all $s \le t$ [so that automatically $\mathscr{R}(t) \ge \widehat{\pi}(t)$], and such that these paths are with high probability in $\varXi(\ell, t)$ for some $\ell \le 2Dt$, and also have a drift to the right at any time s when there are at least two particles at $\widehat{\pi}(s)$. Thus, it will be sufficient to show that every space–time path $\widehat{\pi} \in \cup_{\ell \le 2Dt} \varXi(\ell, t)$ has some particle "near $\widehat{\pi}(s)$ most of the time."

To this end we choose a large integer C_0 and partition space–time $\mathbb{Z} \times [0, \infty)$ into the following blocks of size $\Delta_r := C_0^{6r}$ for some $r > 0$:

$$\mathscr{B}_r(i,k) = [i\Delta_r, (i+1)\Delta_r) \times [k\Delta_r, (k+1)\Delta_r).$$

We call these intervals r-blocks. Next we will define "good" and "bad" r-blocks. There is a standard percolation argument which also partitions space into large blocks which can be good or bad, and then shows that on the one hand the bad blocks do not percolate, and on the other hand that no percolation of bad blocks implies a desired property. In our case the desired property would be that any space–time path $\widehat{\pi} \in \Xi(\ell, t)$ intersects at most ϵt bad r-blocks for a suitable ℓ and for a small ϵ. This is indeed the desired property we are after, but we have not succeeded in simply working with r-blocks for one fixed r, because of the complicated dependence of the configurations in different r-blocks. Instead we work with r-blocks for all r. This is why we say that our proof is based on a multiscale argument.

The notion of being "good" or "bad" depends on the presence of particles in certain sets, so we need to count numbers of particles. We define $N^*(x,t)$ as the number of particles at the space–time point (x,t) in the system which evolves freely, without any B-particles. In this system we start off with $N_{\mathrm{A}}(x,0) = N_{\mathrm{A}}(x,0-)$ particles at x at time 0 and let all these particles perform independent random walks without any interaction. Note that $N^*(x,t) \le N_{\mathrm{A}}(x,t) + N_{\mathrm{B}}(x,t)$. Let $Q_r(x) := [x, x + C_0^r)$. The important counts are

$$U_r(x,v) = \sum_{y \in Q_r(x)} N^*(y,v) = \sum_{y: x \le y < x + C_0^r} N^*(y,v).$$

We call the r-block $B_r(i,k)$ bad if

$$U_r(x,v) < \gamma_r \mu_A C_0^r \tag{12}$$

for some (x,v) with integer v for which $Q_r(x) \times v$ is contained in $\widetilde{\mathscr{B}}_r(i,k)$, and where $\widetilde{\mathscr{B}}_r(i,k) := [(i-3)\Delta_r, (i+4)\Delta_r) \times [(k-1)\Delta_r, (k+1)\Delta_r) \supset \mathscr{B}_r(i,k)$.

For the time being the only important properties are that the γ_r are strictly increasing (but slowly) and satisfy

$$0 < \gamma_0 < \gamma_r < \gamma_\infty \le 1/2, \quad r > 0.$$

Roughly speaking, the bad blocks are blocks in which the number of A-particles in some only-space-like cube of specified size and which is nearby in space–time, is less than half the expected amount. Indeed, it is well known that in our setup each $U_r(x,v)$ has a Poisson distribution of mean $\mu_A C_0^r$. A block is called good if it is not bad.

If a space–time path $\widehat{\pi}$ is in a good r-block at a given time s, then there are a reasonable number of particles within distance C_0^r of $\widehat{\pi}(s)$ at time s, by definition of a good block. We therefore would like to show that "most" space–time paths intersect "few" bad blocks during $[0,t]$. To quantify this statement we define

$$\phi_r(\widehat{\pi}) = \text{ number of bad } r\text{-blocks which intersect the space–time path } \widehat{\pi},$$
$$\Phi_r(\ell) = \sup_{\widehat{\pi}\in\Xi(\ell,t)} \phi_r(\widehat{\pi}).$$

The principal part of the proof is to show that for any choice of $K > 0$ and $\epsilon_0 > 0$ there exists a r_0 such that

$$P\{\Phi_r(\ell) \geq \epsilon_0 C_0^{-6r}(t+\ell) \text{ for some } r \geq r_0, \ell \geq 0\} \leq \frac{2}{t^K}, \tag{13}$$

for all large t. This result has the desired form, because any path $\widehat{\pi}$ spends at most C_0^{6r} time units in a given r-block, and therefore at most $C_0^{6r}\phi_r(\widehat{\pi})$ time units in bad blocks during $[0,t]$. Moreover, as we stated before, we only need to consider space–time paths in $\sum_{\ell\leq 2Dt}\Xi(\ell,t)$. Thus if the property in braces in (13) holds, then any $\widehat{\pi}\in\sum_{\ell\leq 2Dt}\Xi(\ell,t)$ satisfies

$$C_0^{6r}\phi_r(\widehat{\pi}) \leq C_0^{6r} \sup_{\ell\leq zDt} \Phi_r(\ell) \leq \epsilon_0(1+2D)t,$$

and spends at most $\epsilon_0(1+2D)t$ time units in bad blocks (for $r \geq r_0$). For $\epsilon = \epsilon_0(1+2D) < 1/2$ this shows that the paths of interest to us have a drift to the right for at least $t/2$ time units.

This leaves us with the problem of proving (13). This is done by means of a recurrence relation (with random terms) for the Φ_r. Note that each bad r-block has to lie either in a good $(r+1)$-block or in a bad $(r+1)$-block. Since any $(r+1)$-block contains exactly C_0^{12} r-blocks, the number of bad r-blocks which intersect a path $\widehat{\pi}$, and which are contained in a bad $(r+1)$-block (and which necessarily intersects $\widehat{\pi}$) is at most $C_0^{12}\phi_{r+1}(\widehat{\pi}) \leq C_0^{12}\Phi_{r+1}$. A similar estimate holds for the number of bad r-blocks which intersect $\widehat{\pi}$ and which are contained in a good $(r+1)$-block (Fig. 2).

At this stage it may be useful to say a few words about the case of dimension greater than 1. There is no clear analogue of $\mathscr{R}(t)$, or at least none that is helpful. Instead of constructing paths which have a drift to the right at least half the time one now fixes a $x \in \mathscr{C}(C_2t)\cap Z^d$ and tries to construct a space–time path $\lambda(.) = \lambda(., x)$ which has a B-particle at $\lambda(s)$ for all s, and which has a tendency to move toward x. In fact, our $\lambda(s)$ behaves like a (d-dimensional) simple random walk at times s when there is only one particle at $\widehat{\pi}(s)$, but if there are at least two particles at $\lambda(s)$ and a particle jumps away from $\lambda(s.)$ at time s, then the conditional expectation of $\|\lambda(s) - x\|_2$ is smaller than $\|\lambda(s-) - x\|_2$. This will give us a path which with high probability reaches x during $[0,t]$, provided the path has at least two particles "near $\lambda(s)$" at least a positive fraction of the time. In this way all points $x \in \mathscr{C}(C_2t)\cap Z^d$ are reached by the infection during $[0,t]$. From there on there are only minor differences between the cases $d = 1$ and $d > 1$ for Theorem 2.

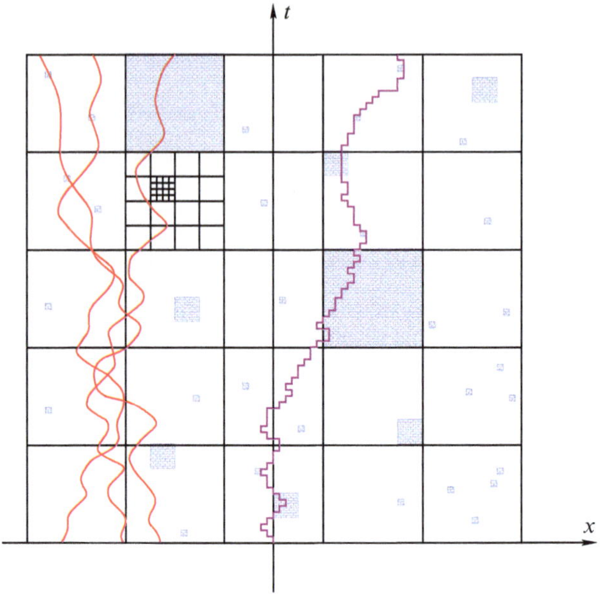

Fig. 2 Bad boxes are gray. Good (= *white*) boxes of scale r may contain bad boxes of scale $r+1$

Remark. In recent work [12] proved an improved upper bound for the velocity for the class of models where A- and B-particles respectively evolve with a diffusion constant $D_B = 1$ and a possibly time dependent jump rate $D_A \geq 0$. Even more generally, A-particles follow some independent stochastic process which also includes long range random walks with drift and various deterministic processes. Assume that the density of A-particles is ρ. Then they get in all dimensions an upper bound of order $\max(\rho, \sqrt{\rho})$ that depends only on ρ and d and not on the specific process followed by A-particles, in particular that does not depend on D_A. They also argue that for $d \geq 2$ or $\rho \geq 1$ this bound can be optimal (in ρ). This leads us to another challenging open problem.

Open Problem. There is numerical evidence and some theoretical arguments in Physics literature (see [26] and references therein) which claim that for given fixed density of A-particles which are moving with the jump rate $D_A > 0$ the limiting velocity depends solely on D_B and not on D_A. Prove or disprove it!

Finally, we turn to the "full" Shape Theorem. Theorem 2, in particular inclusion (6) is a crucial tool for proving the Shape Theorem. We do not know any shortcut which proves the Shape Theorem without much of the development of [19] for (6).

Theorem 3. *If $D_A = D_B > 0$, then there exists a non-random, compact, convex set B_0 such that for all $\epsilon > 0$ almost surely*

$$(1-\epsilon)B_0 \subset \frac{1}{t}B(t) \subset (1+\epsilon)B_0 \quad \text{for all large } t. \tag{14}$$

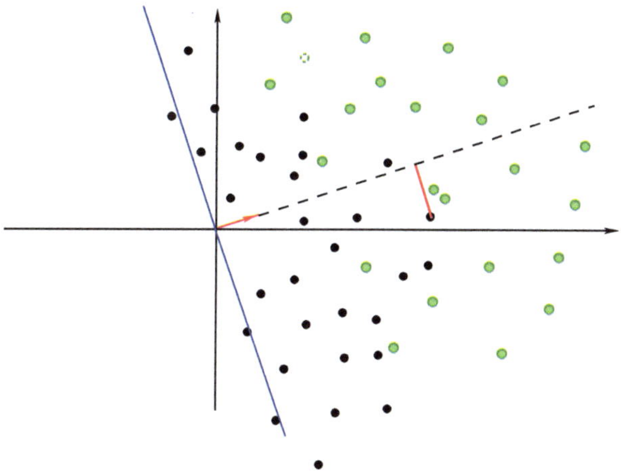

Fig. 3 "Half-space" process starts with the initial configuration of particles restricted to the half-plane. During the dynamics particles move freely and are not restricted to the half-plane. B-particles are black, and projection of the position of "most advanced" B-particle at time t on to the line parallel to chosen vector u determines $H(t,u)$

The origin is an interior point of B_0, and B_0 is invariant under reflections in coordinate hyperplanes and under permutations of the coordinates.

As we already mentioned before, the present model does not have the strict subadditivity properties. But subadditivity still remains the principal ingredient used for the proof of Theorem 3. However, we now use subadditivity only for certain "half-space" processes which approximate the true process. The "half-space" processes are evolving exactly in the same way as the original "full-space" process, but the starting initial conditions are different. We assume that initially particles are located only in half-spaces determined by a chosen direction (Fig. 3).

However even these half-space processes have only approximate superconvolutive properties, but these properties are strong enough to show that for each unit vector u there exists a constant $\lambda(u)$ such that almost surely

$$\lim_{n\to\infty} \frac{1}{t} H(t,u) = \lambda(u), \tag{15}$$

where $H(t,u)$ is basically the maximum of $\langle x, u\rangle$ over all x which have been reached by a B-particle by time t ($\langle x, u\rangle$ is the inner product of x and u; for technical reasons $H(t,u)$ will be calculated in a process in which the starting conditions are slightly different from our original process). Thus the B-particles reach in time t half-spaces in a fixed direction u at a distance which grows linearly in t. Except in dimension 1, it then still requires a considerable amount of technical work to go from this result about the linear growth of the distances of reached half-spaces to the full asymptotic shape result. In particular, the coupling between the two half-space processes clearly relies heavily on the assumption $D_A = D_B$, so that we can assign the same path to a particle in the two processes, even though the types of the particle in the two processes may be different.

Remark. Our proof in [19] shows that the right hand inclusion in (14) remains valid for arbitrary jump-rates of the A- and the B-particles. However, it is still not known whether the left hand inclusion holds in general. The lower bound for $B(t)$ is known only when $D_A = D_B$, or when $D_A = 0$, that is, when the A- and B-particles move according to the same random walk (see [19]).

Open problem. Another interesting open problem is related to renormalization scheme itself. In spite of its apparent robustness our argument requires not only that $D_A = D_B$, but also the initial distribution of particles is a product of i.i.d. Poisson with parameter μ (which is the invariant measure for the system of independent walkers). We do not know how to prove neither a lower bound nor a full Shape Theorem if starting from different initial conditions.

If $D_A \neq D_B$, then we can only prove the upper bound $B(t) \subset \mathscr{C}(C_1 t)$ eventually.

2.2 Phase Transition

In this subsection we will briefly discuss the situation when an "infected" B-particle can recover, that is, all B-particles recuperate (i.e., turn back into A-particles) independently of each other at a rate λ. As before, we assume that we start the system with $N_{\rm A}(x, 0-)$ A-particles at x, and that the $N_{\rm A}(x, 0-)$, $x \in \mathbb{Z}^d$, are i.i.d., mean μ_A Poisson random variables. In addition we start with one additional B-particle at the origin. We show that there is a critical recuperation rate $\lambda_{\rm c} > 0$ such that the B-particles survive (globally) with positive probability if $\lambda < \lambda_{\rm c}$ and die out with probability 1 if $\lambda > \lambda_{\rm c}$.

Before formally stating our theorem we make some comments about the precise formulation of the model, and introduce some notation. First we define for $\eta = A$ or B

$$N_\eta(x, t) = \text{number of } \eta\text{-particles at the space–time point } (x, t).$$

Throughout we write $\mathbf{0}$ for the origin. As stated in the abstract we put $N_{\rm A}(x, 0-)$ A-particles at x just before we start. We then introduce a B-particle at the origin and turn some of the particles at the origin instantaneously to B-particles, so that at time 0 we start with $N_{\rm A}(x, 0) = N_{\rm A}(x, 0-)$ A-particles at $x \neq \mathbf{0}$ and $N_{\rm B}(\mathbf{0}, 0) \in [1, N_{\rm A}(\mathbf{0}, 0-) + 1]$ B-particles at $\mathbf{0}$. However, at any time $t > 0$ an A-particle can turn into a B-particle only if the A-particle itself jumps to the position of a B-particle at t or if some B-particle jumps to the position of the A-particle at time t. Thus, we are not saying that an A-particle turns into a B-particle whenever it coincides with a B-particle. We adopted the rule that a jump is required for the following reason. If we did not make this requirement, then B-particles could effectively not recover at a space-time point (x, t) with several B-particles present. Indeed, if one of them tried to turn back into an A-particle at time t it would immediately become of type B again because it coincided with another B-particle. This creates some sort of singularity in the model which we are unable to handle at the moment (see, however, Remark 3 below). This is the reason for the requirement of a jump for a change from type A to type B at all strictly positive times t. Only at $t = 0$ did we change some A-particles at $\mathbf{0}$ to B-particles because they coincided with a B-particle (even though no jump occurred). The choice of the set of A-particles at $\mathbf{0}$ which is turned into B-particles at time 0 will not influence our arguments. Note that because of the jump requirement there may be particles of both types at a single space–time point.

We say that the infection *survives* if

$$P\{\text{there are some } B\text{-particles at all times}\} > 0. \tag{16}$$

Since there cannot be any B-particles after time t if there are no B-particles at t, it follows that (16) is equivalent to

$$\lim_{t\to\infty} P\{\text{there are some } B\text{-particles at time } t\} > 0. \tag{17}$$

One may even replace $\lim_t$ by $\liminf_{t\to\infty}$ in (17). Note that the survival in (16) or (17) is only global survival. Local survival in its strongest form would say that

$$\liminf_{t\to\infty} P\{N_{\mathrm{B}}(\mathbf{0},t) > 0\} > 0. \tag{18}$$

A weaker form of local survival would be that

$$P\{N_{\mathrm{B}}(\mathbf{0},t) \geq 0 \text{ for some arbitrarily large } t\} > 0. \tag{19}$$

Clearly (18) implies (19), and this, in turn implies (16). We do not know how to prove that either of the forms (18) or (19) of local survival holds if λ is small enough.

The infection is said to *die out* or to *become extinct* if it does not survive, i.e., if

$$P\{\text{there is some (random) } t \text{ such that there are no } B\text{-particles after } t\} = 1. \tag{20}$$

Here is our principal result.

Theorem 4. *There exists* $0 < \lambda_c < \infty$ *such that the infection survives if* $\lambda < \lambda_c$ *and dies out if* $\lambda > \lambda_c$.

Remark. The restriction to only one B-particle at time 0 is for convenience only. The theorem remains valid if we start with any finite number of B-particles at (non-random) positions.

Remark. We already remarked that the theorem does not give local survival if λ is sufficiently small. Neither does it tell us anything about the location of the B-particles as a function of t on the event that the B-particles survive forever.

By a special argument one can show that (19) holds for $d = 1$ and $\lambda < \lambda_c$ on the event that the B-particles survive forever.

Remark. The proof that there is survival for small $\lambda>0$ works even in the case in which an A-particle turns into a B-particle whenever it coincides with a B-particle, that is, if we do not require that the A- or B-particle jumps before reinfection can occur after recuperation of a B-particle.

3 The Stochastic Combustion Process ($D_A = 0$, $D_B > 0$)

In this section we review results concerning the Stochastic Combustion Process. As it was mentioned in the Introduction, this is an interacting particle system which can be interpreted as a model for the burning of a propellant material, where particles represent heat and move as simple random walks which can branch and annihilate. In this section we consider the process being defined on the lattice $\mathbb{Z}^d$. The dynamics is now defined in terms of B-particles (heat particles) and A-particles (propellant material): (a) B-particles move independently as simple symmetric continuous time random walks; (b) A-particles are inert, so they do not move at all; (c) each time a B-particle jumps to a site where there is a A-particle, the A-particle is transformed into a B one and begins to move as a simple symmetric random walk.

There is also a very interesting and important connection between the Combustion Process and the Activated Random Walks (ARW) model broadly studied in physics literature in the context of absorbing state phase transition. If we assume that in the Combustion process B-particles can recover, i.e., again become an A-particle, at rate 1, independently of anything else, then we obtain the well known ARW model. We briefly discuss this case in the last subsection.

3.1 Shape Theorem

For each site $x \in \mathbb{Z}^d$ and time $t \geq 0$, if there is no A-particle at site x at time t, we define $\eta_t(x)$ as the number of B-particles at site x, while $\eta_t(x) := -1$ if there is one A-particle at site x at time t. Consider the stochastic combustion process with an initial condition η_0 where there is one A-particle at each site $x \neq 0$ while one B-particle at site 0. A natural question is to describe the evolution of the set,

$$B_d(t) := \{y \in \mathbb{Z}^d : \eta_t(y) \geq 0\}, \quad t \geq 0.$$

Note that $B_d(0) = \{0\}$. The following result provides a partial answer to such a question. Here, we use the notation $[D] := D \cap \mathbb{Z}^d$ for subsets $D \subset \mathbb{R}^d$.

Theorem 5. *Assume that $B_d(0) = \{0\}$. Then, there is a closed convex bounded subset $B^0_{\mathrm{d}} \subset R^d$, symmetric under permutations of the coordinate axis and with non-empty interior, such that for every $\epsilon > 0$, a.s. eventually in t one has that*

$$[B^0_{\mathrm{d}} t(1-\epsilon)] \subset B_d(t) \subset [B^0_{\mathrm{d}} t(1+\epsilon)].$$

This result was proved in [27] (see also [1]). Let us define the first time site y is visited from the initial configuration $B_d(0) := \{x\}$ (in other words, there is only one B-particle at site x while one A-particle per site $y \neq x$),

$$T_{x,y} := \inf\{t \geq 0 : y \in B_d(t)\}.$$

The proof of Theorem 5 relies on the following crucial subadditivity property,

$$T_{x,y} \leq T_{x,z} + T_{z,y}, \qquad x, y, z \in \mathbb{Z}^d,$$

combined with Kingman's subadditive ergodic theorem. It is however, difficult to prove the required condition $E[T_{0,1}] < \infty$ to apply Kingman's theorem, and most of the work in [27] and [1] is to prove such a condition using different methods, though.

3.2 The Stochastic Combustion Process in Dimension d = 1

Fix some $r \in \mathbb{Z}$. We assume that initially at each site x with $x \geq r+1$ there is one A-particle and no B-particle, while at each site $x \leq r$, there are $\eta(x) \geq 0$ particles and no A-particles. We are interested in the asymptotic behavior of r_t, the position of the right-most visited site by a B-particle at time t. We can prescribe the state space of this system as

$$\mathbb{S}_0 := \{(r, \eta) : r \in \mathbb{Z}, \eta \in \mathbb{N}^{\{\ldots,r-1,r\}}\}.$$

In order to avoid pathological situations (involving for example a super-ballistic behavior of the front), we will consider initial conditions which belong to the set

$$\mathbb{S} := \{(r, \eta) : r \in \mathbb{Z}, \eta \in \mathbb{N}^{\{\ldots,r-1,r\}}, \sum_{x \leq r} e^{\theta(x-r)}\eta(x) < \infty\},$$

where $\theta > 0$ is chosen sufficiently small. Throughout, given an initial condition $\omega \in \mathbb{S}$, we will denote by $\mathbb{P}_\omega$ the law of the stochastic combustion process starting from ω.

3.2.1 Regeneration Times, The Law of Large Numbers and Fluctuations

In [27], a law of large numbers for the front position, with a specific initial condition was proven. Subsequently, in [6], the law of large numbers was generalized according to the following theorem.

Theorem 6. *There exists a deterministic $v > 0$ such that for every initial condition in $\mathbb{S}$ which has at least one B-particle a.s. the following law of large* *numbers is satisfied,*

$$\lim_{t\to\infty} \frac{r_t}{t} = v.$$

Furthermore, in [2] it is proven that the condition required in the definition of $\mathbb{S}$ (that $\sum_{x\le 0} e^{\theta x}\eta(x) < \infty$) is optimal in the following sense: there is a $\theta' > 0$ and a $v' > v > 0$ such that when

$$\sum_{x\le 0} e^{\theta' x}\eta(x) = \infty, \tag{21}$$

one has that a.s.

$$\liminf_{t\to\infty} \frac{r_t}{t} \ge v'.$$

Concerning the fluctuations of the front, the following theorem was established in [6].

Theorem 7. *There exists $\sigma^2 \in (0,\infty)$ non-random and independent of initial conditions in $\mathbb{S}$ such that*

$$\epsilon^{-1/2}(r_{\epsilon^{-1}t} - \epsilon^{-1}tv), \qquad t \ge 0,$$

converges in law as $\epsilon \to 0$ to a Brownian motion with variance σ^2.

The proofs of Theorems 6 and 2 are heavily based on the construction of a regeneration time sequence with useful properties. It is possible to construct sequences of regeneration times which have i.i.d. increments. But the main challenge is to choose a sequence among this group which is nice enough so that one can prove that the first and second moments of the increments are finite. We now describe the construction performed in [6]. Without loss of generality we assume that $r_0 = 0$. We now construct for each $r \ge 0$, a coupling between the process r_{T_r+t}, where

$$T_r := \inf\{t \ge 0 : r_t = r\},$$

and a combustion process $\bar{r}^r_t$ such that $r_o = r$ and the configuration of B-particles at sites $x \le r$ consists of only one particle at site r and none at $x < r$, so that $\bar{r}^r_t \le r_{T_r+t}$ for $t \ge 0$. This coupling will be useful to prove the desired i.i.d. properties of the regeneration times. The idea now is to decouple the effect of the particles behind the front, from the effect of the particles at the front, on the movement. For some reaction–diffusion fronts corresponding to interacting particle system dynamics where the number of particles per site is bounded by some constant, there is a natural way to this (see for example [5] and [16]), where essentially one compares any initial condition of the front with an extremal one, where all the sites behind the front are "full" of particles. Nevertheless, the fact that the number of particles per site is not bounded in the combustion process, makes the above idea useless as presented. The way in which this issue was solved in [6], was introducing a space–time line which separates the effect of the particles at the front from those behind it. This space–time line, corresponds to an imaginary particle traveling at a constant speed which must be chosen smaller than the actual speed v of the front. The point then, is essentially define the first regeneration time, as the first time κ in which the front starting from r_κ and after κ never slows down to the point where an imaginary particle starting from the position r_κ with a constant speed larger than v (fixed) catches it up, and such that all the particles behind the front (at the left of r_κ) never catch up with the imaginary particle. Somehow, this should decouple the effect of "old" particle from the evolution of the front. The above construction can be perfomed, but the main difficulty of the problem requires some new idea: to prove that the moments of the regeneration time increments are finite. The reason why something more is required is that some kind of control is needed on the size of the "cloud" of particles at the left of the front. Indeed, to obtain estimates on the moments, it is necessary to present an explicit construction in terms of stopping times, in a way which is similar to the presentation given by Sznitman and Zerner in [30] for random walks in random environment. What is done in [6] is that the above described definition is slightly modified, so that at each step in the construction with stopping times, there is a good control on the cloud of particle to the left of the front in terms of some exponential norm. Let us now describe this construction. Let

$$U := \inf\{t \geq 0 : \bar{r}_t^r - r < \lfloor \alpha_2 t \rfloor\}.$$

Define also the *exponential density norm* of particles by

$$\phi_z(t) := \sum_{x \in \mathbb{Z}} \mathrm{e}^{\theta(x - r_t)} \eta_z(t, x),$$

where $\eta_z(x, t)$ is the number of particles at time t and site x which originated from a branching at some site $y \leq z$. Now let $L \in \mathbb{N}$ be fixed and

$$W := \inf\{t \geq 0 : \phi_{r-L}(t, r, \eta(0)) \geq \mathrm{e}^{\theta(\lfloor \alpha_1 t \rfloor - (r_t - r))}\}.$$

Note that W is the first time that the exponential density norm of particles which originated from a branching at a site at a distance larger than or equal to L from the initial position of the front, increases beyond $\mathrm{e}^{\theta(\lfloor \alpha_1 t \rfloor - (r_t - r))}\}$. It is possible to perform a construction of the stochastic combustion process so that each X particle originating from a given site z is represented by a continuous time simple symmetric random walk $Z_{z,i}(t)$, where the index i labels all the particles which originate from site z. We say that the X particle originated from site z if initially the particle was at site z, or if the B-particle is created at some time from the branching of an A-particle which was initially at site z. We now define

$$V := \inf\{t \geq 0 : \max_{r - L < z < r} \max_{1 \leq i \leq a} Z_{z,i}(t) > \lfloor \alpha_1 t \rfloor + r\},$$

the first time some of the particles originating from a branching at a site at a distance smaller than L from the initial position of the front, hit the line $\lfloor \alpha_1 t \rfloor + r$.

Let now

$$D := \min\{U, V, W\}.$$

In other words, we have defined D as is the first time that something "bad" happens: the front slow downs too much, corresponding to U; the particles at the left of the front but originally at a distance smaller than L from it, speed up too much; or the particles at distances larger than L from the front and to its left produce a "cloud" whose size grows too much, in the sense that the exponential norm increases above a certain threshold.

Define $U \circ \theta_s$, $V \circ \theta_s$, and $W \circ \theta_s$ as the first times U, V or W happen after time $s \geq 0$, and $D \circ \theta_s := \min\{U \circ \theta_s, V \circ \theta_s, W \circ \theta_s\}$. For each $y \in \mathbb{Z}$, let

$$T_y := \inf\{t \geq 0 : r_t \geq y\},$$

and fix $p \in (0, 1)$. Define for $x \geq r$,

$$J_x := \inf\{j \geq 1 : \phi_{x+(j-1)L}(T_{x+jL}) \leq p \text{ and } m_{x+jL-L^{1/4},x+jL}(T_{x+jL}) \geq aL^{1/4}/2\}, \tag{22}$$

the first trial after the front visits site x, such that the exponential density norm of particles originating at sites at a distance larger than L from the front, decreases to a quantity smaller than p and such that there are sufficiently many particles originating from sites close to the front which are again there at time T_{x+jL} when the front advances L steps.

We define $\mathcal{F}_t$ as the infromation up to time t generated by the random walks $\{Z_{z,i}\}$ used to construct the combustion process (see [6]). Define sequences of $\mathcal{F}_t$-stopping times, $\{S_k : k \geq 0\}$ and $\{D_k : k \geq 1\}$ as follows. $S_0 := 0$, $R_0 := r$, and for $k \geq 0$,

$$S_{k+1} := T_{R_k + J_{R_k} L} \qquad D_{k+1} := D \circ \theta_{S_{k+1}} + S_{k+1}, \qquad R_{k+1} := r_{D_{k+1}}.$$

The $S_k, k \geq 1$ are good times when there is control on the cloud of particles originating from sites far from the front, in the sense that the exponential norm is small enough, and there are enough particles originating from sites close to the front.

For $k \geq 1$, define $U_k := U \circ \theta_{S_k} + S_k$, $V_k := V \circ \theta_{S_k} + S_k$ and $W_k := W \circ \theta_{S_k} + S_k$. Let

$$K := \inf\{k \geq 1 : S_k < \infty, D_k = \infty\},$$

and define the *regeneration time*

$$\kappa := S_K, \tag{23}$$

if $K < \infty$ and $\kappa = \infty$ otherwise. κ is *not* a stopping time.

$\mathcal{G}$, the information up to time κ, is the completion with respect to $\mathbb{P}_w$ of the smallest σ-algebra containing all sets of the form $\{\kappa \leq t\} \cap A$, $A \in \mathcal{F}_t$.

Define the sequence of regeneration times $\kappa_1 \leq \kappa_2 \leq \cdots$ by $\kappa_1 := \kappa$ and for $n \geq 1$

$$\kappa_{n+1} := \kappa_{\rm n} + \kappa(w_{\kappa_{\rm n}+\cdot}),$$

where $\kappa(w_{\kappa_{\rm n}+\cdot})$ is the regeneration time starting from $w_{\kappa_{\rm n}+\cdot}$ and we set $\kappa_{n+1} = \infty$ on $\kappa_{\rm n} = \infty$ for $n \geq 1$. κ_1 is the *first regeneration time* and $\kappa_{\rm n}$ is the *n-th regeneration time*. A crucial property of the regeneration structure that has been defined is the following corollary presented in [6], whose proof is relatively straightforward.

Corollary 1. *Let $w \in \mathbb{S}$. (i) Under $\mathbb{P}_w$, $\kappa_1, \kappa_2 - \kappa_1, \kappa_3 - \kappa_2, \ldots$ are independent, and $\kappa_2 - \kappa_1, \kappa_3 - \kappa_2, \ldots$ are identically distributed with law identical to that of κ_1 under $\mathbb{P}_{a\delta_0}[\cdot|U = \infty]$. (ii) Under $\mathbb{P}_w$, $r_{\cdot\wedge\kappa_1}, r_{(\kappa_1+\cdot)\wedge\kappa_2} - r_{\kappa_1}, r_{(\kappa_2+\cdot)\wedge\kappa_3} - r_{\kappa_2}, \ldots$ are independent, and $r_{(\kappa_1+\cdot)\wedge\kappa_2} - r_{\kappa_1}, r_{(\kappa_2+\cdot)\wedge\kappa_3} - r_{\kappa_2}, \ldots$ are identically distributed with law identical to that of r_{κ_1} under $\mathbb{P}_{a\delta_0}[\cdot|U = \infty]$.*

On the other hand, as previously commented, the most challenging step which renders useful the previous corollary, is the following proposition that was also proved in [6].

Proposition 2. *For every initial data $w \in \mathbb{S}$,*

$$\kappa < \infty, \qquad \mathbb{P}_w - \text{as} \tag{24}$$

Let $a\delta_0$ denote initial data with $r = 0$, $\eta(r) = a$ and $\eta(x) = 0$, $x < 0$. Then

$$\mathbb{E}_{a\delta_0}\left[\kappa^2|U = \infty\right] < \infty \qquad \text{and} \qquad \mathbb{E}_{a\delta_0}\left[r_\kappa^2|U = \infty\right] < \infty. \tag{25}$$

The proof of this result requires some careful and tedious computations which we do not include in this review. Now, a combination of Corollary 1 and Proposition 2, provides an argument to prove Theorem 7.

3.2.2 Large Deviations

In [2], the regeneration time structure presented in the previous subsection, was applied to prove a large deviations principle for the front $r_{\rm t}$. First we need to introduce a condition stronger than the one needed to prove the law of large numbers.

Assumption (G). For all $\theta > 0$

$$\sum_{x \leq 0} \exp(\theta x)\eta(x) < +\infty. \tag{26}$$

We then have the following result proved in [2].

Theorem 8. Large Deviations Principle *There exists a rate function* I*:* $[0,+\infty) \to [0,+\infty)$ *such that, for every initial* ω *condition satisfying* **(G)***,*

$$\limsup_{t\to+\infty} \frac{1}{t} \log \mathbb{P}_\omega \left[\frac{r_t}{t} \in C \right] \leq - \inf_{b\in C} I(b), \qquad \text{for} \quad C \subset [0,+\infty) \quad \text{closed},$$

and

$$\liminf_{t\to+\infty} \frac{1}{t} \log \mathbb{P}_\omega \left[\frac{r_t}{t} \in G \right] \geq - \inf_{b\in G} I(b), \qquad \text{for} \quad G \subset [0,+\infty) \quad \text{open}.$$

Furthermore, I *is identically zero on* $[0,v]$*, positive, convex, and increasing on* $(v,+\infty)$*.*

The proof of the existence of the rate function is based on a straightforward subadditivity argument. Nevertheless, to prove that the zero set of the rate function I is the interval $[0,v]$ it is necessary to use a sophisticated argument which is based on the use of regeneration times. Since the regeneration positions of the stochastic combustion process do not have exponential moments, it is crucial to perform a coupling with a variation of the stochastic combustion process where the random walks have a fixed but small drift towards the right. For this process the regeneration positions do have exponential moments.

In [2], precise estimates for the probability of the slowdown deviations were obtained. Let

$$U(\eta) := \limsup_{x\to-\infty} \frac{1}{\log|x|} \log\left(\sum_{y=0}^{x} \eta(y)\right), \quad u(\eta) := \liminf_{x\to-\infty} \frac{1}{\log|x|} \log\left(\sum_{y=0}^{x} \eta(y)\right),$$

and

$$s(\eta) := \min(1, U(\eta)).$$

For the statement of the following theorem we will write $U, u, ands$ instead of $U(\eta)$, $u(\eta)$ and $s(\eta)$.

Theorem 9. Slowdown deviations estimates. *Let* η *be an initial condition satisfying* **(G)***. Then the following statements are satisfied.*

(a) *For all* $0 \leq c < b < v$*, as* t *goes to infinity,*

$$\mathbb{P}\left[c \leq \frac{r_t}{t} \leq b\right] \geq \exp\left(-t^{s/2+o(1)}\right). \tag{27}$$

(b) *In the special case where* $\eta(x) \geq a$ *for all* $x \leq 0$*, one has that, for every* $0 \leq b < v$*, as* t *goes to infinity,*

$$\mathbb{P}\left[\frac{r_t}{t} \leq b\right] \leq \exp\left(-t^{1/3+o(1)}\right). \tag{28}$$

(c) *When $u < +\infty$, as t goes to infinity,*

$$\exp\left(-t^{U/2+o(1)}\right) \le \mathbb{P}\left[r_t = 0\right] \le \exp\left(-t^{u/2+o(1)}\right). \tag{29}$$

In the case of a homogeneous initial configuration, like $d_- \le \eta(y) \le d_+$ for all $y \le 0$, with $1 \le d_- \le d_+ < +\infty$, or when $(\eta(y))_{y\le 0}$ forms a realization of an i.i.d. family of random variables with positive finite expectation, the above results take a simpler form since $u = U = s = 1$. As a consequence, $\exp(-t^{1/2})$ turns out to be the actual order of magnitude for $\mathbb{P}\left[r_t = 0\right]$, and a lower bound for $\mathbb{P}\left[c \le \frac{r_t}{t} \le b\right]$ when $0 \le c < b < v$.

Open Problem. Extend the above results to higher dimension.

3.3 *Activated Random Walks Model and Absorbing State Phase Transition*

Here we briefly discuss the case when B-particles can recuperate. If, as in the Sect. 2.2, we assume that B-particles recuperate at rate $\lambda > 0$, it brings us to the phenomena of absorbing state phase transition, and this process in Physics literature is known broadly as Activated Random Walk (ARW) model.

Numerical analysis and some general theoretical arguments suggest that the ARW model exhibits a phase transition in the parameters λ – the recuperation rate and μ – the initial density of particles in the system, and that there should be two distinct regimes:

(**a**) Low particle density. There is a phase transition in λ in this case, namely, if λ is large enough, then system locally fixates, i.e., for any finite volume Λ there is almost surely a finite time t_Λ such that after this time there are no B-particles within Λ. If λ is small enough there is no fixation, and we expect that there is a limiting density of active particles in the long-time limit.
(**b**) High particle density. In this case there is no phase transition. For any $\lambda > 0$, the system does not fixate.

In spite of its intuitive transparency, it is remarkably difficult to prove existence of the phase transition. We start with the first basic fact, proved in [29] using the Diaconis–Fulton representation of the model. The Diaconis–Fulton representation provides an Abelian property for the dynamics of the system with finitely many particles, and – what is particularly important – provides monotonicity for the occupation times in μ as well as in λ.

Theorem 10 ([29]). *For $d \geqslant 1$ and any translation-invariant random walk and $\lambda > 0$, there exists $\mu_c \equiv \mu_c(\lambda) \in [0, \infty]$, such that if the initial distribution is i.i.d Poisson with density μ then*

$$P(\text{system locally fixates}) = \begin{cases} 1, & \mu < \mu_c \\ 0, & \mu > \mu_c. \end{cases}$$

Moreover, μ_c is non-decreasing in λ.

For fixed λ the value of $\mu_c(\lambda)$ is not known, however some theoretical arguments suggest, and numerical simulations support, that the following holds:

Conjecture. For any dimension, any random walk, and any $\lambda > 0$,

$$0 < \mu_c(\lambda) < 1.$$

Using Peierls type argument one can show that $\mu_c(\lambda) < +\infty$:

Theorem 11 ([21]). *Consider simple symmetric random walks on $\mathbb{Z}^d, d \geqslant 1$. There exists $\mu_0 < \infty$ such that $\mu_c(\lambda) < \mu_0$ for all λ.*

Recently E. Shellef improved this estimate.

Theorem 12. *Under the same hypotheses,*

$$\mu_c(\lambda) \leqslant 1. \tag{30}$$

However is much more difficult to show that $\mu_c(\lambda) > 0$. So far only the one-dimensional case was understood.

Theorem 13 ([29]). *Consider the Activated Random Walk Model with nearest-neighbor jumps in the one-dimensional lattice $\mathbb{Z}$ with fixed halting rate λ. Suppose the $\eta_0(x)$ are i.i.d. random variables in $\mathbb{N}_0$, having a Poisson distribution with parameter μ. There exists $\mu_c \in \left[\frac{\lambda}{1+\lambda}, 1\right]$ such that the system locally fixates a.s. if $\mu < \mu_c$ and stays active a.s. if $\mu > \mu_c$. In the particular case of $\lambda = +\infty$ we have $\mu_c = 1$.*

Open problem. It remains a challenging problem to prove the analog of the Theorem 13 in dimensions two and more.

For general account on this process and other interesting open problems we refer to [10].

4 Modified Diffusion Limited Aggregation ($D_A > 0, D_B = 0$)

To get a better feeling for the problem which we want to present, one could think of investigating the other extreme case, namely, when $D_B = 0$. This is not an interesting case when taken literally. In this case the infected particles stand still and act as traps for the healthy particles. All what happens with any given A-particle is that it walks around till it coincides with one of the B-particles after which it stands still as well. The infected set $\widetilde{B}(t)$ equals $\widetilde{B}(0)$ at all $t \geq 0$ and the speed at which the infection spreads is 0. To obtain something interesting we have to allow the B-particles to move, at least at some times. The simplest situation is the one-dimensional one, i.e., when $d = 1$. We chose to let the B-particles move one unit to the right, when an A-particle jumps on top of it. According to our rules all A-particles which were one unit to the right of the B-particles are turned into B-particles at the time of this jump. This leads to the model, we will describe now.

We consider the following problem in one-dimensional DLA. At time t we have an "aggregate" consisting of $\mathbb{Z} \cap [0, R(t)]$ (with $R(t)$ a positive integer). We also have $N(i,t)$ particles at $i > R(t)$. All these particles perform independent continuous time symmetric simple random walks until the first time $t' > t$ at which some particle tries to jump from $R(t)+1$ to $R(t)$. The aggregate is then increased to the integers in $[0, R(t')] = [0, R(t)+1]$ (so that $R(t') = R(t)+1$) and all particles which were at $R(t)+1$ at time $t'-$ are removed from the system. The problem is to determine how fast $R(t)$ grows as a function of t if we start at time 0 with $R(0) = 0$ and the $N(i,0)$ i.i.d. Poisson variables with mean $\mu > 0$. What makes this model particularly attractive is that it is conjectured that there is a phase transition for the growth of $R(t)$: we show that if $\mu < 1$, then $R(t)$ is of order $\sqrt{t}$ in a sense which is made precise below, and it is expected that it will grow linearly in t if μ is large enough.

This model is of further interest because it is a one-dimensional version of the celebrated DLA model of Witten and Sander [33]. In this model on $\mathbb{Z}^d$ one again has a growing aggregate $A(t) \subset \mathbb{Z}^d$ and one starts with $A(1) = \{\mathbf{0}\}$ the origin. Usually t is taken to run through the integers and $A(t)$ has cardinality t. $A(t+1)$ is obtained from $A(t)$ by adding one point of $\mathbb{Z}^d$. This added point is the first point of the boundary of $A(t)$ which is reached by a random walker which starts at infinity (see Kesten (1987) for a more precise description). The main difference between the modified DLA model and the DLA model of Witten and Sander is that the latter adds one A-particle to the system at a time, while in the former there are infinitely many A-particles from the start. However, there have been various investigations for related models in which new A-particles are added to the system before all previously released A-particles have reached the boundary of the aggregate and are removed from the system; see for instance Lawler, Bramson and Griffeath [25]. In the physics literature, almost the same model as we discuss here was already studied by simulations in Voss [31]. However, in Voss' paper the A-particles do not perform independent random walks, but the system of A-particles evolves as an exclusion process; moreover, Voss (1984) considers the two-dimensional case. Also, Chayes and Swindle (see [7]) investigated hydrodynamic limits for the one-dimensional case in which the A-particles follow exclusion dynamics. We remark that the particle density in an exclusion process is necessarily at most 1. As we shall see, in our model the case when the particle density μ is less than 1 can be handled much better than the case with $\mu \geq 1$. We have few results in the latter case.

As a side remark we point out that DLA is usually considered in dimension $d > 1$ in which there is a whole new level of difficulty because we do not know how to describe the "shape" of $A(t)$.

Let us now turn to the question about the rate at which $R(t)$ grows. We take $\tau_0 = 0$. As stated we take $R(0) = R(\tau_0) = 0$ and $N(i, 0)$, $i \geq 1$, an i.i.d. sequence of mean μ Poisson random variables. All particles perform independent continuous time simple random walks until they are absorbed by the aggregate. Unless otherwise stated we mean by "simple random walk" a symmetric simple random walk. It is convenient to let the particles continue as a simple random walk even after absorption, by giving the particles also a color, white or black. We start with all particles white, but absorption of the particle by the aggregate is now represented by changing the color of the particle from white to black at the time of its absorption. However, the particle's path is not influenced by its color. After a particle turns black it continues with a continuous time simple random walk path. A black particle has no interaction with any other particle, nor does it influence the motion of $R(\cdot)$. Thus, R is not increased at a time t when a black particle jumps to $R(t)$. In the sequel we shall always use this description of the system with colored particles.

$N(i, t)$ denotes the number of white particles at the space–time point (i, t). We successively define stopping times τ_k and take $R(t) = k$ on the time interval $[\tau_k, \tau_{k+1})$. Moreover, it will follow by induction on k that

$$\text{at time } \tau_k \text{ there are no white particles in } [0, R(\tau_k)] = [0, k]. \tag{31}$$

We take $\tau_0 = \mathbf{0}$. If τ_k and the $N(i, \tau_k)$ have been determined, and $R(\tau_k) = k$ and (31) holds, then we take

$$\tau_{k+1} = \inf\{t > \tau_k : \text{ some white particle jumps to position } R(\tau_k) = k\}. \tag{32}$$

Since the particles perform simple random walk and (31) holds, only white particles at position $k + 1$ at time τ_{k+1}^- can jump to k at time τ_{k+1}. If such a jump occurs, we take $R(\tau_{k+1}) = k + 1$ (i.e.,, $R(\cdot)$ jumps up by 1 at time τ_{k+1}), and we change to black the color of all white particles which were at $R(\tau_k) + 1 = k + 1$ at time $\tau_{k+1}-$ (this includes the particle which jumped to k at τ_{k+1}). It is clear that then (31) with k replaced by $k + 1$ holds, so that we can now define τ_{k+2} etc. It also follows from this description that

$$R(t) = k \text{ for } \tau_k \leq t < \tau_{k+1}. \tag{33}$$

Remark 1. We will not discuss here how our process can be constructed as a Markov process with the strong Markov property. See [22, 23] for details.

Let us now state our results. Throughout $\mathbf{0}$ denotes the origin and $\{S(t)\}_{t \geq 0}$ is a continuous time simple symmetric random walk on $\mathbb{Z}$ with jump rate D. Unless otherwise stated $S(0) = \mathbf{0}$. C_i will denote a constant with value in $(0, \infty)$. Its value may vary from formula to formula. Our first theorem states that for any value of μ, the common expectation of the $N(k, 0)$, it is the case that

$$\limsup_{t \to \infty} \frac{1}{t} R(t) < \infty \text{ a.s.} \tag{34}$$

Theorem 14. *Assume that* $R(0) = \mathbf{0}$ *and that the* $N(i,0), i \geq 1$, *are i.i.d. mean* μ *Poisson variables. Then (34) holds. In fact, there exist constants* $0 < C_i < \infty$ *such that*

$$P\{R(t) > C_1 t\} \leq C_2 \exp[-C_3 t]. \tag{35}$$

Remark 2. Theorem 14 remains valid if the particles perform an asymmetric simple random walk, that is, each jump of the random walk is $+1$ or -1 with probability p_+ and $p_- = 1 - p_+$, respectively. No change in the proof is required for this more general case.

In view of Theorem 14 it is reasonable to conjecture that $\lim_{t\to\infty}(1/t)R(t)$ exists and is constant as One might even assume that this limit is strictly positive, but a quick and quite general argument in the next theorem shows that if $\mu < 1$ "there are not enough particles around" to make $R(t)$ grow linearly with time.

Theorem 15. *Assume that* $\{N(i,0)\}_{i\geq 1}$ *is a stationary ergodic sequence and* $E\{N(i,0)\} = \mu$. *If* $0 < \mu < 1$, *then*

$$\lim_{t\to\infty} \frac{R(t)}{(\log t)^2\sqrt{t}} = 0 \textit{ as} \tag{36}$$

Moreover $R(t)/\sqrt{t},\ t \geq 1$, *is a tight family, i.e.,*

$$P\{R(t) \geq x\sqrt{t}\} \to 0 \textit{ as } x \to \infty, \text{ uniformly in } t \geq 1. \tag{37}$$

If we assume that the initially particles are distributed as i.i.d. Poisson random variables with mean μ, *then (36) can be strengthened to*

$$\limsup_{t\to\infty} \frac{R(t)}{\sqrt{t}} < \infty \textit{ a.s.} \tag{38}$$

Remark 3. One can formulate a d-dimensional analogue of our model and of Theorem 2. In this version one works on $\mathbb{Z}^d$ and at time 0 the aggregate consists of the origin only, while at the site $x \neq \mathbf{0}$ there are $N(x,0)$ particles, with the $N(x,0),\ x \in \mathbb{Z}^d \setminus \{\mathbf{0}\}$ i.i.d. Poisson variables of mean μ. Again all particles perform independent continuous time simple random walks. They all start out as white particles. We denote the aggregate at time t by $A(t)$. If at some time t, a white particle jumps from a site $x \notin A(t-)$ onto the aggregate, then we set $A(t) = A(t-) \cup \{x\}$, and all particles which were at x at time $t-$ are changed to black at time t.

Define an outer radius of the aggregate by:

$$R^{(o,d)} := \sup\{\|x\|_2 : x \in A(t)\},$$

and define an inner radius as:

$$R^{(i,d)}(t) := \inf\{\|x\|_2 : x \notin A(t)\}.$$

The latter is the distance from the origin to the nearest vertex outside $A(t)$. For this model Theorem 14 remains valid. More precisely, (35) and (34) with $R(t)$ replaced by $R^{(o,d)}(t)$ still hold. Theorem 15 has the following analogue: If $\mu < 1$, then

$$\limsup_{t\to\infty} \frac{R^{(i,d)}(t)}{\sqrt{t}} < \infty \text{ a.s.} \tag{39}$$

(Note that (39) is trivially true if there exists a site x_0 which never is occupied by $A(t)$.) We shall not give the proofs of these results here. They are essentially the same as for Theorem 1 and for (38). If we strengthen our assumptions on the

$N(i,0)$, then we can show that in the one-dimensional model $R(t)/\sqrt{t}$ is actually bounded away from 0 in distribution. This holds for all $\mu > 0$.

Theorem 16. *Assume that the $N(i,0), i \geq 1$, are i.i.d. with finite second moment $\mu_2 > 0$. Then for all $\epsilon > 0$ there exists an $\eta = \eta(\epsilon) > 0$ and a $t_0 = t_0(\epsilon)$ such that*

$$P\left\{\frac{R(t)}{\sqrt{t}} > \eta\right\} \geq 1 - \epsilon \quad \text{for all} \quad t \geq t_0. \tag{40}$$

Unfortunately the simple proof of (36) breaks down when $\mu > 1$ and we therefore conjecture that there exists a critical value $\mu_c \geq 1$ such that

$$\lim_{t\to\infty} \frac{1}{t} R(t) \text{ exists and is a.s. a constant which is } \begin{cases} > 0 & \text{if } \mu > \mu_c \\ = 0 & \text{if } \mu < \mu_c. \end{cases} \tag{41}$$

A stronger conjecture would be that

$$\mu_c = 1. \tag{42}$$

Simulations certainly indicate that this is the case, however we have made only little progress towards proving (41), so we pose this as a problem.

Open problem 1. Prove (41), and if this holds, determine μ_c. If one becomes even more ambitious one can ask whether a power laws exist as $\mu \downarrow \mu_c$, and what the critical exponents are. To formulate this problem we have to assume that $\lim_{t\to\infty}(1/t)R(t)$ exists. Let us write $S(\mu)$ for this limit.

Open problem 2. Does

$$\lim_{\mu\downarrow\mu_c} \frac{\log S(\mu)}{\log(\mu - \mu_c))}$$

exist and if so, what is its value?

Fig. 4 DLA clusters with the initial density μ of particles being 0.3, 0.4, 0.6, and 0.68. In the two last cases numerical simulations indicate linear growth

Open problem 3. Does $t^{-1/2}R(t)$ have a limit distribution as $t \to \infty$ when $\mu < 1$?

The obvious approach to proving that $R(t)$ grows linearly in t is to study our system as seen from the right edge of the aggregate. Indeed the collection of positions of the white particles relative to $R(t)$ form a Markov process. Does this Markov process have a non-trivial invariant probability distribution, and if so is the invariant distribution unique? (By non-trivial we mean that we exclude the distribution which puts no particles at all to the right of the aggregate.) On an intuitive level one would like to say that the invariant measure puts at position $R(t) + x$ roughly a Poisson number of particles with mean equal to μ times the probability that a particle at $R(t)+x$ is white. That is, the mean number of particles at $R(t) + x$ should be $\lim_{t\to\infty}\mu\nu(x,t)$, where

$$\nu(x,t) = P\{R(t) + x - S(s) > R(t-s) \text{ for } 0 \le s \le t\}.$$

Actually, all we want to know in first instance is that the density of white particles right in front of $R(t)$ is bounded away from 0 as $t \to \infty$. We want to show that the system does not develop large holes without white particles in front of $R(t)$. To obtain such a result we need some a priori control of $R(t) - R(t-s)$, which we do not know how to control. Tom Kurtz (private communication) showed us that conditionally on the σ-field generated by $\{R(s) : s \le t\}$, the $N(R(t) + x, t)$ have a Poisson distribution with a mean $\mu_x(t)$, and even derived a system of differential equations for the μ_x. Unfortunately this system still involves the unknown random function $R(\cdot)$ in boundary conditions and we have been unable to make use of these differential equations (Fig. 4).

In conclusion we must say that since we were unsuccessful in proving the existence of a non-trivial invariant probability measure for the Markov process of

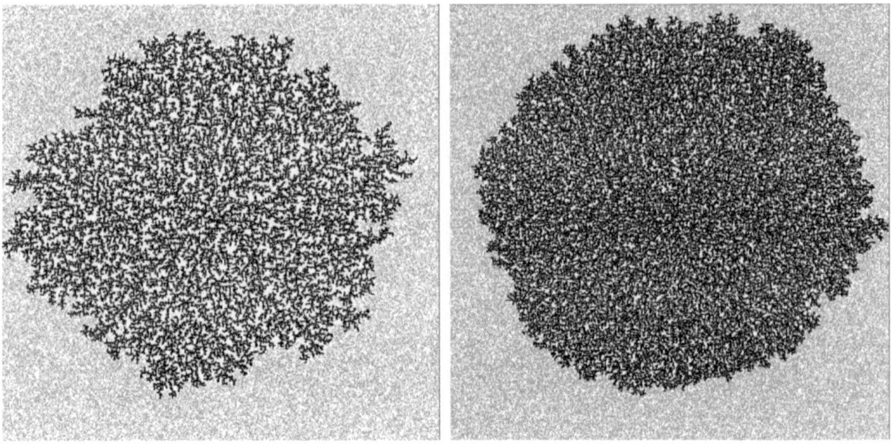

Fig. 5 DLA clusters with the initial density μ of particles being 0.3, 0.4, 0.6, and 0.68. In the two last cases numerical simulations indicate linear growth

the last paragraph, we designed some caricatures of the model. We hope that these caricatures can be regarded as "approximations" to the true model and will help us to treat the true model. These caricatures have some built-in mechanism that makes it more difficult for a large hole to form in front of the aggregate (see [22, 23]).

Open problem. Show that in dimension larger than or equal to two there is linear growth and an asymptotic shape for μ large enough. Actually we conjecture that in dimension $d \geq 2$ the critical density $\mu_c < 1$ (Fig. 5).

Acknowledgements A.F.R and V.S. would like to thank the all organizers of both workshops and in particular to Prof. Wolfgang König for his hospitality. A.F.R. would like to thank the support of FONDECYT grant 1100298.

References

1. Alves, O., Machado, F., Popov, S.: The shape theorem for the frog model Ann. Appl. Probab. **12**, 533–546 (2002)
2. Bérard, J., Ramírez, A.F.: Large deviations of the front in a one dimensional model of $X+Y \to 2X$. In: Bérard, J., Ramírez, A.F. Ann. Probab. **38**(3), 955–1018 (2010)
3. Bramson, M.: Convergence of solutions of the Kolmogorov equation to travelling waves. Mem. Amer. Math. Soc. **44**(285), iv+190 (1983)
4. Bramson, M., Griffeath. D.: On the Williams-Bjerknes Tumour Growth Model II. Math. Proc. Cambridge Philos. Soc. **88**, 339–357 (1980)
5. Comets, F., Quastel, J., Ramírez, A.F.: Fluctuations of the front in a stochastic combustion model , Ann. de l'IHP, Probabilités et Statistiques, Vol. 43. 147–162 (2007)
6. Comets, F., Quastel, J., Ramírez, A.F.: Fluctuations of the Front in a one dimensional model of $X + Y \to 2X$. Trans. Amer. Math. Soc. **361**, 6165–6189 (2009)

7. Chayes, L., Swindle, G.: Hydrodynamic limits for one-dimensional particle systems with moving boundaries. Ann. Probab. **24**(2), 559–598 (1996)
8. Cox, J.T., Durrett, R.: The stepping stone model: new formulas expose old myths, Ann. Appl. Probab. **12**, 1348–1377 (2002)
9. Eden, M.: A two dimensional growth process, in Fourth Berkeley sympos. Math. Statist. Probab. **IV**, 223–239 (1961); In: Neyman, J. (eds.) University of California Press, Berkeley, CA.
10. Dickman, R., Rolla, L.T., Sidoravicius, V.: Activated Random Walkers: Facts, Conjectures and Challanges, J. Stat. Physics. **138**, 126–142 (2010)
11. Garet, O., Marchand, R.: Asymptotic shape for the chemical distance and first-passage percolation in random environment. ESAIM: Probab. Stat. **8**, 169–199 (2004)
12. Gaudilliere, A., Nardi, F.: An upper bound for front propagation velocities inside moving populations. Braz. J. Probab. Stat. **24**, 256–278 (2010)
13. Hammersley, J.M.: Postulates for subadditive processes. Ann. Probab. **2**, 652–680 (1974)
14. Hammersley, J.M., Welsh, D.J.A.: First-passage percolation, subadditive processes, stochastic networks and generalized renewal theory, in Bernoulli, Bayes, Laplace Anniversary Volume. (J. Neyman and L.M. LeCam, eds.), pp. 61–110. Springer, New York (1965)
15. Howard, C.D.: Models of first passage percolation, in Probability on discrete structures (H. Kesten, eds.) pp. 125–173. Springer, New York (2003)
16. Jara, M., Moreno, G., Ramírez, A.F.: Front propagation in an exclusion one-dimensional reactive dynamics. Markov Process. Related Fields **18**, 185–206 (2008)
17. Kesten, H., Sidoravicius, V.: Branching random walk with catalysts, Elec. J. Probab. **8**, paper # 6 (2003)
18. Kesten, H.: Aspects of first passage percolation, in Lecture Notes in Mathematics. vol. 1180, pp. 125–264. Springer, New York (1986)
19. Kesten, H., Sidoravicius, V.: The spread of a rumor or infection in a moving population, Ann. Probab. **33**(6), 2402–2462 (2005)
20. Kesten, H., Sidoravicius, V.: A shape theorem for the spread of an infection, Ann. Math. **167**, 701–766 (2008)
21. Kesten, H., Sidoravicius, V.: A phase transition in a model for the spread of an infection. Illinois J. Math. **50**(3), 547–634 (2006)
22. Kesten, H., Sidoravicius, V.: A problem in one-dimensional diffusion-limited aggregation (DLA) and positive recurrence of Markov chains, Ann. of Probab. **36**(5), 1838–1879 (2008)
23. Kesten, H., Sidoravicius, V.: Positive recurrence of a one-dimensional variant of diffusion limited aggregation. In and out of equilibrium. vol. 2, pp. 429–461. Progr. Probab., **60**, Birkhäuser, Basel (2008)
24. Kingman, J.F.C.: Subadditive processes, in Lecture Notes in Mathematics. vol. 539, pp. 168–223. Springer, New York (1975)
25. Lawler, G.F., Bramson, M., Griffeath, D.: Internal Diffusion Limited Aggregation. Ann. Probab. **20**(4), 2117–2140 (1992)
26. Panja, D.: Effects of fluctuations on propagating fronts, Physics Reports. **393**, 87–174 (2004)
27. Ramírez, A.F., Sidoravicius, V.: Asymptotic behavior of a stochastic combustion growth process, J. Eur. Math. Soc. **6**(3), 293–334 (2004)
28. Richardson, D.: Random growth in a tesselation, Proc. Cambridge Philos. Soc. **74**, 515–528 (1973)
29. Rolla, L., Sidoravicius, V.: Absorbing-state phase transition for stochastic sandpiles and activated random walks. Arxiv:09081152
30. Sznitman, A.S., Zerner, M.: A law of large numbers for random walks in random environment, Ann. Probab. **27**(4), 1851–1869 (1999)
31. Voss, R.F.: Multiparticle fractal aggregation. J. Stat. Phys. **36**(5–6), 861–872 (1984)
32. Wierman, J.C.: The front velocity of the simple epidemic. J. Appl. Probab. **16**, 409–415 (1979)
33. Witten, Jr. T.A., Sander, L.M.: Diffusion-Limited aggregation, a kinetic critical phenomenon. Phys. Rev. Lett. **47**, 1400–1403 (1981)

The Parabolic Anderson Model with Acceleration and Deceleration

Wolfgang König and Sylvia Schmidt

Abstract We describe the large-time moment asymptotics for the parabolic Anderson model where the speed of the diffusion is coupled with time, inducing an acceleration or deceleration. We find a lower critical scale, below which the mass flow gets stuck. On this scale, a new interesting variational problem arises in the description of the asymptotics. Furthermore, we find an upper critical scale above which the potential enters the asymptotics only via some average, but not via its extreme values. We make out altogether five phases, three of which can be described by results that are qualitatively similar to those from the constant-speed parabolic Anderson model in earlier work by various authors. Our proofs consist of adaptations and refinements of their methods, as well as a variational convergence method borrowed from finite elements theory.

MSC 2000: 35K15, 82B44, 60F10, 60K37.

1 Introduction

We consider the solution $u^{(t)}\colon [0,\infty) \times \mathbb{Z}^d \to [0,\infty)$, $t > 0$, to the Cauchy problem for the heat equation with random coefficients and t-dependent diffusion rate,

W. König (✉)
Weierstraß-Institut Berlin, Mohrenstr. 39, 10117 Berlin, Germany
Institut für Mathematik, Technische Universität Berlin, Str. des 17. Juni 136, 10623 Berlin Germany
e-mail: koenig@wias-berlin.de

S. Schmidt
Kompetenzzentrum für klinische Studien, Universität Bremen, Achterstr. 30, 28359 Bremen, Germany
e-mail: sylviaschmidt@math.uni-bremen.de

J.-D. Deuschel et al. (eds.), *Probability in Complex Physical Systems*, Springer Proceedings in Mathematics 11, DOI 10.1007/978-3-642-23811-6_9,

$$\frac{\partial}{\partial s} u^{(t)}(s,z) = \kappa(t)\Delta u^{(t)}(s,z) + \xi(z)u^{(t)}(s,z), \quad s > 0, z \in \mathbb{Z}^d, \tag{1}$$
$$u^{(t)}(0,\cdot) = \mathbb{1}_0,$$

where Δ is the discrete Laplacian,

$$\Delta f(z) = \sum_{x \in \mathbb{Z}^d : |x-z|=1} [f(x) - f(z)],$$

$(\xi(z))_{z\in\mathbb{Z}^d}$ is a field of independent and identically distributed random variables, and $\kappa\colon [0,\infty) \to [0,\infty)$ is a function with $\lim_{t\to\infty} t\kappa(t) = \infty$. Our main goal is to understand the asymptotic behavior as $t \to \infty$ of the expected total mass at time t,

$$U(t) = \sum_{z\in\mathbb{Z}^d} u^{(t)}(t,z).$$

The total mass may be represented in terms of the famous *Feynman–Kac formula*,

$$U(t) = \mathbb{E}_0^{(t)}\Big[\exp\Big\{\int_0^t \xi(X_s)\,\mathrm{d}s\Big\}\Big], \tag{2}$$

where $(X_s)_{s\in[0,\infty)}$ is a random walk with generator $\kappa(t)\Delta$, starting from zero under $\mathbb{E}_0^{(t)}$. Denoting by $\langle\cdot\rangle$ the expectation with respect to the random potential ξ, we will study the logarithmic asymptotics of $\langle U(t)\rangle$ for various choices of the diffusion function $t \mapsto \kappa(t)$.

The model with constant diffusion rate $\kappa(t) \equiv 1$ has been analyzed in [13] and [3] for three important classes of tail distributions of $\xi(0)$, see also [11] for a survey, and [6] for more background. In [15] a classification of all potential distributions into four universality classes was made out such that the qualitative behavior of $\langle U(t)\rangle$ in each of the classes is similar. This classification holds under mild regularity assumptions and depends only on the upper tails of the potential. Heuristically, the main effect in each of these classes is the concentration of the total mass on a so-called intermittent island the size of which is t-dependent and deterministic. The (rescaled) shape of the solution and the potential on this island can be described by a deterministic variational formula. The thinner the tails of the potential distribution are, the larger the islands are, ranging from single sites to large areas, however, still having a radius $\ll t^{1/d}$.

In (1), the diffusion is coupled with time so that it is accelerated if the diffusion function $t \mapsto \kappa(t)$ grows or decelerated if it decreases. Now an interesting competition between the speed of the diffusion and the thickness of the tails of the potential distribution arises: the faster $\kappa(t)$ is, the stronger the flattening effect of the diffusion term is. One rightfully expects that if the speed of this function is not too extreme, then similar formulas should be valid as for constant diffusion rate. Indeed, we will identify a lower critical scale for $\kappa(t)$, which depends on the upper tails of the potential distribution, and marks the threshold below which the mass does not flow unboundedly far away from the origin in the Feynman–Kac formula, see below Assumption 2.1. Then we are in the case of [13]. Furthermore, we will see that – if $\kappa(t)$ is above this lower critical scale – $t^{2/d}$ presents an upper critical scale in the sense that, for $\kappa(t) \ll t^{2/d}$, the main contribution to the total mass comes from extremely high potential values, while for $\kappa(t) \approx t^{2/d}$, it comes from just super-average, but not extreme, values. This is reflected by the fact that the asymptotics can be described in terms of the upper tails of the potential distribution in the former case (then we find the formulas derived in [3] and [15]), but all the details of this distribution are required in the latter. (If the speed is even faster, then, conjecturally, only a rough mean behavior of the potential values will influence the asymptotics.)

The paper is organized as follows. In Sect. 2, we formulate our assumptions on the potential and on the function κ. Then we state our results for the moment asymptotics of $U(t)$ in Sect. 3. Our main result will be the identification of five phases with qualitatively different behavior, which we will describe informally in Sect. 3.1 and rigourously in Sect. 3.2 (for four of them). We will also give a proposition concerning the convergence of a discrete variational formula to the corresponding continuous version, representing one of the main tools used in the proof of the asymptotics. In Sects. 4–6, we give sketches of the proofs of this proposition and of the theorems. The details are rather lengthy and involved; they may be found in the second author's thesis [17].

2 Assumptions and Preliminaries

2.1 Model Assumptions

Let

$$H(t) = \log\langle e^{t\xi(0)}\rangle, \quad t > 0, \tag{3}$$

be the logarithmic moment generating function of $\xi(0)$. We assume $H(t) < \infty$ for all $t > 0$, which is sufficient for the existence of a nonnegative solution of (1) and the finiteness of all its positive moments [12]. Now we recall the discussion on regularity assumptions in [15, Sect. 1.2]. If we assume that $t \mapsto H(t)/t$ is in the de Haan class, then the theory of regularly varying functions provides us with an asymptotic description of H that depends only on two parameters γ and ρ, see [2] and [15, Proposition 1.1]. This leads to the following assumption which will be in force throughout the rest of this paper.

Assumption 2.1. *There exist parameters $\gamma \geq 0$ and $\rho > 0$ and a continuous function $K_H\colon (0,\infty) \to (0,\infty)$, regularly varying with parameter γ, such that, locally uniformly in $y \in [0,\infty)$,*

$$\lim_{t\to\infty} \frac{H(ty) - yH(t)}{K_H(t)} = \rho\widehat{H}(y), \tag{4}$$

where

$$\widehat{H}(y) = \begin{cases} y \log y & \text{if } \gamma = 1, \\ \dfrac{y - y^{\gamma}}{1-\gamma} & \text{if } \gamma \neq 1. \end{cases} \tag{5}$$

The scale function K_H roughly describes the thickness of the potential tails at infinity. As we will see later, the function $t \mapsto K_H(t)/t$ presents a lower critical scale for the diffusion function $\kappa(t)$. The following lemma is a consequence of [2, Theorem 3.6.6].

Lemma 2.2. *Let Assumption* 2.1 *hold.*

(a) *If* $\operatorname{ess\,sup} \xi(0) \in \{0, \infty\}$, *then* H *is regularly varying with index* γ.
(b) *If* $\langle \xi(0) \rangle = 0$, *then* H *is regularly varying with index* $\gamma \vee 1$.

Now we formulate some mild regularity assumptions on the speed function κ.

Assumption 2.3. *The following limits exist:*

$$\lim_{t\to\infty} t\kappa(t) = \infty, \qquad \lim_{t\to\infty} \frac{t\kappa(t)}{K_H(t)} \in [0,\infty], \qquad \lim_{t\to\infty} \frac{\kappa(t)}{t^{2/d}} \in [0,\infty].$$

We also need a scale function $\alpha\colon [0,\infty) \to [0,\infty)$, which will be interpreted as the order of the radius of the relevant island. While we can define $\alpha = 1$ in the results for Phases 1 and 2 of our classification, we will need the following fixed point equation in Phase 3:

$$K_H\Big(\frac{t}{\alpha_t^d}\Big) = \frac{t\kappa(t)}{\alpha_t^{d+2}}. \tag{6}$$

Let us state existence and some important properties of a solution of (6).

Lemma 2.4. *Let* $\kappa(t)$ *be regularly varying with index* $\beta \in (\gamma - 1, 2/d)$. *Then there exists a regularly varying function* α *such that* (6) *holds for all large* t. *Any solution* $\alpha(t) = \alpha_t$ *satisfies* $\lim_{t\to\infty} \alpha_t = \infty$. *Furthermore,* $t/\alpha_t^d \gg 1$ *and* $\alpha_t^x \ll t\kappa(t)$ *for each* $x < d+2$.

Proof. Similar to the proof of [15, Proposition 1.2]. For details, see [17, Lemma 2.1.5]. □

From the assumptions of Theorem 3.1(c) below, we will see that the interval for the index of regular variation for κ is not a hard restriction in Phase 3.

2.2 Variational Formulas

The following variational formulas will play a role in our results. Here, $\mathrm{H}^1(\mathbb{R}^d)$ is the Sobolev space on $\mathbb{R}^d$ and $\mathcal{M}_1(\mathbb{Z}^d)$ is the space of probability measures on $\mathbb{Z}^d$. The inner product on $l^2(\mathbb{Z}^d)$ is denoted by $(\cdot,\cdot)$. All integrals are with respect to

Lebesgue measure. We always have $\rho, \theta > 0$, and $\gamma \in [0,\infty)\setminus\{1\}$.

$$\chi_\gamma^{(\mathrm{B})}(\rho) = \inf_{\substack{g\in \mathrm{H}^1(\mathbb{R}^d)\\ \|g\|_2=1}} \left\{ \int_{\mathbb{R}^d} |\nabla g|^2 + \frac{\rho}{1-\gamma}\int_{\mathbb{R}^d} (g^{2\gamma} - g^2) \right\}, \tag{7}$$

$$\chi^{(\mathrm{AB})}(\rho) = \inf_{\substack{g\in \mathrm{H}^1(\mathbb{R}^d)\\ \|g\|_2=1}} \left\{ \int_{\mathbb{R}^d} |\nabla g|^2 - \rho\int_{\mathbb{R}^d} g^2\log g^2 \right\}, \tag{8}$$

$$\chi^{(\mathrm{DE})}(\rho) = \inf_{p\in\mathcal{M}_1(\mathbb{Z}^d)} \left\{ -\big(\Delta\sqrt{p},\sqrt{p}\big) - \rho\big(p,\log p\big) \right\}, \tag{9}$$

$$\chi_\gamma^{(\mathrm{DB})}(\rho) = \inf_{p\in\mathcal{M}_1(\mathbb{Z}^d)} \left\{ -\big(\Delta\sqrt{p},\sqrt{p}\big) + \frac{\rho}{1-\gamma}\big(p^\gamma - p, 1\big) \right\}, \tag{10}$$

$$\chi_H^{(\mathrm{RWRS})}(\theta) = \inf_{\substack{g\in \mathrm{H}^1(\mathbb{R}^d)\\ \|g\|_2=1}} \left\{ \int_{\mathbb{R}^d} |\nabla g|^2 - \theta\int_{\mathbb{R}^d} H\circ g^2 \right\}. \tag{11}$$

If $\gamma = 0$, then we use the interpretation $\int_{\mathbb{R}^d} g^{2\gamma} = |\mathrm{supp}\, g|$ and $(p^\gamma, 1) = |\mathrm{supp}\, p|$. We sometimes refer to the formulas that are defined in $\mathbb{R}^d$ (i.e., $\chi_\gamma^{(\mathrm{B})}$, $\chi^{(\mathrm{AB})}$, and $\chi_H^{(\mathrm{RWRS})}$) as to "continuous" formulas and to the others as to the "discrete" ones. Clearly, $\chi_\gamma^{(\mathrm{B})}$ and $\chi^{(\mathrm{AB})}$ are the continuous variants of $\chi_\gamma^{(\mathrm{DB})}$ and $\chi^{(\mathrm{DE})}$, respectively. Note that $\chi_\gamma^{(\mathrm{B})}$ is degenerate in the case $\gamma > 1 + 2/d$ (which we do not consider here).

The formulas $\chi^{(\mathrm{DE})}$, $\chi^{(\mathrm{AB})}$, and $\chi_\gamma^{(\mathrm{B})}$ are already known from the study of the parabolic Anderson model for constant diffusion $\kappa(t) \equiv 1$ in three universality classes, see the summary in [15]. Our notation refers to the names of these classes introduced there: "DE" for "double-exponential," "AB" for "almost bounded," and "B" for "bounded." Informally, the functions g^2 and p, respectively, in the formulas have the interpretation of the shape (up to possible rescaling and vertical shifting) of those realizations of the solution $u^{(t)}(t,\cdot)$ that give the overwhelming contribution to the expected total mass $\langle U(t)\rangle$. If the total mass comes from an unboundedly growing island, then a rescaling is necessary, and a continuous formula arises, otherwise a discrete one.

In [16] the existence, uniqueness (up to shift), and some characterizations of the minimizer of $\chi_\gamma^{(\mathrm{B})}$ are shown for $\gamma < 1$, in [15] it is shown that the only minimizer of $\chi^{(\mathrm{AB})}$ is an explicit Gaussian function, and in [13] and [10], the minimizers of $\chi^{(\mathrm{DE})}(\rho)$ are analyzed, which are unique (up to shifts) for any sufficiently large ρ. Formula $\chi_H^{(\mathrm{RWRS})}$ is a rescaling of the Legendre transform of a variational formula which appeared in the study of large deviations for the random walk in random scenery in [9], see (36). Its properties have not been analyzed yet.

However, formula $\chi_\gamma^{(\mathrm{DB})}$ ("DB" refers to "discrete bounded") appears in the study of the parabolic Anderson model for the first time in the present paper. Here are some of its properties.

Proposition 2.5. (a) *For any* $\rho > 0$ *and any* $\gamma \neq 1$ *with* $0 \leq \gamma < \max\{1 + 1/d, 1 + \rho/(2d)\}$, *there exists a minimizer for* $\chi_\gamma^{(\mathrm{DB})}(\rho)$.

(b) *Let* p *be a minimizer for* $\chi_\gamma^{(\mathrm{DB})}(\rho)$. *Then* supp p *is finite if and only if* $\gamma \leq 1/2$. *In the case* $\gamma > 1/2$ *the support of* p *is the whole lattice.*

Proof. See [17, Proposition 2.1.8]. This uses ideas from [14, Lemma 3.2] for the existence and from [10, p. 44] for the size of the support. □

Similarly to the continuous analogue in [15, Proposition 1.16], it is possible to show that $\lim_{\gamma\to 1} \chi_\gamma^{(\mathrm{DB})}(\rho) = \chi^{(\mathrm{DE})}(\rho)$, furthermore we have $\lim_{\rho\to\infty} \chi_\gamma^{(\mathrm{DB})}(\rho) = 2d$.

3 Results

In what follows, we will use the notation $f_t \gg g_t$ if $\lim_{t\to\infty} f_t/g_t = \infty$ and $f_t \asymp g_t$ if $\lim_{t\to\infty} f_t/g_t$ exists in $(0,\infty)$. We will always work under the assumptions made in Sect. 2.1.

3.1 *Five Phases*

Depending on the ratio between the speed $\kappa(t)$ and the critical scales $K_H(t)/t$ and $t^{2/d}$, we make out up to five phases. In the following, we resume heuristically our results for these phases. Recall the Feynman–Kac formula in (2).

Phase 1. $\kappa(t) \ll K_H(t)/t$.
The mass stays in the origin, where the potential takes on its highest value. The expected total mass behaves therefore like $\langle U(t)\rangle \approx \langle u^{(t)}(t,0)\rangle \approx \exp(H(t) - 2dt\kappa(t))$. This includes the single-peak case of [13].

Phase 2. $\kappa(t) \asymp K_H(t)/t$.
The radius of the intermittent island remains bounded in time, and consequently the moment asymptotics are given in terms of a discrete variational formula. Denoting $\kappa_* = \lim_{t\to\infty} t\kappa(t)/K_H(t)$,

$$\lim_{t\to\infty} \frac{1}{t\kappa(t)} \log\langle U(t)\mathrm{e}^{-H(t)}\rangle = -\begin{cases} \chi^{(\mathrm{DE})}(\rho/\kappa_*) & \text{if } \gamma = 1, \\ \chi_\gamma^{(\mathrm{DB})}(\rho/\kappa_*) & \text{if } \gamma \neq 1. \end{cases} \tag{12}$$

While the case $\gamma = 1$ is qualitatively the same as the case of the double-exponential distribution analyzed in [13], the case $\gamma \neq 1$ shows a new effect that was not present for constant diffusion speed $\kappa(t) \equiv 1$. The diffusion is decelerated so strongly that the mass moves only by a bounded amount.

Phase 3. $K_H(t)/t \ll \kappa(t) \ll t^{2/d}$.
The relation between acceleration/deceleration and thickness of potential tails is so strong that the mass flows an unbounded amount of order α_t defined by (6). Since the acceleration is not too strong, the total mass comes from sites of extremely high potential values. Therefore, we get the continuous analogue to (12), but on scale $t\kappa(t)/\alpha_t^2$,

$$\lim_{t\to\infty} \frac{\alpha_t^2}{t\kappa(t)} \log\langle U(t)\exp(-\alpha_t^d H(t\alpha_t^{-d}))\rangle = -\begin{cases} \chi^{(\mathrm{AB})}(\rho) & \text{if } \gamma = 1, \\ \chi_\gamma^{(\mathrm{B})}(\rho) & \text{if } \gamma \neq 1. \end{cases} \tag{13}$$

Hence, for $\gamma = 1$ we are in the almost-bounded case [15] and for $\gamma < 1$ in the bounded case [3]. Note that we can have all $\gamma \in [0, 1 + 2/d)$ here, whereas only the values $\gamma \in [0, 1]$ were allowed in this regime of the constant speed parabolic Anderson model.

Phase 4. $K_H(t)/t \ll \kappa(t) \asymp t^{2/d}$.
As in Phase 3, the mass flows an unbounded distance away from the origin. The acceleration reaches the critical level, such that this distance is of order $t^{1/d}$, which is much larger than in Phase 3. Only so little mass reaches the sites in this large island that the potential is not extremely large here, but only by a bounded amount larger than the mean. Therefore, the characteristic variational formula does not only depend on the tails of the distribution, but on all values of the logarithmic moment generating function H. This regime has strong connections to the large deviation result for a random walk in random scenery model as described in [9].

Phase 5. $\kappa(t) \gg K_H(t)/t$ and $\kappa(t) \gg t^{2/d}$.
The speed is so high that, conjecturally, the values of the potential influence the expected total mass only via their mean, and the diffusion behaves like free Brownian motion with some diffusion constant that depends on the potential distribution. We will not present rigorous results for this phase in the present paper.

Note that, because of regular variation, $K_H(t) = t^{\gamma + o(1)}$. Hence, Phases 3 and 4 can only appear if we have $\gamma \leq 1 + 2/d$. The four universality classes for the constant-diffusion case $\kappa(t) \equiv 1$ are found in Phases 1–3 depending on whether $\gamma = 1$ or $\gamma \neq 1$.

3.2 *Moment Asymptotics*

We now formulate our results. Recall the variational formulas defined in the Sect. 2.2 and set

$$\chi_\gamma^{\mathrm{d}} = \begin{cases} \chi^{(\mathrm{DE})} & \text{if } \gamma = 1, \\ \chi_\gamma^{(\mathrm{DB})} & \text{if } \gamma \neq 1, \end{cases} \qquad \text{and} \qquad \chi_\gamma^{\mathrm{c}} = \begin{cases} \chi^{(\mathrm{AB})} & \text{if } \gamma = 1, \\ \chi_\gamma^{(\mathrm{B})} & \text{if } \gamma \neq 1. \end{cases}$$

Then we have the following result for the first three regimes of our model.

Theorem 3.1 (Phase 1–Phase 3). *Assume* ess sup $\xi(0) \in \{0, \infty\}$.

(a) *If* $\kappa(t) \ll K_H(t)/t$, *then we have for* $t \to \infty$

$$\langle U(t)\rangle = \exp\big(H(t) - 2\mathrm{d}t\kappa(t)(1 + o(1))\big). \tag{14}$$

(b) *If* $\kappa(t) \asymp K_H(t)/t$, *then*

$$\langle U(t)\rangle = \exp\Big(H(t) - t\kappa(t)\chi^{\mathrm{d}}_{\gamma}\Big(\frac{\rho}{\kappa_*}\Big)(1 + o(1))\Big) \tag{15}$$

with $\kappa_* = \lim_{t\to\infty} t\kappa(t)/K_H(t) \in (0, \infty)$.

(c) *Let the assumption of Lemma 2.4 hold, in particular we have* $K_H(t)/t \ll \kappa(t) \ll t^{2/d}$. *Furthermore suppose* $K_H(t) \gg \log t$ *and* $\gamma < 2$. *Then*

$$\langle U(t)\rangle = \exp\Big(\alpha_t^d H\Big(\frac{t}{\alpha_t^d}\Big) - \frac{t\kappa(t)}{\alpha_t^2}\chi^{\mathrm{c}}_{\gamma}(\rho)(1 + o(1))\Big). \tag{16}$$

Note that the assumption ess sup $\xi(0) \in \{0, \infty\}$ is not restrictive, since a shift of the potential would only lead to an additive constant in our results. The assumptions $K_H(t) \gg \log t$ and $\gamma < 2$ in part (c) of the theorem are purely technical, the first one only needed in the case $\gamma = 0$. Since $\gamma < 1 + 2/d$ in the respective phase (which follows from the assumption of Lemma 2.4), $\gamma < 2$ is only a restriction in dimension 1.

Now we come to Phase 4, where we will meet the variational formula $\chi_H^{\mathrm{(RWRS)}}(\theta)$ defined in (11). Since the result will no longer depend on the upper tails of the potential distribution, it will make sense to have an assumption for the expectation of $\xi(0)$ instead of its essential supremum. Again, this is no loss of generality.

Theorem 3.2 (Phase 4). *Assume* $\langle\xi(0)\rangle = 0$ *and* $K_H(t)/t \ll \kappa(t) \asymp t^{2/d}$. *Let* $\gamma \in [0, 1 + 2/d)$, $\gamma < 2$. *Then we have for* $t \to \infty$

$$\langle U(t)\rangle = \exp\Big(-t\kappa^*\chi_H^{\mathrm{(RWRS)}}\Big(\frac{1}{\kappa^*}\Big)(1 + o(1))\Big) \tag{17}$$

with $\kappa^* = \lim_{t\to\infty} \kappa(t)/t^{2/d} \in (0, \infty)$.

3.3 Variational Convergence

We now state a result, which is both important in the proof of Theorem 3.1(c) and of independent interest as a connection between the discrete variational formula $\chi^{\mathrm{d}}_{\gamma}(\rho)$ and its continuous analogue $\chi^{\mathrm{c}}_{\gamma}(\rho)$. In the case $\gamma = 1$, this fact is stated in [15] and is derived without difficulties from an explicit representation of $\chi^{\mathrm{(AB)}}(\rho)$. The proof for the case $\gamma \neq 1$ is much more involved and uses techniques from the theory of finite elements.

Proposition 3.3. *Let $\rho > 0$. As $\kappa \to \infty$, we have*

$$\kappa\chi^{(\mathrm{DE})}\Big(\frac{\rho}{\kappa}\Big) = \chi^{(\mathrm{AB})}(\rho) + \rho\frac{d}{2}\log\kappa + o(1) \tag{18}$$

and for $\gamma \in [0, 1 + 2/d) \setminus \{1\}$

$$\kappa^{1-d\nu}\chi_\gamma^{(\mathrm{DB})}\Big(\frac{\rho}{\kappa}\Big) = \chi_\gamma^{(\mathrm{B})}(\rho) + \rho\frac{1-\kappa^{-d\nu}}{1-\gamma} + o(1) \tag{19}$$

with $\nu = \frac{1-\gamma}{2+d(1-\gamma)}$.

Note that (18) and (19) are consistent, as (18) is a continuous continuation of (19) to $\gamma = 1$. Proposition 3.3 shows that Phases 2 and 3 can be continuously transformed into each other, i.e., the transition between them is actually not a phase transition in the sense of statistical mechanics.

4 Proof of Variational Convergence (Proposition 3.3)

The asymptotics (18) follows from the arguments in [15, p. 313]. To show (19), we remark first that the summand $\frac{\rho}{1-\gamma}$ drops out in both (7) and (10). Therefore (19) is equivalent to

$$\lim_{\kappa\to\infty} \kappa^{1-d\nu} \inf_{p\in\mathcal{M}_1(\mathbb{Z}^d)} \Big\{-(\Delta\sqrt{p}, \sqrt{p}) + \frac{\rho}{\kappa(1-\gamma)}\sum_{z\in\mathbb{Z}^d} p(z)^\gamma\Big\} = \hat{\chi}_\gamma(\rho), \tag{20}$$

where

$$\hat{\chi}_\gamma(\rho) = \inf_{\substack{g\in \mathrm{H}^1(\mathbb{R}^d)\\ \|g\|_2=1}} \Big\{\int_{\mathbb{R}^d} |\nabla g|^2 + \frac{\rho}{1-\gamma}\int_{\mathbb{R}^d} g^{2\gamma}\Big\}.$$

The proof of the upper bound of (20) is standard and we will here only give the idea. To an approximate minimizer g for the infimum in $\hat{\chi}_\gamma(\rho)$ and for small $\varepsilon > 0$, we define a probability measure p_ε by

$$p_\varepsilon(z) = \int_{\varepsilon z+[0,\varepsilon)^d} g(x)^2\,\mathrm{d}x, \quad z \in \mathbb{Z}^d.$$

Assuming that g is smooth and compactly supported, we can make use of Taylor expansions to see that, as $\varepsilon \downarrow 0$,

$$-\varepsilon^{-2}(\Delta\sqrt{p_\varepsilon}, \sqrt{p_\varepsilon}) \to \int_{\mathbb{R}^d} |\nabla g|^2 \quad \text{and} \quad \varepsilon^{d(1-\gamma)}\sum_{z\in\mathbb{Z}^d} p_\varepsilon(z)^\gamma \to \int_{\mathbb{R}^d} g^{2\gamma}.$$

Recall $\gamma < 1 + 2/d$. Putting $\varepsilon = \kappa^{-(1-d\nu)/2} = \kappa^{-1/(2+d(1-\gamma))} \downarrow 0$ as $\kappa \to \infty$, this shows the upper bound.

Let us now turn to the lower bound. This proof is pretty involved and comes in several steps. The principal idea and main arguments are taken from [15, Proof of (5.3)]. However, we could not find an argument for the L^2-normalization of the limit function in their approximation approach, since this involves interchanging integral and limit, which seems to be hard to justify. Hence, we use a different construction. Furthermore, our consideration of $\gamma > 1$ causes some additional difficulties.

We will only treat the case $\gamma > 1$. The structure for $\gamma < 1$ is similar, for details we refer to the proofs of [17, Propositions 3.4.7 and 5.2.1]. We denote $S(p) = -(\Delta\sqrt{p}, \sqrt{p})$.

Step 1. We choose minimizing sequences $\kappa_n \to \infty$ and $(p_n)_n$ from $\mathcal{M}_1(\mathbb{Z}^d)$ for the left hand side of (20). Put $a_n = \kappa_n^{(1-d\nu)/2}$. We now argue that we can assume, without loss of generality, that

$$\sup_{n\in\mathbb{N}} a_n^2 S(p_n) < \infty. \tag{21}$$

For this, we need the following discrete Sobolev inequality.

Lemma 4.1. *Let $\gamma > 1$ with $\gamma(d-2) < d$. There exists a constant $c = c_{d,\gamma}$ such that for all $p \in \mathcal{M}_1(\mathbb{Z}^d)$*

$$\sum_{z\in\mathbb{Z}^d} p(z)^\gamma \le c S(p)^{d(\gamma-1)/2}.$$

Proof. See [17, Lemma 3.2.10] for details; the idea is taken from [7]. □

Now suppose that (21) does not hold. Then, by Lemma 4.1 and because of $d(\gamma-1)/2 < 1$,

$$\lim_{n\to\infty} a_n^2 \Big\{ S(p_n) + \frac{\rho}{a_n^{2+d(1-\gamma)}(1-\gamma)} \sum_{z\in\mathbb{Z}^d} p_n(z)^\gamma \Big\}$$
$$\ge \limsup_{n\to\infty} \Big\{ a_n^2 S(p_n) - \frac{c\rho}{\gamma-1}\big(a_n^2 S(p_n)\big)^{d(\gamma-1)/2} \Big\} = \infty.$$

Since $(p_n)_n$ is a minimizing sequence, the lower bound would now be trivially satisfied. Hence, we can assume (21).

Step 2. We compactify on a box $B_{Ra_n} = [-Ra_n, Ra_n]^d \cap \mathbb{Z}^d$ for $R > 0$. Consider the periodized probability measures

$$p_n^R(z) = \sum_{k\in(2Ra_n+1)\mathbb{Z}^d} p_n(z+k), \quad z \in B_{Ra_n}.$$

In [13, Lemma 1.10], it was shown that $S^{\pi,R}(p_n^R) \le S(p_n)$ in the one-dimensional case, where $S^{\pi,R}$ is the Dirichlet form with periodic boundary condition. This holds as well in higher dimensions, besides we have $\frac{1}{1-\gamma}\sum_{z\in B_{Ra_n}} p_n^R(z)^\gamma \le \frac{1}{1-\gamma}\sum_{z\in\mathbb{Z}^d} p_n(z)^\gamma$ by subadditivity. Therefore it will be sufficient to prove that

$$\liminf_{R\to\infty}\liminf_{n\to\infty} a_n^2\Big\{S^{\pi,Ra_n}\big(p_n^R\big) + \frac{\rho}{1-\gamma}a_n^{-2-d(1-\gamma)}\sum_{z\in B_{Ra_n}}\big(p_n^R(z)\big)^\gamma\Big\} \ge \hat{\chi}_\gamma(\rho). \tag{22}$$

Since $S^{\pi,R}(p_n^R) \le S(p_n)$, (21) implies

$$\sup_{n\in\mathbb{N}} a_n^2 S^{\pi,R}(p_n^R) < \infty. \tag{23}$$

Step 3. Our goal is to construct potential minimizers for $\hat{\chi}_\gamma(\rho)$ that interpolate the values of the rescaled step functions $h_n(x) = \sqrt{a_n^d p_n^R(\lfloor a_n x\rfloor)}$ on the lattice $\{x = z/a_n : z \in B_{Ra_n}\}$. In the present step, we define piecewise linear interpolations $g_n \in \mathrm{H}^1(Q_R^{(n)})$ with $Q_R^{(n)} = [-R, R + a_n^{-1})^d$, which we will slightly modify in Step 4 in order to obtain normalized $\mathrm{H}^1(\mathbb{R}^d)$-functions.

We borrow a technique from finite elements theory, see e.g., [4]. Consider the triangulation

$$Q_R^{(n)} = \bigcup_{z\in B_{Ra_n}}\bigcup_{\sigma\in\mathfrak{S}_d} T_\sigma(z),$$

where $\mathfrak{S}_d$ is the set of permutations of $1,\dots,d$ and $T_\sigma(z)$ is the d-dimensional tetrahedron defined as the convex hull of the points $a_n^{-1}z, a_n^{-1}(z+e_{\sigma(1)}),\dots, a_n^{-1}(z+e_{\sigma(1)}+\dots+e_{\sigma(d)})$, where e_i is the ith unit vector in $\mathbb{R}^d$. Note that the tetrahedra are disjoint up to the boundary. On each tetrahedron $T_\sigma(z)$, we define a function

$$g_{n,z,\sigma}(x) = b_{n,z,\sigma}^{(0)} + \sum_{k=1}^d b_{n,z,\sigma}^{(k)}(a_n x_{\sigma(k)} - z_{\sigma(k)}), \qquad x = (x_1,\dots,x_d) \in T_\sigma(z),$$

where the coefficients are given by

$$b_{n,z,\sigma}^{(0)} = \sqrt{a_n^d p_n^R(z)} = h_n\Big(\frac{z}{a_n}\Big),$$

$$b_{n,z,\sigma}^{(k)} = \sqrt{a_n^d p_n^R(z+e_{\sigma(1)}+\dots+e_{\sigma(k)})} - \sqrt{a_n^d p_n^R(z+e_{\sigma(1)}+\dots+e_{\sigma(k-1)})}$$

for $k = 1,\dots,d$, where p_n^R is continued periodically outside B_{Ra_n}. Then $g_{n,z,\sigma}$ satisfies

$$g_{n,z,\sigma}(x) = h_n(x) \quad \text{for all } x \in T_\sigma(z) \text{ such that } a_n x \in \mathbb{Z}^d.$$

The values of all functions $g_{n,z,\sigma}$ on the common borders of their respective tetrahedra coincide; see [1, Proof of Lemma 2.1] for a detailed argument. Hence, the function $g_n\colon Q_R^{(n)} \to \mathbb{R}$ given by

$$g_n(x) = g_{n,z,\sigma}(x) \quad \text{if } x \in T_\sigma(z)$$

is well-defined and continuous, and $g_n \in \mathrm{H}^1(Q_R^{(n)})$.

A direct calculation for the gradient gives $\partial_{x_{\sigma(k)}} g_n(x) = a_n b_{n,z,\sigma}^{(k)}$ and thus

$$\int_{Q_R^{(n)}} |\nabla g_n|^2 = a_n^2 S^{\pi,R}(p_n^R). \tag{24}$$

Note that by (23) this is bounded in n. Now consider the L^2-norm of g. Because of $|a_n x_{\sigma(k)} - z_{\sigma(k)}| \le 1$ and $b_{n,z,\sigma}^{(k)} = a_n^{-1} \partial_{x_{\sigma(k)}} g_n(x)$ we obtain

$$\|(g_n - h_n)\mathbb{1}_{Q_R^{(n)}}\|_2^2 \le a_n^{-2} \int_{Q_R^{(n)}} \left(\sum_{i=1}^d \frac{\partial}{\partial x_i} g_n(x) \right)^2 \mathrm{d}x.$$

By Jensen's inequality, $(\sum_{i=1}^d c_i)^2 \le d \sum_{i=1}^d c_i^2$. Since $\|h_n \mathbb{1}_{Q_R^{(n)}}\|_2 = 1$, the triangle inequality gives

$$\big|\|g_n \mathbb{1}_{Q_R^{(n)}}\|_2 - 1\big|^2 \le d a_n^{-2} \int_{Q_R^{(n)}} |\nabla g_n|^2, \tag{25}$$

which tends to zero as $n \to \infty$ by (24) and (23).

A similar calculation for the $\mathrm{L}^{2\gamma}$-norm results in

$$\|(g_n - h_n)\mathbb{1}_{Q_R^{(n)}}\|_{2\gamma}^{2\gamma} \le d^\gamma a_n^{-2\gamma} \int_{Q_R^{(n)}} |\nabla g_n|^{2\gamma}.$$

Because of $p_n^R(z) \in [0,1]$, we have $|b_{n,z,\sigma}^{(k)}| \le a_n^{d/2}$ and therefore $|\nabla g_n|^2 \le d a_n^{d+2}$. For $\gamma > 1$, this yields

$$|\nabla g_n(x)|^{2\gamma} = d^\gamma a_n^{(d+2)\gamma} \left(\frac{|\nabla g_n(x)|^2}{d a_n^{d+2}} \right)^\gamma \le d^{\gamma-1} a_n^{2\gamma} a_n^{-2-d(1-\gamma)} |\nabla g_n(x)|^2.$$

Now use triangle inequality to get

$$a_n^{d(\gamma-1)} \sum_{z \in B_{Ra_n}} (p_n^R(z))^\gamma = \|h_n \mathbb{1}_{Q_R^{(n)}}\|_{2\gamma}^{2\gamma} \le \left(\|g_n \mathbb{1}_{Q_R^{(n)}}\|_{2\gamma} + c_n a_n^{\frac{-2-d(1-\gamma)}{2\gamma}} \right)^{2\gamma}, \tag{26}$$

where $c_n = (d^{2\gamma-1} \int_{Q_R^{(n)}} |\nabla g_n|^2)^{1/(2\gamma)}$ is bounded in n.

Step 4. In order to adapt our function g_n to zero boundary conditions, we introduce a cut off function $\Psi_R(x) = \prod_{i=1}^d \psi_R(x_i)$, $x = (x_1, \dots, x_d) \in \mathbb{R}^d$, where $\psi_R = 1$ on $[-R+\sqrt{R}, R-\sqrt{R}]$, $\psi_R = 0$ on $\mathbb{R} \setminus [-R, R]$ and it interpolates linearly in-between. Then $0 \le \psi_R \le 1$ and $|\psi_R'| \le 1/\sqrt{R}$. Let us estimate the relevant terms for the $\mathrm{H}^1(\mathbb{R}^d)$-function $g_n\Psi_R$ (which is zero outside $Q_R = [-R, R]^d$). As for the gradient,

$$\int_{\mathbb{R}^d} \Big(\frac{\partial}{\partial x_i}(g_n\Psi_R)(x)\Big)^2 \,\mathrm{d}x \le \int_{Q_R} \Big(\frac{\partial}{\partial x_i} g_n(x)\Big)^2 \,\mathrm{d}x + \frac{1}{R}\int_{Q_R} g_n(x)^2 \,\mathrm{d}x + \frac{2}{\sqrt{R}}\sqrt{\int_{Q_R} \Big(\frac{\partial}{\partial x_i} g_n(x)\Big)^2 \,\mathrm{d}x}\sqrt{\int_{Q_R} g_n(x)^2 \,\mathrm{d}x},$$

where we used the properties of ψ_R and the Cauchy–Schwarz-inequality. Since all integrals are bounded (recall (23)–(25)), we find a constant $c > 0$ such that for all n and all R

$$\int_{\mathbb{R}^d} |\nabla(g_n\Psi_R)(x)|^2 \,\mathrm{d}x \le \int_{Q_R} |\nabla g_n(x)|^2 \,\mathrm{d}x + \frac{c}{\sqrt{R}}. \tag{27}$$

Our basic tool for estimating the L^2- and $\mathrm{L}^{2\gamma}$-norm of $g_n\Psi_R$ is a variation of the shift lemma [8, Lemma 3.4]. Indeed, using the shift-invariance of the variational problem because of periodic boundary conditions, the mass of a nonnegative function on the boundary $Q_R \setminus Q_{R-\sqrt{R}}$ can, after suitable shifting, be estimated by its total mass on Q_R times the quotient of the volumes. Applying this to $g_n^2 + g_n^{2\gamma}$, we may assume that

$$\int_{Q_R \setminus Q_{R-\sqrt{R}}} (g_n^2 + g_n^{2\gamma}) \le \frac{d}{\sqrt{R}} \int_{Q_R} (g_n^2 + g_n^{2\gamma}).$$

Skipping the details, this leads to

$$\|g_n\Psi_R\|_{2\gamma}^{2\gamma} \ge \Big(1 - \frac{d}{\sqrt{R}}\Big) \|g_n \mathbb{1}_{Q_R^{(n)}}\|_{2\gamma}^{2\gamma} - \frac{c}{\sqrt{R}} \tag{28}$$

and, with use of (25),

$$\big|\|g_n\Psi_R\|_2^2 - 1\big| \le \frac{c}{\sqrt{R}} \tag{29}$$

for a suitable constant, not depending on n or R, which we also denote $c > 0$.

Step 5. Now we put everything together to show (22). We use (24) and (26) and note that $1 < \gamma < 1 + 2/d$ to get

$$\liminf_{n\to\infty} a_n^2 \Big\{ S^{\pi,Ra_n}\big(p_n^R\big) + \frac{\rho}{1-\gamma} a_n^{-2-d(1-\gamma)} \sum_{z\in B_{Ra_n}} \big(p_n^R(z)\big)^\gamma \Big\}$$

$$\geq \limsup_{n\to\infty} \Big(\int_{Q_R^{(n)}} |\nabla g_n|^2 - \frac{\rho}{\gamma-1} \|g_n 1\!\!1_{Q_R^{(n)}}\|_{2\gamma}^{2\gamma} \Big).$$

Next, we plug in (27) and (28) obtaining

$$\liminf_{R\to\infty}\liminf_{n\to\infty} a_n^2 \Big\{ S^{\pi,Ra_n}\big(p_n^R\big) + \frac{\rho}{1-\gamma} a_n^{-2-d(1-\gamma)} \sum_{z\in B_{Ra_n}} \big(p_n^R(z)\big)^\gamma \Big\}$$

$$\geq \limsup_{R\to\infty}\limsup_{n\to\infty} \Big(\int_{\mathbb{R}^d} |\nabla (g_n \Psi_R)|^2 - \frac{\rho}{\gamma-1} \|g_n \Psi_R\|_{2\gamma}^{2\gamma} \Big).$$

With the help of (29), we can replace $g_n\Psi_R$ by its normalized version $g_n\Psi_R/\|g_n\Psi_R\|_2$, which is a candidate for the infimum in $\hat{\chi}_\gamma(\rho)$. This yields the assertion.

5 Proof for Phases 1–3 (Theorem 3.1)

The proof of (a) and (b) is analogous to the proof of [13, Theorem 1.2], therefore we only sketch the idea here and omit all details, like compactification, cutting, or error terms (see [17] for details).

Denote by $\ell_t(z) = \int_0^t 1\!\!1_{\{X_s=z\}}\,\mathrm{d}s$ the local time of the random walk path $(X_s)_{s\in[0,t]}$ with generator $\kappa(t)\Delta$ in the point $z \in \mathbb{Z}^d$. Note that $\sum_{z\in\mathbb{Z}^d} \ell_t(z) = t$. The time change

$$\ell_t(z) = \kappa(t)^{-1} \int_0^{t\kappa(t)} 1\!\!1_{\{X_{s/\kappa(t)}=z\}}\,\mathrm{d}s \tag{30}$$

relates ℓ_t to the local time of the standard random walk with generator Δ, which is given by the integral on the right hand side.

Recall the definition (3) of the cumulant generating function H of $\xi(0)$. Then, rewriting the Feynman–Kac formula (2),

$$\langle U(t)\rangle = \mathbb{E}_0^{(t)}\Big[\exp\Big(\sum_{z\in\mathbb{Z}^d} H(\ell_t(z))\Big)\Big]$$

$$= \mathrm{e}^{H(t)}\mathbb{E}_0^{(t)}\Big[\exp\Big(K_H(t) \sum_{z\in\mathbb{Z}^d} \frac{H(t\frac{\ell_t(z)}{t}) - \frac{\ell_t(z)}{t}H(t)}{K_H(t)}\Big)\Big],$$

we see that we can apply the asymptotics (4) to the normalized local times ℓ_t/t. Heuristically, this gives

$$\langle U(t)\rangle \mathrm{e}^{-H(t)} \approx \mathbb{E}_0^{(t)}\Big[\exp\Big(K_H(t)\sum_{z\in\mathbb{Z}^d}\rho\hat{H}\Big(\frac{\ell_t(z)}{t}\Big)(1+o(1))\Big)\Big], \quad t\to\infty.$$

Denote by $\mathbb{P}_0^{(t)}$ the probability measure related to $\mathbb{E}_0^{(t)}$. Under $\mathbb{P}_0^{(t)}$, the process $(\ell_t/t)_t$ satisfies a large deviation principle on scale $t\kappa(t)$ with rate function $p \mapsto -\big(\Delta\sqrt{p},\sqrt{p}\big)$. This follows via the time change (30) from the LDP for the respective process in the non decelerated/accelerated case, which was shown in [13, Lemma 1.5]. Since $K_H(t) \asymp t\kappa(t)$ in part (b), an application of Varadhan's lemma gives (15). In part (a), one can use the relation $K_H(t) \gg t\kappa(t)$ and the nonpositivity of the function $\hat{H}$ to show that the main contribution on the exponential scale comes from the event that the process $(X_s)_{s\in[0,t]}$ stays in the origin, where the potential takes on its highest value. This leads to formula (14).

The proof of (c) follows mainly the arguments of [15] (who consider only $\gamma = 1$), adapting them to the new scale $t\kappa(t)/\alpha_t^2$. The case $\gamma < 1$ was treated in a similar way in [3], whereas the case $\gamma > 1$ did not appear originally in Phase 3. For convenience, we give a universal derivation for all values $\gamma \in [0, 1+2/d)$.

By an adaption of [15, Proposition 3.4], the rescaled and normalized local times

$$L_t(y) = \frac{\alpha_t^d}{t}\ell_t(\lfloor\alpha_t y\rfloor), \quad y\in\mathbb{R}^d, \tag{31}$$

with α_t defined by (6), satisfy under $\mathbb{P}_0^{(t)}(\,\cdot\,\mathbb{1}_{\{\operatorname{supp} L_t\subseteq Q_R\}})$ a large deviation principle in the weak topology induced by test integrals against continuous functions, where we recall that $Q_R = [-R,R]^d$. The scale of the principle is $t\kappa(t)/\alpha_t^2$ and the rate function is $g^2 \mapsto \int_{\mathbb{R}^d}|\nabla g|^2$ for $g\in \mathrm{H}^1(\mathbb{R}^d)$ with $\operatorname{supp} g\subseteq Q_R$ and $\|g\|_2 = 1$.

For a lower bound, we start again with (2) and insert the indicator on the event $\{\operatorname{supp} L_t\subseteq Q_R\}$, using the notation $\mathbb{E}_{0,R}^{(t)}[\,\cdot\,]$. After transforming

$$\begin{aligned}\langle U(t)\rangle &\geq \mathbb{E}_{0,R}^{(t)}\Big[\exp\Big(\sum_{z\in\mathbb{Z}^d}H(\ell_t(z))\Big)\Big]\\ &= \mathbb{E}_{0,R}^{(t)}\Big[\exp\Big(\alpha_t^d\int_{Q_R}H\Big(\frac{t}{\alpha_t^d}L_t(y)\Big)\,\mathrm{d}y\Big)\Big]\\ &= \mathrm{e}^{\alpha_t^d H(\frac{t}{\alpha_t^d})}\mathbb{E}_{0,R}^{(t)}\Big[\exp\Big(\frac{t\kappa(t)}{\alpha_t^2}\int_{Q_R}\frac{H\big(\frac{t}{\alpha_t^d}L_t(y)\big)-L_t(y)H\big(\frac{t}{\alpha_t^d}\big)}{K_H\big(\frac{t}{\alpha_t^d}\big)}\,\mathrm{d}y\Big)\Big],\end{aligned} \tag{32}$$

we restrict the integral to the part where $L_t(y)\leq M$ for some $M>1$, noting that the integrand on the set $\{L_t(y)>M\}$ is nonnegative because of the convexity of H. Then we apply the locally uniform asymptotics (4). Next, to get rid of the indicator on $\{L_t(y)\leq M\}$, we introduce a Hölder parameter $\eta\in(0,1)$ to separate the expectations over the whole integral and over the difference set $\{L_t(y)>M\}$. The expectation over the rest term can be shown to be negligible on the exponential scale

$t\kappa(t)/\alpha_t^2$ (see [17, pp. 86f]; here we use Lemma 2.2(a) and the assumption that $\gamma < 2$). Finally, we apply the large deviation principle for L_t and Varadhan's lemma; the lower semi-continuity of $g^2 \mapsto \int_{Q_R} \hat{H} \circ g^2$ was proved in [15, Lemma 3.5] for $\gamma = 1$ and can be shown similarly for all positive γ. Summarizing, we obtain for $\gamma > 0$

$$\begin{aligned}
&\liminf_{t\to\infty} \frac{\alpha_t^2}{t\kappa(t)} \log\big(\langle U(t)\rangle \mathrm{e}^{-\alpha_t^d H(t\alpha_t^{-d})}\big)\\
&\geq \liminf_{M\to\infty}\liminf_{t\to\infty} \frac{\alpha_t^2}{t\kappa(t)} \log \mathbb{E}_{0,R}^{(t)}\Big[\exp\Big(\frac{t\kappa(t)}{\alpha_t^2}\int_{Q_R} \rho\hat{H}(L_t(y))\mathbb{1}_{\{L_t(y)\leq M\}}\,\mathrm{d}y\Big)\Big]\\
&\geq \liminf_{t\to\infty} \frac{\alpha_t^2}{t\kappa(t)} \log \mathbb{E}_{0,R}^{(t)}\Big[\exp\Big((1-\eta)\frac{t\kappa(t)}{\alpha_t^2}\int_{Q_R} \rho\hat{H}(L_t(y))\,\mathrm{d}y\Big)\Big]\\
&\geq -\inf_{\substack{g\in \mathrm{H}^1(\mathbb{R}^d)\\ \mathrm{supp}\, g\subseteq Q_R\\ \|g\|_2=1}} \Big\{\int_{Q_R} |\nabla g|^2 - \rho(1-\eta)\int_{Q_R} \hat{H}\circ g^2\Big\}.
\end{aligned}$$

A standard argument shows that the compactified variational formula converges to $\chi_\gamma^{\mathrm{c}}(\rho)$ as $R\to\infty$ and $\eta\downarrow 0$. For the case $\gamma = 0$, we refer to [17, pp. 85f].

Now we prove the upper bound of (16). For technical reasons, we will not work with the large deviation principle, but use a method derived in [5]. First, we compactify with the help of an eigenvalue expansion described in [3] and applied in [15]. Replacing carefully t by $t\kappa(t)$ in their proofs, we find for $R > 0$

$$\frac{\alpha_t^2}{t\kappa(t)}\log\langle U(t)\rangle \leq \frac{C}{R^2} + \frac{\alpha_t^2}{t\kappa(t)}\log\langle U_{4R\alpha_t}(t)\rangle + o(1), \qquad t\to\infty, \tag{33}$$

with some constant $C > 0$, where $U_{R\alpha_t}(t) = \mathbb{E}_{0,R}^{(t)}[\mathrm{e}^{\int_0^t \xi(X_s)\,\mathrm{d}s}]$. Similarly to (32), we can write

$$\langle U_{R\alpha_t}(t)\rangle = \mathrm{e}^{\alpha_t^d H(\frac{t}{\alpha_t^d})}\mathbb{E}_{0,R}^{(t)}\Big[\exp\Big(\frac{t\kappa(t)}{\alpha_t^2}\sum_{z\in B_{R\alpha_t}} \frac{H(\ell_t(z)) - \frac{t}{\alpha_t^d}\ell_t(z) H\big(\frac{t}{\alpha_t^d}\big)}{K_H\big(\frac{t}{\alpha_t^d}\big)}\Big)\Big],$$

where we recall that $B_R = [-R,R]\cap\mathbb{Z}^d$. We split the sum into the part where $\ell_t(z) \leq Mt\alpha_t^{-d}$ and the rest where $\ell_t(z) > Mt\alpha_t^{-d}$ for some $M > 1$, separating the respective expectations with Hölder's inequality. The rest term can again be neglected on the exponential scale $t\kappa(t)/\alpha_t^2$, while an application of (4) in the main term leads to

$$\limsup_{t\to\infty} \frac{\alpha_t^2}{t\kappa(t)} \log\Big(\langle U_{R\alpha_t}(t)\rangle\, \mathrm{e}^{-\alpha_t^d H(\frac{t}{\alpha_t^d})}\Big)$$

$$\le \limsup_{t\to\infty} \frac{\alpha_t^2}{t\kappa(t)} \log \mathbb{E}_{0,R}^{(t)}\Big[\exp\Big(\tilde\rho\, \frac{t\kappa(t)}{\alpha_t^{d+2}} \sum_{z\in B_{R\alpha_t}} \hat H\Big(\frac{\alpha_t^d}{t}\ell_t(z)\Big) 1\!\!1_{\{\ell_t(z)\le M\frac{t}{\alpha_t^d}\}}\Big)\Big], \tag{34}$$

where $\tilde\rho = \rho(1+\eta)$ with the Hölder parameter $\eta\in(0,1)$. Next, we can omit the indicator of the event $\{\ell_t(z)\le Mt\alpha_t^{-d}\}$ noting that the function $\hat H$ is nonnegative on $[1,\infty)$.

We now need the mentioned tool from [5], namely, an explicit description of the local times density, which provides an upper bound on exponential functionals as in (34) in the form of a variational formula: Define

$$G_t(p) = \alpha_t^{-(d+2)} \sum_{z\in\mathbb{Z}^d} \hat H(\alpha_t^d\, p(z))$$

for $p\in\mathcal{M}_1(\mathbb{Z}^d)$. We want to apply [15, Proposition 3.3], which is formulated for the local times related to a random walk with generator Δ. With the help of (30) we can adapt it to our case. This gives

$$\mathbb{E}_{0,R}^{(t)}\Big[\exp\Big(t\kappa(t)\tilde\rho\, G_t\big(\tfrac{\ell_t}{t}\big)\Big)\Big]$$

$$\le \exp\Big(t\kappa(t) \sup_{\substack{p\in\mathcal{M}_1(\mathbb{Z}^d)\\ \mathrm{supp}\, p\subseteq B_{R\alpha_t}}} \{\tilde\rho\, G_t(p) + (\Delta\sqrt{p},\sqrt{p})\}\Big)(2dt\kappa(t))^{|B_{R\alpha_t}|}|B_{R\alpha_t}|$$

$$\le \exp\Big(-\frac{t\kappa(t)}{\alpha_t^2}\chi_t(\tilde\rho)\Big)\, \mathrm{e}^{o(t\kappa(t)/\alpha_t^2)},$$

where we put

$$\chi_t(\tilde\rho) = -\alpha_t^2 \sup_{p\in\mathcal{M}_1(\mathbb{Z}^d)} \{\tilde\rho\, G_t(p) + (\Delta\sqrt{p},\sqrt{p})\}.$$

In the last step, we also used the properties of the scale function α_t mentioned in Lemma 2.4 and the assumption $K_H(t)\gg \log t$.

Now a direct calculation shows that

$$\chi_t(\tilde\rho) = \begin{cases} \alpha_t^2\chi^{(\mathrm{DE})}\Big(\dfrac{\tilde\rho}{\alpha_t^2}\Big) - \tilde\rho\dfrac{d}{2}\log\alpha_t^2 & \text{for } \gamma = 1,\\[2ex] \alpha_t^2\chi_\gamma^{(\mathrm{DB})}\Big(\dfrac{\tilde\rho}{\alpha_t^{2+d(1-\gamma)}}\Big) - \tilde\rho\dfrac{1-\alpha_t^{-d(1-\gamma)}}{1-\gamma} & \text{for } \gamma\ne 1. \end{cases}$$

In both cases, we can apply Proposition 3.3 with $\kappa = \alpha_t^{2+d(1-\gamma)}\to\infty$ for $t\to\infty$, since $\gamma < 1+2/d$. Hence, $\chi_t(\tilde\rho)$ converges to $\chi^{(\mathrm{AB})}(\tilde\rho)$ in the case $\gamma = 1$ and to $\chi_\gamma^{(\mathrm{B})}(\tilde\rho)$ in the case $\gamma\ne 1$, i.e., to $\chi_\gamma^{\mathrm{c}}(\tilde\rho)$ in both cases. In summary, (34) becomes

$$\limsup_{t\to\infty} \frac{\alpha_t^2}{t\kappa(t)} \log\Big(\langle U_{R\alpha_t}(t)\rangle\, \mathrm{e}^{-\alpha_t^d H(\frac{t}{\alpha_t^d})}\Big) \le \limsup_{t\to\infty}(-\chi_t(\tilde\rho)) \le -\chi_\gamma^{\mathrm{c}}(\tilde\rho).$$

By a scaling argument, one can see that $\chi_\gamma^{\mathrm{c}}(\tilde\rho) = \chi_\gamma^{\mathrm{c}}(\rho(1+\eta))$ converges to $\chi_\gamma^{\mathrm{c}}(\rho)$ for $\eta \downarrow 0$. Together with (33), the assertion (16) is thus shown, which finishes the proof of Theorem 3.1. □

6 Proof for Phase 4 (Theorem 3.2)

Phase 4 is characterized by the fact that the space–time scale ratio is constant: $\alpha_t = t^{1/d}$, i.e. $t/\alpha_t^d = 1$. We rescale both local times and potential,

$$L_t(y) = \ell_t(\lfloor \alpha_t y\rfloor) \quad \text{and} \quad \bar\xi_t(y) = \xi(\lfloor \alpha_t y\rfloor), \quad y \in \mathbb{R}^d.$$

Note that because of $\kappa(t) = \kappa^* t^{2/d}(1+o(1))$, the definition of the rescaled (and normalized) local times is asymptotically equivalent to (31), hence we have again an LDP under $\mathbb{P}_0^{(t)}(\,\cdot\, 1\!\!1_{\{\operatorname{supp} L_t \subseteq Q_R\}})$ on scale $t\kappa(t)/\alpha_t^2 \asymp t$ with rate function $g^2 \mapsto \int_{\mathbb{R}^d} |\nabla g|^2$ for $g \in \mathrm{H}^1(\mathbb{R}^d)$ satisfying $\operatorname{supp} g \subseteq Q_R$ and $\|g\|_2 = 1$.

We will frequently make use of arguments from [9], in particular their main result on large deviations for the scalar product $\big(L_t, \bar\xi_t\big)$. The time parameter t in [9] is replaced by $t\kappa(t)$ and our scale function $\alpha_t = t^{1/d}$ corresponds to the scale in [9] at time $t\kappa(t)$, multiplied by $(\kappa^*)^{-1/(d+2)}$. Thus, [9, Theorem 1.3] reads

$$\lim_{t\to\infty} \frac{1}{t} \log \mathbb{P}_0^{(t)} \times \mathrm{Prob}\big(\big(L_t, \bar\xi_t\big) > u\big) = -(\kappa^*)^{d/(d+2)} \Lambda_H^{(\mathrm{RWRS})}(u) \tag{35}$$

for $u > 0$ such that $u \in (\operatorname{supp} \xi(0))^\circ$, where Prob is the probability with respect to the potential and

$$\Lambda_H^{(\mathrm{RWRS})}(u) = \inf_{\substack{g\in \mathrm{H}^1(\mathbb{R}^d)\\ \|g\|_2=1}} \Big\{ \int_{\mathbb{R}^d} |\nabla g(y)|^2\,\mathrm{d}y + \sup_{\beta>0}\Big[\beta u - \int_{\mathbb{R}^d} H(\beta g^2(y))\,\mathrm{d}y\Big]\Big\}.$$

By rescaling and duality, it turns out that the variational problem $\chi_H^{(\mathrm{RWRS})}$ that we wish to find in this proof is essentially the negative Legendre transform of $\Lambda_H^{(\mathrm{RWRS})}(u)$:

$$\sup_{u>0} \big\{\beta u - \Lambda_H^{(\mathrm{RWRS})}(u)\big\} = -\beta^{-2/d} \chi_H^{(\mathrm{RWRS})}(\beta^{1+2/d}), \qquad \beta > 0. \tag{36}$$

Let us come to the lower bound of (17). A transformation of the Feynman–Kac formula (2) gives

$$\langle U(t)\rangle = \langle \mathbb{E}_0^{(t)}\big[\exp\big(t\big(L_t, \bar\xi_t\big)\big)\big]\rangle = \int_{\mathbb{R}} t\mathrm{e}^{ut}\, \mathbb{P}_0^{(t)} \times \mathrm{Prob}\big(\big(L_t, \bar\xi_t\big) > u\big)\,\mathrm{d}u.$$

With the help of (35), we can conclude for fixed $u > 0$ and $\varepsilon > 0$ that

$$\begin{aligned}\langle U(t)\rangle &\geq \varepsilon t e^{(u-\varepsilon)t}\,\mathbb{P}_0^{(t)} \times \mathrm{Prob}\big((L_t, \bar{\xi}_t) > u\big) \\ &= \exp\Big(t\big[u - \varepsilon - (\kappa^*)^{d/(d+2)}\Lambda_H^{(\mathrm{RWRS})}(u)\big](1 + o(1))\Big)\end{aligned}$$

as $t \to \infty$. Now let $\varepsilon \downarrow 0$, take the supremum over all $u > 0$ and use (36) for $\beta = (\kappa^*)^{-d/(d+2)}$ to finish the proof of the lower bound.

For the upper bound, we can first derive an analogue formula to (33) to restrict the support of the local times on a compact box (see [17, Proposition 4.4.3] for details). Therefore, it suffices to consider $U_{R\alpha_t}(t) = \mathbb{E}_{0,R}^{(t)}[\exp(t(L_t, \bar{\xi}_t))]$ for some large $R > 0$ instead of $U(t)$. We will use a similar strategy as in the proof of the upper bound in [9, Theorem 1.3]: In order to be able to apply the LDP for the local times, we need to smooth the scenery, which we can only do after cutting it. For $M > 0$, introduce $\bar{\xi}_t^{(\leq M)} = (\bar{\xi}_t \wedge M) \vee (-M)$ and $\bar{\xi}_t^{(>M)} = (\bar{\xi}_t - M)_+$. Then $\bar{\xi}_t \leq \bar{\xi}_t^{(\leq M)} + \bar{\xi}_t^{(>M)}$. We want to work with the convolution $\bar{\xi}_t^{(\leq M)} \star j_\delta$ with $j_\delta = \delta^{-d} j(\cdot/\delta)$, where $j \geq 0$ is a smooth, rotational invariant, L^1-normalized function supported in Q_1. For brevity, we will not explain in detail how to deal with the remainder terms $\mathbb{E}_{0,R}^{(t)}[\exp(t(L_t, \bar{\xi}_t^{(>M)}))]$ and $\mathbb{E}_{0,R}^{(t)}[\exp(t(L_t, \bar{\xi}_t^{(\leq M)} - \bar{\xi}_t^{(\leq M)} \star j_\delta))]$ (which can be separated from the main term by Hölder's inequality). For the smoothing, one can apply [9, Lemma 3.5], while the cutting is technically involved and follows the proof of [14, (2.12)] (here we need Lemma 2.2(b) and $\gamma < 2$). Let us in the following take for granted that it is enough to show

$$\limsup_{M\to\infty}\limsup_{\delta\downarrow 0}\limsup_{t\to\infty}\frac{1}{t\kappa^*}\log\langle\mathbb{E}_{0,R}^{(t)}\big[\exp\big(t\big(L_t, \bar{\xi}_t^{(\leq M)} \star j_\delta\big)\big)\big]\rangle \leq -\chi_H^{(\mathrm{RWRS})}\Big(\frac{1}{\kappa^*}\Big). \tag{37}$$

Denote $\ell_t^{(\delta)}(z) = \int_{z+[0,1)^d} L_t \star j_\delta(y/\alpha_t)\,\mathrm{d}y$, then by rotational invariance of j, we have $t(L_t, \bar{\xi}_t^{(\leq M)} \star j_\delta) = \sum_{z\in\mathbb{Z}^d}\ell_t^{(\delta)}(z)\xi^{(\leq M)}(z)$, where $\xi^{(\leq M)}(z) = (\xi \wedge M) \vee (-M) \leq \xi(z) \vee (-M)$. Denote by H_M the cumulant generating function of $\xi(0) \vee (-M)$, then it follows

$$\begin{aligned}\langle\mathbb{E}_{0,R}^{(t)}\big[\exp\big(t\big(L_t, \bar{\xi}_t^{(\leq M)} \star j_\delta\big)\big)\big]\rangle &\leq \Big\langle\mathbb{E}_{0,R}^{(t)}\Big[\exp\Big(\sum_{z\in\mathbb{Z}^d}\ell_t^{(\delta)}(z)\xi(z) \vee (-M)\Big)\Big]\Big\rangle \\ &= \mathbb{E}_{0,R}^{(t)}\Big[\exp\Big(\sum_{z\in\mathbb{Z}^d}H_M(\ell_t^{(\delta)}(z))\Big)\Big].\end{aligned}$$

Jensen's inequality for the probability measure $1\!\!1_{\{z+[0,1)^d\}}\,\mathrm{d}y$ yields

$$\exp\Big(\sum_{z\in\mathbb{Z}^d}H_M(\ell_t^{(\delta)}(z))\Big) \leq \exp\Big(t\int_{\mathbb{R}^d}H_M(L_t \star j_\delta(y))\,\mathrm{d}y\Big).$$

Now we are ready to apply Varadhan's lemma to derive

$$\limsup_{t\to\infty}\frac{1}{t\kappa^*}\log\mathbb{E}^{(t)}_{0,R}\big[\mathrm{e}^{t\int_{\mathbb{R}^d}H_M(L_t\star j_\delta(y))\,\mathrm{d}y}\big]$$

$$\le -\inf_{\substack{g\in \mathrm{H}^1(\mathbb{R}^d)\\ \operatorname{supp} g\subseteq Q_R\\ \|g\|_2=1}}\Big\{\int_{\mathbb{R}^d}|\nabla g(y)|^2\,\mathrm{d}y-\frac{1}{\kappa^*}\int_{\mathbb{R}^d}H_M\big(g^2\star j_\delta(y)\big)\,\mathrm{d}y\Big\}.$$

Jensen's inequality for the probability measure j_δ and Fubini's theorem show that we receive an upper bound when omitting the convolution with j_δ. Finally, we can replace H_M by H for $M\to\infty$ by the following consideration. By definition of H_M, we have

$$\mathrm{e}^{H_M(t)}\le \mathrm{e}^{H(t)}+\mathrm{e}^{-tM}\mathrm{Prob}(\xi(0)<-M)$$

for all $t>0$. Using the inequality $\log(1+x)\le x$, we obtain

$$H_M(t)\le H(t)+\mathrm{e}^{-tM-H(t)}\mathrm{Prob}(\xi(0)<-M).$$

Since $H(t)\ge t\langle\xi(0)\rangle=0$ by Jensen's inequality, we can write for all $g\in\mathrm{H}^1(\mathbb{R}^d)$ with $\operatorname{supp} g\subseteq Q_R$

$$\int_{\mathbb{R}^d}H_M\big(g^2(y)\big)\,\mathrm{d}y\le\int_{\mathbb{R}^d}H\big(g^2(y)\big)\,\mathrm{d}y+|Q_R|\mathrm{Prob}(\xi(0)<-M),$$

and the last term tends to zero when $M\to\infty$. Thus, we have arrived at a compactified version of our variational problem $\chi_H^{\mathrm{(RWRS)}}(1/\kappa^*)$, which we can estimate against the whole-space problem. This shows (37) and completes the proof of the theorem. □

References

1. Becker, M., König, W.: Self-intersection local times of random walks: Exponential moments in subcritical dimensions, Probab. Theory Relat. Fields (to appear)
2. Bingham, N.H., Goldie, C.M., Teugels, J.L.: Regular Variation. Cambridge University Press, Cambridge (1987)
3. Biskup, M., König, W.: Long-time tails in the parabolic Anderson model with bounded potential. Ann. Probab. **29**(2), 636–682 (2001)
4. Braess, D.: Finite elements. Theory, fast solvers and applications in elasticity theory (Finite Elemente. Theorie, schnelle Löser und Anwendungen in der Elastizitätstheorie), 4th revised and extended edn. (German). Springer, Berlin (2007)
5. Brydges, D., van der Hofstad, R., König, W.: Joint density for the local times of continuous-time Markov chains. Ann. Probab. **35**(4), 1307–1332 (2007)

6. Carmona, R.A., Molchanov, S.A.: Parabolic Anderson problem and intermittency. Mem. Am. Math. Soc. **518**, 125 (1994)
7. Coudière, Y., Gallouët, T., Herbin, R.: Discrete Sobolev inequalities and L^p error estimates for finite volume solutions of convection diffusion equations. M2AN, Math. Model. Numer. Anal. **35**(4), 767–778 (2001)
8. Donsker, M.D., Varadhan, S.R.S.: Asymptotics for the Wiener sausage. Commun. Pure Appl. Math. **28**, 525–565 (1975)
9. Gantert, N., König, W., Shi, Z.: Annealed deviations of random walk in random scenery. Ann. Inst. Henri Poincaré, Probab. Stat. **43**(1), 47–76 (2007)
10. Gärtner, J., den Hollander, F.: Correlation structure of intermittency in the parabolic Anderson model. Probab. Theory Relat. Fields **114**(1), 1–54 (1999)
11. Gärtner, J., König, W.: The parabolic Anderson model. In: Deuschel, J.-D., et al. (ed.) Interacting Stochastic Systems, pp. 153–179. Springer, Berlin (2005)
12. Gärtner, J., Molchanov, S.A.: Parabolic problems for the Anderson model. I: Intermittency and related topics. Commun. Math. Phys. **132**(3), 613–655 (1990)
13. Gärtner, J., Molchanov, S.A.: Parabolic problems for the Anderson model. II: Second-order asymptotics and structure of high peaks. Probab. Theory Relat. Fields **111**(1), 17–55 (1998)
14. Grüninger, G., König, W.: Potential confinement property of the parabolic Anderson model. Ann. Inst. Henri Poincaré, Probab. Stat. **45**(3), 840–863 (2009)
15. van der Hofstad, R., König, W., Mörters, P.: The universality classes in the parabolic Anderson model. Commun. Math. Phys. **267**(2), 307–353 (2006)
16. Schmidt, B.: On a semilinear variational problem. ESAIM Control Optim. Calc. Var. **17**, 86–101 (2011)
17. Schmidt, S.: Das parabolische Anderson-Modell mit Be- und Entschleunigung (German), PhD thesis, University of Leipzig (2010) and SVH Saarbrücken (2011)

A Scaling Limit Theorem for the Parabolic Anderson Model with Exponential Potential

Hubert Lacoin and Peter Mörters

Abstract The parabolic Anderson problem is the Cauchy problem for the heat equation with random potential and localized initial condition. In this paper, we consider potentials which are constant in time and independent exponentially distributed in space. We study the growth rate of the total mass of the solution in terms of weak and almost sure limit theorems, and the spatial spread of the mass in terms of a scaling limit theorem. The latter result shows that in this case, just like in the case of heavy tailed potentials, the mass gets trapped in a single relevant island with high probability.

1 Introduction and Main Results

1.1 Overview and Background

We consider the heat equation with random potential on the integer lattice $\mathbb{Z}^d$ and study the Cauchy problem with localised initial datum,

$$\begin{aligned} \partial_t u(t,z) &= \Delta u(t,z) + \xi(t,z)\, u(t,z), && \text{for } (t,z) \in (0,\infty) \times \mathbb{Z}^d, \\ \lim_{t \downarrow 0} u(t,z) &= \mathbb{1}_0(z), && \text{for } z \in \mathbb{Z}^d, \end{aligned}$$

H. Lacoin
CEREMADE, Université Paris Dauphine, Paris
e-mail: lacoin@ceremade.dauphine.fr

P. Mörters (✉)
University of Bath, Bath, UK
e-mail: maspm@bath.ac.uk

J.-D. Deuschel et al. (eds.), *Probability in Complex Physical Systems*, Springer Proceedings in Mathematics 11, DOI 10.1007/978-3-642-23811-6_10,

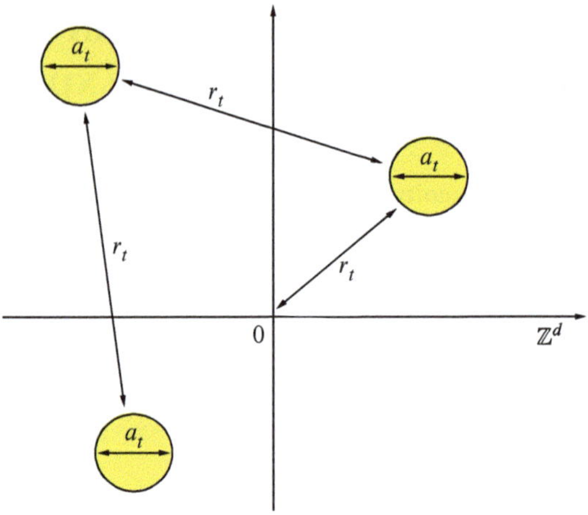

Fig. 1 A schematic picture of intermittency: the mass of the solution is concentrated on relevant islands (*indicated by shaded balls*) with radius of order a_t and distances of order $r_t \gg a_t$

where

$$(\Delta f)(z) = \sum_{y\sim z}[f(y) - f(z)], \qquad \text{for } z \in \mathbb{Z}^d, f : \mathbb{Z}^d \to \mathbb{R},$$

is the discrete Laplacian, and the potential $(\xi(t,z) : t > 0, z \in \mathbb{Z}^d)$ is a random field. This equation is known as the *parabolic Anderson model.*

In the present paper, we assume that the potential field is constant in time and independent, identically distributed in space according to some nondegenerate distribution. Under this hypothesis, the solutions are believed to exhibit *intermittency*, which roughly speaking means that at any late time the solution is concentrated in a small number of *relevant islands* at large distance from each other, such that the diameter of each island is much smaller than this distance, see Fig. 1 for a schematic picture. The relevant islands are located in areas where the potential has favourable properties, e.g., a high density of large potential values. As time progresses, new relevant islands emerge in locations further and further away from the origin at places where the potential is more and more favourable, while old islands lose their relevance. The main aim of the extensive research in this model, which was initiated by Gärtner and Molchanov in [4, 5], is to get a better understanding of the phenomenon of intermittency for various choices of potentials.

Natural questions about the nature of intermittency are the following:

- What is the diameter of the relevant islands? Are they growing in time?
- How much mass is concentrated in a relevant island?
  How big is the potential on a relevant island?
- Where are the relevant islands located? What is the distance of different islands?
- How many relevant islands are there?
- How do new relevant islands emerge? What is the lifetime of a relevant island?

Explicit answers to these questions and, more generally, results on the precise geometry of solutions to the parabolic Anderson model are typically very difficult to obtain. In the related context of Brownian motion among Poissonian obstacles, Sznitman [16] provides methodology to study properties of Brownian paths conditioned on survival, which offer a possible route to the geometry of solutions, at least in the case of bounded potentials. In a seminal paper, Gärtner, König and Molchanov [3] follow a different route to analyse size and position of relevant islands in the case of double exponential potentials. Their results also offer some insight into potentials with heavier tails. In [8] and [11], a complete picture of the geometry of the solutions is given in the case of Pareto distributed potentials, building on the work of [3]. In this case of an extremely heavy tailed potential, it can be shown that, for any $\varepsilon > 0$ at sufficiently late times, there exists a single point carrying a proportion of mass exceeding $1 - \varepsilon$ with probability converging to one. This point constitutes the single relevant island, and very precise results about the location, lifetime and dynamics of this island can be obtained, see also [12] for a survey of this research, and [2] for similar results in the discrete time case.

For more complicated potentials, however, one has to rely on less explicit results. A natural way forward is to investigate the growth rates of the *total mass*

$$U(t) := \sum_{z\in\mathbb{Z}^d} u(t,z)$$

of the solution. If the potential is bounded from above, we define the (quenched) *Lyapunov exponent* as

$$\lambda := \lim_{t\to\infty} L_t \text{ where } L_t := \frac{1}{t}\log U(t),$$

whenever this limit exists in the almost sure sense. If the potential is unbounded, one expects superexponential growth and is interested in an asymptotic expansion of L_t. If the tails of the potential distribution are sufficiently light so that the logarithmic moment generating function

$$H(x) := \log E\mathrm{e}^{x\xi(0)}$$

is finite for all $x \geq 0$, a large deviation heuristics suggests that,we get

$$L_t = \frac{H(\beta_t\alpha_t^{-d})}{\beta_t\alpha_t^{-d}} - \frac{1}{\alpha_t^2}\big(\kappa + o(1)\big), \text{ almost surely as } t\uparrow\infty,$$

where α, β are deterministic scale functions and κ is a deterministic constant. According to the heuristics, the quantity α_t can be interpreted as the diameter of the relevant islands at time t, and the leading term as the size of the potential values on the island. The constant κ is given in terms of a variational problem whose maximiser describes the shape of a vertically shifted and rescaled potential on an island. More details and a classification of light-tailed potentials according to this paradigm are given in [6].

If the potential is such that the moment generating functions do not always exist, this approach breaks down. Indeed, one can no longer expect the leading terms in an expansion of L_t to be deterministic. Instead, one should expect the solutions to be concentrated in islands consisting of single sites and the expansion of L_t to reflect fluctuations in the size of the potential on these sites. One would expect the sites of the islands to be those with the largest potential in some time-dependent centred box and the fluctuations to be similar to those seen in the order statistics of independent random variables. This programme is carried out in detail in [7] for potentials with Weibull (stretched exponential) and Pareto (polynomial) tails. In the present paper, we add the case of standard exponential potentials and present weak (see Theorem 1) and almost sure (see Theorem 2) asymptotic expansions for L_t in this case. These results are taken from the first author's unpublished master thesis [9] and were announced without proof in [7].

Very little has been done so far to get a precise understanding of the number and position of the relevant islands, the very fine results for the Pareto case being the only exception. A natural idea to approach this with somewhat softer techniques is to prove a scaling limit theorem. To this end, we define a probability distribution ν_t on $\mathbb{Z}^d$ associating to each site z a weight proportional to the solution $u(t,z)$, i.e.,

$$\nu_t := \sum_{z\in\mathbb{Z}^d} \frac{u(t,z)}{U(t)}\,\delta(z), \quad \text{for any } t \geq 0,$$

where $\delta(z)$ denotes the Dirac measure concentrated at $z \in \mathbb{R}^d$. For $a > 0$, we also define the distribution of mass at the time t in the scale a as

$$\nu_t^a := \nu_t\big(a\,\cdot\big) = \sum_{z\in\mathbb{Z}^d} \frac{u(t,z)}{U(t)}\,\delta\big(\tfrac{z}{a}\big),$$

which is considered as an element of the space $\mathcal{M}(\mathbb{R}^d)$ of probability measures on $\mathbb{R}^d$. Identifying the scale r_t of the distances between the islands and the origin, intermittency would imply that islands are contracted to points and that $\nu_t^{r_t}$ converges in law to a random probability measure, which is purely atomic with atoms representing intermittent islands and their weights representing the proportion of mass on the islands. In the case of Pareto potentials, such a result follows easily from the detailed geometric picture, see [11, Proposition 1.4], but in principle could be obtained from softer arguments. It therefore seems viable that scaling limit theorems like the above can be obtained for a large class of potentials including some which are harder to analyse because they have much lighter tails.

In Theorem 3 of the present paper, we show that in the case of exponential potentials for $r_t = t/\log\log t$ the random probability measures $\nu_t^{r_t}$ converge in distribution to a point mass in a nonzero random point. In particular, this shows that for exponential potential we also have *only one relevant island*. Moreover, the solution of the parabolic Anderson problem spreads *sublinearly* in space. Our arguments can be adapted to the easier case of Weibull, or stretched exponential, potentials, where there is also only one relevant island but the solution has a *superlinear spread*. These results are new and open up possibilities for further research projects, which we briefly mention in our concluding remarks.

1.2 Statement of Results

We now assume that $(\xi(z): z \in \mathbb{Z}^d)$ is a family of independent random variables with

$$P\big(\xi(z) > x\big) = \mathrm{e}^{-x} \text{ for } x \geq 0.$$

Suppose $(u(t,z): t > 0, z \in \mathbb{Z}^d)$ is the unique nonnegative solution to the parabolic Anderson model with this potential, and let $(U(t): t > 0)$ be the total mass of the solution. We recall that

$$L_t = \frac{1}{t} \log U(t),$$

and first ask for a weak expansion of L_t up to the first nondegenerate random term. This turns out to be the third term in the expansion, which is of constant order. In the following, we use $\Rightarrow$ to indicate convergence in distribution.

Theorem 1 (Weak asymptotics for the growth rate of the total mass). *We have*

$$L_t - d \log t + d \log\log\log t \Rightarrow X,$$

where X has a Gumbel distribution

$$P(X \leq x) = \exp\big\{-2^d \mathrm{e}^{-x+2d}\big\} \qquad \text{for } x \in \mathbb{R}.$$

In an almost sure expansion already the second term exhibits fluctuations.

Theorem 2 (Almost sure asymptotics for the growth rate of the total mass). *Almost surely,*

$$\limsup_{t\uparrow\infty} \frac{L_t - d \log t}{\log\log t} = 1,$$

and

$$\liminf_{t\uparrow\infty} \frac{L_t - d \log t}{\log\log\log t} = -(d+1).$$

Remark 1. Note that neither of these almost sure asymptotics agree with the asymptotics

$$\lim_{t\uparrow\infty} \frac{L_t - d \log t}{\log\log\log t} = -d \quad \text{ in probability,}$$

which follows from Theorem 1. The almost sure results pick up fluctuations on both sides of the second term in the weak expansion, with those above being significantly stronger than those below the mean. This is different in the stretched exponential case studied in [7], where the liminf behaviour coincides with the weak limit behaviour. The limsup behaviour in the exponential case is included in the results of [7] and therefore not proved here.

Recall that the distribution of the mass of the solution at time $t > 0$ and on the scale $a > 0$ is defined as a (random) element of the space $\mathcal{M}(\mathbb{R}^d)$ of probability measures on $\mathbb{R}^d$ by

$$\nu_t^a := \nu_t\,(a\,\cdot) = \sum_{z\in\mathbb{Z}^d} \frac{u(t,z)}{U(t)}\,\delta\big(\tfrac{z}{a}\big).$$

The following theorem is the main result of this paper.

Theorem 3 (Scaling limit theorem). *Defining the sublinear scale function*

$$r_t = \frac{t}{\log\log t},$$

we have

$$\lim_{t\uparrow\infty} \nu_t^{r_t} = \delta(Y)\text{in distribution},$$

where $\delta(x)$ denotes the Dirac measure concentrated in $x \in \mathbb{R}^d$ and Y is a random variable in $\mathbb{R}^d$ with independent coordinates given by standard exponential variables with uniform random sign.

Remark 2. In the case of a Weibull potential with parameter $0 < \gamma < 1$ given by

$$P\big(\xi(z) > x\big) = \mathrm{e}^{-x^\gamma} \text{ for } x \geq 0,$$

a variant of the proof gives convergence of $\nu_t^{r_t}$ for the superballistic scale function

$$r_t = \frac{t\,(\log t)^{\frac{1}{\gamma}-1}}{\log\log t},$$

to a limit measure $\delta(Y)$ where the components of Y are independent exponentially distributed with parameter $d^{1-1/\gamma}$ and uniform sign. Details are left to the reader.

2 Proof of the Main Results

2.1 Overview

The proofs are based on the Feynman–Kac formula

$$u(t,z) = \mathbb{E}\Big[\exp\Big\{\int_0^t \xi(X_s)\,\mathrm{d}s\Big\}\,\mathbb{1}\{X_t = z\}\Big],$$

where $(X_s : s \geq 0)$ is a continuous-time simple random walk on $\mathbb{Z}^d$ started at the origin and the probability $\mathbb{P}$ and expectation $\mathbb{E}$ refer only to this walk and not to the potentials. Recall that $(X_s : s \geq 0)$ is the Markov process generated by the discrete Laplacian Δ featuring in the parabolic Anderson problem. It is shown in [4] that the Feynman–Kac formula gives the unique solution to the parabolic Anderson problem under a moment condition on the potential, which is satisfied in the exponential case. By summing over all sites, the Feynman–Kac formula implies that the total mass is given by

$$U(t) = \mathbb{E}\Big[\exp\Big\{ \int_0^t \xi(X_s)\, \mathrm{d}s \Big\} \Big].$$

An analysis of this formula allows us to approximate $L_t = \frac{1}{t} \log U(t)$ almost surely from above and below by variational problems for the potential. These variational problems have the structure that one optimizes over all sites $z \in \mathbb{Z}^d$ the difference between the potential value $\xi(z)$, corresponding to the reward for spending time in the site, and a term corresponding to the cost of getting to the site, which is going to infinity when $z \to \infty$ and thus ensure that the problem is well-defined.

We can use the result for the lower bound given in [7, Lemmas 2.1 and 2.3]. Here and throughout this paper we use $|\cdot|$ to denote the ℓ^1-norm on $\mathbb{R}^d$.

Lemma 1 (Lower bound on L_t). *Let*

$$\underline{N}(t) := \max_{z \in \mathbb{Z}^d} \Big\{ \xi(z) - \frac{|z|}{t} \log \xi(z) \Big\},$$

then, almost surely, for all sufficiently large t, we have

$$L_t \geq \underline{N}(t) - 2d + o(1).$$

The appearance of $\xi(z)$ in the cost term can be explained by the fact that part of the cost arises from the fact that the optimal paths leading to z spend a positive proportion of the overall time traveling to the site and therefore miss out on the optimal potential value for some considerable time, see Sect. 1.3 in [7] for a heuristic derivation of this formula.

The corresponding upper bound will be our main concern here.

Lemma 2 (Upper bound on L_t). *For any $c > 0$, let*

$$\overline{N}_c(t) := \max_{t/(\log t)^2 \leq |z| \leq t \log t} \Big\{ \xi(z) - \frac{|z|}{t} \big(\log\log|z| + c \big) \Big\}.$$

Then, for any $\varepsilon > 0$ there exists $c = c(\varepsilon) > 0$ such that, almost surely, for all sufficiently large t, we have

$$L_t \leq \overline{N}_{c(\varepsilon)}(t) - 2d + \varepsilon + o(1).$$

This lemma will be proved in two steps: We first remove paths that do not make an essential contribution from the average in the Feynman–Kac formula using an ad-hoc approach, see Lemma 7 and Lemma 8. Then we use the properties of the remaining paths to refine the argument and get an improved bound, see Proposition 1.

The variational problems for the upper and lower bound can then be studied using an extreme value analysis, which follows along the lines of [7]. It turns out that the weak and almost sure asymptotics of the two problems coincide up to the accuracy required to prove Theorem 1 and Theorem 2.

For the proof of the scaling limit we need to give an upper bound on the growth rate of the contribution of all those paths ending in a site at distance more than δr_t, for some $\delta > 0$, from the site with the largest potential among those sites that can be reached by some path with the same number of jumps. This bound needs to be strictly better than the lower bound on the overall growth rate. To this end, in a first step, we again use Lemma 7 and Lemma 8 to eliminate some paths using ad-hoc arguments. In the second step, we remove paths that never hit the site with largest potential that is within their reach. This is done on the basis of the gap between the largest and the second largest value for the variational problem in the upper bound. In the third step, it remains to analyse the contribution of paths that hit the optimal site but then move away by more than δr_t. Again, it turns out that the rate of growth of the contribution of these paths is strictly smaller than the lower bound on the growth rate of the total mass. Proposition 1 is set up in such a way that it can deal with both the second and third step. We conclude from this that the solution is concentrated in a single island of diameter at most δr_t around the optimal site. An extreme value analysis characterizes the location of the optimal site and concludes the proof of Theorem 3.

The remainder of the paper is structured as follows: In Sect. 2.2, we give some notation and collect auxiliary results from [7]. Section 2.3 contains the required upper bounds and constitutes the core of the proof. Section 2.4 studies the variational problem arising in the upper bound. Using these approximations, we complete the proof of Theorem 2 in Sect. 2.5 and of Theorem 1 in Sect. 2.6. The proof of the scaling limit theorem, Theorem 3, is completed in Sect. 2.7.

2.2 Auxiliary Results

Let $B_{\mathrm{r}} = \{|z| \leq r\}$ be the ball of radius r centred at the origin in $\mathbb{Z}^d$. The number l_r of points in B_{r} grows asymptotically like r^d. More precisely, there exists a constant κ_d such that, $\lim_{r\to\infty} l_r r^{-d} = \kappa_d$. We define $M_{\mathrm{r}} = \max_{|z|\leq r} \xi(z)$ to be the maximal value of the potential on B_{r}. The behaviour of M_{r} is described quite accurately in [7, Lemma 4.1], which we restate now.

Lemma 3 (Bounds for M_{r}). *Let $\delta \in (0,1)$ and $c > 0$. Then, almost surely,*

$$M_{\mathrm{r}} \le d\,\log r + \log\log r + (\log\log r)^{\delta} \text{ for all sufficiently large } r,$$
$$M_{\mathrm{r}} \ge d\,\log r - (1+c)\log\log\log r \qquad \text{for all sufficiently large } r.$$

In particular, for any pair of constant c_1 and c_2 satisfying $c_1 < d < c_2$, we have

$$c_1 \log r \le M_{\mathrm{r}} \le c_2 \log r \quad \text{for all sufficiently large } r.$$

Let $M_{\mathrm{r}}^{(i)}$ denote the i-th biggest value taken by the potential in the ball of radius r centered at the origin. The next lemma gives us estimates for upper order statistics for the potential.

Lemma 4 (Rough asymptotic behaviour for upper order statistics). *Let $0 < \beta < 1$ be a fixed constant. Then, almost surely,*

$$\lim_{n\to\infty} \frac{M_n^{(\lfloor n^\beta \rfloor)}}{\log n} = d - \beta.$$

Proof. Recalling that l_{n} is the number of points in a ball of radius n in $\mathbb{Z}^d$, we get

$$P\left(M_n^{(\lfloor n^\beta \rfloor)} \le x\right) = \sum_{i=0}^{\lfloor n^\beta \rfloor -1} \binom{l_{\mathrm{n}}}{i} \mathrm{e}^{-xi}\,(1-\mathrm{e}^{-x})^{l_{\mathrm{n}}-i}. \tag{1}$$

We fix $\varepsilon > 0$ and infer that

$$\begin{aligned} P\left(M_n^{(\lfloor n^\beta \rfloor)} \le (d-\beta-\varepsilon)\log n\right) &\le \sum_{i=0}^{\lfloor n^\beta \rfloor} \left(l_{\mathrm{n}} n^{-d+\beta+\varepsilon}\right)^i \left(1-n^{-d+\beta+\varepsilon}\right)^{l_{\mathrm{n}}-n^\beta} \\ &\le \left(n^\beta+1\right)\left((\kappa_d+o(1))n^{\beta+\varepsilon}\right)^{n^\beta} \exp\left[-(\kappa_d+o(1))n^{\beta+\varepsilon}\right] \\ &= \exp\left[-n^{\beta+\varepsilon}(\kappa_d+o(1))\right]. \end{aligned}$$

Since this sequence is summable, we can use the Borel–Cantelli lemma to obtain the lower bound. Similarly, for the upper bound, we use (1) to get

$$P\left(M_n^{(\lfloor n^\beta \rfloor)} \ge (d-\beta+\varepsilon)\log n\right) \le \sum_{i=\lfloor n^\beta \rfloor}^{l_{\mathrm{n}}} \binom{l_{\mathrm{n}}}{i} n^{-(d-\beta+\varepsilon)i}. \tag{2}$$

We now use a rough approximation for the binomial coefficient, namely

$$\binom{l_{\mathrm{n}}}{i} \le \frac{(l_{\mathrm{n}})^i}{i!} \le \left(\frac{e l_{\mathrm{n}}}{i}\right)^i,$$

when i is big enough. Combining this with (2) and using that the first term in the ensuing sum is the largest, we obtain, for all sufficiently large n,

$$P\left(M_n^{(\lfloor n^\beta \rfloor)} \geq (d-\beta+\varepsilon)\log n\right) \leq \sum_{i=\lfloor n^\beta \rfloor}^{l_n} \left(\frac{el_n}{i n^{d-\beta+\varepsilon}}\right)^i \leq l_n \left(\frac{el_n}{n^{d+\varepsilon}}\right)^{n^\beta} \leq \mathrm{e}^{-n^\beta}.$$

Using the Borel–Cantelli lemma again we obtain an upper bound, completing the proof of our statement.

Let $0 < \sigma < \rho < \frac{1}{2}$ be some fixed constants. We define

$$k_n = \lfloor n^\sigma \rfloor \qquad \text{and} \qquad m_n = \lfloor n^\rho \rfloor$$

Combining Lemma 3 and Lemma 4, we get the following result.

Lemma 5. *For any constant $c > 0$, for all sufficiently large n, we have*

(i) $M_n^{(1)} - M_n^{(k_n)} > (\sigma - c)\log n$;
(ii) $M_n^{(k_n)} - M_n^{(m_n)} > (\rho - \sigma - c)\log n$.

Finally, we use Lemma 3 to give a lower bound for $\underline{N}(t)$.

Lemma 6 (Eventual lower bound for $\underline{N}(t)$). *For any small $\varepsilon > 0$, we have*

$$\underline{N}(t) \geq d\log t - (d+1+\varepsilon)\log\log\log t,$$

for all sufficiently large t, almost surely.

Proof. Using Lemma 3 we get, for any fixed $c > 0$ and $c_2 > d$,

$$\begin{aligned}\underline{N}(t) &\geq \max_{r>0}\left[M_r - \frac{r}{t}\log M_r\right]\\ &\geq \max_{r>0}\left[d\log r - (1+c)\log\log\log r - \frac{r}{t}\log\log r - \frac{r}{t}\log c_2\right],\end{aligned}$$

if the maximum of the expression in the square brackets (which we denote by $f_t(r)$) is attained at a point r_t, large enough so that both bounds of Lemma 3 hold.

The solution $r = r_t$ of $f_t'(r) = 0$ satisfies

$$\frac{d}{r} = \frac{\log\log r}{t}\big(1 + o(1)\big).$$

Writing $r_t = t\varphi(r_t)$, where $\varphi(r) = d(\log\log r)^{-1}(1+o(1))$ we get that

$$\log\varphi(r) = -\log\log\log r + \log d + o(1), \tag{3}$$

and hence $\log r_t = \log t + \log\varphi(r_t) = \log t + o(\log r_t)$, which implies $\log r_t / \log t = 1 + o(1)$. Note that this implies $r_t \to \infty$ as $t \to \infty$, which justifies *a posteriori* the application of Lemma 3. Combining this with (3) we get,

$$\begin{aligned} f(r_t) &= d(\log(t\varphi(r_t))) - (1+c)\log\log\log r_t - \varphi(r_t)(\log\log r_t + \log c_2) \\ &= d\log t - (1+d+c)\log\log\log t + O(1). \end{aligned}$$

2.3 Upper Bounds

We start by showing ad hoc bounds for the growth rates of the contribution of certain families of paths. These can be compared to the lower bound for the growth rate of $U(t)$ showing that the paths can be be neglected. For a path $(X_s : s \geq 0)$ on the lattice $\mathbb{Z}^d$, we denote by J_t the number number of jumps up to time t. Recall that $M_n^{(k)}$ denotes the kth largest potential value on the sites $z \in \mathbb{Z}^d$ with $|z| \leq n$.

Lemma 7. *Fix* $0 < \sigma < \frac{1}{2}$ *and* $k_n = n^{\sigma}$. *Let*

$$U_2(t) = \mathbb{E}\Bigg[\exp\Bigg\{\int_0^t \xi(X_s)\mathrm{d}s\Bigg\}\mathbb{1}\Bigg\{\tfrac{t}{(\log t)^2} \leq J_t \leq t\log t,\ \max_{0\leq s\leq t}\xi(X_s) \leq M_{J_t}^{(k_{J_t})}\Bigg\}\Bigg].$$

Then

$$\lim_{t\uparrow\infty}\frac{1}{t}\log\frac{U_2(t)}{U(t)} = -\infty.$$

Proof. Simply replacing $\xi(X_s)$ in the integral by the maximum, we get

$$\begin{aligned} U_2(t) &= \sum_{t/(\log t)^2 \leq n \leq t\log t} \mathbb{E}\Bigg[\exp\Bigg\{\int_0^t \xi(X_s)\mathrm{d}s\Bigg\}\mathbb{1}\{J_t = n,\ \max_{0\leq s\leq t}\xi(X_s) \leq M_n^{(k_n)}\}\Bigg] \\ &\leq \sum_{t/(\log t)^2 \leq n \leq t\log t} \mathrm{e}^{tM_n^{(k_n)}}\,\mathbb{P}(J_t = n) \leq \max_{t/(\log t)^2 \leq n \leq t\log t} \mathrm{e}^{tM_n^{(k_n)}}. \end{aligned}$$

By Lemma 4, we have $M_n^{(k_n)} = (d-\sigma)\log n + o(\log n)$ and hence

$$\frac{1}{t}\log U_2(t) \leq (d-\sigma)\log t + o(\log t),$$

so that the result follows by comparison with Lemma 1 and Lemma 6.

Lemma 8. *Let*

$$U_3(t) = \mathbb{E}\Bigg[\exp\Bigg\{\int_0^t \xi(X_s)\mathrm{d}s\Bigg\}\Bigg(\mathbb{1}\{J_t < \tfrac{t}{(\log t)^2}\} + \mathbb{1}\{J_t > t\log t\}\Bigg)\Bigg].$$

Then

$$\lim_{t\uparrow\infty} \frac{1}{t} \log \frac{U_3(t)}{U(t)} = -\infty.$$

Proof. We first show that almost surely,

$$\frac{1}{t} \log U_3(t) \le \max_{n<t/(\log t)^2} \left\{ M_n - \frac{n}{t} \log \frac{n}{2det} \right\} - 2d + o(1). \tag{4}$$

Indeed, we have

$$\begin{aligned} U_3(t) &\le \sum_{\substack{\{n<t/(\log t)^2\}\\ \cup\{n>t\log t\}}} \mathrm{e}^{tM_n} \mathbb{P}(J_t = n) = \sum_{\substack{\{n<t/(\log t)^2\}\\ \cup\{n>t\log t\}}} \mathrm{e}^{tM_n} \frac{(2\mathrm{d}t)^n \mathrm{e}^{-2\mathrm{d}t}}{n!} \\ &\le \sum_{\substack{\{n<t/(\log t)^2\}\\ \cup\{n>t\log t\}}} \exp\big(tM_n - 2\mathrm{d}t + n\log(2\mathrm{d}t) - \log(n!)\big). \end{aligned} \tag{5}$$

To estimate $n!$ we use Stirling's formula,

$$n! = \sqrt{2\pi n} \left(\frac{n}{e}\right)^n \mathrm{e}^{\delta(n)}, \qquad \text{with } \lim_{n\uparrow\infty} \delta(n) = 0.$$

Fixing some $\varepsilon > 0$ we know from Lemma 3, that $M_n \le (d+\varepsilon)\log n$ for all sufficiently large n, so for t large enough, we obtain for all $n > t \log t$,

$$\begin{aligned} tM_n - 2\mathrm{d}t + n\log(2\mathrm{d}t) - \log(n!) &\le t(d+\varepsilon)\log n - n\log \tfrac{n}{2\mathrm{ed}t} - \delta(n) \\ &\le t(d+\varepsilon)\log n \left(1 - \tfrac{1+o(1)}{(d+\varepsilon)} \log\left(\tfrac{\log t}{2ed}\right) + o(1)\right) \\ &\le -2\log n, \end{aligned}$$

by noticing that $n \mapsto \frac{n}{t\log n} \log \frac{n}{2\mathrm{ed}t}$ is decreasing on $(t\log t, \infty)$. Hence, almost surely,

$$\sum_{n>t\log t} \exp\left(tM_n - 2\mathrm{d}t + n\log(2\mathrm{d}t) - \log(n!)\right) = o(1),$$

so that using (5) the following upper bound for U_3

$$\begin{aligned} U_3(t) &\le \frac{t}{(\log t)^2} \max_{n<t/(\log t)^2} \exp\left(tM_n - 2\mathrm{d}t + n\log(2\mathrm{d}t) - \log(n!)\right) + o(1) \\ &\le \frac{t}{(\log t)^2} \max_{n<t/(\log t)^2} \exp\big(tM_n - 2\mathrm{d}t - n\log \tfrac{n}{2\mathrm{ed}t} + o(t)\big) + o(1), \end{aligned}$$

and hence (4) follows. As a second step we show that

$$\frac{1}{t}\log U_3(t) \le d\log t - (2d-1)\log\log t + o(\log\log t). \tag{6}$$

Recall that $r \mapsto M_{\rm r}$ is a non-decreasing function and check that

$$r \longmapsto \frac{r}{t}\log\frac{r}{2det} \text{ is decreasing on } (0, 2det),$$

hence, replacing r in the bracket by $t/(\log t)^2$

$$\max_{r<t/(\log t)^2}\left[M_{\rm r} - \frac{r}{t}\log\frac{r}{2det}\right] = M_{t/(\log t)^2} + o(1).$$

By Lemma 3 we have $M_{\rm r} \le d\log r + \log\log r + o(\log\log r)$ for all sufficiently large r, we get, for t large enough

$$\max_{r<t/(\log t)^2}\left[M_{\rm r} - \frac{r}{t}\log\frac{r}{2det}\right] \le d\log t - (2d-1)\log\log t + o(\log\log t), \tag{7}$$

and combining (4) and (7), we have proved (6). Using Lemma 1 and Lemma 6,

$$\begin{aligned}\frac{1}{t}\log\frac{U_3(t)}{U(t)} &\le \frac{1}{t}\log U_3(t) - \underline{N}(t) - 2d + o(1)\\ &\le -(2d-1)\log\log t + o(\log\log t) \to -\infty,\end{aligned}$$

and hence, our statement is proved.

The following versatile upper bound is the main tool in the proof of all our theorems and will be used repeatedly. Note, for example, that, together with Lemmas 7 and 8 it implies Lemma 2 if the parameters in (ii) are chosen as $k = 1$ and $\delta = 0$.

Proposition 1. *For a path $(X_s : s \ge 0)$ on the lattice $\mathbb{Z}^d$, we denote by $J_{\rm t}$ the number of jumps up to time t. We denote by $M_n^{(k)}$ the* kth *largest potential value on the sites $z \in \mathbb{Z}^d$ with $|z| \le n$, and let $Z_n^{(k)}$ be the site where this maximum is attained. Further fix $0 < \sigma < \frac{1}{2}$ and let $k_n = \lfloor n^\sigma \rfloor$ and $a_t \downarrow 0$.*

(a) For $n \in \mathbb{N}$, let

$$U_1^{(n)}(t) = \mathbb{E}\left[\exp\left\{\int_0^t \xi(X_s)\,{\rm d}s\right\}\mathbb{1}\{J_{\rm t} = n\}\,\mathbb{1}\{\max_{0\le s\le t}\xi(X_s) > M_n^{(k_n)}\}\right].$$

Then, for all $\varepsilon > 0$ there exists $C_\varepsilon > 0$ such that uniformly for all $ta_t \le n \le t\log t$,

$$\frac{1}{t}\log U_1^{(n)}(t) \le M_n^{(1)} - \frac{n}{2t}\big(\log\log n - C_\varepsilon\big) + \varepsilon - 2d + o(1) \quad \textit{as } t \uparrow \infty.$$

(b) For fixed $\delta \geq 0$ and $k, n \in \mathbb{N}$ let

$$U_1^{(\delta,k,n)}(t) = \mathbb{E}\Bigg[\exp\left\{ \int_0^t \xi(X_s)\,\mathrm{d}s \right\} \\ \mathbb{1}\{J_t = n\}\, \mathbb{1}\left\{ \sup_{0\leq s\leq t} \xi(X_s) \notin \{M_n^{(1)}, \ldots, M_n^{(k-1)}\} \right\} \\ \mathbb{1}\big\{Z_n^{(k)} \in \{X_s : 0 \leq s \leq t\}, |X_t - Z_n^{(k)}| \geq \delta r_t \big\} \Bigg].$$

Then, almost surely,

$$\text{uniformly in } k \leq k_n \text{ and } \tfrac{t}{(\log t)^2} \leq n \leq t a_t,$$

we have that

$$\frac{1}{t} \log U_1^{(\delta,k,n)}(t) \leq M_n^{(k)} - \frac{|Z_n^{(k)}|}{t} \log\log |Z_n^{(k)}| - 2d - \delta + o(1) \quad \text{as } t \uparrow \infty.$$

The first step in the proof is to integrate out the waiting times of the continuous time random walk paths. The following fact taken from [7] helps with this.

Lemma 9. *Let $\eta_0, \ldots, \eta_n$ be fixed real numbers attaining their maximum only once, i.e., there is an index $0 \leq k \leq n$ with $\eta_k > \eta_i$ for all $i \neq k$. Then, for all $t > 0$,*

$$\int_{\mathbb{R}_+^n} \exp\left\{ \sum_{i=0}^{n-1} t_i \eta_i + \left(t - \sum_{i=0}^{n-1} t_i\right) \eta_n \right\} \mathbb{1}\left\{ \sum_{i=0}^{n-1} t_i < t \right\} \mathrm{d}t_0 \ldots \mathrm{d}t_{n-1} \leq e^{t\eta_k} \prod_{i\neq k} \frac{1}{\eta_k - \eta_i}.$$

Proof. First, we prove the result for the case $k = n$, i.e., $\eta_n > \eta_i$ for all $i < n$. We have

$$\int_{\mathbb{R}^n} \exp\left\{ \sum_{i=0}^{n-1} t_i \eta_i + \left(t - \sum_{i=0}^{n-1} t_i\right) \eta_n \right\} \mathbb{1}\left\{ \sum_{i=0}^{n-1} t_i < t, t_i \geq 0 \forall i \leq n-1 \right\} \mathrm{d}t_0 \ldots \mathrm{d}t_{n-1}$$

$$= \mathrm{e}^{t\eta_n} \int_{\mathbb{R}_+^n} \exp\left\{ \sum_{i=0}^{n-1} t_i (\eta_i - \eta_n) \right\} \mathbb{1}\left\{ \sum_{i=0}^{n-1} t_i < t \right\} \mathrm{d}t_0 \ldots \mathrm{d}t_{n-1}$$

$$\leq \mathrm{e}^{t\eta_n} \int_{\mathbb{R}_+^n} \exp\left\{ \sum_{i=0}^{n-1} t_i (\eta_i - \eta_n) \right\} \mathrm{d}t_0 \ldots \mathrm{d}t_{n-1} = \mathrm{e}^{t\eta_n} \prod_{i<n} \frac{1}{\eta_n - \eta_i}.$$

Now we show that any permutation of the indices does not change the value of the integral above and this will be sufficient to prove the statement. First, it is obvious that transposition of i and j does not change the integral if $i, j \leq n-1$. Now we consider the case of a transposition τ on j and n, where $j < n$. We change variables such that $t_i' = t_i$ if $i \neq j, i \leq n-1$ and $t_j' = t - \sum_{i=0}^{n-1} t_i$, and get

$$\int_{\mathbb{R}_+^n} \exp\left\{\sum_{i=0}^{n-1} t_i \eta_i + \left(t - \sum_{i=0}^{n-1} t_i\right)\eta_n\right\} \mathbb{1}\left\{\sum_{i=0}^{n-1} t_i < t\right\} dt_0 \dots dt_{n-1}$$

$$= \int_{\mathbb{R}^n} \exp\left\{\sum_{i=0}^{n-1} t_i' \eta_{\tau(i)} - \left(t - \sum_{i=0}^{n-1} t_i'\right)\eta_{\tau(n)}\right\}$$

$$\times \mathbb{1}\left\{\sum_{i=0}^{n-1} t_i' < t, t_i' \geq 0 \forall i \leq n-1\right\} dt_0' \dots dt_{n-1}',$$

which completes the proof.

For the proof of Proposition 1 (b) denote by

$$\mathcal{P}^{(\delta,k,n)} = \Big\{ y = (y_0, \dots, y_n)\colon y_0 = 0,\ |y_{i-1} - y_i| = 1,$$

$$\{y_0, \dots, y_n\} \cap \{Z_n^{(1)}, \dots, Z_n^{(k-1)}\} = \emptyset,\ Z_n^{(k)} \in \{y_0, \dots, y_n\},\ |y_n - Z_n^{(k)}| \geq \delta r_t \Big\}$$

the set of all "good" paths and let (τ_i) be a sequence of independent, exponentially distributed random variables with parameter $2d$.

Denote by E the expectation with respect to (τ_i). We have

$$U_1^{(\delta,k,n)}(t) = \sum_{y \in \mathcal{P}^{(\delta,k,n)}} (2d)^{-n} \mathsf{E}\Bigg[\exp\left\{\sum_{i=0}^{n-1} \tau_i \xi(y_i) + \left(t - \sum_{i=0}^{n-1} \tau_i\right)\xi(y_n)\right\}$$

$$\times \mathbb{1}\left\{\sum_{i=0}^{n-1} \tau_i < t, \sum_{i=0}^{n} \tau_i > t\right\}\Bigg]. \tag{8}$$

In the further proof, we apply Lemma 9 to the values of the potential ξ along a path y. However, to do so we need the maximum of ξ along the path y to be attained only once. Therefore, we have to modify the potential along the path slightly.

We fix $y \in \mathcal{P}^{(\delta,k,n)}$ and let

$$i(y) = \min\left\{i \in \{0, \dots, n\}\colon y_i = Z_n^{(k)}\right\},$$

be the index of the first instant where the maximum of the potential over the path is attained. Now we define a slight variation of ξ on y in the following way. Fix $\varepsilon > 0$

and define $\xi^y\colon \{0,\dots,n\} \to \mathbb{R}$ by $\xi_i^y = \xi(y_i)$ if $i \neq i(y)$, and $\xi_{i(y)}^y = \xi(y_{i(y)}) + \varepsilon$. We obtain, using $\xi(y_i) \le \xi_i^y$, that

$$\begin{aligned}
&\mathsf{E}\Bigg[\exp\Bigg\{\sum_{i=0}^{n-1}\tau_i\xi(y_i) + \Bigg(t-\sum_{i=0}^{n-1}\tau_i\Bigg)\xi(y_n)\Bigg\}\mathbb{1}\Bigg\{\sum_{i=0}^{n-1}\tau_i < t, \sum_{i=0}^{n}\tau_i > t\Bigg\}\Bigg]\\
&\le \mathsf{E}\Bigg[\exp\Bigg\{\sum_{i=0}^{n-1}\tau_i\xi_i^y + \Bigg(t-\sum_{i=0}^{n-1}\tau_i\Bigg)\xi_n^y\Bigg\}\mathbb{1}\Bigg\{\sum_{i=0}^{n-1}\tau_i < t, \sum_{i=0}^{n}\tau_i > t\Bigg\}\Bigg]\\
&= (2d)^{n+1}\int_{\mathbb{R}_+^d}\exp\Bigg\{\sum_{i=0}^{n-1}t_i\xi_i^y + \Bigg(t-\sum_{i=0}^{n-1}t_i\Bigg)\xi_n^y\Bigg\}\\
&\qquad \mathbb{1}\Bigg\{\sum_{i=0}^{n-1}t_i < t, \sum_{i=0}^{n}t_i > t\Bigg\}\,\mathrm{e}^{-2d\sum_{i=0}^{n}t_i}\,\mathrm{d}t_0\dots\mathrm{d}t_{n-1}\mathrm{d}t_n\\
&= (2d)^n\mathrm{e}^{-2dt}\int_{\mathbb{R}_+^d}\exp\Bigg\{\sum_{i=0}^{n-1}t_i\xi_i^y + \Bigg(t-\sum_{i=0}^{n-1}t_i\Bigg)\xi_n^y\Bigg\}\mathbb{1}\Bigg\{\sum_{i=0}^{n-1}t_i < t\Bigg\}\,\mathrm{d}t_0\dots\mathrm{d}t_{n-1}\\
&\le (2d)^n\mathrm{e}^{-2dt}\mathrm{e}^{\xi_{i(y)}^y t}\prod_{i\neq i(y)}\frac{1}{\xi_{i(y)}^y - \xi_i^y},
\end{aligned} \tag{9}$$

where the last line follows from Lemma 9. Using the definition of our function ξ^y, we get

$$\begin{aligned}
\mathrm{e}^{\xi_{i(y)}^y t}\prod_{i\neq i(y)}\frac{1}{\xi_{i(y)}^y - \xi_i^y} &= \mathrm{e}^{(\xi(y_{i(y)})+\varepsilon)t}\prod_{i\neq i(y)}\frac{1}{\varepsilon + \xi(y_{i(y)}) - \xi(y_i)}\\
&\le \mathrm{e}^{(\xi(y_{i(y)})+\varepsilon)t}\varepsilon^{-n}\prod_{(\xi(y_{i(y)})-\xi(y_i))>1}\frac{1}{\xi(y_{i(y)}) - \xi(y_i)}.
\end{aligned} \tag{10}$$

Next recall that ρ is fixed, and $m_n = \lfloor n^\rho \rfloor$. Let

$$G_n = \{Z_n^{(1)},\dots,Z_n^{(m_n)}\} \subset \{z \in \mathbb{Z}^d\colon |z| \le n\},$$

and call the complement G_n^{c} the set of sites with very low potential. Note that there are at least $|Z_n^{(k)}| + \lfloor \delta r_t \rfloor - m_n$ points in the path y that belong to G_n^{c}. Hence, there are at least

$$|Z_n^{(k)}| + \lfloor \delta r_t \rfloor - m_n,$$

terms in the product in the left hand side of (10) that are smaller than

$$\big(M_n^{(k_n)} - M_n^{(m_n)}\big)^{-1},$$

provided this is less than 1. Combining this with (8), (9) and (10), we get

$$\begin{aligned}U_1^{(\delta,k,n)}(t) &\leq \sum_{y\in\mathscr{P}(\delta,k,n)} \varepsilon^{-n}\mathrm{e}^{\left(M_n^{(k)}+\varepsilon-2d\right)t}\left(M_n^{(k_n)}-M_n^{(m_n)}\right)^{-|Z_n^{(k)}|-\lfloor\delta r_t\rfloor+m_n}\\ &\leq (2d)^n\varepsilon^{-n}\mathrm{e}^{\left(M_n^{(k)}+\varepsilon-2d\right)t}\left(\tfrac{\rho-\sigma}{2}\log n\right)^{-|Z_n^{(k)}|-\lfloor\delta r_t\rfloor+m_n}.\end{aligned}$$

Taking the log of the above and defining $C_\varepsilon := \log(\frac{2d}{\varepsilon}) - \log(\frac{\rho-\sigma}{2})$ we get

$$\begin{aligned}\frac{1}{t}\log U_1^{(\delta,k,n)}(t) &\leq \tfrac{n}{t}\log\tfrac{2d}{\varepsilon} + M_n^{(k)} - 2d + \varepsilon - \tfrac{1}{t}\left(|Z_n^{(k)}| + \lfloor\delta r_t\rfloor - m_n\right)\\ &\qquad \times\log\left(\tfrac{\rho-\sigma}{2}\log n\right)\\ &\leq M_n^{(k)} - \tfrac{1}{t}|Z_n^{(k)}|\log\log|Z_n^{(k)}| - 2d + \varepsilon + \tfrac{n}{t}C_\varepsilon - \delta\,\tfrac{\log\log n}{\log\log t} + o(1),\end{aligned}$$

where we use that $|Z_n^{(k)}| + \lfloor\delta r_t\rfloor \leq n$. Observing that $\log\log n \geq (1+o(1))\log\log t$ and $\frac{n}{t}C_\varepsilon = o(1)$, uniformly for all n in the given range, concludes the proof of (b).

To prove part (a) we show that regardless of the distance travelled by the path, it hits a site with very low potential in every other step. Recall that a set H of vertices of $\mathbb{Z}^d$ is *totally disconnected* if there is no pair of vertices $(x, y) \in H^2$ such that $|x - y| = 1$.

Lemma 10. *Almost surely, for sufficiently large n, the set G_n is totally disconnected.*

Proof. We prove the statement for $d \geq 2$ first. If i and j are distinct integers in $\{1,\dots,m_n\}$, the random pair of points $(Z_n^{(i)}, Z_n^{(j)})$ is uniformly distributed over all possible pairs of points in the ball of radius n. As no vertex has more than $2d$ neighbours, we have $P(Z_n^{(i)} \sim Z_n^{(j)}) \leq 2d/l_n$. Summing over all possible pairs i, j we get

$$P\left(G_n \text{ not totally disconnected}\right) \leq \sum_{i<j} P\left(Z_n^{(i)} \sim Z_n^{(j)}\right) \leq \binom{m_n}{2}\frac{2d}{l_n} \leq Cn^{2\rho-d}. \tag{11}$$

for some constant C. Since $\rho < \frac{1}{2}$ and $d \geq 2$ we can apply the Borel–Cantelli lemma and obtain the result. We now prove the the same result when $d = 1$. We introduce a new quantity

$$m_n' = \lfloor n^{\rho'}\rfloor \text{ with } \rho < \rho' < \tfrac{1}{2}.$$

Let G_n' be the set of the m_n' vertices in the ball of radius n where the biggest values of ξ are taken, and let p_n be the biggest integer power of 2, which is less than n. Note that, by (11), the set G_{p_n}' is totally disconnected for all sufficiently large n.

We now prove that

$$G_n \subseteq G'_{2p_n} \quad \text{for all sufficiently large } n. \tag{12}$$

For this it suffices to show that at least m_n points of G'_{2p_n} are in the ball of radius n. Indeed, if we assume this and also that $G_n \nsubseteq G'_{2p_n}$ we can find a vertex z_0 satisfying, $|z_0| \leq n$, $z_0 \in G'_{2p_n}$ and $z_0 \notin G_n$. This implies that every $z \in G_n$ satisfies $\xi(z) > \xi(z_0)$, because G_n is the set where the largest values of ξ are achieved. Then, because $z_0 \in G'_{2p_n}$, we have

$$G_n \subseteq \{\xi(z) > \xi(z_0)\} \cap B_n \subseteq \{\xi(z) > \xi(z_0)\} \cap B_{2p_n} \subseteq G'_{2p_n},$$

which leads to a contradiction to our assumption.

In fact, we will prove the slightly stronger statement that there are at least m_{2p_n} vertices of G'_{2p_n} in the ball of radius p_n, and we will now write p instead of p_n. We write

$$G'_{2p} = \{a'_0, \ldots, a'_{m'_{2p}-1}\},$$

where a'_i is the vertex where $\xi(a'_i) = M_{2p}^{(i+1)}$ and introduce

$$X = (X_i)_{0 \leq i \leq m'_{2p}-1} \text{ with } X_i = \mathbb{1}_{\{|a'_i| \leq p\}} \text{ and } |X| = \sum_{i=0}^{m'_{2p}-1} X_i.$$

Observing that $m'_{2p} = o(p)$ and that G'_{2p} is uniformly distributed over all possible ordered sets and recalling that the box of radius p contains $2p+1$ vertices, it is easy to see that for p big enough,

$$P\big(X_j = 1 \,\big|\, X_i = x_i, \forall i < j\big) < \tfrac{3}{4} \text{ and } P\big(X_j = 0 \,\big|\, X = x_i, \forall i < j\big) < \tfrac{3}{4},$$

for all $j \leq m'_{2p} - 1$ and for all fixed $(x_0, \ldots, x_{j-1}) \in \{0,1\}^j$. Hence

$$\begin{aligned}
P\big(|X| < m_{2p}\big) &= \sum_{i=0}^{m_{2p}-1} \sum_{|x|=i} P(X = x) \\
&\leq \sum_{i=0}^{m_{2p}-1} \binom{m'_{2p}}{i} \left(\frac{3}{4}\right)^{m'_{2p}} \leq m_{2p} \left(m'_{2p}\right)^{m_{2p}-1} \left(\frac{3}{4}\right)^{m'_{2p}} \\
&= \exp\Big(-m'_{2p} \log(4/3) + (m_{2p} - 1) \log m'_{2p} + \log m_{2p}\Big) \\
&= \mathrm{e}^{-(2p)^{\rho'}(1+o(1))} \leq \mathrm{e}^{-n^{\rho'}} \quad \text{as } n \leq 2p_n.
\end{aligned}$$

Using the Borel–Cantelli lemma we can prove (12), which implies the statement.

We define the set of paths $\mathscr{P}_n$ to be

$$\mathscr{P}_n = \Big\{ y = (y_0, \dots, y_n) \colon y_0 = 0,\ |y_{i-1} - y_i| = 1,$$
$$\{y_0, \dots, y_n\} \cap \{Z_n^{(1)}, \dots, Z_n^{(k_n-1)}\} \neq \emptyset \Big\},$$

so that

$$U_1^{(n)}(t) = \sum_{y \in \mathscr{P}_n} (2d)^{-n} \mathsf{E}\Bigg[\exp\Big\{ \sum_{i=0}^{n-1} \tau_i \xi(y_i) + \Big(t - \sum_{i=0}^{n-1} \tau_i\Big) \xi(y_n) \Big\}$$
$$\times \mathbb{1}\Big\{ \sum_{i=0}^{n-1} \tau_i < t, \sum_{i=0}^{n} \tau_i > t \Big\} \Bigg].$$

We can now argue similarly as for part (b) but using this time the fact that for any path in $\mathscr{P}_n$ the number of step out of G_n is at least $\lfloor n/2 \rfloor$. More precisely,

$$U_1^{(n)}(t) \leq \sum_{y \in \mathscr{P}_n} \varepsilon^{-n} \mathrm{e}^{(M_n^{(1)} + \varepsilon - 2d)t} \big(M_n^{(k_n)} - M_n^{(m_n)}\big)^{-\lfloor n/2 \rfloor},$$

and taking the log of the above and defining $C_\varepsilon := 2\log\big(\frac{2d}{\varepsilon}\big) - \log\big(\frac{\rho-\sigma}{2}\big)$, we get

$$\begin{aligned}\frac{1}{t} \log U_1^{(n)}(t) &\leq \tfrac{n}{t} \log \tfrac{2d}{\varepsilon} + M_n^{(1)} - 2d + \varepsilon - \tfrac{1}{t} \lfloor n/2 \rfloor \log\big(\tfrac{\rho-\sigma}{2} \log n\big) \\ &= M_n^{(1)} - \tfrac{n}{2t}\big(\log\log n - C_\varepsilon\big) - 2d + \varepsilon + o(1),\end{aligned}$$

which concludes the proof of (a).

2.4 Analysis of the Variational Problem

We use the point process framework established in [7, Sect. 2.2] adapting the approach of [13, Chap. 3]. We only give an outline of the framework and sketched proofs here, see [7, Sect. 2.2] for more details.

Observe that $\mu(dy) := e^{-y}\, dy$ is a Radon measure on $G := (-\infty, \infty]$. For any $z \in \mathbb{Z}^d$, $x \in \mathbb{R}$ and $r > 0$, we have

$$r^d\, P\big(\xi(z) - d \log r \geq x\big) = r^d\, e^{-d \log r - x} = \mathrm{e}^{-x} = \mu([x, \infty]).$$

Define, for any $q, \tau > 0$ the set $H_\tau^q := \{(x, y) \in \dot{\mathbb{R}}^d \times G \colon y \geq q|x| + \tau\}$, where $\dot{\mathbb{R}}^d$ is the one-point compactification of $\mathbb{R}^d$. As in [7, Lemma 4.3], we see that the point process

$$\zeta_r = \sum_{z\in\mathbb{Z}^d} \delta\big((\tfrac{z}{r}, \xi(z) - d\log r)\big),$$

converges in law to the Poisson process ζ with intensity $\mathrm{Leb}_d \otimes \mu$ in the sense that, for any pairwise disjoint compact sets $K_1, \ldots, K_n \subset H_\tau^q$ with $\mathrm{Leb}_{d+1}(\partial K) = 0$, we have that $(\zeta_r(K_1), \ldots, \zeta_r(K_n))$ converge in law to

$$\bigotimes_{i=1}^n \mathrm{Poiss}\big(\mathrm{Leb}_d \otimes \mu(K_i)\big).$$

We further note that for $|z| = t^{1+o(1)}$, we have

$$\psi_t(z) := \xi(z) - \frac{|z|}{t}\log\log|z| = \xi(z) - \frac{|z|}{r_t}\big(1 + o(1)\big).$$

As in [7, Lemma 4.4] applied to $T_t(z,x) := (z, x - |z|)$, we infer from this the convergence of the point process

$$\varpi_t := \sum_{z\in\mathbb{Z}^d} \delta\big((\tfrac{z}{r_t}, \psi_t(z) - d\log r_t)\big),$$

in law to a Poisson process ϖ with intensity

$$\big(\mathrm{Leb}_d \otimes \mu\big) \circ T_t^{-1} = \mathrm{e}^{-|z|-y}\,\mathrm{d}z\,\mathrm{d}y,$$

where now the compact sets $K_1, \ldots, K_n$ can be chosen from the set $H_\tau := \dot{\mathbb{R}}^{d+1} \setminus (\mathbb{R}^d \times (-\infty, \tau))$. The form of these and the previous domains and in particular the use of the compactification, ensure that we can use these convergence results to analyse the right hand side of the final formula in Proposition 1.

Lemma 11. *Let $X_t^{(1)}$ and $X_t^{(2)}$ be the sites corresponding to the largest and second largest value of $\psi_t(z)$, $z \in \mathbb{Z}^d$. Then, $\psi_t(X_t^{(1)}) - \psi_t(X_t^{(2)})$ converges in law to a standard exponential random variable.*

Proof. Using careful arguments in the convergence step we obtain, for any $a \geq 0$,

$$\begin{aligned}
&P\big(\psi_t(X_t^{(1)}) - \psi_t(X_t^{(2)}) \geq a\big)\\
&\quad= \sum_y P\big(\varpi_t\big(\mathbb{R}^d \times (y,\infty)\big) = 0, \varpi_t(\mathbb{R}^d \times \{y\}) = 1, \varpi_t\big(\mathbb{R}^d \times (y-a, y)\big) = 0\big)\\
&\quad\to \int P\big(\varpi\big(\mathbb{R}^d \times (y,\infty)\big) = 0\big) P\big(\varpi\big(\mathbb{R}^d \times (y-a, y)\big) = 0\big) e^{-y}\,\mathrm{d}y\\
&\quad= \int \exp(-e^{-y+a}) e^{-y}\,\mathrm{d}y = e^{-a}.
\end{aligned}$$

Lemma 12. *Let $X_t^{(1)}$ be the site corresponding to the largest value of $\psi_t(z)$, $z \in \mathbb{Z}^d$. Then, $X_t^{(1)}/r_t$ converges in law to a random variable in $\mathbb{R}^d$ with coordinates given by independent standard exponential variables with uniform random signs.*

Proof. As above we obtain, for any $A \subset \mathbb{R}^d$ Borel with $\mathrm{Leb}_d(\partial A) = 0$,

$$P\left(\frac{X_t^{(1)}}{r_t} \in A\right) = \sum_y P\big(\varpi_t\big(\mathbb{R}^d \times (y,\infty)\big) = 0, \varpi_t\big(A \times \{y\}\big) = 1\big)$$

$$\to \int_A \mathrm{d}z \int \mathrm{d}y\, \mathrm{e}^{-|z|-y}\, P\big(\varpi\big(\mathbb{R}^d \times (y,\infty)\big) = 0\big)$$

$$= \int_A \mathrm{d}z \int \mathrm{d}y\, \exp(-e^{-y})\mathrm{e}^{-y-|z|} = \int_A 2^{-d}\, \mathrm{e}^{-|z|}\, \mathrm{d}z.$$

Observe that this implies that the limit variable has the given distribution.

2.5 Proof of the Almost Sure Asymptotics

Note that combining Lemma 1 and Lemma 6 establishes the almost sure lower bound for the liminf result in Theorem 2. To find a matching upper bound, recall from Lemma 2 that, for sufficiently large t,

$$L_t \le N_\varepsilon(t) - 2d + \varepsilon$$

for $N_\varepsilon(t) := \overline{N}_{c(\varepsilon)}(t)$. We now approximate the distribution of $N_\varepsilon(t)$.

Lemma 13 (Approximation for the distribution of $N_\varepsilon(t)$). *Let $b_t \uparrow \infty$, then*

$$\log\big(P\big(N_\varepsilon(t) \le b_t\big)\big) = -\mathrm{e}^{-b_t} r_t^d 2^d\, (1 + o(1))\,.$$

Proof. Observe that

$$P\,(N_\varepsilon(t) \le b_t) = \prod_{t/(\log t)^2 \le |z| \le t\log t} F\left(b_t + \frac{|z|}{t}\,(\log\log|z| - C_\varepsilon)\right).$$

The values which $|z|$ can take are such that $\log\log|z| = \log\log t + o(1)$ uniformly for all z, and since $b_t \to \infty$, we have,

$$\log\big(P\big(N_\varepsilon(t) \le b_t\big)\big)$$

$$= \sum_{t/(\log t)^2 \le |z| \le t\log t} \log\left(1 - \exp\big(-b_t - \tfrac{|z|}{t}\big(\log\log t - C_\varepsilon + o(1)\big)\big)\right)$$

$$= -(1+o(1)) \sum_{t/(\log t)^2 \le |z| \le t \log t} \mathrm{e}^{-b_t - \frac{|z|}{r_t}(1+o(1))}$$

$$= -\mathrm{e}^{-b_t} r_t^d \,(1+o(1)) \int_{\mathbb{R}^d} e^{-|x|(1+o(1))} \mathbb{1}_{\{\log\log t/(\log t)^2 \le |x| \le \log t \,\log\log t\}} \,\mathrm{d}x$$

To obtain our final result, we apply the dominated convergence theorem to the integral, which converges to 2^d.

We are now ready to prove the upper bound. We consider a sequence of times $t_n := \exp(n^2)$ for which $N_\varepsilon(t_n)$ are independent random variables, in order to use Borel–Cantelli.

Lemma 14 (Upper bound for lower envelope of $N_\varepsilon(t_n)$). *For any small $c > 0$, almost surely there are infinitely many n such that*

$$N_\varepsilon(t_n) \le d \log t_n - (1+d-c)\log\log\log t_n.$$

Proof. Note that $(N_\varepsilon(t_n))_{n \ge N}$ is a sequence of independent variables if N is large enough. To see this, it suffices to notice that the different $(N_\varepsilon(t_n))_{n\ge N}$ depend on the values of the potential on disjoints areas. Indeed

$$\frac{t_{n+1}}{(\log t_{n+1})^2} = \frac{\exp\left(n^2+2n+1\right)}{(n+1)^4} > n^2 \exp\left(n^2\right) = t_n \log t_n \text{ for all large } n.$$

Now we use Lemma 13 with $b_t = d \log t - (1+d-c)\log\log\log t$ and we get,

$$\log\left(P\left(N_\varepsilon(t_n) \le b_{t_n}\right)\right) = -2^d \left(\log\log t_n\right)^{1-c} (1+o(1)) \ge -\log n,$$

for all sufficiently large n. Hence, the sum over the probabilities diverges and we obtain our result by applying the converse of the Borel–Cantelli lemma.

2.6 *Proof of the Weak Asymptotics*

To prove Theorem 1, we show that the upper and lower bounds we found earlier for L_t both satisfy the required limit statement. We first state the result of [7, Proposition 4.12], which describes the limit result for the lower bound $\underline{N}(t)$.

Lemma 15 (Weak asymptotics for $\underline{N}(t)$). *As t tends to infinity,*

$$\underline{N}(t) - d \log t + d \log\log\log t \Rightarrow X, \quad \text{where} \quad P(X \le x) = \exp\left(-2^d e^{-x}\right).$$

Next, we check the analogous limit theorem for the upper bound $N_\varepsilon(t)$ and thus complete the proof of Theorem 1.

Lemma 16 (Weak asymptotics for $N_\varepsilon(t)$). *As t tends to infinity,*

$$N_\varepsilon(t) - d\log t + d\log\log\log t \Rightarrow X, \quad \textit{where} \quad P(X \le x) = \exp\big(-2^d e^{-x}\big).$$

Proof. Fix $x \in \mathbb{R}$ and apply Lemma 13 with $b_t = d\log t - d\log\log\log t + x$ to get

$$\log\big(P\big(N_\varepsilon(t) - d\log t + d\log\log\log t \le x\big)\big) = -e^{-x}2^d\ (1+o(1)),$$

which proves our result.

2.7 *Proof of the Scaling Limit Theorem*

We recall that $X_t^{(k)}$ $(k = 1,\ 2)$ is the site at which

$$\psi_t(z) = \xi(z) - \frac{|z|}{t}\log\log|z|$$

takes its kth largest value. Fix $\delta > 0$ and write

$$U(t) = U_1(t) + U_2(t) + U_3(t) + U_4(t) + U_5(t) + U_6(t),$$

where U_2 and U_3 were defined in Lemma 7, resp. Lemma 8, and

$$U_1(t) = \mathbb{E}\Bigg[\exp\Bigg\{\int_0^t \xi(X_s)\,\mathrm{d}s\Bigg\}\,\mathbb{1}\Bigg\{\tfrac{2t}{(\log t)^2} \le J_t \le ta_t,\ \max_{0\le s\le t}\xi(X_s) > M_{J_t}^{(kJ_t)}\Bigg\}$$
$$\mathbb{1}\big\{X_t^{(1)} \in \{X_s : 0 \le s \le t\},\ |X_t - X_t^{(1)}| \le \delta r_t\big\}\Bigg],$$

$$U_4(t) = \mathbb{E}\Bigg[\exp\Bigg\{\int_0^t \xi(X_s)\,\mathrm{d}s\Bigg\}\,\mathbb{1}\big\{ta_t < J_t \le t\log t,\ \max_{0\le s\le t}\xi(X_s) > M_{J_t}^{(kJ_t)}\big\}\Bigg],$$

$$U_5(t) = \mathbb{E}\Bigg[\exp\Bigg\{\int_0^t \xi(X_s)\,\mathrm{d}s\Bigg\}\,\mathbb{1}\big\{\tfrac{2t}{(\log t)^2} \le J_t \le ta_t,\ \max_{0\le s\le t}\xi(X_s) > M_{J_t}^{(kJ_t)}\big\}$$
$$\mathbb{1}\big\{X_t^{(1)} \notin \{X_s : 0 \le s \le t\}\big\}\Bigg],$$

$$U_6(t) = \mathbb{E}\Bigg[\exp\Bigg\{\int_0^t \xi(X_s)\,\mathrm{d}s\Bigg\}\,\mathbb{1}\big\{\tfrac{2t}{(\log t)^2} \le J_t \le ta_t,\ \max_{0\le s\le t}\xi(X_s) > M_{J_t}^{(kJ_t)}\big\}$$
$$\mathbb{1}\big\{X_t^{(1)} \in \{X_s : 0 \le s \le t\},\ |X_t - X_t^{(1)}| > \delta r_t\big\}\Bigg].$$

Observe that our result follows if the contributions of $U_i(t)$ for $i = 2, \ldots, 6$ to the total mass are negligible, as $U_1(t)$ only contributes to the mass distributed on points close to $X_t^{(1)}$ on the r_t scale.

Lemma 17. *Suppose $a_t \downarrow 0$ and $a_t \log\log t \to \infty$. Then we have, in probability,*

$$\lim_{t\uparrow\infty} \frac{U_4(t)}{U(t)} = \lim_{t\uparrow\infty} \frac{U_5(t)}{U(t)} = \lim_{t\uparrow\infty} \frac{U_6(t)}{U(t)} = 0.$$

Proof. For the first statement we use Proposition 1 (a) to see that

$$\frac{1}{t} \log U_4(t) \leq \sup_{n \geq t a_t} \left\{ M_n^{(1)} - \frac{n}{2t}\left(\log\log n - C_\varepsilon\right) \right\} + \varepsilon - 2d + o(1).$$

By Lemmas 12 and 11, the limit of the right hand side is strictly smaller than the growth rate of $U(t)$, proving that the first limit in the statement equals zero.

Using Proposition 1 (b) with $\delta = 0$ and summing over all $1 \leq k \leq t^\sigma$ with $X_t^{(1)} \neq Z_n^{(k)}$, and over all n with $2t/(\log t)^2 \leq n \leq t a_t$ we get

$$\frac{1}{t} \log U_5(t) \leq \max_{z \setminus \{X_t^{(1)}\}} \psi_t(z) - 2d + o(1) = \psi_t(X_t^{(2)}) - 2d + o(1) \quad \text{in probability.}$$

By Lemma 11, we find $\varepsilon > 0$ such that, with a probability arbitrarily close to one

$$\frac{1}{t} \log U_5(t) \leq \psi_t(X_t^{(1)}) - 2d - \varepsilon + o(1),$$

and a comparison with the lower bound $\underline{N}(t)$ for the growth rate of $U(t)$ proves the second result.

For the third statement, we use Proposition 1 (b) with the choice of $\delta > 0$ from the statement. Summing over all $1 \leq k \leq t^\sigma$ and n with $2t/(\log t)^2 \leq n \leq t a_t$ we get, as above,

$$\frac{1}{t} \log U_6(t) \leq \psi_t(X_t^{(1)}) - 2d - \delta + o(1).$$

We can now argue as before that this rate is strictly smaller than the lower bound $\underline{N}(t)$ for $U(t)$, proving the final statement.

We can now complete the proof of Theorem 3. By definition, we have

$$1 \geq \liminf_{t\uparrow\infty} \nu_t\{z \in \mathbb{Z}^d \mid |z - X_t^{(1)}| \leq \delta r_t\} \geq \liminf_{t\uparrow\infty} \frac{U_1(t)}{U(t)} = 1 - \limsup_{t\uparrow\infty} \sum_{j=2}^{6} \frac{U_j(t)}{U(t)}.$$

Combining Lemmas 7, 8 and 17 we see that the limsup is zero, so that we get

$$\lim_{t\uparrow\infty} \nu_t\{z \in \mathbb{Z}^d \mid |z - X_t^{(1)}| \leq \delta r_t\} = 1 \quad \text{in probability.}$$

Combining this with the convergence of $X_t^{(1)}/r_t$ given in Lemma 12 and recalling that $\delta > 0$ was arbitrary concludes the proof.

3 Concluding Remarks

It would be interesting to study scaling limit theorems for potentials with lighter tails and thus shed further light on the number of relevant islands in these cases.

The techniques of the present paper appear suitable to treat cases where the relevant islands are single sites, which is the case for potentials heavier than the double-exponential distributions. For the double-exponential distribution itself and lighter tails, arguments related to classical order statistics of i.i.d. random variables need to be replaced by eigenvalue order statistics for the random Schrödinger operator $\Delta + \xi$ on $\ell^2(\mathbb{Z}^d)$, making the problem much more complex. Work in an advanced state of progress by Biskup and König [1] deals with the double-exponential case and strongly hints at localization in a single island of finite size in this and other cases of unbounded potentials.

For *bounded* potentials, the question of the number of relevant islands and the formulation of a scaling limit theorem at present seems wide open and constitutes an attractive research project. Sznitman in [14] discusses an "elliptic version" of the Anderson problem, describing Brownian paths in a Poissonian potential conditioned to reach a remote location. Sznitman's technique of enlargement of obstacles, described in [16], offers a possible approach to the scaling limit theorem, leading in [15] to a study of fluctuations of the principal eigenvalues of the operator $\Delta + \xi$ and moreover an analysis of variational problems somewhat similar to those that we expect to arise in the proof of a scaling limit theorem.

In the light of our result and this discussion, it would be of particular interest to know whether there at all exist potentials which lead to more than one relevant island, and if so, to find the nature and location of the transition between phases of one and several islands.

Acknowledgements: Special thanks are due to the organizers of the *Workshop on Random Media*, in celebration of Jürgen Gärtner's 60th birthday, which provided an ideal forum for discussing the problems raised in this paper. The first author acknowledges the support of ERC grant PTRELSS. The second author is grateful for the support of EPSRC through an Advanced Research Fellowship.

References

1. Biskup, M., König, W.: Eigenvalue order statistics for random Schrödinger operators with doubly exponential tails. In preparation (2010)
2. Caravenna, F., Carmona, P., Pétrélis, N.: The discrete-time parabolic Anderson model with heavy-tailed potential. Preprint arXiv:1012.4653v1[math.PR] (2010)

3. Gärtner, J., König, W., Molchanov, S.: Geometric characterisation of intermittency in the parabolic Anderson model. Ann. Probab. **35**, 439–499 (2007)
4. Gärtner, J., Molchanov, S.: Parabolic problems for the Anderson model. I. Intermittency and related topics. Commun. Math. Phys. **132**, 613–655 (1990)
5. Gärtner, J., Molchanov, S.: Parabolic problems for the Anderson model. II. Second-order asymptotics and structure of high peaks. Probab. Theory Relat. Fields **111**, 17–55 (1998)
6. van der Hofstad, R., König, W., Mörters, P.: The universality classes in the parabolic Anderson model. Commun. Math. Phys. **267**, 307–353 (2006)
7. van der Hofstad, R., Mörters, P., Sidorova, N.: Weak and almost sure limits for the parabolic Anderson model with heavy tailed potentials. Ann. Appl. Probab. **18**, 2450–2494 (2008)
8. König, W., Lacoin, H., Mörters, P., Sidorova, N.: A two cities theorem for the parabolic Anderson model. Ann. Probab. **37**, 347–392 (2009)
9. Lacoin, H.: Calcul d'asymptotique et localization p.s. pour le modèle parabolique d'Anderson. Mémoire de Magistère, ENS, Paris (2007)
10. Mörters, P.: The parabolic Anderson model with heavy-tailed potential. In: Surveys in Stochastic Processes. Proceedings of the 33rd SPA Conference in Berlin, 2009. Edited by J. Blath, P. Imkeller, and S. Roelly. pp. 67–85. EMS Series of Congress Reports (2011)
11. Mörters, P., Ortgiese, M., Sidorova, N.: Ageing in the parabolic Anderson model. Annales de l'Institut Henri Poincaré: Probab. et Stat. **47**, 969–1000 (2011)
12. Mörters, P.: The parabolic Anderson model with heavy-tailed potential. To appear in: 'Surveys in Stochastic Processes'. J. Blath et al. (eds.) EMS Conference Reports (2011)
13. Resnick, S.I.: Extreme values, regular variation, and point processes. Springer Series in OR and Financial Engineering. Springer, New York (1987)
14. Sznitman, A.-S.: Crossing velocities and random lattice animals. Ann. Probab. **23**, 1006–1023 (1995)
15. Sznitman, A.-S.: Fluctuations of principal eigenvalues and random scales. Commun. Math. Phys. **189**, 337–363 (1997)
16. Sznitman, A.-S.: Brownian motion, obstacles and random media. Springer, New York (1998)

Part II
Self-Interacting Random Walks and Polymers

The Strong Interaction Limit of Continuous-Time Weakly Self-Avoiding Walk

David C. Brydges, Antoine Dahlqvist, and Gordon Slade

Abstract The strong interaction limit of the discrete-time weakly self-avoiding walk (or Domb–Joyce model) is trivially seen to be the usual strictly self-avoiding walk. For the continuous-time weakly self-avoiding walk, the situation is more delicate, and is clarified in this paper. The strong interaction limit in the continuous-time setting depends on how the fugacity is scaled, and in one extreme leads to the strictly self-avoiding walk, in another to simple random walk. These two extremes are interpolated by a new model of a self-repelling walk that we call the "quick step" model. We study the limit both for walks taking a fixed number of steps and for the two-point function.

Dedicated to Erwin Bolthausen and Jürgen Gärtner on the occasion of their 65th and 60th birthday celebration

1 Domb–Joyce Model: Discrete Time

The discrete-time weakly self-avoiding walk, or Domb–Joyce model [6], is a useful adaptation of the strictly self-avoiding walk that continues to be actively studied [1].

D.C. Brydges (✉)
Department of Mathematics, University of British Columbia, Vancouver, BC, Canada V6T 1Z2
e-mail: db5d@math.ubc.ca

A. Dahlqvist
LPMA-UPMC (Paris 6), Case courrier 188 LPMA, 4 Place Jussieu, 75252 PARIS Cedex 05, France
e-mail: antoine.dahlqvist@upmc.fr

G. Slade
Department of Mathematics, University of British Columbia, Vancouver, BC, Canada V6T 1Z2,
e-mail: slade@math.ubc.ca

J.-D. Deuschel et al. (eds.), *Probability in Complex Physical Systems*,
Springer Proceedings in Mathematics 11, DOI 10.1007/978-3-642-23811-6_11,

It is defined as follows. For simplicity, we restrict attention to the nearest-neighbor model on $\mathbb{Z}^d$, although a more general formulation is easy to obtain.

Let $d \geq 1$ and $n \geq 0$ be integers, and let $\mathscr{W}_{\mathrm{n}}$ denote the set of nearest-neighbor walks in $\mathbb{Z}^d$, of length n, which start from the origin. In other words, $\mathscr{W}_{\mathrm{n}}$ consists of sequences $Y = (Y_0, Y_1, \ldots, Y_n)$ with $Y_i \in \mathbb{Z}^d$, $Y_0 = 0$, $|Y_{i+1} - Y_i| = 1$ (Euclidean distance). Let $\mathscr{S}_{\mathrm{n}}$ denote the set of nearest-neighbor self-avoiding walks in $\mathscr{W}_{\mathrm{n}}$; these are the walks with $Y_i \neq Y_j$ for all $i \neq j$. Let c_n denote the cardinality of $\mathscr{S}_{\mathrm{n}}$. For $Y \in \mathscr{W}_{\mathrm{n}}$ and $x \in \mathbb{Z}^d$, let $n_x = n_x(Y) = \sum_{i=0}^{n} \mathbb{1}_{Y_i = x}$ denote the number of visits to x by Y. The Domb–Joyce model is the measure on $\mathscr{W}_{\mathrm{n}}$ which assigns to a walk $Y \in \mathscr{W}_{\mathrm{n}}$ the probability

$$P_{g,n}^{\mathrm{DJ}}(Y) = \frac{1}{c_n^{\mathrm{DJ}}(g)} \mathrm{e}^{-g \sum_{x \in \mathbb{Z}^d} n_x(Y)(n_x(Y)-1)}, \tag{1}$$

where g is a positive parameter and

$$c_n^{\mathrm{DJ}}(g) = \sum_{Y \in \mathscr{W}_{\mathrm{n}}} e^{-g \sum_{x \in \mathbb{Z}^d} n_x(Y)(n_x(Y)-1)}. \tag{2}$$

The Domb–Joyce model interpolates between simple random walk and self-avoiding walk. Indeed, the case $g = 0$ corresponds to simple random walk by definition, and also

$$\lim_{g \to \infty} \mathrm{e}^{-g \sum_{x \in \mathbb{Z}^d} n_x(Y)(n_x(Y)-1)} = \mathbb{1}_{Y \in \mathscr{S}_{\mathrm{n}}}, \tag{3}$$

and hence

$$\lim_{g \to \infty} P_{g,n}^{\mathrm{DJ}}(Y) = \frac{1}{c_n} \mathbb{1}_{Y \in \mathscr{S}_{\mathrm{n}}}. \tag{4}$$

This shows that the strong interaction limit of the Domb–Joyce model is the uniform measure on $\mathscr{S}_{\mathrm{n}}$. (For an analogous result for weakly self-avoiding lattice trees, which is more subtle than for self-avoiding walks, see [2].)

A standard subadditivity argument (see, e.g., [10, Lemma 1.2.2]) implies that the limits

$$\mu(g) = \lim_{n \to \infty} c_n^{\mathrm{DJ}}(g)^{1/n}, \qquad \mu = \lim_{n \to \infty} c_n^{1/n}, \tag{5}$$

exist and obey $c_n^{\mathrm{DJ}}(g) \geq \mu(g)^n$ and $c_n \geq \mu^n$ for all n. The number of walks that take steps only in the positive coordinate directions is d^n, and such walks are self-avoiding, so $c_n \geq d^n$. Also, it follows from (2) that if $0 \leq g < g_0$ then $(2d)^n \geq c_n^{\mathrm{DJ}}(g) \geq c_n^{\mathrm{DJ}}(g_0) \geq c_n \geq d^n$, and hence $2d \geq \mu(g) \geq \mu(g_0) \geq \mu \geq d$. In particular, by monotonicity, $\lim_{g \to \infty} \mu(g)$ exists in $[\mu, 2d]$. If we take the limit $g \to \infty$ in the inequality $c_n^{\mathrm{DJ}}(g) \geq \mu(g)^n \geq \mu^n$, we obtain $c_n \geq (\lim_{g \to \infty} \mu(g))^n \geq \mu^n$. Taking nth roots and then the limit $n \to \infty$ then gives

$$\lim_{g \to \infty} \mu(g) = \mu. \tag{6}$$

Let $\mathscr{W}_{\rm n}(x)$ denote the subset of $\mathscr{W}_{\rm n}$ consisting of walks that end at $x \in \mathbb{Z}^d$. Let $\mathscr{S}_{\rm n}(x) = \mathscr{S}_{\rm n} \cap \mathscr{W}_{\rm n}(x)$, and let $c_n(x)$ denote the cardinality of $\mathscr{S}_{\rm n}(x)$. Let

$$c_{n,g}^{\rm DJ}(x) = \sum_{Y \in \mathscr{W}_{\rm n}(x)} {\rm e}^{-g \sum_{\tilde{x} \in \mathbb{Z}^d} n_{\tilde{x}}(Y)(n_{\tilde{x}}(Y)-1)}. \tag{7}$$

Let $z \geq 0$. The two-point functions of the Domb–Joyce and self-avoiding walk models are defined as follows:

$$G_{g,z}^{\rm DJ}(x) = \sum_{n=0}^{\infty} c_{n,g}^{\rm DJ}(x) z^n, \qquad G_z(x) = \sum_{n=0}^{\infty} c_n(x) z^n. \tag{8}$$

These series converge for $z < \mu(g)^{-1}$ and $z < \mu^{-1}$, respectively. Presumably they converge also for $z = \mu(g)^{-1}$ and $z = \mu^{-1}$ but this is a delicate question that is unproven except in high dimensions (in fact, the decay of the two-point function with $z = \mu^{-1}$ is known in some cases [4, 8, 9]). The following proposition shows that the strong interaction limit of $G_{g,z}^{\rm DJ}(x)$ is $G_z(x)$.

Proposition 1. *For* $z \in [0, \mu^{-1})$ *and* $x \in \mathbb{Z}^d$,

$$\lim_{g \to \infty} G_{g,z}^{\rm DJ}(x) = G_z(x). \tag{9}$$

Proof. Fix $z \in [0, \mu^{-1})$. By (6), if g_0 is sufficiently large then $z < \mu(g_0)^{-1}$. Thus, since $c_n^{\rm DJ}(g)$ is nonincreasing in g, there are $r < 1$ and $C > 0$ such that $c_n^{\rm DJ}(g)z^n \leq c_n^{\rm DJ}(g_0)z^n \leq C r^n$ for all n, uniformly in $g \geq g_0$. Thus, for all $g \geq g_0$,

$$G_{g,z}^{\rm DJ}(x) \leq \sum_{x \in \mathbb{Z}^d} G_{g,z}^{\rm DJ}(x) = \sum_{n=0}^{\infty} c_n^{\rm DJ}(g) z^n \leq \frac{C}{1-r} < \infty. \tag{10}$$

By (3), $\lim_{g \to \infty} c_{n,g}^{\rm DJ}(x) = c_n(x)$, and the desired result then follows by dominated convergence. □

2 The Continuous-Time Weakly Self-Avoiding Walk

Our goal is to study the analogues of (4) and Proposition 1 for the continuous-time weakly self-avoiding walk. The continuous-time model is a lattice version of the Edwards model [7]. It has been useful in particular due to its representation in terms of functional integrals [5] that have been employed in renormalization group analyses.

2.1 Fixed-Length Walks

We first consider the case of fixed-length walks in which a fixed number n of steps is taken by the walk. We will find that the strong interaction limit depends on how an auxiliary parameter ρ is scaled, where e^{ρ} plays the role of a fugacity. The scaling is parametrized by $a \in [-\infty, \infty]$. The case $a = \infty$ leads to the strictly self-avoiding walk, the case $a = -\infty$ leads to simple random walk, and the interpolating cases, $a \in (-\infty, \infty)$, define a new model of a self-repelling walk that we call the "quick step" model.

Let X denote the continuous-time Markov process with state space $\mathbb{Z}^d$, in which uniformly random nearest-neighbor steps are taken after independent Exp(1) holding times. Let $\mathbb{E}$ denote expectation for this process started at 0. We distinguish between the continuous-time walk X and the sequence of sites visited during its first n steps, which we typically denote by $Y \in \mathscr{W}_{\mathrm{n}}$. Conditioning on the first n steps of X to be Y is denoted by $\mathbb{E}(\cdot \mid Y)$.

For fixed-length walks, the continuous-time weakly self-avoiding walk is the measure $Q_{g,\rho,n}$ on $\mathscr{W}_{\mathrm{n}}$ defined as follows. Here ρ is a real parameter at our disposal, which we allow to depend on $g > 0$. Let T_{n} denote the time of the $(n+1)^{\mathrm{st}}$ jump of X, and let $L_{x,n}(X) = \int_0^{T_{\mathrm{n}}} \mathbb{1}_{X(s)=x} \mathrm{d}s$ denote the local time at x up to time T_{n}. By definition, $\sum_{x\in\mathbb{Z}^d} L_{x,n} = T_{\mathrm{n}}$. For $Y \in \mathscr{W}_{\mathrm{n}}$, let

$$Q_{g,\rho,n}(Y) = \frac{1}{Z_n(g,\rho)} \mathbb{E}\left(e^{-g\sum_x L_{x,n}^2 + \rho\sum_x L_{x,n}} \mid Y\right), \tag{11}$$

where

$$Z_n(g,\rho) = \sum_{Y\in\mathscr{W}_{\mathrm{n}}} \mathbb{E}\left(e^{-g\sum_x L_{x,n}^2 + \rho\sum_x L_{x,n}} \mid Y\right). \tag{12}$$

For $a \in \mathbb{R}$ and $m \in \mathbb{N}$, let

$$I_m(a) = \int_{-a}^{\infty} \frac{(a+u)^{m-1}}{(m-1)!} \mathrm{e}^{-u^2}\, \mathrm{d}u. \tag{13}$$

Proposition 2. *Let $\alpha = \alpha(g,\rho) = \frac{1}{2} g^{-1/2}(\rho - 1)$, and let $\rho = \rho(g)$ be chosen in such a way that $a = \lim_{g\to\infty} \alpha(g,\rho(g))$ exists in $[-\infty, \infty]$. Let $n \geq 1$ and $Y \in \mathscr{W}_{\mathrm{n}}$. Then*

$$\lim_{g\to\infty} Q_{g,\rho(g),n}(Y) = \begin{cases} \frac{1}{Z_a} \prod_{x\in Y} e^{a^2} I_{n_x(Y)}(a) & \text{if } a \in (-\infty,\infty), \\ \frac{1}{c_n} \mathbb{1}_{Y\in\mathscr{S}_{\mathrm{n}}} & \text{if } a = \infty, \\ \frac{1}{(2d)^n} & \text{if } a = -\infty, \end{cases} \tag{14}$$

where Z_a is a normalization constant, and the product over x is over the distinct vertices visited by Y.

Proof. As before, we write $n_x = n_x(Y)$ for the number of times that x is visited by Y. Thus $\sum_x n_x = n + 1$ is the number of vertices visited by Y (with multiplicity). Since the sum of m independent Exp(1) random variables has a Gamma$(m, 1)$ distribution, we have

$$\mathbb{E}\left(\mathrm{e}^{-g\sum_x L_{x,n}^2+\rho\sum_x L_{x,n}} \mid Y\right) = \prod_{x\in Y}\int_0^\infty \frac{s_x^{n_x-1}}{(n_x-1)!}\mathrm{e}^{-s_x}\mathrm{e}^{-gs_x^2+\rho s_x}\,\mathrm{d}s_x, \tag{15}$$

where the product is over the *distinct* vertices visited by Y. We make the changes of variables $t_x = g^{1/2}s_x$ and then $u_x = t_x - \alpha$. After completing the square, this leads to

$$\mathbb{E}\left(e^{-g\sum_x L_{x,n}^2+\rho\sum_x L_{x,n}} \mid Y\right) = g^{-(n+1)/2}\prod_{x\in Y} e^{\alpha^2} I_{n_x}(\alpha). \tag{16}$$

Case $a \in (-\infty,\infty)$: the quick step model. Suppose that $\alpha \to a \in (-\infty,\infty)$ as $g \to \infty$. In this case, by the continuity of $I_m(a)$ in a,

$$\mathbb{E}\left(\mathrm{e}^{-g\sum_x L_{x,n}^2+\rho(g)\sum_x L_{x,n}} \mid Y\right) \sim g^{-(n+1)/2}\prod_{x\in Y} e^{a^2} I_{n_x}(a), \tag{17}$$

and thus

$$\lim_{g\to\infty} Q_{g,\rho(g),n}(Y) = \frac{1}{Z_a}\prod_{x\in Y} e^{a^2} I_{n_x(Y)}(a) \qquad (\alpha \to a \in (-\infty,\infty)). \tag{18}$$

Case $a = \infty$: limit is uniform on $\mathscr{S}_\mathrm{n}$. Suppose that $\alpha \to \infty$ as $g \to \infty$. In this case, since α is nonzero we can use (16) to write

$$\mathbb{E}\left(e^{-g\sum_x L_{x,n}^2+\rho\sum_x L_{x,n}} \mid Y\right)$$
$$= (g^{-1/2}e^{\alpha^2})^{n+1}(\alpha e^{-\alpha^2})^{n+1-|Y|}\prod_{x\in Y}\int_{-\alpha}^\infty \frac{(1+u_x/\alpha)^{n_x-1}}{(n_x-1)!}e^{-u_x^2}\,\mathrm{d}u_x, \tag{19}$$

where $|Y|$ denotes the number of distinct vertices visited by Y. Since the factor $(\alpha e^{-\alpha^2})^{n+1-|Y|}$ goes to zero unless Y is self-avoiding, in which case the factor is equal to 1 and $n_x = 1$ for the vertices visited by Y, and since also

$$\lim_{\alpha\to\infty}\int_{-\alpha}^\infty e^{-u_x^2}\,\mathrm{d}u_x = \sqrt{\pi}, \tag{20}$$

this gives

$$\mathbb{E}\left(e^{-g\sum_x L_{x,n}^2+\rho(g)\sum_x L_{x,n}} \mid Y\right) \sim (g^{-1/2}e^{\alpha^2}\sqrt{\pi})^{n+1}\mathbb{1}_{Y\in\mathscr{S}_\mathrm{n}}. \tag{21}$$

When we take the normalization into account we find that

$$\lim_{g\to\infty} Q_{g,\rho(g),n}(Y) = \frac{1}{c_n} \mathbb{1}_{Y\in\mathscr{S}_n} \qquad (\alpha \to \infty). \tag{22}$$

Case $a = -\infty$: limit is uniform on $\mathscr{W}_n$. Suppose that $\alpha \to -\infty$ as $g \to \infty$. We will show that, for $m \geq 1$,

$$e^{\alpha^2} I_m(\alpha) \sim (-2\alpha)^{-m} \quad \text{as } \alpha \to -\infty. \tag{23}$$

With (16), this claim implies that

$$\mathbb{E}\left(e^{-g\sum_x L_{x,n}^2 + \rho(g)\sum_x L_{x,n}} \mid Y\right) \sim g^{-(n+1)/2} \prod_{x\in Y} (-2\alpha)^{-n_x} = (-2\alpha g^{-1/2})^{n+1}. \tag{24}$$

Since the right-hand side is independent of Y, this proves that the limiting measure is uniform on $\mathscr{W}_n$, as required. Finally, to prove (23), we set $b = -\alpha$ and obtain

$$\begin{aligned}(2b)^m e^{b^2} I_m(-b) &= (2b)^m e^{b^2} \int_b^\infty \frac{(-b+u)^{m-1}}{(m-1)!} e^{-u^2}\, du \\ &= \int_0^\infty \frac{u^{m-1}}{(m-1)!} e^{-(u/(2b))^2 - u}\, du.\end{aligned} \tag{25}$$

By dominated convergence, as $b \to \infty$, the integral on the right-hand side approaches 1 because it becomes the integral over the $\Gamma(m, 1)$ probability density function. □

Proposition 2 shows that the case $\alpha \to \infty$ leads to the uniform measure on self-avoiding walks, whereas $\alpha \to -\infty$ leads to simple random walk. These two extremes are interpolated by the quick step walk, for $\alpha \to a \in (-\infty, \infty)$ (e.g., $a = 0$ if $|\rho| = o(g^{1/2})$ or $a = c$ if $\rho \sim 2cg^{1/2}$). The name "quick step walk" is intended to convey the idea that the large g limit of the continuous-time walk should be dominated by quickly moving continuous-time walks. In fact, when $\rho = 2ag^{1/2}$, by completing the square the weight $\mathrm{e}^{-\sum_x (gL_{x,n}^2 - \rho L_{x,n})}$ can be rewritten as $\mathrm{e}^{\sum_x [-(g^{1/2}L_{x,n} - a)^2 + a^2]}$. Thus, walks with smaller $L_{x,n}$ receive larger weight, and this effect grows in importance as $g \to \infty$.

The particular case of Proposition 2 for the choice

$$\rho(g) = (2g\log(g/\pi))^{1/2}, \tag{26}$$

which corresponds to $a = \infty$, was proved previously in [3].

For the case $a = 0$, evaluation of $I_{n_x(Y)}(0)$ in (18) gives

$$\lim_{g\to\infty} Q_{g,\rho(g),n}(Y) = \frac{1}{Z_0} \prod_{x\in Y} \frac{\Gamma(n_x(Y)/2)}{2\Gamma(n_x(Y))} \qquad (\alpha \to 0). \tag{27}$$

Large values of n_x are penalised under this limiting probability, so this is a model of a self-repelling walk. It is an interesting question whether the quick step walk is in the same universality class as the self-avoiding walk, for $a \in (-\infty, \infty)$. We do not have an answer to this question.

2.2 Two-Point Function

Now we show that when ρ is chosen carefully, depending on g, the two-point function for the continuous-time weakly self-avoiding walk converges, as $g \to \infty$, to the two-point function of the strictly self-avoiding walk. The two-point function of the continuous-time weakly self-avoiding walk can be written in two equivalent ways. This is discussed in a self-contained manner in [5], and we summarise the situation as follows.

The version of the two-point function that we will work with is written in terms of a modified Markov process $X = X(t)$, whose definition depends on a choice of $\delta \in (0, 1)$. The state space is $\mathbb{Z}^d \cup \{\partial\}$, where ∂ is an absorbing state called the cemetery. When X arrives at state x it waits for an Exp(1) holding time and then jumps to a neighbor of x with probability $(2d)^{-1}(1 - \delta)$ and jumps to the cemetery with probability δ. The holding times are independent of each other and of the jumps. The two-point function is defined, for $x \in \mathbb{Z}^d$, to be

$$G^{\mathrm{CT}}_{g,\rho}(x) = \frac{1}{\delta}\mathbb{E}^{(\delta)}\left(e^{-g\sum_{v\in\mathbb{Z}^d} L_v^2 + \rho\zeta}\mathbb{1}_{X(\zeta^-)=x}\right), \tag{28}$$

where we leave implicit the dependence of G^{CT} on δ, where $\mathbb{E}^{(\delta)}$ denotes expectation with respect to the modified process, and where ρ is any real number for which the expectation is finite. The random number of steps taken by X before jumping to the cemetery is denoted η, and the independent sequence of holding times will be denoted $\sigma_0, \sigma_1, \ldots, \sigma_\eta$.

A special case of the conclusions of [5, Sect. 3.2] (there with $d_x = 1$ and $\pi_{x,\partial} = \delta$ for all x, and restricted to finite state space) is the equivalent formula

$$G^{\mathrm{CT}}_{g,\rho}(x) = \int_0^\infty \mathbb{E}\left(e^{-g\sum_{v\in\mathbb{Z}^d} L_{v,T}^2}\mathbb{1}_{X(T)=x}\right)\mathrm{e}^{(\rho-\delta)\mathrm{T}}\mathrm{d}T, \tag{29}$$

where now X is the original continuous-time Markov process X without cemetery state, and $\mathbb{E}$ denotes its expectation when started from the origin of $\mathbb{Z}^d$. Here $L_{v,T} = \int_0^T \mathbb{1}_{X(s)=v}\mathrm{d}s$ is the local time of X at $v \in \mathbb{Z}^d$ up to time T. We will work with (28) rather than (29).

As in Proposition 2, we write $\alpha = \alpha(g,\rho) = \frac{1}{2}g^{-1/2}(\rho - 1)$. Throughout this section, we mainly choose $\rho = \rho(g)$ in such a way that

$$\lim_{g\to\infty} g^{-1/2}\mathrm{e}^{\alpha^2(g,\rho(g))} = p \in [0, \infty). \tag{30}$$

For example, (30) holds for $p > 0$ when $\rho(g) = 2[g \log(p\sqrt{g})]^{1/2}$, which is a choice closely related to that in (26). Note that $\lim_{g\to\infty} \rho(g) = \infty$ when $p > 0$. It is natural to consider $\rho \to \infty$, because if ρ is fixed to a value such that $G^{\mathrm{CT}}_{g_0,\rho}(x) < \infty$ for some $g_0 > 0$, then by dominated convergence $\lim_{g\to\infty} G^{\mathrm{CT}}_{g,\rho}(x) = 0$. The conclusion of Proposition 3 shows that this trivial behavior persists even when $\rho(g) \to \infty$ in such a way that $p = 0$.

Given $p \in [0,\infty)$, let

$$z = (2d)^{-1}(1-\delta)p\sqrt{\pi}. \tag{31}$$

The following proposition shows that, under the scaling (30), the strong interaction limit of the continuous-time weakly self-avoiding walk two-point function is the two-point function of the strictly self-avoiding walk defined in (8).

Proposition 3. *Let $\delta \in (0,1)$, $z \in [0,\mu^{-1})$, and $x \in \mathbb{Z}^d$. Suppose that* (30) *holds with the value of $p \in [0,\infty)$ specified by z via* (31)*. Then,*

$$\lim_{g\to\infty} G^{\mathrm{CT}}_{g,\rho(g)}(x) = p\sqrt{\pi}G_z(x). \tag{32}$$

The proof of Proposition 3 uses three lemmas, and we discuss these next. For $m \in \mathbb{N}$ and $\alpha > 0$, let

$$J_m(\alpha) = \int_{-\alpha}^{\infty} \frac{(1+u/\alpha)^{m-1}}{(m-1)!} \mathrm{e}^{-u^2}\, du. \tag{33}$$

Lemma 1. *Given any $\epsilon > 0$ there exists $A_0 > 0$ such that for all $\alpha \geq A \geq A_0$ and $m \geq 1$,*

$$J_m(\alpha) \leq (1+\epsilon)J_m(A). \tag{34}$$

Proof. For $m \geq 2$, $J_m(\alpha)$ is a nonincreasing function of $\alpha \in (0,\infty)$ because

$$\begin{aligned}
\frac{dJ_m(\alpha)}{d\alpha} &= -\frac{1}{(m-2)!}\int_{-\alpha}^{\infty} \frac{u}{\alpha^2}(1+u/\alpha)^{m-2}\mathrm{e}^{-u^2}\, du \\
&= -\frac{1}{(m-2)!}\Bigg[\int_{\alpha}^{\infty} \frac{u}{\alpha^2}(1+u/\alpha)^{m-2}\mathrm{e}^{-u^2}\, du \\
&\quad + \int_0^{\alpha} \frac{u}{\alpha^2}[(1+u/\alpha)^{m-2} - (1-u/\alpha)^{m-2}]\mathrm{e}^{-u^2}\, du\Bigg] \\
&\leq 0,
\end{aligned} \tag{35}$$

(note that in the first line the contribution from differentiating the limit of integration vanishes), and thus (34) holds even with $\epsilon = 0$. Consider the remaining case $m = 1$.

Since J_1 is increasing and $\lim_{\alpha\to\infty} J_1(\alpha) = \sqrt{\pi}$ (see (20)), given any $\epsilon > 0$ there exists $A_0 > 0$ such that if $\alpha \geq A \geq A_0$ then $1 \leq J_1(\alpha)/J_1(A) \leq 1+\epsilon$. □

Recall that η is the random number of steps taken by X before jumping to the cemetery state. For $x \in \mathbb{Z}^d$, let

$$w_n(g,\rho;x) = \frac{1}{\delta}\mathbb{E}^{(\delta)}[e^{-g\sum_v L_v^2+\rho\zeta}\mathbb{1}_{X(\zeta^-)=x}\mathbb{1}_{\eta=n}], \tag{36}$$

$$w_n(g,\rho) = \frac{1}{\delta}\mathbb{E}^{(\delta)}[e^{-g\sum_v L_v^2+\rho\zeta}\mathbb{1}_{\eta=n}]. \tag{37}$$

Let $w_n(g;x) = w_n(g,\rho(g);x)$ and $w_n(g) = w_n(g,\rho(g))$ with $\rho(g)$ chosen according to (30).

Lemma 2. *Suppose that* (30) *holds with* $p > 0$, *and let* z *be given by* (31). *Then for* $n \geq 0$ *and* $x \in \mathbb{Z}^d$,

$$\lim_{g\to\infty} w_n(g;x) = p\sqrt{\pi}c_n(x)z^n. \tag{38}$$

Proof. Given that $\eta = n$, let $Y \in \mathscr{W}_n(x)$ denote the sequence of jumps made by X before landing in the cemetery, and let $|Y|$ denote the cardinality of the range of Y. By conditioning on Y and using (19), we see that, as $g \to \infty$,

$$\begin{aligned} w_n(g;x) &= [(2d)^{-1}(1-\delta)]^n (g^{-1/2}e^{\alpha^2})^{n+1} \sum_{Y\in\mathscr{W}_n(x)} (\alpha e^{-\alpha^2})^{n+1-|Y|} \prod_{v\in Y} J_{n_v}(\alpha) \\ &\sim [(2d)^{-1}(1-\delta)]^n p^{n+1} \sum_{Y\in\mathscr{W}_n(x)} (\alpha e^{-\alpha^2})^{n+1-|Y|} \prod_{v\in Y} J_{n_v}(\alpha), \end{aligned} \tag{39}$$

where the product is over the distinct vertices visited by Y and $|Y|$ denotes the number of such vertices. It suffices to show that, for any $Y \in \mathscr{W}_n(x)$,

$$\lim_{g\to\infty} (\alpha e^{-\alpha^2})^{n+1-|Y|} \prod_{v\in Y} J_{n_v}(\alpha) = \mathbb{1}_{Y\in\mathscr{S}_n}\pi^{(n+1)/2}. \tag{40}$$

Since $p > 0$, we have $\alpha \to \infty$, and so $\alpha e^{-\alpha^2} \to 0$. Therefore, the above limit is zero unless $n+1 = |Y|$, which corresponds to $Y \in \mathscr{S}_n$; the product over v remains bounded as $\alpha \to \infty$ and poses no difficulty. Since $J_1(\alpha) \to \sqrt{\pi}$ as in (20), the result follows. □

Lemma 3. *Suppose that* (30) *holds with* $p \in (0,\infty)$, *and let* z *be specified by* (31). *Let*

$$\mu(g,\rho) = \limsup_{n\to\infty} w_n(g,\rho)^{1/n}. \tag{41}$$

Then

$$\limsup_{g\to\infty} \mu(g,\rho(g)) \leq z\mu. \tag{42}$$

Proof. Let $L_{x,[i,j]} = \sum_{k=i}^{j} \sigma_k \mathbb{1}_{Y_k=x}$, where the σ_k are the exponential holding times. Let $\mathbb{E}_y^{(\delta)}$ denote the expectation for the process started in state y instead of state 0. For integers $n \geq 1$ and $m \geq 1$, an elementary argument using the strong Markov property leads to

$$\begin{aligned}
w_{n+m}(g,\rho) &\leq \frac{1}{\delta}\mathbb{E}^{(\delta)}[\mathrm{e}^{-g\sum_x L_{x,[0,n]}^2+\rho\sum_x L_{x,[0,n]}} \\
&\quad \times \mathrm{e}^{-g\sum_x L_{x,[n+1,n+m]}^2+\rho\sum_x L_{x,[n+1,n+m]}} \mathbb{1}_{\eta=n+m}] \\
&= \sum_y \mathbb{E}^{(\delta)}[\mathrm{e}^{-g\sum_x L_{x,[0,n]}^2+\rho\sum_x L_{x,[0,n]}} \mathbb{1}_{Y_{n+1}=y}] \\
&\quad \times \frac{1}{\delta}\mathbb{E}_y^{(\delta)}[\mathrm{e}^{-g\sum_x L_x^2+\rho\sum_x L_x} \mathbb{1}_{\eta=m-1}] \\
&= \frac{1-\delta}{\delta}\mathbb{E}^{(\delta)}[\mathrm{e}^{-g\sum_x L_{x,[0,n]}^2+\rho\sum_x L_{x,[0,n]}} \mathbb{1}_{\eta=n}] w_{m-1}(g,\rho) \\
&\leq w_n(g,\rho) w_{m-1}(g,\rho). \qquad (43)
\end{aligned}$$

It is straightforward to adapt the proof of [10, Lemma 1.2.2] to obtain from this approximate subadditivity the equality

$$\mu(g,\rho) = \inf_{n\geq 1} w_n(g,\rho)^{1/(n+1)}. \qquad (44)$$

Then, we have

$$w_n(g,\rho)^{1/(n+1)} \geq \mu(g,\rho). \qquad (45)$$

We let $g \to \infty$ in the above inequality, with $\rho(g)$ chosen as in (30); note that $\alpha \to \infty$ since $p > 0$. By Lemma 2, for $n \geq 0$,

$$\lim_{g\to\infty} w_n(g) = p\sqrt{\pi} c_n z^n. \qquad (46)$$

By (45), this gives

$$(p\sqrt{\pi} c_n)^{1/(n+1)} z^{n/(n+1)} \geq \limsup_{g\to\infty} \mu(g,\rho(g)). \qquad (47)$$

Now we take $n \to \infty$ to get

$$\mu z \geq \limsup_{g\to\infty} \mu(g,\rho(g)), \qquad (48)$$

as required. □

Proof of Proposition 3. We consider separately the cases $p > 0$ and $p = 0$.

Case $p > 0$. We write $\rho = \rho(g)$. By (28), and by (36) with $\rho = \rho(g)$,

$$G^{\mathrm{CT}}_{g,\rho}(x) = \sum_{n=0}^{\infty} w_n(g;x). \tag{49}$$

By Lemma 2, the result of taking the limit $g \to \infty$ under the summation gives the desired result

$$p\sqrt{\pi}\sum_{n=0}^{\infty} c_n(x)z^n, \tag{50}$$

and it suffices to justify the interchange of limit and summation. For this, we will use dominated convergence. Since $w_n(g;x) \le w_n(g)$, it suffices to find a $g_0 > 0$ and a summable sequence B_n such that, for $g \ge g_0$ and $n \in \mathbb{N}_0$,

$$w_n(g;x) \le B_n. \tag{51}$$

This will follow if we show the stronger statement that for large g

$$w_n(g) \le B_n. \tag{52}$$

Since $z\mu < 1$, there exists $\epsilon > 0$ such that $(1+\epsilon)^2(\mu z + \epsilon) < 1$. Since $g^{-1/2}e^{\alpha^2} \to p > 0$, there is a (large) g_0 such that if $g \ge g_0$ then $g^{-1/2}e^{\alpha^2} \le g_0^{-1/2}e^{\alpha_0^2}(1+\epsilon)$, where α_0 is the value of α corresponding to $g = g_0$; also $\alpha e^{-\alpha^2} \le \alpha_0 e^{-\alpha_0^2}$. Therefore, by (39), and by Lemma 1 (increasing g_0 if necessary),

$$\begin{aligned} w_n(g) &= [(2d)^{-1}(1-\delta)]^n (g^{-1/2}\mathrm{e}^{\alpha^2})^{n+1} \sum_{Y\in\mathscr{W}_n} (\alpha \mathrm{e}^{-\alpha^2})^{n+1-|Y|} \prod_{v\in Y} J_{n_v}(\alpha) \\ &\le [(2d)^{-1}(1-\delta)]^n (g_0^{-1/2}\mathrm{e}^{\alpha_0^2}(1+\epsilon)^2)^{n+1} \sum_{Y\in\mathscr{W}_n} (\alpha_0 \mathrm{e}^{-\alpha_0^2})^{n+1-|Y|} \prod_{v\in Y} J_{n_v}(\alpha_0) \\ &= (1+\epsilon)^{2(n+1)} w_n(g_0). \end{aligned} \tag{53}$$

We set $B_n = (1+\epsilon)^{2(n+1)} w_n(g_0)$. Then

$$\limsup_{n\to\infty} B_n^{1/n} = (1+\epsilon)^2 \mu(g_0, \rho(g_0)) \le (1+\epsilon)^2(z\mu + \epsilon) < 1, \tag{54}$$

by taking g_0 larger if necessary and applying Lemma 3. Therefore $\sum_n B_n$ converges, and the proof is complete for the case $p > 0$.

Case $p = 0$. We will prove that

$$\lim_{g\to\infty} \sum_{n=0}^{\infty} w_n(g) = 0. \tag{55}$$

By (49), this is more than sufficient. We again write $\rho = \rho(g)$. By conditioning on Y and using (16), for $n \geq 0$ we have

$$w_n(g) = [(2d)^{-1}(1-\delta)]^n \sum_{Y \in \mathscr{W}_n} \prod_{x \in Y} g^{-n_x/2} e^{\alpha^2} I_{n_x}(\alpha). \tag{56}$$

The change of variables $s = a + u$ in (13) gives, for $m \geq 1$,

$$\begin{aligned} e^{\alpha^2} I_m(\alpha) &= e^{\alpha^2} \int_0^\infty \frac{s^{m-1}}{(m-1)!} e^{-(s-\alpha)^2} \, ds \\ &\leq e^{\alpha^2} \int_0^\infty \frac{s^{m-1}}{(m-1)!} e^{-s} \left(\sup_{s \in \mathbb{R}} e^{s-(s-\alpha)^2} \right) ds = e^{\alpha^2+\alpha+1/4}. \end{aligned} \tag{57}$$

Let $\epsilon > 0$. Since $g^{-1/2} e^{\alpha^2} \to p = 0$, we can find $g(\epsilon)$ such that for $g \geq g(\epsilon)$ and $m \geq 2$,

$$g^{-1/2} e^{\alpha^2} \sqrt{\pi} \leq \epsilon, \qquad g^{-m/2} e^{\alpha^2+\alpha+1/4} \leq \epsilon^m. \tag{58}$$

Henceforth, we assume that $g \geq g(\epsilon)$. By (57),

$$g^{-m/2} e^{\alpha^2} I_m(\alpha) \leq \epsilon^m \quad \text{for } m \geq 2. \tag{59}$$

For $m = 1$, we obtain an upper bound by extending the range of the integral in the first line of (57) to the entire real line, whereupon it evaluates to $\sqrt{\pi}$. Thus, by (58), $g^{-1/2} e^{\alpha^2} I_1(\alpha) \leq \epsilon$. By (56) and the fact that the number of walks in $\mathscr{W}_n$ is $(2d)^n$, for $n \geq 0$ we then have

$$w_n(g) \leq [(2d)^{-1}(1-\delta)]^n \sum_{Y \in \mathscr{W}_n} \prod_{v \in Y} \epsilon^{n_v} = (1-\delta)^n \epsilon^{n+1}. \tag{60}$$

(The case $n = 0$ corresponds to $m = 1$ because the number of visits to state 0 is $n_0 = 1$.) Therefore $\limsup_{g \to \infty} \sum_{n=0}^\infty w_n(g) = O(\epsilon)$. Since ϵ is arbitrary, this proves (55), and the proof is complete. □

Acknowledgements The work of DB and GS was supported in part by NSERC of Canada.

References

1. Bolthausen, E., Ritzmann, C., Rubin, F.: Work in progress.
2. Borgs, C., Chayes, J.T., van der Hofstad, R., Slade, G.: Mean-field lattice trees. Ann. Combinatorics **3**, 205–221 (1999)
3. Bovier, A., Felder, G., Fröhlich, J.: On the critical properties of the Edwards and the self-avoiding walk model of polymer chains. Nucl. Phys. B **230**, [FS10] 119–147 (1984)

4. Brydges, D., Slade, G.: Renormalisation group analysis of weakly self-avoiding walk in dimensions four and higher. In: Proceedings of the International Congress of Mathematicians (Hyderabad 2010), Volume 4, pp. 2232–2257, eds. R. Bhatia et al., World Scientific, Singapore (2011)
5. Brydges, D.C., Imbrie, J.Z., Slade, G.: Functional integral representations for self-avoiding walk. Probab. Surveys **6**, 34–61 (2009)
6. Domb, C., Joyce, G.S.: Cluster expansion for a polymer chain. J. Phys. C: Solid State Phys. **5**, 956–976 (1972)
7. Edwards, S.F.: The statistical mechanics of polymers with excluded volume. Proc. Phys. Soc. London **85**, 613–624 (1965)
8. Hara, T.: Decay of correlations in nearest-neighbor self-avoiding walk, percolation, lattice trees and animals. Ann. Probab. **36**, 530–593 (2008)
9. Hara, T., van der Hofstad, R., Slade, G.: Critical two-point functions and the lace expansion for spread-out high-dimensional percolation and related models. Ann. Probab. **31**, 349–408 (2003)
10. Madras, N., Slade, G.: The Self-Avoiding Walk. Birkhäuser, Boston (1993)

Copolymers at Selective Interfaces: Settled Issues and Open Problems

Francesco Caravenna, Giambattista Giacomin, and Fabio Lucio Toninelli

Abstract We review the literature on the localization transition for the class of polymers with random potentials that goes under the name of *copolymers near selective interfaces*. We outline the results, sketch some of the proofs and point out the open problems in the field. We also present in detail some alternative proofs that simplify what one can find in the literature.

2010 Mathematics Subject Classification: 60K35, 82B41, 82B44.

1 Copolymers and Selective Solvents

1.1 A Basic Model

In [14], T. Garel, D. A. Huse, S. Leibler and H. Orland introduced a simple model in order to look into how *the statistical behavior of macromolecules can be strongly affected by randomness in the physico-chemical properties of their constituents.*

F. Caravenna (✉)
Dipartimento di Matematica e Applicazioni, Università degli Studi di Milano-Bicocca, via Cozzi 53, 20125 Milano, Italy
e-mail: francesco.caravenna@unimib.it

G. Giacomin
Université Paris Diderot (Paris 7) and Laboratoire de Probabilités et Modèles Aléatoires (CNRS U.M.R. 7599)

U.F.R. Mathématiques, Case 7012 (Site Chevaleret), 75205 Paris cedex 13, France
e-mail: giacomin@math.jussieu.fr

F.L. Toninelli
Ecole Normale Supérieure de Lyon, Laboratoire de Physique and CNRS, UMR 5672, 46 Allée d'Italie, 69364 Lyon Cedex 07, France
e-mail: fabio-lucio.toninelli@ens-lyon.fr

J.-D. Deuschel et al. (eds.), *Probability in Complex Physical Systems*, Springer Proceedings in Mathematics 11, DOI 10.1007/978-3-642-23811-6_12,

They aimed at a special class of macromolecules of linear type, the *random hydrophilic-hydrophobic copolymers*, in a medium of water and oil, separated by an interface. Such a polymer chain is just made up of monomers that differ for their affinity for water (or oil). The affinity is reduced to a real parameter that we call *charge*: the charge of the j-th monomer is denoted in [14] by ζ_j and, in mathematical terms, $\{\zeta_j\}_{j=1,2,\dots}$ is an IID sequence of Gaussian random variables, with given mean and variance. In order both to conform with the mathematical literature and to generalize slightly the problem we will write ζ_j as $\omega_j + h$, where $h \in \mathbb{R}$ (or $h \geq 0$ as we will do next) and $\omega = \{\omega_j\}_{j=1,2,\dots}$ is an IID sequence of random variables (often referred to as the *disorder*) such that

$$\mathrm{M}(t) := \mathbb{E}\left[\exp\left(t\omega_1\right)\right] < \infty, \tag{1}$$

for every $t \in \mathbb{R}$ and such that $\mathbb{E}\omega_1 = 0$, $\mathbb{E}\omega_1^2 = 1$. Apart for the larger class of charges that we allow and for notations, the Hamiltonian of the polymer model set forth in [14] is

$$H_{N,\omega,h}(S) := \sum_{n=1}^{N} (\omega_n + h)\,\mathrm{sign}(S_n), \tag{2}$$

where N is the length of the polymer and $S = \{S_0, S_1, \dots\}$ is a simple symmetric random walk trajectory ($S_0 = 0$, $\{S_{n+1} - S_n\}_{n=0,1,\dots}$ is a sequence of independent identically distributed symmetric random variables that take only values ± 1: the law of S is denoted by $\mathbf{P}$ and we stress that ω and S are independent).

We invite the reader to have a look from now at Fig. 1 for the directed polymer interpretation of the trajectories of the model. A small detail to deal with is sign(0): $\mathrm{sign}(S_n)$ should be read as $\mathrm{sign}(S_{n-1})$ when $S_n = 0$ and this convention is particularly natural in directed polymer terms, because $\mathrm{sign}(S_n)$ is $+1$ (-1) if the nth monomer is in the upper (lower) half plane, that is in oil (water), see Fig. 1. Still to conform with most of the mathematical literature on copolymers, the inverse temperature is denoted by $\lambda(\geq 0)$ instead of the more customary β, so that the Boltzmann factor that defines the polymer model of length N is $\exp(\lambda H_{N,\omega,h}(S))$. We are interested in the *quenched* system so we underline the very different nature of the two sources of randomness: ω is chosen once for all at the beginning of the experiment (the hydrophilic or hydrophobic character of the monomers does not change, while the chain fluctuates).

At a superficial level the effect of the charges on the polymer is quite intuitive: for $\lambda > 0$ positively charged monomers ($\omega_n + h > 0$, that is hydrophobic monomers) prefer lying in the upper half-plane (oil) and the opposite is true for the negatively charged ones. But for large N these *energetically favorable* trajectories become more and more atypical for $\mathbf{P}$ since placing the monomers in their preferred solvent strongly reduces the fluctuation freedom of the chain. We are therefore dealing with an energy-entropy competition that in the limit $N \to \infty$ leads to a localization-delocalization transition: localization arises when energy prevails and the polymer

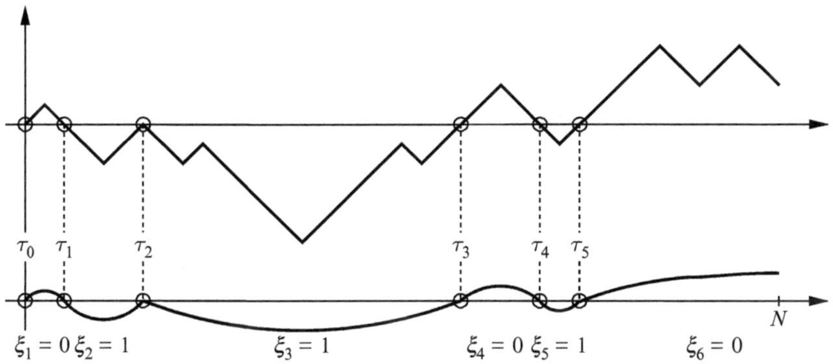

Fig. 1 The upper part of the figure shows a copolymer configuration: each *bond*, or segment, $[(n-1, S_{n-1}), (n, S_n)]$ of the random walk trajectory represents a monomer, so that sign(S_n) should be read as $+1$ (resp. -1) if this monomer is in the upper (*resp. lower*) half-plane. Note that the Hamiltonian of the copolymer, cf. (2), does not depend on the details of S within an excursion, but only on the length and sign of the excursion. This naturally leads to the generalized model introduced in Sect. 1.2 (*lower part of the figure*) in terms of a general discrete renewal process τ and a sequence of independent signs ξ. It is mathematically useful to use the variable Δ_n, that in the random walk case is just $(1 - \text{sign}(S_n))/2$, *i.e.* the indicator function that the copolymer is below the axis. Note in fact that, by (2), $H_{N,\omega,h}(S) = -2\sum_{n=1}^{N}(\omega_n + h)\Delta_n + c_N(\omega)$, with $c_N(\omega) = \sum_{n=1}^{N}(\omega_n + h)$ which does not depend on S (or Δ), therefore we can drop it without changing the polymer measure (*that is what we do in* (4))

sticks to the oil–water interface, visiting thus both oil and water, while delocalization corresponds to the case in which the polymer prefers to stay away from the interface. We will come back to this with much more details, but we anticipate that this entropy-energy competition turns out to be rather challenging: the phase diagram of the model is for the moment only partially understood and sound conjectures (or even only convincing physical heuristics) are lacking on several fundamental issues. This may appear rather surprising in view of the very simple nature of the model and of the fact, at the heart of the motivation of [14], that it represents one of the simplest instances of a general mechanism that plays a crucial role in a variety of extremely important phenomena (protein folding, to name one).

1.2 The (General) Copolymer Model

As argued in Fig. 1 and its caption, the basic copolymer model does not depend on all the details of S, but just on its zero level set, which is a renewal set, and on the sign of the excursions, that is simply an independent fair coin tossing sequence. It is therefore natural, and at times really helpful, to look at the following generalized framework. Let us consider a discrete renewal process $\tau = \{\tau_n\}_{n\geq 0}$ on the non-negative integers $\mathbb{N} \cup \{0\}$, *i.e.*, a sequence of random variables such that $\tau_0 = 0$

and $\{\tau_{j+1} - \tau_j\}_{j=0,1,\ldots}$ is an IID sequence of positive integer-valued variables, with marginal law satisfying

$$K(n) := \mathbf{P}(\tau_1 = n) \overset{n\to\infty}{\sim} \frac{c_K}{n^{1+\alpha}}, \tag{3}$$

where $\alpha > 0$ and $c_K > 0$ (we write $f_n \sim g_n$ for $f_n/g_n \to 1$). Since we are dealing with renewal processes it is important to stress that $\mathbf{P}(\tau_1 < \infty) = 1$, that is $\sum_{n=1}^{\infty} K(n) = 1$, so that τ is a *persistent* renewal. More generally, one could replace the constant c_k in (3) by a slowly varying function $L(n)$, but we stick for simplicity to the purely polynomial asymptotic behavior (3). It is well-known that the first return time to zero of the simple symmetric random walk $\inf\{n > 0 : S_n = 0\}$ satisfies (3) (restricted to the even integers, due to the usual periodicity issue) with $\alpha = \frac{1}{2}$. In particular, the basic model presented in the previous subsection is a special case of the generalized copolymer model we are defining, as it will be clear in a moment.

Remark 1. It is practical to switch freely from looking at τ as a sequence to considering it a *random set*, so for example $|\tau \cap [0, N]|$ is the number of renewals up to N, or $n \in \tau$ is the event that there exists j such that $\tau_j = n$. For a comprehensive references on renewal processes see for example [3].

The renewal τ identifies the polymer-interface contacts: we still need to know whether the excursion is above or below the axis. For this let $\xi = \{\xi_n\}_{n\in\mathbb{N}}$ denote an IID sequence of $B(1/2)$ variables (that is $\mathbf{P}(\xi_n = 0) = \mathbf{P}(\xi_n = 1) = \frac{1}{2}$) independent of τ, that we still call signs. Starting from the couple (τ, ξ) we build a new sequence $\Delta = \{\Delta_n\}_{n\in\mathbb{N}}$ by setting $\Delta_n = \sum_{j=1}^{\infty} \xi_j \, \mathbf{1}_{(\tau_{j-1},\tau_j]}(n)$, in analogy with the simple random walk case: the signs Δ_n are constant between the points in τ and they are determined by ξ.

By *copolymer model* we mean the probability law $\mathbf{P}_{N,\omega} = \mathbf{P}_{N,\omega,\lambda,h}$ for the sequence Δ defined by

$$\frac{d\mathbf{P}_{N,\omega}}{d\mathbf{P}}(\Delta) := \frac{1}{Z_{N,\omega}} \exp\left(-2\lambda \sum_{n=1}^{N} \Delta_n(\omega_n + h)\right), \tag{4}$$

where $N \in \mathbb{N}$, $\lambda \geq 0$, $h \in \mathbb{R}$ (but we can and will assume $h \geq 0$ without loss of generality) and $\omega = \{\omega_n\}_{n\in\mathbb{N}}$ has been introduced in § 1. The partition function $Z_{N,\omega} = Z_{N,\omega,\lambda,h}$ is given by

$$Z_{N,\omega} := \mathbf{E}\left[\exp\left(-2\lambda \sum_{n=1}^{N} \Delta_n(\omega_n + h)\right)\right]. \tag{5}$$

In order to emphasize the value of α in (3), we will sometimes speak of a α-*copolymer model*, but $\mathbf{P}_{N,\omega}$ depends on the full distribution $K(\cdot)$, not only on α.

1.3 The Free Energy: Localization and Delocalization

We introduce the free energy of the copolymer by

$$\mathrm{F}(\lambda, h) := \lim_{N\to\infty} \mathrm{F}_N(\lambda, h), \qquad \text{where} \qquad \mathrm{F}_N(\lambda, h) := \frac{1}{N}\mathbb{E}\left[\log Z_{N,\omega,\lambda,h}\right]. \quad (6)$$

The existence of such a limit follows by a standard argument based on super-additivity, see for example [15] or [16, Ch. 4], where it is also proved that

$$\mathrm{F}(\lambda, h) = \lim_{N\to\infty}\frac{1}{N}\log Z_{N,\omega,\lambda,h}, \qquad \mathbb{P}(\mathrm{d}\omega)\text{-a.s. and in } L^1(\mathbb{P}). \quad (7)$$

Equations (6)–(7) are telling us that the limit in (7) does not depend on the (typical) realization of ω, however it does depend on $\mathbb{P}$, that is on the law of ω_1, as well as on the inter-arrival law $K(\cdot)$. This should be kept in mind, even if we omit $\mathbb{P}$ and $K(\cdot)$ from the notation $\mathrm{F}(\lambda, h)$. Let us point out from now that $\mathrm{F}(\lambda, \cdot)$ and $\mathrm{F}(\cdot, h)$ are convex functions, since they are limits of convex functions. As a matter of fact $\mathrm{F}(\cdot,\cdot)$ is only separately convex because of the choice of the parametrization, but it is straightforward to see that $(\lambda, h) \mapsto \mathrm{F}(\lambda, h/\lambda)$ is convex.

Remark 2. It is sometimes useful to consider the *constrained* partition function $Z^{\mathtt{c}}_{N,\omega} = Z^{\mathtt{c}}_{N,\omega,\lambda,h}$ of the model, defined by

$$Z^{\mathtt{c}}_{N,\omega,\lambda,h} := \mathbf{E}\left[\exp\left(-2\lambda\sum_{n=1}^{N}(\omega_n + h)\Delta_n\right)\mathbf{1}_{\{N\in\tau\}}\right], \quad (8)$$

which differs from (5) only by the *boundary condition* factor $\mathbf{1}_{\{N\in\tau\}}$. It is a standard fact [16, Remark 1.2] that for all N, λ, h we have

$$Z^{\mathtt{c}}_{N,\omega,\lambda,h} \le Z_{N,\omega,\lambda,h} \le C\, N\, Z^{\mathtt{c}}_{N,\omega,\lambda,h}, \quad (9)$$

where C is a positive constant. In particular, the free energy $\mathrm{F}(\lambda, h)$ does not change if $Z_{N,\omega,\lambda,h}$ is replaced by $Z^{\mathtt{c}}_{N,\omega,\lambda,h}$ in (6) and (7). Furthermore, since $N \mapsto \mathbb{E}(\log Z^{\mathtt{c}}_{N,\omega,\lambda,h})$ is a real super-additive sequence, we can write

$$\mathrm{F}(\lambda, h) = \lim_{N\to\infty}\frac{1}{N}\mathbb{E}(\log Z^{\mathtt{c}}_{N,\omega,\lambda,h}) = \sup_{N\in\mathbb{N}}\frac{1}{N}\mathbb{E}(\log Z^{\mathtt{c}}_{N,\omega,\lambda,h}). \quad (10)$$

A crucial observation is:

$$\mathrm{F}(\lambda, h) \ge 0 \text{ for every } \lambda, h \ge 0. \quad (11)$$

This follows by restricting the expectation in (5) to the event $\{\tau_1 > N, \xi_1 = 0\}$, on which we have $\Delta_1 = 0, \ldots, \Delta_N = 0$, hence we obtain $Z_{N,\omega} \geq \frac{1}{2}\mathbf{P}(\tau_1 > N)$ and it suffices to observe that $N^{-1} \log \mathbf{P}(\tau_1 > N)$ vanishes as $N \to \infty$, thanks to (3). Notice that the event $\{\tau_1 > N, \xi_1 = 0\}$ corresponds to the set of trajectories that never visit the lower half plane, therefore the right hand side of (11) may be viewed as the contribution to the free energy given by these trajectories.

Based on this, we say that $(\lambda, h) \in \mathscr{D}$ (*delocalized regime*) if $\mathrm{F}(\lambda, h) = 0$, while $(\lambda, h) \in \mathscr{L}$ (*localized regime*) if $\mathrm{F}(\lambda, h) > 0$. This may look at first as a cheap way to escape from the real localization/delocalization issue, that is inherently linked to the path properties of the measure $\mathbf{P}_{N,\omega}$, but it is not the case. Notice in fact that, if $h \mapsto \mathrm{F}(\lambda, h)$ is differentiable (which fails at most for a countable number of values of h, by convexity), by differentiating (7) and by convexity arguments we have

$$-\frac{1}{2\lambda}\frac{\partial}{\partial h}\mathrm{F}(\lambda, h) = \lim_{N\to\infty} \mathbf{E}_{N,\omega}\left[\frac{\mathscr{N}_N}{N}\right], \qquad \mathbb{P}\text{-a.s.}, \tag{12}$$

where

$$\mathscr{N}_N := \sum_{n=1}^{N} \Delta_n, \tag{13}$$

is just the total number of monomers in the lower half-plane, that is in water (cf. Fig. 1). Therefore if (λ, h) is chosen in the interior of $\mathscr{D}$, where $\mathrm{F} \equiv 0$, the polymer visits water with null density ($\mathscr{N}_N/N \to 0$). On the other hand, if (λ, h) is in $\mathscr{L}$, the polymer puts a positive density, precisely $-(1/(2\lambda))\partial\mathrm{F}(\lambda, h)/\partial h \in (0, 1)$, of monomers in water and the rest, still a positive density, in oil.

We will deal below, *cf.* Sect. 5, with sharper results on path behavior, but the elementary observation we have just made shows that the definition we have set forth of localization and delocalization is far from being artificial. As a matter of fact, it is the natural physical definition, and in fact it has been used already in [14], while in the mathematical literature was first introduced by [8].

1.4 The Phase Diagram

Convexity and the evident monotonicity of $\mathrm{F}(\lambda, \cdot)$ put strong a priori constraints on the phase diagram: let us go through this before going toward sharper questions. We can set

$$h_c(\lambda) := \sup\{h : \mathrm{F}(\lambda, h) > 0\}, \tag{14}$$

and the monotonicity of $\mathrm{F}(\lambda, \cdot)$ guarantees that $(\lambda, h) \in \mathscr{L}$ if and only if $h < h_c(\lambda)$, namely that $h_c(\cdot)$ is the critical curve. Let us derive a number of elementary properties of $h_c(\cdot)$.

The fact that $h_c(\lambda) < \infty$ for every λ follows by the standard *annealed bound*:

$$\mathbb{E} \log Z_{N,\omega} \le \log \mathbb{E} Z_{N,\omega} = \log \mathbf{E}\mathbb{E}\left[\exp\left(-2\lambda \sum_{n=1}^{N} \Delta_n(\omega_n + h)\right)\right]$$
$$= \log \mathbf{E} \exp\left((\log \mathrm{M}(-2\lambda) - 2\lambda h) \sum_{n=1}^{N} \Delta_n\right), \quad (15)$$

so that $\mathrm{F}(\lambda, h) \le 0$ (hence, recall (11), $\mathrm{F}(\lambda, h) = 0$) if $h \ge \log \mathrm{M}(-2\lambda)/(2\lambda)$, and $\log \mathrm{M}(-2\lambda)/(2\lambda) < \infty$ for every λ by (1).

Remark 3. The exponential of the rightmost term in (15) is the partition function of the *annealed* model associated to our quenched model. The free energy of the annealed model is rather trivial: it is in fact an elementary exercise to see that

$$\lim_{N\to\infty} \frac{1}{N} \log \mathbf{E} \exp\left((\log \mathrm{M}(-2\lambda) - 2\lambda h) \sum_{n=1}^{N} \Delta_n\right) = (\log \mathrm{M}(-2\lambda) - 2\lambda h)_+, \quad (16)$$

where $a_+ := a\mathbf{1}_{a>0}$. The annealed model has therefore a (de)localization transition too: its critical curve is $h_c^{\mathrm{ann}}(\lambda) := \log \mathrm{M}(-2\lambda)/(2\lambda)$ and we have just remarked that $h_c(\lambda) \le h_c^{\mathrm{ann}}(\lambda)$ (this inequality on the critical curves is also referred to as annealed bound). It can be noticed also that the annealed free energy is not C^1 at criticality, that is the transition is of first order. Let us stress that the annealed free energy looses a lot of details of the original model (in particular: no trace of $K(\cdot)$!).

We have also $h_c(\lambda) > 0$ as soon as $\lambda > 0$, but this is not a trivial statement. In fact this means showing that $\mathrm{F}(\lambda, 0) > 0$ for every $\lambda > 0$: in more dramatic terms, if the interaction does not select, on the average, a preferred solvent ($h = 0$), the polymer is localized even at arbitrarily weak coupling, a result established first in [28]. We skip the proof of this fact (in Sect. 2 the proof of stronger results is sketched) and we simply observe that, together with the annealed upper bound, it implies that $h_c(\lambda) \longrightarrow 0$ as $\lambda \searrow 0$.

At this point convexity can be used in a very profitable way: since $\{(\lambda, y) : \mathrm{F}(\lambda, y/\lambda) \le 0)\}$ is a convex set, its lower boundary $\lambda \mapsto \lambda h_c(\lambda)$ is a convex function. So we can write $h_c(\lambda) = g(\lambda)/\lambda$, with $g(\cdot)$ convex such that $g(\lambda) = o(\lambda)$ as $\lambda \searrow 0$. This directly implies in particular continuity of $h_c(\cdot)$ and, with a little bit more of work, also the fact that $h_c(\cdot)$ is strictly increasing [5].

Quite a bit of effort has been put into pinning down the value of $h_c(\cdot)$. Figure 2 sums up the results that are known on $h_c(\cdot)$ and, in particular, the content of

Theorem 1. *For every $\lambda > 0$ the following explicit bounds hold:*

$$\frac{1}{2\lambda/(1+\alpha)} \log \mathrm{M}\left(-2\lambda/(1+\alpha)\right) \le h_c(\lambda) < \frac{1}{2\lambda} \log \mathrm{M}\left(-2\lambda\right) = h_c^{\mathrm{ann}}(\lambda), \quad (17)$$

where the left inequality is strict when $\alpha \ge 0.801$ (at least for λ small).

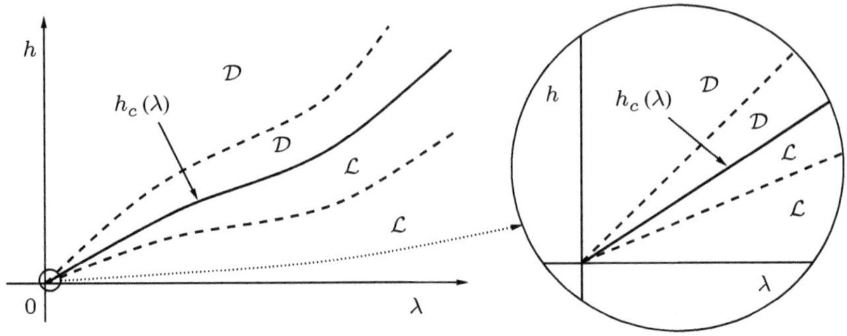

Fig. 2 On the left there is a sketch of the phase diagram and the critical curve $\lambda \mapsto h_c(\lambda)$. The localized (*resp. delocalized*) regime corresponds to (λ, h) lying strictly below (*resp. above*) the critical curve. Explicit upper and lower bounds on $h_c(\cdot)$ are known, cf. Theorem 1, and they are schematically drawn as dashed lines. On the right one finds a zoom of the region near the origin, where the critical curve is close to a straight line and essentially all the relevant information close to the origin (*weak coupling regime*) is encoded in the slope of this line: while the details of the critical curve do depend on the law of the disorder ω and on the details of $K(\cdot)$, the slope depends only on α. This is an important universality feature to which Sect. 4 is devoted

The lower bound in (17) is proved in [5]. The strict inequality in the upper bound in (17) was first proved in [32] to hold for large λ and then extended to every $\lambda > 0$ in [6]. In Sect. 3 we give an alternative, more direct proof.

Highlighted in Fig. 2 is the small λ behavior of $h_c(\lambda)$. In fact, as more extensively explained in Sect. 4, for $\alpha \in (0, 1)$ we have $h_c(\lambda) \overset{\lambda \searrow 0}{\sim} m_\alpha \lambda$, with $m_\alpha > 0$ depending only on α. The slope m_α is therefore a universal feature of the model: it does not depend on the details of the disorder sequence ω and of the underlying renewal τ. The proof of such a result goes through showing that for λ and h small the free energy of the copolymer is close to the free energy of a suitable continuum polymer model.

It becomes therefore quite relevant to get a hold of the value of m_α (at least for $\alpha \in (0, 1)$). As a matter of fact, from (17) one directly extracts

$$\frac{1}{1+\alpha} \leq m_\alpha \leq 1, \qquad \text{for every } \alpha > 0. \tag{18}$$

but this result can be sharpened to

$$\max\left(\frac{1}{2}, \frac{g(\alpha)}{\sqrt{1+\alpha}}, \frac{1}{1+\alpha}\right) \leq m_\alpha < 1, \tag{19}$$

where $g(\cdot)$ is a continuous function (of which we have an expression in terms of the primitive of an explicit function) such that $g(\alpha) = 1$ for $\alpha \geq 1$ and for which one can show that $g(\alpha)/\sqrt{1+\alpha} > 1/(1+\alpha)$ for $\alpha \geq 0.801$ (by evaluating $g(\cdot)$ numerically one can extend this result to $\alpha \geq 0.65$). Since the existence of m_α is not guaranteed for $\alpha \geq 1$, to be precise both in (18) and in (19) m_α should be replaced by the inferior and superior limits of $h_c(\lambda)/\lambda$.

Remark 4. For sake of conciseness we have left aside the $\alpha = 0$, that would require replacing the power law behavior (3) with a regularly varying behavior (allowing, in particular, the presence of logarithmic multiplicative corrections). The bounds that we have just presented directly generalize [16, Ch. 6] and for $\alpha = 0$ the three terms in (17) coincide.

We would like to stress that Theorem 1 and the bounds (19) show that some claims in the physical literature are wrong. Notably in [14, 34] it is claimed, for $\alpha = 1/2$, that $h_c(\cdot)$ coincides with $h_c^{(\mathrm{ann})}(\cdot)$ (and (19) shows that even the weaker claim that $h_c(\lambda) \stackrel{\lambda \searrow 0}{\sim} h_c^{(\mathrm{ann})}(\lambda)$ is not correct). In [24, 30] it is claimed that, still for $\alpha = 1/2$ (or, at least, for what we call "basic model," but nothing but the value of $K(n)$ for large n seems to be used in [24, 30]), the inequality in the left-hand side of (17) is an equality. Theorem 1 falls short of proving that also this claim is false (even if it suggests it). A numerical study [10] lead for the basic model, complemented by a careful statistical analysis using concentration inequalities, strongly suggests that the lower bound in (17) is strict and that the critical curve is somewhat *halfway* between the lower and the upper bound. Moreover for all positive values of α, λ, and ε one can find a $K(\cdot) = K_{\lambda,\varepsilon}(\cdot)$ that satisfies (3) and such that the α-copolymer (the one based on the inter-arrival law $K_{\lambda,\varepsilon}(\cdot)$, not an arbitrary one!) is such that $h_c(\lambda) > h_c^{\mathrm{ann}}(\lambda) - \varepsilon$. In fact fix $\lambda > 0$ and $\varepsilon > 0$ and consider an auxiliary model with $\alpha = 0$, or with a very small value of α if you want to avoid slowly varying functions: call $K_0(\cdot)$ the inter-arrival law for such a model, for which we know that $h_c(\cdot) = h_c^{\mathrm{ann}}(\cdot)$ (or they are very close, if α is not precisely zero). By the consequence (10) of super-additivity we see that we can find N such that $\mathbb{E} \log Z^{\mathrm{c}}_{N,\omega,\lambda,h} > 0$ for $h > h_c^{\mathrm{ann}}(\lambda) - \varepsilon$. But the same holds for any model such that $K(n) := K_0(n)$ for $n \leq N$, so, in particular, for some α-copolymer models and the claim is proved.

What the previous results and discussion expose is first of all that $h_c(\cdot)$, unlike m_α (at least for $\alpha < 1$), does not depend only on α: in fact, if it were the case, the argument we have just outlined would imply $h_c(\cdot) = h_c^{\mathrm{ann}}(\cdot)$ for $\alpha > 0$, which is in contrast with Theorem 1. But it exposes also the lack of a convincing heuristic theory predicting the value of the slope of the critical curve at the origin (let us leave alone predicting what $h_c(\cdot)$ is). In this sense we consider that capturing the value of m_α, for $\alpha \in (0, 1)$, is an important open problem. Since computing explicitly quenched quantities may be really out of reach, here are two sub-problems that are open:

- ⋆ Show that $m_{1/2} > 2/3$;
- ⋆ Is $m_\alpha = 1/(1 + \alpha)$ for some $\alpha > 0$?

Finally, we point out that a *reduced (simplified) copolymer model* was introduced in [5], taking inspiration from the approach in [24]. The original hope was that this simplified model could catch the main features of the original copolymer model.

However, this does not seem to be the case, since it has been shown [7, 32] that for the reduced model one has $m_\alpha = 1/(1+\alpha)$ for all $\alpha \in (0,1)$.

1.5 The Critical Behavior and a Word About Pinning Models

Claims can be found in the physical literature about the critical behavior of this model (at least in the original set-up, $\alpha = 1/2$, *e.g.* [11, 22, 24, 34]), but these claims do not always agree with each other, apart for the fact that they all claim, not surprisingly, a smoothing effect of disorder. A rigorous result available on this issue has been proved in [20]: the transition of the general copolymer model is smooth (the derivative of the free energy vanishes at least linearly when the critical point is approached, hence it is at least Lipschitz continuous at the critical point), in contrast with the annealed case (where the the derivative of the free energy has a discontinuity at criticality, *cf.* Remark 3): for every $\lambda > 0$ there exists $c(\lambda) < \infty$ such that

$$\mathrm{F}(\lambda, h_c(\lambda) + \delta) \leq (1+\alpha)c(\lambda)\delta^2 \tag{20}$$

for every $\delta > 0$. The result was obtained under some technical conditions on the disorder law, which are satisfied for instance in the case of Gaussian or bounded charges. One can of course wonder whether the critical behavior of the copolymer model depends or not on α (and on λ?) and how, but once again, the substantial lack of sound physical predictions is quite disappointing. A natural open question however is:

⋆ Can one improve (20), in the sense of replacing the exponent 2 with a larger value?

A somewhat deeper insight into this issue can be achieved by considering also another class of models, as we explain next.

The bound (20) in fact coincides with the one available for disordered pinning models. Pinning models are a close companion to the copolymer, since the Boltzmann factor of a pinning model is

$$\exp\left(\sum_{n=1}^{N}(\beta\omega_n + h)\delta_n\right), \tag{21}$$

where $\delta_n = \mathbf{1}_{n\in\tau}$, that is $\mathbf{1}_{S_n=0}$ in the random walk set-up (ω is chosen as before, so that $\beta\omega_n + h$ is a random variable of mean $h \in \mathbb{R}$ and variance β^2). Therefore in this case the polymer has an interaction with the environment only when it touches the oil–water interface (or simply when it touches the x axis, usually called *defect line*, since the model does not depend on the sign of the excursions). It is well known that pining models exhibit a localization transition too and they can be dealt, to a certain extent, with similar techniques [16]. However, in the end, there are considerable differences, but let us try to single them out and see what they suggest for copolymers:

1. The annealed pinning model is much richer than the annealed copolymer (*cf.* Remark 3). In particular, the annealed free energy does depend on $K(\cdot)$ and the critical behavior depends on α: the transition is continuous (that is, the free energy is C^1) as soon as $\alpha \leq 1$ and it becomes smoother and smoother as α approaches 0 [16]. Harris criterion (see references in [16, 17]) gives a precise prediction on what to expect for systems for which the annealed system has a transition that is sufficiently smooth (for the pinning case the criterion boils down to $\alpha < 1/2$): essentially it says that quenched and annealed systems have the same critical behavior and it gives a precise prediction of the shift in the critical point due to the disorder (*irrelevant disorder regime*). At the same time it suggests/predicts that for $\alpha > 1/2$ *disorder is relevant*, even if arbitrarily weak. This scenario has now been made rigorous, see [1, 17] and references therein, for pinning models. The crucial point for us is however the fact that the free energy of the annealed copolymer model is not differentiable at the critical point and therefore, in the Harris sense, disorder is always relevant.
2. Understanding critical phenomena when disorder is relevant is a major challenge and the possible scenarios set forth in the physical literature are quite intriguing, but very challenging and, at times, controversial (see *e.g.* [25, 35] and references therein). In this sense also the question that we have raised about improving (20) acquires particular importance.
3. When $\alpha > 1$, also the annealed pinning model free energy is not C^1, and the critical curve has been identified with no more precision than for the copolymer model. In fact the annealed pinning critical curve (again, a curve separating localized and delocalized regimes, in the (β, h) plane) behaves like $-\beta^2/2$ when β is small and the quenched critical curve is in $[-c_+\beta^2, -c_-\beta^2]$ for β small (with explicit values of the constants $0 < c_- < c_+ < 1/2$, cf. [12]). This is absolutely parallel to the fact that for the copolymer model $h_c(\lambda) \in [c_-\lambda, c_+\lambda]$, as one reads for example out of (18), (19).

Finally, it is natural to wonder what happens when a pinning interaction is added to the copolymer model, that is when not only the solvents are selective, but something special goes on at the interface (thus taking into account for example the lack of sharpness of the interface or the fact that impurities could be trapped at the interface). There are works on this model, often called *copolymer with adsorption* (see for example [27, 29, 36]), but the understanding is very limited: we refer to [16, § 6.3.2] for a detailed overview on this issue.

1.6 Organisation of the Chapter

The rest of the chapter is devoted to going deeper into various results that we have stated, by giving either a sketch of arguments of proof, or alternative proofs and results that complement what can be found in the literature. More precisely:

- In Sect. 2 we give a sketch of the proof on the lower bound in (17).
- In Sect. 3 we give an alternative proof of the upper bound in (17).
- In Sect. 4 we discuss the universality features of the copolymer model in the weak coupling regime, i.e., for small values of λ, h.
- Finally, Sect. 5 is devoted to the description of the available results on the path properties of the copolymer model.

2 Localization Estimates

The aim of this section is to give a sketch of the proof of the lower bound in Theorem 1, that is of the left inequality in (17), as well as of the left inequality in (19).

The key-phrase for the approaches in this section is: *rare stretch strategies*. The idea, inspired by the renormalization group approach in [24], is to restrict the partition function to polymer trajectories that can visit the lower half-plane only when there is a stretch of monomers that are *particularly, and anomalously, hydrophilic*. To do this we introduce an intermediate scale ℓ (large, but fixed) and look at the sequence of charges in blocks of ℓ charges at a time. We will consider two strategies (A and B) and, for simplicity, we will assume $\omega_1 \sim \mathcal{N}(0,1)$:

A: The jth block is good if $\sum_{n=(j-1)\ell+1}^{j\ell}(\omega_n + h) \leq -m\ell$, with m a positive value to be chosen below. For ℓ large, the probability that a given block is good is very small, about $\exp(-\ell(h+m)^2/2)$, so that good blocks are typically separated by a distance of about $\exp(\ell(h+m)^2/2)$.

B: The jth block is good if $\sum_{n=(j-1)\ell+1}^{j\ell}(\omega_n + h) = o(\ell)$. For ℓ large, the probability that a given block is good is again very small for $h > 0$, about $\exp(-\ell h^2/2)$, so that good blocks in this case are typically separated by a distance of about $\exp(\ell h^2/2)$.

Given a sequence ω of charges, the good blocks are identified by the rules just given and we introduce a set of polymer trajectories $\Omega^A_{N,\omega}$, resp. $\Omega^B_{N,\omega}$, for the strategy A, resp. B, defined as follows. $\Omega^A_{N,\omega}$ is the set of trajectories that stay in the upper half plane except in presence of good blocks, when they stay in the lower half plane, see the upper part of Fig. 3. The set of polymer trajectories $\Omega^B_{N,\omega}$ still includes all trajectories that stay in the upper half plane except in presence of good blocks, but when there is a good block the only restriction is that the polymer has to touch the oil–water interface just before every good block and it has to touch it again at the end of the block, see the lower part of Fig. 3.

The estimates are now just based on observing that

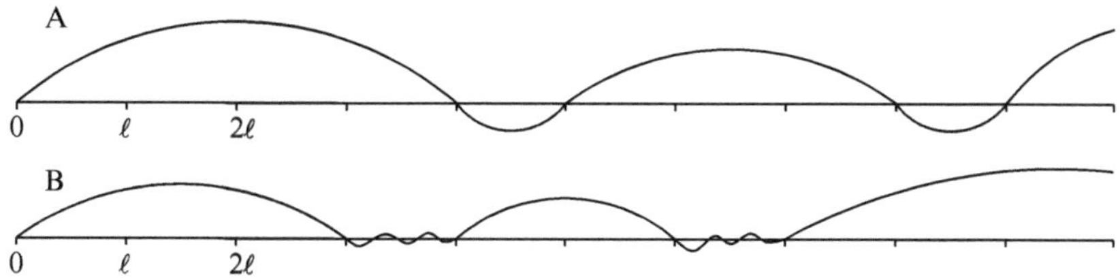

Fig. 3 The lower bound strategies presented in this section are two and they are both based on selecting some *good blocks*, just by looking at the environment. They are actually blocks in which the environment is atypically negative: in strategy A we really select blocks in which the empirical average of the charges $\omega_n + h$ is smaller that a value $-m < 0$ (selected in the end, in order to maximize the gain) and in strategy B we just aim at an empirical average close to zero (a rare event anyway, for $h > 0$). Then for strategy A we make a lower bound on the partition function by visiting the lower half plane if and only if a block is good and by insisting that in a good block the walk stays in the lower half plane. In strategy B we target again the good blocks, but we put no constraint on the walk in the good blocks

$$
\begin{aligned}
Z_{N,\omega} &\geq \mathbf{E}\left[\exp\left(-2\lambda\sum_{n=1}^{N}\Delta_n(\omega_n+h)\right);\, \Omega^A_{N,\omega}\right] \\
&= \frac{K(i_1\ell)}{2}\exp\left(-2\lambda\sum_{n=i_1\ell+1}^{(i_1+1)\ell}(\omega_n+h)\right)\frac{K(\ell)}{2} \\
&\quad\times\frac{K((i_2-i_1)\ell)}{2}\exp\left(-2\lambda\sum_{n=i_2\ell+1}^{(i_2+1)\ell}(\omega_n+h)\right)\frac{K(\ell)}{2}\times\ldots
\end{aligned}
\tag{22}
$$

where the first good block is the i_1th and so on. By using the definition of $\Omega^A_{N,\omega}$ we see that each term $-2\lambda\sum_{n=i_j\ell+1}^{(i_j+1)\ell}(\omega_n+h)$ is bounded below by $2\lambda m\ell$ and we notice that the right-hand side of (22) is a product of terms that, typically and to leading order, are just the same term, because the distance of good block is about $\exp(\ell(h+m)^2/2)$. Therefore this product of terms will give origin to an exponential growth in N (localization!) if

$$
-\frac{1+\alpha}{2}\ell(h+m)^2 + 2\lambda\ell m + o(\ell) > 0 \tag{23}
$$

(we have of course used (3)). If we now optimize the choice of m we readily see that this condition is met if $h < \lambda/(1+\alpha)$, for ℓ sufficiently large. Therefore $h_c(\lambda) \geq \lambda/(1+\alpha)$, which is the lower bound in (17) for the special case of Gaussian charges (the extension to general disorder is straightforward).

Strategy B exploits a factorization similar to (22) but this time the contribution in the good blocks is of about $\exp(\mathrm{F}(\lambda,0)\ell)$, since in the good blocks the empirical average of the charges is zero, so the charges are essentially centered and for large ℓ the random variables in a good block have a distribution that is close to centered

IID standard Gaussian variables (in the complete proof, *cf.* [6, Sect. 2], this step is performed via a change of measure argument that makes rigorous the heuristics presented here: the argument is slightly more involved for non Gaussian charges). The analog of (23) in this case becomes

$$-\frac{1+\alpha}{2}\ell h^2 + \mathrm{F}(\lambda,0)\ell + o(\ell). \tag{24}$$

At this point we need to estimate $\mathrm{F}(\lambda,0)$. For example if one can show that $\mathrm{F}(\lambda,0) \geq c\lambda^2$ for come $c > 0$ (say, for $\lambda \leq \lambda_0$) then for the same values of λ one would have that $h_c(\lambda) \geq \lambda\sqrt{2c/(1+\alpha)}$. This is the basic idea leading to the (middle) lower bound in (19). Of course the work is now on estimating c. We will not go into this issue which, ultimately, is a refinement of the result in [28] and we refer to [6, Sect. 2] for details. But we point out that:

- As explained in Sect. 4, one can show that $\mathrm{F}(\lambda,0)/\lambda^2$ has a positive limit that can be expressed in term of the free energy of a continuum polymer;
- In order to improve on the lower bound in (18) one needs $\sqrt{2c/(1+\alpha)} > 1/(1+\alpha)$, that is $c > 1/(2+2\alpha)$. This can be established, as recalled just below (19), for $\alpha \geq 0.65$. If we were to improve on the lower bound in (18) with this strategy for $\alpha = 1/2$ we would need to show $c > 1/3$, but numerical estimations suggest that $\lim_{\lambda\searrow 0}\mathrm{F}(\lambda,0)/\lambda^2$ is smaller (probably by little) than $1/3$, so it is very likely that this strategy (barely) fails to establish that the lower bound in (18) can be made strict for $\alpha = 1/2$. The interest on this issue is because it would prove that the claims in [24, 30] are not correct.

3 Delocalization Estimates

In this section we address the upper bound in Theorem 1, that is the right inequality in (17). We will actually present (in full) an argument that is substantially easier than the one originally used [6], even if it has the drawback to work only for $\alpha \in (0,1)$. This argument is still based on the fractional moment method (first used in the copolymer context in [32] to show that the upper bound in (17) holds for λ large and unbounded disorder), but it avoids the change of measure argument used in [6]. The change of measure argument in [6, 12] is an important technique, as well as its refinement in [33] that leads to the upper bound in (19), but we will not discuss these techniques here.

3.1 Fractional Moment Method: The General Principle

We consider the fractional moment method in its most elementary application. For the constrained partition function (8) we can write

$$Z^{\mathrm{c}}_{N,\omega,\lambda,h} = \sum_{k=1}^{N} \sum_{0=t_0<t_1<\ldots<t_k=N} \prod_{i=1}^{k} \frac{1+\mathrm{e}^{-2\lambda(\sum_{t_{i-1}<j\le t_i}\omega_j + h(t_i-t_{i-1}))}}{2} K(t_i-t_{i-1}). \tag{25}$$

For $\gamma \in [0,1]$, from the inequality $(a+b)^\gamma \le a^\gamma + b^\gamma$, valid for all $a,b \ge 0$, we obtain the upper bound

$$\mathbb{E}((Z^{\mathrm{c}}_{N,\omega,\lambda,h})^\gamma) \le \sum_{k=1}^{N} \sum_{0=t_0<t_1<\ldots<t_k=N} \prod_{i=1}^{k} \widetilde{K}_{\gamma,\lambda,h}(t_i-t_{i-1}), \tag{26}$$

where we define

$$\widetilde{K}_{\gamma,\lambda,h}(n) := \frac{1+\mathrm{e}^{-(2\lambda\gamma h - \log \mathrm{M}(-2\lambda\gamma))n}}{2^\gamma} K(n)^\gamma. \tag{27}$$

We also set

$$\Sigma(\gamma,\lambda,h) := \sum_{n\in\mathbb{N}} \widetilde{K}_{\gamma,\lambda,h}(n). \tag{28}$$

Assume that $\Sigma(\gamma,\lambda,h) \le 1$. By classical renewal theory [3], the right-hand side of (26) equals the probability that a renewal process with step probability (or sub-probability, if $\Sigma(\gamma,\lambda,h) < 1$) $\widetilde{K}_{\gamma,\lambda,h}(\cdot)$ passes through N; in particular, it is bounded by 1. Then

$$\begin{aligned} \mathrm{F}(\lambda,h) &= \lim_{N\to\infty} \frac{1}{N}\mathbb{E}(\log Z^{\mathrm{c}}_{N,\omega,\lambda,h}) = \lim_{N\to\infty} \frac{1}{\gamma N}\mathbb{E}(\log (Z^{\mathrm{c}}_{N,\omega,\lambda,h})^\gamma) \\ &\le \lim_{N\to\infty} \frac{1}{\gamma N} \log \mathbb{E}((Z^{\mathrm{c}}_{N,\omega,\lambda,h})^\gamma) = 0, \end{aligned}$$

whence $\mathrm{F}(\lambda,h) = 0$ by (11).

This means that, when there exists $\gamma \in [0,1]$ such that $\Sigma(\gamma,\lambda,h) \le 1$, it follows that $(\lambda,h) \in \mathcal{D}$. This allows to give explicit estimates on the delocalized region. Note that for $\gamma = 1$ we find the *annealed delocalized regime*, that we have already introduced: in fact $\Sigma(1,\lambda,h) \le 1$ when $h > h^{\mathrm{ann}}_{\mathrm{c}}(\lambda) := \log \mathrm{M}(-2\lambda)/(2\lambda)$. Since for $\gamma \in [0, 1/(1+\alpha)]$ one sees immediately that $\Sigma(\gamma,\lambda,h) = +\infty$, the interesting range is $\gamma \in (1/(1+\alpha), 1)$.

3.2 Fractional Moment Method: Application

Let us define for $\lambda > 0$

$$\overline{h}(\lambda) := \inf\{h > 0 : \exists \gamma \in [0,1] \text{ such that } \Sigma(\gamma,\lambda,h) < 1\}. \tag{29}$$

Proposition 1. *For $\alpha \in (0,1)$ we have $h_c(\lambda) \le \overline{h}(\lambda) < h_c^{\mathrm{ann}}(\lambda)$ for every $\lambda > 0$.*

Proof. We have just remarked that $\Sigma(\gamma,\lambda,h) \le 1$ implies $(\lambda,h) \in \mathscr{D}$, therefore $h_c(\lambda) \le \overline{h}(\lambda)$. It remains to show that $\overline{h}(\lambda) < h_c^{\mathrm{ann}}(\lambda)$ for every $\lambda > 0$, that is, for $\epsilon > 0$ sufficiently small we can choose $\gamma \in [0,1]$ such that $\Sigma(\gamma,\lambda,h_c^{\mathrm{ann}}(\lambda)-\epsilon) < 1$. Note that for $\gamma \in (0,1)$

$$\frac{\partial\Sigma}{\partial\gamma}(\gamma,\lambda,h_c^{\mathrm{ann}}(\lambda)) = \sum_{n\in\mathbb{N}} \left(1+\mathrm{e}^{-(\gamma\log\mathrm{M}(-2\lambda)-\log\mathrm{M}(-2\lambda\gamma))n}\right)\left(\log\frac{K(n)}{2}\right)\left(\frac{K(n)}{2}\right)^{\gamma}$$
$$-2\lambda\left[(\log\mathrm{M})'(-2\lambda)+\frac{\log\mathrm{M}(-2\lambda)}{2\lambda}\right]\sum_{n\in\mathbb{N}}\frac{\mathrm{e}^{-(\gamma\log\mathrm{M}(-2\lambda)-\log\mathrm{M}(-2\lambda\gamma))n}}{2^{\gamma}}\,n\,K(n)^{\gamma}.$$

By the strict convexity of $\log\mathrm{M}(\cdot)$ and the fact that $\log\mathrm{M}(0)=0$,

$$(\log\mathrm{M})'(-2\lambda) < \frac{\log\mathrm{M}(0)-\log\mathrm{M}(-2\lambda)}{2\lambda} = -\frac{\log\mathrm{M}(-2\lambda)}{2\lambda}. \tag{30}$$

Recalling our assumption (3), for $\alpha \in (0,1)$ we have $\sum_{n\in\mathbb{N}} nK(n) = \infty$, therefore by Fatou's lemma

$$\frac{\partial\Sigma}{\partial\gamma}(1^-,\lambda,h_c^{\mathrm{ann}}(\lambda)) := \lim_{\gamma\uparrow 1}\frac{\partial\Sigma}{\partial\gamma}(\gamma,\lambda,h_c^{\mathrm{ann}}(\lambda)) = +\infty. \tag{31}$$

Since $\Sigma(1,\lambda,h_c^{\mathrm{ann}}(\lambda)) = 1$, it follows that $\Sigma(1-\eta,\lambda,h_c^{\mathrm{ann}}(\lambda)) < 1$, for $\eta > 0$ small enough. By continuity, $\Sigma(1-\eta,\lambda,h_c^{\mathrm{ann}}(\lambda)-\epsilon) < 1$ for ϵ small enough, and the proof is completed. □

4 Continuum Model and Weak Coupling Limit

In this section we explain in some detail the universality feature sketched in Fig. 2 and its caption. The idea is that at weak coupling the details of the model, that is the law of the renewal beyond the exponent α and the law of the disorder, are inessential and a suitable continuum model captures the leading behavior of the (large class of) discrete models we consider.

As we already remarked, it is convenient to look at the renewal process $\tau = \{\tau_k\}_{k\ge 0}$ as a random subset of $[0,\infty)$. It follows from our assumption (3) that the rescaled random set $\epsilon\tau = \{\epsilon\tau_k\}_{k\ge 0}$ converges in distribution as $\epsilon \searrow 0$ toward a limit random set $\widetilde{\tau}^{\alpha}$, the so-called α-stable regenerative set (we refer to [13] for more details; cf. also [9]). This is a random closed subset of $[0,\infty)$ which is scale-invariant ($c\widetilde{\tau}^{\alpha}$ has the same law as $\widetilde{\tau}^{\alpha}$, for every $c > 0$), has zero Lebesgue measure and no isolated points. For $\alpha = \frac{1}{2}$ we have the representation $\widetilde{\tau}^{\alpha} = \{t \in [0,\infty) : B_t = 0\}$, where B is Brownian motion.

The complementary set $(\widetilde{\tau}^\alpha)^c$, being open, is the countable union of disjoint open intervals $\{I_n\}_{n\in\mathbb{N}}$. We can then define a continuous-time process $\widetilde{\Delta}^\alpha = \{\widetilde{\Delta}_t^\alpha\}_{t\in[0,\infty)}$, which is constant on each I_n and takes the value 0 or 1, decided by fair coin tossing: more precisely, in analogy with the discrete case, we set $\Delta_t := \sum_{n\in\mathbb{N}} \xi_n \mathbf{1}_{I_n}(t)$ where $\{\xi_n\}_{n\in\mathbb{N}}$ are i.i.d. $B(\frac{1}{2})$ random variables. For $\alpha = \frac{1}{2}$ we have the representation $\widetilde{\Delta}_t^\alpha = \mathbf{1}_{\{B_t<0\}}$ with B a Brownian motion. In general, $\widetilde{\Delta}^\alpha$ may be viewed as the limit in distribution of the rescaled discrete process $\{\Delta_{\lfloor t/\epsilon\rfloor}\}_{t\in[0,\infty)}$ as $\epsilon \searrow 0$.

Let now $(\beta = \{\beta_t\}_{t\in[0,\infty)}, \mathbb{P})$ be a Brownian motion, independent of $(\widetilde{\Delta}^\alpha, \mathbf{P})$. We proceed for a moment in a somewhat informal way: as $a \searrow 0$, $a^{-1}\omega_{\lfloor t/a^2\rfloor}$ converges toward the white noise $\mathrm{d}\beta_t/\,\mathrm{d}t$ and $\Delta_{\lfloor t/a^2\rfloor}$ converges toward $\widetilde{\Delta}_t$, therefore

$$a\lambda \sum_{n=1}^{N/a^2} (\omega_n + ah)\Delta_n = \lambda \int_0^N (a^{-1}\omega_{\lfloor t/a^2\rfloor} + h)\Delta_{\lfloor t/a^2\rfloor}\,\mathrm{d}t \approx \lambda \int_0^N (\mathrm{d}\beta_t + h\,\mathrm{d}t)\widetilde{\Delta}_t^\alpha.$$

This hints at introducing a continuum partition function

$$\widetilde{Z}_{t,\beta} = \widetilde{Z}_{t,\beta,\lambda,h} := \mathbf{E}\left[\exp\left(-2\lambda \int_0^t \widetilde{\Delta}_s^\alpha(\mathrm{d}\beta_s + h\,\mathrm{d}s)\right)\right], \tag{32}$$

so that, recalling (5), one should have $Z_{N/a^2,\omega,a\lambda,ah} \approx \widetilde{Z}_{N,\beta,\lambda,h}$ for a small.

We now turn to precise statements. One can show that the definition (32) of the continuum partition function is well-posed, for $\mathbb{P}$-a.e. β, and one introduces the corresponding continuum free energy $\widetilde{\mathrm{F}}^\alpha(\lambda,h)$ in the usual way:

$$\widetilde{\mathrm{F}}_\alpha(\lambda,h) = \lim_{t\to\infty} \frac{1}{t}\,\mathbb{E}\log \widetilde{Z}_{t,\beta,\lambda,h}. \tag{33}$$

The existence of this limit and the fact that it is self-averaging (i.e., the expectation $\mathbb{E}$ can be dropped) require a much longer and technical proof than the discrete counterpart, cf. [9].

Also the continuum free energy is non-negative: $\widetilde{\mathrm{F}}_\alpha(\lambda,h) \geq 0$ for all $\lambda, h \geq 0$, as one can easily check. The localized and delocalized regimes can therefore be defined in analogy with the discrete case, namely $(\lambda,h) \in \widetilde{\mathscr{L}}$ (resp. $(\lambda,h) \in \widetilde{\mathscr{D}}$) if $\widetilde{\mathrm{F}}_\alpha(\lambda,h) > 0$ (resp. $\widetilde{\mathrm{F}}_\alpha(\lambda,h) = 0$), and it is easily shown that they are separated by a critical curve: $\widetilde{\mathscr{L}} = \{(\lambda,h) : h < \widetilde{h}_c^\alpha(\lambda)\}$ and $\widetilde{\mathscr{D}} = \{(\lambda,h) : h \geq \widetilde{h}_c^\alpha(\lambda)\}$. There is however a major simplification with respect to the discrete case: the scaling properties of the processes $\widetilde{\Delta}^\alpha$ and β yield easily $\widetilde{\mathrm{F}}_\alpha(a\lambda, ah) = a^2\widetilde{\mathrm{F}}_\alpha(\lambda,h)$ for all $\lambda, h, a \geq 0$, therefore the critical curve is a straight line: $\widetilde{h}_c^\alpha(\lambda) = m_\alpha\lambda$ for some m_α.

We can finally come back to the rough consideration $Z_{N/a^2,\omega,a\lambda,ah} \approx \widetilde{Z}_{N,\beta,\lambda,h}$, that was discussed above. This can be made precise in the form of the following theorem.

Theorem 2. *For an arbitrary discrete α-copolymer model we have*

$$\lim_{a\searrow 0}\frac{1}{a^2}\mathrm{F}(a\lambda, ah) \;=\; \widetilde{\mathrm{F}}_\alpha(\lambda, h), \qquad \forall \lambda, h \geq 0. \tag{34}$$

Moreover

$$\lim_{\lambda\searrow 0}\frac{h_c(\lambda)}{\lambda} = m_\alpha. \tag{35}$$

This result was first proved in [8] in the special case of the basic model of Sect. 1, i.e., for the discrete copolymer model based on the simple random walk on $\mathbb{Z}$, corresponding to $\alpha = \frac{1}{2}$ (in [18] one can find an argument to relax the assumption in [8] of binary charges and in [27] the case with *adsorption* is treated, *cf.* the end of Sect. 1.5). The generalization to arbitrary α-copolymer models, with general disorder distribution, is in [9].

Note that (34) yields directly the existence of the limit as $\lambda \searrow 0$ of $\mathrm{F}(\lambda, 0)/\lambda^2$, that was anticipated in Sect. 2, as well as the fact that this limit coincides with $\widetilde{\mathrm{F}}_\alpha(1, 0) > 0$. We also point out that (35) is not a direct consequence of (34).

The importance of Theorem 2 relies in its *universality* content: for any fixed $\alpha \in (0, 1)$ there is a single continuum model that captures the behavior of all discrete α-copolymer models for small values of λ and h. In other words, the differences among these models become irrelevant in the weak coupling limit. From this viewpoint, the slope m_α of the continuum critical curve is an extremely interesting object: improving the known bounds (19) would most probably mean a substantial improvement in the understanding of the phase transition in this class of models.

5 Path Properties

Up to now we have discussed the localization-delocalization transition only in terms of free energy. A complementary, and equally interesting, point of view is that of looking at path properties. In other words, how does the typical (under $\mathbf{P}_{N,\omega}$, for typical ω) polymer trajectory look like? The bottom-line of the picture which has emerged up to this day is the following. In the localized region the polymer makes order of N excursions between the two half-planes; the lengths of such excursions are $O(1)$ and their distribution has an exponential tail. In the delocalized region, on the other hand, the number of monomers in the unfavorable solvent (i.e. in the lower half plane) is not only sub-linear in N (this information can be obtained immediately from the fact that the free energy is zero there, cf. (12)) but it actually does not exceed $O(\log N)$, with high probability. In the following, we discuss this picture in a bit more detail.

5.1 The Localized Phase

This subsection is extracted from [19], to which we refer for additional results, concerning for instance the exponential tail of the length of the polymer excursions between the two solvents. Path properties in the localized phase have been studied also in [2, 4].

Let $\mathbf{M}_N := \max_{i:\tau_i \le N}(\tau_i - \tau_{i-1})$ be the length of the longest polymer excursion between the two solvents. The following result says that in the localized region correlations decay exponentially fast, and the longest excursion is of order $\log N$:

Theorem 3. *Let $(\lambda, h) \in \mathscr{L}$. There exist constants c_1, c_2 such that, for every pair of bounded local functions A, B of τ we have*

$$\sup_N \mathbb{E}\left[|\mathbf{E}_{N,\omega}(AB) - \mathbf{E}_{N,\omega}(A)\mathbf{E}_{N,\omega}(B)|\right] \le c_1 \|A\|_\infty \|B\|_\infty e^{-c_2 d(A,B)}, \qquad (36)$$

where $d(A, B)$ denotes the distance between the supports of A and B.

Moreover, for every $\epsilon \in (0, 1)$ the following holds in $\mathbb{P}$-probability:

$$\lim_{N\to\infty} \mathbf{P}_{N,\omega}\left(\frac{1-\epsilon}{\mu(\lambda,h)} \le \frac{\mathbf{M}_N}{\log N} \le \frac{1+\epsilon}{\mu(\lambda,h)}\right) = 1, \qquad (37)$$

where μ is defined as

$$\mu(\lambda, h) = -\lim_{N\to\infty} \frac{1}{N} \log \mathbb{E} \frac{1 + \mathrm{e}^{-2\lambda \sum_{n=1}^{N}(\omega_n + h)}}{Z_{N,\omega}}. \qquad (38)$$

The existence of the limit (38), together with the bounds $0 < \mu(\lambda, h) \le \mathrm{F}(\lambda, h)$ in the localized phase ($\mu(\lambda, h) \ge 0$ in general, and this is seen in the same way as for $\mathrm{F}(\lambda, h) \ge 0$), is proved in [19], where one can also find an argument showing that $\mu(\lambda, h) < \mathrm{F}(\lambda, h)$ under suitable (but not too restrictive) assumptions on the law of the charges.

As a simple consequence of the exponential decay of correlations one can prove that:

1. The free energy is infinitely differentiable (in both λ and h) in the localized region $\mathscr{L}$.
2. For every bounded local observable A the limit $\lim_{N\to\infty} \mathbf{E}_{N,\omega}(A)$ exists $\mathbb{P}(\mathrm{d}\omega)$ almost surely and is reached exponentially fast.

As expected, the rate of exponential decay of the correlation functions (or inverse correlation length), i.e. c_2 in (36), tends to zero as $h \nearrow h_c(\lambda)$. In general, it is a very interesting open problem to understand the relation between this and the way the free energy vanishes close to the critical point. In [31], for the special case where $K(\cdot)$ is the law of the first return to zero of the symmetric simple random walk on $\mathbb{Z}$, it was proved that the best constant c_2 coincides with $\mu(\lambda, h)$ defined in (38).

There would be much to say about $\mu(\lambda,h)$ and $\mathrm{F}(\lambda,h)$ and we prefer to refer the reader to the introduction of [21] where this issue is treated in detail for pinning models. Here, in a simplistic way, we just point out that Theorem 3 and the bounds mentioned just after it are telling us in particular that $\mu(\lambda,h)$ is as good as $\mathrm{F}(\lambda,h)$ for detecting the localization transition. But:

- ⋆ Is it true that $\log\mu(\lambda,h) \sim \log \mathrm{F}(\lambda,h)$ as $h \searrow h_c(\lambda)$? In view of the discussion on disorder relevance in Sect. 1.5, we expect that this is not the case and establishing such a result would be very interesting.

5.2 The Delocalized Phase

Recall the definition $\mathcal{N}_N = \sum_{n=1}^N \Delta_n$ in (13). The following theorem shows that, strictly inside the delocalized region, $\mathcal{N}_N$ is typically at most of order $\log N$:

Theorem 4 ([18]). *For any $\delta > 0, \lambda > 0$ there exist $c > 0, q > 0$ such that for every $N \in \mathbb{N}$*

$$\mathbb{E}\,\mathbf{P}_{N,\omega,\lambda,h_c(\lambda)+\delta}(\mathcal{N}_N \geq n) \leq e^{-cn} \qquad \forall n \geq q \log N. \tag{39}$$

This result was proved in [18] under the assumption that the disorder law $\mathbb{P}$ satisfies a concentration inequality of sub-Gaussian type. This holds for instance in the case of Gaussian or bounded charges, and more generally whenever the distribution of ω_1 satisfies a Log-Sobolev inequality.

Here we give a simpler argument, inspired by [23], which works under the general assumptions of Sect. 1 on the disorder law and gives the weaker statement

$$\mathbb{E}\,\mathbf{E}_{N,\omega,\lambda,h_c(\lambda)+\delta}(\mathcal{N}_N) \leq \frac{c}{2\lambda\delta}\log N \tag{40}$$

for some constant c. The same argument also shows that at the critical point $\mathcal{N}_N$ is typically at most of order $\sqrt{N\log N}$:

$$\mathbb{E}\,\mathbf{E}_{N,\omega,\lambda,h_c(\lambda)}(\mathcal{N}_N) \leq c'\sqrt{N\log N} \tag{41}$$

for some other constant c'.

Recalling (5), for all $h_0, h_1 \geq 0$ we can write

$$Z_{N,\omega,\lambda,h_0} := Z_{N,\omega,\lambda,h_1} \cdot \mathbf{E}_{N,\omega,\lambda,h_1}\left[\exp\left(2\lambda(h_1-h_0)\mathcal{N}_N\right)\right]. \tag{42}$$

Restricting Z_{N,ω,λ,h_1} on the event $\{\tau_1 > N, \zeta_1 = 0\}$, we obtain the bound $Z_{N,\omega,\lambda,h_1} \geq \frac{1}{2}\sum_{n>N} K(n) \sim (const.)N^{-\alpha}$, by (3). Applying Jensen's inequality we obtain

$$\begin{aligned}\frac{1}{N}\mathbb{E}\left[\log Z_{N,\omega,\lambda,h_0}\right] &\geq \frac{1}{N}\mathbb{E}\left[\log Z_{N,\omega,\lambda,h_1}\right] + \frac{2\lambda(h_1-h_0)}{N}\mathbb{E}\,\mathbf{E}_{N,\omega,\lambda,h_1}\left[\mathcal{N}_N\right] \\ &\geq -c_1\frac{\log N}{N} + \frac{2\lambda(h_1-h_0)}{N}\mathbb{E}\,\mathbf{E}_{N,\omega,\lambda,h_1}\left[\mathcal{N}_N\right],\end{aligned} \tag{43}$$

where $c_1 \in (0,\infty)$. By (9) we have $\log Z^{\mathrm{c}}_{N,\omega,\lambda,h} \geq \log Z_{N,\omega,\lambda,h} - c_2 \log N$, for some constant c_2, therefore by (10) we can write, for some $c \in (0,\infty)$,

$$\mathrm{F}(\lambda,h_0) \geq \sup_{N\in\mathbb{N}}\left\{\frac{2\lambda(h_1-h_0)}{N}\mathbb{E}\,\mathbf{E}_{N,\omega,\lambda,h_1}\left[\mathcal{N}_N\right] - c\,\frac{\log N}{N}\right\}. \tag{44}$$

- Take $h_0 = h_c(\lambda)$ and $h_1 = h_c(\lambda)+\delta$ with $\delta>0$. Since $\mathrm{F}(\lambda,h_0)=0$, from (44) we obtain (40).
- Now take $h_0 = h_c(\lambda)-\delta$, with $\delta>0$, and $h_1 = h_c(\lambda)$. Since $\mathrm{F}(\lambda,h_c(\lambda)-\delta) \leq c_2\delta^2$ for some $c_2 = c_2(\lambda) \in (0,\infty)$ (cf. (20)), it follows again from (44) that

$$\limsup_{N\to\infty}\frac{\mathbb{E}\,\mathbf{E}^{\mathrm{c}}_{N,\omega,\lambda,h_c(\lambda)}\left[\mathcal{N}_N\right]}{\sqrt{N\log N}} \leq C, \qquad \text{where} \quad C := \frac{\sqrt{c_1\,c_2(\lambda)}}{\lambda} \in (0,\infty), \tag{45}$$

whence (41).

Remark 5. Recalling (27), let $\lambda, h > 0$ and $\gamma \in [0,1]$ be chosen such that $\Sigma(\gamma,\lambda,h) < 1$, that is, $\widetilde{K}_{\gamma,\lambda,h}(\cdot)$ is a sub-probability kernel on $\mathbb{N}_0$. Since $\widetilde{K}_{\gamma,\lambda,h}(\cdot) \sim cK(N)^\gamma \sim c'N^{-\gamma(1+\alpha)}$ as $N\to\infty$ for some constants $c,c'>0$, by (27) and (3), it is a basic result in renewal theory that the right-hand side of (26) is asymptotically equivalent to $c''\widetilde{K}_{\gamma,\lambda,h}(N)$ as $N\to\infty$ for some constant $c''>0$, cf. [16, Theorem A.4]. Therefore we have for all $N\in\mathbb{N}$

$$\mathbb{E}((Z^{\mathrm{c}}_{N,\omega,\lambda,h})^\gamma) \leq C_1\widetilde{K}_{\gamma,\lambda,h}(N) \leq C_1\,K(N)^\gamma, \tag{46}$$

where here and in the sequel C_i denotes a generic positive constant. Recalling (9) and (3), for the original (non constrained) partition function we have

$$\mathbb{E}\left((Z_{N,\omega,\lambda,h})^\gamma\right) \leq C_2\,N^\gamma\,K(N)^\gamma \leq C_3 N^{-\gamma\alpha}. \tag{47}$$

This relation can be exploited to improve Theorem 4, showing that with high probability $\mathcal{N}_N$ is of order 1. More precisely, from the bound $Z_{N,\omega,\lambda,h} \geq \frac{1}{2}\sum_{n>N}K(n) \sim (const.)N^{-\alpha}$ and (42) it follows that for any $\delta>0$

$$\mathbf{E}_{N,\omega,\lambda,h+\delta}\left[\exp\left(2\lambda\delta\mathcal{N}_N\right)\right] = \frac{Z_{N,\omega,\lambda,h}}{Z_{N,\omega,\lambda,h+\delta}} \leq C_4\,N^\alpha\,Z_{N,\omega,\lambda,h}, \tag{48}$$

hence applying (47) we obtain

$$\mathbb{E}\left[\left(\mathbf{E}_{N,\omega,\lambda,h+\delta}\left[\exp\left(2\lambda\delta\mathcal{N}_N\right)\right]\right)^\gamma\right] \leq C_4^\gamma\,N^{\alpha\gamma}\,\mathbb{E}\left((Z_{N,\omega,\lambda,h})^\gamma\right) \leq C_5. \tag{49}$$

Recalling that $\gamma \in [0,1]$, by Markov's inequality we can write

$$\mathbb{E}\,\mathbf{P}_{N,\omega,\lambda,h+\delta}(\mathcal{N}_N \geq n) \leq \mathbb{E}\,\left(\mathbf{P}_{N,\omega,\lambda,h+\delta}(\mathcal{N}_N \geq n)^\gamma\right) \\ \leq \mathrm{e}^{-2\lambda\delta\gamma n}\,\mathbb{E}\,[(\mathbf{E}_{N,\omega,\lambda,h+\delta}\,[\exp{(2\lambda\delta\mathcal{N}_N)}])^\gamma] \leq C_5\,\mathrm{e}^{-2\lambda\delta\gamma n}.$$

Summarizing, we have shown that whenever relation (47) holds true, there exists a constant $C > 0$ such that for every $\delta > 0$

$$\mathbb{E}\,\mathbf{P}_{N,\omega,\lambda,h+\delta}(\mathcal{N}_N \geq n) \leq C\,e^{-2\lambda\delta\gamma n}, \qquad \forall n \in \mathbb{N}. \tag{50}$$

We stress that relation (47) holds true in particular for every λ, h with $h > \overline{h}(\lambda)$ (with a suitable choice of $\gamma \in [0,1]$, recall (29) and Proposition 1), hence also *below* the annealed critical curve. This is therefore an improvement of (1.12) in [18].

We point out that delocalization properties were also studied in [4]. However the nature of the delocalized phase, in the pathwise sense, is still very little understood and, notably, almost sure results are lacking (see however [26]). For example:

- ⋆ Is it true that, if (λ, h) is in the interior of $\mathcal{D}$, for every $\varepsilon > 0$ we have $\lim_{N\to\infty}\mathbf{P}_{N,\omega}(\Delta_n = 0 \text{ for every } n \in [\varepsilon N, N] \cap \mathbb{N}) = 0$ $\mathbb{P}(\mathrm{d}\omega)$-almost surely?

Acknowledgements We gratefully acknowledge the support of the University of Padova (F.C. under grant CPDA082105/08) and of ANR (G.G. and F.L.T. under grant SHEPI).

References

1. Alexander, K., Zygouras, N.: Quenched and annealed critical points in polymer pinning models. Comm. Math. Phys. **291**, 659–689 (2008)
2. Albeverio, S., Zhou, X.Y.: Free energy and some sample path properties of a random walk with random potential. J. Statist. Phys. **83**, 573–622 (1996)
3. Asmussen, S.: Applied probability and queues. (2nd edn), Applications of Mathematics **51**, Springer, New York (2003)
4. Biskup, M., den Hollander, F.: A heteropolymer near a linear interface. Ann. Appl. Probab. **9**, 668–687 (1999)
5. Bodineau, T., Giacomin, G.: On the localization transition of random copolymers near selective interfaces. J. Statist. Phys. **117**, 801–818 (2004)
6. Bodineau, T., Giacomin, G., Lacoin, H., Toninelli, F.L.: Copolymers at selective interfaces: new bounds on the phase diagram. J. Statist. Phys. **132**, 603–626 (2008)
7. Bolthausen, E., Caravenna, F., de Tilière, B.: The quenched critical point of a diluted disordered polymer model. Stoch. Process. Appl. **119**, 1479–1504 (2009)
8. Bolthausen, E., den Hollander, F.: Localization transition for a polymer near an interface. Ann. Probab. **25**, 1334–1366 (1997)
9. Caravenna, F., Giacomin, G.: The weak coupling limit of disordered copolymer models. Ann. Probab. **38**, 2322–2378 (2010)
10. Caravenna, F., Giacomin, G., Gubinelli, M.: A numerical approach to copolymers at selective interfaces. J. Statist. Phys. **122**, 799–832 (2006)

11. Causo, M.S., Whittington, S.G.: A Monte Carlo investigation of the localization transition in random copolymers at an interface. J. Phys. A: Math. Gen. **36**, L189–L195 (2003)
12. Derrida, B., Giacomin, G., Lacoin, H., Toninelli, F.L.: Fractional moment bounds and disorder relevance for pinning models. Commun. Math. Phys. **287**, 867–887 (2009)
13. Fitzsimmons, P.J., Fristedt, B., Maisonneuve, B.: Intersections and limits of regenerative sets. Z. Wahrscheinlichkeitstheorie verw. Gebiete **70**, 157–173 (1985)
14. Garel, T., Huse, D.A., Leibler, S., Orland, H.: Localization transition of random chains at interfaces. Europhys. Lett. **8**, 9–13 (1989)
15. den Hollander, F.: Random Polymers, École d'Été de Probabilités de Saint-Flour XXXVII–2007, Lecture Notes in Mathematics **1974**, Springer (2009)
16. Giacomin, G.: Random Polymer Models. Imperial College Press, World Scientific (2007)
17. Giacomin, G., Lacoin, H., Toninelli, F.L.: Marginal relevance of disorder for pinning models. Commun. Pure Appl. Math. **63**, 233–265 (2010)
18. Giacomin, G., Toninelli, F.L.: Estimates on path delocalization for copolymers at selective interfaces. Probab. Theory Relat. Fields **133**, 464–482 (2005)
19. Giacomin, G., Toninelli, F.L.: The localized phase of disordered copolymers with adsorption. ALEA **1**, 149–180 (2006)
20. Giacomin, G., Toninelli, F.L.: Smoothing effect of quenched disorder on polymer depinning transitions. Commun. Math. Phys. **266**, 1–16 (2006)
21. Giacomin, G., Toninelli, F.L.: On the irrelevant disorder regime of pinning models. Ann. Probab. **37**, 1841–1873 (2009)
22. Habibzadah, N., Iliev, G.K., Martin, R., Saguia, A., Whittington, S.G.: The Order of the Localization Transition for a Random Copolymer. J. Phys. A: Math. Gen. **39**, 5659–5667 (2006)
23. Lacoin, H.: The martingale approach to disorder irrelevance for pinning models. Electron. Comm. Probab. **15**, 418–427 (2010)
24. Monthus, C.: On the Localization of Random heteropolymers at the interface between two selective solvents. Eur. Phys. J. B **13**, 111–130 (2000)
25. Monthus, C., Garel, T.: Delocalization transition of the selective interface model: distribution of pseudo-critical temperatures. J. Stat. Mech. P12011 (2005)
26. Mourrat, J.-C.: On the delocalized phase of the random pinning model, Séminaire de Probabilités (to appear)
27. Pétrélis, N.: Copolymer at selective interfaces and pinning potentials: weak coupling limits. Ann. Instit. H. Poincaré **45**, 175–200 (2009)
28. Sinai, Ya.G.: A random walk with a random potential. Theory Probab. Appl. **38**, 382–385 (1993)
29. Soteros, C.E., Whittington, S.G.: The statistical mechanics of random copolymers. J. Phys. A: Math. Gen. **37**, R279–R325 (2004)
30. Stepanow, S., Sommer, J.-U., Erukhimovich, I.Ya.: Localization transition of random copolymers at interfaces. Phys. Rev. Lett. **81**, 4412–4416 (1998)
31. Toninelli, F.L.: Critical properties and finite-size estimates for the depinning transition of directed random polymers. J. Stat. Phys. **126**, 1025–1044 (2007)
32. Toninelli, F.L.: Disordered pinning models and copolymers: beyond annealed bounds. Ann. Appl. Probab. **18**, 1569–1587 (2008)
33. Toninelli, F.L.: Coarse graining, fractional moments and the critical slope of random copolymers. Electron. J. Probab. **14**, 531–547 (2009)
34. Trovato, A., Maritan, A.: A variational approach to the localization transition of heteropolymers at interfaces. Europhys. Lett. **46**, 301–306 (1999)
35. Vojta, T., Sknepnek, R.: Critical points and quenched disorder: From Harris criterion to rare regions and smearing. Phys. Stat. Sol. B **241**, 2118–2127 (2004)
36. Whittington, S.G.: Randomly coloured self-avoiding walks: adsorption and localization. Markov Proc. Rel. Fields **13**, 761–776 (2007)

Some Locally Self-Interacting Walks on the Integers

Anna Erschler, Bálint Tóth, and Wendelin Werner

Abstract We study certain self-interacting walks on the set of integers that choose to jump to the right or to the left randomly but influenced by the number of times they have previously jumped along the edges in the finite neighbourhood of their current position (in this paper, typically, we will discuss the case where one considers the neighbouring edges and the next-to-neighbouring edges). We survey a variety of possible behaviours, including some where the walk is eventually confined to an interval of large length. We also focus on certain "asymmetric" drifts, where we prove that with positive probability, the walks behave deterministically on large scale and move like $n \mapsto c\sqrt{n}$ or like $n \mapsto c \log n$.

1 Introduction

A locally self-interacting random walk on the set of integers is a sequence of integer-valued random variables $(X_n, n \geq 0)$ with $|X_{n+1} - X_n| = 1$ for all $n \geq 0$, which can be defined inductively as follows: One initializes a "local time profile" by first choosing some function L_0 that associates with each edge e of the lattice $\mathbb{Z}$ a real number $L_0(e)$ (the set E of unoriented edges can be identified to the set $\mathbb{Z} + 1/2$ or

A. Erschler
Laboratoire de Mathématiques, Université Paris-Sud, 91405 Orsay cedex, France
e-mail: anna.erschler@math.u-psud.fr;

B. Tóth (✉)
Institute of Mathematics, Technical University Budapest, Egry József u. 1, 1111 Budapest, Hungary
e-mail: balint@math.bme.hu

W. Werner
Laboratoire de Mathématiques, Université Paris-Sud, 91405 Orsay cedex, France

DMA, ENS, 45 rue d'Ulm, 75230 Paris cedex 05, France
e-mail: wendelin.werner@math.u-psud.fr

J.-D. Deuschel et al. (eds.), *Probability in Complex Physical Systems*, Springer Proceedings in Mathematics 11, DOI 10.1007/978-3-642-23811-6_13,

to the set of couples $\{x, x+1\}$ where $x \in \mathbb{Z}$) and one chooses the starting point X_0. The simplest choice is of course to set this initial function L_0 to be equal to zero on all edges, and to start at the origin (i.e. $X_0 = 0$) with probability one.

The law of the walker is then described via some function R from $\mathbb{R}^E$ to $[0, 1]$. This function describes the probability of jumping to the right in terms of the local time profile "seen from the walker at its current position". More precisely, one defines inductively, for each $n \geq 0$, X_{n+1} to be equal to either $X_n + 1$ or $X_n - 1$ in such a way that

$$P(X_{n+1} = X_n + 1 \mid X_0, \dots, X_n) = R((\ell_n(e), e \in E)),$$

where

$$\ell_n(\cdot) = L_n(X_n + \cdot)$$

is the local time profile seen from the particle at time n. At each step, one updates this local time profile as follows: $e_{n+1} = \{X_n, X_{n+1}\}$ denotes the edge of the $(n+1)$th jump and

$$L_{n+1}(e) = L_n(e) + 1_{\{e=e_{n+1}\}}$$

is the new local time profile. When L_0 is identically zero, then $L_n(e)$ denotes the number of times the walk has jumped on the edge e before time n. This procedure clearly describes completely the law of the family $(X_n, n \geq 0)$.

Similar definitions work of course on other graphs than $\mathbb{Z}$ (for instance on $\mathbb{Z}^d$). In this paper, we will, however, restrict our discussion to the one-dimensional case. One can also use local times on sites instead of edges, but the latter case turns out to be often easier to handle.

Depending on the function R, the walker can be of self-repelling type (it prefers to go into the direction it has less visited in the past), self-attracting, or more complicated (it can be for instance repelled by the visits to the neighbouring edge, but attracted by its past visits to further edges).

When the function R is "cylindric", i.e. when $R((\ell(e), e \in E))$ is a function on the values $\ell(e_1), \dots, \ell(e_K)$ for a finite set of edges $e_1, \dots, e_K$, the interaction can be described as "local". In other words, for some constant K', the past trajectory $(X_0, X_1, \dots, X_n)$ influences the outcome of $X_{n+1} - X_n$ only via the values of $L_n(e)$ for the edges in the K'-neighbourhood of X_n. Note that in this cylindric case, if there is a large-scale limit to the process $(X_n, n \geq 0)$, its evolution will be "purely" local as the finite range becomes scaled down.

Such locally self-interacting walks provide a particular class of random evolutions of functions as the sequence of functions $(\ell_n(\cdot))_{n\geq 0}$ form a Markov process. It is in fact very natural to study the evolutions where the function R is also height-invariant, i.e. the particular case where

$$R((\ell(e), e \in E)) = R((l + \ell(e), e \in E))$$

for all $(\ell(\cdot))$ and all real l. In other words, the walk is sensitive to the local discrete gradient of ℓ but not to its actual height. This excludes for instance the family of "linear" reinforcements (see for instance [3] for a survey of these other types of reinforcements). Another way to view these evolutions is to define, for all $x \in \mathbb{Z}$ and $n \geq 0$, the local gradient

$$\eta_n(x) = \nabla \ell_n(x) := \ell_n(x + 1/2) - \ell_n(x - 1/2)$$

and to say that $R((\ell))$ is in fact a function $\tilde{R}$ of (η).

In [6], a case where $R((\ell))$ was just a function of $\eta(0) := \ell(1/2) - \ell(-1/2)$ was studied. Loosely speaking, this walk (sometimes referred to as true self-avoiding walk or true self-repelling walk – TSRW) is driven by the negative gradient of the local time at its actual position, i.e. the walk prefers to go in the direction it has visited less often in the past. It turns out (see [6, 7]) that a non-trivial scaling shows up, as $X_n/n^{2/3}$ converges in law to some non-trivial random variable. In fact, the entire limiting process can be described [8] (it is called the true self-repelling motion – TSRM). It corresponds to a random and local evolution of functions, that is neither a deterministic partial differential equation nor constructed via a white-noise driven stochastic partial differential equation.

This raises the questions whether other such processes exist, whether the previous scaling limit is "stable" (do all similar discrete models have the same limiting behaviour? What properties of the discrete model are actually captured by this scaling limit?), and provides a motivation to study self-interacting random walks associated with more general functions R.

The family of models that we will focus on in this paper are those where the self-interaction is governed by a linear combination of the values of the local time. In other words, we let

$$D((\ell)) = \sum_{e \in E} a_e \ell(e),$$

where all but finitely many a_e's are zero, and we choose the local "drift" at step n according to the value of $D((\ell_n))$. The fact that this evolution in height-invariant corresponds to the condition

$$\sum_{e \in E} a_e = 0.$$

Then, we take

$$R((\ell)) = F(D((\ell)))$$

for some increasing function F such that $F(-y) = 1 - F(y)$. In other words, if D is (very) positive (respectively, negative), then the walk will tend to jump to the right (respectively, to the left), so that D indeed plays the role of a local drift. A particularly natural choice for F turns out to be

$$F(y) = \frac{e^y}{e^y + e^{-y}}$$

and we will in fact restrict our discussion to this case in this paper.

In order to start investigating the variety of possible behaviours, one can for instance start to look at the case where $R(\ell)$ is a function of $(\ell(-3/2), \ell(-1/2), \ell(1/2), \ell(3/2))$. A first natural choice is to pick two real numbers a and b, and to drive the walker by the linear combination

$$D^{a,b}((\ell)) = D^{a,b}((\ell)) := a\ell(-3/2) + b\ell(-1/2) - b\ell(1/2) - a\ell(3/2). \quad (1)$$

Note that these are the choices such that both height-invariance (because the sum of the terms $a + b - b - a$ is zero) and *left–right symmetry* (the walks (X_n) and $(-X_n)$ have the same law because the coefficients in front of $\ell(e)$ and $\ell(-e)$ are opposite) hold.

Note that when b is positive and $a = 0$, this is the previously mentioned TSRW with TSRM as scaling limit. When $b > 0$ and $|a|$ is very small, one can therefore view this process as a perturbation of the TSRW. At the other end of the picture, when b is negative, it is easy to check that the walk will (with positive probability) eventually jump back and forth on one single edge (the more it jumps on it, the more it is attracted by it) which is not a particularly interesting case. In Sect. 2, we will describe various possible asymptotic behaviours, depending on the values of a and b. We shall see that intuition can sometimes be misleading; repellance by second neighbours does not always have the same effect as repellance by immediate neighbours as it can create traps. One interesting feature that we will mention is the existence (when $b > 0 > a$) of a phase transition between stuck walks (that can eventually remain in a finite interval) and the non-stuck ones (with almost surely an infinite range), depending on whether $-a > b/3$ or not.

A second natural possibility is to break the left–right symmetry, and to drive the walker by a linear combination such as

$$-\ell(-3/2) + \ell(-1/2) + \ell(1/2) - \ell(3/2)$$

which will be among those studied in Sect. 3 (where, at least with positive probability, one has $X_n \sim f(n)$ as n tends to infinity, for some deterministic function f that tends to infinity).

Section 2 is more of survey-type presenting results proved in [1], as well as heuristics and conjectures, while Sect. 3 will contain original results with actual proofs.

Throughout the paper, we will mention results that hold with positive probability, like "with positive probability, the walk will eventually remain stuck on a single edge". Clearly, one then implicitly conjectures that this is the generic behaviour of the walk (i.e. that the corresponding asymptotic event will hold almost surely), but in general, proving such almost sure results for self-interacting walks is not a straightforward task (and it requires different arguments), see for instance [4] for the linearly reinforced walk. We will not focus on these questions in this paper.

We would like to emphasize here that (despite the "phase diagram" that we will draw at the end of the next section) the aim of this paper is not to give a full classification of the possible asymptotic behaviour of all these cylindric self-interacting walks. Our goal is rather to point out certain phenomenological features, to shed some light on how sensitive (or not) these walks are with respect to their microscopic definition, and what feature is kept in the scaling limit.

A Warm-Up As a warm-up to maybe help (if it does not help the reader, then she/he should not hesitate to skip this paragraph as it will not be essential later on) the following analysis, let us make some non-rigorous analogy with Fibonacci-type sequences and their related polynomials. Recall that our models are "driven" by a linear combination of the type

$$D((\ell)) = \sum_{k=k_0}^{k_1} a_{k+1/2}\ell(k+1/2),$$

where $a_{k_0} \neq 0 \neq a_{k_1}$. If a walk is in some local equilibrium state (it does not "shoot in one direction", it is somewhere in the middle of its past range, and the values of ℓ are all very large), then one could expect that the value of D is rather small at most of the sites (at least compared to the total number of steps). After renormalization by the number of steps, this would indicate that

$$\sum_{k=k_0}^{k_1} a_{k+1/2}\ell_n(j+k+1/2) << \ell_n(1/2)$$

i.e. that $j \mapsto F_j = \ell_n(j+1/2)/\ell_n(1/2)$ follows some Fibonacci-type progression. As we know, the behaviour of such progressions is controlled by the values (and position with respect to the unit circle) of the roots of the corresponding polynomial $P(x)$ where

$$P(x)x^{k_0} = \sum_{k \geq k_0} a_{k+1/2}x^k. \tag{2}$$

The fact that the dynamics is height-invariant corresponds to the case where 1 is a root of P. Hence, it is in a way more appropriate to describe the self-interaction by its polynomial (in the case where the self-interaction is left–right symmetric, it does indeed characterize the self-interaction), and we will use this when we will illustrate some of our simulations. Note that the function $\theta \mapsto P(e^{i\theta})$ defined for $\theta \in [-\pi, \pi)$ is actually the Fourier transform of the sequence (a_k). Thus, the qualitative behaviour of the walk will depend on the type of the Fibonacci polynomial $P(x)$, or, what is equivalent of the Fourier transform of the sequence of coefficients (a_k). For example, positive definiteness of this sequence plays an essential role.

2 Survey of Left–Right Symmetric Cases

During most of this section, the setup is the one described in the introduction, where the walk is "driven" by $D^{a,b}((\ell))$ defined as in (1), i.e. when we are considering the next-to-neighbouring interactions that are both height-invariant and left–right symmetric (we will comment on interactions with larger range at the end of the section). The meaning of a and b is as follows:

- If b is positive (respectively, negative), then the walk is repelled (respectively, attracted) by its previous visits to the neighbouring edge.
- a plays the same role as b but for the next-to-neighbouring edges.
- $|a|$ and $|b|$ describe the intensity of these self-interactions.

As we shall see, in most of these cases, the qualitative asymptotic behaviour of the walk will depend on the signs of a and b, and on the ratio between $|a|$ and $|b|$.

We can divide the parameter space into several classes. Let us first notice that in the case where $b < 0$, then with positive probability, the walk X_n will forever jump back and forth along one single edge (between 0 and 1, say). This is easily checked via the Borel–Cantelli lemma, using the fact that $F(y)$ decays rapidly to 0 as $y \to -\infty$. Similarly, when $b = 0$ and a is negative, then with positive probability, the walk will be stuck on a set of two edges (say, it will visit only the three sites 0, 1 and 2). These are the *purely self-attractive cases.*

2.1 *When* $b > 0$ *and* $-b/3 < a < b$*: The TRSM Regime?*

Let us first very briefly recall here the case where $b > 0$ and $a = 0$: The walker is then driven by the negative gradient of the local time at its current position. As we have already mentioned, this is the so-called true self-repelling walk TSRW.[1] A sample is depicted on Fig. 1. This scaling limit was clarified in [6] and [8]; it is the true self-repelling motion (TRSM).

It is quite natural to expect that this TSRW scaling behaviour is stable under perturbation, i.e. when choosing $a \neq 0$, but sufficiently small (in absolute value). There are some good reasons to guess that in the parameter range $a \in (-b/3, b)$ (and for $b > 0$), the long-time asymptotic behaviour and scaling limit of the walk are similar to the TSRW case. Indeed, besides numerical evidence, one can note that:

1. In this parameter range, we can identify a stationary and ergodic distribution of the "environment seen by the random walker" process. This distribution is qualitatively similar to the TSRW case, the difference being that instead of a

[1] In this terminology originating in the physics literature, "true" refers to the fact that this is a true walk – the distributions of the first n steps is consistent with the distribution of the first $n + 1$ steps – and this is not always the case for measures on self-avoiding or self-repelling walks; the term "myopic" random walk has also been used.

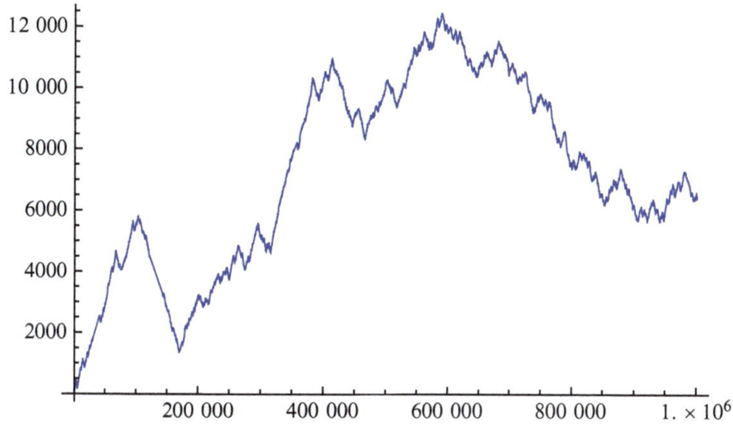

Fig. 1 TSRW trajectory: $a = 0$, $b = 1$, $P(x) = (x - 1)$

product measure we get an exponentially mixing Gibbs measure. For details, see Sect. 2.5.

2. In this stationary and ergodic regime, using the variational method initiated in [2] and exploited in [5] for similar but continuous space–time models, one could very likely obtain superdiffusive lower bounds of the type $E(X_n^2) \geq Cn^{5/4}$ for the asymptotic variance of the displacement of the random walker. For details of similar computations in continuous space–time, see [5].

2.2 When $b > 0$ and $a < -b/3$: The Stuck Case

In this regime, the competition between the nearest edge self-repellence due to $b > 0$ and next-to-nearest edge self-attractiveness due to $a < 0$ is won by the latter because $|a|$ is large enough, and the random walker is eventually trapped in an interval of finite length. The length of the trapping interval increases to infinity as the ratio $|a|/b$ decreases to the critical value $1/3$. More precisely, let us define for all positive integer k,

$$A_k := 1 + 2\cos\left(\frac{2\pi}{k+2}\right).$$

This is an increasing sequence ($A_1 = 0$, $A_2 = 1$, $A_4 = 2$ and $\lim_{n\to\infty} A_n = 3$). In [1], the following theorem is proved:

Theorem 1 ([1]). *Suppose that $a < 0 < b$.*

- *If $b/|a| \in (A_k, A_{k+1})$ for some $k \geq 1$, then with positive probability, the walk remains stuck on a set of $k+2$ consecutive sites (and visits all these sites infinitely often).*

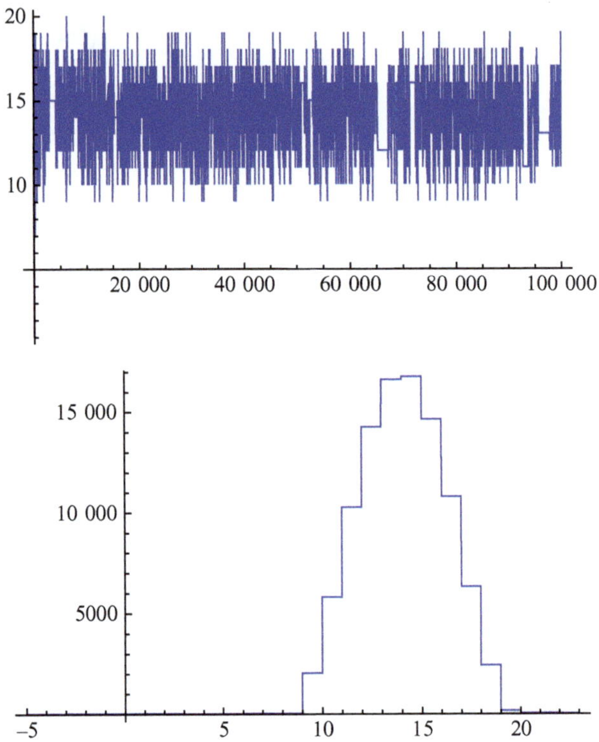

Fig. 2 A walk ($a = -1.1$, $b = 3$) stuck on 11 sites: The trajectory and the local time

- *If* $b/|a| > A_k$, *then almost surely, the walk does not get stuck on a set with less than* $k + 2$ *sites.*

For instance, for some values of the parameter $|a|$ larger but very close to $b/3$, the walk can remain stuck on a set of exactly 10^9 sites. On the other hand, when $0 < -a \leq b/3$, then the walk does almost surely not get stuck on a finite set.

For a proof of Theorem 1 and more refined details on the asymptotic behaviour of the walk in this particular trapping regime, see [1].

2.3 *When* $b \geq 0$ *and* $a > b$*: The Slow Phase?*

In this regime, nearest and next-to-nearest edges are all repelling. Nevertheless, these effects do not just always add up to produce a TSRM-type asymptotic behaviour. The second neighbours can produce traps that are difficult to get out from.

In order to illustrate this, let us describe a scenario of building up traps, for the extreme case where $b = 0$ and $a > 0$ (see a sample on Figure 3). The random

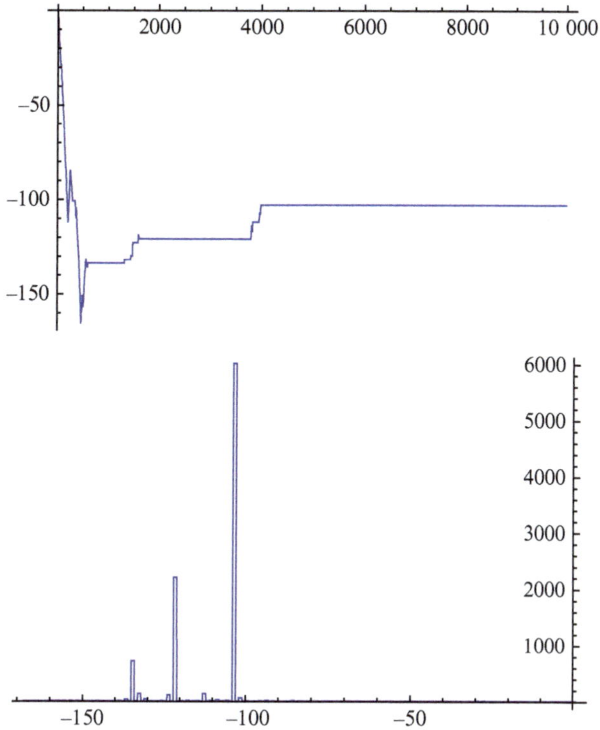

Fig. 3 The trapping phenomenon for $a = 1$ and $b = 0$: The trajectory and the local time

walk starts by jumping back and forth on the edge e_0 between 0 and 1, a large even number N of times. This happens with probability $2^{-N} = e^{-c_1 N}$. Then, in some $O(1)$ number of steps it moves to site -4. (Note that due to the large number of visits at edge e_0, the walker does not have a chance to return from site -1.) Now, it performs yet again a number of $N + o(N)$ back and forth jumps on the edge e_{-4} (between sites -4 and -3). This happens with probability $e^{-c_2(N+o(N))}$. After all this, in $O(1)$ number of steps the walker lands on edge e_{-2} (between sites -2 and -1). Now, due to the $N + o(N)$ number of visits at edges e_0 and e_{-4} the walker does not have a chance to get away from edge e_{-2} before jumping $e^{aN+o(N)}$ times back and forth on this edge.

Hence, we see that while it is very unlikely to build up a trap (and it therefore typically takes a long time to build up one), once it is built and if e^a is large enough, then it takes even much longer to escape from it. In fact, the previously described strategy to build such traps is certainly not the optimal one, and the probability to create one is in fact smaller, so that for all positive a, it takes much longer to escape from the trap than to create it. This slowing-down mechanism is reminiscent of that of random walks in random environment, and it indicates that there is no scaling limit process with continuous trajectories.

2.4 The Two Critical Cases

When b is positive, we do not have precise results at this point for the two borderline cases where $a = -b/3$ and $a = b$.

- $a = -b/3$ is the borderline case between the conjectural TSRM regime (where the walk should scale like $n^{2/3}$) and the sticky regime (when the walk is eventually stuck to a finite interval); see a sample on Figures 4 and 5. We have seen that the walk does almost surely not stay in a finite interval, but its asymptotic behaviour is not really well understood. Naive scaling arguments (and the stationary measure computations mentioned at the very end of the next

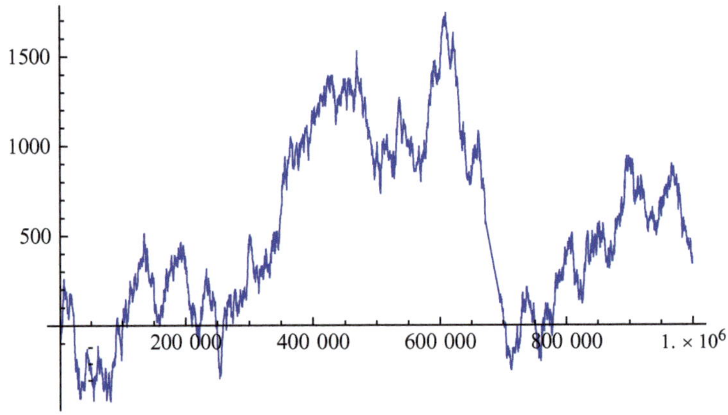

Fig. 4 The "third derivative walk", i.e. $P(x) = (1-x)^3$: The trajectory

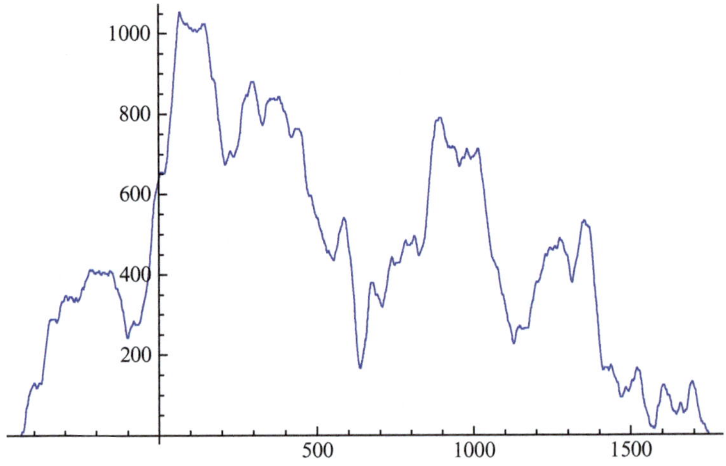

Fig. 5 The third derivative walk: The (smoother) local time profile

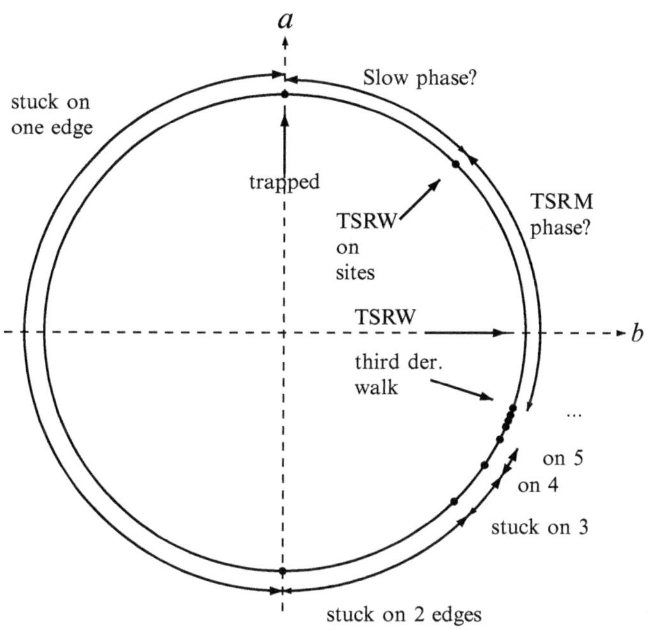

Fig. 6 The phase diagram in terms of (b, a) on the unit circle

section) would at first suggest that the non-degenerate scaling of the walk should be like $n^{2/5}$. But it can be rigorously proved that in a stationary regime there exists a diffusive lower bound on the asymptotic variance of the displacement, and intuitively, one might guess diffusive behaviour (the stationary local time regime – see also Fig. 5 – and the walk displacement might have different scaling exponents and not interact in the large-scale limit).

- The case where $a = b$ can be interpreted as the TSRW with self-repulsion defined in terms of occupation times on sites rather than edges. Its scaling behaviour is expected to be similar to the TSRW on edges (see [1] for references).

Figure 6 sums up the rough phase diagram of the asymptotic behaviour of the walk depending on the values of (b, a) on the unit circle.

2.5 Stationary Measures for the Cases Where $b > 0$ and $-b/3 < a < b$

We now quickly survey some computations related to the existence of stationary measures for the current environment (i.e. the local time fluctuations) as seen from the random walk in the conjectural "TSRM" regime. We will work in the more general context where the dynamics is driven by a height-invariant linear function of the local time, i.e. where

$$R((\ell)) = e^D/(e^D + e^{-D})$$

with

$$D((\ell)) = \tilde{D}((\eta)) = \sum_x \alpha(x)\eta(x),$$

where all but finitely many $\alpha(x)$'s are zero. We will write $R((\ell)) = p(\eta)$.

It is convenient to work with an initial profile η_0 that belongs to the set

$$\Omega = \{\eta \ : \ \text{ for all } x, \ \eta(x) + 1_{x=0} \in 2\mathbb{Z}\}.$$

It is immediate to check that this parity condition will be conserved for all $n \geq 0$ (the opposite parity convention, i.e. $\eta + 1 \in \Omega$ would also be preserved) so that $(\eta_n, n \geq 0)$ is a Markov chain on Ω. We can also restrict ourselves to the family of η's in Ω with finite norm $\|\eta\| = \sum_{x\in\mathbb{Z}} 2^{-|x|}|\eta(x)|$.

We will use the following notation. If $X_{n+1} - X_n = 1$, i.e. X jumps to the right, we write $\eta_{n+1} = \mathscr{R}\eta_n$, and if it jumps to the left, we write $\eta_{n+1} = \mathscr{L}\eta_n$. In other words, for all x,

$$\mathscr{R}\eta(x) = \eta(x+1) - \delta_0(x) + \delta_{-1}(x), \quad \mathscr{L}\eta(x) = \eta(x-1) + \delta_0(x) - \delta_{+1}(x).$$

The Markov chain is described by the fact that (given η_n) η_{n+1} is equal to $\mathscr{R}\eta_n$ with conditional probability $p(\eta_n)$, and that $\eta_{n+1} = \mathscr{L}\eta_n$ otherwise. The transition operator of the Markov process is therefore:

$$Pf(\eta) = p(\eta)f(\mathscr{R}\eta) + (1 - p(\eta))f(\mathscr{L}\eta).$$

As often, it is convenient to view this Markov chain as the jump chain associated with a continuous-time Markov chain on the same state space. This continuous-time Markov chain $(\tilde{\eta}_t, t \geq 0)$ jumps to the right with intensity $\mathrm{d}t \times \exp\{\tilde{D}(\tilde{\eta}_t)\}$ and to the left with intensity $\mathrm{d}t \times \exp\{-\tilde{D}(\tilde{\eta}_t)\}$.

The particular choice for the dynamics leads us naturally to look for a stationary distribution for $\tilde{\eta}_t$ of "discrete Gaussian type". We are going to suppose in the present section that:

- $(\alpha(\cdot))$ is even (i.e. $\alpha(x) = \alpha(-x)$ for all x). This corresponds to the left–right symmetry mentioned in the Introduction.
- The self-interaction has finite spatial range (i.e. all but finitely many $\alpha(x)$'s are equal to zero)
- It is positive definite in the sense that for all $(\lambda(x), x \in \mathbb{Z})$ such that all of them but a finite positive number is equal to zero,

$$\sum_{x,y\in\mathbb{Z}} \lambda(x)\lambda(y)\alpha(y-x) > 0 \tag{3}$$

(this implies in particular that $\alpha(0) > 0$).

Note that in the case where $D = D^{a,b}$ and $b > 0$, this last condition precisely means that $-b/3 < a < b$.

Under these assumptions, there exists a unique Gibbs measure π_0 on Ω formally given by

$$d\pi_0(\eta) \sim \exp\{-\frac{1}{2}\sum_{x,y}\eta(x)\eta(y)\alpha(y-x)\}.$$

The proper, rigorous definition of the Gibbs measure π_0 is in terms of conditional specifications on finite intervals, given fixed boundary conditions. This Gibbs measure is well understood, and its correlations decay exponentially fast with distance and the zero-one law holds for tail events.

Proposition 1. *Under the previous assumptions, the probability measure π_0 is stationary and ergodic for the Markov process $\tilde{\eta}_t$.*

The proof is straightforward and based on the following observation:

$$\frac{\mathrm{d}\pi_0(\mathscr{R}\eta)}{\mathrm{d}\pi_0(\eta)} = \exp\{-(\alpha(1)-\alpha(0)) - \sum_{x\in\mathbb{Z}}(\alpha(x)-\alpha(x-1))\eta(x)\} = \frac{\exp\{\tilde{D}(\eta)\}}{\exp\{\tilde{D}(\mathscr{R}\eta)\}}$$

and similar formulae when one replaces $\mathscr{R}$ by $\mathscr{L}$. It proves that in fact the measure π_0 is invariant under the process with "right-jumps" only, and under the process with "left-jumps" only.

Ergodicity also follows from the above observation. The Dirichlet form

$$\mathscr{D}(f) = \int_{\Omega} f(\eta)(I-P)f(\eta)\mathrm{d}\pi_0(\eta)$$

is easily computed. Using the above observation, we get

$$2\mathscr{D}(f) = \int_{\Omega} \big(p(\eta)(f(\eta)-f(\mathscr{R}\eta))^2 + (1-p(\eta))(f(\eta)-f(\mathscr{L}\eta))^2\big)\mathrm{d}\pi_0(\eta).$$

Hence, since $0 < p(\eta) < 1$ almost surely, it follows that $\mathscr{D}(f) = 0$ if and only if almost surely

$$f(\mathscr{R}\eta) = f(\eta) = f(\mathscr{L}\eta).$$

But it is a standard fact that the above almost sure identities hold if and only if f is tail measurable. Hence, by the zero-one law mentioned before, f is almost surely constant. This proves ergodicity.

Also, one can note that the mean waiting time of the continuous-time Markov chain at η is given by

$$\tau(\eta) := e^{\tilde{D}(\eta)} + e^{-\tilde{D}(\eta)},$$

and that $Z := \pi_0(1/\tau)$ is finite. From this, it follows that one can define a probability measure π on Ω by

$$\frac{\mathrm{d}\pi}{\mathrm{d}\pi_0}(\eta) = \frac{1}{Z \times \tau(\eta)}$$

and that:

Corollary 1. *This probability measure π is stationary for the discrete chain (η_n).*

Hence, we see that in this regime, and under these well-chosen starting distributions, the large-scale behaviour of the local-time profile is of Brownian type.

We can note that in the borderline case where $b > 0$ and $a = -b/3$ (i.e. the discrete "third derivative" case), the previous stationary distribution analysis can be adapted to show that the discrete gradient of $\tilde{\eta}_t$ (i.e. the discrete second derivative of the local time profile) has a stationary measure of discrete Gaussian type. More precisely, we see in this interesting case that if we define for all $e \in E$,

$$\nabla\eta(e) = \eta(e + 1/2) - \eta(e - 1/2),$$

then the process $(\nabla\tilde{\eta}_t, t \geq 0)$ is a Markov process, and that it has a stationary distribution given by i.i.d. discrete Gaussians.

For details of similar computations of stationary and ergodic measures of the environment seen by the moving particle, in continuous space-time setup, and their consequences see also [5].

2.6 Some Comments

To conclude this section, let us very quickly list a few observations:

- In the setting of the TSRW, it was natural to describe the dynamics of the walker by saying that it was driven by an approximate negative gradient of the local time at its current position, and the dynamics of its scaling limit could also be described in this way. Our previous analysis, however, shows that this type of description has to be handled with care. The case $\{a = 1, b = 0\}$ seems at first to also correspond to a negative discrete gradient, but its large-scale behaviour is completely different.

- In fact, in the general case, what seems more relevant to classify the asymptotic behaviour of the walk is the quadratic form appearing in (3) or alternatively the study of the roots of $P(x)$. This is also apparent in the study of the stuck walks in [1]. Note also that the borderline cases correspond to (multiples of) the polynomials $(1-x)^3$ and $(1-x)^2(1+x)$. In the general case (with larger interaction-range), the phase diagram is seemingly more complicated, with different connected components of the parameter space corresponding to similar asymptotic behaviours.
- A particular role is then played by the "odd" derivatives, i.e. more precisely, the polynomials $(1-x)^{2n+1}$. Then, the discrete nth derivative of $\tilde{\eta}$ (and therefore of η) has a stationary measure of discrete Gaussian type.

3 Some Cases Without Left–Right Symmetry

Let us stress that (as opposed to the cases studied in the previous section) the examples studied in the present section break the left–right symmetry, i.e. the laws of $(X_n, n \geq 0)$ and $(-X_n, n \geq 0)$ are different even if one chooses L_0 to be symmetric.

3.1 Setup and Statement

In this section, we will first focus on the case where $D = -\Delta/2$, where

$$\Delta((\ell(e), e \in E)) = \ell(-3/2) - \ell(-1/2) - \ell(1/2) + \ell(3/2).$$

We then choose

$$R((\ell(e), e \in E)) = \frac{e^{-\Delta/2}}{e^{\Delta/2} + e^{-\Delta/2}},$$

which penalizes jumps in the direction of this second derivative. The precise form of the function R is not really important for what follows. For instance, our results clearly remain true if we only assume in addition of locality and height-invariance, that

- $R((-\ell)) = 1 - R((\ell))$.
- $R((\ell))$ is a function of Δ goes to 0 very quickly as $\Delta \to +\infty$.

These conditions can in fact also be relaxed.

Note that $\Delta = \eta(1) - \eta(-1)$ is a discrete and symmetric version of the second derivative of the local time, but this feature is not essential in what follows.

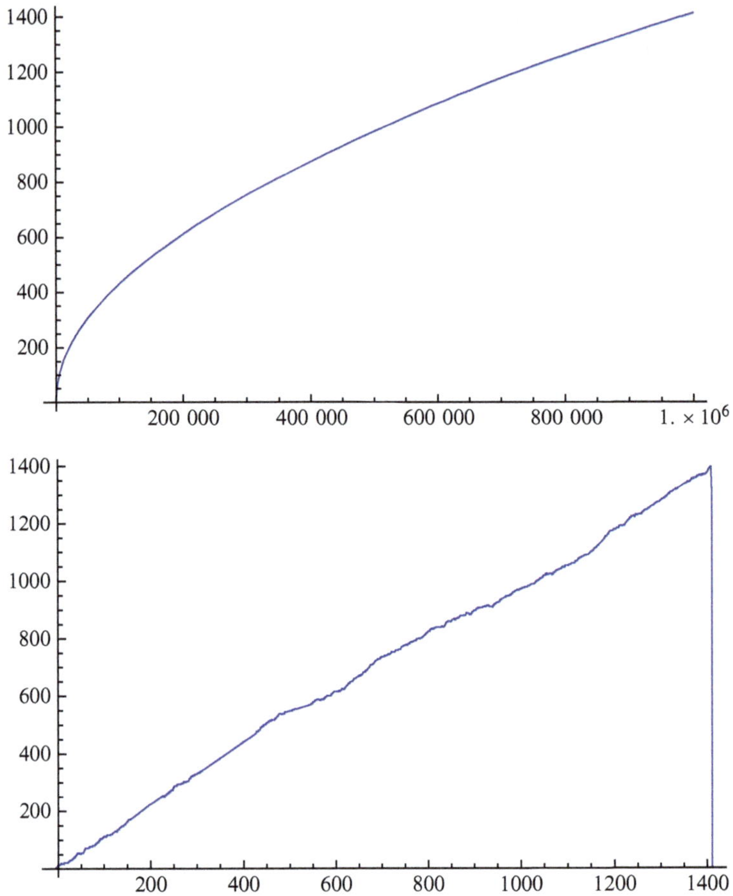

Fig. 7 The second derivative walk: The trajectory and the local time

Intuitively speaking, the walk will tend to move to the right if the local time profile is locally concave and to the left if it is locally convex.

The main result of the present section is the following: Suppose that the initial profile is identically zero (again, this is not really necessary – we only need this initial profile not to be too "wild"). Then:

Proposition 2. *With positive probability,* $X_n \sim \sqrt{2n}$ *as* $n \to \infty$.

On large scale and on a positive fraction of the probability space, the walk becomes almost deterministic, but non-ballistic, and the scaling is of Brownian type. As we shall see, the proof gives in fact a more precise description of the behaviour of $(X_n, n \geq 0)$ on this event of positive probability. See Fig. 7 for a sample of this walk.

Let us immediately stress that the proof will in fact show that the results remain true if one replaces Δ by any function $\tilde{\Delta}$ of the type

$$\tilde{\Delta}((\ell(e), e \in E)) = \ell(-3/2) - \ell(-1/2) - \ell(1/2) + \sum_{k \in E,\ k \geq 3/2} a_k \ell(k).$$

The facts that Δ is a local version of the second derivative, that it is translation-invariant and that it is "anti-symmetric" are therefore not so crucial after all. In fact, our conclusion holds true if we replace Δ by any local function $\hat{\Delta}((\ell(e), e \in E)) = \sum_{e \in E} a_e \ell(e)$ such that

- For some $e_0 < 0$, $a_e = 0$ for all $e \leq e_0$.
- $\sum_{e<0, e \in E} a_e = 0$
- $a_{-1/2}$ and $a_{1/2}$ are both negative.

This for instance includes discrete pth derivatives for "even" p's such as

$$\hat{\Delta}((\ell)) = -\ell(-5/2) + 3\ell(-3/2) - 2\ell(-1/2) - 2\ell(1/2) + 3\ell(3/2) - \ell(5/2)$$

that corresponds to a discrete forth derivative.

In the "continuous limit", the occupation time density profile of the motion $t \mapsto \sqrt{2t}$ has a discontinuity at its local position (i.e. at time t, it is equal to $x 1_{x \in [0, \sqrt{2t}]}$), and the size of this jump is growing with time. Hence, one could a priori naively expect that the walk would not slow down, but on the contrary speed up when this gap widens. This can be explained by the fact that the self-interacting random walk therefore captures some feature of the discrete local time profile that is not visible in the continuous limit.

3.2 The Scenario

The coming three subsections are devoted to the proof of Proposition 2. Let us first describe in plain words a possible behaviour for our walk X (we shall then prove that indeed, this behaviour occurs with a positive probability): The walk X starts at the origin, jumps to its right i.e. $X_1 = 1$, and it forever remains positive i.e. $X_n > 0$ for all positive times. In fact, when $n \to \infty$, X_n goes off to $+\infty$. For each $n \geq 0$, denote $S_n = \max(X_m, m \leq n)$ the past maximum of the walk. In our scenario, $X_n \geq S_n - 2$ for all n. In other words, for each n, either X_n is equal to its past maximum S_n, or it is equal to $S_n - 1$ or $S_n - 2$.

We see that we can therefore decompose the set of times as follows: Define for each $x \geq 0$ the hitting time τ_x of x by X, and let $I_x = [\tau_{x+1}, \tau_{x+2})$. Then, in our scenario, during each of the intervals I_x, the walk jumps back and forth on the edges between $x - 1$ and $x + 1$. When it stops doing so, it first jumps to $x + 2$, and then jumps back and forth on the edges between x and $x + 2$ and so on.

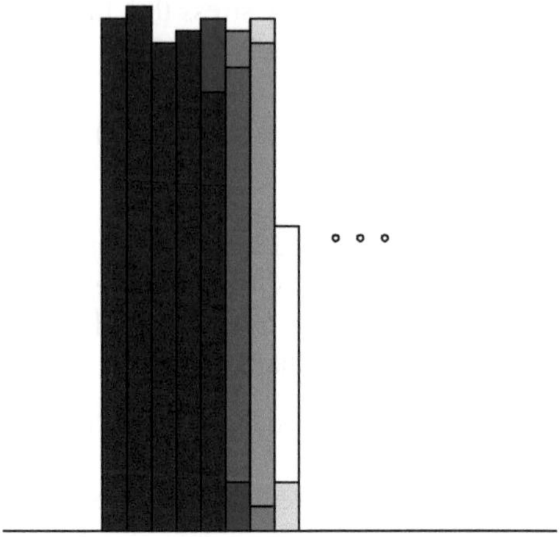

Fig. 8 The local time profile in our scenario: Darker cells are constructed first

Imagine for a moment that this has happened for a while and that just before τ_x (for a large x), i.e. when $n = \tau_x - 1$, the following loosely defined event $G(x)$ is true: The walker has visited each of the edges $\{x-2, x-1\}$, $\{x-3, x-2\}$, and $\{x-4, x-3\}$ many times, and the number of times it has visited these three edges remain however comparable (in the sense that the difference between these three numbers of visits is small). In particular, immediately before it chooses to jump to x for the first time, the value of $\Delta((\ell_n)) = \ell_n(-3/2) - \ell_n(-1/2)$ is rather close to zero (mind that at that moment $X_n = x - 1$, and that $\ell_n(1/2) = \ell_n(3/2) = 0$ because the walk has not visited x yet) so that the probability to indeed jump from $x-1$ to x is therefore not too small.

Then, the walk arrives at x for the first time. At that moment, the two edges to its right have not yet been visited, and the edge between x and $x-1$ (to its left) has been visited only once (because the walker arrives at x for the first time). On the other hand, the edge between $x-1$ and $x-2$ has been visited a lot of times. Hence, the walk will (very likely) jump back to $x-1$. Once it is back at $x-1$, the probability to jump to x or to $x-2$ at that moment is neither very small nor very close to 1 (because the situation can not drastically change because of the two previous jumps). Note that if the walk then jumps to $x-2$, the value of Δ will then be very small, so that the walk will want to jump immediately back to $x-1$. Hence, the walk is (with high probability) trapped between $x-2$ and x for a while. How long does this happen? Well, one should note that each time the walk jumps on $\{x-2, x-1\}$ or on $\{x-1, x\}$, it will increase the chance to jump to the right next time it is at $x-1$. Hence, after a short while, the walk will in fact only jump back and forth between $x-1$ and x. This will be the case until the number of times at which it has jumped on $\{x-1, x\}$ starts to be comparable with the number of times at which it has jumped on $\{x-2, x-1\}$, because the walk will then have a significant chance to move to $x+2$. Hence, we end up in a situation where $G(x+1)$ holds.

One important point in this scenario is that in the end, the walk will tend to jump a little more often on $\{x, x+1\}$ than on $\{x-1, x\}$. So, by a law of large numbers type argument, the total number of visits of the edge $\{x-1, x\}$ will turn out to grow "linearly" in x as $x \to \infty$. This will a posteriori justify our assumption (in the definition of $G(x)$) that at the time τ_x, the walk has visited $\{x-2, x-1\}$ and $\{x-3, x-2\}$ many times.

3.3 Auxiliary Sequences

We now start to turn the previous ideas into a rigorous proof. Let us define a family of independent random variables $(\xi^x_{j,k})$, where $j \in \mathbb{Z}/2$, $k \geq 1$, $x \in \mathbb{Z}$. For each j, k and x, the law of the random variable is the following:

$$P(\xi^x_{j,k} = +1) = 1 - P(\xi^x_{j,k} = -1) = \frac{e^{-j}}{e^j + e^{-j}}.$$

We can then define (deterministically) our random walk X using this family of random variables as follows. Let $X_0 = 0$ and for each $n \geq 0$, choose inductively

$$X_{n+1} = X_n + \xi^x_{\delta,k},$$

where n is the kth time at which $X_n = x$ and $\Delta_n = 2\delta$. Note that when $X_n \geq 2$, then Δ_n is necessarily even (because $\ell_n(-3/2)$ and $\ell_n(-1/2)$ are odd, while $\ell_n(1/2$ and $\ell_n(3/2)$ are even)). We will for the time being forget about this coupling between X and the ξ's, and just list a few simple observations concerning this family $(\xi^x_{j,k})$ of random variables.

A particular role will be played by the random variables $\xi^x_{j,k}$ when $x \geq 0$, $k = 1$ and j is an integer. We denote them by ξ^x_j (and drop the subscript k). Let us insist on the fact that ξ^x_j are not defined for semi-integer j's (as opposed to the $\xi^x_{j,k}$'s). Let us also keep in mind that for a given k and x, when j is very large, the probability that $\xi^x_{j,k} = -1$ and $\xi^x_{-j,k} = +1$ is very close to 1.

We now define certain events:

- Let A denote the event that for all x, all $j > \sqrt{|x|}/100$ and all $k \leq 100x^2$,

$$\xi^x_{j,k} = -\xi^x_{-j,k} = -1.$$

- Note that almost surely, for each x, the number of positive (respectively, negative) j's for which ξ^x_j is positive (respectively, negative) is finite (it follows immediately from the definition that it has a finite expectation). B is the event that for all $x \geq 0$,

$$\sum_{j>0} \left(1_{\xi^x_j = 1} + 1_{\xi^x_{-j} = -1} \right) \leq \sqrt{x}/100.$$

- For all $x \geq 0$, let us define

$$M^x = \sum_{j \geq 0} 1_{\xi_j^x = 1} - \sum_{j < 0} 1_{\xi_j^x = -1}.$$

There is no problem with the definition of M^x for the same reason as above. Note that because of symmetry, $E(M^x) = P(\xi_0^x = 1) = 1/2$. The law of large numbers therefore ensures that almost surely

$$\lim_{y \to \infty} y^{-1} \sum_{x=0}^{y} M^x = 1/2.$$

We define the event C that for all $y \geq 1$, $\sum_{x=0}^{y} M^x \in [y/4, 4y]$.

A simple application of the Borel–Cantelli Lemma enables to prove that:

Lemma 1. *With positive probability, $A \cap B \cap C$ holds.*

The proof is elementary and safely left to the reader.

3.4 The Coupling

In order to simplify our notations, we will assume that at time zero, the initial profile is such that $\ell_0(-1/2) = \ell_0(-3/2) = 1$ and that ℓ_0 is zero otherwise. Then, in our "good scenario", the Δ_n's will always remain even (because $\ell_n(-3/2)$ and $\ell_n(-1/2)$ are both odd, whereas $\ell_n(1/2)$ and $\ell_n(3/2)$ are both even).

This assumption is not a problem. If we start with ℓ_0 being identically equal to 0, we let X move twice to the right, and at this time $n = 2$, the situation is exactly the previous one. Since we will prove asymptotic results about X with positive probability, our theorem will follow immediately from the results with this particular initial profile.

Recall that our walk $(X_n, n \geq 0)$ is defined deterministically from the $\xi_{j,k}^x$'s as follows: $X_0 = 0$ and for each $n \geq 0$,

$$X_{n+1} = X_n + \xi_{j,k}^x,$$

where $x = X_n$, $j = \Delta_n/2$ and k is the cardinal of the set

$$\{m \leq n \; : \; X_m = x \text{ and } \Delta_m = 2j\}.$$

Clearly, this definition ensures that one uses never the same $\xi_{j,k}^x$ twice, and that indeed

$$P(X_{n+1} = X_n + 1 \mid X_0, \ldots, X_n) = \frac{e^{-\Delta_n/2}}{e^{\Delta_n/2} + e^{-\Delta_n/2}}.$$

We are going to assume that $A \cap B \cap C$ holds, and we will show that our "good scenario" holds as well. Note that there is nothing probabilistic in the following arguments. For this, we will inductively prove that for each $x \geq 0$, certain events $E(x)$ hold. Define for each $x \geq 1$, the stopping time $\sigma_x = \sigma(x)$ as the first time at which the edge $\{x-1, x\}$ has been crossed more than $x/8$ times and the walk is at $x-1$. We say that the event $E(x)$ holds if the following four conditions hold:

1. The walk did not visit $x+1$ before σ_x.
2. On $[\sigma_{x-1}, \sigma_x]$, the walk did only visit the three sites $x-2, x-1$ and x.
3. The last $x/10$ jumps of the walk before σ_x were all on the edge between $x-1$ and x.
4. At the time $n = \sigma_x$, the two quantities $\ell_n(-1/2)$ and $\ell_n(-3/2)$ differ by not more than $\sqrt{x}$, and they are both larger than $x/6$ and smaller than $50x$.

Assume that $X_1 = 1$, $X_2 = 2$, and that $E(3), \ldots, E(x)$ do hold (and that $A \cap B \cap C$ hold as well).

For simplicity, let us first assume that ξ_j^x are all equal to $+1$ when $j \leq 0$ and to -1 when $j > 0$ (we will then later see what difference it makes if for finitely many values of j this is not the case). Because $E(x)$ holds, it implies that at the ca. $x/10$ first times at which the walk has been at x before time σ_x, the value of Δ_n at those times was greater than $x/20$ (because $\ell_n(-1/2)$ was small, $\ell_n(-3/2)$ was large and $\ell_n(1/2) = \ell_n(3/2) = 0$). Furthermore, the value of Δ_n at the time $n = \sigma(x) = \sigma$ is very negative (recall that $X_{\sigma(x)} = x-1$, that at this time $\ell_\sigma(-3/2)$ and $\ell_\sigma(-1/2)$ are comparable, while $\ell_\sigma(1/2)$ is greater than $x/8$ and $\ell_n(3/2) = 0$). Hence, our assumption on the $\xi_{j,k}^x$'s ensures that $X_{\sigma+1} = x$, $X_{\sigma+2} = x-1$ and $X_{\sigma+3} = x$ (note that the k used in those steps cannot exceed the number of times one visited the sites). After these two steps, the value of Δ_n (when the walk is at $x-1$) has decreased by 2, i.e. $\Delta_{\sigma+2} = \Delta_\sigma - 2$, whereas $\Delta_{\sigma+3} = \Delta_{\sigma+1} - 2$ if

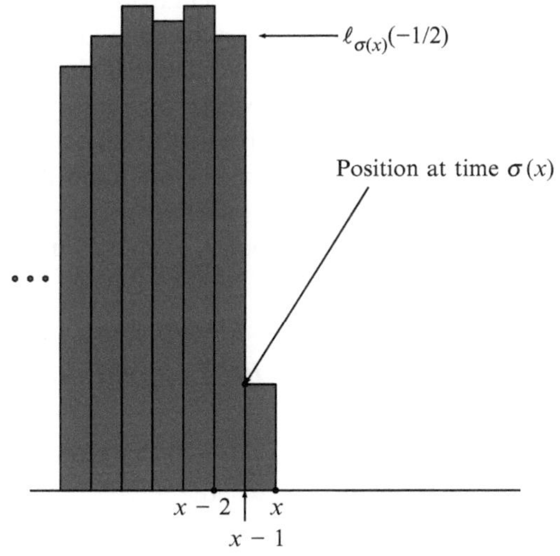

Fig. 9 The local time profile at $\sigma(x)$

$X_{\sigma+2} = x$. Hence, we see that the walk will have to jump back and forth a number of times on the edge between $x-1$ and x; this will stop being the case because at some point, the walk will be at x and will meet a $\xi^x_{j,k}$ that is equal to one. Because before that, the Δ's (at the visiting times of x) have been decreasing, we see that this can only happen when one uses a $k=1$, i.e. when one uses $\xi^x_0 = 1$ (this is the ξ^x_j that is equal to one, for which j is the largest).

From that moment (call it τ) onwards, the walk will start jumping back and forth on the edge between x and $x+1$ (this is because when it is at x, it will use ξ^x_j's for negative j's, and when it is at $x+1$, it will use ξ^{x+1}_j for very large values of j). This will happen until the time σ_{x+1}. At that time, $\ell_{\sigma(x+1)}(-1/2) = \ell_{\sigma(x)}(-1/2)$, i.e. the number of times at which the walk jumped on $\{x-1,x\}$ is equal to the number of times the walk jumped on $\{x-2,x-1\}$, which ensures that in this very particular situation, $E(x+1)$ indeed holds.

What is the difference due to those ξ^x_j's that are equal to -1 if $j \geq 0$ or to $+1$ if $j \leq 0$? The first remark is that all these particular j's will indeed be used in our process (note that Δ is decreasing two by two at each visit of x, and that it starts from a very positive value and become very negative, and therefore has to use all ξ^x_j's for $j \in [-\sqrt{x}, \sqrt{x}]$). Each time it meets a positive j such that $\xi^x_j = +1$, it jumps to the right instead of jumping to the left. At the end of the day (i.e. at $\sigma(x+1)$) this will mean that $\ell(-1/2)$ will be diminished by two. Conversely, if it meets a non-negative j such that $\xi^x_j = -1$, this will add 2 to the number of jumps on $\{x-1,x\}$. Hence, we see that in our "real" case,

$$\ell_{\sigma(x+1)}(-1/2) = \ell_{\sigma(x)}(-1/2) + 2M^x$$

and that $E(x+1)$ will still hold because of our assumptions on the sum of the M^x's.

As a consequence, we see that on the event $A \cap B \cap C$ of positive probability, all $E(x)$'s hold. This implies in particular that for each x, the walk will not come back to the site $x-1$ after $\sigma(x+1)$ i.e. that $\ell_{\sigma(x+1)}(-1/2)$ is in fact the total number of times the walk does jump on the edge $\{x-1,x\}$.

Recall that $E(M^x) = 1/2$ and that almost surely, $\sum_{x=0}^{y}(2M^x) \sim y$ as $y \to \infty$, so that on our event of positive probability, $\sum_{x=0}^{y} \ell_{\sigma(x)}(-1/2) \sim y^2/2$. It follows that on our event of positive probability, $\sigma(x) \sim x^2/2$ so that $n \sim (X_n)^2/2$ (recall that between $\sigma(x)$ and $\sigma(x+1)$ we know that X_n is equal to $x-1$, x or $x+1$) and $X_n \sim \sqrt{2n}$ as $n \to \infty$.

This concludes the proof of Proposition 2.

3.5 An Example with Logarithmic Behaviour

We now focus on an example that exhibits yet another possible asymptotic behaviour for the walk: $D = -\Delta^{\#}$ where

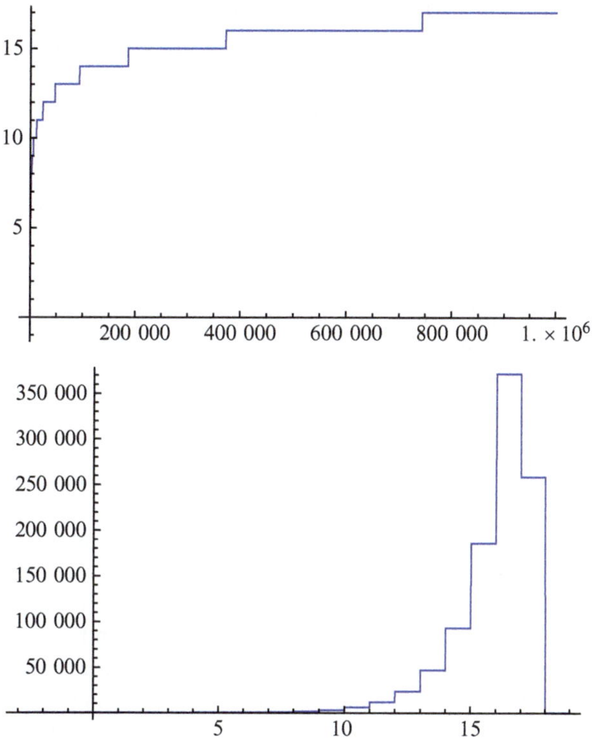

Fig. 10 The logarithmic walk: The trajectory and the local time

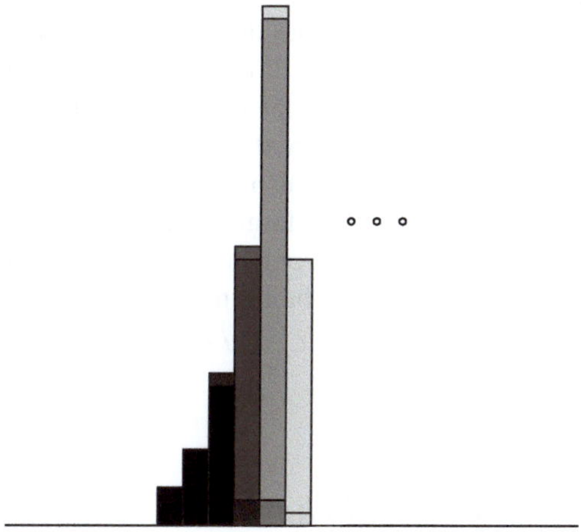

Fig. 11 The local time profile in the good scenario: Darker cells are constructed first

$$\Delta^{\#}((\ell(e), e \in E)) = 2\ell(-3/2) - \ell(-1/2) - \ell(1/2) = -2\eta(-1) - \eta(0)$$

and we then choose F as before. A naive first guess would be that this walk is self-attractive (it is "driven" by the positive gradient of its local time) and that it should get stuck. However:

Proposition 3. *With positive probability,* $X_n \sim (\log n)/(\log 2)$ *as* $n \to \infty$.

See Fig. 10 for a sample of this walk. Again, the proof shows that this behaviour is valid for a wider class of self-interactions. In fact, it is quite similar to the case (with Δ) studied in the previous subsections. The main difference is that (in the "good" scenario that we will describe) the walk will visit approximatively twice more the edge $e+1$ than the edge e for large e. Since the proof is otherwise almost identical to the previous case, we will only describe in plain words this "scenario", and leave the details of the proof to the interested reader.

Suppose that for some particular large time n_x:

- The walk is at its past maximum: $X_{n_x} = \max_{m \le n_x} X_m$. We call this site $x = X_{n_k}$, and for all $n \ge 0$, we define U_n, V_n and W_n to be the respective values of L_{n_x+n} at $x-5/2$, $x-3/2$ and $x-1/2$.
- The values of U_0, V_0 and W_0 satisfy:

$$U_0 \in [(9/10) \times V_0/2, (11/10) \times V_0/2] \text{ and } W_0 \in [V_0/2, V_0].$$

We also suppose that the value of V_0 is very large. Note that under these assumptions, it very likely that $X_{n_x+1} = x-1$ and $X_{n_x+2} = x$, i.e. that the walk will jump back and forth along the edge $x-1/2$ between x and $x-1$, because $-2U_0+V_0+W_0$ is very large, while $-2V_0+W_0$ is negative and has a large absolute value. A quick analysis shows that (with high probability) the walk will jump back and forth on this edge until the negative drift at x stops being huge, and the walk will then jump to $x+1$ for the first time. When this happens, this means that at this time, $2V_n$ and W_n are comparable. Then, for some time, it will jump on the three sites $x-1$, x and $x+1$, but while doing so, the drift at x (i.e. that it feels when it is at x) grows fast, so that it will quickly be forced to jump along the edge between x and $x+1$ only. We then let n_{x+1} the first time n at which $L_n(x+1/2)$ is greater than $L_n(x-1/2)/2$, and note that n_{x+1} satisfies the same conditions as those we required for n_x. Note also that once n_x is found, the scenario is very likely to hold until n_{x+1} when V_0 is large (its conditional probability goes to 1 very quickly with as V_0 gets larger). From this, it follows easily that with positive probability, the "good scenario" will be infinitely repeated (Fig. 11). Then, clearly the total number of visits to the edge $y+1/2$ grows like $2^{y+o(y)}$ as $y \to \infty$, and the proposition follows.

3.6 Ballistic Behaviour

To conclude, let us finally mention a much less surprising possible asymptotic behaviour, namely the ballistic one. The following pictures (Fig. 12) correspond to a trajectory and the corresponding local time for the case where $D = -\Delta^b$ where

$$\Delta^b((\ell(e), e \in E)) = -\ell(3/2) - \ell(-1/2) + 2\ell(1/2) = 2\eta(1) + \eta(0)$$

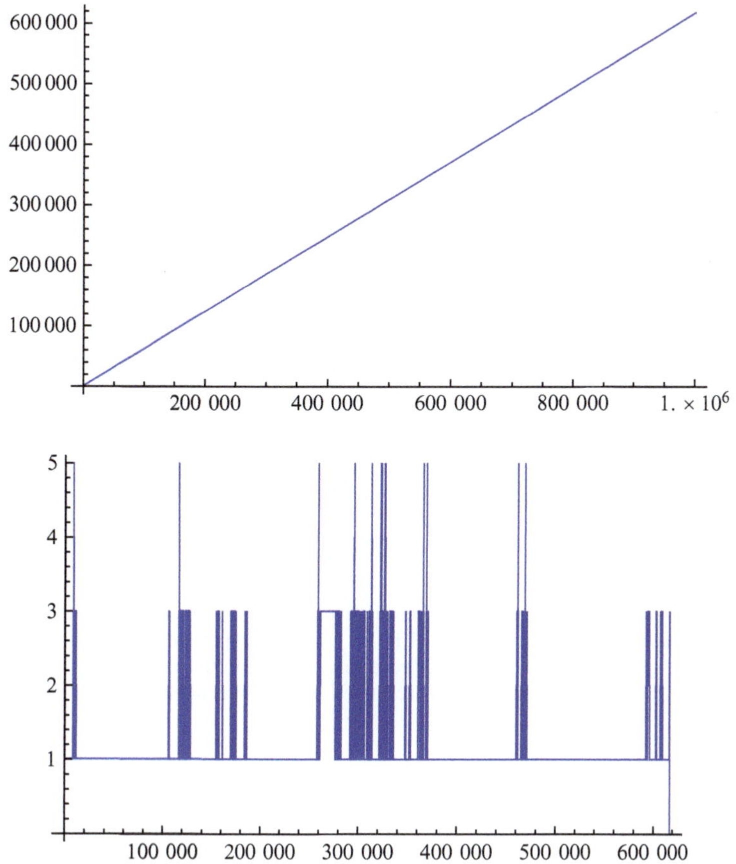

Fig. 12 A ballistic walk: The trajectory and the local time

that could seem at first glance to be again one of the walks driven (like TSRW) by the negative gradient of η. This example, like many others in this paper, illustrates how sensitive the discrete model is to little shifts in the definition of the driving dynamics. Note that a proof of the ballistic behaviour of this walk would require somewhat different techniques than for the previous cases, for instance because the sequence $(\max_{m \le n}(X_m - X_n), n \ge 0)$ will not almost surely be a bounded sequence.

4 Some Open Questions

Let us conclude with a list of (possibly accessible) open problems related to the questions that we have just discussed:

1. Prove the $2/3$ scaling behaviour of the walk in the "TSRM regime" (i.e. where $b > 0$ and $a \in (-b/3, b)$ and the stationary distribution makes sense), or even

the convergence to TSRM. Or construct another related Markovian model (for instance with another function R) where one can prove this?
2. Describe in some detail the "actual" (i.e. the one that actually dominates) trapping strategy in the "trapped case" $b = 0$ and $a = 1$.
3. Get some (even partial) description of the dynamics in the "slow phase". Does the qualitative behaviour depend only on b/a?
4. In the case where the walk goes deterministically to infinity, are other scaling behaviours than ct, $c\sqrt{t}$ and $c \log t$ possible?
5. Improve some of the results to almost sure statements (instead of "with positive probability").
6. Is it possible that for some choice of the parameters in such self-interacting random walks with finite range, the qualitative asymptotic behaviour is actually not almost sure (i.e. can there be two different asymptotic behaviours, each occurring with positive probability?).
7. Do qualitatively really new asymptotic behaviours arise when one considers larger (but finite) self-interaction ranges?

Other questions more directly related to the "stuck case" are listed in [1].

Acknowledgements BT thanks the kind hospitality of Ecole Normale Supérieure, Paris, where part of this work was done. The research of BT is partially supported by the Hungarian National Research Fund, grant no. K60708. WW's research was supported in part by ANR-06-BLAN-00058. The cooperation of the authors is facilitated by the French–Hungarian bilateral mobility grant Balaton/02/2008.

References

1. Erschler, A., Tóth, B., Werner, W.: Stuck walks. Probability Theory and related fields http://ariv.org/abs/1011.1103 (to appear, 2010)
2. Landim, C., Quastel, J., Salmhofer, M., Yau, H.-T.: Superdiffusivity of asymmetric exclusion process in dimensions one and two. Commun. Math. Phys. **244**, 455–481 (2004)
3. Pemantle, R.: A survey of random processes with reinforcement. Prob. Surv. **4**, 1–79 (2007)
4. Tarrès, P.: VRRW on $\mathbb{Z}$ eventually gets stuck at a set of five points. Ann. Prob. **32**, 2650–2701 (2004)
5. Tarrès, P., Tóth, B., Valkó, B.: Diffusivity bounds for 1d Brownian polymers. Ann. Probab. http://arxiv.org/abs/0911.2356 (to appear, 2010)
6. Tóth, B.: True self-avoiding walk with bond repulsion on Z: Limit theorems. Ann. Prob. **23**, 1523–1556 (1995)
7. Tóth, B.: Self-interacting random motions – A Survey. In: Revesz, P., Tóth, B. (eds.) Random Walks – A Collection of Surveys. Bolyai Society Mathematical Studies, vol. 9, pp. 349–384 (1999)
8. Tóth, B., Werner, W.: The true self-repelling motion. Prob. Theor. Relat. Fields, **111**, 375–452 (1998)

Stretched Polymers in Random Environment

Dmitry Ioffe and Yvan Velenik

Abstract We survey recent results and open questions on the ballistic phase of stretched polymers in both annealed and quenched random environments.

This paper is dedicated to Erwin Bolthausen
on the occasion of his 65th birthday

1 Introduction

Stretched polymers or drifted random walks in random potentials could be considered either in their own right or as a more sophisticated and physically more realistic version of directed polymers. Indeed, directed polymers were introduced in [15] as an effective SOS-type model for domain walls in Ising model with random ferromagnetic interactions [11]. Thus, directed polymers do not have overhangs or self-intersections, whereas models of stretched polymers do not impose such constraints and in this respect, resemble "real" Ising interfaces.

Obviously stretched polymers inherit all the pending questions which are still open for their directed counterparts; in particular, a general mathematical description of the strong disorder regime is still missing. It is perhaps unreasonable to expect that these issues would be easier to settle in the stretched context. However, attempts to analyze the model give rise to other more amenable issues of an intrinsic

D. Ioffe (✉)
Department of Industrial Engineering and Management, Technion, Technion City, Haifa 32000, Israel
e-mail: ieioffe@ie.technion.ac.il

Y. Velenik
Department of Mathematics, University of Geneva, 1211 Genève 4, Switzerland
e-mail: Yvan.Velenik@unige.ch

J.-D. Deuschel et al. (eds.), *Probability in Complex Physical Systems*, Springer Proceedings in Mathematics 11, DOI 10.1007/978-3-642-23811-6_14,

interest. To start with, even the annealed model is nontrivial in the stretched case. Furthermore, both quenched and annealed models of stretched polymers exhibit a sub-ballistic to ballistic transition in terms of the pulling force which leads to a rich morphology of the corresponding phase diagram to be explored. Finally, models of stretched polymers do not have a natural underlying martingale structure, which rules out an immediate application of martingale techniques which played such a prominent role in the analysis of directed polymers (see e.g. [2, 4, 7, 22] as well as the review [5] and references therein). It should be noted, however, that both an adjustment of the martingale approach (as based on, e.g., [20] – see also Sect. 5.3 below) and nonmartingale methods developed in the directed context [17,21,25,26] continue to be relevant tools for the stretched models as well.

In this paper, we try to summarize the current state of knowledge about stretched polymers. A large deviation level investigation of the model was initiated in the continuous context by Sznitman ([23] and references therein) and then adjusted to the discrete setup in [8, 19]. The case of high temperature discrete Wiener sausage with drift was addressed in [24]. The existence of weak disorder in higher dimensions has been established first for on-axis directions in [10] and then extended to arbitrary directions in [28]. The main input of the latter work was a proof of a certain mass-gap condition for the (conjugate – see below) annealed model at high temperatures. In fact, the mass-gap condition in question holds for a general class of off-critical self-interacting polymers in attractive potentials at all temperatures [13], which leads to a complete Ornstein-Zernike level analysis of the off-critical annealed case. In the quenched case, such an analysis paves the way to a refined description of what we call below the very weak disorder regime [12, 14], which yields a stretched counter-part of the results of [2]. Finally, the approach of [26] and the fractional moment method of [17] were adjusted in [29] for a study of strong disorder in low ($d = 2, 3$) dimensions.

The paper is organized as follows: The rest of Sect. 1 is devoted to a precise mathematical definition of the model and to an explanation of the key notions. The large deviation level theory is exposed in Sect. 2. Path decomposition as described in Sect. 3 is in the heart of our approach. It justifies an effective directed structure of stretched polymers in the ballistic regime. In the annealed case, it leads to a complete description of off-critical ballistic models, which is the subject of Sect. 4. The remaining Sect. 5 (Weak disorder) and Sect. 6 (Strong disorder) are devoted to a description of the rather incomplete state of knowledge for the quenched models in the ballistic regime.

1.1 Class of Models

1.1.1 Polymers

A polymer $\gamma = (\gamma_0, \ldots, \gamma_n)$ is a nearest-neighbor trajectory on the integer lattice $\mathbb{Z}^d$. Unless stressed otherwise, γ_0 is always placed at the origin. The length of the polymer is $|\gamma| = n$ and its spatial extension is $X(\gamma) = \gamma_n - \gamma_0$. In the most general case, neither the length nor the spatial extension are fixed.

1.1.2 Random Environment

The random environment is a collection $\{V(x)\}_{x\in\mathbb{Z}^d}$ of nondegenerate nonnegative i.i.d. random variables which are normalized by $0 \in \mathrm{supp}(V)$. In the sequel, we shall tacitly assume that $\mathrm{supp}(V)$ is bounded. Limitations and extensions (e.g., $\mathbb{E}V^d < \infty$ or existence of traps $\{V(x) = \infty\}$) will be discussed separately in each particular case. Probabilities and expectations with respect to the environment are denoted with bold letters $\mathbb{P}$ and $\mathbb{E}$. The underlying probability space is denoted as $(\Omega, \mathscr{F}, \mathbb{P})$.

1.1.3 Weights

The reference measure $P(\gamma) \stackrel{\Delta}{=} (2d)^{-|\gamma|}$ is given by simple random walk weights. The most general polymer weights we are going to consider are quantified by three parameters:

- The inverse temperature $\beta \geq 0$;
- The external pulling force $h \in \mathbb{R}^d$;
- The mass per step $\lambda \geq 0$.

The random quenched weights are given by

$$q_{\lambda,h}^{\beta}(\gamma) \stackrel{\Delta}{=} \exp\Big\{h \cdot X(\gamma) - \lambda\,|\gamma| - \beta \sum_{1}^{|\gamma|} V(\gamma_i)\Big\} P(\gamma). \tag{1}$$

In the sequel, we shall drop the index β from the notation, and we shall drop the indices λ or h whenever they equal zero. The corresponding deterministic annealed weights are given by

$$a_{\lambda,h}(\gamma) \stackrel{\Delta}{=} \mathbb{E}q_{\lambda,h}(\gamma) = \exp\left\{h \cdot X(\gamma) - \lambda\,|\gamma| - \Phi_\beta(\gamma)\right\} P(\gamma), \tag{2}$$

where $\Phi_\beta(\gamma) \stackrel{\Delta}{=} \sum_x \phi_\beta\big(\ell_\gamma(x)\big)$, with $\ell_\gamma(x)$ denoting the local time (number of visits) of γ at x, and

$$\phi_\beta(\ell) = -\log \mathbb{E}\mathrm{e}^{-\beta\ell V}. \tag{3}$$

Note that the annealed potential is attractive, in the sense that $\phi_\beta(\ell + m) \leq \phi_\beta(\ell) + \phi_\beta(m)$.

1.1.4 Path Measures and Conjugate Ensembles

There are two natural types of ensembles to be considered: Those with fixed polymer length $|\gamma| = n$, and those with fixed spatial extension $X(\gamma) = x$ or, alternatively, with fixed $h \cdot X(\gamma) = N$. Accordingly, we define the quenched partition functions by

$$Q_\lambda(x) \stackrel{\Delta}{=} \sum_{X(\gamma)=x} q_\lambda(\gamma), \quad Q_\lambda(N) \stackrel{\Delta}{=} \sum_{h \cdot x = N} Q_\lambda(x) \text{ and } Q_n(h) \stackrel{\Delta}{=} \sum_{|\gamma|=n} q_h(\gamma), \tag{4}$$

and use $A_\lambda(x) \stackrel{\Delta}{=} \mathbb{E} Q_\lambda(x)$, $A_\lambda(N)$ and $A_n(h)$ to denote their annealed counterparts.

Of particular interest is the case of a polymer with fixed length $|\gamma| = n$. We define the corresponding quenched and annealed path measures by

$$\mathbb{Q}_n^h(\gamma) \stackrel{\Delta}{=} \mathbf{1}_{\{|\gamma|=n\}} \frac{q_h(\gamma)}{Q_n(h)} \text{ and } \mathbb{A}_n^h(\gamma) \stackrel{\Delta}{=} \mathbf{1}_{\{|\gamma|=n\}} \frac{a_h(\gamma)}{A_n(h)}. \tag{5}$$

Following (4), the probability distributions $\mathbb{Q}_\lambda^x, \mathbb{A}_\lambda^x, \mathbb{Q}_\lambda^N$ and $\mathbb{A}_\lambda^N$ are defined in the obvious way.

1.2 Ballistic and Sub-Ballistic Phases

In both the quenched and the annealed setups, there is a competition between the attractive potential and the pulling force h: For small values of h the attraction wins and the polymer is sub-ballistic, whereas it becomes ballistic if h is large enough. These issues were investigated on the level of Large Deviations first in the continuous context of drifted Brownian motion among random obstacles in [23] and then for the models we consider here in [8, 19]. Such large deviation analysis, however, overlooks the detailed sample-path structure of polymers and, in particular, does not imply law of large numbers or even existence of limiting spatial extension (speed). The law of large numbers in the annealed case was established in [13] together with other more refined analytic properties of annealed polymer measures in the ballistic regime (see Definition 1 below). As is explained below, in the regime of weak disorder the annealed law of large numbers implies the quenched law of large numbers with the same limiting macroscopic spatial extension. We record all these facts as follows:

Theorem 1. *There exist compact convex sets* $\mathbf{K}^a \subseteq \mathbf{K}^q$ *with nonempty interiors,* $0 \in \mathrm{int}\mathbf{K}^a$*, such that:*

1. If $h \in \mathrm{int}\mathbf{K}^a$*, respectively* $h \in \mathrm{int}\mathbf{K}^q$*, then, for any* $\epsilon > 0$*,*

$$\lim_{n\to\infty} \mathbb{A}_n^h\Big(\Big|\frac{X(\gamma)}{n}\Big| > \epsilon\Big) = 0, \textit{ respectively } \lim_{n\to\infty} \mathbb{Q}_n^h\Big(\Big|\frac{X(\gamma)}{n}\Big| > \epsilon\Big) = 0 \ \mathbb{P}\text{-a.s.}, \tag{6}$$

exponentially fast in n [8, 19].

2. *If $h \notin \mathbf{K}^a$, then there exists $v = v^a(h, \beta) \neq 0$, such that, for any $\epsilon > 0$,*

$$\lim_{n\to\infty} \mathbb{A}_n^h\Big(\big|\frac{X(\gamma)}{n} - v\big| > \epsilon\Big) = 0, \tag{7}$$

exponentially fast in n [8, 13, 19].

3. *If $h \notin \mathbf{K}^q$, then there exists a compact set $0 \notin \mathscr{M}_h$ such that*

$$\liminf_{n\to\infty} \mathbb{Q}_n^h\Big(\mathrm{d}\big(\frac{X(\gamma)}{n}, \mathscr{M}_h\big)\Big) = 0 \ \mathbb{P}\text{-a.s.}, \tag{8}$$

exponentially fast in n [8, 19]. Furthermore, if the dimension $d \geq 4$, then for any $h \neq 0$ fixed the set $\mathscr{M}_h = \{v^a(h, \beta)\}$ as soon as β is sufficiently small.

As described in the next subsection, $\mathbf{K}^a$ and $\mathbf{K}^q$ are support sets for certain Lyapunov exponents (norms).

Definition 1. The phases corresponding to $h \notin \mathbf{K}^a$ and, respectively, $h \notin \mathbf{K}^q$ are called ballistic. For $* \in \{a, q\}$ the drifts h are called sub-critical (respectively critical and super-critical) if $h \in \mathrm{int}\mathbf{K}^*$ (respectively $h \in \partial\mathbf{K}^*$ and $h \notin \mathbf{K}^*$).

A general (i.e., without an assumption of weak disorder) characterization of the set $\mathscr{M}_h$ is given in Lemma 2. The question whether quenched models in the ballistic phase, or equivalently, quenched models at super-critical drifts satisfy a law of large numbers is open with an exception of the small noise higher dimensional case (see Lemma 11 below).

Remark 1. The above theorem and its far reaching refinements hold for a large class of annealed models with attractive interactions. In particular, no moment assumptions on V are needed. For instance, both the set $\mathbf{K}^a$ is defined and the corresponding results hold in the case of pure traps $V \in \{0, \infty\}$.

The critical cases $h \in \partial\mathbf{K}^a$ and, of course, $h \in \partial\mathbf{K}^q$ are open. It is easy to see that $\mathbf{K}^a \subset \mathbf{K}^q$ for sufficiently low temperatures. It is, however, an open question (which depends on dimension $d = 2, 3$ or $d \geq 4$) whether the sets of critical drifts coincide for moderate or small values of β.

The subcritical case $h \in \mathrm{int}\mathbf{K}^a$ has been worked out by Sznitman in the context of drifted Brownian motion among random obstacles [23]; see also [1] for some results in the case of random walks.

1.3 Lyapunov Exponents

The quenched and annealed Lyapunov exponents are defined via

$$\mathfrak{q}_\lambda(x) \triangleq -\lim_{N\to\infty} \frac{1}{N} \log Q_\lambda(\lfloor Nx \rfloor) \ \text{ and } \ \mathfrak{a}_\lambda(x) \triangleq -\lim_{N\to\infty} \frac{1}{N} \log A_\lambda(\lfloor Nx \rfloor). \tag{9}$$

Theorem 2. *Both $\mathfrak{q}_\lambda$ and $\mathfrak{a}_\lambda$ are defined for all $\lambda \geq 0$. Moreover, for every $\lambda \geq 0$, $\mathfrak{q}_\lambda \geq \mathfrak{a}_\lambda$ and both are equivalent norms on $\mathbb{R}^d$: there exist $c_\lambda^1, c_\lambda^2 \in (0,\infty)$ such that*

$$c_\lambda^1 |x| \leq \mathfrak{a}_\lambda(x) \leq \mathfrak{q}_\lambda(x) \leq c_\lambda^2 |x|. \tag{10}$$

In particular, $\mathfrak{q}_\lambda$ and $\mathfrak{a}_\lambda$ are support functions,

$$\mathfrak{q}_\lambda(x) = \max_{h \in \partial \mathbf{K}_\lambda^q} h \cdot x \quad \text{and} \quad \mathfrak{a}_\lambda(x) = \max_{h \in \partial \mathbf{K}_\lambda^a} h \cdot x.$$

of compact convex sets $\mathbf{K}_\lambda^a \subseteq \mathbf{K}_\lambda^q$ with non-empty interior containing 0.

Remark 2. The annealed Lyapunov exponent is always defined. The proof of the existence of the (nonrandom) quenched Lyapunov exponent in [19] is based on subadditive ergodic theorem and requires an $\mathbb{E}V^d < \infty$ assumption. This was relaxed to $\mathbb{E}V < \infty$ in [16]. However, no moment assumptions (apart from $\mathbb{P}(V = \infty)$ being small) are needed to justify existence of quenched Lyapunov exponents in the very weak disorder case in higher dimensions [14].

The sets $\mathbf{K}_\lambda^a$ and $\mathbf{K}_\lambda^q$ can be described equivalently as the unit balls for the polar norms

$$\mathfrak{q}_\lambda^*(h) = \max_{x \neq 0} \frac{h \cdot x}{\mathfrak{q}_\lambda(x)} \quad \text{and, accordingly,} \quad \mathfrak{a}_\lambda^*(h) = \max_{x \neq 0} \frac{h \cdot x}{\mathfrak{a}_\lambda(x)}.$$

The set $\mathbf{K}^a$, respectively $\mathbf{K}^q$, in Theorem 1 is given by $\mathbf{K}_0^a$, respectively $\mathbf{K}_0^q$.

1.4 Very Weak, Weak, and Strong Disorder

Given $\lambda \geq 0$ and $\beta \geq 0$, we say that the disorder is weak if $\mathfrak{a}_\lambda = \mathfrak{q}_\lambda$ and strong otherwise. Note that this definition is slightly different from the one employed in the directed case [5]. The condition of being *very weak* is of a technical nature. It means that the dimension is $d \geq 4$ and that, given either a fixed value of $h \neq 0$ or of $\lambda > 0$, the inverse temperature β is sufficiently small. More precisely, we need a validity of (52) below, which enables a fruitful L_2-type control of partition functions and related quantities. In particular, the disorder is weak if it is very weak [10, 14, 28] and, furthermore, in the regime of very weak disorder, both a $\mathbb{P}$-a.s. LLN and a $\mathbb{P}$-a.s. CLT hold for the limiting macroscopic spatial extension [12, 14]. As we explain in Sect. 5.1, the LLN is inherited by quenched models in the weak disorder regime. However, contrary to the directed case [6], it is not known whether CLT holds whenever the ratio between quenched and annealed partition functions stays bounded away from zero. Furthermore, it is not known whether, in $d \geq 4$, the disorder is weak for all $\lambda > 0$ as soon as β is small. In particular, proving that $\mathbf{K}^a = \mathbf{K}^q$ for small β remains an open problem.

Under mild assumptions on the potential V ($\{x : V(x) = 0\}$ does not percolate and $\lim_{\beta\to\infty} \log \mathbb{E}e^{-\beta V}/\beta = 0$), it is easy to see [29] that, for a given λ, the disorder is strong as soon as β is large enough. Such a result is well-known even in the original context of the Ising model with random interactions [27]. It was recently proved [29] that in $d = 2, 3$ the disorder is strong for any $\lambda > 0$ and $\beta > 0$; a short proof of the case $d = 2$ is given in Sect. 6.

Furthermore, the approach of Vargas [26] was adjusted in [29] in order to show that in the regime of strong disorder quenched conjugate measures necessarily contain macroscopic atoms.

2 Large Deviations

The following result holds under the presumably technical assumption that $\mathbb{E}V^d < \infty$ in the quenched case, but in full generality in the annealed case.

Theorem 3. *For any $h \in \mathbb{R}^d$, the rescaled spatial polymer extension $X(\gamma)/n$ satisfies large deviation principles (with speed n) under both $\mathbb{A}_n^h$ and, $\mathbb{P}$-a.s., under $\mathbb{Q}_n^h$ with the corresponding (nonrandom) rate functions J_a^h and J_q^h given by*

$$J_a^h(v) = \max_\lambda \{\mathfrak{a}_\lambda(v) - \lambda\} + (\Lambda_a(h) - h \cdot v),$$
$$J_q^h(v) = \max_\lambda \{\mathfrak{q}_\lambda(v) - \lambda\} + \big(\Lambda_q(h) - h \cdot v\big), \tag{11}$$

where $\Lambda_a(h) \triangleq \lim_{n\to\infty} \frac{1}{n} \log A_n(h)$ and $\Lambda_q(h) \triangleq \lim_{n\to\infty} \frac{1}{n} \log Q_n(h)$.

Let us explain Theorem 3: The following lemma shows that the sets $\mathbf{K}_\lambda^a$ and $\mathbf{K}_\lambda^q$ can be characterized as domains of convergence of certain power series.

Lemma 1. *1. For every $\lambda \geq 0$, if $h \in \mathrm{int}\mathbf{K}_\lambda^a$ or, equivalently, if $\mathfrak{a}_\lambda^*(h) < 1$, then*

$$\sum_x e^{h\cdot x} A_\lambda(x) < \infty. \tag{12}$$

2. For every $\lambda > 0$, if $h \notin \mathbf{K}_\lambda^a$ or, equivalently, if $\mathfrak{a}_\lambda^*(h) > 1$, then there exists $\alpha = \alpha(h) > 0$ such that, all n sufficiently large one can find $y = y_n$ satisfying:

$$e^{h\cdot y} A_{\lambda,n}(y) \triangleq \sum_{X(\gamma)=y, |\gamma|=n} e^{h\cdot y} a_\lambda(\gamma) \geq e^{\alpha n}. \tag{13}$$

A completely analogous statement holds $\mathbb{P}$-a.s. in the quenched case.

We sketch the proof of Lemma 1 at the end of the section. For the moment, let us assume its validity. For any $f \in \mathbb{R}^d$,

$$\mathrm{e}^{-\lambda n} A_n(f) = \sum_x \mathrm{e}^{f\cdot x} A_{\lambda,n}(x).$$

Hence, by (12), $\limsup_{n\to\infty} \frac{1}{n} \log A_n(f) \leq \lambda$ whenever $f \in \mathrm{int}\mathbf{K}_\lambda^a$, whereas (13) implies that $\liminf_{n\to\infty} \frac{1}{n} \log A_n(f) \geq \lambda$ for any $f \notin \mathbf{K}_\lambda^a$. It is easy to see that the strict inclusion $\mathbf{K}_\lambda^a \subset \mathbf{K}_{\lambda'}^a$ holds for any $0 \leq \lambda < \lambda'$. Furthermore,

$$\mathbf{K}_\lambda^a = \bigcap_{\lambda'>\lambda} \mathbf{K}_{\lambda'}^a = \overline{\bigcup_{\lambda'<\lambda} \mathbf{K}_{\lambda'}^a}.$$

Finally, since $\lim_{\ell\to\infty} \phi_\beta(\ell)/\ell = 0$, it is always the case that

$$\liminf_{n\to\infty} \frac{1}{n} \log A_n(f) \geq 0.$$

Putting all these observations together, we deduce that, for any $f \in \mathbb{R}^d$,

$$\Lambda_a(f) = \lim_{n\to\infty} \frac{1}{n} \log A_n(f) = \begin{cases} \lambda, & \text{if } f \in \partial\mathbf{K}_\lambda^a, \\ 0, & \text{if } f \in \mathbf{K}^a. \end{cases} \tag{14}$$

Similarly, since $0 \in \mathrm{supp}(V)$,

$$\Lambda_q(f) = \lim_{n\to\infty} \frac{1}{n} \log Q_n(f) = \begin{cases} \lambda, & \text{if } f \in \partial\mathbf{K}_\lambda^q, \\ 0, & \text{if } f \in \mathbf{K}^q. \end{cases} \tag{15}$$

Obviously, the distribution of $X(\gamma)/n$ is exponentially tight under both $\mathbb{A}_n^h$ and $\mathbb{Q}_n^h$. It follows that the annealed large deviation principle is satisfied with the rate function

$$\sup_f \{f\cdot v - \Lambda_a(h+f)\} + \Lambda_a(h) = \sup_f \{f\cdot v - \Lambda_a(f)\} + (\Lambda_a(h) - h\cdot v)\,.$$

The latter is easily seen to coincide with J_a in (11), using (14) and $\mathfrak{a}_\lambda(v) = \max_{f\in\partial\mathbf{K}_\lambda^a} v\cdot f$. The quenched case is dealt with in the same way. □

2.1 Ramifications for Ballistic Behavior

The assertion of Theorem 1 is now straightforward. Set $\mathfrak{a} \equiv \mathfrak{a}_0$ and $\mathfrak{q} \equiv \mathfrak{q}_0$.

(1) Since $\mathfrak{a}^*(h)\mathfrak{a}(v) \geq h\cdot v$, we infer that, for any $v \neq 0$,

$$J_a^h(v) \geq (1-\mathfrak{a}^*(h))\mathfrak{a}(v) > c(h)\,|v| > 0, \tag{16}$$

whenever $h \in \mathrm{int}\mathbf{K}^a$. The same argument also applies in the quenched case.

Note that formula (11) readily implies that $J_a^h(0) = \Lambda_a(h)$, respectively, $J_q^h(0) = \Lambda_q(h)$. In particular, $J_a^h(0) = 0$ (respectively $J_a^h(0) = 0$) whenever $h \in \partial\mathbf{K}^a$ (respectively, $h \in \partial\mathbf{K}^q$).

On the other hand, in the ballistic case of super-critical drifts $h \in \partial\mathbf{K}_\lambda^a$ or, respectively, $h \in \partial\mathbf{K}_\lambda^q$, for some $\lambda > 0$, the value of the corresponding rate functions at zero is strictly positive (and is equal to λ).

(2) As we shall explain in more details in Sects. 4.2 and 4.3, in the annealed case the control is complete: Outside $\mathbf{K}^a$ the function $\Lambda_a(\cdot)$ is locally analytic and Hess$[\Lambda_a]$ is nondegenerate. Consequently, there is a unique minimum $v = \nabla\Lambda_a(h)$ of J_a^h for any supercritical $h \notin \mathbf{K}^a$.

(3) Following Flury [9], zeroes of the quenched rate function can be described as follows.

Lemma 2. *Let $\mu > 0$ and $h \in \partial\mathbf{K}_\mu^q$. Then the set $\mathcal{M}_h \triangleq \left\{v : J_q^h(v) = 0\right\}$ in (8) can be characterized as follows:*

$$v \in \mathcal{M}_h \Longleftrightarrow \begin{cases} \mathfrak{q}_\mu(v) = h \cdot v, \\ \frac{\mathrm{d}^-}{\mathrm{d}\lambda}\Big|_{\lambda=\mu}\mathfrak{q}_\lambda(v) \geq 1 \geq \frac{\mathrm{d}^+}{\mathrm{d}\lambda}\Big|_{\lambda=\mu}\mathfrak{q}_\lambda(v). \end{cases} \tag{17}$$

In particular, $\mathcal{M}_h = \{v\}$ is a singleton if and only if $\partial\mathbf{K}_\mu^q$ is smooth at h and $\mathfrak{q}_\lambda(v)$ is smooth at μ.

Proof. By (11),

$$v \in \mathcal{M}_h \Longleftrightarrow \max_\lambda \{\mathfrak{q}_\lambda - \lambda\} + (\mu - h \cdot v) = 0.$$

The choice $\lambda = \mu$ implies that $\mathfrak{q}_\mu(v) \leq h \cdot v$. Since $h \in \partial\mathbf{K}_\mu^q$, the first condition in the rhs of (17) follows. Consequently, for any λ,

$$\mathfrak{q}_\lambda(v) - \mathfrak{q}_\mu(v) \leq \lambda - \mu.$$

Since $\mathfrak{q}_\lambda$ is concave in λ, both right- and left-derivatives are defined and the second condition in the rhs of (17) follows as well. □

As will be explained in Sect. 5, the existence of a unique minimizer $v = \nabla\Lambda_q(h) = \nabla\Lambda_a(h)$ of the quenched rate function easily follows from the corresponding annealed statement in the weak disorder regime. Moreover, an almost-sure CLT can be established when the disorder is very weak; see Sect. 5.2.

2.2 Proof of Lemma 1

The annealed case is easy. Since the potential is attractive, the Lyapunov exponent A_λ is superadditive. Hence, the second limit in (9) is well-defined and, in addition,

$$A_\lambda(x) \le \mathrm{e}^{-\mathfrak{a}_\lambda(x)}. \tag{18}$$

Since $h \cdot x \le \mathfrak{a}_\lambda^*(h)\mathfrak{a}_\lambda(x)$, the bound (12) follows from (18) and (10) and holds for all subcritical drifts $h \in \mathrm{int}\mathbf{K}_\lambda^a$.

In the supercritical case $\mathfrak{a}_\lambda^*(h) > 1$, pick a unit vector x satisfying $h \cdot x = \mathfrak{a}_\lambda^*(h)\mathfrak{a}_\lambda(x)$. Then,

$$A_\lambda(mx)\mathrm{e}^{mh\cdot x} \ge \exp\left\{\frac{m(\mathfrak{a}_\lambda^*(h)-1)\mathfrak{a}_\lambda(x)}{2}\right\},$$

for all m sufficiently large. Obviously, only paths with $|\gamma| \ge m$ can contribute to $A_\lambda(mx)$. On the other hand, for any $\lambda > 0$ one can ignore paths with $|\gamma| \ge c_\lambda m$ for some c_λ sufficiently large. It follows that one can find $\alpha > 0$, $n_0 > 0$ and y_0 such that

$$\mathrm{e}^{h\cdot y_0} A_{\lambda,n_0}(y_0) \ge \mathrm{e}^{2\alpha n_0}.$$

In view of subadditivity, the target (13) follows by setting $n = kn_0 + r$ and iterating.

The quenched case is slightly more involved. Under suitable assumptions on V (e.g., boundedness of supp(V) or $\mathbb{E}V < \infty$), the existence of

$$\mathfrak{q}_\lambda(x) = -\lim_{N\to\infty}\frac{1}{N}\log Q_\lambda(\lfloor Nx\rfloor) = -\lim_{N\to\infty}\frac{1}{N}\mathbb{E}\log Q_\lambda(\lfloor Nx\rfloor) \tag{19}$$

follows from the subadditive ergodic theorem [16, 19].

In order to mimic the proofs of (12) and (13), one needs to apply concentration inequalities in order to control fluctuations of the random quantities on the lhs of (19) around their expectations. This is done in [19], under the assumption of $\mathbb{E}V^d < \infty$. The speed of convergence of the expectations on the rhs of (19) is under control exactly as in the annealed case. □

3 Geometry of Typical Polymers

3.1 Skeletons of Paths

Let $\lambda > 0$ and $x \in \mathbb{Z}^d$ be a distant point. Our characterization of the path measures $\mathbb{Q}_x^\lambda$ and $\mathbb{A}_x^\lambda$ hinges upon a renormalization construction. In the sequel, U_λ denotes the unit ball in either the quenched ($\mathfrak{q}_\lambda$) or the annealed ($\mathfrak{a}_\lambda$) norms. We choose a large scale K and use the dilated shifted balls $KU_\lambda(u) \overset{\Delta}{=} u + KU_\lambda$ for a coarse-grained decomposition of paths $\gamma \in \mathscr{D}(x) \overset{\Delta}{=} \{\gamma : 0 \mapsto x\}$ (see Sect. 2.2 of [13]),

$$\gamma = \gamma_1 \cup \eta_1 \cup \gamma_2 \cup \ldots \cup \eta_m \cup \gamma_{m+1}.. \tag{20}$$

This decomposition enjoys properties (a)–(d) below:

(a) For $i = 1, \dots, m$, the paths γ_i are of the form $\gamma_i : u_{i-1} \mapsto v_i \in \partial K U_\lambda(u_{i-1})$.
(b) The last path $\gamma_{m+1} : u_m \mapsto x$.
Given a set G, let us say that a path γ with endpoints u and v is in $\mathscr{D}_G(u; v)$ if $\gamma \setminus v \subseteq G$. Define $G_1 = KU_\lambda$ and $G_i = KU_\lambda(u_{i-1}) \setminus \left(\cup_{j<i} KU_\lambda(u_{j-1})\right)$.
(c) $\gamma_i \in \mathscr{D}_{G_i}(u_{i-1}, v_i)$.
(d) The paths η_i are of the form $\eta_i : v_i \mapsto u_i$ and $\eta_i \cap G_{j-1} = \varnothing$ for any $j < i$.

Definition 2. The set $\hat{\gamma}_K \triangleq (0 = u_0, v_1, u_1, \dots, v_m, u_m = x)$ is called the K-skeleton of γ. We say that $\underline{\gamma} = (\gamma_1, \dots, \gamma_{m+1}) \sim \hat{\gamma}_K$ and $\underline{\eta} = (\eta_1, \dots, \eta_m) \sim \hat{\gamma}_K$ if they satisfy Conditions (a)–(d) above.

The collection $\underline{\eta}$ is called the hairs of $\hat{\gamma}_K$. In the sequel we shall concentrate on controlling the geometry of the skeletons. The geometry of hairs is, for every $\lambda > 0$ fixed, controlled by a crude comparison with killed random walks, and we refer to Sect. 2.2 of [13] for the corresponding arguments.

It follows from Condition (c) that the paths γ_i are pairwise disjoint. Consequently,

- In the annealed case, $\Phi_\beta(\gamma_1 \cup \dots \cup \gamma_{m+1}) = \sum_i \Phi_\beta(\gamma_i)$. As a result,

$$\sum_{\underline{\gamma} \sim \hat{\gamma}_K} a_\lambda(\gamma_1 \cup \dots \cup \gamma_{m+1}) = \prod A_\lambda(u_{i-1}; v_i \big| G_i), \tag{21}$$

  with the obvious notation $A_\lambda(u; v \big| G) \triangleq \sum_{\gamma \in \mathscr{D}_G(u;v)} a_\lambda(\gamma)$.
- In the quenched case,

$$\sum_{\underline{\gamma} \sim \hat{\gamma}_K} q_\lambda(\gamma_1 \cup \dots \cup \gamma_{m+1}) = \prod Q_\lambda(u_{i-1}; v_i \big| G_i), \tag{22}$$

  and the variables $Q_\lambda(u_{i-1}; v_i \big| G_i) \triangleq \sum_{\gamma \in \mathscr{D}_{G_i}(u_{i-1};v_i)} q_\lambda(\gamma)$ are jointly independent.

3.2 Annealed Models

Let $\lambda > 0$ and $h \in \partial \mathbf{K}_\lambda^a$ such that $h \cdot x = \mathfrak{a}_\lambda(x)$. Observe first that, by the very definition of $\mathfrak{a}_\lambda$,

$$A_\lambda(x) \geq \mathrm{e}^{-\mathfrak{a}_\lambda(x)(1+o(1))}. \tag{23}$$

Now, let $\hat{\gamma}_K = (u_0 = 0, v_1, u_1, \dots, v_m, u_m = x)$ be a K-skeleton. On the one hand, (21) and (18) imply that

$$\log \sum_{\underline{\gamma} \sim \hat{\gamma}_K} a_\lambda(\underline{\gamma}) \leq -Km.$$

On the other hand, independently of the scale K,

$$\log \sum_{\underline{\eta} \sim \hat{\gamma}_K} \mathrm{e}^{-\lambda|\underline{\eta}|} P(\underline{\eta}) \leq c_1(\lambda) m. \tag{24}$$

Notice that $\mathfrak{a}_\lambda(u_i - u_{i-1}) = K + O(1)$ by construction. We deduce that

$$\begin{aligned}\mathbb{A}_x^\lambda(\hat{\gamma}_K) &\leq \exp\left\{-(1-\epsilon_K)\sum_{i=1}^m \mathfrak{a}_\lambda(u_i - u_{i-1}) - \mathfrak{a}_\lambda(x) + \mathrm{o}(1)|x|\right\} \\ &\leq \exp\left\{-(1-\epsilon_K)\sum_{i=1}^m \big(\mathfrak{a}_\lambda(u_i - u_{i-1}) - h\cdot(u_i - u_{i-1})\big) + \mathrm{o}(1)|x|\right\} \\ &= \exp\left\{-(1-\epsilon_K)\sum_{i=1}^m \mathfrak{s}_a^h(u_i - u_{i-1}) + \mathrm{o}(1)|x|\right\},\end{aligned}$$

where we have introduced the (annealed) surcharge function $\mathfrak{s}_a^h(y) \stackrel{\Delta}{=} \mathfrak{a}_\lambda(y) - h\cdot y$, and ϵ_K can be chosen arbitrarily small, provided that K is chosen large enough. Defining the (annealed) surcharge of a skeleton $\hat{\gamma}_K$ by

$$\mathfrak{s}_a^h(\hat{\gamma}_K) \stackrel{\Delta}{=} \sum_{i=1}^m \mathfrak{s}_a^h(u_i - u_{i-1}),$$

we finally obtain the following fundamental surcharge inequality (see [13] for details):

Lemma 3. *For every small $\epsilon > 0$, there exists $K_0(d,\beta,\lambda,\epsilon)$ such that*

$$\mathbb{A}_x^\lambda\big(\mathfrak{s}_a^h(\hat{\gamma}_K) > 2\epsilon|x|\big) \leq \mathrm{e}^{-\epsilon|x|},$$

uniformly in $x \in \mathbb{Z}^d$, $h \in \partial\mathbf{K}_\lambda^a$ such that $h\cdot x = \mathfrak{a}_\lambda(x)$, and scales $K > K_0$.

3.3 Quenched Models

In the quenched case, (logarithms of) partition functions are random quantities and we need to control both the averages and the fluctuations.

Lemma 4. *For any $\lambda > 0$, there exists $c = c(\lambda) > 0$ such that*

$$\mathbb{P}\big(|\log Q_\lambda(x) - \mathbb{E}\log Q_\lambda(x)| \geq t\big) \leq \exp\left\{-c\frac{t^2}{|x|}\right\}, \tag{25}$$

uniformly in $|x|$ large enough.

Proof. We follow Flury [10], although working with the conjugate λ-ensemble helps. For a given realization $\underline{\mathfrak{v}} = \{\mathfrak{v}_x\}$ of the environment, define

$$F_\lambda^x[\underline{\mathfrak{v}}] \triangleq \log \sum_{X(\gamma)=x} \mathrm{e}^{-\lambda|\gamma| - \sum_y \mathfrak{v}_y \ell_\gamma(y)} P(\gamma) \text{ and } \mathbb{Q}_\lambda^{x,\underline{\mathfrak{v}}}(\gamma) \triangleq \frac{\mathrm{e}^{-\lambda|\gamma| - \sum_y \mathfrak{v}_y \ell_\gamma(y)} P(\gamma)}{\mathrm{e}^{F_\lambda^x[\underline{\mathfrak{v}}]}}.$$

Since $\lambda \geq 0$ and the entries of $\underline{\mathfrak{v}}$ are non-negative, there exists $c = c(\lambda)$ such that

$$\mathbb{Q}_\lambda^{x,\underline{\mathfrak{v}}}\Big(\sum_z \ell_\gamma^2(z)\Big) \leq c\,|x|\,. \tag{26}$$

In order to see this, given $z \in \mathbb{Z}^d$, define the set of loops

$$\mathscr{L}_z \triangleq \{\eta : z \mapsto z \,:\, \ell_\eta(z) = 1\}\,.$$

Evidently,

$$\sum_{\eta \in \mathscr{L}_z} Q_\lambda(\eta) \leq \mathrm{e}^{-\lambda}.$$

Now, any path $\gamma \in \mathscr{D}(x)$ with $\ell_\gamma(z) = n$ has a well-defined decomposition

$$\gamma = \gamma_0 \cup \eta_1 \cup \cdots \cup \eta_{n-1},$$

with $\ell_{\gamma_0}(z) = 1$ and $\eta_i \in \mathscr{L}_z$. It follows that

$$\mathbb{Q}_\lambda^{x,\underline{\mathfrak{v}}}\left(\ell_\gamma(z)^2 \middle| \ell_z(\gamma) > 0\right) \leq \sum_n n^2 \mathrm{e}^{-\lambda(n-1)} \triangleq c_1(\lambda), \tag{27}$$

uniformly in the realizations $\underline{\mathfrak{v}}$ of the environment. Consequently,

$$\mathbb{Q}_\lambda^{x,\underline{\mathfrak{v}}}\Big(\sum_z \ell_\gamma^2(z)\Big) \leq c_1(\lambda) \sum_z \mathbb{Q}_\lambda^{x,\underline{\mathfrak{v}}}\big(\ell_\gamma(z) > 0\big) \leq c_1 \mathbb{Q}_\lambda^{x,\underline{\mathfrak{v}}}\left(|\gamma|\right) \leq c_2\,|x|\,.$$

The last inequalllity above is straightforward since we assume that $\lambda > 0$ and that the distribution of random environment has bounded support.

At this stage, we infer that $F_\lambda^x[\cdot]$ is Lipschitz: Given two realizations of the environment $\underline{\mathfrak{w}}$ and $\underline{\mathfrak{v}}$, define $\mathfrak{v}^t_\cdot \triangleq t\mathfrak{w}_\cdot + (1-t)\mathfrak{v}_\cdot$. Then,

$$F_\lambda^x[\underline{\mathfrak{w}}] - F_\lambda^x[\underline{\mathfrak{v}}] = \int_0^1 \mathbb{Q}_\lambda^{x,\underline{\mathfrak{v}}^t}\Big(\sum_z (\mathfrak{w}_z - \mathfrak{v}_z)\ell_\gamma(z)\Big)\mathrm{d}t \leq \sqrt{c\,|x|} \cdot \|\underline{\mathfrak{w}} - \underline{\mathfrak{v}}\|_2.$$

Indeed,

$$\mathbb{Q}_\lambda^{x,\underline{\mathfrak{v}}}\left(\sqrt{\sum_z \ell_\gamma(z)^2}\right) \leq \sqrt{\mathbb{Q}_\lambda^{x,\underline{\mathfrak{v}}}\left(\sum_z \ell_\gamma(z)^2\right)}$$

and (26) applies. Since $F_\lambda^x[\cdot]$ is convex and, as we have checked above, Lipschitz, (25) follows from concentration inequalities on product spaces (see, e.g., [18, Corollary 4.10]) . □

Lemma 4 leads to a lower bound on the random partition function $Q_\lambda(x)$. Define

$$\hat{\epsilon}(r) \triangleq - \min_{\mathfrak{q}_\lambda(z)\geq r} \left\{ \frac{\mathfrak{q}_\lambda(z) + \mathbb{E}\log Q_\lambda(z)}{\mathfrak{q}_\lambda(z)} \right\}.$$

By subadditivity, $\hat{\epsilon}(r)$ is non-negative, and $\lim_{r\to\infty}\hat{\epsilon}(r) = 0$. By (25),

$$\mathbb{P}\left(Q_\lambda(x) \leq \mathrm{e}^{-\mathfrak{q}_\lambda(x)(1+\hat{\epsilon}(|x|)+t)} \right) \leq \exp\left\{-ct^2|x|\right\}. \tag{28}$$

We may thus assume that there exists $\epsilon(r) \to 0$ such that

$$\log Q_\lambda(x) \geq -\mathfrak{q}_\lambda(x)\left(1 + \epsilon(|x|)\right), \tag{29}$$

$\mathbb{P}$-a.s. for all $|x|$ sufficiently large.

The lower bound (29) is used to control the geometry of the skeletons $\hat{\gamma}_K$. Namely,

$$\log Q_\lambda(\hat{\gamma}_K) = \log \sum_{\underline{\eta}\sim\hat{\gamma}_K} q_\lambda(\underline{\eta}) + \sum_{i=1}^{m} \log Q_\lambda\left(u_{i-1}; v_i \middle| G_i\right). \tag{30}$$

A comparison with the simple random walk killed at the constant rate $\lambda > 0$ reveals that the following bounds hold uniformly in the realizations of the environment:

$$\log \sum_{\underline{\eta}\sim\hat{\gamma}_K} q_\lambda(\underline{\eta}) \leq c_2(\lambda)m \quad\text{and}\quad \log Q_\lambda\left(u_{i-1}; v_i \middle| G_i\right) \leq -c_3(\lambda)K.$$

It follows that we may restrict our attention to *moderate* trunks with at most $m \leq c_4|x|/K$ vertices. Consequently, the first term in (30) is at most of order $c_2c_4|x|/K$.

Assuming that $m \leq c_4|x|/K$, let us focus on the second term in (30). To simplify notations, we shall describe it as a random variable $F_{\hat{\gamma}_K} \triangleq \sum_i \log Q_\lambda\left(u_{i-1}; v_i \middle| G_i\right)$. First of all, since both u_i and v_i belong to $\partial K U_\lambda(u_{i-1})$,

$$\mathbb{E}F_{\hat{\gamma}_K} \leq -\sum \mathfrak{q}_\lambda(v_i - u_{i-1}) = -\sum \mathfrak{q}_\lambda(u_i - u_{i-1}) + O\left(\frac{|x|}{K}\right).$$

Lemma 5. *For any $\lambda > 0$, there exists $c = c(\lambda) > 0$ such that*

$$\mathbb{P}\left(\left|F_{\hat{\gamma}_K} - \mathbb{E}F_{\hat{\gamma}_K}\right| \geq t\right) \leq \exp\left\{-\frac{t^2}{|x|}\right\}, \tag{31}$$

uniformly in $|x|$ large enough, in renormalization scales K and in moderate skeletons $\hat{\gamma}_K$.

The proof of this lemma is similar to the proof of Lemma 4 and we shall sketch it below. The size of the scale K is not essential for the proof. It is essential, however, for an efficient use of the lemma: Assuming that (31) holds, we choose $1 \gg \delta \gg \sqrt{\log K/K}$. By (31),

$$\mathbb{P}\left(\left|F_{\hat{\gamma}_K} - \mathbb{E}F_{\hat{\gamma}_K}\right| \geq \delta\,|x|\right) \leq \exp\left\{-c_5\delta^2\,|x|\right\},$$

for any moderate trunk $\hat{\gamma}_K$. Since there are at most $\exp\left\{c_6\frac{\log K}{K}\,|x|\right\}$ moderate trunks, we conclude that

Lemma 6. *For any $\delta > 0$, there exists a finite scale K such that*

$$F_{\hat{\gamma}_K} \leq -\sum_i \mathfrak{q}_\lambda(u_i - u_{i-1}) + \delta\,|x|\,, \tag{32}$$

$\mathbb{P}$-a.s. for all $|x|$ large enough (and all the corresponding moderate skeleton trunks of paths $\gamma \in \mathscr{D}(x)$).

Proof (of Lemma 5). Introduce the following notation: Given a realization $\underline{\mathfrak{v}}_i$ of the environment on G_i, let $\mathbb{Q}_\lambda^{\underline{\mathfrak{v}}_i}\left(\cdot\middle|G_i\right)$ be the corresponding probability distribution on the set of paths $\mathscr{D}_{G_i}(u_{i-1}, v_i)$. In this notation,

$$F_{\hat{\gamma}_K}(\underline{\mathfrak{w}}) - F_{\hat{\gamma}_K}(\underline{\mathfrak{v}}) = \sum_{i=1}^m \int_0^1 \mathbb{Q}_\lambda^{\underline{\mathfrak{v}}_i^t}\left(\sum_{z\in G_i}\ell_{\gamma_i}(z)\,(\mathfrak{w}_z - \mathfrak{v}_z)\;\middle|\; G_i\right),$$

where $\underline{\mathfrak{v}}^t = t\underline{\mathfrak{w}} + (1-t)\underline{\mathfrak{v}}$. The conclusion follows as in the proof of Lemma 4. □

We can now proceed as in the annealed case and introduce the (quenched) surcharge of a skeleton $\hat{\gamma}_K$,

$$\mathfrak{s}_q^h(\hat{\gamma}_K) \triangleq \sum_{i=1}^m \big(\mathfrak{q}_\lambda(u_i - u_{i-1}) - h\cdot(u_i - u_{i-1})\big).$$

We then obtain the following quenched version of the surcharge inequality:

Lemma 7. *For every small $\epsilon > 0$, there exists $K_0(d,\beta,\lambda,\epsilon)$ such that, $\mathbb{P}$-a.s.,*

$$\mathbb{Q}_x^\lambda\big(\mathfrak{s}_q^h(\hat{\gamma}_K) > 2\epsilon|x|\big) \leq \mathrm{e}^{-\epsilon|x|},$$

uniformly in sufficiently large $x \in \mathbb{Z}^d$, $h \in \partial\mathbf{K}_\lambda^q$ such that $h\cdot x = \mathfrak{q}_\lambda(x)$, and scales $K > K_0$.

3.4 Irreducible Decomposition and Effective Directed Structure

The surcharge inequalities of Lemmas 3 and 7 pave the way to a detailed analysis of the structure of typical paths, as they reduce probabilistic estimates to purely geometric ones. We only describe here the resulting picture, but details can be found in [13].

Let $\lambda > 0$ and $h \in \partial \mathbf{K}_\lambda^a$. Let us fix $\delta \in (0,1)$. We define the forward cone by

$$Y_\delta^>(h) = \left\{ y \in \mathbb{Z}^d \ : \ \mathfrak{s}_a^h(y) < \delta \mathfrak{a}_\lambda(y) \right\},$$

and the *backward cone* by $Y_\delta^<(h) = -Y_\delta^>(h)$. Given $\gamma = (\gamma_0, \dots, \gamma_n) : 0 \to x$, we say that γ_k is a cone point of γ if

$$\gamma \subset \left(\gamma_k + Y_\delta^>(h)\right) \cup \left(\gamma_k + Y_\delta^<(h)\right).$$

The next theorem shows that typical paths have a positive density of cone-points.

Theorem 4. *[13] Let $\#_{\text{cone}}(\gamma)$ be the number of cone points of γ. There exist c, $C > 0$ and $\delta' > 0$, depending only on d, β, δ, and λ, such that*

$$\mathbb{A}_\lambda^x \left(\#_{\text{cone}}(\gamma) < c|x|\right) \leq \mathrm{e}^{-C|x|}, \tag{33}$$

uniformly in all sufficiently large $x \in \mathbb{Z}^d$ satisfying $\mathfrak{s}_a^h(x) \leq \delta' \mathfrak{a}_\lambda(x)$.

Remark 3. By (14) and Theorem 3, there exist $\alpha \in [1, \infty)$ and $c' > 0$ such that

$$1 \asymp \mathrm{e}^{-\lambda n} A_n(h) = \sum_{\substack{\alpha^{-1} n \leq |x| \leq \alpha n \\ \mathfrak{s}_a^h(x) \leq \delta' \mathfrak{a}_\lambda(x)}} A_{\lambda,n}(x) \mathrm{e}^{h \cdot x} + \mathrm{o}\left(\mathrm{e}^{-c' n}\right), \tag{34}$$

as n becomes large. It follows that sets of paths which are uniformly exponentially improbable under the $\mathbb{A}_\lambda^x$-measures will remain exponentially improbable under $\mathbb{A}_n^h$. In particular, (33) implies that there exist $c, C > 0$, depending only on d, β, δ, and h such that

$$\mathbb{A}_n^h \left(\#_{\text{cone}}(\gamma) < cn\right) \leq \mathrm{e}^{-Cn}, \tag{35}$$

uniformly in n sufficiently large.

With the help of (35), we can decompose typical ballistic paths into a string of irreducible pieces. A path $\gamma = (\gamma_0, \dots, \gamma_n)$ is said to be backward irreducible if γ_n is the only cone point of γ. Similarly, γ is said to be forward irreducible if γ_0 is the only cone-point of γ. Finally, γ is said to be irreducible if γ_0 and γ_n are the only cone points of γ. We denote by $\mathscr{F}^>(y)$, $\mathscr{F}^<(y)$ and $\mathscr{F}(y)$ the corresponding sets of irreducible paths connecting 0 to y.

In view of (35), we can restrict our attention to paths γ possessing at least $c'n$ cone-points, at least when n is sufficiently large. We can then unambiguously decompose γ into irreducible subpaths:

$$\gamma = \omega_{>} \cup \omega_1 \cup \cdots \cup \omega_m \cup \omega_{<}. \tag{36}$$

We thus have the following expression

$$\mathrm{e}^{-\lambda n} A_n(h) = \sum_{m \geq c'n} \sum_{\omega_{>} \in \mathscr{F}^{>}} \sum_{\omega_1,\ldots\omega_m \in \mathscr{F}} \sum_{\omega_{<} \in \mathscr{F}^{<}} a_{\lambda,h}(\gamma)\mathbf{1}_{\{|\gamma|=n\}} + O\big(\mathrm{e}^{-Cn}\big). \tag{37}$$

Observe now that the weight $a_{\lambda,h}(\gamma)$ of a path γ can be nicely factorized over its irreducible components (see (36)):

$$a_{\lambda,h}(\gamma) = a_{\lambda,h}(\omega_{>})\, a_{\lambda,h}(\omega_{<}) \prod_{i=1}^{m} a_{\lambda,h}(\omega_i),$$

Similarly, Lemma 7 implies:

Theorem 5. *Let $\lambda > 0$ and $h \in \partial\mathbf{K}_{\lambda}^{q}$. There exist $c, C > 0$, depending only on d, β, δ, and h such that*

$$\mathbb{Q}_n^h\big(\#_{\text{cone}}(\gamma) < cn\big) \leq \mathrm{e}^{-Cn}, \tag{38}$$

$\mathbb{P}$-a.s., uniformly in n sufficiently large. In particular, using the same notation (36) *for the irreducible decomposition of γ,*

$$\mathrm{e}^{-\lambda n} Q_n(h) = \sum_{m \geq c'n} \sum_{\omega_{>} \in \mathscr{F}^{>}} \sum_{\omega_1,\ldots\omega_m \in \mathscr{F}} \sum_{\omega_{<} \in \mathscr{F}^{<}} q_{\lambda,h}(\gamma)\mathbf{1}_{\{|\gamma|=n\}} + O\big(\mathrm{e}^{-Cn}\big). \tag{39}$$

$\mathbb{P}$-a.s., for all n large enough.

3.5 Basic Partition Functions

Let us say that a path $\gamma : 0 \to x$ is cone confined if $\gamma \subset Y_{\delta}^{>}(h) \cap (x + Y_{\delta}^{<}(h))$. Let $\mathscr{T}(x) \subset \mathscr{D}(x)$ be the collection of all cone confined paths leading from 0 to x, and let $\mathscr{F}(x) \subset \mathscr{T}(x)$ be the collection of all irreducible cone-confined paths. Clearly, $\omega_{>} = \omega_{<} = \varnothing$ in the irreducible decomposition (36) of paths $\gamma \in \mathscr{T}(x)$. Let $\lambda > 0$ and $h \in \partial\mathbf{K}_{\lambda}^{q}$. We define the following in general unnormalized quenched partition functions,

$$t_{x,n}^{\omega} \triangleq \sum_{\gamma \in \mathscr{T}(x)} \mathbf{1}_{\{|\gamma|=n\}} q_{\lambda,h} \text{ and } f_{x,n}^{\omega} \triangleq \sum_{\gamma \in \mathscr{F}(x)} \mathbf{1}_{\{|\gamma|=n\}} q_{\lambda,h}. \tag{40}$$

Their annealed counterparts are denoted by $\mathbf{t}_{x,n} \triangleq \mathbb{E}t^{\omega}_{x,n}$ and $\mathbf{f}_{x,n} \triangleq \mathbb{E}f^{\omega}_{x,n}$. As we shall see below, $\{\mathbf{f}_{x,n}\}$ is normalized – it is a probability distribution,

$$\sum_n \sum_x \mathbf{f}_{x,n} \triangleq \sum_n \mathbf{f}_n = 1.$$

with exponentially decaying tails. The tails are exponential by Theorem 4. The probabilistic normalization is explained in Sect. 4.1.

For $n \geq 1$, let $\mathbf{t}_n \triangleq \sum_x \mathbf{t}_{x,n}$ and $\mathbf{f}_n \triangleq \sum_x \mathbf{f}_{x,n}$, and set $\mathbf{t}_0 \triangleq 1$. The irreducible decomposition (36) of paths imply the following renewal-type relations for $\mathbf{t}_n$ and for $t^{\omega}_{x,n}$:

$$\mathbf{t}_n = \sum_{m=0}^{n-1} \mathbf{t}_m \mathbf{f}_{n-m}, \qquad t^{\omega}_{x,n} = \sum_{m=0}^{n-1} \sum_y t^{\omega}_{y,m} f^{\theta_y \omega}_{x-y,n-m}. \tag{41}$$

For the rest of the paper, we shall work mainly with the above basic ensembles of paths. All the results can be routinely extended (as in, e.g.,[14]) to general ensembles by summing out over the paths $\omega_<$ and $\omega_>$ paths in the irreducible decomposition (36), their weights being exponentially decaying.

4 The Annealed Model

4.1 Asymptotics of $\mathbf{t}_n = \sum_x \mathbf{t}_{x,n}$

Annealed asymptotics are not related to the strength of disorder and hold for all values of $\beta \geq 0$. Neither do they require any moment assumptions on V.

Lemma 8. *Let $\lambda > 0$ and $h \in \partial\mathbf{K}^a_\lambda$. There exists $\kappa = \kappa(\lambda, h)$ such that*

$$\lim_{n\to\infty} \mathbf{t}_n = \left\{ \sum_n n\mathbf{f}_n \right\}^{-1} \triangleq \frac{1}{\kappa} \tag{42}$$

exponentially fast.

Proof. For $|u| \leq 1$, define the generating functions

$$\hat{\mathbf{t}}[u] \triangleq \sum_{n=0}^{\infty} u^n \mathbf{t}_n \text{ and } \hat{\mathbf{f}}[u] \triangleq \sum_{n=1}^{\infty} u^n \mathbf{f}_n.$$

It follows from Theorem 4 that the second series converges on some disc $\mathbb{D}_{1+\nu} = \{u \in \mathbb{C} : |u| < 1 + \nu\}$. The first series blows up at any $\mathbb{R} \ni u > 1$. Since by (41), $\hat{\mathbf{t}}[u] = \left(1 - \hat{\mathbf{f}}[u]\right)^{-1}$, it follows that $\hat{\mathbf{f}}[1] = 1$, which is the probabilistic normalization mentioned above. We identify κ in (42) as $\kappa = \hat{\mathbf{f}}'[1]$. It then follows from the renewal relation (41) that

$$\hat{\mathbf{t}}[u] = \frac{1}{1-\hat{\mathbf{f}}[u]} = \frac{1}{\kappa(1-u)} + \frac{\hat{\mathbf{f}}[u]-\hat{\mathbf{f}}[1]-\kappa(u-1)}{\kappa(1-\hat{\mathbf{f}}[u])(1-u)} \stackrel{\Delta}{=} \frac{1}{\kappa(1-u)} + \Delta[u]. \quad (43)$$

Since the function Δ is analytic on some disc $\mathbb{D}_{1+\nu'}$, the claim follows from Cauchy's formula. □

4.2 *Geometry of* $\mathbf{K}_\lambda^a$*, Annealed LLN and CLT*

Let $\lambda > 0$ and $h \in \partial \mathbf{K}_\lambda^a$. In the ballistic phase of the annealed model, the CLT is obtained on the level of a local limit description: Given $z \in \mathbb{C}$, let us try to find $\mu = \mu(z) \in \mathbb{C}$ such that

$$\mathbf{F}(z,\mu) \stackrel{\Delta}{=} \sum_{n,x} \mathrm{e}^{-\mu n + z\cdot x}\mathbf{f}_{x,n} = 1. \quad (44)$$

Since $\{\mathbf{f}_n\}$ has exponential decay, the implicit function theorem implies that

Lemma 9. *There exist* $\delta, \eta > 0$ *and an analytic function* μ *on* $\mathbb{D}_\delta^d$ *such that*

$$\left\{(z,\mu) \in \mathbb{D}_\delta^d \times \mathbb{D}_\eta \;:\; \mathbf{F}(z,\mu) = 1\right\} = \left\{(z,\mu) \in \mathbb{D}_\delta^d \times \mathbb{D}_\eta \;:\; \mu = \mu(z)\right\}. \quad (45)$$

Moreover, $\mathrm{Hess}[\mu](0)$ *is nondegenerate.*

If z is real and $|z|$ is sufficiently small, then $\mu(z) = \Lambda_a(h+z) - \lambda$. Indeed, if $|z|$ is small, then the leading contribution to $\sum_n \mathrm{e}^{-(\mu+\lambda)n} A_n(h+z)$ is still coming from

$$\sum_n \sum_x \mathbf{t}_{x,n}\, e^{-\mu n + z\cdot x}$$

By (41) and (44), $\mu = \mu(z)$ describes the radius of convergence of the latter series, whereas $\mu + \lambda = \Lambda_a(h+z)$ descibes the radius of convergence of the former one.

Therefore, $\partial \mathbf{K}_\lambda^a$ is locally given by the level set $\{h + z \,:\, \mu(z) = 0\}$. In addition Λ_a inherits analyticity and nondegeneracy properties of μ:

$$\nabla\mu(0) = \nabla\Lambda_a(h) \stackrel{\Delta}{=} v = v^a(h,\beta) \text{ and define } \mathrm{Hess}[\mu](0) = \mathrm{Hess}[\Lambda_a](h) \stackrel{\Delta}{=} \Xi^{-1}. \quad (46)$$

Given $z \in \mathbb{D}_\delta$ as above, define

$$\mathbf{f}_{x,n}(z) \stackrel{\Delta}{=} \mathbf{f}_{x,n}\mathrm{e}^{-\mu(z)n + z\cdot x} \text{ and, respectively, } \mathbf{t}_{x,n}(z) \stackrel{\Delta}{=} \mathbf{t}_{x,n}\mathrm{e}^{-\mu(z)n + z\cdot x}.$$

Set $\mathbf{f}_n(z) = \sum_x \mathbf{f}_{x,n}(z)$, $\mathbf{t}_n(z) = \sum_x \mathbf{t}_{x,n}(z)$ and $\kappa(z)^{-1} = \sum_n n\mathbf{f}_n(z)$. Literally repeating the derivation of (42), we infer that there exists $\alpha > 0$ such that, uniformly in $z \in \bar{\mathbb{D}}_\delta^d$,

$$\left| \mathbf{t}_n(z) - \frac{1}{\kappa(z)} \right| \leq \mathrm{e}^{-\alpha n}. \tag{47}$$

By Cauchy's formula, $\nabla \log \mathbf{t}_n(z) = \mathrm{O}(1)$. Therefore,

$$\mathrm{O}\left(\frac{1}{n}\right) = \frac{1}{n}\nabla \log \mathbf{t}_n(0) = -\nabla\mu(0) + \frac{1}{n}\sum_x x\frac{\mathbf{t}_{x,n}}{\mathbf{t}_n} = -v + \frac{1}{n}\sum_x x\frac{\mathbf{t}_{x,n}}{\mathbf{t}_n}. \tag{48}$$

(48) is an annealed law of large numbers which, in particular, identifies v as the limiting macroscopic spatial extension.

Next, the following form of the annealed CLT holds: for any $\alpha \in \mathbb{R}^d$,

$$\begin{aligned}\frac{\mathbf{S}_n(\alpha)}{\mathbf{t}_n} &\triangleq \sum_x \frac{\mathbf{t}_{x,n}}{\mathbf{t}_n} \exp\left\{ i\alpha \cdot \frac{x - nv}{\sqrt{n}} \right\} = \frac{\mathbf{t}_n\left(\frac{i\alpha}{\sqrt{n}}\right)}{\mathbf{t}_n} \exp\left\{ n\mu\left(\frac{i\alpha}{\sqrt{n}}\right) - nv \cdot \frac{i\alpha}{\sqrt{n}} \right\} \\ &= \exp\left\{ -\frac{1}{2}\Xi^{-1}\alpha\cdot\alpha \right\} \left(1 + O\left(n^{-1/2}\right)\right),\end{aligned} \tag{49}$$

with the second asymptotic equality holding uniformly in α on compact subsets of $\mathbb{R}^d$.

4.3 Local Limit Theorem for the Annealed Polymer

In this subsection, we shall explain the local limit picture behind (48) and (49). Recall that

$$\begin{aligned}\mathbf{t}_{x,n} &= \sum_{m=0}^{n-1}\sum_y \mathbf{t}_{y,m}\mathbf{f}_{x-y,n-m} \\ &= \sum_{N\geq 1}\ \sum_{m_1,\dots,m_N\geq 1}\ \sum_{y_1,\dots,y_N\in\mathbb{Z}^d} \mathbf{1}_{\left\{\sum m_i = n, \sum y_i = x\right\}} \prod_{i=1}^N \mathbf{f}_{y_i - y_{i-1},m_i},\end{aligned} \tag{50}$$

where we have set, for convenience, $y_0 = 0$. As explained above, the weights $\mathbf{f}_{y,m}$ form a probability distribution on $\mathbb{Z}^d \times \mathbb{N}$,

$$\sum_{y\in\mathbb{Z}^d, m\in\mathbb{N}} \mathbf{f}_{y,m} = 1,$$

and decay exponentially both in y and m. Let us consider an i.i.d. sequence of random vectors $(Y_k, M_k)_{k\geq 1}$ whose joint distribution $\mathrm{P}_{\mathrm{eff}}$ is given by these weights. Then, (50) can be expressed as

$$\mathbf{t}_{x,n} = \sum_{N\geq 1} \mathrm{P}_{\mathrm{eff}}\left(\sum_{i=1}^{N}(Y_i, M_i) = (x,n)\right).$$

Consequently, sharp asymptotics for $\mathbf{t}_{x,n}$ readily follow from a local limit analysis of the empirical mean of the i.i.d. random vectors $(Y_k, M_k)_{k\geq 1}$ with exponential tails. In this way, one obtains the following sharp asymptotics for the extension of an annealed polymer, covering all possible deviation scales.

Theorem 6. *[13] Suppose that $h \notin \mathbf{K}^a$. Let $v_h = \nabla\Lambda_a(h)$. Then, for some small enough $\epsilon > 0$, the rate function J_h^a is real analytic and strictly convex on the ball $B_\epsilon(v_h) \stackrel{\Delta}{=} \{u : |u - v_h| < \epsilon\}$ with a nondegenerate quadratic minimum at v_h. Moreover, there exists a strictly positive real analytic function G on $B_\epsilon(v_h)$ such that*

$$\mathbb{A}_n^h\Big(\frac{X(\gamma)}{n} = u\Big) = \frac{G(u)}{\sqrt{n^d}}\,\mathrm{e}^{-nJ_h^a(u)}\big(1 + o(1)\big), \tag{51}$$

uniformly in $u \in B_\epsilon(v) \cap \frac{1}{n}\mathbb{Z}^d$.

Remark 4. We would like to note that a local limit result for a particular instance of the annealed model (discrete Wiener sausage with a fixed nonzero drift at small β) was obtained in [24]. We are grateful to Erwin Bolthausen for sending us a copy of this work.

5 Weak Disorder

In this section, we focus on the supercritical quenched models in the weak disorder regime. Accordingly, we consider higher dimensional models on $\mathbb{Z}^d$ with $d \geq 4$. Let us say that the weak disorder holds at (h, β) if there exists $\lambda > 0$ such that $h \in \partial\mathbf{K}_\lambda^a$ and the disorder is weak in the conjugate ensemble at (λ, β), that is the Lyapunov exponents coincide; $\mathfrak{a}_\lambda(\cdot) \equiv \mathfrak{q}_\lambda(\cdot)$. In particular, $\lambda = \Lambda_a(h) = \Lambda_q(h)$.

5.1 LLN at Supercritical Drifts

An important, albeit elementary, observation is that, in this regime, events of exponentially small probability under the annealed measure $\mathbb{A}_n^h$ are also exponentially unlikely under the quenched measure $\mathbb{Q}_n^h$.

Lemma 10. *Assume that weak disorder holds, $\Lambda_a(h) = \Lambda_q(h)$. Let E be a path event such that $\mathbb{A}_n^h(E) \le \mathrm{e}^{-cn}$ for some constant $c > 0$ and all n large enough. Then there exists $c' > 0$ such that, $\mathbb{P}$-a.s.,*

$$\mathbb{Q}_n^h(E) \le \mathrm{e}^{-c'n},$$

for all n large enough.

Proof. We note that, since $\Lambda_a(h) = \Lambda_q(h)$, it follows from Markov's inequality that, for all n large enough,

$$\mathbb{P}\big(\mathbb{Q}_n^h(E) > \mathrm{e}^{-c'n}\big) = \mathbb{P}\big(Q_n(h;E) > Q_n(h)\mathrm{e}^{-c'n}\big)$$
$$\le \mathbb{P}\left(Q_n(h;E) > A_n(h)\mathrm{e}^{-\frac{1}{2}c'n}\right) \le \mathbb{A}_n^h(E)\mathrm{e}^{+\frac{1}{2}c'n} \le \mathrm{e}^{-\left(c-\frac{1}{2}c'\right)n},$$

where $Q_n(h;E)$ denotes the quenched partition function restricted to paths in E. The conclusion now follows from Borel–Cantelli. □

Recall that a pulling force $h \in \partial\mathbf{K}_\lambda^q$ is called supercritical if $\lambda > 0$. Using Lemma 10, it is very easy to prove that, in the supercritical weak disorder regime, the quenched model satisfies LLN, and that the polymer has the same limiting macroscopic extension under the quenched and annealed path measures.

Lemma 11. *Assume that weak disorder holds, $\lambda = \Lambda_a(h) = \Lambda_q(h) > 0$. Let $v = \nabla\Lambda_a(h)$ be the macroscopic extension of the polymer under the annealed path measure. Then, for any $\epsilon > 0$,*

$$\lim_{n\to\infty} \mathbb{Q}_n^h\left(\left|\frac{X(\gamma)}{n} - v\right| > \epsilon\right) = 0, \qquad \mathbb{P}\text{-a.s.},$$

exponentially fast in n.

Proof. The claim immediately follows from (7) and Lemma 10. □

5.2 Very Weak Disorder

Recall that we use notation $(\Omega, \mathscr{F}, \mathbb{P})$ for the (product) probability space generated by the random environment. For the rest of this section, we consider the regime of very weak disorder, which should be understood in the following sense: we fix either $\lambda > 0$ or $h \ne 0$ and then, for β sufficiently small, we pick the remaining parameter (h or λ) according to $h \in \partial\mathbf{K}_\lambda^q$.

The regime of very weak disorder is quantified in terms of the following upper bound :

Lemma 12. *Fix an external force $h \neq 0$. Then, for all β small enough, the random weights* (40) *(with $\lambda = \lambda(h,\beta)$ being determined by $h \in \partial \mathbf{K}^a_\lambda$) satisfy: There exist $c_1, c_2 < \infty$ such that, uniformly in x_1, x_2, m_1, m_2, ℓ and in all cylindrical sub-σ-algebras $\mathscr{F}_1$, of $\mathscr{F}$,*

$$\begin{aligned}&\left|\mathbb{E} t^{\omega}_{x_1,\ell} t^{\omega}_{x_2,\ell} \mathbb{E}\left(f^{\theta_{x_1}\omega}_{m_1} - \mathbf{f}_{m_1} \mid \mathscr{F}_1\right) \mathbb{E}\left(f^{\theta_{x_2}\omega}_{m_2} - \mathbf{f}_{m_2} \mid \mathscr{F}_1\right)\right| \\ &\quad \leq \frac{c_1 \mathrm{e}^{-c_2(m_1+m_2)}}{\ell^d} \exp\left\{-c_2\left(|x_1 - x_2| + \frac{|x_1 - \ell v|^2}{\ell}\right)\right\},\end{aligned} \tag{52}$$

where $v = \nabla\lambda(h)$.

Although (52) looks technical, it has a transparent intuitive meaning: the expressions on the rhs are just local limit bounds for a couple of independent annealed polymers with exponential penalty for disagreement at their end-points. In the regime of very weak disorder, the interaction between polymers does not destroy these asymptotics. The proof will be given elsewhere [12]. Closely related upper bounds were already derived in [14].

5.3 Convergence of Partition Functions

As mentioned above, the rescaled quenched partition functions satisfy the following multidimensional renewal relation:

$$t^{\omega}_{z,n} = \sum_{m=0}^{n-1} \sum_{x} t^{\omega}_{x,m} f^{\theta_x \omega}_{z-x,n-m} \text{ and } t^{\omega}_{n} = \sum_{z} t^{\omega}_{z,n}. \tag{53}$$

Theorem 7. *In the regime of very weak disorder,*

$$\lim_{n\to\infty} t^{\omega}_n = \frac{1}{\kappa}\left(1 + \sum_{x,y} t^{\omega}_x \left(f^{\theta_x \omega}_{y-x} - \mathbf{f}_{y-x}\right)\right) \stackrel{\Delta}{=} \frac{1}{\kappa} s^{\omega} \in (0,\infty), \tag{54}$$

$\mathbb{P}$-a.s. and in $L_2(\Omega)$.

Proof. We rely on an expansion similar to the one employed by Sinai and rewrite (53) as

$$t^{\omega}_{z,n} = \mathbf{t}_{z,n} + \sum_{l+m+r=n} \sum_{x,y} t^{\omega}_{x,l} \left(f^{\theta_x \omega}_{y-x,m} - \mathbf{f}_{y-x,m}\right) \mathbf{t}_{z-y,r}. \tag{55}$$

This is just a resummation based on the following identity: Let $0 < \ell_1, \dots, \ell_k$ and let $x_1, \dots x_k \in \mathbb{Z}^d$. Set $x_0 = 0$. Then,

$$\prod_1^k f_{x_j - x_{j-1}, \ell_j}^{\theta_{x_{j-1}}\omega} = \prod_1^k \mathbf{f}_{x_j - x_{j-1}, l_j} + \left(f_{x_1,\ell_1}^{\omega} - \mathbf{f}_{x_1,\ell_1}\right) \prod_2^k \mathbf{f}_{x_j - x_{j-1}, l_j}$$
$$+ f_{x_1,\ell_1}^{\omega} \left(f_{x_2 - x_1, \ell_2}^{\theta_{x_1}\omega} - \mathbf{f}_{x_2 - x_1, \ell_2}\right) \prod_3^k \mathbf{f}_{x_j - x_{j-1}, l_j}$$
$$+ \cdots + \prod_1^{k-1} f_{x_j - x_{j-1}, \ell_j}^{\theta_{x_{j-1}}\omega} \left(f_{x_k - x_{k-1}, \ell_k}^{\theta_{x_{k-1}}\omega} - \mathbf{f}_{x_k - x_{k-1}, \ell_k}\right).$$

(55) implies,

$$\begin{aligned} t_n^{\omega} &= \mathbf{t}_n + \sum_{l+m+r=n} \sum_x t_{x,l}^{\omega} \left(f_m^{\theta_x \omega} - \mathbf{f}_m\right) \mathbf{t}_r \\ &= \mathbf{t}_n + \frac{1}{\kappa} \sum_{l+m \le n} \sum_x t_{x,l}^{\omega} \left(f_m^{\theta_x \omega} - \mathbf{f}_m\right) + \sum_{l+m+r=n} \sum_x t_{x,l}^{\omega} \left(f_m^{\theta_x \omega} - \mathbf{f}_m\right) \left(\mathbf{t}_r - \frac{1}{\kappa}\right) \\ &= \frac{1}{\kappa} s_n^{\omega} + (\mathbf{t}_n - 1/\kappa) + \epsilon_n^{\omega}, \end{aligned} \tag{56}$$

where

$$s_n^{\omega} = 1 + \sum_{l+m \le n} \sum_x t_{x,l}^{\omega} \left(f_m^{\theta_x \omega} - \mathbf{f}_m\right), \tag{57}$$

and the correction term ϵ_n^{ω} is given by

$$\epsilon_n^{\omega} = \sum_{l+m+r=n} \sum_x t_{x,l}^{\omega} \left(f_m^{\theta_x \omega} - \mathbf{f}_m\right) \left(\mathbf{t}_r - \frac{1}{\kappa}\right). \tag{58}$$

We claim that, $\mathbb{P}$-a.s.,

$$\lim_{n \to \infty} s_n^{\omega} = s^{\omega} \text{ and } \sum_n \mathbb{E}(\epsilon_n^{\omega})^2 < \infty. \tag{59}$$

The assertion of Theorem 7 follows. □

The main input for proving (59) is the upper bound of (52) and the following maximal inequality of McLeish.

Maximal Inequality. Let $X_1, X_2, \dots$ be a sequence of zero mean and square integrable random variables. Let also $\{\mathscr{F}_k\}_{-\infty}^{\infty}$ be a filtration of σ-algebras. Suppose that we have chosen $\epsilon > 0$ and numbers $a_1, a_2, \dots$ in such a way that

$$\mathbb{E}\left(\mathbb{E}\left(X_\ell \,|\mathscr{F}_{\ell-m}\right)^2\right) \le \frac{a_\ell^2}{(1+m)^{1+\epsilon}} \text{ and } \mathbb{E}\left(X_\ell - \mathbb{E}\left(X_\ell \,|\mathscr{F}_{\ell+m}\right)\right)^2 \le \frac{a_\ell^2}{(1+m)^{1+\epsilon}}, \tag{60}$$

for all $\ell = 1, 2, \ldots$ and $m \geq 0$. Then there exists $K = K(\epsilon) < \infty$ such that, for all $n_1 \leq n_2$,

$$\mathbb{E} \max_{n_1 \leq m \leq n_2} \Big(\sum_{n_1}^{m} X_\ell \Big)^2 \leq K \sum_{n_1}^{n_2} a_\ell^2. \tag{61}$$

In particular, if $\sum_\ell a_\ell^2 < \infty$, then $\sum_\ell X_\ell$ converges $\mathbb{P}$-a.s.

Proof of (59)**.** The difficult part of s_n^ω in (57) is

$$\sum_{\ell \leq n} \sum_x t_{x,\ell}^\omega \left(f^{\theta_x \omega} - 1 \right) \stackrel{\Delta}{=} \sum_{\ell=1}^{n} X_\ell. \tag{62}$$

To simplify the exposition, let us consider the case of an on-axis external force $h = h\mathsf{e}_1$. By lattice symmetries, the mean displacement $v = \nabla\lambda(h) = v\mathsf{e}_1$. At this stage, let us define the hyperplanes $\mathscr{H}_m^- \stackrel{\Delta}{=} \{x \,:\, x \cdot \mathsf{e}_1 \leq mv\}$ and the σ-algebras

$$\mathscr{F}_m \stackrel{\Delta}{=} \sigma \left(V(x) \,:\, x \in \mathscr{H}_m^- \right).$$

Then,

$$\mathbb{E}\left(X_\ell \mid \mathscr{F}_{\ell-m} \right) = \sum_{x \in \mathscr{H}_{\ell-m}^-} t_x^\omega \mathbb{E}\left(f^{\theta_x\omega} - 1 \mid \mathscr{F}_{\ell-m} \right).$$

Consequently,

$$\mathbb{E}\Big(\mathbb{E}\left(X_\ell \mid \mathscr{F}_{\ell-m} \right)^2 \Big) \leq \sum_{x,x' \in \mathscr{H}_{\ell-m}^-} \left| \mathbb{E} t_x^\omega t_{x'}^\omega \mathbb{E}\left(f^{\theta_x\omega} - 1 \mid \mathscr{F}_{\ell-m} \right) \mathbb{E}\left(f^{\theta_x\omega} - 1 \mid \mathscr{F}_{\ell-m} \right) \right|.$$

The following notation is convenient: We say that $a_\ell \lesssim b_\ell$ if there exists a constant $c > 0$ such that $a_\ell \lesssim c b_\ell$ for all ℓ . In this language, using (52), we bound the latter expression (for m large enough) by

$$\begin{aligned} &\lesssim \frac{1}{\ell^d} \sum_{x \in \mathscr{H}_{\ell-m}^-} \mathrm{e}^{-c_2|x - v\ell|^2/\ell} \sum_{x'} \mathrm{e}^{-c_2|x'-x|} \lesssim \frac{1}{\ell^d} \int_{|y| > \frac{m|v|}{2}} \mathrm{e}^{-c_2|y|^2/\ell} \mathrm{d}y \\ &= \frac{1}{\ell^d} \int_{\frac{m|v|}{2}}^{\infty} r^d \mathrm{e}^{-c_2 r^2/l} \mathrm{d}r \lesssim \frac{1}{\ell^{(d+1)/2}} \mathrm{e}^{-c_3 m^2/\ell}. \end{aligned} \tag{63}$$

Noting that, for any ϵ fixed,

$$\frac{\mathrm{e}^{-c_3 m^2/\ell}}{\ell^{1/2+\epsilon}} \lesssim \frac{1}{(1+m)^{1+\epsilon}},$$

we conclude that

$$\mathbb{E}\left(\mathbb{E}\left(X_\ell \mid \mathscr{F}_{\ell-m} \right) \right)^2 \lesssim \frac{1}{\ell^{(d-1)/2-\epsilon}} \frac{1}{(1+m)^{1+\epsilon}}. \tag{64}$$

Similarly, the main contribution to $X_\ell - \mathbb{E}\left(X_\ell \mid \mathscr{F}_{\ell+m}\right)$ comes from

$$\sum_{x \in \mathscr{H}^+_{\ell+m}} t^{\omega}_{x,\ell}\left(f^{\theta_x \omega} - 1\right).$$

By a completely similar computation,

$$\mathbb{E}\left(X_\ell - \mathbb{E}\left(X_\ell \mid \mathscr{F}_{\ell+m}\right)\right)^2 \lesssim \frac{1}{\ell^{(d-1)/2-\epsilon}} \cdot \frac{1}{(1+m)^{1+\epsilon}}. \tag{65}$$

Therefore, (60) applies with $a_\ell^2 \cong \ell^{-(d-1)/2-\epsilon}$. Since $d \geq 4$, we rely on (61) and deduce (59).

5.4 Quenched CLT

One possible strategy for proving a $\mathbb{P}$-a.s CLT would be to try to adjust a powerful approach by Bolthausen–Sznitman [3] which was developed in the context of ballistic RWRE. It appears, however, that a direct work on generating functions goes through. Let us introduce

$$\mathscr{S}^{\omega}_n(\alpha) \stackrel{\Delta}{=} \sum_z \mathbf{t}_{z,n} \mathrm{e}^{i\alpha\cdot(z-nv)/\sqrt{n}}.$$

The asymptotics of $\mathbf{S}_n(\alpha) \stackrel{\Delta}{=} \mathbb{E}\mathscr{S}^{\omega}_n(\alpha)$ are given in (49). Using (56), we can write

$$\mathscr{S}^{\omega}_n(\alpha) = \mathbf{S}_n(\alpha) + \sum_{l+m+r=n} \sum_{x,y,z} t^{\omega}_{x,l}\left(f^{\theta_x\omega}_{y-x,m} - \mathbf{f}_{y-x,m}\right) \mathbf{t}_{z-y,r} \mathrm{e}^{i\alpha\cdot(z-nv)/\sqrt{n}}. \tag{66}$$

Define $\alpha^r_n \stackrel{\Delta}{=} \alpha\sqrt{r/n}$ and

$$G^{\omega}_m(\alpha) \stackrel{\Delta}{=} \sum_y \mathrm{e}^{(y-mv)\cdot i\alpha}\left(f^{\omega}_{y,m} - \mathbf{f}_m\right). \tag{67}$$

We can rewrite (66) as

$$\begin{aligned}
\mathscr{S}^{\omega}_n(\alpha) &= \mathbf{S}_n(\alpha) + \sum_{l+m+r=n} \mathbf{S}_r\left(\alpha^r_n\right) \sum_x t^{\omega}_{x,\ell} \mathrm{e}^{(x-\ell v)\cdot i\frac{\alpha}{\sqrt{n}}} G^{\theta_x\omega}_m\left(\frac{\alpha}{\sqrt{n}}\right) \\
&= \mathbf{S}_n(\alpha)\left(1 + \sum_{\ell+m\leq n} \sum_x t^{\omega}_{x,\ell}\left(f^{\theta_x\omega}_m - \mathbf{f}_m\right)\right)
\end{aligned}$$

$$
\begin{aligned}
&+ \sum_{l+m+r=n} \left(\mathbf{S}_r(\alpha_n^r) - \mathbf{S}_n(\alpha)\right) \sum_x t_{x,\ell}^\omega \left(f_m^{\theta_x\omega} - \mathbf{f}_m\right) \\
&+ \sum_{l+m+r=n} \mathbf{S}_r(\alpha_n^r) \sum_x t_{x,\ell}^\omega \left(G_m^{\theta_x\omega}\left(\frac{\alpha}{\sqrt{n}}\right) - G_m^{\theta_x\omega}(0)\right) \\
&+ \sum_{l+m+r=n} \mathbf{S}_r(\alpha_n^r) \sum_x t_{x,\ell}^\omega \left(\mathrm{e}^{(x-\ell v)\cdot i\alpha/\sqrt{n}} - 1\right) G_m^{\theta_x\omega}\left(\frac{\alpha}{\sqrt{n}}\right) \\
&\triangleq \mathbf{S}_n(\alpha) s_n^\omega + \sum_{i=1}^{3} \hat{\mathscr{E}}_n^i(\omega),
\end{aligned} \tag{68}
$$

where s_n^ω is as in (57).

Theorem 8. *For every $\alpha \in \mathbb{R}^d$, the correction terms $\hat{\mathscr{E}}_n^i(\omega)$ in (68) satisfy*

$$
\text{For } i = 1, 2, 3, \quad \lim_{n\to\infty} \hat{\mathscr{E}}_n^i(\omega) = 0\,, \ \mathbb{P}\text{-a.s. and in } L_2(\Omega). \tag{69}
$$

The proof of Theorem 8 is technical and will appear elsewhere [12]. In view of (49) and (54), the convergence in (69) implies that

$$
\lim_{n\to\infty} \frac{\mathscr{S}_n^\omega(\alpha)}{t_n^\omega} = \exp\left\{-\frac{1}{2}\Xi^{-1}\,\alpha\cdot\alpha\right\}, \tag{70}
$$

$\mathbb{P}$-a.s. for every $\alpha \in \mathbb{R}^d$ fixed.

6 Strong Disorder

In this section, we do not impose any moment assumptions on the environment $\{V(x)\}$. Even the case of traps (i.e., when $\mathbb{P}(V = \infty) > 0$) is not excluded. We still need that $\mathbb{P}(V \neq 0, \infty) > 0$. Without loss of generality, we shall assume that $\mathbb{P}(V \in (0, 1]) > 0$. Under this sole assumption, the environment is always strong in two dimensions in the following sense.

Theorem 9. *Let $d = 2$ and $\beta, \lambda > 0$. There exists $c = c(\beta, \lambda) > 0$ such that the following holds: For any $x \in \mathbb{S}^1$ define $x_n = \lfloor nx \rfloor$. Then,*

$$
\limsup_{n\to\infty} \frac{1}{n} \log \frac{Q_\lambda(x_n)}{A_\lambda(x_n)} < -c. \tag{71}
$$

In particular, $\mathfrak{a}_\lambda < \mathfrak{q}_\lambda$ whenever $\mathfrak{q}_\lambda$ is well defined.

Remark 5. As in [17] and, subsequently, [29] proving strong disorder in dimension $d = 3$ is a substantially more delicate task.

Let us explain Theorem 9: By the exponential Markov's inequality (and Borel–Cantelli) it is sufficient to prove that there exist $c' > 0$ and $\alpha > 0$ such that

$$\mathbb{E}\left(\frac{Q_\lambda(x_n)}{A_\lambda(x_n)}\right)^\alpha \leq \mathrm{e}^{-c'n}. \tag{72}$$

6.1 Normalization

In order to facilitate the notation, we shall proceed with an on-axis case $x = (0, x)$. Let $h = (0, h) \in \partial \mathbf{K}_\lambda^a$ be unambiguously defined by the relation $\mathfrak{a}_\lambda(x) = h \cdot x$. We shall explore

$$\frac{1}{n}\log\frac{Q_{\lambda,h}(x_n)}{A_{\lambda,h}(x_n)} \stackrel{\Delta}{=} \frac{1}{n}\log\frac{\mathrm{e}^{h\cdot x_n} Q_\lambda(x_n)}{\mathrm{e}^{h\cdot x_n} A_\lambda(x_n)}.$$

Since the annealed Lyapunov exponent $\mathfrak{a}_\lambda$ is well-defined, we can rely on the logarithmic equivalence $A_{\lambda,h}(x_n) \asymp 1$. Note that, for any family of paths Γ_n,

$$A_{\lambda,h}\left(X(\gamma) = x_n; \Gamma_n\right) \stackrel{\Delta}{=} A_{\lambda,h}\left(x_n; \Gamma_n\right) = \mathbb{E}Q_{\lambda,h}\left(x_n; \Gamma_n\right).$$

Consequently, by the exponential Markov inequality and Borel–Cantelli Lemma, we can ignore the families Γ_n for which $A_{\lambda,h}\left(X(\gamma) = x_n; \Gamma_n\right) \leq \mathrm{e}^{-c|x_n|}$.

6.2 Reduction to Basic Partition Functions

In particular, we can restrict attention to paths γ which have at least two cone points. With a slight abuse of notation,

$$Q_{\lambda,h}\left(x_n\right) = \sum_{y,z} f_>^\omega(y) t_{z-y}^{\theta_y\omega} f_<^{\theta_z\omega}(x_n - z),$$

where $f_>^\omega$ and $f_<^{\theta_z\omega}$ are the weights of the initial and the final irreducible pieces $\omega_>$ and $\omega_<$ in the decomposition (36). The left and right irreducible partition functions satisfy $\mathbb{E}f_>^\omega(u), \mathbb{E}f_<^\omega(u) \leq \mathrm{e}^{-c|u|}$. Consequently, for fractional powers $\alpha \in (0, 1)$,

$$\mathbb{E}\left(Q_{\lambda,h}\left(x_n\right)\right)^\alpha \leq \sum_{y,z} \mathrm{e}^{-c|y|}\mathbb{E}\left(t_{z-y}^\omega\right)^\alpha \mathrm{e}^{-c|x_n - z|}.$$

As a result, we need to check that

$$\limsup_{u\to\infty}\frac{1}{|u|}\mathbb{E}\left(t_u^\omega\right)^\alpha < 0. \tag{73}$$

In its turn, (73) is routinely implied by the following statement ((75) below): Let r_N^{ω} be the partition function of N irreducible steps:

$$r_N^{\omega} = \sum_x r_{x,N}^{\omega} \triangleq \sum_x \sum_{u_1,\cdots,u_{N-1}} f_{u_1}^{\omega} f_{u_2-u_1}^{\theta_{u_1}\omega} \cdots f_{x-u_{N-1}}^{\theta_{u_{N-1}}\omega}. \tag{74}$$

Then,

$$\limsup_{N\to\infty} \frac{1}{N} \log \mathbb{E} \left(r_N^{\omega}\right)^{\alpha} < 0. \tag{75}$$

Note by the way that by the very definition of the irreducible decomposition any trajectory γ which contributes to a $\underline{u} = (u_0, u_1, \ldots, u_N)$-term in (74) is confined to the set $D(\underline{u}) = \cup_{\ell} D(u_{\ell-1}, u_{\ell})$ where the diamond shape $D(u_{\ell-1}, u_{\ell}) \triangleq \left(u_{\ell-1} + Y_{\delta}^{>}(h)\right) \cap \left(u_{\ell} + Y_{\delta}^{<}(h)\right)$.

6.3 Fractional Moments

Following Lacoin [17], (75) follows once we show that there exist N and $\alpha \in (0,1)$ such that

$$\mathbb{E} \sum_x \left(r_{x,N}^{\omega}\right)^{\alpha} < 1. \tag{76}$$

Pick K sufficiently large and ϵ small, and consider

$$A_N = \{0, \ldots, KN\} \times \{-N^{1/2+\epsilon}, \ldots, N^{1/2+\epsilon}\} \subset \mathbb{Z}^2.$$

Since $\mathbb{E}\left(r_{x,N}^{\omega}\right)^{\alpha} \leq \left(\mathbb{E} r_{N,x}^{\omega}\right)^{\alpha} \triangleq r_{N,x}^{\alpha}$, annealed estimates enable us to restrict attention to $x \in A_N$. Furthermore, since, as was explained in Sect. 4.3, r_N is a partition function which corresponds to an effective random walk with an on-axis drift and exponential tails we may restrict attention only to the effective trajectories $\underline{u}$ which satisfy $D(\underline{u}) \subseteq A_N$. By the confinement property of the irreducible decomposition we may therefore restrict attention to microscopic polymer configurations γ which stay inside A_N.

At this stage, we shall modify the distribution of the environment inside A_N in the following way: The modified law of the environment $\tilde{\mathbb{P}}$ is still product and, for every $x \in A_N$,

$$\frac{\mathrm{d}\tilde{\mathbb{P}}}{\mathrm{d}\mathbb{P}}(V) \triangleq \mathrm{e}^{-\delta_N (V\wedge 1) + g(\delta_N)}, \quad \text{where } \mathrm{e}^{-g(\delta)} = \mathbb{E}\mathrm{e}^{-\delta(V\wedge 1)}.$$

From Hölder's inequality,

$$\mathbb{E}\left(r_{x,N}^{\omega}\right)^{\alpha} \leq \left(\tilde{\mathbb{E}}\left(\frac{\mathrm{d}\tilde{\mathbb{P}}}{\mathrm{d}\mathbb{P}}\right)^{1/(1-\alpha)}\right)^{1-\alpha} \left(\tilde{\mathbb{E}}\, r_{x,N}^{\omega}\right)^{\alpha}.$$

Now, the first term is

$$\left(\tilde{\mathbb{E}}\left(\frac{\mathrm{d}\tilde{\mathbb{P}}}{\mathrm{d}\mathbb{P}}\right)^{1/(1-\alpha)}\right)^{1-\alpha} = \left(\tilde{\mathbb{E}}\ \exp\{(\delta_N(V\wedge 1) - g(\delta_N))/(1-\alpha)\}\right)^{(1-\alpha)|A_N|}.$$

However, the first order terms in δ_N cancel,

$$\begin{aligned}\tilde{\mathbb{E}}\ \exp\left\{\frac{\delta_N(V\wedge 1) - g(\delta_N)}{1-\alpha}\right\} &= \mathbb{E}\ \exp\left\{\frac{\alpha}{1-\alpha}\left(\delta_N(V\wedge 1) - g(\delta_N)\right)\right\} \\ &= \exp\left\{-g\left(-\frac{\alpha}{1-\alpha}\delta_N\right) - \frac{\alpha}{1-\alpha}g\left(\delta_N\right)\right\} \leq \exp\left\{\frac{\alpha}{(1-\alpha^2)^2}\delta_N^2\right\}.\end{aligned} \tag{77}$$

On the other hand (recall that $\mathscr{F}$ is the set of irreducible paths),

$$\tilde{\mathbb{E}}r^{\omega}_{x,N} \leq \tilde{\mathbb{E}}r^{\omega}_N = \left(\tilde{\mathbb{E}}\sum_{\gamma\in\mathscr{F}} q_{\lambda,h}(\gamma)\right)^N \stackrel{\Delta}{=} \tilde{f}(\delta_N)^N.$$

It is straightforward to check that $\tilde{f}'(0) < 0$. As a result, $\tilde{\mathbb{E}}r^{\omega}_N \leq \mathrm{e}^{-c\delta_N}$.
We are now ready to specify the choice of δ_N. We want to have simultaneously

$$\delta_N^2\,|A_N| \ll \delta_N N \quad \text{and} \quad \delta_N N \gg N^{\epsilon}.$$

The choice $\delta_N = N^{-1/2-2\epsilon}$ with $\epsilon \in (0, 1/3)$ qualifies, and (76) follows. □

Acknowledgements The research of DI was supported by the Israeli Science Foundation (grant No. 817/09). YV is partially supported by the Swiss National Science Foundation. The authors would like to thank the anonymous referee for a very careful reading and useful remarks.

References

1. Antal, P.: Enlargement of obstacles for the simple random walk. Ann. Probab. **23**(3), 1061–1101 (1995)
2. Bolthausen, E.: A note on diffusion of directed polymers in random environment. Comm. Math. Phys. **123**(4), 529–534 (1989)
3. Bolthausen, E., Sznitman, A.-S.: On the static and dynamic points of view for certain random walks in random environment. Methods Appl. Anal. **9**(3), 345–375 (2002)
4. Carmona, P., Hu, Y.: On the partition function of a directed polymer in a random environment. Probab. Theory Rel. Fields **124**(3), 431–457 (2002)
5. Comets, F., Yoshida, N.: Proabilistic analysis of directed polymers in random environment: a review. Stochastic analysis of large scale interacting systems. Adv. Stud. Pure. Math. **39**, 115–142 (2004)

6. Comets, F., Yoshida, N.: Directed polymers in random environment are diffusive at weak disorder. Ann. Probab. **34**(5), 1746–1770 (2006)
7. Comets, F., Shiga, T., Yoshida, N.: Directed polymers in a random environment: path localization and strong disorder. Bernoulli **9**(4), 705–723 (2003)
8. Flury, M.: Large deviations and phase transition for random walks in random nonnegative potentials. Stochastic Process. Appl. **117**(5), 596–612 (2007)
9. Flury, M.: A note on the ballistic limit of random motion in a random potential. Electron. Commun. Probab. **13**, 393–400 (2008)
10. Flury, M.: Coincidence of Lyapunov exponents for random walks in weak random potentials. Ann. Probab. **36**(4), 1528–1583 (2008)
11. Huse, D.A., Henley, C.: Pinning and roughening of domain walls in Ising systems due to random impurities. Phys. Rev. Lett. **54**(25), 2708–2711 (1985)
12. Ioffe, D., Velenik, Y.: In preparation
13. Ioffe, D., Velenik, Y.: Ballistic phase of self-interacting random walks. In: Analysis and Stochastics of Growth Processes and Interface Models, pp. 55–79. Oxford University Press, Oxford (2008)
14. Ioffe, D., Velenik, Y.: Crossing random walks and stretched polymers at weak disorder. Ann. Probab; arXiv:1002.4289 (to appear, 2012)
15. Kardar, M.: Roughening by impurities at finite temperatures. Phys. Rev. Lett. **55**(26), 2923 (1985)
16. Kosygina, E., Mountford, T., Zerner, M.: Lyapunov exponents of Green's functions for random potentials tending to zero. Prob. Theory Rel. Fields. **150**(1–2), 43–59 doi:10.1007/s00440-010-0266-y
17. Lacoin, H.: New bounds for the free energy of directed polymers in dimension $1+1$ and $1+2$. Comm. Math. Phys. **294**(2), 471–503 (2010)
18. Ledoux, M.: The concentration of measure phenomenon, vol. 89 of Mathematical Surveys and Monographs. American Mathematical Society, Providence, RI (2001)
19. Martin, P., Zerner, W.: Directional decay of the Green's function for a random nonnegative potential on $\mathbf{Z}^d$. Ann. Appl. Probab. **8**(1), 246–280 (1998)
20. McLeish, D.L.: A maximal inequality and dependent strong laws. Ann. Probab. **3**(3), 829–839 (1975)
21. Sinai, Y.G.: A remark concerning random walks with random potentials. Fund. Math **147**, 173 –180 (1995)
22. Song, R., Zhou, X.Y.: A remark on diffusion of directed polymers in random environments. J. Statist. Phys. **85**(1–2), 277–289 (1996)
23. Sznitman, A.-S.: Brownian motion, obstacles and random media. Springer Monographs in Mathematics. Springer, Berlin (1998)
24. Trachsler, M.: Phase Transitions and Fluctuations for Random Walks with Drift in Random Potentials. PhD thesis, Universität Zürich (1999)
25. Vargas, V.: A local limit theorem for directed polymers in random media: the continuous and the discrete case. Ann. Inst. H. Poincaré Probab. Statist. **42**(5), 521–534 (2006)
26. Vargas, V.: Strong localization and macroscopic atoms for directed polymers. Probab. Theory Related Fields **138**(3–4), 391–410 (2007)
27. Wouts, M.: Surface tension in the dilute Ising model. The Wulff construction. Comm. Math. Phys. **289**(1), 157–204 (2009)
28. Zygouras, N.: Lyapounov norms for random walks in low disorder and dimension greater than three. Probab. Theory Related Fields **143**(3–4), 615–642 (2009)
29. Zygouras, N.: Strong disorder in semidirected random polymers. preprint, arXiv:1009.2693 (2010)

Part III
Branching Processes

Multiscale Analysis: Fisher–Wright Diffusions with Rare Mutations and Selection, Logistic Branching System

Donald A. Dawson and Andreas Greven

Abstract We study two types of stochastic processes, first a mean-field spatial system of interacting Fisher–Wright diffusions with an inferior and an advantageous type with rare mutation (inferior to advantageous) and selection and second a mean-field spatial system of supercritical branching random walks with an additional death rate, which is quadratic in the local number of particles. The former describes a standard two-type population under selection, mutation and the latter model describes a population under scarce resources causing additional death at high local population intensity. Geographic space is modelled by $\{1, \ldots, N\}$. The first process starts in an initial state with only the inferior type present or an exchangeable configuration and the second one with a single initial particle. This material is a special case of the theory developed in [8] and describes the results of Section 7 therein.

We study the behaviour in two time windows, first between time 0 and T and second after a large time when in the Fisher–Wright model the rare mutants succeed, respectively, in the branching random walk the particle population reaches a positive spatial intensity. It is shown that asymptotically as $N \to \infty$ the second phase for both models sets in after time $\alpha^{-1} \log N$, if N is the size of geographic space and N^{-1} the rare mutation rate and $\alpha \in (0, \infty)$ depends on the other parameters. We identify the limit dynamics as $N \to \infty$ in both time windows and for both models as a nonlinear Markov dynamic (McKean–Vlasov dynamic), respectively, a corresponding random entrance law from time $-\infty$ of this dynamic.

D.A. Dawson (✉)
School of Mathematics and Statistics, Carleton University, Ottawa K1S 5B6, Canada
e-mail: ddawson@math.carleton.ca

A. Greven
Department Mathematik, Universität Erlangen-Nürnberg, Bismarckstraße 1 1/2, D-91054 Erlangen, Germany
e-mail: greven@mi.uni-erlangen.de

J.-D. Deuschel et al. (eds.), *Probability in Complex Physical Systems*, Springer Proceedings in Mathematics 11, DOI 10.1007/978-3-642-23811-6_15,

Finally, we explain that the two processes are just two sides of the very same coin, a fact arising from a new form of duality, in particular the particle model generates the genealogy of the Fisher–Wright diffusions with selection and mutation. We discuss the extension of this duality in relation to a multitype model with more than two types.

1 Motivation and Background

We study here features of the longtime behaviour of two models for the stochastic evolution of populations, the classical (mean-field) spatial version of a system of interacting Fisher–Wright diffusions with selection and mutation, on the one hand, and a logistic spatial branching particle model, on the other hand. Both populations live on the geographic space $\{1, \dots, N\}$. We shall later explain the mathematical relation between them. We describe now the two models and the two main new mathematical methods shortly.

Fisher–Wright model. This process comes from population genetics and models a population of individuals of two types evolving under migration, resampling, selection and mutation. It is the many individual limit of a discrete model. Migration here means individuals move in geographic space, resampling that pairs are replaced by an offspring pair each choosing a parent at random and adopting the parents type, mutation is a spontaneous change of type of an individual and under selection the choice of the parent in the resampling event is biased according to the parents' fitness. For more information on this model, compare [3, 14].

Here, we are particularly concerned with a situation where we have an inferior type and an advantageous type. The case we are interested in is that the mutation rate from inferior to advantageous is very small so that in finite time we expect $O(1)$-many mutations in the complete collection of sites as the number of sites gets large. We want to follow the population through the *emergence* and *fixation* of the whole population in the advantageous type.

Logistic branching model. Here, we consider a population of particles (all of the same type) which migrate in space, have offspring at a certain rate s, die with a certain rate (all particles act independently) but here the risk of death is at a rate increasing with the population size of the site and being zero for one particle. The latter mechanism induces an interaction of the families, which therefore do not evolve anymore independently of each other.

Here for this model we want to see how the population starting from one particle spreads and eventually colonizes the whole space as time evolves, meaning that a positive spatial intensity is reached and a local equilibrium situation arises where locally the process neither becomes extinct nor grows and becomes infinitely large as both N and $t \to \infty$. This is in contrast to the behaviour of classical branching models with their survival versus extinction dichotomy in finite geographic space and reflects the limited resources in a given colony.

McKean–Vlasov equation and random entrance laws. The techniques we use to study the questions raised above for two models is the mean-field limit, where we choose migration to occur according to the uniform distribution on $\{1, \dots, N\}$ and rare mutation having rate N^{-1} and where we let $N \to \infty$. As limit dynamics a nonlinear Markov process of the type first introduced by McKean [22] arises, i.e. an evolution with a generator where the parameters depend on the current state of the process *and* on the current law. The transition probabilities solve the *McKean–Vlasov equation* which has a similar structure to the equations introduced by Vlasov to describe the dynamics of a plasma consisting of charged particles with long-range interaction. A similar scenario holds for the branching particle system.

The first rigorous and systematic analysis of mean-field models and resulting McKean–Vlasov dynamics is Gärtner's fundamental paper in 1988 [16], which established the existence and uniqueness of weak solutions for a general class of McKean–Vlasov equations and the associated non-linear martingale problems but under the condition that the diffusion matrix is *strictly* positive definite. Hence, it does not cover the case dealt with in this paper due to the fact that for the Fisher–Wright diffusions the diffusion function vanishes at the boundary. The first mean-field limit result for neutral Fisher–Wright models is in [4, 5] and with selection and mutation in [7].

We must also extend here the methodology of the McKean–Vlasov equation further in order to describe the limiting behaviour in different time windows. In order to also consider a late time window we need first of all to introduce the notion of an entrance law from time $-\infty$ (our time parameter has the form $T_N + t,\ t \in \mathbb{R}$ and $T_N \to \infty$ as $N \to \infty$) but since in the initial time phase some randomness is involved (rare mutation, respectively, very small early particle intensity) we even have to work with *random entrance laws* to the McKean–Vlasov equation.

Duality. The mathematical structure which relates our two models is the fact that they are in duality, i.e. expectations of certain functionals under one dynamic are given by expectations of appropriate functionals under the other dynamic with the time direction reversed. This will be explained in detail later on. For the use of duality in population models, compare [9] and [2].

The duality relation which we present here is a special case of a broader new duality theory for Fleming–Viot models with selection and mutation, which allows a historical and genealogical interpretation and which has been developed in [8] covering much more general situations than we can discuss here. For further extensions, see [13]. More information on the particle system can be found in [26].

Remark 1. In the framework just sketched, it is possible to analyse the features of the population as described above. If one wants to adapt a more realistic model for geographic space, the mean-field limit has to be replaced by the *hierarchical mean-field limit* for which the present analysis is a key ingredient (see [2, 6, 9] for more on this technique). Then it is possible to study asymptotically two-dimensional geographic space via its approximations by the hierarchical group of order N and $N \to \infty$. This also allows us to investigate the question of *universality* of the behaviour. All this is carried out in [8] and we refer the interested reader to this paper.

1.1 Outline

In Sect. 2, we shall present the Fisher–Wright model; in Sect. 3, the logistic branching model and in Sect. 4 the connection between both via duality.

2 The Fisher–Wright Model with Rare Mutation and Selection: Behaviour in Two Time Windows

Here, we introduce the Fisher–Wright model, some relevant time windows, state its properties in two time windows and consider the McKean–Vlasov limit for the model in five separate subsections.

2.1 A Two-Type Mean-Field Diffusion Model and Its Description

We study a population with two types, type 1 of low fitness and type 2 of high fitness, where initially all the population is of the inferior type but by a rare mutation the advantageous type appears and spreads with time. The population is described specifying the proportion of the two types (here type 1 is the inferior) in every spatial colony.

Formally we look at a process $(X^N(t))_{t\geq 0}$ with $N \in \mathbb{N}$ of the following form:

$$X^N(t) = ((x_1^N(i,t), x_2^N(i,t)); i = 1, \dots, N), \tag{1}$$

$$x_1^N(i,0) = 1, \quad i = 1, \dots, N. \tag{2}$$

Let $c, m, d, s, L \in (0, \infty)$ be parameters. Then X^N satisfies the well-known SSDE:

$$\begin{aligned} dx_1^N(i,t) &= c(\bar{x}_1^N(t) - x_1^N(i,t))dt - s x_1^N(i,t) x_2^N(i,t)dt - \frac{m}{L} x_1^N(i,t)dt \\ &\quad + \sqrt{d \cdot x_1^N(i,t) x_2^N(i,t)}\,dw_1(i,t), \end{aligned} \tag{3}$$

$$\begin{aligned} dx_2^N(i,t) &= c(\bar{x}_2^N(t) - x_2^N(i,t))dt + s x_1^N(i,t) x_2^N(i,t)dt + \frac{m}{L} x_1^N(i,t)dt \\ &\quad + \sqrt{d \cdot x_1^N(i,t) x_2^N(i,t)}\,dw_2(i,t), \end{aligned} \tag{4}$$

where w_2 defined in terms of w_1, namely $w_2(i,t) = -w_1(i,t)$ and $\{(w_1(i,t))_{t\geq 0}, i = 1, \dots, N\}$ are i.i.d. Brownian motions. Furthermore,

$$\bar{x}_\ell^N(t) = \frac{1}{N} \widehat{x}_\ell^N(t) \text{ with } \ell = 1, 2, \quad \widehat{x}_\ell^N(t) = \sum_{i=1}^N x_\ell^N(i,t). \tag{5}$$

Note that $x_2(i,t)$ will be positive for $t > 0$ with probability 1. Later we will use the parameter of mutation strength and size of geographic space satisfying

$$L = N. \tag{6}$$

We can study this system in various ways, *locally* by looking at K tagged sites

$$(x_\ell^N(1,t),\dots,x_\ell^N(K,t)), \quad \ell = 1,2 \tag{7}$$

or *globally* using the concept of the empirical measure of the complete population:

$$Xi_N(t) := \frac{1}{N}\sum_{i=1}^{N} \delta_{(x_1^N(i,t),x_2^N(i,t))} \in \mathscr{P}(\mathscr{P}(\{1,2\})), \tag{8}$$

and the empirical measure process of either type:

$$\Xi_N(t,\ell) := \frac{1}{N}\sum_{i=1}^{N} \delta_{x_\ell^N(i,t)} \in \mathscr{P}([0,1]),\ \ell = 1,2. \tag{9}$$

Note that since $x_1^N(i,t) + x_2^N(i,t) = 1$ it suffices in the case of two types to know one component of the pair in (9), the other is then determined by this condition. Then for two types we can effectively replace $\mathscr{P}(\mathscr{P}(\{1,2\}))$ by $\mathscr{P}([0,1])$ as states of the empirical measure.

However since very soon rare mutants appear somewhere in space they are present for $t > 0$, but of course the set of sites where they are visible, i.e. exceed a given threshold $\delta > 0$, is *sparse*, we have to find a way to describe this small subset where the advantageous type appears. One (global) way is to consider the process

$$\widehat{x}_\ell^N(t) = \sum_{i=1}^{N} x_\ell^N(i,t) \text{ with } \ell = 1,2, \tag{10}$$

but this way we lose the internal structure of the droplet of sites with advantageous types of substantial mass. In order to keep track of the sparse set of sites at which nontrivial mass appears, we will give a random label to each site and define the following *atomic-measure-valued process*.

We assign independent of the process a point $a(j)$ randomly in $[0,1]$ to each site $j \in \{1,\dots,N\}$, that is, we introduce

$$\{a(j), \quad j = 1,\dots,N\} \text{ which is a collection of i.i.d. uniform random variables on } [0,1]. \tag{11}$$

We then associate with our process and a realization of the random labels a measure-valued process on $\mathscr{P}([0,1])$, which we denote by $(\gimel_t^{N,m})_{t\geq 0}$, where

$$\gimel_t^{N,m} = \sum_{j=1}^{N} x_2^N(j,t)\delta_{a(j)}. \tag{12}$$

The process $(\gimel_t^{N,m})_{t\geq 0}$ describes the essential features of the advantageous population.

2.2 *Two Time Windows for the Spread of the Advantageous Type*

The first question now is how the advantageous type develops in finite time and the second on what time scale does the advantageous type take over the whole population. This means that we want to identify a time $T_{N,L}^{\varepsilon}$ for which the advantageous type has first positive spatial intensity, i.e. $\bar{x}_2(T_{N,L}^{\varepsilon}) \geq \varepsilon$ for a threshold ε we think of as very small. This time we call the time of *emergence*. A key role is played by space and the fact that the dynamic is random and not deterministic as in the scenario looked at by [1]. So we look at the cases $N = 1$, N large and L small, $d = 0, d > 0$.

Case $N = 1$ (nonspatial).

Here we find the time scales in which we emerge ($T_{N,L} = T_L$) as $L \to \infty$ to be

$$O(s^{-1}\log L) \text{ for } d = 0, \tag{13}$$

$$O(L) \text{ for } d > 0. \tag{14}$$

This qualitative difference is essentially due to the fact that in the stochastic model the diffusion can hit 0 by sheer random effects, while in the deterministic model the advantageous type expands exponentially fast leading to a $\log L$ time scale.

Case $N \gg 1$ (large spatial model).

Here we take $L = N(T_{N,L} = T_N)$ and then we find also in the deterministic case the time scale $s^{-1}\log N$, but now in the stochastic model, $d > 0$, we have again a time period $\alpha^{-1}\log N$ but the constant α will turn out to be strictly smaller than s. In the sequel, we analyse the latter case in more detail and identify the constant α in terms of the model parameters.

Remark 2. The relation $L = N^{-1}$ is appropriate if one considers the mean-field model which in the hierarchical mean-field limit is the potential-theoretic analogue of two-dimensional space as $N \to \infty$, see [8] and for which all three mechanisms migration, mutation and selection are important.

The task is now to describe the system in the limit $N \to \infty$ in two time windows:

- Times of order 1 after starting: $(X^N(t))_{t \geq 0}$
- Times $\alpha^{-1} \log N$ after starting: $(X^N(\alpha^{-1} \log N + t)^+)_{t \in \mathbb{R}}$.

In the first window, the evolution of the small advantageous droplet is of primary interest, in other words the process $\gimel^{N,m}$, while in the second case when the droplet covers a positive proportion of the whole space, the system is best described by the empirical measure Ξ_N (see 8) or the tagged sample see (7).

2.3 *The Early Time Window as* $N \to \infty$

Here, we want to describe (1) the evolution of the atomic measure-valued process $\gimel^{N,m}$ in the limit $N \to \infty$ over times in some finite interval and (2) the limit $N \to \infty$ of the dynamic of the empirical measure $\Xi_N(t, 1)$.

(1) *Droplet evolution*

During the early times, there will be a finite random number of sites where the advantageous type has mass exceeding some prescribed $\varepsilon > 0$. Therefore as $N \to \infty$ we expect a limiting evolution of $(\gimel_t^N)_{t \geq 0}$. First of all, we can show the following where we denote by $\mathcal{M}_a(E)$ the space of atomic measures on the topological space E.

Proposition 1. *(Limiting droplet dynamic)*
As $N \to \infty$

$$\mathcal{L}[(\gimel_t^{N,m})_{t \geq 0}] \underset{N \to \infty}{\Longrightarrow} \mathcal{L}[(\gimel_t^m)_{t \geq 0}], \tag{15}$$

in the sense of convergence of continuous $\mathcal{M}_a([0,1])$*-valued processes where the set of atomic measures* $\mathcal{M}_a([0,1])$ *is equipped with the weak atomic topology.* □

Remark 3. The weak atomic topology on $\mathcal{M}_a([0,1])$ is introduced in [15] and we do not expand on this here.

To identify the limit evolution $\gimel^m$, we need a bit of classical excursion theory (compare [24] Sect. 3, [19]).

Lemma 1. *(Single site: entrance and excursion laws)*

(a) Let $c > 0, d > 0,\ s > 0$ *and let* W *be Brownian motion. Consider the diffusion*

$$\mathrm{d}x(t) = -cx(t)\mathrm{d}t + sx(t)(1 - x(t))\mathrm{d}t + \sqrt{d \cdot x(t)(1 - x(t))}\mathrm{d}w(t). \tag{16}$$

Then 0 *is an exit boundary with a* σ*-finite entrance law from state* 0 *at time 0, denoted*

$$\mathbb{Q} = \mathbb{Q}^{c,d,s} \text{ a measure on } W_0, \tag{17}$$

$$W_0 := \{w \in C([0,\infty),\mathbb{R}^+),\ w(0)=0,\ w(t)>0\, for\, 0<t<\zeta$$
$$for\ some\ \zeta \in (0,\infty)\}. \tag{18}$$

(b) Denote by P^ε the law of the process started with $w(0)=\varepsilon$ and $\varepsilon>0$, then:

$$\mathbb{Q}(\cdot) = \lim_{\varepsilon\to 0} \frac{P^\varepsilon(\cdot)}{S(\varepsilon)}, \tag{19}$$

where $S(\cdot)$ is the scale function of the diffusion (16), defined by the relation,

$$P_\varepsilon(T_\eta < \infty) = \frac{S(\varepsilon)}{S(\eta)}, \qquad 0<\varepsilon<\eta<\infty, \qquad T_\eta = \inf(t|x(t)=\eta). \tag{20}$$

For the diffusion in (16), S is given by the initial value problem (cf. [25], V28)):

$$S(0)=0, \frac{\mathrm{d}S}{\mathrm{d}x} = \frac{\mathrm{e}^{-2sx}}{(1-x)^{2c}}, \left(\text{so that } \lim_{\varepsilon\to 0}\frac{S(\varepsilon)}{\varepsilon}=1\right). \tag{21}$$

(c) The measure $\mathbb{Q}$ satisfies, for any $\eta>0, \zeta$ as in (18),

$$\mathbb{Q}(\{w : \zeta(w) > \eta\}) < \infty, \tag{22}$$

$$\mathbb{Q}(\sup_t(w(t)) > \eta) = \mathbb{Q}(T_\eta < \infty) = \frac{1}{S(\eta)} \longrightarrow \infty\ as\ \eta \to 0, \tag{23}$$

$$\int_0^1 x\mathbb{Q}(\sup_t(w(t)) \in \mathrm{d}x) = \infty. \tag{24}$$

□

Now the limit ($N\to\infty$) droplet dynamic $\gimel^m$ can be identified as follows.

Proposition 2. *(A continuous atomic-measure-valued Markov process)*
Let $N(ds, da, \mathrm{d}u, \mathrm{d}w)$ be a Poisson random measure on (recall (18) for W_0)

$$[0,\infty)\times[0,1]\times[0,\infty)\times W_0, \tag{25}$$

with intensity measure

$$\mathrm{d}s\,\mathrm{d}a\,\mathrm{d}u\,\mathbb{Q}(\mathrm{d}w), \tag{26}$$

where $\mathbb{Q}$ is the single site excursion law defined in (19) in Lemma 1). Then the following two properties hold.

(a) The stochastic integral equation for $(\gimel^m_t)_{t\ge 0}$ given as

$$\gimel^m_t = \int_0^t \int_{[0,1]} \int_0^{q(s,a)} \int_{W_0} w(t-s)\delta_a N(\mathrm{d}s,\mathrm{d}a,\mathrm{d}u,\mathrm{d}w), \quad t\ge 0, \tag{27}$$

where $q(s,a)$ denotes the non-negative predictable function

$$q(s,a) := (m + c\gimel^m_{s-}([0,1])), \tag{28}$$

has a unique continuous $\mathscr{M}_a([0,1])$-valued solution with law which equals $\mathscr{L}[(\gimel^m_t)_{t\geq 0}]$ defined by (15).

(b) The process $(\gimel^m_t)_{t\geq 0}$ has the following properties:

- *Branching property (two pieces of mass evolve independently)*
- *The mass of each atom observed from the time of its creation follows an excursion from zero generated from the excursion law $\mathbb{Q}$ (see (19)),*
- *New excursions are produced at time t at rate*

$$m + c\gimel^m_t([0,1]). \tag{29}$$

- *Each new excursion produces an atom located at a point $a \in [0,1]$ chosen according to the uniform distribution on $[0,1]$,*
- *At each t and $\varepsilon > 0$ there are at most finitely many atoms of size $\geq \varepsilon$.* □

(2) *McKean–Vlasov equation for limiting empirical measure.*

We now turn to the global description of the complete population by the empirical measure. In the limit $N \to \infty$, the evolution of the empirical measure in a finite time window is given by the McKean–Vlasov limit of our system, but of course it is trivial, that is, totally concentrated on type 1 if the given initial state has this property. This is however different at late times. Consider therefore the above system (1)-(4) of N interacting sites with type space $\{1,2\}$ starting at time $t = 0$ from a product measure (that is, i.i.d. initial values at the N sites). The *basic McKean–Vlasov limit theorem* (cf. [7], Theorem 9) says that if we start initially in an i.i.d. distribution, then

$$\{\Xi_N(t)\}_{0\leq t\leq T} \underset{N\to\infty}{\Longrightarrow} \{\mathscr{L}_t\}_{0\leq t\leq T}, \tag{30}$$

where the $\mathscr{P}(\mathscr{P}(\{1,2\})$-valued *deterministic* path $\{\mathscr{L}_t\}_{0\leq t\leq T}$ is the law of a *nonlinear* Markov process solving a *forward* equation, namely the unique weak solution of the *McKean–Vlasov equation*:

$$\frac{d\mathscr{L}_t}{\mathrm{d}t} = (L_t^{\mathscr{L}_t})^* \mathscr{L}_t, \tag{31}$$

where for $\pi \in \mathscr{P}(\mathscr{P}(\{1,2\}))$, L^π is given by the generator of the process given by the evolution of type 1 in (33) below, (and $\pi = m(t)$) and the $*$ indicates the adjoint of an operator mapping from a dense subspace of $C_b(E,\mathbb{R})$ into $C_b(E,\mathbb{R})$ w.r.t. the pairing of $\mathscr{P}(E)$ and $C_b(E,\mathbb{R})$ given by the integral of the function with respect to the measure.

As pointed out above in (9), in the special case of the type set $\{1,2\}$, we can simplify by considering only the frequency of type 2 and by reformulating (31) living on $\mathscr{P}(\mathscr{P}(\{1,2\}))$ in terms of $\mathscr{L}_t(2) \in \mathscr{P}[0,1]$. This we now carry out.

Namely we note that given the mean-curve

$$m(t) = \int_{[0,1]} y\, \mathscr{L}_t(2)(\mathrm{d}y), \tag{32}$$

the process $(\mathscr{L}_t(2))_{t \geq 0}$ is the *law* of the solution of (i.e. the unique weak solution) the SDE:

$$\mathrm{d}y(t) = c(m(t) - y(t))\mathrm{d}t + sy(t)(1 - y(t))\mathrm{d}t + \sqrt{\mathrm{d}y(t)(1 - y(t)}\mathrm{d}w(t), \tag{33}$$

with w being standard BM. Then informally $(\mathscr{L}_t)_{t \geq 0}$ corresponds to the solution of the nonlinear diffusion equation. Namely for $t > 0$, $\mathscr{L}_t(2)(\cdot)$ is absolutely continuous and for

$$\mathscr{L}_t(2)(\mathrm{d}x) = u(t,x)\mathrm{d}x \; \in \mathscr{P}([0,1]), \tag{34}$$

the evolution equation of the probability density $u(t,\cdot)$ (the forward equation) is given by:

$$\begin{aligned} \frac{\partial}{\partial t} u(t,x) = & -c \frac{\partial}{\partial x} \left\{ \left[\int_{[0,1]} y u(t,y)\mathrm{d}y - x \right] u(t,x) \right\} \\ & - s \frac{\partial}{\partial x}(x(1-x)u(t,x)) + \frac{d}{2} \frac{\partial^2}{\partial x^2}(x(1-x)u(t,x)). \end{aligned} \tag{35}$$

We have the following basic property for the McKean–Vlasov equation.

Proposition 3. *(McKean–Vlasov equation and its solution)*

(a) Given the initial state $\mu_0 \in \mathscr{P}([0,1])$ there exists a unique solution

$$\mathscr{L}_t(2)(\mathrm{d}x) = \mu_t(\mathrm{d}x), \quad t \geq t_0, \tag{36}$$

to (31) with initial condition $\mathscr{L}_{t_0}(2) = \mu_0$.

(b) If $s > 0$ and $\int_{[0,1]} x\mu_{t_0}(\mathrm{d}x) > 0$, then this solution satisfies:

$$\lim_{t \to \infty} \mathscr{L}_t(2)(\mathrm{d}x) = \delta_1(\mathrm{d}x). \tag{37}$$

□

2.4 The Late Time Window as $N \to \infty$

In the late time window, we see (global) *emergence* and then *fixation* of the advantageous type.

(1) *Emergence times.*

Here we begin by studying emergence by identifying the time of emergence and of fixation of the advantageous type as follows.

Proposition 4. *(Macroscopic emergence and fixation times)*
(a) (Emergence-time)

There exists a constant α with:

$$0 < \alpha < s, \tag{38}$$

such that if $T_N = \frac{1}{\alpha}\log N$, then for $t \in \mathbb{R}$ and asymptotically as $N \to \infty$, type 2 is present at times $T_N + t$, i.e. there exists a $\varepsilon > 0$ such that for every i,

$$\liminf_{N\to\infty} P[x_2^N(i, T_N + t) > \varepsilon] > 0, \tag{39}$$

and type 2 is not present earlier, namely for $1 > \varepsilon > 0$:

$$\lim_{t\to-\infty} \limsup_{N\to\infty} [P(x_1^N(i, T_N + t) < 1 - \varepsilon] = 0. \tag{40}$$

(b) (Fixation time)

After emergence the fixation occurs in times $O(1)$ as $N \to \infty$, i.e. for any $\varepsilon > 0$, $i \in \mathbb{N}$

$$\lim_{t\to\infty} \limsup_{N\to\infty} P[x_1^N(i, T_N + t) > \varepsilon] = 0. \tag{41}$$

□

Corollary 1. *(Emergence and fixation times of spatial density)*
The relations (39), (40) and (41) hold for $\bar{x}_2^N$, respectively, $\bar{x}_1^N$ as well. □

We can identify the parameter α as follows from the droplet growth behaviour (recall (15)).

Proposition 5. *(Long-time growth behaviour of $\mathfrak{J}_t^m$)*
Assume that $m > 0$. Then the following growth behaviour of $\mathfrak{J}^m$ holds.

(a) There exists an α^* such that the following limit exists

$$\lim_{t\to\infty} \mathrm{e}^{-\alpha^* t} E[\mathfrak{J}_t^m([0,1])] \in (0, \infty), \tag{42}$$

with (here α is from (38))

$$\alpha^* = \alpha, \text{ where } \alpha \text{is given below in (92)} \tag{43}$$

$$e^{-\alpha t}\mathfrak{J}_t^m([0,1]) \underset{t\to\infty}{\Longrightarrow} \mathscr{W}^* \text{in probability}, \ 0 < \mathscr{W}^* < \infty \text{a.s.} \tag{44}$$

(b) The growth factor in the exponential is truly random:

$$0 < Var[\lim_{t\to\infty} \mathrm{e}^{-\alpha t}\mathfrak{J}_t^m([0,1])] < \infty. \tag{45}$$

□

Remark 4. The random variable $\mathscr{W}^*$ reflects the growth of $\mathfrak{J}_t^m([0,1])$ in the beginning, as is the case in a supercritical branching process and hence $\mathscr{E}^* = \alpha^{-1}\log \mathscr{W}^*$ can be viewed as the random time shift of that exponential $e^{\alpha t}$, which matches the total mass of $\mathfrak{J}_t^m$ for large t.

Remark 5. Even though one might think that for small mass of the advantageous type this expands at the rate e^{st}, this is not the case due to the stochastic effects leading to a subtle interplay between the parameters s, d and c resulting in $\alpha^* = \alpha < s$.

(2) *Fixation dynamic.*

We now understand the pre-emergence situation and the time of emergence and fixation. In order to describe the whole dynamics of *macroscopic fixation*, we consider the limiting distributions of the empirical measure-valued processes in a second time window $(\alpha^{-1}\log N + t)^+$, $t \in \mathbb{R}$. Define (we suppress the truncation in the notation below)

$$\Xi_N^{\log,\alpha}(t) := \frac{1}{N}\sum_{i=1}^N \delta_{\left(x_1^N\left(i,\frac{\log N}{\alpha}+t\right),x_2^N\left(i,\frac{\log N}{\alpha}+t\right)\right)},$$

$$t \in \mathbb{R}, (\Xi_N^{\log,\alpha}(t) \in \mathscr{P}(\mathscr{P}(\{0,1\}))), \tag{46}$$

and then the two empirical marginals are given as

$$\Xi_N^{\log,\alpha}(t,\ell) := \frac{1}{N}\sum_{i=1}^N \delta_{x_\ell^N(i,\frac{\log N}{\alpha}+t)}, \ell = 1,2 \text{ and } t \in \mathbb{R}, \quad (\Xi_N^{\log,\alpha}(t,\ell) \in \mathscr{P}([0,1])). \tag{47}$$

Note that for each t and given ℓ the latter is a random measure on $[0,1]$. Furthermore, we have the representation of the empirical mean of type 2 as follows:

$$\bar{x}_2^N\left(\frac{\log N}{\alpha}+t\right) = \int_{[0,1]} x\ \Xi_N^{\log,\alpha}(t,2)(\mathrm{d}x). \tag{48}$$

Since we consider the limits of systems observed in the interval $const \cdot \log N + [-\frac{T}{2}, \frac{T}{2}]$ with T any positive number, that is setting $t_0(N) = \text{const}\cdot\log N - \frac{T}{2}$, we need to identify entrance laws for the process from $-\infty$ (by considering $T \to \infty$) out of the state concentrated on type 1 with certain additional properties.

The next main result is on the fixation stage in which $\Xi_N^{\log,\alpha}$ converges as $N \to \infty$ and the resulting limit can be explicitly identified as a random McKean–Vlasov entrance law from $-\infty$, a concept now we explain.

Definition 1. (Entrance law from $t = -\infty$)

In the two-type case, we say that a probability measure-valued function $\mathscr{L} : \mathbb{R} \to \mathscr{P}([0,1])$, is an entrance law at $-\infty$ starting from type 1 if $(\mathscr{L}_t)_{t\in(-\infty,\infty)}$

is such that $\mathscr{L}_t$ solves the McKean–Vlasov equation (31) and $\mathscr{L}_t \to \delta_1$ as $t \to -\infty$. □

We will indeed establish that the emergence of rare mutants gives rise to "random" solutions of the McKean–Vlasov dynamics. In particular, we will show that the limiting empirical measures at times of the form $C \log N + t$ are *random* probability measures on $[0, 1]$ and therefore given by sequences of $[0, 1]$-valued truly exchangeable random variables which are *not* i.i.d., that is, the exchangeable σ-algebra is not trivial. This means that the limiting empirical mean turns out to be a *random variable* and this is the driving term due to migration for the local evolution of a site in the McKean–Vlasov limit. Therefore, both the random driving term and the non-linearity of the evolution equation come seriously into play. However, once we condition on the exchangeable σ-algebra, we then get for the further evolution again a deterministic limiting equation for the empirical measures, namely the McKean–Vlasov equation. The reason for this is the fact that conditioned on the exchangeable σ-algebra we obtain an asymptotically (as $N \to \infty$) i.i.d. configuration to which the classical convergence theorem applies. Using the Feller property of the system, we get our claim. This leads to the task of identifying an entrance law in terms of a *random initial condition at time* $-\infty$.

The above discussion shows that we need to introduce the notion of a truly random McKean–Vlasov entrance law from $-\infty$.

Definition 2. (Random entrance laws of McKean–Vlasov from $t = -\infty$)
We say that the probability measure-valued process $\{\mathscr{L}(t)\}_{t\in\mathbb{R}}$ is a *random solution of the McKean–Vlasov equation* (31) if

- $\{\mathscr{L}_t : t \in \mathbb{R}\}$ is a.s. a solution to (31), that is, for every t_0 the distribution of $\{\mathscr{L}_t : t \geq t_0\}$ conditioned on $\mathscr{F}_{t_0} = \sigma\{\mathscr{L}_s : s \leq t_0\}$ is given by $\delta_{\{\mu_t\}_{t\geq t_0}}$, where μ_t is a solution of the McKean–Vlasov equation with $\mu_{t_0} = \mathscr{L}_{t_0}$,
- The time t marginal distributions of $\{\mathscr{L}_t : t \in \mathbb{R}\}$ are truly random. □

The emergence of the advantageous type and the subsequent evolution to fixation in this type is characterized as follows.

Proposition 6. *(Asymptotic macroscopic emergence-fixation process)*

(a) For each $-\infty < t < \infty$ the empirical measures converge weakly to a random measure:

$$\mathscr{L}[\{\Xi_N^{\log,\alpha}(t,\ell)\}] \underset{N\to\infty}{\Longrightarrow} \mathscr{L}[\{\mathscr{L}_t(\ell)\}] = P_t^\ell \in \mathscr{P}(\mathscr{P}([0,1])), \text{for } \ell = 1, 2. \tag{49}$$

In addition, we have path convergence:

$$w - \lim_{N\to\infty} \mathscr{L}[(\Xi_N^{\log,\alpha}(t))_{t\in\mathbb{R}}] = P \in \mathscr{P}[C((-\infty,\infty), \mathscr{P}(\mathscr{P}(\{0,1\})))]. \tag{50}$$

A realization of P is denoted $(\mathscr{L}_t)_{t\in\mathbb{R}}$, respectively, its marginal processes $(\mathscr{L}_t(1))_{t\in\mathbb{R}}$, $(\mathscr{L}_t(2))_{t\in\mathbb{R}}$.

(b) *The process $(\mathcal{L}_t)_{t\in\mathbb{R}}$ describes the emergence and fixation dynamics, that is, for $t \in \mathbb{R}$, and $\varepsilon > 0$,*

$$\lim_{t\to-\infty} \text{Prob}[\mathcal{L}_t(2)((\varepsilon,1]) > \varepsilon] = 0, \tag{51}$$

$$\lim_{t\to\infty} \text{Prob}[\mathcal{L}_t(2)([1-\varepsilon,1]) < 1-\varepsilon] = 0, \tag{52}$$

with

$$\mathcal{L}_t(2)((0,1)) > 0 \quad , \quad \forall t \in \mathbb{R},\ a.s. \tag{53}$$

(c) *The limiting dynamic in (50) is identified as follows:*
The probability measure P in (50) is such that the canonical process is a random solution (recall Definition 2) and entrance law from time $-\infty$ to the McKean–Vlasov (31).

(d) *The limiting dynamic in (50) satisfies with α as in (38):*

$$\mathcal{L}[\mathrm{e}^{\alpha|t|} \int_{[0,1]} x\mathcal{L}_t(2)(\mathrm{d}x)] \Rightarrow \mathcal{L}[{}^*\mathcal{W}] \text{as } t \to -\infty, \tag{54}$$

and we explicitly identify the random element generating P in (50), namely P arises from random shift of a deterministic path:

$$P = \mathcal{L}[\tau_{{}^*\mathcal{E}}\mathcal{L}^*], \quad {}^*\mathcal{E} = (\log {}^*\mathcal{W})/\alpha, \quad \tau_r \textit{ is the time-shift of path by } r, \tag{55}$$

where $\mathcal{L}^*$ is the unique and deterministic entrance law of the McKean–Vlasov equation (31) with projection $\mathcal{L}_t^*(2)$ on the type 2 coordinate satisfying:

$$\mathrm{e}^{\alpha|t|} \int_{[0,1]} x\mathcal{L}_t^*(2)(\mathrm{d}x) \longrightarrow 1, \text{as } t \to -\infty. \tag{56}$$

The random variable ${}^*\mathcal{W}$ satisfies

$$0 < {}^*\mathcal{W} < \infty \text{ a.s.}, \quad E[{}^*\mathcal{W}] < \infty, \quad 0 < Var({}^*\mathcal{W}) < \infty. \tag{57}$$

(e) *We have for $s_N \to \infty$ with $s_N = o(\log N)$ the approximation property for the growth behaviour of the limit dynamic by the finite N model, namely:*

$$\mathcal{L}[\mathrm{e}^{\alpha s_N} \bar{x}_2^N \left(\frac{\log N}{\alpha} - s_N\right)] \underset{N\to\infty}{\Longrightarrow} \mathcal{L}[{}^*\mathcal{W}]. \tag{58}$$

□

How does this emergence in (58) relate to the droplet growth? We have

Proposition 7. *(Microscopic emergence and evolution: droplet formation)*
The total type-2 mass $\beth^{N,m}_{t_N}([0,1])$ *grows at exponential rate* α,

$$\mathscr{L}\left[\beth^{N,m}_{t_N}([0,1])e^{-\alpha t_N}\right] \underset{N\to\infty}{\Longrightarrow} \mathscr{L}[\mathscr{W}^*], \; \textit{for } t_N \uparrow \infty \textit{ with } t_N-(\alpha^{-1}\log N) \longrightarrow -\infty;, \tag{59}$$

$$\mathscr{L}[{}^*\mathscr{W}] = \mathscr{L}[\mathscr{W}^*]. \tag{60}$$

□

The basic information needed to identify the fixation process as done above is the following property of the McKean–Vlasov equation and possible random entrance laws.

Proposition 8. *(Random entrance laws)*

(a) There exists a solution $(\mathscr{L}^{**}_t(2))_{t\in\mathbb{R}}$ *to (31) satisfying the conditions:*

$$\lim_{t\to-\infty} \mathscr{L}^{**}_t(2) = \delta_0, \tag{61}$$

$$\lim_{t\to\infty} \mathscr{L}^{**}_t(2) = \delta_1$$

$$\int_{[0,1]} x\mathscr{L}^{**}_0(2,\mathrm{d}x) = \frac{1}{2}.$$

This solution is called an entrance law from $-\infty$ *with mean* $\frac{1}{2}$ *at* $t=0$.

(b) We can obtain a solution in (a) such that:

$$\exists\, \alpha \in (0,s) \textit{ and } A_0 \in (0,\infty) \text{such that } \lim_{t\to-\infty} \mathrm{e}^{\alpha|t|}\int_{[0,1]} x\mathscr{L}^{**}_t(2,\mathrm{d}x) = A_0. \tag{62}$$

(c) The solution of (31) also satisfying (62) for prescribed A_0 *is unique and if* $A_0 \in (0,\infty)$ *then* α *is necessarily uniquely determined.*

For any deterministic solution

$$\{\mathscr{L}_t,\; t\in\mathbb{R}\} \tag{63}$$

to (31) with

$$0 \le \limsup_{t\to-\infty} \mathrm{e}^{\alpha|t|}\int_{[0,1]} x\mathscr{L}_t(2,\mathrm{d}x) < \infty, \tag{64}$$

the limit $A = \lim_{t\to-\infty} e^{\alpha|t|}\int_{[0,1]} x\mathscr{L}_t(2,\mathrm{d}x)$ *exists.*

If $A>0$, *then* $\{\mathscr{L}_t, t\in\mathbb{R}\}$ *is given by a time shift of the unique* $\{\mathscr{L}^{**}_t, t\in\mathbb{R}\}$ *singled out in (62), namely*

$$\mathscr{L}_t = \mathscr{L}^{**}_{t+\tau}, \quad \tau = \alpha^{-1}\log\frac{A}{A_0}. \tag{65}$$

For future reference, we define $(\mathcal{L}_t^*)_{t \in \mathbb{R}}$ to be the unique solution satisfying

$$\lim_{t \to -\infty} e^{\alpha |t|} \int_{[0,1]} x \mathcal{L}_t^*(2, dx) = 1 \textit{ for some } \alpha \in (0, s). \tag{66}$$

(d) *Any random solution $(\mathcal{L}_t)_{t \in \mathbb{R}}$ to (31) such that*

$$\limsup_{t \to -\infty} e^{\alpha |t|} E\left[\int_{[0,1]} x \mathcal{L}_t(2, dx)\right] < \infty, \quad \liminf_{t \to -\infty} e^{\alpha |t|} \left[\int_{[0,1]} x \mathcal{L}_t(2, dx)\right] > 0 \text{ a.s.}, \tag{67}$$

is a random time shift of $(\mathcal{L}_t^{**})_{t \in \mathbb{R}}$ and of $(\mathcal{L}_t^*)_{t \in \mathbb{R}}$. □

Example 1. Let $\mathcal{L}_t^*$ be a solution satisfying (35), (62) and A be a true real-valued random variable. Then $\{\mathcal{L}_{t-\tau}^*\}_{t \in \mathbb{R}}$ with τ as in (65) is a truly random solution. This can also be viewed as saying that we have a solution with an exponential growth factor A which is truly random.

3 A Logistic Branching Random Walk and Its Growth

We define here a logistic branching population, take the mean-field limit $N \to \infty$, study the expansion of the droplet of occupied sites, determine the late time window and finally the behaviour in a late time window (i.e. shift of observation time interval and size of space tend to infinity) in four subsections.

3.1 The Logistic Branching Particle Model

We now consider a particle system on the geographic space $\{1, \ldots, N\}$ and its occupation numbers, which are specified by an element of the state space

$$(\mathbb{N}_0)^{\{1,\ldots,N\}}. \tag{68}$$

Here individual particles are subject to three mechanisms, they:

- *Migrate* according to a continuous time (rate c) random walk with uniform step distribution.
- *Die* at rate $d(k-1)$ if they occupy a site i and if we have k particles at site i.
- Give *birth* to one new particle at rate s at the same site.

If we observe only the occupation numbers at sites, this uniquely defines a strong Markov pure jump process on our state space, which we denote by

$$(\eta_t^N)_{t\geq 0}. \tag{69}$$

Remark 6. Note that the mean production rate (mean growth rate) of this model in state k is

$$sk - d(k(k-1)), \tag{70}$$

which is a concave function f with $f(0) = 0$, $f(1) = s$, $f(k) > 0$ for $k \leq k_0$ and $f(k) < 0$ for $k > k_0$. This production rate can be interpreted as reflecting limited local resources for a population.

Remark 7. We can view this process as a supercritical branching random walk (supercriticality parameter s) with an additional linear death rate $d(k-1)$ per individual, thus inducing an interaction between families. The process is also called a coalescing branching random walk.

Due to the quadratic death rate the population can only expand indefinitely by having individuals move to vacant sites. When the space is filled the expansion of the population is replaced by an equilibrium situation. How to make this precise?

We want to study this particle system for $N \to \infty$ in finite time windows, one early starting at time 0 and the other time window beginning at a late time when space fills up with particles. Here, we start with one particle and determine the late time window by asking when does the population develop a positive spatial intensity even as $N \to \infty$. Then in particular we focus on the influence of the death rate d and the migration rate c on the speed of spatial spread. The goal then is to establish that the population grows exponentially fast with a rate α which is positive but strictly less than the birth rate s as long as we have no collisions.

Remark 8. Note that if $d = 0$, then in fact we have a supercritical branching process at exponential rate s and it would take time $\frac{1}{s} \log N$ to develop positive spatial intensity.

If we have only one site we have a classical birth and death process, in the spatial model we have a collection of such processes. Since the death rate is zero if we have only one particle at a site, and furthermore because the death rate is quadratic, we have a positive recurrent Markov process on the state space $\mathbb{N}_0^N$ with a unique equilibrium law denoted:

$$\pi^N = \pi^N_{c,s,d}. \text{ Write } \pi_{c,s,d} \text{ for } \pi^1_{c,s,d}. \tag{71}$$

Given the initial state with finitely many particles, new sites can be occupied by migration and the population can grow till the jumps can only hit already occupied sites and as a result a population intensity on the whole space can develop and a local equilibrium forms, resulting in a global equilibrium density.

3.2 The Early Time Window as $N \to \infty$

As $N \to \infty$ the geographic space expands to $\mathbb{N}$ and the migrating particles eventually do not hit occupied sites in a fixed time interval if we start with finitely many initial particles. We call this the collision-free regime in contrast to the collision regime arising at late times, when sites interact again by migration. More precisely, we want to establish that, as $N \to \infty$, we get as limit dynamic a collection of birth and death processes with emigration at rate c and immigration at rate c from a reservoir with the current intensity in the total population. This is carried out as follows.

Consider a branching random walk with birth and death as before but now with geographic space $\mathbb{N}$, so that the state space becomes

$$(\mathbb{N}_0)^{N}, \tag{72}$$

and migration changes to:

- Emigration out of any component at rate c to the unoccupied site of lowest index as long as there is more than one particle.
- Immigration at rate $c \cdot \iota$ into every colony, with $\iota \in [0, \infty)$.

Here if we start the system in an exchangeable initial state, we choose

$$\iota = \lim_{N \to \infty} \left(N^{-1} \sum_{i=1}^{N} \eta(i) \right). \tag{73}$$

Later in larger time scales we will obtain a ι which is time dependent and arises from the law at a tagged site. For $\iota = 0$, we obtain what we call the *collision-free process*.

The strong Markov process defined by the above McKean–Vlasov dynamic is denoted

$$(\eta_t^{\iota})_{t \geq 0} \quad (\text{ resp. } (\eta_t)_{t \geq 0} \text{ if } \iota = 0). \tag{74}$$

This process has for every value $\iota \in [0, \infty)$ a unique equilibrium

$$\Pi^{\iota} = \bigotimes \pi_{c,s,d}^{(\iota)}, \tag{75}$$

where $\pi_{c,s,d}^{\iota}$ is the single site equilibrium. Let r be any map which acts on paths and permutes the locations so that $1, 2, \ldots, |\{i | \eta_t^N(i) > 0\}|$ are all occupied. We prove

Proposition 9. *(Convergence to ι-process, collision-free process)*
If we start in an i.i.d. distribution with $E[\eta_t^N(i)] = \iota$, then

$$\mathscr{L}[(\eta_t^N)_{t \geq 0}] \underset{N \to \infty}{\Longrightarrow} \mathscr{L}[(\eta_t^{(\iota)})_{t \geq 0}]. \tag{76}$$

If we start with one initial particle at site 1 and if $t_N = o(\log N)$, *then*

$$\mathscr{L}[(r \circ \eta_t^N)_{t \geq 0}] \underset{N \to \infty}{\Longrightarrow} \mathscr{L}[(\eta_t)_{t \geq 0}]. \tag{77}$$

□

3.3 The Droplet Expansion and Crump–Mode–Jagers Processes

As time gets large we expect the populations of particles to occupy more and more sites (droplet), such that the overall population expands as time grows, i.e. we have a growing *droplet of occupied sites*. This droplet growth is expected to be exponential and as time grows to become (at least if the further evolution is viewed globally) more and more deterministic due to a law of large numbers effect. Here, we have to distinguish a time window, which is late but where particles essentially always move to new unoccupied sites if they migrate and a later phase where spatial intensity builds up and collisions play a role. This subsection handles the first regime.

If we establish an exponential growth behaviour, we would be able to determine the position of the late time window in which the population density becomes positive as time and N tend to infinity. How can we make all this rigorous?

To analyse this time window, an important concept is that of a CMJ-process (Crump–Mode–Jagers process, see [21]), which allows us to describe the number of occupied sites in the process $(\eta_t)_{t \geq 0}$ starting from finitely many particles occupying all sites $1, \dots, k$ with $k \geq 1$. Let

$$K_t = \#\{i \in \mathbb{N} | \eta_t(i) > 0\}, \tag{78}$$

$$\zeta_t = \{\zeta_t(i), \quad i \in 1, \dots, K_t\} \quad , \quad \zeta_t(i) = r \circ \eta_t(i), \quad i = 1, \dots, K_t. \tag{79}$$

We will find that K, ζ are processes which have the structure of a CMJ-process, that is, the process of *occupied sites* is a type of generalised branching process. We begin by recalling this concept.

1. *Crump–Mode–Jagers process*

The CMJ-process models individuals in a branching population whose dynamics is as follows. Individuals can die or give birth to new individuals based on the following ingredients:

- Individuals have a lifetime (possibly infinite).
- For each individual an independent realization of a point process $\xi(t)$ starting at the birth time specifying the times at which the individual gives birth to new individuals.
- Different individuals act independently.
- The process of birth times is not concentrated on a lattice.

The process might be growing exponentially and then we want to determine its exponential growth rate. The corresponding *Malthusian parameter*, $\alpha > 0$ is obtained as the unique solution of

$$\int_0^\infty e^{-\alpha t} \mu(dt) = 1 \text{ where } \mu([0,t]) = E[\xi([0,t])], \tag{80}$$

with

$$\xi(t) \text{ counting the number of births of a single individual up to time } t, \tag{81}$$

(see for example [20], [23] (1.4)).

If we know the Malthusian parameter, we need to know that it is actually equal to the almost sure growth rate of the population. It is known for a CMJ-process $(K_t)_{t\geq 0}$ that (Proposition 1.1 and Theorem 5.4 in [23]) the following basic growth theorem holds. If

$$E[X \log(X \vee 1)] < \infty, \quad \text{where } X = \int_0^\infty e^{-\alpha t} d\xi(t), \tag{82}$$

then

$$\lim_{t\to\infty} \frac{K_t}{e^{\alpha t}} = W, \text{ a.s. and in } L_1, \tag{83}$$

where W is a random variable which has two important properties, namely

$$W > 0, \ a.s. \text{ and } \ E[W] < \infty. \tag{84}$$

The empirical distribution of individuals of a certain age in the growing population converges as $t \to \infty$ to a *stable age distribution*

$$\mathscr{U}(\infty, du) \text{ on } [0,\infty), \tag{85}$$

according to Corollary 6.4 in [23], if condition 6.1 therein holds. The condition 6.1 in [23] or (3.1) in [21] requires that (with μ as in (80)):

$$\int_0^\infty e^{-\beta t} \mu(dt) < \infty \quad \text{for some } \beta < \alpha. \tag{86}$$

2. *Application of the CMJ-theory to* (K_t, ζ_t)

The $(\zeta_t)_{t\geq 0}$ we have introduced is a CMJ where individuals are the occupied sites and satisfy all the conditions posed above. In addition, we have more structure, namely the birth process is determined from an *internal state* of the individual (here a site) which follows a Markovian evolution (here our $(\zeta_t(i))_{t\geq 0}$). This allows to obtain some stronger statements as follows.

Define

$$(\zeta_0(t,i))_{t\geq 0}, \text{ as the process } (\zeta_t(i))_{t\geq 0} \text{ if } i \text{ gets occupied at time } 0. \tag{87}$$

In our case, we then have

$$\mu([0,t]) = c \int_0^t E[\zeta_0(s) 1_{(\zeta_0(s) \geq 2)}] \mathrm{d}s. \tag{88}$$

Define the random measure

$$\mathscr{U}(t, \mathrm{d}u, j) = \#\{\text{ sites of size } j \text{ with birth time } t - \mathrm{d}u\} K_t^{-1}. \tag{89}$$

It follows that the random measure $\mathscr{U}(t,\cdot,\cdot)$ converges to a *deterministic* object, the *stable age and size* distribution, i.e.

$$\mathscr{U}(t,\cdot,\cdot) \Longrightarrow \mathscr{U}(\infty,\cdot,\cdot), \text{ as } t \to \infty \text{ in law}, \quad \mathscr{U}(\infty,\cdot,\mathbb{N}) = \mathscr{U}(\infty,\cdot) \text{ from (85).} \tag{90}$$

We can obtain from this the following representation of the Malthusian parameter α.

Given site i let $\tau_i \geq 0$ denote the time at which a migrant (or initial particle) first occupies it. Noting that we can verify Condition 5.1 in [23] we have that

$$\lim_{t\to\infty} \frac{1}{K_t} \sum_{i=1}^{K_t} \zeta_{\tau_i}(t - \tau_i) = \int_0^\infty E[\zeta_0(u)] \mathscr{U}(\infty, \mathrm{d}u) = B \text{ (a constant), a.s.,} \tag{91}$$

by Corollary 5.5 of [23]. The constant B in (91) is in our case given by the average number of particles per occupied site and the growth rate α arises from this quantity neglecting single occupation. Namely define

$$\alpha = c \sum_{j=2}^\infty j \mathscr{U}(\infty, [0,\infty), j) < \infty, \qquad \gamma = c\, \mathscr{U}(\infty, [0,\infty), 1). \tag{92}$$

Then the mean occupation number of an occupied site is given by

$$B = \frac{\alpha + \gamma}{c} > 0. \tag{93}$$

Furthermore, the average birth rate of new sites (by arrival of a migrant at an unoccupied site) at time t (in the process in the McKean–Vlasov dual) is equal to

$$\alpha = c\, B - \gamma. \tag{94}$$

Here, a remarkable point is the relation between α and s. Recall s is the parameter of supercriticality in the branching part of the internal state dynamics. The action of the quadratic death part and the role of migration lead to a number α with

$$0 < \alpha < s, \tag{95}$$

since the fraction of the supercritical branching process with supercriticality parameter s, which die by the quadratic death rate is positive.

3.4 Time Point of Emergence as $N \to \infty$

If we take the collision-free model $(\eta_t)_{t\geq 0}$ and we observe at time $T_N(t) = \alpha^{-1}\log N + t$ the number of individuals or the number of occupied sites, both normalized by N, then these quantities satisfy (recall (93), (83)) that there exists a random variable W such that

$$\left(\frac{1}{N}\sum_{i=1}^{K_{T_N(t)}} \eta_{T_N(t)}(i)\right) \underset{N\to\infty}{\longrightarrow} B e^{\alpha(t+\alpha^{-1}\log W)},$$

$$(K_{T_N(t)}/N)_{t\in\mathbb{R}} \underset{N\to\infty}{\longrightarrow} \left(e^{\alpha(t+\alpha^{-1}\log W)}\right)_{t\in\mathbb{R}}, \tag{96}$$

with $0 < W < \infty$ a.s. and $Var(W) > 0$. Hence in particular a positive intensity develops in space and on the occupied sites the intensity converges:

$$\left(\frac{1}{K_{T_N(t)}}\sum_{i=1}^{K_{T_N(t)}} \eta_{T_N(t)}(i)\right) \underset{N\to\infty}{\longrightarrow} B, \quad \forall\, t \in \mathbb{R}. \tag{97}$$

In particular in this time window η^N and η differ significantly, but one can show that still:

$$\mathcal{L}\left[(\{r\circ\eta^N_{T_N+t}(1),\dots,(r\circ\eta^N_{T_N+t}(k)\})_{t\geq 0}\right] \Longrightarrow \mathcal{L}\left[(\eta_t(1),\dots,\eta_t(k))_{t\geq 0}\right],$$
$$\text{as } N\to\infty, \tag{98}$$

provided $T_N - \alpha^{-1}\log N \to -\infty$ as $N \to \infty$. In fact even though the CMJ-approximation breaks down at times $\alpha \log N + t$ we can still prove emergence occurs at time $\alpha^{-1}\log N$, since η^N develops a positive intensity meaning the total population is comparable to N, which as $t \to -\infty$ is asymptotically equivalent to the r.h.s. of (96). Indeed, time $\alpha^{-1}\log N$ separates the collision-free droplet growth from the emergence.

Proposition 10. *(Emergence of positive intensity)*
We have for $N \to \infty$ and $\nu = \nu((t_N)_{N\in\mathbb{N}})$:

$$\mathcal{L}\left[N^{-1}\sum_{i=1}^{N}\eta^N_{t_N}(i)\right] \underset{t\to\infty}{\Longrightarrow} \begin{cases} \delta_0, & \text{if } t_N - \alpha^{-1}\log N \to -\infty \\ \nu,\ \nu((0,\infty)) = 1, & \text{if } \lim\limits_{N\to\infty}(t_N - \alpha^{-1}\log N) = t > -\infty. \end{cases} \tag{99}$$

□

3.5 The Late Time Window as $N \to \infty$

To treat the time window $\alpha^{-1}\log N + t$ and to obtain the limiting emergence-equilibration dynamics based on *random entrance laws* of the *McKean–Vlasov equation*, we need some ingredients.

We consider the number of sites occupied at time t denoted K_t^N and the corresponding measure-valued process on $[0,\infty) \times \mathbb{N}_0$ giving the unnormalized number of sites of a certain age u in $[a,b)$ and occupation size j:

$$\Psi^N(t,[a,b),j) = \int_{(t-b)}^{(t-a)} 1_{(K_u^N > K_{u-}^N)} 1_{(\zeta_u^N(t)=j)} dK_u^N, \tag{100}$$

where $\zeta_u^N(t)$ denotes the occupation number at time t of the site born at time u, that is, a site first occupied the last time at time u, which is therefore exactly of age $t-u$ at time t.

The *normalized empirical age and size distribution* among the occupied sites is defined as:

$$U^N(t,[a,b),j) = \frac{1}{K_t^N}\Psi^N(t,[a,b),j), \quad t \geq 0,\ j \in \{1,2,3,\dots\}. \tag{101}$$

Denote now for convenience the number of sites:

$$u^N(t) := K_t^N. \tag{102}$$

We have obtained with (101), (102) a pair which is $\mathbb{N} \times \mathscr{P}([0,\infty) \times \mathbb{N})$-valued and which describes our particle system completely provided we consider individuals and sites as exchangeable labels. This pair is denoted

$$(u^N(t), U^N(t,\cdot,\cdot))_{t\geq 0}. \tag{103}$$

In the interval $\alpha^{-1}[\log N - \log\log N, \log N + T]$ the process $(u^N(t))_{t\geq 0}$ increases by one respectively decreases by one at rates

$$\alpha_N(t)\left(1 - \frac{u^N(t)}{N}\right)u^N(t), \text{ respectively } \gamma_N(t)\frac{(u^N(t))^2}{N}, \tag{104}$$

where $\alpha_N(t), \gamma_N(t)$ are defined:

$$\alpha_N(t) = c\int_0^t \sum_{j=2}^{\infty} jU^N(t,ds,j)ds, \quad \gamma_N(t) = c\int_0^t U^N(t,ds,1). \tag{105}$$

The above rates of change of $U^N(t,\cdot,\cdot)$ follow directly from the dynamics of the particle system η^N.

Our goal is now to study the behaviour of (u^N, U^N) at times $\alpha^{-1}\log N + t$ and to show that this follows a limiting fixation dynamics. We start the system with k particles at ℓ distinct sites and write k, ℓ as superscript. We need the time-shifted quantities:

$$\widetilde{u}^{N,k,\ell}(t) = u^{N,k,\ell}\left(\left(\frac{\log N}{\alpha} + t\right) \vee 0\right), \; t \in (-\infty, T], \quad \widetilde{u}^{N,k,\ell}\left(-\frac{\log N}{\alpha}\right) = \ell \tag{106}$$

$$\widetilde{U}^{N,k,\ell}(t) = U^{N,k,\ell}\left(\left(\frac{\log N}{\alpha} + t\right) \vee 0\right), \; t \in (-\infty, T], \quad \widetilde{U}^{N,k,\ell}\left(-\frac{\log N}{\alpha}\right) = \delta_{(k,0)}. \tag{107}$$

We now formulate first a limit result for an initial time t_0, which is fixed, then later consider the case as above where the initial time tends to $-\infty$ as $N \to \infty$ (Proposition 14).

Proposition 11. *(Convergence to a colonization-equilibration dynamic in the $N \to \infty$ limit)*

Assume that for some $t_0 \in \mathbb{R}$ as $N \to \infty$, $(\frac{1}{N}\widetilde{u}^{N,k,\ell}(t_0), \widetilde{U}^{N,k,\ell}(t_0))$ converges in law to the pair $(u(t_0), U(t_0))$ automatically contained in $[0,\infty) \times L_1(\mathbb{N}, \nu)$.

Then as $N \to \infty$

$$\mathscr{L}\left[\left(\frac{1}{N}\widetilde{u}^{N,k,\ell}(t), \widetilde{U}^{N,k,\ell}(t,\cdot,\cdot)\right)_{t \geq t_0}\right] \Longrightarrow \mathscr{L}\left[(u^{k,\ell}(t), U^{k,\ell}(t,\cdot,\cdot))_{t \geq t_0}\right], \tag{108}$$

in law on pathspace, where the r.h.s. is supported on the solution of the nonlinear *system (110) and (111) below corresponding to the initial state $(u(t_0), U(t_0))$. (Note that the mechanism of the limit dynamics does not depend on k or ℓ, but the initial state at t_0 does.)* □

We obtain the limiting system (u, U) as follows. We specify the pair

$$(u, U) = (u(t), U(t))_{t \in \mathbb{R}}, \text{ with } (u(t), U(t)) \in \mathbb{R}^+ \times \mathscr{P}_1(\mathbb{R}^+ \times \mathbb{N}), \tag{109}$$

by the (coupled) system of nonlinear *forward* evolution equations:

$$\frac{du(t)}{dt} = \alpha(t)(1 - u(t))u(t) - \gamma(t)u^2(t), \tag{110}$$

$$\begin{aligned}\frac{\partial U(t, dv, j)}{\partial t} &= -\frac{\partial U(t, dv, j)}{\partial v} \\ &\quad + s(j-1)1_{\{j \neq 1\}}U(t, dv, j-1) - sjU(t, dv, j) \\ &\quad + \frac{d}{2}(j+1)jU(t, dv, j+1) - \frac{d}{2}j(j-1)1_{\{j \neq 1\}}U(t, dv, j)\end{aligned} \tag{111}$$

$$
\begin{aligned}
&+c(j+1)U(t,\mathrm{d}v,j+1)-cjU(t,\mathrm{d}v,j)1_{\{j\neq 1\}}\\
&-cu(t)U(t,\mathrm{d}v,1)1_{\{j=1\}}\\
&+u(t)(\alpha(t)+\gamma(t))[1_{\{j\neq 1\}}U(t,\mathrm{d}v,j-1)-U(t,\mathrm{d}v,j)]\\
&+(1-u(t))\alpha(t)1_{\{j=1\}}\delta_0(\mathrm{d}v)\\
&-\Big(\alpha(t)(1-u(t))-\gamma(t)u(t)\Big)U(t,\mathrm{d}v,j).
\end{aligned}
$$

We have to constrain the state space to guarantee the r.h.s. above is well defined. Therefore, set ν to be the measure on $\mathbb{N}$ given by

$$\nu(j)=1+j^2 \tag{112}$$

and consider $\mathbb{R}\otimes L^1(\nu,\mathbb{N})$ as a basic space for the analysis. Then the (110), (111) have the following properties.

Proposition 12. *(Uniqueness of the pair (u,U))*

(a) The pair of equations (110)–(111), given an initial state satisfying

$$(u(t_0),U(t_0,\cdot,\cdot))\in\mathbb{R}^+\times\mathscr{P}_1(\mathbb{R}^+\times\mathbb{N}),\ U(t_0,\mathbb{R}^+,\cdot)\in L^1(\mathbb{N},\nu), \tag{113}$$

at time t_0, has a unique solution $(u(t),U(t))_{t\geq t_0}$ with values in $\mathbb{R}^+\times\mathscr{P}_1(\mathbb{R}^+\times\mathbb{N})$ such that $U(t,\mathbb{R}^+,\cdot)\in L^1_+(\mathbb{N},\nu)$, and which satisfies $U(t,\mathbb{R}^+\times\mathbb{N})\equiv 1$.

(b) There exists a solution (u,U) with time parameter $t\in\mathbb{R}$ for every $A\in(0,\infty)$ with values in $\mathbb{R}^+\otimes\mathscr{P}_1(\mathbb{R}^+\times\mathbb{N})$, such that (here $\Longrightarrow$ denotes weak convergence)

$$u(t)\mathrm{e}^{-\alpha t}\to A \text{ as } t\to-\infty, \tag{114}$$

$$U(t)\underset{t\to-\infty}{\Longrightarrow}\mathscr{U}(\infty). \tag{115}$$

The law $\mathscr{U}(\infty)$ is the stable age and size distribution of the CMJ-process corresponding to the particle process $(K_t,\zeta_t)_{t\geq 0}$ given by the McKean–Vlasov dual process η, as defined in (79).

(c) Given any solution (u,U) of (110)–(111) for $t\in\mathbb{R}$ with values in the space $\mathbb{R}\otimes L^1(\mathbb{N},\nu)$ satisfying

$$u(t)\geq 0,\quad \limsup_{t\to-\infty}\mathrm{e}^{-\alpha t}u(t)<\infty, \tag{116}$$

then u satisfies

$$A=\lim_{t\to-\infty}\mathrm{e}^{-\alpha t}u(t) \tag{117}$$

exists and the solution satisfying this for given A is unique. Furthermore, U satisfies

$$U(t)\Longrightarrow\mathscr{U}(\infty)\text{as } t\to-\infty. \tag{118}$$

□

Remark 9. Potential limits arising from (u^N, U^N) satisfy (116).

Remark 10. Note that looking at the form of the equation we see that a solution indexed by $\mathbb{R}$ remains a solution if we make a time shift. This corresponds to the different possible values for the growth constant A in (114). In particular, the entrance law from 0 at time $-\infty$ is unique up to the time shift.

We can now identify the behaviour of (u^N, U^N) for $N \to \infty$ in terms of the early growth behaviour of the droplet of colonized sites as follows.

Proposition 13. *(Identification of colonization-equilibrium dynamics)*
The limits $(u^{k,\ell}(t), U^{k,\ell}(t,\cdot,\cdot)))_{t\geq t_0}$, in (108) can be represented as the unique solution of the nonlinear system (110) and (111) satisfying

$$\lim_{t\to-\infty} \mathrm{e}^{-\alpha t} u^{k,\ell}(t) = W^{k,\ell}, \quad \lim_{t\to-\infty} U^{k,\ell}(t) = \mathscr{U}(\infty), \tag{119}$$

with $W^{k,\ell}$ having the law of the random variable appearing as the scaling (by $\mathrm{e}^{-\alpha t}$) limit $t \to \infty$ of the CMJ-process $K_t^{k,\ell}$ started with k particles at each of ℓ sites. □

We can now ask, what happens if we consider times $\alpha^{-1}\log N + t_N$ with $t_N \to \infty$. Then we reach a global stable state with law $\widetilde{\Pi}^{\iota^*}_{c,d,s}$ ($\Pi^{\iota^*}_{c,d,s}$ conditioned to be strictly positive). In particular

Proposition 14. *(Equilibrium population)*
We have

$$N^{-1}u^N(\alpha^{-1}\log N + t_N) \underset{N\to\infty}{\longrightarrow} \iota^* \quad , \quad U^N(\alpha^{-1}\log N + t_N) \underset{N\to\infty}{\Longrightarrow} \pi^{\iota^*}_{c,d,s}, \tag{120}$$

where ι^* satisfies the self consistency relation:

$$\sum_{k=1}^{\infty} k\pi^{(\iota)}_{c,d,s}(\{k\}) = \iota. \tag{121}$$

□

4 The Duality Relation

In this section, we relate the processes from Sects. 2 and 3 with each other by duality and we discuss the extension of the duality to more than two types.

4.1 A Classical Duality Formula

The key tool in relating the two processes we have introduced is duality. A *duality* between two Markov processes X and Y on state spaces E and E' w.r.t. a duality function $H : E \times E' \longmapsto \mathbb{R}$ such that $\{H(\cdot, y), y \in E'\}$ is measure-determining holds, if

$$E[H(X_t, Y_0)|X_0] = E[H(X_0, Y_t)|Y_0], \qquad X_0 \in E, Y_0 \in E'. \tag{122}$$

All the dualities we shall discuss fit this framework. Recall for example the classical relation between a single Fisher–Wright diffusion and a quadratic death process.

Consider $(X_t)_{t\geq 0}$ solving

$$\mathrm{d}X_t = \sqrt{d \cdot X_t(1 - X_t)}\mathrm{d}W_t \tag{123}$$

and $(D_t)_{t\geq 0}$ being the $\mathbb{N}$-valued death process

$$n \to n - 1 \text{ at rate } d \cdot \binom{n}{2}. \tag{124}$$

Then a generator calculation shows that with initial states X_0 and $D_0 = k$ (we explain more on the background below) meaning $E = [0, 1], E' = \mathbb{N}$ and with $H(x, n) = x^n$:

$$E_{X_0}[(X_t)^k] = E_k[X_0^{D_t}]. \tag{125}$$

The analogous relation can be formulated for our mean-field spatial model including mutation and selection. Define for the process starting with k-particles at each of ℓ sites:

$$\Pi_t^{N,k,\ell} = u^N(t) \sum_{j=1}^{\infty} j U^N(t, \mathbb{R}^+, j), \tag{126}$$

where $(u^N(t), U^N(t, \mathbb{R}^+, \mathbb{N}))$ are given by (103). Then in [8] we get by a generator calculation the duality formula (compare also the explanation following (150)):

$$E[x_1^N(i,t)] = E\left[\exp\left(-\frac{m}{N}\int_0^t \Pi_u^{N,1,1}\mathrm{d}u\right)\right], \quad i \in \{1, \dots, N\}. \tag{127}$$

Hence if we start initially with $x_1^N(i, 0) = 1; i = 1, \dots, N$ we see that in order to observe a mean which is less than one we need an occupation measure of the population from the branching coalescing random walk which is of order N. Due to the exponential growth, this amounts to $\Pi^{N,1,1}$ growing to order N. Since we proved the latter behaves at times $\alpha^{-1} \log N + t$ for very negative t approximately like

$$W \exp(\alpha(\alpha^{-1} \log N + t)) = \mathrm{e}^{\alpha t} \cdot WN;, \tag{128}$$

the advantageous type emerges since $E[x_1^N(1, \alpha^{-1} \log N + t)]$ as $N \to \infty$ is strictly between 0 and 1 for $t \in \mathbb{R}$. By the analysis of moments we actually can prove the results on the Fisher–Wright diffusion model stated in Sect. 2 from the results of the coalescing particles model stated in Sect. 3. In Sect. 7 of [8], [11] methods are developed to turn this idea into rigorous mathematics.

4.2 The Genealogy and Duality

The formula (125) can be understood on a deeper level, since the dual process $(D_t)_{t\geq 0}$ can be interpreted in terms of the genealogy of the Fisher–Wright diffusion model. This will also exhibit the role of selection a bit better.

For that purpose, the death process $(D_t)_{t\geq 0}$ has to be enriched to the Kingman coalescent which is a partition-valued process. This means its states are the partitions of the set $\{1,\ldots,k\}$ starting in $\{\{1\},\ldots,\{k\}\}$, where partition elements coalesce at rate d independent of each other (and in the spatial case perform continuous time random walks at rate c). If we consider a Fisher–Wright diffusion, it can be viewed as the limit of the individual based Moran model taking the population size to infinity. In this model individuals migrate according to random walks, mutate according to the kernel $M(\cdot,\cdot)$ and two individuals are replaced at random times by two new individuals such that one of the two old individuals have two offspring and the other no offspring with probabilities depending in the fitness of the two dying individuals.

Pick then from the population at time t a *random sample* of exactly k individuals and look at their genealogy. Suppose that the fitness function is constant. This genealogy then has the same law as the genealogy of the Kingman coalescent with the genealogical distance between two individuals defined to be the first time they are in the same partition element. Given this genealogy we can calculate the probability, that all individuals in the sample have the same type, say one, which is x^k if x is the current (time t) frequency of type one, in terms of the Kingman coalescent getting the duality relation (125). See [13, 17, 18] for more details on the genealogical processes.

If we include selection and if we follow the tagged sample from the population back, note that one of the individuals might interact with the rest of the population by a selection event (note now individuals under selection and mutation are *not* exchangeable (as in the neutral case) so that the outside individual does not have the same statistical properties as the one already in the sample). What results from this event, however, now depends on the current types of the two involved individuals. Therefore, we have to also follow the other individual further back. This means we expect that each time a selection event occurs our tagged sample has to be enriched by a further particle. This is reflected in the birth of new individuals in the dual particle model.

In order to handle multiple types and mutation, the point now is that the dual process has to be complemented by a function-valued part. The basic idea behind this is explained in the next subsection and has many applications (see [8] and [13].

4.3 The Dual for General Type Space

In the multitype situation with more than 2 types, the picture is more subtle than for the case of two types. We next sketch the basic new ideas from [8] which are used for this purpose.

4.3.1 A Multitype Model

The process considered in Sect. 2 can be defined for general type space $\mathbb{I}$ (a subset of $[0,1]$) as $(\mathscr{P}(\mathbb{I}))^S$-valued process ($S$ = a finite or countable geographic space) and is called the interacting Fleming–Viot process with selection and mutation, see [7], (which becomes a multitype Fisher–Wright diffusion in the case of finitely many types and the model of Sect. 2 for two types). Then types have a fitness given by a function and mutation occurs via a jump kernel which are denoted, respectively, by

$$\chi : \mathbb{I} \longrightarrow \mathbb{R}^+,\ 0 \le \chi \le 1,\ M(i, dj) \text{ a } \mathbb{I} \times \mathbb{I}\text{-probability transition kernel.} \tag{129}$$

The generator of the Fleming–Viot diffusion is a second order differential operator. Recall that for functions F on signed measures x

$$\frac{\partial F(x)}{\partial x}[y] = \lim_{\varepsilon \to 0} \left(\frac{F(x + \varepsilon y) - F(x)}{\varepsilon} \right). \tag{130}$$

Note that it suffices to consider $y = \delta_u, \quad u \in \mathbb{I}$ since finite atomic measures are weakly dense in $\mathscr{P}(\mathbb{I})$. The generator of the nonspatial Fleming–Viot process acts on monomials F of order n with test function $f \in C_b(\mathbb{I})$ defined as

$$F(x) = \int f(u_1, \dots, u_n) x(\mathrm{d}u_1) \cdots x(\mathrm{d}u_n), \quad x \in \mathscr{P}(\mathbb{I}), \tag{131}$$

as follows, with setting $Q_x(\mathrm{d}u, \mathrm{d}v) = x(\mathrm{d}u)\delta_u(\mathrm{d}v) - x(u)x(\mathrm{d}v)$:

$$\begin{aligned}
(\mathrm{GF})(x) = {} & s \int_{\mathbb{I}} \left\{ \frac{\partial F(x)}{\partial x}(u) \left(\chi(u) - \int_{\mathbb{I}} \chi(w) x(\mathrm{d}w) \right) \right\} x(\mathrm{d}u) \\
& + m \int_{\mathbb{I}} \left\{ \int_{\mathbb{I}} \frac{\partial F(x)}{\partial x}(v) M(u, \mathrm{d}v) - \frac{\partial F(x)}{\partial x}(u) \right\} x(\mathrm{d}u) \\
& + d \int_{\mathbb{I}} \int_{\mathbb{I}} \frac{\partial^2 F(x)}{\partial x \partial x}(u, v) Q_x(\mathrm{d}u, \mathrm{d}v), \qquad x \in \mathscr{P}(\mathbb{I}).
\end{aligned} \tag{132}$$

In the spatial case, a corresponding drift term from migration appears as well:

$$\sum_{i,j\in S} a(i,j)\int\left(\frac{\partial F(x)}{\partial x_j}(u)-\frac{\partial F(x)}{\partial x_i}(u)\right)x_i(\mathrm{d}u). \tag{133}$$

In [10] a dual for a Fleming–Viot process with mutation and selection is introduced in order to show that the process is well defined by its martingale problem. In [8], a new dual for the spatial case was developed which makes possible the study of the long-time behaviour and the genealogy of the system. We now explain this. In order to introduce the main ideas, we first consider the case with no mutation and the special case $N=1$ and then the general case.

4.3.2 The Dual with Selection

Let the mutation rate be zero and let $|S|=1$, so that the state space is $\mathscr{P}(\mathbb{I})$ and we only have resampling and selection. Consider the class of functions

$$F((n,f),x) := \left[\int_{\mathbb{I}}\dots\int_{\mathbb{I}} f(u_1,\dots,u_n)x(\mathrm{d}u_1)\dots x(\mathrm{d}u_n)\right], \tag{134}$$

for all $n\in\mathbb{N}$, $f\in L_\infty((\mathbb{I})^n,\mathbb{R})$ and $x\in\mathscr{P}(\mathbb{I})$. Note that given a random probability measure X on $\mathbb{I}$, the collection

$$\{E[F((n,f),X)], n\in\mathbb{N}, f\in L_\infty((\mathbb{I})^n,\mathbb{R})\} \tag{135}$$

uniquely characterizes the law of X.

The class given above contains the functions which we will use for our dual representations. In fact it suffices to take smaller, more convenient sets of test functions f:

$$f(u_1,\dots,u_n)=\prod_{i=1}^{n} f_i(u_i), \quad f_i\in L_\infty(\mathbb{I}). \tag{136}$$

If we consider the case when $\mathbb{I}$ is finite, it would even suffice to take functions

$$f_j(u)=1_{\{j\}}(u) \text{ or } f_j(u)=1_{A_j}(u), \tag{137}$$

where 1_j is the indicator function of $j\in\mathbb{I}$.

The *function-valued dual processes* $(\eta_t,\mathscr{F}_t)_{t\geq 0}$ and $(\eta_t,\mathscr{F}_t^+)_{t\geq 0}$ are constructed from the following four ingredients:

- N_t the number of *individuals* present in the dual process which is a non-decreasing $\mathbb{N}$-valued process with $N_0=n$, the number of initially tagged individuals, and $N_t\geq n$.

- $\zeta_t = \{1, \ldots, N_t\}$ is an *ordered particle system.*
- π_t: is a *partition* $(\pi_t^1, \ldots, \pi_t^{|\pi_t|})$ of ζ_t, i.e. an ordered family of subsets, where the *index* of a partition element is the smallest element of the partition element.
- $\mathscr{F}_t$ is for given states of π_t, ζ_t, N_t a function in $L_\infty(\mathbb{I}^{|\pi_t|})$, which is obtained from a function in $L_\infty(\mathbb{I}^{N_t})$ by setting variables equal, which are corresponding to one and the same partition element and $\mathscr{F}_t$ changes further driven selection (see below).

Definition 3. (Evolution of $(\eta, \mathscr{F})$ and $(\eta, \mathscr{F}^+)$)

(a) The process η is driven by coalescence at rate d of every pair of partition elements and by the birth of a new individual at rate s, which forms its own partition element.
(b) Conditioned on the process η the evolution of $\mathscr{F}$ respectively $\mathscr{F}^+$ is as follows:

- The coalescence mechanism: If a coalescence of two partition elements occurs, then the corresponding variables of $\mathscr{F}_t$ are set equal to the variable indexing the partition element, i.e. for $\mathscr{F}_{t-} = g$ we have the transition (here $\widehat{u}_j$ denotes an omitted variable)

$$g(u_1, \ldots, u_i, \ldots, u_j, \ldots, u_m) \longmapsto \widehat{g}(u_1, \ldots, \widehat{u}_j, \ldots, u_m) \qquad (138)$$
$$= g(u_1, \ldots, u_i, \ldots, u_{j-1}, u_i, u_{j+1}, \ldots, u_m),$$

so that the function changes from an element of $L_\infty(\mathbb{I}^m)$ to one of $L_\infty(\mathbb{I}^{m-1})$.
- The selection mechanisms:

  - Feynman–Kac. For $\mathscr{F}_t$, if a birth occurs in the process (ζ_s) due to the partition element to which the element i of the basic set belongs, then for $\mathscr{F}_{t-} = g$ the following transition occurs from an element in $L_\infty((\mathbb{I})^m)$ to elements in $L_\infty((\mathbb{I})^{m+1})$:

$$g(u_1, \ldots, u_m) \longmapsto \chi(u_i) g(u_1, \ldots, u_m) - \chi(u_{m+1}) g(u_1, \ldots, u_m). \quad (139)$$

  - Non-negative. For $\mathscr{F}_t^+$, the transition (139) is replaced by (provided that χ satisfies $0 \le \chi \le 1$):

$$g(u_1, \ldots, u_m) \longmapsto \widehat{g}(u_1, \ldots, u_{m+1}) = (\chi(u_i) + (1 - \chi(u_{m+1})))$$
$$\times g(u_1, \ldots, u_m). \qquad (140)$$

□

Now we can obtain two different duality relations, the first below working in all cases and a second below working in a large subclass of models.

Proposition 15. *(Duality relation – signed with Feynman–Kac dual)*
Let $(X_t)_{t\ge 0}$ be a solution of the Fleming–Viot martingale problem with finite type

space $\mathbb{I}$*, fitness function* χ *and selection rate* s *with* $X_0 = x \in \mathscr{P}(\mathbb{I})$*. Let* $\mathscr{F}_0 = f \in L_\infty((\mathbb{I})^n)$ *for some* $n \in \mathbb{N}$*.*

Assume that t_0 *is such that:*

$$E\left(\exp\left(s\int_0^{t_0}|\pi_r|\mathrm{d}r\right)\right) < \infty. \tag{141}$$

Then for $0 \le t \le t_0$*,* $(\eta_t, \mathscr{F}_t)$ *is the Feynman–Kac dual of* (X_t)*, that is:*

$$E[F((\eta_0, f), X_t)] = E_{(\eta_0,\mathscr{F}_0)}\left\{\left[\exp\left(s\int_0^t|\pi_r|\mathrm{d}r\right)\right]\right. \tag{142}$$

$$\left.\times\left[\int_{\mathbb{I}}\cdots\int_{\mathbb{I}}\mathscr{F}_t(u_1,\ldots,u_{|\pi_t|})x(\mathrm{d}u_1)\ldots x\left(\mathrm{d}u_{|\pi_t|}\right)\right]\right\}, \tag{143}$$

where the initial state $(\eta_0, \mathscr{F}_0)$ *is for* $n \in \mathbb{N}$ *chosen given by*

$$\begin{aligned}\pi_0 &= [(\{1\},\{2\},\ldots,\{n\})], \\ \mathscr{F}_0 &= f \in C_b(\mathbb{I}^n).\end{aligned} \tag{144}$$

□

Remark 11. In [7], it was shown that there exists $t_0 > 0$ for which (141) is satisfied.

The disadvantage of the dual above is the exponential term together with the signed function. This involves the interplay of a cancelation effect and the exponential growth factor which is often hard to analyse as $t \to \infty$. The key observation is that if the fitness function χ is a bounded function then we can obtain the following duality relation that does *not* involve a Feynman–Kac factor and preserves the positivity of functions:

Proposition 16. *(Duality relation – non-negative)*

With the notation and assumptions as in Proposition 15 (except with (139) replaced by (140)) we get for χ *with*

$$0 \le \chi \le 1, \tag{145}$$

that for all $t \in [0, \infty)$ *we have:*

$$E[F((\eta_0, f), X_t)] = E_{(\eta_0,\mathscr{F}_0^+)}\left[\int_{\mathbb{I}}\cdots\int_{\mathbb{I}}\mathscr{F}_t^+(u_1,\ldots,u_{|\pi_t|})x_1(\mathrm{d}u_1)\cdots x_{(|\pi_t|)}(\mathrm{d}u_{|\pi_t|})\right]. \tag{146}$$

Moreover, $\mathscr{F}_t^+$ *is always non-negative if* $\mathscr{F}_0^+$ *is.* □

4.3.3 The Dual with Migration, Selection and Mutation

We now consider the case of (3), (4) including selection and migration but for the moment setting $m = 0$. The partition elements of the dual now have *locations* given by

$$\xi_t : \pi_t \longrightarrow S^{|\pi_t|}, \quad S = \{1, \dots, N\}. \tag{147}$$

The corresponding additional *dual migration* dynamics is as follows. At rate c, each partition element can jump from its current location to a randomly chosen point in $\{1, \dots, N\}$. In the limiting case ($N \to \infty$), the particle always migrates to an empty site which for convenience we can take to be the smallest unoccupied site $n \in \mathbb{N}$. This then results in precisely the logistic branching particle model, respectively, the Crump–Mode–Jagers process described in Sect. 3.

The duality relation is now given by

$$E[F((\eta_0, f), X_t)] = E_{(\eta_0, \mathscr{F}_0^+)}\left[\int_{\mathbb{I}} \cdots \int_{\mathbb{I}} \mathscr{F}_t^+(u_1, \dots, u_{|\pi_t|}) x_{\xi_t(1)}(du_1) \cdots x_{\xi_t(|\pi_t|)}(du_{|\pi_t|})\right], \tag{148}$$

where $X_0 = (x_1, \dots, x_n),\ x_i \in \mathscr{P}(\mathbb{I}),\ i = 1, \dots, N,\ \eta_0 = (\zeta_0, \pi_0, \xi_0)$.

The mutation can now be incorporated by adding a corresponding transition of $\mathscr{F}$, at rate m for every variable the operation acting on $g \in (L^\infty(\mathbb{I}))^n$ by

$$g(u_1, u_2, \dots, u_n) \longrightarrow \int g(u_1, \dots, v, u_{i+1}, \dots, u_n) M(u_i, dv). \tag{149}$$

Then the relation (146) holds with mutation.

Example of application. Now consider the case with $m > 0$, $\mathbb{I} = \{1, 2\}$ as in Sect. 1 and $\mathscr{F}_0 = 1_1$. The effect of rare mutation from type 1 to type 2 results in the dual in the transition

$$1_1 \to 0 \quad \text{at rate } \frac{m}{N}. \tag{150}$$

Returning to our duality relation, we now see that the particles in our process η stand for possible individuals, represented by the factor 1_1 which could by mutation represent a possible line through which the advantageous type can enter the population. More precisely, if $\mathscr{F}_0 = 1_1$, then we can check that as time increases we have an increasing number, $\Pi_u^{N,1,1}$ of such factors and any of these can undergo the rare mutation transition.

Therefore,

$$V_t^N = \frac{m}{N} \int_0^t \Pi_u^{N,1,1} du \tag{151}$$

represents the *hazard function* for a rare mutation to occur and therefore $E[1 - \exp(-V_t^N)]$ is the mean of $x_2^N(t)$. Hence, the growth of the spatial intensity in the logistic branching particle model relates to the emergence of the rare mutant population.

As a simple application of this dual, we can now verify (13), (14) when there is only one site. In the case $d = 0$, we have a linear birth process with birth rate s and therefore V_t^N grows like e^{st} and therefore the first rare mutation occurs after a time of order $O\left(\frac{\log L}{s}\right)$. On the other hand when $d > 0$ the process does not grow indefinitely but approaches an equilibrium. In this case, V_t^N grows only in a linear fashion and therefore requires time of order $O(L)$.

We note that since in the process, respectively, the dual, time has to be read forward, respectively, backward, from t, the rare mutation jumps occuring in times $\alpha^{-1}\log N + t$ for the dual after the expansion of the population up to time $\alpha^{-1}\log N$, correspond to rare mutations in the original process occurring at times between 0 and t and then growing and eventually taking over after time $\alpha^{-1}\log N$.

4.4 Outlook on Set-Valued Duals

If we consider $|\mathbb{I}| < \infty$, then we can enrich the dual (by introducing different types of births, compare Sect. 4.4 and Sect. 8 in [8]) such that if we use for $\mathscr{F}_0^+$ *products of indicator functions* we obtain under the evolution *sums of products of indicator functions*, where the dynamic of the different summands is coupled by the transition occuring in the underlying process η. This can be used to introduce a *set-valued dual*.

Here, we briefly describe the main idea carried out in Sect. 8 of [8], [12] on how to describe the dual based on a set-valued process but where we introduce the order of individuals and we use a change in the coupling between summands.

To explain the main idea, we again consider the case $\mathbb{I} = \{1, 2\}$, $N = 1$. Now consider the case

$$\mathscr{F}_0^+ = 1_2. \tag{152}$$

Then at the random time τ at which one selection operation occurs we have

$$\mathscr{F}_\tau = 1 \otimes 1_2 + 1_1 \otimes 1_2. \tag{153}$$

We can now regard this as defining a subset of $\mathbb{I}^2$. If we now couple the transitions in the two summands in a different way, namely instead of (153) we write (note that we can do since the dual expression depends only on the marginal law of the summands not the joint law)

$$1_2 \otimes 1 + 1_1 \otimes 1_2, \tag{154}$$

then we can ensure that the summands continue to correspond to *disjoint* subsets of $\mathbb{I}^{\mathbb{N}}$ and therefore we obtain a dual process with values in subsets of $\mathbb{I}^{\mathbb{N}}$.

A much more complicated set-valued dual can be constructed for the general multitype Fisher–Wright diffusion with mutation, selection and migration. The key point is to first take the non-negative function-valued dual driven by the particle system we introduced and then to introduce the *order of factors* and an appropriate *coupling* of its decomposition into a set of summands. Then we replace the transition (140) by using the indicator of level sets A of the fitness function and writing:

$$g(u_1, \dots, u_n) \longmapsto 1_A(u_i) g(u_1, \dots, u_m) \\ + (1 - 1_A)(u_i) g(u_1, \dots, u_{i-1}, u_{m+1}, u_{i+1}, \dots, u_m). \quad (155)$$

This allows us to obtain a duality relation for general finite type space and additive selection with a bounded fitness function.

This duality relation can be interpreted in terms of the *ancestral lines of a tagged sample* of individuals picked from the time-t population, since the dual now gives a decomposition in disjoint events for the ancestral lines and genealogical tree for a tagged sample of n individuals from the time t population from which we can read off current types and genealogical distance of the tagged sample.

References

1. Bürger, R.: The Mathematical Theory of Selection, Recombination, and Mutation, Wiley (2001)
2. Cox, J.T., Dawson, D.A., Greven, A.: Mutually catalytic super branching random walks: Large finite systems and renormalization analysis. Memoirs of the AMS, **171**, No. 809, 2nd of 4 numbers (2004)
3. Dawson, D.A.: Measure-valued Markov Processes. In: École d'Été de Probabilités de Saint Flour XXI, Lecture Notes in Mathematics, **1541**, pp. 1–261. Springer (1993)
4. Dawson, D.A., Greven, A.: Multiple time scale analysis of interacting diffusions. Probab. Theory Rel. Fields, **95**, 467–508 (1993)
5. Dawson, D.A., Greven, A.: Hierarchical models of interacting diffusions: multiple time scale phenomena. Phase transition and pattern of cluster-formation. Probab. Theory Rel. Fields, **96**, 435–473 (1993)
6. Dawson, D.A., Greven, A.: Multiple space-time scale analysis for interacting branching models. Electron. J. Probab. **1**(14), 1–8 (1996)
7. Dawson, D.A., Greven, A.: Hierarchically interacting Fleming-Viot processes with selection and mutation: Multiple space time scale analysis and quasi equilibria. Electron. J. Probab. **4**(4), 1–81 (1999)
8. Dawson, D., Greven, A.: On the effects of migration in spatial Fleming-Viot models with selection and mutation, in preparation (2011)
9. Dawson, D.A., Greven, A., Vaillancourt, J.: Equilibria and Quasi-equilibria for infinite systems of Fleming-Viot processes. Trans. Am. Math. Soc. **347**(7), 2277–2360 (1995)
10. Dawson, D.A., Kurtz, T.G.: Applications of duality to measure-valued diffusions. Springer Lecture Notes in Control and Inf. Sci. **42**, 177–191 (1982)
11. Dawson, D.A., Greven, A.: Invasion by rare mutants in a spatial two-type system with selection, arXiv:1104.0253v1[math.PR] (2011)
12. Dawson, D.A., Greven, A.: Duality for spatially interacting Fleming-Viot processes with mutation and selection, arXiv:1104.1099v1[math.PR] (2011)
13. Depperschmidt, A., Greven, A., Pfaffelhuber, P.: Tree-valued Fleming-Viot dynamics with mutation and selection, Submitted to Annals of Probab. (2010)
14. Ethier, S.N., Kurtz, T.: *Markov Processes. Characterization and Convergence*, Paperback reprint in 2005, Wiley, New York (1986)

15. Ethier, S.N., Kurtz, T.G.: Convergence to Fleming-Viot processes in the weak atomic topology. Stoch. Process Appl. **54**, 1–27 (1994)
16. Gärtner, J.: On the McKean–Vlasov limit for interacting diffusions. Math. Nachr. **137**, 197–248 (1988)
17. Greven, A., Pfaffelhuber, P., Winter, A.: Convergence in distribution of random metric measure spaces (The Λ-coalescent measure tree). PTRF **145**(1), 285–322 (2009)
18. Greven, A., Pfaffelhuber, P., Winter, A.: Tree-valued resampling dynamics: Martingale problems and applications, submitted to PTRF 2010.
19. Hutzenthaler, M.: The virgin island model. Electr. J. Probab. **14**, 1117–1161 (2009)
20. Jagers, P.: Stability and instability in population dynamics. J. Appl. Probab. **29**, 770–780 (1992)
21. Jagers, P., Nerman, O.: The growth and composition of branching populations. Adv. in Appl. Probab. **16**, 221–259 (1984)
22. McKean, H.P. Jr.: A class of Markov processes associated with nonlinear parabolic equations. Proc. N.A.S., USA **56**, 1907–19 (1966)
23. Nerman, O.: On the convergence of supercritical general (C-M-J) branching processes. Zeitschrift f. Wahrscheinlichkeitsth. verw. Gebiete **57**, 365–395 (1981)
24. Pitman, J., Yor, M.: A decomposition of Bessel bridges. Z. Wahr. verw. Geb. **59**, 425–457 (1982)
25. Rogers, L.C.G., Williams, D.: Diffusions, Markov processes and martingales. **2**, Wiley (1987)
26. Schirmeier, F.: A spatial population model in separating time windows. Master thesis, Department Mathematik, Erlangen (2010)

Properties of States of Super-α-Stable Motion with Branching of Index $1 + \beta$

Klaus Fleischmann, Leonid Mytnik, and Vitali Wachtel

Abstract It has been well known for a long time that the measure states of the process in the title are absolutely continuous at any fixed time provided that the dimension of space is small enough. However, besides the very special case of one-dimensional continuous super-Brownian motion, properties of the related density functions were not well understood. Only in 2003, Mytnik and Perkins [21] revealed that in the Brownian motion case and if the branching is discontinuous, there is a dichotomy for the densities: Either there are continuous versions of them or they are locally unbounded. We recently showed that the same type of fixed time dichotomy holds also in the case of discontinuous motion. Moreover, the continuous versions are locally Hölder continuous, and we determined the optimal index for them. Finally, we determine the optimal index of Hölder continuity at given space points which is strictly larger than the optimal index of local Hölder continuity.

AMS 2010 Subject Classification. Primary 60J80; Secondary 60G57.

K. Fleischmann (✉)
Weierstrass Institute for Applied Analysis and Stochastics, Leibniz Institute in Forschungsverbund Berlin e.V., Mohrenstr. 39, D–10117 Berlin, Germany
e-mail: fleischm@wias-berlin.de

L. Mytnik
Faculty of Industrial Engineering and Management, Technion Israel Institute of Technology, Haifa 32000, Israel
e-mail: leonid@ie.technion.ac.il; http://ie.technion.ac.il/leonid.phtml

V. Wachtel
Mathematical Institute, University of Munich, Theresienstrasse 39, D–80333 Munich, Germany
e-mail: wachtel@mathematik.uni-muenchen.de

J.-D. Deuschel et al. (eds.), *Probability in Complex Physical Systems*, Springer Proceedings in Mathematics 11, DOI 10.1007/978-3-642-23811-6_16,

1 Model: Super-α-Stable Motion with Branching of Index $1 + \beta$

The process in the title, sometimes also called (α, d, β)-superprocess, is a finite measure-valued process $X = \{X_t : t \geq 0\}$ describing the evolution of populations of infinitesimally small individuals/particles. The process can be constructed as a limit of branching particle systems, where the particles move independently according to symmetric α-stable motions in Euclidean space R^d, and additionally they branch according to a branching mechanism in the domain of attraction of a stable law of index $1+\beta$. Here $\alpha \in (0,2]$ and $1+\beta \in (1,2]$. Of course, for the very special case of $\alpha = 2$ and $\beta = 1$ we obtain the famous continuous super-Brownian motion in R^d.

A convenient description of X can be given via the log-Laplace transition functional, which is determined by the log-Laplace equation

$$\frac{\mathrm{d}}{\mathrm{d}t} u_t = \boldsymbol{\Delta}_\alpha u_t + a u_t - b u_t^{1+\beta}, \quad t > 0, \tag{1}$$

with fixed constants $a \in \mathsf{R}$ and $b > 0$. Parameter a is responsible for the growth rate of the total mass process of X, and b can be seen as a scaling constant. Note that the branching is critical if $a = 0$. The fractional Laplacian $\boldsymbol{\Delta}_\alpha := -(-\boldsymbol{\Delta})^{\alpha/2}$ describes a symmetric stable motion in R^d of index $\alpha \in (0,2]$, whereas the two other terms at the right-hand side of (1) reflect the continuous-state branching of index $1 + \beta \in (1,2]$. To be more specific, the log-Laplace transition functional of the homogeneous measure-valued Markov process X is defined as

$$\log \mathbf{E}_\mu \exp \langle X_t, -\varphi \rangle = \big\langle \mu, -u(t,\cdot) \big\rangle, \quad t > 0. \tag{2}$$

Here $\mu \in \mathscr{M}_\mathrm{f}$ (the set of finite measures on R^d), $\varphi \geq 0$ is a test function, and the nonnegative function $u = \big\{u(s,x) : s > 0, x \in \mathsf{R}^d\big\}$ solves the log-Laplace integral equation

$$\begin{aligned} u(s,x) = & \int_{\mathsf{R}^d} \mathrm{d}y \; p_s^\alpha(y-x)\, \varphi(y) \\ & + \int_0^s \mathrm{d}r \int_{\mathsf{R}^d} \mathrm{d}y \; p_{s-r}^\alpha(y-x) \Big[a u(r,y) - b\big(u(r,y)\big)^{1+\beta} \Big], \end{aligned} \tag{3}$$

which is the mild form of the log-Laplace equation (1) with initial condition $u(0+,\cdot) = \varphi$. Also, p^α describes the transition kernel of the particles' α-stable motion.

This rather general model was introduced by Iscoe in his thesis 1980 at Carleton University, published in [10, 11], and investigated later by many authors. From the beginning, one of the central issues was the question of the nature of the states of the process X.

2 Dichotomy of States at Fixed Times

Since 1988 (see Fleischmann [4]) it is known that at any fixed time $t > 0$, with probability one, the measure $X_t = X_t(\mathrm{d}x)$ is absolutely continuous with respect to the Lebesgue measure, provided that the dimension d is sufficiently small: $d < \frac{\alpha}{\beta}$. To be more precise, in [4] it was assumed that $a = 0$, but $a \neq 0$ requires just the obvious changes. On the other hand, in all higher dimensions $d \geq \alpha/\beta$, the states are singular a.s. The singularity statement was proved in [4] only in the critical dimension, and for the case of $d > \alpha/\beta$ it follows from Theorem 7.3.4 of Dawson [1].

3 Absolutely Continuous States

From now on assume $d < \frac{\alpha}{\beta}$, that is, for any time t, the measure $X_t(\mathrm{d}x)$ has a density function $x \mapsto X_t(x)$, which by a slight abuse of notation is denoted by the same symbol X_t as the corresponding measure. How to characterize the density and what are its properties? In the very special case of one-dimensional continuous super-Brownian motion ($\alpha = 2, d = 1 = \beta$) it is well known that a jointly continuous density field $\{X_t(x) : t > 0, x \in \mathsf{R}\}$ exists and satisfies a stochastic partial differential equation (SPDE); see Konno and Shiga [17] as well as Reimers [23]. However, it took a long time to make some progress in the case of $\beta < 1$. For the Brownian case $\alpha = 2$ (and $a = 0$, that is, critical branching), it was proved in Mytnik [20] that a version of the density $\{X_t(x) : t > 0, x \in \mathsf{R}^d\}$ of the measure $X_t(\mathrm{d}x)\mathrm{d}t$ exists that satisfies – in a weak sense – the following SPDE:

$$\frac{\partial}{\partial t}X_t(x) = \boldsymbol{\Delta} X_t(x) + \left(X_{t-}(x)\right)^{1/(1+\beta)} \dot{L}(t,x), \tag{4}$$

where $\dot{L}$ is a $(1+\beta)$-stable noise without negative jumps. Of course, this is a counterpart of the SPDE result for the continuous super-Brownian motion mentioned above.

From now on assume that $\beta < 1$. In other words, we restrict our discussion to the case of a discontinuous branching mechanism and ask for properties of the density function at fixed times.

3.1 *Dichotomy of Density Functions*

The particular case of $\alpha = 2$ was first treated in [21], where regularity and irregularity properties of the density at fixed times t had been revealed. More precisely, it was shown that:

- These densities have continuous versions if $d = 1$.

- They are locally unbounded on open sets of positive $X_t(\mathrm{d}x)$-measure in all higher dimensions.

However the case of discontinuous motion ($\alpha < 2$) was not considered in [21].

Here, we take care of the general case of $\alpha \leq 2$, that is, we include also discontinuous underlying motions. We show that the same type of fixed time dichotomy still holds, and this is our first result, taken from Fleischmann et al. [5].

Theorem 1 (Fixed time dichotomy of density function). *Recall that* $\alpha \leq 2$, $\beta < 1$, *and* $d < \frac{\alpha}{\beta}$. *Fix an initial state* $X_0 = \mu \in \mathcal{M}_\mathrm{f}$ *and any time* $t > 0$.

(a) (Continuity) *If* $d = 1$ *and* $1 + \beta < \alpha$, *then a.s. there exists a continuous version, say* $\widetilde{X}_t$, *of the density function of the measure* $X_t(\mathrm{d}x)$.

(b) (Local unboundedness) *If* $d > 1$ *or* $1 + \beta \geq \alpha$, *then a.s., for all open* $U \subseteq \mathsf{R}^d$,

$$\|X_t\|_U := \operatorname*{ess\,sup}_{x \in U} X_t(x) = \infty \text{ whenever } X_t(U) > 0. \tag{5}$$

The proof of (b) is rather technical, heavily uses ideas of [21], and roughly goes as follows. Let U be a fixed open ball. One first shows that on the event $\{X_t(U) > 0\}$ there are always sufficiently "big" jumps of X that occur in U close to time t. Then with the help of properties of solutions of the log-Laplace equation one is able to show that the "big" jumps are large enough to ensure the unboundedness of the density at time t in U. Loosely speaking, the density is getting unbounded in the proximity of "big" jumps. Finally, the exceptional set concerning the a.s. statement in (b) can be chosen uniformly in U since each U contains a non-empty ball with rational center and radius.

As for (a), the continuity of the density is verified via the Kolmogorov criterion. Note that besides the continuity, this criterion gives also some Hölder exponent of continuity. This immediately raises the question of determining the optimal Hölder index for the density and this question is addressed in the next section. Here, we just mention that in the case of one-dimensional continuous super-Brownian motion ($\alpha = 2, d = 1 = \beta$), the densities are locally Hölder continuous (in the spatial variable) for any index $\eta < \frac{1}{2}$, and the bound $\frac{1}{2}$ is moreover optimal.

3.2 Local Hölder Continuity of Continuous Density Functions

Here is our next result, again taken from [5].

Theorem 2 (Local Hölder continuity). *Fix* $X_0 = \mu \in \mathcal{M}_\mathrm{f}$ *and* $t > 0$. *Suppose* $d = 1$ *and* $1 + \beta < \alpha$.

(a) (Local Hölder continuity) *For each* $\eta < \eta_\mathrm{c} := \frac{\alpha}{1+\beta} - 1$, *with probability one, the continuous version* $\widetilde{X}_t$ *of the density function* X_t *is locally Hölder continuous*

of index η :

$$\sup_{x_1,x_2\in K,x_1\neq x_2} \frac{\left|\widetilde{X}_t(x_1)-\widetilde{X}_t(x_2)\right|}{|x_1-x_2|^{\eta}} < \infty, \quad \text{for all } K\subset \mathsf{R} \text{ compact.} \tag{6}$$

(b) (Optimality of η_c*)* *With probability one, for any open* $U\subseteq \mathsf{R}$,

$$\sup_{x_1,x_2\in U,x_1\neq x_2} \frac{\left|\widetilde{X}_t(x_1)-\widetilde{X}_t(x_2)\right|}{|x_1-x_2|^{\eta_c}} = \infty \quad \text{whenever } X_t(U)>0. \tag{7}$$

However, how is the optimal index η_c related to the optimal index $\frac{1}{2}$ in the excluded boundary case of continuous super-Brownian motion?

3.3 Some Transition Curiosity

Suppose for the moment that $\alpha=2$, and let $\beta\uparrow 1$. Then

$$\eta_c = \frac{2}{1+\beta}-1 \downarrow 0 \neq \frac{1}{2}. \tag{8}$$

That is, we have some surprising discontinuity while passing to the boundary case of continuous super-Brownian motion. How to understand this phenomenon? An explanation can be given using the notion of Hölder continuity at a point. The latter notion is recalled in the next section, and related results concerning our process are given.

3.4 Hölder Continuity at a Given Point

Recall that a function f is Hölder continuous with index $\eta\in(0,1)$ *at a point* x_0 if there exists a neighborhood $U(x_0)$ such that

$$\left|f(x)-f(x_0)\right| \leq C\,|x-x_0|^{\eta}, \quad x\in U(x_0). \tag{9}$$

The *optimal* Hölder index, say $H(x_0)$, of f at the point x_0 is defined as the supremum over all such η. Of course, there exist functions f where $H(x_0)$ indeed depends on x_0. Clearly, the optimal index of local Hölder continuity in a domain is determined by the infimum of $H(x_0)$ over points x_0 in the domain.

This phenomenon of difference of optimal index of local Hölder continuity from optimal Hölder index at some points can be observed in our case of the continuous densities of superprocesses with discontinuous branching ($\beta<1$). Here is our result, presented in Fleischmann et al. [6].

Theorem 3 (Hölder continuity at a given point). *Fix* $X_0 = \mu \in \mathcal{M}_{\mathrm{f}}$ *and* $t > 0$ *as well as* $x_0 \in \mathsf{R}$. *Suppose* $d = 1$ *and* $1+\beta < \alpha$.

(a) (Hölder continuity at a given point) *For each* $\eta > 0$ *satisfying*

$$\eta < \bar{\eta}_{\mathrm{c}} := \min\left\{\frac{1+\alpha}{1+\beta} - 1,\, 1\right\}, \tag{10}$$

with probability one the continuous version $\widetilde{X}_t$ *of the density is Hölder continuous of order* η *at the point* x_0. *That is, for all neighborhoods* $U(x_0)$ *of* x_0,

$$\sup_{x\in U(x_0),\, x\neq x_0} \frac{\left|\widetilde{X}_t(x) - \widetilde{X}_t(x_0)\right|}{|x-x_0|^{\eta}} < \infty. \tag{11}$$

(b) (Optimality of $\bar{\eta}_{\mathrm{c}}$*)* *If additionally* $\beta > (\alpha-1)/2$, *then* $\bar{\eta}_{\mathrm{c}}$ *is optimal. That is, with probability one for all neighborhoods* $U(x_0)$ *of* x_0,

$$\sup_{x\in U(x_0),\, x\neq x_0} \frac{\left|\widetilde{X}_t(x) - \widetilde{X}_t(x_0)\right|}{|x-x_0|^{\bar{\eta}_{\mathrm{c}}}} = \infty \quad \text{whenever } X_t(x_0) > 0. \tag{12}$$

Note that, in fact,

$$\bar{\eta}_{\mathrm{c}} > \eta_{\mathrm{c}}\,. \tag{13}$$

The above results imply that for the density $\widetilde{X}_t$ the optimal Hölder index H varies from point to point. The optimal local Hölder index $\eta_{\mathrm{c}} = \frac{\alpha}{1+\beta} - 1$ equals the infimum of H over an open domain. Therefore, there have to be (random) points x_0 in the domain with $H(x_0)$ arbitrary close or equal to η_{c}. On the other hand by Theorem 3 there are also points x_0 with the optimal Hölder index $H(x_0) = \bar{\eta}_{\mathrm{c}} > \eta_{\mathrm{c}}$, and hence we can conclude that H varies from point to point.

Heuristically the reason for the fact that there exist points in an open domain with different Hölder indexes is as follows. It can be seen from the proofs that the Hölder index at a point is highly influenced by relatively "big" jumps of the superprocess that occur close to time t in the proximity of the point. Therefore when we choose any fixed point in space, the size of the "biggest" jump close to it may be, and in fact is, much smaller than the "biggest" jump somewhere in an open domain which in turn influences the index of continuity at some exceptional random point in the domain. This consequently implies that the local Hölder index of continuity in an open domain is smaller than the modulus of continuity at a fixed point.

Now note that, if $\alpha = 2$, then as $\beta \uparrow 1$,

$$\bar{\eta}_{\mathrm{c}} = \left(\frac{3}{1+\beta} - 1\right) \wedge 1 \downarrow \frac{1}{2}\,. \tag{14}$$

That is, for Hölder continuity at fixed points we have continuity for the optimal index $\bar{\eta}_c$ while approaching the boundary case of $\beta = 1$, whereas earlier we observed discontinuity of η_c. Heuristically this can be explained as follows. We conjecture that the fact that the optimal Hölder index of continuity at fixed points equals $\bar{\eta}_c$ implies that, with probability one, the optimal Hölder index $H(x)$ of continuity at $X_t(\mathrm{d}x)$-a.e. point x equals $\bar{\eta}_c$, and moreover the Lebesgue measure of the points with optimal Hölder index of continuity η_c equals 0. On the other hand for continuous super-Brownian motion we have that almost surely $H(x) = \frac{1}{2}$ for all x, and so we see that the continuity at the boundary case $\beta = 1$ holds for the optimal Hölder index $\bar{\eta}_c$ that describes the modulus of continuity at a.e. point and not just at exceptional points.

3.5 Some Open Problems

We would like to list here some open problems.

At the first sight, in Theorem 3(b) there is the additional assumption $\beta > (\alpha - 1)/2$. But note that the opposite case $\beta \leq (\alpha-1)/2$ implies that $\bar{\eta}_c = 1$. Therefore the optimality of $\bar{\eta}_c$ follows here automatically from the definition of $H(x_0)$. Our first conjecture deals with the finer analysis of the case $\beta < (\alpha - 1)/2$. Note that we exclude here the boundary case of $\beta = (\alpha - 1)/2$.

Conjecture 1 (Lipschitz). Let $\beta < (\alpha - 1)/2$. Then at any given point x_0, with probability one, the density function $\widetilde{X}_t$ is Lipschitz continuous at x_0. ◇

Next we turn to the topic of so-called *multifractal spectrum* of random functions and measures. It has attracted attention already for a while and has been studied for example by the following authors: Dembo et al. [2], Durand [3], Hu and Taylor [9], Klenke and Mörters [16], Le Gall and Perkins [18], Mörters and Shieh [19], and Perkins and Taylor [22]. The multifractal spectrum of singularities that describe the Hausdorff dimension of sets of different Hölder exponents of functions was investigated for deterministic and random functions in Jaffard [12]–[14] as well as in Jaffard and Meyer [15].

Based on experience concerning the mentioned papers, we have the following conjectures in our situation:

Conjecture 2 (Multifractal spectrum). Fix $X_0 = \mu \in \mathcal{M}_\mathrm{f}$ and $t > 0$. Let $d = 1$ and $1 + \beta < \alpha$.

(a) (Full spectrum) We conjecture that for any $\eta \in (\eta_c, \bar{\eta}_c)$ with probability one there are (random) points $x_0 \in \mathsf{R}$ such that the optimal Hölder index $H(x_0)$ of $\widetilde{X}_t$ at x_0 is exactly η.

(b) (Hausdorff dimension) For $\eta \in (\eta_c, \bar{\eta}_c)$, let $D(\eta)$ denote the Hausdorff dimension of the (random) set $\{x_0 : H(x_0) = \eta\}$. We conjecture that

$$\lim_{\eta \downarrow \eta_c} D(\eta) = 0 \quad \text{and} \quad \lim_{\eta \uparrow \bar{\eta}_c} D(\eta) = 1 \quad \text{a.s.}$$

◇

That function $\eta \mapsto D(\eta)$ reveals the multifractal spectrum concerning the optimal Hölder index H for the densities of superprocesses with branching of index $1+\beta<\alpha$ and it is definitely worth studying.

We would also like to mention here that we got ideas to study regularity properties of the densities of (α, d, β)-superprocesses when we had been dealing with super-α-stable motions, say X^0, with *Neveu's branching mechanism.* This process is defined via the log-Laplace equation

$$\frac{\mathrm{d}}{\mathrm{d}t} u_t = \boldsymbol{\Delta}_\alpha u_t + a u_t - u_t \log u_t, \quad t > 0, \tag{15}$$

$0<\alpha\leq 2$, and consequently formally corresponds to the earlier excluded boundary case of $\beta=0$. By the way, (15) is interesting in itself since the nonlinear term has a local non-Lipschitz property. Despite the fact that the branching mechanism has infinite expectation here, the process X^0 exists and was constructed in Fleischmann and Sturm [7]. It was also shown there, that the process is immortal and propagates mass instantaneously everywhere in space, opposed, for instance, to supercritical super-Brownian motions with finite expectation; see [7, Proposition 16]. The large-scale behavior of X^0 is also not at all typical for supercritical spatial branching processes. In fact, in Fleischmann and Wachtel [8, Theorem 1] it was shown that X_t^0 normalized by its total masses $X_t^0(\mathsf{R}^d)$, with time t speeded up by a factor k, and contracted in space by $k^{1/\alpha}$, converges as $k \uparrow \infty$ toward a measure-valued process describing a single atom of mass one which fluctuates in macroscopic time according to an α-stable process.

What else can be expected concerning the nature of states of X^0? Here, we have the following conjectures about fixed time state properties.

Conjecture 3 (Superprocess with Neveu's branching mechanism). Fix $X_0^0 = \mu \in \mathscr{M}_\mathrm{f}$ and $t>0$.

(a) (Absolute continuity) In all dimensions, with probability one, the measure $X_t^0(\mathrm{d}x)$ is absolutely continuous.

(b) (Dichotomy of density functions) If $d=1$ and $\alpha>1$, with probability one there exists a continuous version, say $\widetilde{X}_t^0$, of the density function X_t^0 of the measures $X_t^0(\mathrm{d}x)$. On the other hand, if $d>1$ or $\alpha\leq 1$, we have local unboundedness of X_t^0.

For the remaining statements, suppose $d=1$ and $\alpha>1$.

(c) (Optimal local Hölder continuity) Write $\eta_\mathrm{c}^0 := \alpha - 1$. Then the continuous density $\widetilde{X}_t^0$ is locally Hölder continuous of every index $\eta < \eta_\mathrm{c}^0$. Moreover, η_c^0 is optimal.

(d) (Lipschitz continuity at a given point) Fix $x_0 \in \mathsf{R}$. Then a.s. $\widetilde{X}_t^0$ is Lipschitz continuous at x_0. ◇

Note that all these conjectures are based on our results above together with Conjecture 1, by letting formally $\beta \downarrow 0$.

By the way, as in the earlier case of (α, d, β)-superprocesses with $\beta > 0$, the verification of existence of density functions should be done along the lines of proving existence of mild fundamental solutions to the log-Laplace equation (15). Such solutions should exist despite the nonlocal Lipschitz property in the branching mechanism.

4 Main Tools to get the Hölder Statements

Clearly, a standard procedure to get an optimal Hölder index of continuity is via Kolmogorov's criterion by using "high" moments. This, for instance, can be done in the case of one-dimensional continuous super-Brownian motion ($\alpha = 2$, $d = 1 = \beta$) to show local Hölder continuity in the space variable of any index smaller than $\frac{1}{2}$, see the estimates in the proof of Corollary 3.4 in Walsh [24]. But in our $\beta < 1$ case, "high" moments do not exist, and it turns out that we cannot use this method for the entire range of parameters α, β. Hence we have to go deeply into the jump structure of the superprocess to obtain the needed estimates.

Actually, the starting point for all of our Hölder proofs is the well-known martingale decomposition of the (α, d, β)-superprocess X, valid for any α, d, β; see, e.g., [5, Lemma 1.6]:

For all sufficiently smooth bounded test functions $\varphi \geq 0$ on R^d and $t \geq 0$,

$$\langle X_t, \varphi\rangle = \langle \mu, \varphi\rangle + \int_0^t \mathrm{d}s\ \langle X_s, \boldsymbol{\Delta}_\alpha \varphi\rangle + M_t(\varphi) + a\ I_t(\varphi), \tag{16}$$

with discontinuous martingale

$$t \mapsto M_t(\varphi) := \int_{(0,t]\times\mathsf{R}^d\times\mathsf{R}_+} \tilde{N}\big(\mathrm{d}(s,x,r)\big)\, r\, \varphi(x) \tag{17}$$

and increasing process

$$t \mapsto I_t(\varphi) := \int_0^t \mathrm{d}s\ \langle X_s, \varphi\rangle. \tag{18}$$

Here $\tilde{N} := N - \hat{N}$, where $N\big(\mathrm{d}(s,x,r)\big)$ is a Poisson random measure on $(0,\infty) \times \mathsf{R}^d \times (0,\infty)$ describing all the jumps $r\delta_x$ of X at times s at sites x of size r, which are the only discontinuities of the process X. Moreover,

$$\hat{N}\big(\mathrm{d}(s,x,r)\big) = \varrho\ \mathrm{d}s\ X_s(\mathrm{d}x)\, r^{-2-\beta}\mathrm{d}r \tag{19}$$

is the compensator of N, where $\varrho := b\,(1+\beta)\beta/\Gamma(1-\beta)$ with Γ denoting the Gamma function.

Recall that under our assumption $d < \frac{\alpha}{\beta}$, for fixed $t > 0$, the random measure $X_t(\mathrm{d}x)$ is absolutely continuous a.s. From the Green function representation related to (16), see, e.g., [5, (1.9)], we obtain the following representation of a version of the density function of $X_t(\mathrm{d}x)$ (see, e.g., [5, (1.12)]):

$$\begin{aligned} X_t(x) &= \int_{\mathsf{R}^d} \mu(\mathrm{d}y)\, p_t^\alpha(x-y) + \int_{(0,t]\times\mathsf{R}^d} M\big(\mathrm{d}(s,y)\big)\, p_{t-s}^\alpha(x-y) \\ &\quad + a \int_{(0,t]\times\mathsf{R}^d} I\big(\mathrm{d}(s,y)\big)\, p_{t-s}^\alpha(x-y) \\ &=: Z_t^1(x) + Z_t^2(x) + Z_t^3(x), \quad x \in \mathsf{R}^d. \end{aligned} \tag{20}$$

Here $M\big(\mathrm{d}(s,y)\big)$ is the martingale measure related to (17) and $I\big(\mathrm{d}(s,y)\big)$ the random measure related to (18).

It is easy to see that the deterministic function Z_t^1 is locally Lipschitz continuous. It is also not difficult to show that Z_t^3 is a.s. locally Lipschitz, and hence we focus our attention on the main term Z_t^2 involving the martingale measure M. Note that Z_t^2 is the most difficult term to analyze. Here, the starting point is that the random increment $Z_t^2(x_1) - Z_t^2(x_2)$ can be represented as the difference of the values of two spectrally positive $(1+\beta)$-stable processes L^1, L^2 at some random times T_+, T_-, respectively. Recall that per definition L is a spectrally positive $(1+\beta)$-stable process, if it is an R-valued time-homogeneous process with independent increments and with Laplace transform given by

$$\mathbf{E}\,\mathrm{e}^{-\lambda L(t)} = \mathrm{e}^{t\lambda^{1+\beta}}, \quad \lambda, t \geq 0. \tag{21}$$

Consequently, there is a representation

$$Z_t^2(x_1) - Z_t^2(x_2) = L^1(T_+) - L^2(T_-). \tag{22}$$

Here, the random times $T_\pm$ are given by

$$T_\pm := \int_0^t \mathrm{d}s \int_{\mathsf{R}} X_s(\mathrm{d}y)\, \big(p_{t-s}^\alpha(x_1-y) - p_{t-s}^\alpha(x_2-y)\big)_\pm^{1+\beta} \tag{23}$$

with $\pm$ referring to the positive and negative parts.

It follows from (22) that the Hölder continuity can be destroyed by "big" values of the processes L^1 and L^2. Now, it is known from the standard theory of spectrally positive stable processes that "big" values are due to "big" positive jumps. Thus, to prove the Hölder continuity, one needs to control all the jumps of the processes L^1, L^2 by time $T_\pm$. More precisely, we show in the proof that there are no jumps, which can destroy the Hölder continuity of order η smaller than the critical index η_c or $\bar{\eta}_\mathrm{c}$.

The more complicated parts are the optimality proofs for the indexes. To prove the optimality of η_{c} we show that there exists a sequence of "big" jumps of X that occur close to time t in the considered domain U in Theorem 2(b), and these jumps indeed destroy the local Hölder continuity of index η_{c}. But the existence of such a sequence is not sufficient for the proof of optimality. We need additionally to show that the influence of "big" jumps of one of the stable processes L^1, L^2 cannot be compensated by "big" jumps of the other one.

To prove the optimality of $\bar{\eta}_{\mathrm{c}}$, additionally the "big" jumps have to be found in the vicinity of the fixed x_0. Moreover, values of "big" jumps in Theorem 3(b) are of a smaller order than those in Theorem 2(b), because it is more likely to have "big" jumps in a domain than in a vicinity of a fixed point. This creates additional technical difficulties in the proof of Theorem 3(b).

Now, we would like to explain a bit the occurrence of the critical value $\eta_{\mathrm{c}} = \frac{\alpha}{1+\beta} - 1$ for local Hölder continuity in Theorem 2 and we will skip the discussion on $\bar{\eta}_{\mathrm{c}}$ which goes along similar lines.

Our first observation is that, up to a probability error of $\varepsilon \in (0, \frac{1}{1+\beta})$, all the jumps $\Delta M(s,y)$ of the martingale measure $M\big(\mathrm{d}(s,y)\big)$ at times $s<t$ are bounded by $(t-s)^{\frac{1}{1+\beta}-\varepsilon}$, that is,

$$\mathbf{P}\Big(\Delta M(s,y) \leq c\,(t-s)^{\frac{1}{1+\beta}-\varepsilon} \text{ for all } s<t \text{ and } y \in \mathsf{R}\Big) \geq 1-\varepsilon, \tag{24}$$

see [5, Lemma 2.14]. However, it follows from (20) and (22) that the jumps ΔL^1 and ΔL^2 of L^1 and L^2, respectively, generated by the jumps $\Delta M(s,y)$ do not exceed

$$\Delta M(s,y) \sup_{y \in \mathsf{R}} \big|p^{\alpha}_{t-s}(x_1-y) - p^{\alpha}_{t-s}(x_2-y)\big|. \tag{25}$$

Hence, from (24) and an estimate for α-stable kernels (see [5, Lemma 2.1]) we obtain the following bound

$$\begin{aligned} &c\,(t-s)^{\frac{1}{1+\beta}-\varepsilon} \sup_{y\in\mathsf{R}} \big|p^{\alpha}_{t-s}(x_1-y) - p^{\alpha}_{t-s}(x_2-y)\big| \\ &\leq c\,(t-s)^{\frac{1}{1+\beta}-\varepsilon} \frac{|x_1-x_2|^{\delta}}{(t-s)^{\delta/\alpha+1/\alpha}}, \qquad 0<\delta\leq 1, \end{aligned} \tag{26}$$

for the jumps ΔL^1 and ΔL^2. If now $\delta = \frac{\alpha}{1+\beta} - 1 - \alpha\varepsilon = \eta_{\mathrm{c}} - \alpha\varepsilon$ (for sufficiently small ε), then

$$\Delta L^1, \Delta L^2 \leq c\,|x_1-x_2|^{\eta_{\mathrm{c}}-\alpha\varepsilon}. \tag{27}$$

But if the jumps of a spectrally positive stable process are not "big", then the process values cannot be "big" as well. Consequently,

$$\mathbf{P}\Big(L^1, L^2 \leq c\,|x_1-x_2|^{\eta_{\mathrm{c}}-\alpha\varepsilon}\Big) \geq 1-\varepsilon. \tag{28}$$

In view of (22), the latter implies

$$\mathbf{P}\Big(\big|Z_t^2(x_1) - Z_t^2(x_2)\big| \leq c\,|x_1 - x_2|^{\eta_c - \alpha\varepsilon}\ \forall x_1, x_2 \in K\Big) \geq 1 - \varepsilon, \tag{29}$$

(with K a compact), which gives the Hölder continuity of Z_t^2 of any exponent smaller than η_c.

To show the optimality of η_c we first prove that there exists a sequence (s_n, y_n, r_n) such that

$$s_n \uparrow t,\ \ y_n \in (-1,1),\ \ \Delta M(s_n, y_n) = r_n \geq (t - s_n)^{\frac{1}{1+\beta}} \log \frac{1}{t - s_n}. \tag{30}$$

Using again a representation as in (22), with corresponding spectrally positive stable processes and random times indexed by n, we have

$$Z_t^2(y_n) - Z_t^2\big(y_n + (t - s_n)^{1/\alpha}\big) = L_n^1(T_{n,+}) - L_n^2(T_{n,-}). \tag{31}$$

One can see that the jump of L_n^1 generated by $\Delta M(s_n, y_n)$ is bounded from below by

$$r_n\Big(p_{t-s_n}^{\alpha}(0) - p_{t-s_n}^{\alpha}\big((t - s_n)^{1/\alpha}\big)\Big) \geq \big(p_1^{a}(0) - p_1^{a}(1)\big)\,(t - s_n)^{\frac{1}{1+\beta} - \frac{1}{\alpha}} \log \frac{1}{t - s_n}.$$

Now "big" jumps of a spectrally positive stable process lead to "big" values of the process, that is,

$$L_n^1(T_{n,+}) \geq c\ (t - s_n)^{\frac{1}{1+\beta} - \frac{1}{\alpha}} \log \frac{1}{t - s_n}. \tag{32}$$

Since the probability of having another "big" jump is small, one has

$$L_n^2(T_{n,-}) \leq c\,(t - s_n)^{\frac{1}{1+\beta} - \frac{1}{\alpha}}. \tag{33}$$

As a result we have

$$\begin{aligned} \frac{Z_t^2(y_n) - Z_t^2\big(y_n + (t - s_n)^{1/\alpha}\big)}{\big((t - s_n)^{1/\alpha}\big)^{\eta_c}} &= \frac{L_n^1(T_{n,+}) - L_n^2(T_{n,-})}{\big((t - s_n)^{1/\alpha}\big)^{\eta_c}} \\ &\geq c\ \frac{(t - s_n)^{\frac{1}{1+\beta} - \frac{1}{\alpha}} \log \frac{1}{t - s_n}}{(t - s_n)^{\frac{1}{1+\beta} - \frac{1}{\alpha}}} = c\ \log \frac{1}{t - s_n} \underset{n\uparrow\infty}{\longrightarrow} \infty. \end{aligned} \tag{34}$$

In other words, Z_t^2 is not Hölder continuous of index η_c.

Acknowledgements We thank an anonymous referee for a careful reading of our manuscript. This work was supported by the German Israeli Foundation for Scientific Research and Development, Grant No. G-807-227.6/2003. Moreover, Mytnik was partially supported by ISF grant and Wachtel by GIF Young Scientists Program Grant No. G-2241-2114.6/2009.

References

1. Dawson, D.A.: Infinitely divisible random measures and superprocesses. In: Stochastic analysis and related topics (Silivri, 1990), volume 31 of Progr. Probab., pp. 1–129. Birkhäuser Boston, Boston, MA (1992)
2. Dembo, A., Peres, Y., Rosen, J., Zeitouni, O.: Thick points for planar Brownian motion and the Erdös-Taylor conjecture on random walk. Acta Math. **186**, 239–270 (2001)
3. Durand, A.: Singularity sets of Lévy processes. Probab. Theor. Relat. Fields. **143**(3–4), 517–544 (2009)
4. Fleischmann, K.: Critical behavior of some measure-valued processes. Math. Nachr. **135**, 131–147 (1988)
5. Fleischmann, K., Mytnik, L., Wachtel, V.: Optimal local Hölder index for density states of superprocesses with $(1+\beta)$-branching mechanism. Ann. Probab. **38**(3), 1180–1220 (2010)
6. Fleischmann, K., Mytnik, L., Wachtel, V.: Hölder index at a given point for density states of super-α-stable motion of index $1+\beta$. J. Theor. Probab. **24**(1), 66–92 (2011)
7. Fleischmann, K., Sturm, A.: A super-stable motion with infinite mean branching. Ann. Inst. Henri Poincaré Probab. Stat. **40**(5), 513–537 (2004)
8. Fleischmann, K., Wachtel, V.: Large scale localization of a spatial version of Neveu's branching process. Stoch. Proc. Appl. **116**(7), 983–1011 (2006)
9. Hu, X., Taylor, S.J.: Multifractal structure of a general subordinator. Stoch. Process. Appl. **88**, 245–258 (2000)
10. Iscoe, I.: A weighted occupation time for a class of measure-valued critical branching Brownian motions. Probab. Theor. Relat. Fields. **71**, 85–116 (1986)
11. Iscoe, I.: Ergodic theory and a local occupation time for measure-valued critical branching Brownian motion. Stochastics **18**, 197–243 (1986)
12. Jaffard, S.: The multifractal nature of Lévy processes. Probab. Theor. Relat. Fields. **114**, 207–227 (1999)
13. Jaffard, S.: On Lacunary wavelet series. Ann. Appl. Probab. **10**(1), 313–329 (2000)
14. Jaffard, S.: Wavelet techniques in multifractal analysis. Proc. Symp. Pure Math. **72**(2), 91–151 (2004)
15. Jaffard, S., Meyer, Y.: Wavelet methods for pointwise regularity and local oscillations of functions. Memb. Am. Math. Soc. **123**, 587 (1996)
16. Klenke, A., Mörters, P.: The multifractal spectrum of Brownian intersection local time. Ann. Probab. **33**, 1255–1301 (2005)
17. Konno, N., Shiga,T.: Stochastic partial differential equations for some measure-valued diffusions. Probab. Theor. Relat. Fields. **79**, 201–225 (1988)
18. Le Gall, J.-F., Perkins, E.A.: The Hausdorff measure of the support of two-dimensional super-Brownian motion. Ann. Probab. **23**(4), 1719–1747 (1995)
19. Mörters, P., Shieh, N.R.: On the multifractal spectrum of the branching measure on a Galton-Watson tree. J. Appl. Probab. **41**, 1223–1229 (2004)
20. Mytnik, L.: Stochastic partial differential equation driven by stable noise. Probab. Theor. Relat. Fields. **123**, 157–201 (2002)
21. Mytnik, L., Perkins, E.: Regularity and irregularity of $(1+\beta)$-stable super-Brownian motion. Ann. Probab. **31**(3), 1413–1440 (2003)
22. Perkins, E.A., Taylor, S.J.: The multifractal structure of super-Brownian motion. Ann. Inst. H. Poincaré Probab. Stat. **34**(1), 97–138 (1998)
23. Reimers, M.: One-dimensional stochastic partial differential equations and the branching measure diffusion. Probab. Theor. Relat. Fields. **81**, 319–340 (1989)
24. Walsh, J.B.: An introduction to stochastic partial differential equations. volume 1180 of Lecture Notes Math., pp. 265–439. École d'Été de Probabilités de Saint-Flour XIV–1984, Springer, Berlin, (1986)

Part IV
Miscellaneous Topics in Statistical Mechanics

A Quenched Large Deviation Principle and a Parisi Formula for a Perceptron Version of the GREM

Erwin Bolthausen and Nicola Kistler

Dedicated to Jürgen Gärtner on the occasion of his 60th birthday.

Abstract We introduce a perceptron version of the Generalized Random Energy Model, and prove a quenched Sanov-type large deviation principle for the empirical distribution of the random energies. The dual of the rate function has a representation through a variational formula, which is closely related to the Parisi variational formula for the SK-model.

1 Introduction

There has been important progress in the mathematical study of mean-field spin glasses over the last 10 years. By results of Guerra [5] and Talagrand [7], the free energy of the Sherrington–Kirkpatrick model is known to be given by the formula predicted by Parisi [4]. Furthermore, the description of the *high* temperature is remarkably accurate, see [6] and references therein. On the other hand, results for the Gibbs measure at *low* temperature are more scarce and are restricted to models with a simpler structure, like Derrida's generalized random energy model, the GREM, [2] and [3], the nonhierarchical GREMs [1] and the p-spin model with large p [6]. To put on rigorous ground, the full Parisi picture remains a major

E. Bolthausen (✉)
Institut für Mathematik, Universität Zürich, Winterthurerstrasse 190, CH-8057 Zürich
e-mail: eb@math.uzh.ch

N. Kistler
Institut für Angewandte Mathematik, Rheinische Friedrich-Wilhelms-Universität Bonn, Endenicher Allee 60, 53115 Bonn, Germany
e-mail: nkistler@uni-bonn.de

J.-D. Deuschel et al. (eds.), *Probability in Complex Physical Systems*, Springer Proceedings in Mathematics 11, DOI 10.1007/978-3-642-23811-6_17,

challenge, and even more so in view of its alleged universality, at least for mean-field models.

We introduce here a model which hopefully sheds some new light on the issue.

In this paper, we derive the free energy, which can be analyzed by large deviation techniques. The limiting free energy turns out to be given by a Gibbs variational formula, which can be linked to a Parisi-type formula by a duality principle, so that it becomes evident why an infimum appears in the latter. This duality also gives an interesting interpretation of the Parisi order parameter in terms of the sequence of inverse of temperatures associated with the extremal measures from the Gibbs variational principle.

In a forthcoming paper, we will give a full description of the Gibbs measure in the thermodynamic limit in terms of the Ruelle cascades.

2 A Perceptron Version of the GREM

Let $\{X_{\alpha,i}\}_{\alpha\in\Sigma_N,1\leq i\leq N}$ be random variables which take values in a Polish space S equipped with the Borel σ-field $\mathcal{S}$, and defined on a probability space $(\Omega,\mathcal{F},\mathbb{P})$. We write $\mathcal{M}_1^+(S)$ for the set of probability measures on $(S,\mathcal{S})$, which itself is a Polish space. Σ_N is exponential in size, typically $|\Sigma_N| = 2^N$. It is assumed that all $X_{\alpha,i}$ have the same distribution μ, and that for any fixed $\alpha\in\Sigma_N$, the collection $\{X_{\alpha,i}\}_{1\leq i\leq N}$ is independent. It is, however, not assumed that they are independent for different α. The perceptron Hamiltonian is defined by

$$-H_{N,\omega}(\alpha) \stackrel{\text{def}}{=} \sum_{i=1}^{N} \phi\left(X_{\alpha,i}(\omega)\right), \tag{1}$$

where $\phi : S\to\mathbb{R}$ is a measurable function. One may allow that the index set for i is rather $\{1,\ldots,[aN]\}$ with a some positive real number, but for convenience, we always stick to $a=1$ here. The case which is best investigated (see [6]) takes for α spin sequences: $\alpha=(\sigma_1,\ldots,\sigma_N)\in\{-1,1\}^N$, $S=\mathbb{R}$, and the $X_{\alpha,i}$ are centered Gaussians with

$$\mathbb{E}\left(X_{\alpha,i}X_{\alpha',i'}\right) = \delta_{i,i'}\frac{1}{N}\sum_{j=1}^{N}\sigma_j\sigma_j'. \tag{2}$$

This is closely related to the SK-model, and is actually considerably more difficult. The model has been investigated by Talagrand [6], but a full Parisi formula for the free energy is lacking.

The Hamiltonian (1) can be written in terms of the empirical measure

$$L_{N,\alpha} \stackrel{\text{def}}{=} \frac{1}{N}\sum_{i=1}^{N}\delta_{X_{\alpha,i}}, \tag{3}$$

i.e.,

$$-H_{N,\omega}(\alpha) = N \int \phi(x) L_{N,\alpha}(\mathrm{d}x).$$

The quenched free energy is the almost sure limit of

$$\frac{1}{N} \log \sum_{\alpha} \exp[-H_{N,\omega}(\alpha)],$$

and it appears natural to ask whether this free energy can be obtained by a quenched Sanov-type large deviation principle for $L_{N,\alpha}$ in the following form:

Definition 2.1. We say that $\{L_N\}$ satisfies a **quenched large deviation principle** (in short QLDP) with good rate function $J : \mathcal{M}_1^+(S) \to [-\infty, \infty)$, provided the level sets of J are compact, and for any weakly continuous bounded map $\Phi : \mathcal{M}_1^+(S) \to \mathbb{R}$, one has

$$\lim_{N\to\infty} \frac{1}{N} \log \sum_{\alpha\in\Sigma_N} \exp[N\Phi(L_{N,\alpha})] = \log 2 + \sup_{\nu\in\mathcal{M}_1^+(S)} [\Phi(\nu) - J(\nu)], \ \mathbb{P}\text{-a.s.}$$

The annealed version of such a QLDP is just Sanov's theorem:

$$\begin{aligned}\lim_{N\to\infty} \frac{1}{N} \log \sum_{\alpha} \mathbb{E} \exp[N\Phi(L_{N,\alpha})] &= \log 2 + \lim_{N\to\infty} \frac{1}{N} \log \mathbb{E} \exp[N\Phi(L_{N,\alpha})] \\ &= \log 2 + \sup_{\nu} (\Phi(\nu) - H(\nu|\mu)),\end{aligned}$$

where $H(\nu|\mu)$ is the usual relative entropy of ν with respect to μ, the latter being the distribution of the $X_{\alpha,i}$:

$$H(\nu|\mu) \stackrel{\text{def}}{=} \begin{cases} \int \log \frac{\mathrm{d}\nu}{\mathrm{d}\mu}\, \mathrm{d}\nu & \text{if } \nu \ll \mu \\ \infty & \text{otherwise} \end{cases}.$$

There is no reason to believe that $H(\nu|\mu) = J(\nu)$.

Conjecture 2.2. *The empirical measures $\{L_{N,\alpha}\}$ with (2) satisfy a QLDP.*

We do not know how this conjecture could be proved, nor do we have a clear picture what J should be in this case. The only support we have for the conjecture is that it is true in a perceptron version of the GREM, a model we are now going to describe.

For $n \in \mathbb{N}$, $\alpha = (\alpha_1, \ldots, \alpha_n)$ with $1 \le \alpha_k \le 2^{\gamma_i N}$, $\sum_k \gamma_k = 1$, and $1 \le i \le N$, let

$$X_{\alpha,i} = \left(X^1_{\alpha_1,i}, X^2_{\alpha_1,\alpha_2,i}, \ldots, X^n_{\alpha_1,\alpha_2,\ldots,\alpha_n,i}\right),$$

where the X^j are independent, taking values in some Polish Space $(S, \mathcal{S})$ with distribution μ_j. For notational convenience, we assume that the $\gamma_i N$ are all integers. Put

$$\Gamma_j \stackrel{\text{def}}{=} \sum_{k=1}^{j} \gamma_j.$$

We assume that all the variables in the bracket are independent. The $X_{\alpha,i}$ take values in S^n. The distribution is

$$\mu \stackrel{\text{def}}{=} \mu_1 \otimes \cdots \otimes \mu_n.$$

The empirical measure $L_{N,\alpha}$ is defined by (3), which is a random element in $\mathcal{M}_1^+(S^n)$. n is fixed in all we are doing.

Given a measure $\nu \in \mathcal{M}_1^+(S^n)$, and $1 \le j \le n$, we write $\nu^{(j)}$ for its marginal on the first j coordinates. We define subsets $\mathcal{R}_j$ of $\mathcal{M}_1^+(S^n)$, $1 \le j \le n$ by

$$\mathcal{R}_j \stackrel{\text{def}}{=} \left\{\nu \in \mathcal{M}_1^+(S^n) : H\left(\nu^{(j)} \mid \mu^{(j)}\right) \le \Gamma_j \log 2\right\}.$$

We will also consider the sets

$$\mathcal{R}_j^= \stackrel{\text{def}}{=} \left\{\nu \in \mathcal{M}_1^+(S^n) : H\left(\nu^{(j)} \mid \mu^{(j)}\right) = \Gamma_j \log 2\right\}.$$

For $\nu \in \mathcal{M}_1^+(S^n)$, let

$$J(\nu) = \begin{cases} H(\nu \mid \mu) \text{ if } \nu \in \bigcap_{j=1}^n \mathcal{R}_j \\ \infty \qquad \text{otherwise} \end{cases}.$$

It is evident that J is convex and has compact level sets.

Our first main result is:

Theorem 2.3. *$\{L_{N,\alpha}\}$ satisfies a QLDP with rate function J.*

For the rest of this section, we will focus on linear functionals, $\Phi(\nu) = \int \phi(x)\nu(dx)$, for a bounded continuous function $\phi : S^n \to \mathbb{R}$. For a probability measure ν on S^n, we set

$$\text{Gibbs}(\phi, \nu) \stackrel{\text{def}}{=} \int \phi(x)\nu(dx) - H(\nu \mid \mu),$$

and define the Legendre transform of J by

$$J^*(\phi) \stackrel{\text{def}}{=} \sup_\nu \left[\int \phi(x)\nu(dx) - J(\nu)\right] = \sup\left\{\text{Gibbs}(\phi, \nu) : \nu \in \bigcap\nolimits_{j=1}^n \mathcal{R}_j\right\}.$$

whenever the a.s.-limit exists. As a corollary of Theorem 2.3, we have

Corollary 2.4. *Assume that $\phi : S \to \mathbb{R}$ is bounded and continuous.*

$$\lim_{N\to\infty} \frac{1}{N} \log \sum_{\alpha} \exp\left[\sum\nolimits_{i=1}^{N} \phi\left(X_{\alpha,i}\right)\right] = J^{*}(\phi) + \log 2, \text{ a.s.}$$

We next discuss a dual representation of $J^{*}(\phi)$. Essentially, this comes up by investigating which measures solve the variational problem. Remark that without the restrictions $\nu \in \bigcap_{j=1}^{n} \mathcal{R}_j$, we would simply get

$$\mathrm{d}\nu = \frac{\mathrm{e}^{\phi}\mathrm{d}\mu}{\int \mathrm{e}^{\phi}\mathrm{d}\mu},$$

as the maximizer.

Let Δ be the set of sequences $\mathbf{m} = (m_1, \ldots, m_n)$ with $0 < m_1 \le m_2 \le \cdots \le m_n \le 1$. For $\mathbf{m} \in \Delta$, and $\phi : S^n \to \mathbb{R}$ bounded, we define recursively functions ϕ_j, $0 \le j \le n$, $\phi_j : S^j \to \mathbb{R}$, by

$$\phi_n \overset{\text{def}}{=} \phi, \tag{4}$$

$$\phi_{j-1}\left(x_1, \ldots, x_{j-1}\right) \overset{\text{def}}{=} \frac{1}{m_j} \log \int \exp\left[m_j \phi_j\left(x_1, \ldots, x_{j-1}, x_j\right)\right] \mu_j\left(\mathrm{d}x_j\right). \tag{5}$$

ϕ_0 is just a real number, which we denote by $\phi_0(\mathbf{m})$.

Remark that if some of the m_i agree, say $m_k = m_{k+1} = \cdots = m_l$, $k < l$, then ϕ_{k-1} is obtained from ϕ_l by

$$\phi_{k-1}(x_1, \ldots, x_{k-1}) = \frac{1}{m_k} \log \int \exp\left[m_k \phi_l(x_1, \ldots, x_{k-1}, x_k, \ldots, x_l)\right] \prod_{j=k}^{l} \mu_j\left(\mathrm{d}x_j\right).$$

In particular, if all the m_i are 1, then

$$\phi_0 = \log \int \exp[\phi]\, \mathrm{d}\mu.$$

This latter case corresponds to the "replica symmetric" situation. Put

$$\mathrm{Parisi}(\mathbf{m}, \phi) \overset{\text{def}}{=} \sum\nolimits_{i=1}^{n} \frac{\gamma_i \log 2}{m_i} + \phi_0(\mathbf{m}) - \log 2 \tag{6}$$

Theorem 2.5. *Assume that $\phi : S \to \mathbb{R}$ is bounded and continuous. Then*

$$J^{*}(\phi) = \inf_{\mathbf{m}\in\Delta} \mathrm{Parisi}(\mathbf{m}, \phi). \tag{7}$$

The expression for $J^{*}(\phi)$ in this theorem is very similar to the Parisi formula for the SK-model. Essentially the only difference is the first summand, which in the SK-case is a quadratic expression. In our case (in contrast to the still open situation in the SK-model), we can prove that the infimum is uniquely attained, as we will discuss below.

The derivation of the theorem from Corollary 2.4 is done by identifying first the possible maximizers in the variational formula for $J^*(\phi)$. They belong to a family of distributions, parametrized by **m**. The maximizer inside this family is then obtained by minimizing **m** according to (7), and one then identifies the two expressions. The procedure is quite standard in large deviation situations.

Two conventions: C stands for a generic positive constant, not necessarily the same at different occurences. If there are inequalities stated between expressions containing N, it is tacitely assumed that they are valid maybe only for large enough N.

3 Proofs

3.1 The Gibbs Variational Principle: Proof of Theorem 2.3

If $A \in \mathcal{S}$, we put $H(A \mid \mu) \stackrel{\text{def}}{=} \inf_{\nu \in A} H(\nu \mid \mu)$. If S is a Polish Space, and $\mathcal{S}$ its Borel σ-field, then it is well known that $\nu \to H(\nu \mid \mu)$ is lower semicontinuous in the weak topology. This follows from the representation

$$H(\nu \mid \mu) = \sup_{u \in \mathcal{U}} \left[\int u \, \mathrm{d}\nu - \log \int \mathrm{e}^u \mathrm{d}\mu \right], \tag{8}$$

where $\mathcal{U}$ is the set of bounded continuous functions $S \to \mathbb{R}$.

For $(S, \mathcal{S}), (S', \mathcal{S}')$ two Polish Spaces, and $\nu \in \mathcal{M}_1^+(S \times S')$. If $\mu \in \mathcal{M}_1^+(S)$, $\mu' \in \mathcal{M}_1^+(S')$ we have

$$H\left(\nu \mid \mu \otimes \mu'\right) = H\left(\nu^{(1)} \mid \mu\right) + H\left(\nu \mid \nu^{(1)} \otimes \mu'\right), \tag{9}$$

where $\nu^{(1)}$ is the first marginal of ν on S.

Lemma 3.1. *$H(\nu \mid \nu^{(1)} \otimes \mu')$ is a lower semicontinuous function of ν in the weak topology.*

Proof. Applying (8) to

$$H(\nu \mid \nu^{(1)} \otimes \mu') = \sup_{u \in \mathcal{U}} \left[\int u \mathrm{d}\nu - \log \int \mathrm{e}^u d \left(\nu^{(1)} \otimes \mu'\right) \right],$$

where $\mathcal{U}$ denotes the set of bounded continuous functions $S \times S' \to \mathbb{R}$. For any fixed $u \in \mathcal{U}$, both functions $\nu \to \int u \, \mathrm{d}\nu$ and $\nu \to \log \int \mathrm{e}^u d\left(\nu^{(1)} \otimes \mu'\right)$ are continuous, and from this the desired semicontinuity property follows. □

We will need the following "relative" version of Sanov's theorem. Consider three independent sequences of i.i.d. random variables $(X_i), (Y_i), (Z_i)$, taking values in three Polish spaces S, S', S'', and with laws μ, μ', μ''. We consider the empirical processes

$$L_N \stackrel{\text{def}}{=} \frac{1}{N}\sum_{i=1}^{N}\delta_{(X_i,Y_i)},\ R_N \stackrel{\text{def}}{=} \frac{1}{N}\sum_{i=1}^{N}\delta_{(X_i,Z_i)}.$$

The pair (L_N, R_N) takes values in $\mathcal{M}_1^+(S\times S')\times\mathcal{M}_1^+(S\times S'')$.

Lemma 3.2. *The sequence (L_N, R_N) satisfies an LDP with rate function*

$$J(\nu,\theta) = \begin{cases} H\left(\nu^{(1)} \mid \mu\right) + H\left(\nu \mid \nu^{(1)}\otimes\mu'\right) + H\left(\theta \mid \theta^{(1)}\otimes\mu''\right), & \text{if } \nu^{(1)} = \theta^{(1)} \\ \infty & \text{otherwise.}\end{cases}$$

Proof. We apply the Sanov theorem to the empirical measure

$$M_N = \frac{1}{N}\sum_{i=1}^{N}\delta_{(X_i,Y_i,Z_i)} \in \mathcal{M}_1^+(S\times S'\times S'').$$

We use the two natural projections $p : S\times S'\times S'' \to S\times S'$ and $q : S\times S'\times S'' \to S\times S''$. Then $(L_N, R_N) = M_N(p,q)^{-1}$, and by continuous projection, we get that (L_N, R_N) satisfies a good LDP with rate function

$$J'(\nu,\theta) = \inf\left\{H(\rho \mid \mu\otimes\mu'\otimes\mu'') : \rho p^{-1} = \nu, \rho q^{-1} = \theta\right\}.$$

It only remains to identify this rate function with the function J given above.

Clearly $J'(\nu,\theta) = \infty$ if $\nu^{(1)} \neq \theta^{(1)}$. Therefore, assume $\nu^{(1)} = \theta^{(1)}$. If we define $\hat{\rho}(\nu,\theta) \in \mathcal{M}_1^+(S\times S'\times S'')$ to have marginal $\nu^{(1)} = \theta^{(1)}$ on S, and the conditional distribution on $S'\times S''$ given the first projection is the product of the conditional distributions of ν and θ, then applying twice (9), we get

$$H(\hat{\rho} \mid \mu\otimes\mu'\otimes\mu'') = H\left(\nu^{(1)} \mid \mu\right) + H\left(\nu \mid \nu^{(1)}\otimes\mu'\right) + H\left(\theta \mid \theta^{(1)}\otimes\mu''\right),$$

and therefore $J \geq J'$.

To prove the other inquality, consider any ρ satisfying $\rho p^{-1} = \nu, \rho q^{-1} = \theta$. We want to show that $J(\nu,\theta) \leq H(\rho \mid \mu\otimes\mu'\otimes\mu'')$. For that, we can assume that the right-hand side is finite. Then

$$H\left(\rho \mid \mu\otimes\mu'\otimes\mu''\right) = H\left(\rho \mid \hat{\rho}(\nu,\theta)\right) + \int d\rho \log\frac{d\hat{\rho}(\nu,\theta)}{d\left(\mu\otimes\mu'\otimes\mu''\right)}.$$

The first summand is ≥ 0, and the second equals

$$\int \mathrm{d}\hat{\rho}(\nu,\theta)\log\frac{\mathrm{d}\hat{\rho}(\nu,\theta)}{d\left(\mu\otimes\mu'\otimes\mu''\right)} = J(\nu,\theta).$$

So, we have proved that

$$J(\nu,\theta) \leq H\left(\rho \mid \mu \otimes \mu' \otimes \mu''\right),$$

for any ρ satisfying $\rho p^{-1} = \nu, \rho q^{-1} = \theta$. □

We now step back to the setting of Theorem 2.3: For $j = 1, \ldots, n$, we have sequences $\left\{X^j_{\alpha_1,\ldots,\alpha_j,i}\right\}$ of independent random variables with distribution μ_j. We emphasize that henceforth $\mu = \mu_1 \otimes \cdots \otimes \mu_n$ and $\mu^{(j)}$ will denote the marginal on the first k components. Moreover, for $\alpha = (\alpha_1, \ldots, \alpha_n)$, we write $\alpha^{(j)} = (\alpha_1, \ldots, \alpha_j)$ and set

$$L^{(j)}_{N,\alpha^{(j)}} = \frac{1}{N}\sum_{i=1}^{N} \delta_{\left(X^1_{\alpha_1,i}, X^2_{\alpha_1,\alpha_2,i}, \ldots, X^j_{\alpha_1,\ldots,\alpha_j,i}\right)},$$

for $j \leq n$, which is the marginal of $L_{N,\alpha}$ on S^j. With the notation

$$X^{(j)}_{\alpha,i} \stackrel{\text{def}}{=} \left(X^1_{\alpha_1,i}, \ldots, X^j_{\alpha_1,\ldots,\alpha_j,i}\right),$$

$$\hat{X}^{(j)}_{\alpha,i} \stackrel{\text{def}}{=} \left(X^{j+1}_{\alpha_1,\ldots,\alpha_{j+1},i}, \ldots, X^n_{\alpha_1,\ldots,\alpha_n,i}\right),$$

we can write

$$L_{N,\alpha} \stackrel{\text{def}}{=} \frac{1}{N}\sum_{i=1}^{N} \delta_{\left(X^{(j)}_{\alpha,i}, \hat{X}^{(j)}_{\alpha,i}\right)}. \tag{10}$$

For $A \subset \mathcal{M}^+_1(S^n)$, we put $M_N(A) \stackrel{\text{def}}{=} \#\{\alpha : L_{N,\alpha} \in A\}$.

Lemma 3.3. *Assume $\nu \in \mathcal{M}^+_1(S^n)$ satisfies $H(\nu \mid \mu) < \infty$, and let V be an open neighborhood of ν, and $\varepsilon > 0$. Then there exists an open neighborhood U of ν, $U \subset V$, and $\delta > 0$ such that*

$$\mathbb{P}\Big[M_N(U) \geq \exp\left[N\left(\log 2 - H(\nu \mid \mu) + \varepsilon\right)\right]\Big] \leq \mathrm{e}^{-\delta N}.$$

Proof. If $B_r(\nu)$ denotes the open r-ball around ν in one of the standard metrics, e.g. the Prohorov metric, then by the semicontinuity property of the relative entropy, one has

$$H(B_r(\nu) \mid \mu) \uparrow H(\nu \mid \mu)$$

as $r \downarrow 0$. We can choose a sequence $r_k > 0, r_k \downarrow 0$ with $H(B_{r_k}(\nu) \mid \mu) = H(\mathrm{cl}\left(B_{r_k}(\nu)\right) \mid \mu) \uparrow H(\nu \mid \mu)$. Given $\varepsilon > 0$, and V, we can find k such that

$$H(B_{r_k}(\nu) \mid \mu) = H(\mathrm{cl}\left(B_{r_k}(\nu)\right) \mid \mu) \geq H(\nu \mid \mu) - \varepsilon/4,$$

and $B_{r_k}(\nu) \subset V$. By Sanov's theorem, we therefore get

$$\mathbb{P}\Big[L_{N,\alpha} \in B_{r_k}(\nu)\Big] \leq \exp\left[N(-H(\nu \mid \mu) + \varepsilon/2)\right],$$

and therefore

$$\mathbb{E}\Big[M_N\left(B_{r_k}(\nu)\right)\Big] \leq \exp\left[N(\log 2 - H(\nu \mid \mu) + \varepsilon/2)\right].$$

By the Markov inequality, the claim follows by taking $\delta = \varepsilon/3$. □

Lemma 3.4. *Assume $\nu \in \mathcal{M}_1^+(S^n)$ satisfies $H\left(\nu^{(j)} \mid \mu^{(j)}\right) > \Gamma_j \log 2$ for some $j \leq n$, and let V be an open neighborhood of ν. Then there is an open neighborhood U of ν, $U \subset V$ and $\delta > 0$ such that*

$$\mathbb{P}\big[M_N(U) \neq 0\big] \leq \mathrm{e}^{-\delta N},$$

for large enough N.

Proof. As in the previous lemma, we choose a neighborhood U' of $\nu^{(j)}$ in S^j such that $H(\mathrm{cl}\,(U') \mid \mu^{(j)}) = H(U' \mid \mu^{(j)}) > \Gamma_j \log 2 + \eta$, for some $\eta > 0$. Then we put

$$U \stackrel{\mathrm{def}}{=} \left\{\nu \in \mathcal{M}_1^+(S^n) : \nu \in V, \nu^{(j)} \in U'\right\}.$$

If $L_{N,\alpha} \in U$ then $L_{N,\alpha}^{(j)} \in U'$,

$$\begin{aligned}
\mathbb{P}\left[\exists\alpha : L_{N,\alpha} \in U\right] &\leq \mathbb{P}\left[\exists\alpha : L_{N,\alpha}^{(j)} \in U'\right] \\
&\leq 2^{\Gamma_j N}\mathbb{P}\left[L_{N,\alpha}^{(j)} \in U'\right] \\
&\leq 2^{\Gamma_j N}\exp\left[-NH\left(\mathrm{cl}\left(U'\right) \mid \mu^{(j)}\right) + N\eta/2\right] \\
&\leq 2^{\Gamma_j N}\exp\left[-N\Gamma_j \log 2 - N\eta/2\right] = \mathrm{e}^{-N\eta/2}.
\end{aligned}$$

This proves the claim. □

Lemma 3.5. *Assume that $\nu \in \mathcal{M}_1^+(S^n)$ satisfies $H\left(\nu^{(j)} \mid \mu^{(j)}\right) < \Gamma_j \log 2$ for all j, and let V be an open neighborhood of ν, and $\varepsilon > 0$. Then there exists an open neighborhood U of ν, $U \subset V$, and a $\delta > 0$ such that*

$$\mathbb{P}\Big[M_N(U) \leq \exp\left[N\left(\log 2 - H(\nu \mid \mu) - \varepsilon\right)\right]\Big] \leq \mathrm{e}^{-\delta N}.$$

Proof. We claim that we can find U as required, and some $\delta > 0$, such that

$$\mathrm{var}\left[M_N(U)\right] \leq \mathrm{e}^{-2N\delta}\left\{\mathbb{E}\left[M_N(U)\right]\right\}^2. \tag{11}$$

From this estimate, we easily get the claim: From Sanov's theorem, we have for any $\chi > 0$

$$\mathbb{E}M_N(U) = 2^N \mathbb{P}\left(L_{N,\alpha} \in U\right) \geq \exp\left[N\left(\log 2 - H(\nu \mid \mu) - \chi\right)\right]. \tag{12}$$

Using this, we get by taking $\chi = \varepsilon/2$

$$\begin{aligned}
&\mathbb{P}\left(M_N(U) \leq \mathrm{e}^{N(\log 2 - H(\nu|\mu) - \varepsilon)}\right) \\
&= \mathbb{P}\left(M_N(U) - \mathbb{E}M_N(U) \leq \mathrm{e}^{-N\varepsilon/2}\mathrm{e}^{N(\log 2 - H(\nu|\mu) - \varepsilon/2)} - \mathbb{E}M_N(U)\right) \\
&\leq \mathbb{P}\left(M_N(U) - \mathbb{E}M_N(U) \leq \left(\mathrm{e}^{-N\varepsilon/2} - 1\right)\mathbb{E}M_N(U)\right) \\
&\leq \mathbb{P}\left(M_N(U) - \mathbb{E}M_N(U) \leq -\frac{1}{2}\mathbb{E}M_N(U)\right) \\
&\leq \mathbb{P}\left(\left|M_N(U) - \mathbb{E}M_N(U)\right| \geq \frac{1}{2}\mathbb{E}M_N(U)\right) \\
&\leq 4\frac{\mathrm{var}\left[M_N(U)\right]}{\left\{\mathbb{E}M_N(U)\right\}^2} \leq 4\mathrm{e}^{-2N\delta} \leq \mathrm{e}^{-\delta N}.
\end{aligned}$$

So it remains to prove (11). We first claim that for any j

$$\begin{aligned}
&\lim_{r\to 0} \inf\nolimits_{\rho,\theta\in \mathrm{cl}B_r(\nu):\rho^{(j)}=\theta^{(j)}} \left\{H(\rho \mid \mu) + H\left(\theta \mid \theta^{(j)} \otimes \hat{\mu}^{(j)}\right)\right\} \\
&\quad = H(\nu \mid \mu) + H\left(\nu \mid \nu^{(j)} \otimes \hat{\mu}^{(j)}\right),
\end{aligned} \tag{13}$$

where $\hat{\mu}^{(j)} \stackrel{\mathrm{def}}{=} \mu_{j+1} \otimes \cdots \otimes \mu_n$. The inequality $\leq$ is evident by taking $\rho = \theta = \nu$, and the opposite follows from the semicontinuity properties: One gets that for a sequence (ρ_n, θ_n) with $\rho_n^{(j)} = \theta_n^{(j)}$ and $\rho_n, \theta_n \to \nu$, we have

$$\begin{aligned}
\liminf_{n\to\infty} H\left(\rho_n \mid \mu\right) &\geq H(\nu \mid \mu), \\
\liminf_{n\to\infty} H\left(\theta_n \mid \theta_n^{(j)} \otimes \hat{\mu}^{(j)}\right) &\geq H\left(\nu \mid \nu^{(j)} \otimes \hat{\mu}^{(j)}\right),
\end{aligned}$$

the first inequality by the standard semicontinuity, and the second by Lemma 3.1. This proves (13).

Choose $\eta > 0$ such that $H\left(\nu^{(j)} \mid \mu^{(j)}\right) < \Gamma_j \log 2 - \eta$, for all $1 \leq j \leq n$. By (13), we may choose r small enough such that $\mathrm{cl}B_r(\nu) \subset V$, and for all $1 \leq j \leq n$,

$$\begin{aligned}
&\inf\nolimits_{\rho,\theta\in \mathrm{cl}B_r(\nu):\rho^{(j)}=\theta^{(j)}} \left\{H(\rho \mid \mu) + H\left(\theta \mid \theta^{(j)} \otimes \hat{\mu}^{(j)}\right)\right\} \\
&\quad \geq H(\nu \mid \mu) + H\left(\nu \mid \nu^{(j)} \otimes \hat{\mu}^{(j)}\right) - \eta/2 \\
&\quad = 2H(\nu \mid \mu) - H\left(\nu^{(j)} \mid \mu^{(j)}\right) - \eta/2 \\
&\quad \geq 2H(\nu \mid \mu) - \Gamma_j \log 2 + \eta/2.
\end{aligned}$$

For two indices α, α' we write $q(\alpha, \alpha') \overset{\text{def}}{=} \max\left\{j : \alpha^{(j)} = \alpha'^{(j)}\right\}$ with $\max \emptyset \overset{\text{def}}{=} 0$. Then

$$
\begin{aligned}
\mathbb{E}M_N^2(U) &= \sum_{j=0}^{n} \sum_{\alpha,\alpha' : q(\alpha,\alpha')=j} \mathbb{P}\left[L_{N,\alpha} \in U, L_{N,\alpha'} \in U\right] \\
&= \sum_{\alpha,\alpha' : q(\alpha,\alpha')=0} \mathbb{P}\left[L_{N,\alpha} \in U\right] \mathbb{P}\left[L_{N,\alpha'} \in U\right] \\
&\quad + \sum_{j=1}^{n} \sum_{\alpha,\alpha' : q(\alpha,\alpha')=j} \mathbb{P}\left[L_{N,\alpha} \in U, L_{N,\alpha'} \in U\right] \\
&\leq \mathbb{E}[M_N(\mathrm{cl}U)]^2 + \\
&\quad + \sum_{j=1}^{n} \sum_{\alpha,\alpha' : q(\alpha,\alpha')=j} \mathbb{P}\left[L_{N,\alpha} \in \mathrm{cl}U, L_{N,\alpha'} \in \mathrm{cl}U\right].
\end{aligned}
$$

We write the empirical measure in the form (10), and use Lemma 3.2. For any $1 \leq j \leq n$, we have

$$
\begin{aligned}
&\sum_{\alpha,\alpha' : q(\alpha,\alpha')=j} \mathbb{P}\left[L_{N,\alpha} \in \mathrm{cl}\, U, L_{N,\alpha'} \in \mathrm{cl}\, U\right] \\
&\quad = 2^{\Gamma_j N} 2^{(1-\Gamma_j)N} \left(2^{(1-\Gamma_j)N} - 1\right) \mathbb{P}\left[L_{N,\alpha} \in \mathrm{cl}\, U, L_{N,\alpha'} \in \mathrm{cl}\, U\right],
\end{aligned}
$$

where on the right-hand side α, α' is an arbitrary pair with $q(\alpha, \alpha') = j$. Using Lemma 3.2, we have

$$
\begin{aligned}
&\mathbb{P}\left[L_{N,\alpha} \in \mathrm{cl}\, U,\ L_{N,\alpha} \in \mathrm{cl}\, U\right] \\
&\quad \leq \exp\Bigg[- N \inf_{\rho,\theta \in \mathrm{cl}\, U, \rho^{(j)}=\theta^{(j)}} \Big\{ H\left(\rho^{(j)} \mid \mu^{(j)}\right) \\
&\qquad + H\left(\rho \mid \rho^{(j)} \otimes \hat{\mu}^{(j)}\right) + H\left(\theta \mid \theta^{(j)} \otimes \hat{\mu}^{(j)}\right) \Big\} + \frac{N\eta}{4} \Bigg] \\
&\quad = \exp\left[-N \inf_{\rho,\theta \in \mathrm{cl}\, U, \rho^{(j)}=\theta^{(j)}} \left\{ H(\rho \mid \mu) + H\left(\theta \mid \theta^{(j)} \otimes \hat{\mu}^{(j)}\right) \right\} + \frac{N\eta}{4} \right] \\
&\quad \leq 2^{\Gamma_j N} \exp\left[-2NH(\nu \mid \mu) - \frac{N\eta}{4} \right],
\end{aligned}
$$

and thus

$$\sum_{\alpha,\alpha':q(\alpha,\alpha')=j} \mathbb{P}\left[L_{N,\alpha} \in \operatorname{cl} U,\ L_{N,\alpha} \in \operatorname{cl} U\right] \le 2^{2N} \exp\left[-2NH(\nu \mid \mu) - \frac{N\eta}{4}\right].$$

Combining, we obtain by taking $\chi = \eta/16$ in (12)

$$\operatorname{var}[M_N(U)] \le 2^{2N} \exp\left[-2NH(\nu \mid \mu) - \frac{N\eta}{4}\right] \le \mathrm{e}^{-N\eta/8}\mathbb{E}[M_N(U)]^2,$$

which proves our claim. □

Proof of Theorem 2.3 We set

$$\mathcal{G} \stackrel{\text{def}}{=} \left\{\nu \in \mathcal{M}_1^+(S^n) : H\left(\nu^{(j)} \mid \mu^{(j)}\right) \le \Gamma_j \log 2,\ j = 1, \dots, n\right\},$$

which is a compact set.

Step 1. We first prove the lower bound. By compactness of $\mathcal{G}$ and the semicontinuity of H, there exists $\nu_0 \in \mathcal{G}$ such that

$$\sup_{\nu \in \mathcal{G}} \{\Phi(\nu) - H(\nu \mid \mu)\} = \Phi(\nu_0) - H(\nu_0 \mid \mu).$$

We set $\nu_\lambda \stackrel{\text{def}}{=} (1-\lambda)\nu_0 + \lambda\mu$ for $0 < \lambda < 1$. By convexity of $H(\nu \mid \mu)$ in ν, we see that $H\left(\nu_\lambda^{(j)} \mid \mu^{(j)}\right) < \Gamma_j \log 2$ for all $1 \le j \le n$. Furthermore, $\nu_\lambda \to \nu_0$ weakly as $\lambda \to 0$, and $\Phi(\nu_\lambda) \to \Phi(\nu_0)$, $H(\nu_\lambda \mid \mu) \to H(\nu_0 \mid \mu)$.

Given $\varepsilon > 0$ we choose $\lambda > 0$ such that

$$\Phi(\nu_\lambda) - H(\nu_\lambda \mid \mu) \ge \Phi(\nu_0) - H(\nu_0 \mid \mu) - \varepsilon.$$

By the continuity of Φ and Lemma 3.5, we find a neighborhood U of ν_λ, and $\delta > 0$ such that

$$\Phi(\theta) - \Phi(\nu_\lambda) \le \varepsilon,\ \theta \in U,$$

and

$$\mathbb{P}\left[M_N(U) \le 2^N \exp[-NH(\nu_\lambda \mid \mu) - N\varepsilon]\right] \le \mathrm{e}^{-\delta N},$$

Then with probability greater than $1 - \mathrm{e}^{-\delta N}$,

$$\begin{aligned}
Z_N &= 2^{-N} \sum_\alpha \exp[N\Phi(L_{N,\alpha})] \\
&\ge 2^{-N} \sum_{\alpha:L_{N,\alpha}\in U} \exp[N\Phi(L_{N,\alpha})] \\
&\ge \exp[N\Phi(\nu_\lambda) - N\varepsilon] \exp[-NH(\nu_\lambda \mid \mu) - N\varepsilon] \\
&\ge \exp\left[N \sup_{\nu \in \mathcal{G}} \{\Phi(\nu) - H(\nu \mid \mu)\} - 3N\varepsilon\right].
\end{aligned}$$

By Borel–Cantelli, we therefore get, as ε is arbitrary,

$$\liminf_{N\to\infty} \frac{1}{N} \log Z_N \geq \sup_{\nu\in\mathcal{G}} \{\Phi(\nu) - H(\nu \mid \mu)\}$$

almost surely.

Step 2. We prove the upper bound. Let again $\varepsilon > 0$ and set

$$\overline{\mathcal{G}} \stackrel{\text{def}}{=} \{\nu : H(\nu \mid \mu) \leq \log 2\}.$$

If $\nu \in \mathcal{G}$ we choose $r_\nu > 0$ such that $|\Phi(\theta) - \Phi(\nu)| \leq \varepsilon$, $\theta \in B_{r_\nu}(\nu)$ and

$$\mathbb{P}\left[M_N(B_{r_\nu}(\nu)) \geq 2^N \exp\left[-NH(\nu \mid \mu) + N\varepsilon\right]\right] \leq \mathrm{e}^{-N\delta_\nu},$$

for some $\delta_\nu > 0$ and large enough N (using Lemma 3.3). If $\nu \in \overline{\mathcal{G}} \setminus \mathcal{G}$ we choose r_ν such that $|\Phi(\theta) - \Phi(\nu)| \leq \varepsilon$, $\theta \in B_{r_\nu}(\nu)$, and

$$\mathbb{P}\left[M_N(B_{r_\nu}(\nu)) \neq 0\right] \leq \mathrm{e}^{-N\delta_\nu}, \tag{14}$$

again for large enough N (and by Lemma 3.4). As $\overline{\mathcal{G}}$ is compact, we can cover it by a finite union of such balls, i.e.

$$\overline{\mathcal{G}} \subset U \stackrel{\text{def}}{=} \bigcup_{j=1}^{m} B_{r_j}(\nu_j),$$

where $r_j \stackrel{\text{def}}{=} r_{\nu_j}$. We also set $\delta \stackrel{\text{def}}{=} \min_j \delta_{\nu_j}$. We then estimate

$$Z_N \leq 2^{-N} \sum_{l=1}^{m} \sum_{\alpha : L_{N,\alpha} \in B_{r_l}(\nu_l)} \exp\left[N\Phi(L_{N,\alpha})\right] + 2^{-N} \sum_{\alpha : L_{N,\alpha} \notin U} \exp\left[N\Phi(L_{N,\alpha})\right]. \tag{15}$$

we first claim that almost surely the second summand vanishes provided N is large enough, i.e. that there is no α with $L_{N,\alpha} \notin U$. By Sanov's theorem, we have

$$\limsup_{N\to\infty} \frac{1}{N} \log \mathbb{P}\left[L_{N,\alpha} \notin U\right] \leq -\inf\nolimits_{\nu\notin U} H(\nu \mid \mu) < -\log 2.$$

Therefore, almost surely, there is no α with $L_{N,\alpha} \notin U$, and therefore the second summand in (15) vanishes for large enough N, almost surely. The same applies to those summands in the first part for which $\nu_l \notin \mathcal{G}$, using (14). We therefore have, almost surely, for large enough N,

$$Z_N \leq 2^{-N} \sum_{l : \nu_l \in \mathcal{G}} \sum_{\alpha : L_{N,\alpha} \in B_{r_l}(\nu_l)} \exp\left[N\Phi(L_{N,\alpha})\right]$$

$$\leq e^{N\varepsilon} \sum_{l:\nu_l \in \mathcal{G}} \exp\left[N\Phi(\nu_l)\right] M_N(B_{r_l}(\nu_l))$$

$$\leq e^{2N\varepsilon} \sum_{l:\nu_l \in \mathcal{G}} \exp\left[N\Phi(\nu_l)\right] \exp\left[-NH(\nu_l \mid \mu)\right]$$

$$\leq e^{2N\varepsilon} m \exp\left[N \sup_{\nu \in \mathcal{G}} \left\{\Phi(\nu) - H(\nu \mid \mu)\right\}\right].$$

As ε is arbitrary, we get

$$\limsup_{N\to\infty} \frac{1}{N} \log Z_N \leq \sup_{\nu \in \mathcal{G}} \left[\Phi(\nu) - H(\nu \mid \mu)\right].$$

This finishes the proof of Theorem 2.3. □

3.2 *The Dual Representation. Proof of the Theorem 2.5*

We define a family $\mathcal{G}(\phi) = \{G_{\phi,\mathbf{m}}\}$ of probability distributions on S^n, which depend on the parameter $\mathbf{m} = (m_1, \ldots, m_n) \in \Delta$. The probability measure $G = G_{\phi,\mathbf{m}}$ is described by a "starting" measure γ on S, and for $2 \leq j \leq n$ Markov kernels K_j from S^{j-1} to S, so that G is the semidirect product

$$G = \gamma \otimes K_2 \otimes \cdots \otimes K_n.$$

$$\gamma(\mathrm{d}x) \stackrel{\mathrm{def}}{=} \frac{\exp\left[m_1\phi_1(x)\right]\mu_1(\mathrm{d}x)}{\exp\left[m_1\phi_0\right]},$$

$$K_j\left(\mathbf{x}^{(j-1)}, \mathrm{d}x_j\right) \stackrel{\mathrm{def}}{=} \frac{\exp\left[m_j\phi_j\left(\mathbf{x}^{(j)}\right)\right]\mu_j\left(\mathrm{d}x_j\right)}{\exp\left[m_j\phi_{j-1}\left(\mathbf{x}^{(j-1)}\right)\right]},$$

where we write $\mathbf{x}^{(j)} \stackrel{\mathrm{def}}{=} \left(x_j, \ldots, x_j\right)$. Remember the definition of the function $\phi_j : S^j \to \mathbb{R}$ in (4), (5). It should be remarked that these objects are defined for all $\mathbf{m} \in \mathbb{R}^n$, and not just for $\mathbf{m} \in \Delta$. We also write

$$G^{(j)} \stackrel{\mathrm{def}}{=} \gamma \otimes K_2 \otimes \cdots \otimes K_j,$$

which is the marginal of G on S^j. In order to emphasize the dependence on $\mathbf{m}$, we occasionally will write $\phi_{j,\mathbf{m}},\ \gamma_{\mathbf{m}},\ K_{j,\mathbf{m}}$ etc.

We remark that by a simple computation

$$\int H\left(K_j\left(\mathbf{x}^{(j-1)}, \cdot\right) \mid \mu_j\right) G^{(j-1)}\left(\mathrm{d}\mathbf{x}^{(j-1)}\right) \tag{16}$$

$$= m_j \left[\int \phi_j \mathrm{d}G^{(j)} - \int \phi_{j-1} \mathrm{d}G^{(j-1)} \right].$$

$\phi_j, \ldots, \phi_n$ do not depend on m_j, but $\phi_0, \ldots, \phi_{j-1}$ do. Differentiating the equation

$$\mathrm{e}^{m_{r+1}\phi_r} = \int \mathrm{e}^{m_{r+1}\phi_{r+1}} \mathrm{d}\mu_{r+1},$$

with respect to m_j, we get for $0 \leq r \leq j-2$

$$\frac{\partial \phi_r \left(\mathbf{x}^{(r)}\right)}{\partial m_j} = \int \frac{\partial \phi_{r+1} \left(\mathbf{x}^{(r)}, x_{r+1}\right)}{\partial m_j} K_{r+1} \left(\mathrm{d}\mathbf{x}^{(r)}, x_{r+1}\right), \tag{17}$$

and for $r = j-1$

$$\phi_{j-1} \mathrm{e}^{m_j \phi_j} + m_j \frac{\partial \phi_{j-1}}{\partial m_j} \mathrm{e}^{m_j \phi_j} = \int \phi_j \mathrm{e}^{m_j \phi_j} \mathrm{d}\mu_j,$$

i.e.

$$\frac{\partial \phi_{j-1}}{\partial m_j} \left(\mathbf{x}^{(j)}\right) = \frac{1}{m_j} \left[\int \phi_j \left(\mathbf{x}^{(j-1)}, x_j\right) K_j \left(\mathbf{x}^{(j-1)}, \mathrm{d}x_j\right) - \phi_{j-1} \left(\mathbf{x}^{(j-1)}\right) \right].$$

Combining that with (16), (17), we get

$$\begin{aligned} \frac{\partial \phi_{0,\mathbf{m}}}{\partial m_j} &= \frac{1}{m_j} \left[\int \phi_j \mathrm{d}G^{(j)} - \int \phi_{j-1} \mathrm{d}G^{(j-1)} \right] \qquad (18) \\ &= \frac{1}{m_j^2} \int H \left(K_j \left(\mathbf{x}^{(j-1)}, \cdot\right) \mid \mu_j\right) G^{(j-1)} \left(\mathrm{d}\mathbf{x}^{(j-1)}\right). \end{aligned}$$

Theorem 2.5 is immediate from the following result:

Proposition 3.6. *Assume that $\phi : S^n \to \mathbb{R}$ is bounded and continuous. Then there is a unique measure ν maximizing* Gibbs (ν, ϕ) *under the constraint $\nu \in \bigcap_{j=1}^n \mathcal{R}_j$. This measure is of the form $\nu = G_{\phi,\mathbf{m}}$, where $\mathbf{m}$ is the unique element in Δ minimizing (7). For this $\mathbf{m}$, we have*

$$\text{Gibbs}\,(G, \phi) = \text{Parisi}\,(\phi, \mathbf{m}). \tag{19}$$

Proof. From strict convexity of the relative entropy, and the fact that $\bigcap_{j=1}^n \mathcal{R}_j$ is compact and convex, it follows that there is a unique maximizer ν of Gibbs (ν, ϕ) under this constraint.

Also, a straightforward application of Hölder's inequality shows that Parisi $(\phi, \mathbf{m})$ is a strictly convex function in the variables $1/m_j$. Therefore, it follows that there is a uniquely attained minimum of Parisi $(\phi, \mathbf{m})$ as a function of

$\mathbf{m} \in \Delta$. This minimizing $\mathbf{m} = (m_1, \ldots, m_n)$, we can be split into subblocks of equal values: There is a number K, $0 \leq K \leq n$, and indices $0 < j_1 < j_2 < \cdots < j_K \leq n$ such that

$$\begin{aligned}0 &< m_1 = \cdots = m_{j_1} < m_{j_1+1} = \cdots = m_{j_2} \\ &< m_{j_2+1} \cdots < m_{j_{K-1}+1} = \cdots = m_{j_K} \\ &< m_{j_K+1} = \cdots m_n = 1.\end{aligned}$$

$K = 0$ just means that all $m_i = 1$. If $j_K = n$, then all m_i are < 1. We write $G = G_{\phi,\mathbf{m}}$.

From (18), we immediately have

$$\frac{\partial \,\text{Parisi}\,(\phi, \mathbf{m})}{\partial m_j} = \frac{1}{m_j^2}\left[\int H\left(K_j\left(\mathbf{x}^{(j-1)}, \cdot\right) \mid \mu_j\right) G^{(j-1)}\left(d\mathbf{x}^{(j-1)}\right) - \gamma_j \log 2\right]. \tag{20}$$

Set $d_j \stackrel{\text{def}}{=} \int H\left(K_j\left(\mathbf{x}^{(j-1)}, \cdot\right) \mid \mu_j\right) G_{\mathbf{m}}^{(j-1)}\left(d\mathbf{x}^{(j-1)}\right)$.We use (20) and the minimality of Parisi $(\phi, \cdot)$ at $\mathbf{m}$. We can perturb $\mathbf{m}$ by moving a whole block $m_{j_r+1} = \cdots = m_{j_{r+1}}$ up and down locally, without leaving Δ, provided it is not the possibly present block of values 1. This leads to

$$\sum_{i=j_r+1}^{j_{r+1}} d_i = \log 2 \sum_{i=j_r+1}^{j_{r+1}} \gamma_i.$$

Furthermore, we can always move first parts of blocks, say $m_{j_r+1} = \cdots = m_k$, $k \leq j_{r+1}$ locally down, without leaving Δ, so that we get

$$\sum_{i=j_r+1}^{j_k} d_i \leq \log 2 \sum_{i=j_r+1}^{j_k} \gamma_i.$$

These two observations imply

$$G \in \bigcap_{j=1}^{n} \mathcal{R}_j \cap \bigcap_{r=1}^{K} \mathcal{R}_{j_r}^{=}. \tag{21}$$

We next prove

$$\text{Gibbs}\,(\nu, \phi) \leq \text{Gibbs}\,(G, \phi) \tag{22}$$

for any $\nu \in \bigcap_{j=1}^{n} \mathcal{R}_j$.

We first prove the case $n = 1$. If $m < 1$, then

$$H\,(G \mid \mu) = \log 2 \geq H\,(\nu \mid \mu),$$

by (21) and the assumption $\nu \in \mathcal{R}_1$. Therefore, in any case

$$\begin{aligned}\text{Gibbs}(G,\phi) - \text{Gibbs}(\nu,\phi) &\geq \int \phi \mathrm{d}G - \frac{1}{m} H(G \mid \mu) \\ &\quad - \left[\int \phi \mathrm{d}\nu - \frac{1}{m} H(\nu \mid \mu)\right] \\ &= \frac{1}{m} H(\nu \mid G) \geq 0.\end{aligned}$$

The general case follows by a slight extension of the above argument. Put

$$D_k \stackrel{\text{def}}{=} \int \phi_k \mathrm{d}G^{(k)} - \frac{1}{m_{k+1}} H\left(G^{(k)} \mid \mu^{(k)}\right) - \int \phi_k \mathrm{d}\nu^{(k)} + \frac{1}{m_{k+1}} H\left(\nu^{(k)} \mid \mu^{(k)}\right),$$

$D_0 \stackrel{\text{def}}{=} 0$, $D_n = \text{Gibbs}(G,\phi) - \text{Gibbs}(\nu,\phi)$. We prove $D_{k-1} \leq D_k$ for all k, so that the claim follows. Remark that as above in the $n = 1$ case, if $m_k < m_{k+1}$, then $H\left(G^{(k+1)} \mid \mu^{(k+1)}\right) = \Gamma_k \log 2$, and therefore, in any case

$$\begin{aligned}D_k &\geq \int \phi_k \mathrm{d}G^{(k)} - \frac{1}{m_k} H\left(G^{(k)} \mid \mu^{(k)}\right) - \int \phi_k \mathrm{d}\nu^{(k)} + \frac{1}{m_k} H\left(\nu^{(k)} \mid \mu^{(k)}\right) \\ &= \int \phi_{k-1} \mathrm{d}G^{(k-1)} - \frac{1}{m_k} H\left(G^{(k-1)} \mid \mu^{(k-1)}\right) - \int \phi_k \mathrm{d}\nu^{(k)} + \frac{1}{m_k} H\left(\nu^{(k)} \mid \mu^{(k)}\right).\end{aligned}$$

As

$$\begin{aligned}&H\left(\nu^{(k)} \mid \mu^{(k)}\right) - m_k \int \phi_k \mathrm{d}\nu^{(k)} + m_k \int \phi_{k-1} \mathrm{d}\nu^{(k-1)} \\ &= H\left(\nu^{(k-1)} \mid \mu^{(k-1)}\right) + \int \log \frac{\nu^{(k)}\left(\mathrm{d}x_k \mid \mathbf{x}^{(k-1)}\right) \mathrm{e}^{m_k \phi_{k-1}\left(\mathbf{x}^{(k-1)}\right)}}{\mu_k\left(\mathrm{d}x_k\right) \mathrm{e}^{m_k \phi_k\left(\mathbf{x}^{(k)}\right)}} \nu^{(k)}\left(\mathrm{d}\mathbf{x}^{(k)}\right) \\ &\geq H\left(\nu^{(k-1)} \mid \mu^{(k-1)}\right),\end{aligned}$$

(22) is proved.

(21) and (22) identify $G = G_{\phi,\mathbf{m}}$ as the unique maximizer of $G(\cdot,\phi)$ under the constraint $\bigcap_{j=1}^n \mathcal{R}_j$.

The identification (19) comes by a straightforward computation. □

Acknowledgements E. Bolthausen is supported in part by the grant No 200020$_1$25247/1 of the Swiss Science Foundation. N. Kistler is partially supported by the German Research Council in the SFB 611 and the Hausdorff Center for Mathematics.

References

1. Bolthausen, E., Kistler, N.: On a nonhierarchical version of the generalized random energy model. II. Ultrametricity. Stoch. Proces. Appl. **119**, 2357–2386 (2009)
2. Bovier, A., Kurkova, I.: Derrida's generalized random energy models I & II. Annals de l'Institut Henri Poincaré **40**, 439–495 (2004)
3. Derrida, B.: A generalization of the random energy model that includes correlations between the energies. J. Phys. Lett. **46**, 401–407 (1985)
4. Mézard, M., Parisi, G., Virasoro, M.: Spin Glass theory and beyond, World Scientific, Singapore (1987)
5. Guerra, F.: Broken replica symmetry bounds in the mean field spin glass model. Comm. Math. Phys. **233**, 1–12 (2003)
6. Talagrand, M.: Mean field models of spin glasses, Vol I, II. Ergebnisse der Mathematik und ihrer Grenzgebiete. 3. Folge. vol. 54 & 55. Springer (2011)
7. Talagrand, M.: The Parisi Formula. Ann. Math. **163** (2006)

Metastability: From Mean Field Models to SPDEs

Anton Bovier

Abstract Kramer's equation of a diffusion in a double well potential has been the pardigm for a metastable system since 1940. The theme of this note is to partially explain, why and in what sense this is a good model for metastable systems. In the process, I review recent progress in a variety of models, ranging from mean field spin systems to stochastic partial differential equations.

1 Introduction

Metastability is in essence the dynamical signature of a first order phase transition in statistical mechanics. In equilibrium statistical mechanics, a first order phase transition is said to occur if a systems is very sensitive to the change of a parameter (resp. boundary conditions), in the sense that an extensive variable (e.g., density or magnetization) shows a discontinuity as functions of some intensive variable (e.g., pressure or magnetic field), in the thermodynamics limit. Dynamically, for a finite system, this fact manifests itself in that as the parameter is varied across the phase transition line, the system will remain a considerable (and mostly random) amount of time in the "wrong phase" before suddenly changing into the true equilibrium phase (in other words, the sensitive variable will change its value as a function of time with a random delay).

Metastability is a very widespread phenomenon that occurs in a large variety of systems, both natural and artificial. In many instances, it has important effects that are crucial for the proper functioning of the system and there has been great interest in understanding metastability in quantitative terms over at least the last

A. Bovier (✉)
Institut für Angewandte Mathematik, Rheinische Friedrich-Wilhelms-Universität, Endenicher Allee 60, 53115 Bonn, Germany
e-mail: bovier@uni-bonn.de

J.-D. Deuschel et al. (eds.), *Probability in Complex Physical Systems*,
Springer Proceedings in Mathematics 11, DOI 10.1007/978-3-642-23811-6_18,

century. Most metastable systems of practical relevance are many-body systems whose dynamics is very hard to analyze, both analytically and numerically. This is particularly true with respect to metastability, due to the very long time scales that are involved.

One of the first *mathematical models* for metastability was proposed in 1940 by Hendrik Anthony Kramers [25]. It consists of the simple, one-dimensional diffusion equation

$$dX_t = b(X_t)dt + \sqrt{2\epsilon}dB_t, \tag{1}$$

where $b(x) = -V'(x)$, with $V(x)$ a *double well potential*, i.e., a function with two local minima that tends to infinity at $\pm\infty$, and B_t is Brownian motion. In fact, this equation emerged as a special case of the more general equations he considered, namely

$$\mu^{-1}X_t'' = -X_t' + b(X_t) + \sqrt{2\epsilon}B_t', \tag{2}$$

in the limit $\mu \uparrow \infty$. Thus, Kramers' (1) can be seen as the equation of motion of a particle moving under the influence of a gradient force and a random force with friction in the limit where the friction becomes infinitely strong.

Kramers' (1) has become the paradigm of metastability. Kramers had been able to solve all interesting questions in the context of this model. In particular, he derived the so-called *Kramers-formula* for the average transition time, $\mathbb{E}_a\tau_b$, from a minimum at a, via the maximum, z^*, to the minimum, b, as

$$\mathbb{E}_a\tau_b = \frac{2\pi}{\sqrt{V''(a)V''(z^*)}}\exp\left(\epsilon^{-1}\left(V(z^*) - V(a)\right)\right)(1 + o(1)). \tag{3}$$

The multidimensional generalization of this formula is attributed to Eyring and called *Eyring-Kramers formula* (see also [39]). Note that Eyring's so-called *reaction rate theory* [22] is based on quantum mechanical considerations and quite different from the classical theory of Kramers, although it appears to have the idea of interpreting V as a restricted (quantum mechanical) free energy in it. For a historical discussion, see the recent paper by Pollak and Talkner [35].

The question I want to discuss in this paper is how one may understand that indeed this simple equation can reflect quite properly the metastable behavior of the complex dynamics of many-body systems.

Before entering into any further discussion, we need to talk about the dynamics of many-body systems we want to discuss.

1.1 Stochastic Ising Models

To analyse in any sense of rigour, the microscopic dynamics of many-body systems as given by Newton's laws or even by the many-body Schrödinger equation is analytically beyond today's technology. A reasonable compromise, to which I will stick here, are stochastic Ising models. Here, the setting is the following: we consider a state space $\mathscr{S}_\Lambda \equiv \{-1,+1\}^\Lambda$, with $\Lambda \subset \mathbb{Z}^d$ a finite subset of a lattice. A configuration $\sigma \in \mathscr{S}_\Lambda$ describes the values of the magnetic moments of the atoms placed at the sites of Λ, in the magnetic interpretation. Alternatively, one my think of a *lattice gas* in which case the variables $(1+\sigma_i)/2$ represent the number of particles on the site i.

The interaction of the model is described by a *Hamiltonian*, $H_\Lambda : \mathscr{S}_\Lambda \to \mathbb{R}$, which is a real valued function on configuration space representing the *energy* of a configuration.

By *Glauber dynamics*, I will mean a Markov chain (in continuous or discrete time) with transitions rates $p(\sigma,\sigma')$ that are reversible with respect to the *Gibbs measure*, $\mu_{\beta,\Lambda}$, given by[1]

$$\mu_{\beta,\Lambda}(\sigma) \equiv \frac{1}{Z_{\beta,\Lambda}} \exp\left(-\beta H_\Lambda(\sigma)\right). \tag{4}$$

I will assume the dynamics to be local, in the sense that $p(\sigma,\sigma')$ are non-zero only if σ and σ' differ in at most one site. One may of course think of many similar, but different, situations.

In this context, the extensive variable one would like to consider is the *magnetisation*,

$$m_\Lambda(\sigma) \equiv \frac{1}{|\Lambda|}\sum_{i\in\Lambda}\sigma_i. \tag{5}$$

Under reasonable assumptions on the Hamiltonian, the random variable m_Λ satisfies a *large deviation principle* under the Gibbs measure, i.e.

$$\mu_{\beta,\Lambda}\left(m_\Lambda(\sigma) = m\right) \sim \exp\left(-\Lambda\beta f_\beta(m)\right), \tag{6}$$

where the rate function f_β is called the *free energy*. A first order phase transition occurs when f_β is not strictly convex. Can we interpret Kramers' equation as an approximation of the behavior of $m_\Lambda(\sigma(t))$, when $\sigma(t)$ is a Glauber dynamics for our model? This is loosely speaking the issue around which this note will turn.

2 The Curie–Weiss Model

There is a simple model where all works well, the Curie–Weiss model of a ferromagnet. Here the Hamiltonian is very simple, namely

$$H_\Lambda(\sigma) \equiv -\frac{|\Lambda|}{2}\left(m_\Lambda(\sigma)\right)^2 = -\frac{1}{2|\Lambda|}\sum_{i,j\in\Lambda}\sigma_i\sigma_j. \tag{7}$$

[1] Note that I will not seriously enter the discussion of infinite volume dynamics and so I also do not enter the formalism of infinite volume Gibbs measures.

Since the form of Λ does not enter here, let us fix $\Lambda = \{1,\dots,N\}$. Let us for definiteness opt for a discrete time dynamics and fix the transition rates as *Metropolis rates*

$$p(\sigma,\sigma') = \begin{cases} N^{-1}\exp\left(-[H_\Lambda(\sigma') - H_\Lambda(\sigma)]_+\right), & \text{if } \|\sigma-\sigma'\|_1 = 2, \\ 0, & \text{if } \|\sigma-\sigma'\|_1 > 2, \\ 1 - \sum_{\eta\neq\sigma} p(\sigma,\eta), & \text{if } \sigma = \sigma'. \end{cases} \tag{8}$$

Now let us look at the time evolution of $m_\Lambda(t) \equiv m_\Lambda(\sigma(t))$. Clearly, in each step it can only increase or decrease by $2/N$, and one easily checks that the probability of increasing resp. decreasing depends only on the value of H_Λ at the starting configuration and on the number of -1's resp. $+1$'s present in the configuration σ. But these are known once $m_\Lambda(\sigma)$ is given. In other words,

$$\mathbb{P}\left[m_\Lambda(t+1) = m' | \mathcal{F}_t\right] = r(m_\Lambda(\sigma(t)), m'), \tag{9}$$

is a function of $m_\Lambda(t)$ only. From this, one deduces readily that $m_\Lambda(t)$ is itself a Markov chain with transition rates $r(m,m')$ on the state space

$$\Gamma_N \equiv \{-1, -1+2/N, \dots, 1-2/N, 1\}, \tag{10}$$

which is reversible with respect to the measure

$$\mathbb{Q}_{\beta,\Lambda}(m) \equiv \mu_{\beta,\Lambda}\left(m_\Lambda(\sigma) = m\right). \tag{11}$$

Now it is well known that

$$\mathbb{Q}_{\beta,\Lambda}(m) \sim \exp\left(-\beta N f_\beta(m)\right) \tag{12}$$

with

$$f_\beta(m) = -\frac{m^2}{2} + \beta^{-1} I(m), \tag{13}$$

where

$$I(m) \equiv \frac{1+m}{2}\ln(1+m) + \frac{1-m}{2}\ln(1-m), \tag{14}$$

is Cramér's entropy function. f_β is a double well whenever $\beta > 1$. Thus, $m_N(t)$ is a random walk with reversible measure (close to) $\exp(-\beta N f_\beta(m))$ on a lattice with spacing $2/N$ in $[-1,1]$; this is quite close to the diffusion equation of Kramers if we chose $V(x) = \beta f_\beta(x)$ and $\epsilon = (\beta N)^{-1}$. So in the dynamics of the Curie–Weiss model, Kramers' equation can be interpreted as a diffusion approximation of the actual dynamics of the magnetization! This is definitely a strong point in favor of Kramers' ideas.

The weak point of this observation is that it is very unstable under modifications. We are using the full permutation symmetry of this special model which is necessary to ensure that $m_\Lambda(t)$ is even a Markov process.

Let us mention that the knowledge of the behavior of $m_\Lambda(t)$ a priori does not answer all questions on the dynamics of $\sigma(t)$. This issue has been addressed quite recently by Levin et al. [27].

There are a number of generalized mean field models that permit a similar reduction to a *multi-dimensional* diffusive Markov chain, see, e.g., [10].

3 Large Deviations

The mention of the word metastability often triggers the immediate reaction to think about *large deviations*. This is undoubtedly due to the seminal work of Freidlin and Wentzell [23], that pioneered the rigorous analysis of stochastic dynamics exhibiting metastability through the use of large deviations on path space. It is also the basis of the so-called *pathwise approach* to metastability, that was initiated by Cassandro, Galves, Olivieri, and Vares [14] in 1984. The recent monograph on metastability by Olivieri and Vares [34] gives an in-depth overview from this angle.

Let us look at this in a slightly abstract way. Let us assume that we are working with a family, X^ϵ, of Markov processes on a state space $\mathcal{S}$, which we may assume to be a complete separable normed space. Let us denote by Γ the set of all paths, $\gamma : [0,T] \to \mathcal{S}$, with T arbitrary. We may naturally equip Γ with the supremum norm inherited from the norm on $\mathcal{S}$. By a large deviation principle on path space we mean that we have a nonnegative, lower semicontinuous function, $I : \Gamma \to \mathbb{R}_+$, with compact level sets, such that for some ϵ small, for any set $A \subset \Gamma$,

$$- \inf_{\gamma \in \mathrm{int} A} I(\gamma) \le \epsilon \ln \mathbb{P}\left(X^\epsilon \in A\right) + o(1) \le - \inf_{\gamma \in \mathrm{cl} A} I(\gamma). \tag{15}$$

In a way more speaking for us is the essentially equivalent formulation that for any small enough $\delta > 0$, and any $\gamma \in \Gamma$,

$$\epsilon \ln \mathbb{P}\left(\|X^\epsilon - \gamma\| \le \delta\right) = -I(\gamma) + o(1). \tag{16}$$

In the presence of such a formulation, metastability will arise if we can identify two (or more) sets $A, B \subset \mathcal{S}$, such that

$$\inf_{\gamma : A \to B} I(\gamma) > 0, \tag{17}$$

and

$$\inf_{\gamma : B \to A} I(\gamma) > 0, \tag{18}$$

whereas there exists paths $\gamma : A \to A$ and $\gamma' : B \to B$, of striclty positive lenght, such that

$$I(\gamma) = I(\gamma') = 0. \tag{19}$$

Here, the notation $\gamma : A \to B$ means the set of curves (of arbitrary length) that begin in A and end in B.

Clearly in such a situation one can the state space into two parts, one containing A and the other B, such that the process will stay exponentially (in $1/\epsilon$) long in one of the parts before going to the other; this is clearly what we understand by metastable behavior. Also, the minimizer in the variational problems (17) and (18) are clearly the most likely strategies to realize the unlikely transitions, in the sense that with probability tending to one, the process conditioned to move from A to B in finite time[2] will do this by remaining in an arbitrarily small neighborhood of a minimizer γ; if the minimizers are unique, then they represent the optimal transition strategy.

In this latter case, we may again see a confirmation of the one-dimensional model of Kramer: it suggests that we may replace the entire process by one that is confined into an arbitrarily thin, properly chosen, tube around the optimal path and obtain the same result. But there are two difficulties: first, one would have to identify this path, and second, the equivalence would hold only on the level of precision that is given by the large deviation theory.

Before commenting on this latter aspect, let us comment on the question of where a large deviation principle on path space can be expected.

3.1 Diffusions with Small Diffusivity

The multidimensional analog of Kramers' equation was the main example in the original work of Wentzell and Freidlin [23]. In that case, the rate function is given by

$$I(\gamma) = \frac{1}{2}\int_0^T \|\dot{\gamma}(t) - b(\gamma(t))\|^2 \, dt. \tag{20}$$

The rate, ϵ, is simply the ϵ from the coefficient in front of the Brownian motion. We see that zero-action curves are only those that follow the drift field, b, almost all the time. Thus, the analysis of the vector field b provides the full picture of metastable states. In the case when b is the gradient of a potential function V, this analysis boils down to the analysis of the valley structure of the landscape given by V.

3.2 Jump Processes Under Rescaling

Markov jump processes with non-heavy-tailed increments in finite dimensional spaces will often satisfy a large deviation principle under suitable rescaling of space and time, e.g.,

[2] I am not very careful with time here. We may assume that our sets are big enough so that the optimal connecting paths do so in ϵ-independent time.

$$X^{\epsilon}(t) \equiv \epsilon X_{\epsilon^{-1}t}. \tag{21}$$

One may then expect a large deviation principle for X^{ϵ}, with rate function of the form

$$I(\gamma) = \int_0^{\mathrm{T}} \mathscr{L}(\gamma(t), \dot{\gamma}(t))\mathrm{d}t, \tag{22}$$

with a certain *Lagrange function* that can be computed as a Legendre transform of the log-moment generating function of the laws of the increments of the process X. We see that the rate, ϵ, arises here from the rescaling of the process.

3.3 Markov Processes with Exponentially Small Transition Probabilities

Markov processes on finite state space where some transitions occur with probabilities that are exponentially small in some parameter were considered first by Freidlin and Wentzell [23] as they occur naturally as effective processes describing the jumps of a metastable systems between its metastable states (or "cycles" as they were called). They found renewed interest in the context of stochastic dynamics of Ising type models in the limit as the inverse temperature, β, tends to infinity. It is clear that in that case, all moves that will increase the value of the energy, H_{Λ}, will have an exponentially small probability. These models were intensely studied in recent years by various groups, such as Catoni and Cerf [15], Cerf and Ben Arous [2], and Neves and Schonmann [30, 31], Olivieri, Scoppola, den Hollander, Nardi, etc. [20, 21, 32, 33]. In this situation, individual microscopic path can realize the minimizers of the variational principles, since path entropy plays not role in comparison to the probabilities of individual paths. One should view this as a very singular and atypical situation.

3.4 Large Deviations by Massive Entropy Production

The stochastic Ising models we said we are interested in do not fall into any of the settings above. The dimension of state space is very high, and individual paths have not only very small probability, but even very small probability to stay "close" so a prescribed path. Thus, in order to get to a large deviation description, we must seriously "lump" paths together to transform entropy into probability. One way to do this we have seen at work in the Curie–Weiss model. Passing to the variables $m_{\Lambda}(t) = m_{\Lambda}(\sigma(t))$ identifies all path that follow the same magnetisation pattern. Typically, one path in m-space will correspond to exponentially many paths in σ-space. Through this map we obtain a large deviation principle in the sense of our second example above. Of course in the Curie–Weiss model, all is quite simple again due to the fact that the Hamiltonian depends only on the variables m_{Λ}. In the general case, the computation of the rate function will be a far more difficult problem, to say the least.

The choice of the map used in the entropy production procedure leaves of course a lot of freedom. One can in fact think of any coarse graining method familiar from equilibrium statistical mechanics, such as block-spin averages over boxes of some "mesoscopic" size. In principle this appears robust, but hard to carry out in practice. The most impressive example where this was done remains the seminal paper [36] by Schonmann and Shlosman on the two-dimensional Ising model under Glauber dynamics.

Another class of models that have been considered in this sense are again mean field type interaction diffusions in $\mathbb{R}^d$. Here, one considers a system of n stochastic differential equations,

$$\mathrm{d}X_k = \sigma(X_k)\mathrm{d}B_k + b(X_k, \nu_n(\mathbf{X})\mathrm{d}r, \tag{23}$$

where B_k are independent d-dimensional Brownian motions and $\nu_n(\mathbf{X})$ is the empirical measure

$$\nu_n(\mathbf{X}) \equiv n^{-1} \sum_{k=1}^{n} \delta_{X_k}. \tag{24}$$

Dawson and Gärtner [18,19,24] proved that the trajectories of the empirical measure converge do solutions of the McKean–Vlasov equation

$$\frac{\mathrm{d}}{\mathrm{d}t}\mu(t) = \mathscr{L}(\mu(t))^* \mu(t), \tag{25}$$

where $\mathscr{L}(\mu)^*$ is the adjoint generator of the diffusion (23) with $\nu_n(\mathbf{X})$ replaced by μ. They also proved a large deviation principle for the trajectories of the empirical measure and studied the metastable behavior of the this system.

Similar results were also obtained for a class of spin systems with long-range interactions by Comets [16].

4 Limitations of the Large Deviation Approach and Alternatives

The large deviation method is certainly very versatile and may, in principle at least, be employed in any model context. Its main drawback is the limited precision it makes available. Indeed, all physical quantities, such as escape probabilities and transition times are computed up to multiplicative errors of the form $\exp(\pm\delta/\epsilon)$, with δ arbitrarily small but independent of ϵ. Partly this is due to the fact that we obtain too much information: we localize the optimal path and then calculate the probability that the process follows that path, then in reality we are interested in much less, e.g., the law of an exit time. On the other hand, even in the one-dimensional case, we fail to see the fine details of the behavior of the process near the critical saddle points that are crucial for the precise behavior of the relevant probabilities.

Since we are mostly interested in reversible Markov chains, the approach via potential theory and capacity estimates presents a convenient alternative. I have presented this at length in various occasions [7, 8] and thus give just a very short sketch of the key elements here.

I consider a general Markov chain (in discrete time for definiteness) with discrete state space $\mathcal{S}$ and transition matrix P. I like to call $P - \mathbb{1} \equiv L$ the generator. Note that the entire formalism carries over (under some suitable regularity conditions) to the continuous setting, of course.

For two disjoint sets $A, B \subset S$, the equilibrium potential, $h_{A,B}$, is the harmonic function, i.e., the solution of the equation

$$(Lh_{A,B})(\sigma) = 0, \quad \sigma \notin A \cup B, \tag{26}$$

with boundary conditions

$$h_{A,B}(\sigma) = \begin{cases} 1, & \text{if } \sigma \in A \\ 0, & \text{if } \sigma \in B \end{cases}. \tag{27}$$

The *equilibrium measure* is the function

$$e_{A,B}(\sigma) \equiv -(Lh_{A,B})(\sigma) = (Lh_{B,A})(\sigma), \tag{28}$$

which clearly is nonvanishing only on A and B. The capacity, $\mathrm{cap}(A, B)$ is defined as

$$\mathrm{cap}(A, B) \equiv \sum_{\sigma \in A} \mu(\sigma) e_{A,B}(\sigma). \tag{29}$$

By the discrete analog of the first Green's identity, we get that alternatively,

$$\mathrm{cap}(A, B) = \frac{1}{2} \sum_{\sigma,\sigma' \in S} \mu(\sigma) p(\sigma, \sigma')[h_{A,B}(\sigma) - h_{A,B}(\sigma')]^2 = \Phi(h_{A,B}), \tag{30}$$

where the right-hand side is The functional appearing on the left-hand sides of these relations is called the *Dirichlet form* or *energy*. As a consequence of the *maximum principle*, the function $h_{A,B}$ is the unique minimizer of Φ with boundary conditions (27), which implies the *Dirichlet principle*:

$$\mathrm{cap}(A, B) = \inf_{h \in \mathcal{H}_{A,B}} \Phi(h), \tag{31}$$

where $\mathcal{H}_{A,B}$ denotes the space of functions satisfying (27).

An important observation is that equilibrium potentials and equilibrium measures also determine the Green's function. In fact (see, e.g., [8, 11]),

$$h_{A,B}(\sigma) = \sum_{\sigma' \in A} G_{S\setminus B}(\sigma, \sigma') e_{A,B}(\sigma'). \tag{32}$$

Equation (32) can be used to give the following representation for mean hitting times

$$\sum_{\sigma \in A} \mu(\sigma) e_{A,B}(\sigma) \mathbb{E}_\sigma \tau_B = \sum_{\sigma' \in S} \mu(\sigma') h_{A,B}(\sigma'), \tag{33}$$

or, after normalizing the left-hand side to be an expectation,

$$\mathbb{E}_{\nu_{A,B}} \tau_B = \frac{1}{\mathrm{cap}(A, B)} \sum_{\sigma' \in S} \mu(\sigma') h_{A,B}(\sigma'). \tag{34}$$

The point to retain for us is that estimates in hitting times can be obtained once we have control over capacities and the equilibrium potential. Note that of course, knowing the equilibrium potential alone is good enough, since we hen can get the capacity by just plugging it into the Dirichlet form. The point, however, is that it is easier to estimate the capacity then to find the equilibrium potential: we will see why.

Note that computing the equilibrium potential amounts to solving a boundary value problem for a finite difference operator, which is the discrete analogue of solving a boundary value problem for an elliptic pde. The only case when this is easily doable is when $\mathcal{S}$ has the structure of a one-dimensional set and the transition matrix connects only nearest neighbors. In that case equation (26) can be solved by recursion and we obtain an explicit solution it terms of a sum. This is analogous to the case of a one-dimensional diffusion, where we can solve the boundary value problem in terms of an explicit integral. This is the second important fact that we will keep in mind.

5 Capacity Estimates

The first pleasant surprise is that the Dirichlet principle is perfectly suited for the idea of (imperfect) lumping or coarse graining. Let m map $\mathcal{S}$ to some lower dimensional space, Γ. Let us for simplicity assume that two sets A, B are adapted to the map m in the sense that $A = m^{-1}(m(A))$, and likewise for B.

Then we have the following obvious bound:

$$
\begin{aligned}
\operatorname{cap}(A,B) &= \inf_{h\in\mathscr{H}_{A,B}} \frac{1}{2}\sum_{\sigma,\sigma'\in\mathscr{S}} \mu(\sigma)p(\sigma,\sigma')\left[h(\sigma)-h(\sigma')\right]^2 \\
&\leq \inf_{u\in\mathscr{G}_{m(A),m(B)}} \frac{1}{2}\sum_{\sigma,\sigma'\in\mathscr{S}} \mu(\sigma)p(\sigma,\sigma')\left[u(m(\sigma))-u(m(\sigma'))\right]^2 \\
&= \inf_{u\in\mathscr{G}_{m(A),m(B)}} \frac{1}{2}\sum_{x,x'\in\Gamma} \left[u(x)-u(x')\right]^2 \sum_{\sigma\in m^{-1}(x)} \mu(\sigma) \sum_{\sigma'\in m^{-1}(x')} p(\sigma,\sigma') \\
&\equiv \inf_{u\in\mathscr{G}_{m(A),m(B)}} \frac{1}{2}\sum_{x,x'\in\Gamma} \mathscr{Q}(x)r(x,x')\left[u(x)-u(x')\right]^2 \\
&\equiv \operatorname{CAP}(m(A),m(B)).
\end{aligned} \tag{35}
$$

with

$$
r(x,x') \equiv \frac{1}{\mathscr{Q}_{\beta,N}[\omega](\mathbf{x})} \sum_{\sigma\in m^{-1}(x)} \mu(\sigma) \sum_{\sigma'\in m^{-1}(x')} p(\sigma,\sigma'). \tag{36}
$$

Here,

$$
\mathscr{H}_{A,B} \equiv \{h : S \to [0,1] : \forall\sigma\in A, h(\sigma)=1, \forall\sigma\in B, h(\sigma)=0\} \tag{37}
$$

and

$$
\mathscr{G}_{m(A),m(B)} \equiv \{u : \Gamma \to [0,1] : \forall x\in m(A),\ u(x)=1, \forall \mathbf{x}\in m(B),\ u(x)=0\}. \tag{38}
$$

Thus, the map introduces always new transition rates and a new Dirichlet form, thus a new, “lumped,” Markov process. Equality in the above relation holds if and only if the equilibrium potential corresponding to the original chain is in fact a function of the new variables m only. This is exactly the case when lumping in the original sense works, i.e., when the image of the chain under the map m is Markov.

But note now that we are in a much better shape as before. In particular, we understand that the quality of the upper bound will not depend on the global quality of the approximation of the equilibrium potential by a function depending only on $m(\sigma)$, but only on the quality of this approximation in the region of phase space where the main contribution to the capacity is expected to come from. In practice, we tend to believe that the best way to get a good estimate for the capacity is to find a good mapping, m, and then to find an almost optimal solution for the new Dirichlet form. Of course, to justify such a belief, we must in the end have a way to prove a lower bound.

5.1 Random Path Representation and Lower Bounds on Capacities

The canonical way to get lower bounds on capacities, first exploited in [6] and then in [9], is to exploit a dual variational representation of capacities in terms of flows, due to Berman and Konsowa [4].

It will be convenient to think of the quantities $\mu(\sigma)p_N(\sigma,\sigma')$ as *conductances*, $c(\sigma,\sigma')$, associated to the edges $e = (\sigma,\sigma')$ of the graph of allowed transitions of our dynamics.

Definition 1. Given two disjoint sets, $A, B \subset S$, a nonnegative, cycle free *unit flow*, f, from A to B is a function $f : \mathscr{E} \to \mathbb{R}_+ \cup \{0\}$, such that the following conditions are verified:

(i) If $f(e) > 0$, then $f(-e) = 0$;
(ii) f satisfies *Kirchoff's law*, i.e., for any vertex $a \in S \setminus (A \cup B)$,

$$\sum_b f(b,a) = \sum_d f(a,d); \tag{39}$$

(iii)

$$\sum_{a\in A}\sum_b f(a,b) = 1 = \sum_a \sum_{b\in B} f(a,b); \tag{40}$$

(iv) Any path, γ, from A to B such that $f(e) > 0$ for all $e \in \gamma$, is self-avoiding.

We denote the space of non-negative, cycle free unit flows from A to B by $\mathbb{U}_{A,B}$.

An important unit flow is the *harmonic flow*, associated to the equilibrium potential, $h_{A,B}$. It is defined as

$$f^*(a,b) \equiv \frac{1}{\mathrm{cap}(A,B)} c(a,b)\,(h_{A,B}(a) - h_{A,B}(b))_+ \,. \tag{41}$$

One can easily verify, using that $h_{A,B}$ is a harmonic function, f^* is a non-negative unit flow.

The key observation is that *any* $f \in \mathbb{U}_{A,B}$ gives rise to a *lower bound* on the capacity $\mathrm{cap}(A,B)$, and that this bound becomes sharp for the harmonic flow. To see this, we construct from f a stopped Markov chain $\mathbb{X} = (\mathbb{X}_0,\dots,\mathbb{X}_\tau)$ as follows: For each $a \in S \setminus B$ define $F(a) = \sum_b f(a,b)$.

We define the initial distribution of our chain as $\mathbb{P}^f(a) = F(a)$, for $a \in A$, and zero otherwise. The transition probabilities are given by

$$q^f(a,b) = \frac{f(a,b)}{F(a)}, \tag{42}$$

for $a \notin B$, and the chain is stopped on arrival in B.

Thus, given a trajectory $\mathscr{X} = (a_0, a_1, \ldots, a_r)$ with $a_0 \in A$, $a_r \in B$ and $a_\ell \in S \setminus (A \cup B)$ for $\ell = 0, \ldots, r-1$,

$$\mathbb{P}^f(\mathbb{X} = \mathscr{X}) = \frac{\prod_{\ell=0}^{r-1} f(e_\ell)}{\prod_{\ell=0}^{r-1} F(a_\ell)}, \tag{43}$$

where $e_\ell = (a_\ell, a_{\ell+1})$ and we use the convention $0/0 = 0$. Note that, with the above definitions, the probability that $\mathbb{X}$ passes through an edge e is

$$\mathbb{P}^f(e \in \mathbb{X}) = \sum_{\mathscr{X}} \mathbb{P}^f(\mathscr{X}) \mathbb{1}_{\{e \in \mathscr{X}\}} = f(e). \tag{44}$$

Consequently, we have a partition of unity,

$$\mathbb{1}_{\{f(e)>0\}} = \sum_{\mathscr{X}} \frac{\mathbb{P}^f(\mathscr{X}) \mathbb{1}_{\{e \in \mathscr{X}\}}}{f(e)}. \tag{45}$$

We are ready now to derive our f-induced lower bound: For every function h with $h|_A = 0$ and $h|_B = 1$,

$$\begin{aligned} \frac{1}{2} \sum_e c(e) \left(\nabla_e h\right)^2 &\geq \sum_{e: f(e)>0} c(e) \left(\nabla_e h\right)^2 \\ &= \sum_{\mathscr{X}} \sum_{e \in \mathscr{X}} \mathbb{P}^f(\mathscr{X}) \frac{c(e)}{f(e)} \left(\nabla_e h\right)^2 . \end{aligned}$$

As a result, interchanging the minimum and the sum,

$$\begin{aligned} \operatorname{cap}(A, B) \geq \sum_r \sum_{\mathscr{X}=(a_0,\ldots,a_r)} \mathbb{P}^f(\mathscr{X}) & \min_{h(a_0)=0,\, h(a_r)=1} \sum_0^{r-1} \frac{c(a_\ell, a_{\ell+1})}{f(a_\ell, a_{\ell+1})} \\ & \times \left(h(a_{\ell+1}) - h(a_\ell)\right)^2 \\ = \sum_{\mathscr{X}} \mathbb{P}^f(\mathscr{X}) & \left[\sum_{e \in \mathscr{X}} \frac{f(e)}{c(e)} \right]^{-1} . \end{aligned} \tag{46}$$

Since for the equilibrium flow, f^*,

$$\sum_{e \in \mathscr{X}} \frac{f^*(e)}{c(e)} = \frac{1}{\operatorname{cap}(A, B)}, \tag{47}$$

with $\mathbb{P}^{f^*}$-probability one, the bound (46) is sharp.

Thus, we have proven the following result from [4]:

Proposition 1. *Let $A, B \subset S$. Then, with the notation introduced above,*

$$\operatorname{cap}(A,B) = \sup_{f \in \mathbb{U}_{A,B}} \mathbb{E}^f \left[\sum_{e \in \mathscr{X}} \frac{f(e)}{c(e)} \right]^{-1}. \tag{48}$$

One should note that the rather intricate character of the lower bound through these flows is particular to the discrete graph structure of discrete models. In the case of diffusions, this trivializes to the replacement of gradients by optimally chosing directional derivatives in the Dirichlet form.

So these are our basic tools. The next step is to show that all will work on the level of the mesoscopic approximations.

5.2 *Capacity Estimates for Mesoscopic Chains and the Return of $d = 1$*

We have said before that we believe that by choosing good coarse grainings we will get good upper bounds. The can make this more precise: under reasonable assumptions, we can find optimal upper bounds among the class of test-functions that depend only on the coarse grained variables.

Now we must be more specific. We will assume that our coarse graining goes through functions

$$m : \mathscr{S}_\Lambda \to \Gamma \subset \mathbb{R}^n \tag{49}$$

where in principle we may allow n to depend on Λ.

We will also assume that the induced equilibrium measure on m,

$$\mathbb{Q}(m) \equiv \mu \circ m^{-1}, \tag{50}$$

will be of the form

$$\mathbb{Q}(m) \sim \exp\left(-NF_\Lambda(m)\right), \tag{51}$$

where of course F_Λ depends on everything, but is of order unity. Computing F will be hard problem in equilibrium statistical mechanics, well known from the theory of renormalisation. There are instances, however, when this has been achieved completely [26]. The parameter N can be made large depending on the choice of the coarse graining (in fact, it corresponds to the volume of blocks in the case of block spin variables).

On the level of such a *mesoscopic* description, metastable states correspond to local minima of the function F. One will then expect that the essential contributions from the Dirichlet form small neighborhoods of the *essential saddle* points over which two such minima can be connected. Saddle points may in some cases be numerous, in particular if they correspond to localized structures which then, by symmetry, may entail volume factors.

Basically, we expect to be able to identify a neighborhood, D, of the essential saddle points with the following properties:

(i) F is well approximated by a quadratic[3] function, $\widehat{F}$, on D, in the sense that on B, $|F(m) - \widehat{F}(m)| = o(1/N)$.
(ii) The contributions from D^c to the Dirichlet form can be neglected.

The latter request may sound strange: for part of the complement of D, one may be able to show this simply because the integral over $\exp(-NF)$ becomes negligible. However, this only concerns the region where $F(m) > F(z^*)$ ($F(z^*)$ being the value of F at the saddle. To control the rest, we must effectively know that the harmonic potential is almost a constant.

Let us postpone the discussion how point (ii) can be verified for a moment. The first point to note is that, if $\widehat{F}$ is quadratic, with possibly some zero-eigenvalues corresponding to symmetries, we should expect that D extends in the directions of nonzero eigenvalues by something of order $N^{-1/2} \ln N$. This means that under plausible continuity assumptions on the transition rates $r(m, m')$, we can assume that without introducing significant errors, we can replace the Dirichlet form on D by the simplified one (assuming also the rates are such that only one coordinate can be changed by an amount of order $1/N$, which is reasonable if they are derived from Glauber dynamics),

$$\widehat{\Phi}(h) \equiv \sum_m \sum_{\ell=1}^n r_\ell \exp\left(-\frac{N}{2}(m, Am)\right) [h(m + e_\ell/N) - h(m)]^2 . \tag{52}$$

This, on the other hand is very close to its continuum approximation,

$$\widetilde{\Phi}(h) \equiv \int \exp\left(-\frac{N}{2}(x, Ax)\right) (\nabla h(x), \nabla h(x))_r \, dx, \tag{53}$$

where $(f, g)_r \equiv \sum_{\ell=1}^n f_\ell g_\ell r_\ell$. The point is that for such a functional, one can readily find a family of harmonic functions. Namely, a harmonic function for (53) solves

$$\sum_\ell r_\ell \left(\partial_\ell h(x) - (e_\ell, Ax)\partial_\ell h(x)\right) = 0. \tag{54}$$

Now, let $f(t)$ be a solution of $f'' - \gamma N t f' = 0$. Set $h(x) = f((x, v))$, for some vector v. Inserting this ansatz into (54) yields

$$\sum_\ell r_\ell v_\ell \left(v_\ell \gamma(x, v) - (e_\ell, Ax)\right) = 0. \tag{55}$$

In the case $r_\ell = 1$, we see that this is satisfied if v is an eigenvector of A with eigenvalue γ; in the general case, v should satisfy

[3] This may be general forms, see [3] for examples.

$$r_\ell \sum_k A_{\ell k} v_k = \gamma v_\ell. \tag{56}$$

The point is that the solutions of this equation form an orthonormal basis for the inner product $(\cdot,\cdot)_r$. Choosing such a vector with negative γ (this will be unique), we can construct f that goes from 0 to 1. Using this, to construct a sufficiently good approximation to the minimizer everywhere to get the desired bound on the capacity is just patchwork.

This is basically good news. On the level of the mesoscopic model, we can expect to get sharp results. Thus, we may hope to get to sharp estimates for the full model by just refining the coarse graining. The only nontrivial example so far where this programme has been pushed through and shown to be successful is the Random Field Curie Weiss model [6]. The problems in verifying the result lies in the application of the Berman Konsowa principle, i.e., in the construction of an optimal flow for the lower bound. We do not yet have a canonical way of doing this and this is clearly still place where more work will be needed.

6 Stochastic Partial Differential Equations

From the point of view of the discussion before, a natural class of intermediate models to study are stochastic partial differential equations, to be seen, e.g., as heuristic diffusion limits of block spin approximations. However, spde's with small noise arise in any other modeling contexts and are certainly interesting models in their own right.

The simplest spde of interest is the stochastic Allen–Cahn equation,

$$\mathrm{d}X(x,t) = \frac{\gamma}{4\pi^2}\Delta X(x,t)\mathrm{d}t - \nabla f(X(x,t))\mathrm{d}t + \sqrt{2\epsilon}\mathrm{d}B(x,t), \tag{57}$$

with $x \in D \subset \mathbb{R}^d$ and $t \in \mathbb{R}_+$. f is a double well potential,

$$f(s) = \frac{x^4}{4} - \frac{x^2}{2}. \tag{58}$$

In the case when $d = 1$, the noise term B can be chosen as space–time white noise. It is well-known that in that case this equation admits classical solutions [17]. In higher dimensions this is not the case, and some regularization of the noise term is needed.

Equation (57) arises as the limit of a system of ordinary stochastic differential equations (for simplicity, we write only the case of one spatial dimension)

$$\mathrm{d}X^N(t) = -\nabla F_{\gamma,N}(X^N(t))\mathrm{d}t + \sqrt{2\epsilon}\mathrm{d}B^N(t), \tag{59}$$

where $B^N(t)$ is an N-dimensional vector of Brownian motions and

$$F_{\gamma,N}(X) = \frac{1}{N} \sum_{i \in \Lambda_N} \left(\frac{1}{4} X_i^4 - \frac{1}{2} X_i^2 + N^2 \frac{\gamma}{2\pi^2} (X_{i+1} - X_i)^2 \right) \tag{60}$$

Then $\widehat{X}^N(x,t) = X_i^N(tN)$, for $x \in [i/N, (i+1)/N)$ converges to a solution of (57), as N goes to zero.

For any finite N, one can now use the results of [12] and obtain precise formulae for the mean hitting time of, e.g., a neighborhood of the minimum $X \equiv 1$ when starting from $X \equiv -1$. In the simpler case, when $\gamma > 1$, there are only three critical points ± 1 and the saddle 0, and one gets (see [1])

$$\mathbb{E}_{-1}[\tau_{B_+}] = \frac{2\pi e^{\frac{N}{4\epsilon}} \sqrt{|\det(\nabla^2 F_{\gamma,N}(O))|}}{\sqrt{\det(\nabla^2 F_{\gamma,N}(-1))}} (1 + O(\sqrt{\epsilon} |\ln \epsilon|^3)). \tag{61}$$

The determinants can be computed explicitly and one gets

$$\frac{\sqrt{|\det(\nabla^2 F_{\gamma,N}(O))|}}{\sqrt{\det(\nabla^2 F_{\gamma,N}(I_-))}} = \left[1 - \frac{3}{2+2\gamma}\right]^{\frac{e(N)}{2}} \prod_{k=1}^{\lfloor \frac{N-1}{2} \rfloor} \left[\frac{\gamma N^2 \sinh^2(k\pi/N)/\pi^2 - 1}{\gamma N^2 \sinh^2(k\pi/N)/\pi^2 + 2} \right], \tag{62}$$

where $e(N) = 1$ if N is even and 0 if N is odd. One readily sees that this latter expression converges, as it should, to

$$V(\mu) = \prod_{k=1}^{+\infty} \left[\frac{\gamma k^2 - 1}{\gamma k^2 + 2} \right]. \tag{63}$$

This fact will be quite general and holds also for smaller values of γ, see e.g. [28].

The main issue is thus to prove the uniformity of the error estimates in N. This may look difficult in view of our discussion above on capacity estimates, which required uniformly good quadratic approximations on a neighborhood of the saddle point. The point is, however, that the the second derivative of the function $F_{\gamma,N}$ as an operator has eigenvalues of order k^2. Hence, the invariant measure concentrates very rapidly in the directions of the higher eigenvalues, making the problem *effectively* finite dimensional.

7 Open Issues

To conclude this brief review, let me mention some of the open issues that I feel need to be addressed in the coming years.

7.1 Initial Distributions and Regularity Theory

The key equation linking capacities to physical quantities is (34). While as an equation it is perfectly true, it is useful only if the sets A, B are chosen "not too small." Otherwise, both numerator and denominator will be excessively small, and the calculation of the ratio becomes an issue of subtle second order corrections, which will in most situations be next to impossible to achieve.

In many examples, this does not cause to much of a problem. Functions such as $\mathbb{E}_x \tau_B$ should be relatively constant on "small" sets A of interest. This can be proven either by analytic means (see [12] using Hölder estimates in the setting of diffusion processes, or coupling arguments, see [29] for the case of stochastic (partial) differential equations, or [5] for a stochastic spin system. In the last case, the absence of contracting drifts on the level of the microscopic variables made the analysis already rather cumbersome and rather model dependent. It would clearly be desirable to have more universally applicable tools at our disposal.

A similar issue arises in the context of spectral theory. In a number of contexts (finite dimensional diffusions [13], (essentially) finite state space [11]), there is a very sharp link between metastable states and small eigenvalues of of the generator. One would clearly expect this to be true in much more generality. The method used in [13], essentially an application of an old idea of Wentzell [37,38] is again based on regularity properties, this time of eigenfunction of the generator. One should again expect this to be true in more generality, but a good theory seems to be missing so far.

7.2 Canonical Constructions of Flows

In [6], we have proven a lower bound on capacities by constructing a specific *microscopic flow* for the Berman–Konsowa principle. The fact that this worked out well was due to self-averaging resp. homogenization effects, and both the construction and the proofs relied quite heavily on specific properties of the model. Whenever we want to control a model of microscopic spin- or particle dynamics, we will have to be able to do the same. Is there some generic way of doing and proving this? This is clearly one of the most pressing issues to be resolved in the coming years.

Acknowledgements This notes summarizes thoughts that have come up through extensive work on metastability with numerous people. The entire subject started with an intensive collaboration with Michael Eckhoff, Véronique Gayrard, and Markus Klein. These early works set the stage for the potential theoretic approach. More recently I collaborated with Florent Barret, Alessandra Bianchi, Alessandra Faggionato, Frank den Hollander, Dima Ioffe, Francesco Manzo, Sylvie Méléard, Francesca Nardi, and Cristian Spitoni, on various special issues and models. I thank all of them for sharing their thoughts and insights.

The huge project on metastability was possible also only due to the excellent working conditions I had from 1992 on at the WIAS. This was in in more than one way due to Jürgen Gärtner to whom I am deeply grateful.

One person deserves spatial thanks: Erwin Bolthausen handled our first paper [10] on metastability as Editor-in-Chief of PTRF, and through an extensive correspondence, that paper was finally published.

These notes were written while I was holding a Lady Davis Visiting Professorship at the Technion, Haifa. I thank the William Davidson Faculty of Industrial Engineering and Management and in particular Dmitry Ioffe for their kind hospitality. Much of the work reported on here was also supported by a grant from the German-Israeli Foundation (GIF).

Financial support from the German Research Council (DFG) through SFB 611 and the Hausdorff Center for Mathematics is gratefully acknoledged.

References

1. Barret, F., Bovier, A., Méléard, S.: Uniform estimates for metastable transition times in a coupled bistable system. Elect. J. Probab. **15**, 323–345 (2010)
2. Ben Arous, G., Cerf, R.: Metastability of the three-dimensional Ising model on a torus at very low temperatures. Electron. J. Probab. **1**(10), p. 55 (1996)
3. Berglund, N., Gentz, B.: Anomalous behavior of the Kramers rate at bifurcations in classical field theories. J. Phys. A **42**(5), 052001, 9 (2009)
4. Berman, K.A., Konsowa, M.H.: Random paths and cuts, electrical networks, and reversible Markov chains. SIAM J. Discrete Math. **3**(3), 311–319 (1990)
5. Bianchi, A., Bovier, A., Ioffe, D.: Pointwise estimates and exponential laws in metastable systems via coupling methods. preprint, SFB 611, Bonn University (2009) (to appear in Ann. Probab)
6. Bianchi, A., Bovier, A., Ioffe, D.: Sharp asymptotics for metastability in the random field Curie-Weiss model. Electron. J. Probab. **14**(53), 1541–1603 (2009)
7. Bovier, A.: Metastability: a potential theoretic approach. In: International Congress of Mathematicians. Vol. III, pp. 499–518. Eur. Math. Soc., Zürich (2006)
8. Bovier, A.: Metastability. In: Methods of contemporary mathematical statistical physics, volume 1970 of Lecture Notes in Math., pp. 177–221. Springer, Berlin (2009)
9. Bovier, A., den Hollander, F., Spitoni, C.: Homogeneous nucleation for Glauber and Kawasaki dynamics in large volumes and low temperature. Ann. Probab. **38**, 661–713 (2010)
10. Bovier, A., Eckhoff, M., Gayrard, V., Klein, M.: Metastability in stochastic dynamics of disordered mean-field models. Probab. Theor. Relat. Field. **119**(1), 99–161 (2001)
11. Bovier, A., Eckhoff, M., Gayrard, V., Klein, M.: Metastability and low lying spectra in reversible Markov chains. Comm. Math. Phys. **228**(2), 219–255 (2002)
12. Bovier, A., Eckhoff, M., Gayrard, V., Klein, M.: Metastability in reversible diffusion processes. I. Sharp asymptotics for capacities and exit times. J. Eur. Math. Soc. (JEMS) **6**(4), 399–424 (2004)
13. Bovier, A., Gayrard, V., Klein, M.: Metastability in reversible diffusion processes. II. Precise asymptotics for small eigenvalues. J. Eur. Math. Soc. (JEMS) **7**(1), 69–99 (2005)
14. Cassandro, M., Galves, A., Olivieri, E., Vares, M.E.: Metastable behavior of stochastic dynamics: a pathwise approach. J. Statist. Phys. **35**(5–6), 603–634 (1984)
15. Catoni, O., Cerf, R.: The exit path of a Markov chain with rare transitions. ESAIM Probab. Stat. **1**, 95–144 (1995/97)
16. Comets, F.: Nucleation for a long range magnetic model. Ann. Inst. H. Poincaré Probab. Statist. **23**(2), 135–178 (1987)
17. Da Prato, G., Zabczyk, J.: Stochastic equations in infinite dimensions, volume 44 of Encyclopedia of Mathematics and its Applications. Cambridge University Press, Cambridge (1992)

18. Dawson, D.A., Gärtner, J.: Large deviations from the McKean-Vlasov limit for weakly interacting diffusions. Stochastics **20**(4), 247–308 (1987)
19. Dawson, D.A., Gärtner, J.: Long time behaviour of interacting diffusions. In Stochastic calculus in application (Cambridge, 1987), volume 197 of Pitman Res. Notes Math. Ser., pp. 29–54. Longman Sci. Tech., Harlow (1988)
20. den Hollander, F., Nardi, F.R., Olivieri, E., Scoppola, E.: Droplet growth for three-dimensional Kawasaki dynamics. Probab. Theor. Relat. Field. **125**(2), 153–194 (2003)
21. den Hollander, F., Olivieri, E., Scoppola, E.: Metastability and nucleation for conservative dynamics. J. Math. Phys. **41**(3), 1424–1498 (2000); Probabilistic techniques in equilibrium and nonequilibrium statistical physics.
22. Eyring, H.: The activated complex in chemical reactions. J. Chem. Phys **3**, 107 (1935)
23. Freidlin, M.I., Wentzell, A.D.: Random perturbations of dynamical systems, volume 260 of Grundlehren der Mathematischen Wissenschaften [Fundamental Principles of Mathematical Sciences]. Springer, New York (1984); Translated from the Russian by Joseph Szücs.
24. Gärtner, J.: On the McKean-Vlasov limit for interacting diffusions. Math. Nachr. **137**, 197–248 (1988)
25. Kramers, H.A.: Brownian motion in a field of force and the diffusion model of chemical reactions. Physica **7**, 284–304 (1940)
26. Külske, C.: On the Gibbsian nature of the random field Kac model under block-averaging. J. Stat. Phys. **104**(5–6), 991–1012 (2001)
27. Levin, D.A., Luczak, M.J., Peres, Y.: Glauber dynamics for the mean-field Ising model: cut-off, critical power law, and metastability. Probab. Theor. Relat. Field. **146**(1–2), 223–265 (2010)
28. Maier, R., Stein, D.: Droplet nucleation and domain wall motion in a bounded interval. Phys. Rev. Lett. **87**, 270601–1–270601–4 (2001)
29. Martinelli, F., Olivieri, E., Scoppola, E.: Small random perturbations of finite- and infinite-dimensional dynamical systems: unpredictability of exit times. J. Stat. Phys. **55**(3–4), 477–504 (1989)
30. Neves, E.J., Schonmann, R.H.: Critical droplets and metastability for a Glauber dynamics at very low temperatures. Comm. Math. Phys. **137**(2), 209–230 (1991)
31. Neves, E.J., Schonmann, R.H.: Behavior of droplets for a class of Glauber dynamics at very low temperature. Probab. Theor. Relat. Field. **91**(3–4), 331–354 (1992)
32. Olivieri, E., Scoppola, E.: Markov chains with exponentially small transition probabilities: first exit problem from a general domain. I. The reversible case. J. Statist. Phys. **79**(3–4), 613–647 (1995)
33. Olivieri, E., Scoppola, E.: Markov chains with exponentially small transition probabilities: first exit problem from a general domain. II. The general case. J. Stat. Phys. **84**(5–6), 987–1041 (1996)
34. Olivieri, E., Vares, M.E.: Large deviations and metastability, volume 100 of Encyclopedia of Mathematics and its Applications. Cambridge University Press, Cambridge (2005)
35. Pollak, E., Talkner, P.: Reaction rate theory: What it was, where is it today, and where is it going? Chaos **15**, 026116 (2005)
36. Schonmann, R.H., Shlosman, S.B.: Wulff droplets and the metastable relaxation of kinetic Ising models. Comm. Math. Phys. **194**(2), 389–462 (1998)
37. Ventcel′, A.D.: Formulas for eigenfunctions and eigenmeasures that are connected with a Markov process. Teor. Verojatnost. i Primenen. **18**, 3–29 (1973)
38. Ventcel′, A.D.: The asymptotic behavior of the first eigenvalue of a second order differential operator with a small parameter multiplying the highest derivatives. Teor. Verojatnost. i Primenen. **20**(3), 610–613 (1975)
39. Weidenmüller, H.A., Zhang, J.S.: Stationary diffusion over a multidimensional potential barrier: a generalization of Kramers' formula. J. Stat. Phys. **34**(1–2), 191–201 (1984)

Hydrodynamic Limit for the $\nabla\varphi$ Interface Model via Two-Scale Approach

Tadahisa Funaki

Abstract We prove the hydrodynamic limit for the Ginzburg–Landau $\nabla\varphi$ interface model based on a two-scale approach recently introduced by Grunewald et al. [6] under the assumptions of the strict convexity of the coarse-grained Hamiltonian and the logarithmic Sobolev inequality for canonical Gibbs measures. In particular, strictly convex potentials satisfy these assumptions.

1 Introduction

We apply the method of [6] called the two-scale approach to the proof of the hydrodynamic limit for the Ginzburg–Landau $\nabla\varphi$ interface model for a certain class of interaction potentials. Our hope is that this could extend the result of [5] for strictly convex potentials to some nonconvex ones, combining with a recent work of Cotar et al. [2], which showed the strict convexity of the corresponding surface tension, see Remark 1.2 below. One of the advantages of the method of [6] is that the local ergodicity, which is usually shown via the so-called one-block estimate in the framework of the celebrated entropy method due to Guo et al. [7], follows implicitly and, in particular, it is unnecessary to characterize the family of all reversible measures of the dynamics. Compared with the case discussed in [6], the Gibbs measures have long correlations in our case and this causes a difficulty in analysis.

T. Funaki (✉)
Graduate School of Mathematical Sciences, University of Tokyo, Komaba, Tokyo 153-8914, Japan
e-mail: funaki@ms.u-tokyo.ac.jp

J.-D. Deuschel et al. (eds.), *Probability in Complex Physical Systems*, Springer Proceedings in Mathematics 11, DOI 10.1007/978-3-642-23811-6_19,

1.1 Setting

Let $N, M, K \in \mathbb{N}$ be given and fulfill the relation $N = MK$. The microscopic system is defined on the lattice torus $\Gamma_N = \{1, 2, \dots, N\}^d \equiv (\mathbb{Z}/N\mathbb{Z})^d$ of size N. Let Γ_N^* be the set of all directed bonds $b = \langle x, y\rangle$ in Γ_N. Let $\Lambda_K = [1, K]^d \cap \mathbb{Z}^d$ and $\Lambda_K(i) = \Lambda_K + Ki$ for $i \in \mathbb{Z}^d$; $\Lambda_K(i)$ are called boxes of mesoscopic size. We imbed $\Lambda_K(i)$ in Γ_N by considering the elements $x \in \mathbb{Z}^d$ modulo N componentwisely. Let $\Lambda_K^*(i)$ be the set of all directed bonds $b = \langle x, y\rangle$ in $\Lambda_K(i)$, i.e., all b's with $x, y \in \Lambda_K(i)$. We denote the set of all $b = \langle x, y\rangle$ with $x \in \Lambda_K(i)$ and $y \in \Lambda_K(j)$ for $i \neq j$, i.e., bonds crossing two different mesoscopic boxes, by $\Gamma_{N,K,\partial}^*$. Similarly as before, $\Lambda_K^*(i)$ and $\Gamma_{N,K,\partial}^*$ are imbedded in Γ_N^*. Thus, we have the decompositions:

$$\Gamma_N = \bigcup_{i\in\Gamma_M} \Lambda_K(i) \quad \text{and} \quad \Gamma_N^* = \left(\bigcup_{i\in\Gamma_M} \Lambda_K^*(i)\right) \bigcup \Gamma_{N,K,\partial}^*.$$

Let $X \equiv X_N = \mathbb{R}^{\Gamma_N}$ and $Y \equiv Y_M = \mathbb{R}^{\Gamma_M}$ be the configuration spaces of all microscopic height variables $\phi = \{\phi(x)\}_{x\in\Gamma_N}$ and all mesoscopic ones $\psi = \{\psi_i\}_{i\in\Gamma_M}$, respectively. These two Euclidean spaces are equipped with the inner products:

$$\phi \underset{X}{\cdot} \tilde{\phi} \equiv \langle \phi, \tilde{\phi}\rangle_X = \sum_{x\in\Gamma_N} \phi(x)\tilde{\phi}(x),$$

and

$$\psi \underset{Y}{\cdot} \tilde{\psi} \equiv \langle \psi, \tilde{\psi}\rangle_Y = \frac{1}{M^d} \sum_{i\in\Gamma_M} \psi_i\tilde{\psi}_i,$$

respectively, for $\phi = \{\phi(x)\}_{x\in\Gamma_N}, \tilde{\phi} = \{\tilde{\phi}(x)\}_{x\in\Gamma_N} \in X_N$ and $\psi = \{\psi_i\}_{i\in\Gamma_M}, \tilde{\psi} = \{\tilde{\psi}_i\}_{i\in\Gamma_M} \in Y_M$. The scaling operator $P = P_{N,K} : X_N \to Y_M$ is defined by

$$\psi_i = (P\phi)_i = \frac{1}{K^d} \sum_{x\in\Lambda_K(i)} \frac{1}{N}\phi(x), \quad i \in \Gamma_M.$$

In other words, we take the block spin average of the corresponding macroscopic height variables $\phi(x)/N$ over a box $\Lambda_K(i)$ of mesoscopic size. Its dual operator $P^t = P_{N,K}^t : Y_M \to X_N$, defined by $\langle P\phi, \psi\rangle_Y = \langle \phi, P^t\psi\rangle_X$, is given by

$$(P^t\psi)(x) = \frac{1}{N^{d+1}}\psi_{[x]},$$

where $i = [x] \in \Gamma_M$ is equivalent to "$x \in \Lambda_K(i)$". In particular, the relation

$$P \cdot N^{d+2} P^t = i\mathrm{d}_Y, \tag{1}$$

holds.

1.2 The Ginzburg–Landau $\nabla\varphi$ Interface Model

Let $V : \mathbb{R} \to \mathbb{R}$ be a symmetric C^2-function and we assume that $\|V''\|_\infty < \infty$ and $V(\eta)$ grows at least quadratically as $|\eta| \to \infty$, that is, there exist $c, C > 0$ such that $V(\eta) \geq c\eta^2 - C$ for every $\eta \in \mathbb{R}$. The time evolution of the microscopic height $\phi_t = \{\phi_t(x)\}_{x\in\Gamma_N} \in X_N$ of the interface is governed by the stochastic differential equations

$$\mathrm{d}\phi_t(x) = -\sum_{y\in\Gamma_N, |x-y|=1} V'(\phi_t(x) - \phi_t(y))\mathrm{d}t + \sqrt{2}\mathrm{d}w_t(x), \quad x \in \Gamma_N, \tag{2}$$

where $\{w_t(x)\}_{x\in\Gamma_N}$ is a collection of independent Brownian motions, see [4, 5].

The measure $\mu(\mathrm{d}\phi) = \mu^N(\mathrm{d}\phi)$ on X_N defined by

$$\mu(\mathrm{d}\phi) = \mathrm{e}^{-H(\phi)}\mathrm{d}\phi,$$

is reversible under the dynamics ϕ_t, where $\mathrm{d}\phi = \prod_{x\in\Gamma_N} \mathrm{d}\phi(x)$ is the Lebesgue measure on X and

$$H(\phi) = \frac{1}{2} \sum_{b=\langle x,y\rangle \in \Gamma_N^*} V(\phi(x) - \phi(y)). \tag{3}$$

Note that μ is invariant under the translation: $\phi \mapsto \phi + a$ for every $a \in \mathbb{R}$ and therefore it is an unnormalizable (infinite) measure. The probability density with respect to $\mathrm{d}\mu^N$ of the distribution of the time-changed process $\phi_{N^2 t} \in X_N$, denoted by $f = f^N(t, \phi)$, satisfies the forward equation:

$$\frac{\partial f}{\partial t} = N^2 e^H D \cdot (e^{-H} Df), \tag{4}$$

where $D = \{\partial/\partial\phi(x)\}_{x\in\Gamma_N}$ and $\cdot$ is the inner product in the space X.

Let $\bar{\mu} := \mu \circ P^{-1}$ be the (infinite) measure on Y, which is the image measure of μ under the scaling operator P, and decompose μ into a skew product:

$$\mu(\mathrm{d}\phi) = \mu(\mathrm{d}\phi|\psi)\,\bar{\mu}(\mathrm{d}\psi), \tag{5}$$

where $\mu(\mathrm{d}\phi|\psi) := \mu(\mathrm{d}\phi|P\phi = \psi)$ is a probability measure on X for each $\psi \in Y$. Note that such probability measure $\mu(\mathrm{d}\phi|\psi)$ exists. In fact, for each $a \in \mathbb{R}$, let $X^a = \{\phi \in X; (P\phi)_0 = a\}$ and define a probability measure μ^a on X^a by $\mu^a(d\phi^{(a)}) := e^{-H(\phi)} d\phi^{(a)}/Z$, where $d\phi^{(a)}$ is the Lebesgue measure on X^a and Z is the normalizing constant, which is independent of a. Let $Y^a = \{\psi \in Y; \psi_0 = a\}$ and define a probability measure $\bar{\mu}^a = \mu^a \circ P^{-1}$ on Y^a. Then since both μ^a and $\bar{\mu}^a$ are probability measures, one can define the conditional probability $\mu^a(d\phi^{(a)}|\psi) := \mu^a(\cdot|P\phi = \psi)$ for each a and $\psi \in Y^a$, and it holds $\mu^a(d\phi^{(a)}) = \mu^a(d\phi^{(a)}|\psi)\bar{\mu}^a(d\psi)$. It is now clear that, once we define $\mu(d\phi|\psi) := \mu^a(d\phi|\psi^{(a)})$ under the decomposition of $\psi \in Y$ into $(a, \psi^{(a)})$ with $a = \psi_0$ and $\psi^{(a)} \in Y^a$, then we have the identity (5).

We introduce a coarse-grained Hamiltonian $\bar{H}(\psi) = \bar{H}^N(\psi)$ by

$$\bar{H}(\psi) = -\frac{1}{N^d} \log\left(\frac{d\bar{\mu}}{d\psi}\right),$$

where $d\psi = \prod_{i \in \Gamma_M} d\psi_i$ is the Lebesgue measure on Y. In other words,

$$d\bar{\mu} = e^{-N^d \bar{H}(\psi)} d\psi,$$

and

$$\frac{d\bar{\mu}}{d\psi} = \int_{\{P\phi = \psi\}} e^{-\frac{1}{2}\sum_{b \in \Gamma_N^*} V(\phi(x) - \phi(y))} d\phi,$$

where $d\phi$ is the Lebesgue measure on the affine space $\{\phi \in X; P\phi = \psi\}$.

1.3 Main Result

Let $\bar{D} \equiv \bar{D}_M = \{M^d \partial/\partial\psi_i\}_{i \in \Gamma_M}$ be the gradient acting on functions $F = F(\psi)$ of ψ under the metric structure of Y, i.e., $\bar{D}F \underset{Y}{\cdot} \tilde{\psi} := DF \cdot \tilde{\psi}$ $(= \sum_{i \in \Gamma_M} \partial F/\partial\psi_i \tilde{\psi}_i$ in the Euclidean inner product) for every $\tilde{\psi} \in Y$. For $\psi = \{\psi_i\}_{i \in \Gamma_M} \in Y$, we define its macroscopic (right and discrete) derivatives $\nabla_M \psi \in Y^d$ by $\nabla_M \psi(i) = \{M(\psi_{i+e_\alpha} - \psi_i)\}_{\alpha=1}^d \in \mathbb{R}^d$, or we sometimes regard $\nabla_M \psi \in \mathbb{R}^{\Gamma_M^*}$ by defining $\nabla_M \psi(\langle i, j \rangle) = M(\psi_i - \psi_j)$ for $\langle i, j \rangle \in \Gamma_M^*$.

Let $h(t) = h^{N,M}(t) \in Y_M$ be the solution of the ODE (at the mesoscopic level):

$$\frac{dh}{dt} = -\bar{D}_M \bar{H}^N(h), \tag{6}$$

and set for $t \geq 0$

$$\Theta(t) = \int_{X_N} \frac{1}{N^d} \left|\frac{1}{N}\phi - N^{d+1} P^t h(t)\right|_X^2 f(t, \phi)\, \mu^N(d\phi),$$

where $f(t,\phi) = f^N(t,\phi)$ is the solution of the forward equation (4). This is the microscopic expression of the L^2-distance between the scaled process defined from ϕ_{N^2t} and $h(t)$. The essentially same quantity was introduced in (5.1) of [5] and also in [1].

We are now at the position to state our main result. The conditions, which we require, are summarized in the following assumption.

Assumption A. (0) The potential V satisfies the condition stated at the beginning of Sect. 1.2.

(1) (Uniform strict convexity of coarse-grained Hamiltonian for K sufficiently large). There exists $K_0 \geq 1$ and $c_- > 0$ such that

$$\left(\bar{D}\bar{H}^N(\psi^1) - \bar{D}\bar{H}^N(\psi^2)\right) \underset{Y}{\cdot} (\psi^1 - \psi^2) \geq c_-|\nabla_M \psi^1 - \nabla_M \psi^2|^2_{Y^d},$$

for every $\psi^1, \psi^2 \in Y_M$ and $K \geq K_0$.

(2) (Logarithmic Sobolev inequality for $\mu(\mathrm{d}\phi|\psi)$ uniformly in ψ). There exists a constant $\rho = \rho_{N,K} > 0$, which is independent of ψ, such that

$$\int \Phi(f)\,\mu(\mathrm{d}\phi|\psi) - \Phi\left(\int f\,\mu(\mathrm{d}\phi|\psi)\right) \leq \frac{1}{2\rho}\int \frac{|Df|^2_X}{f}\mu(\mathrm{d}\phi|\psi)$$

holds for every $f = f(\phi) > 0$, where $\Phi(f) = f \log f$, and the constant ρ satisfies the condition that $\rho_{N,K}\sqrt{K}M \to \infty$ as $N, M, K \to \infty$.

(3) (Assumptions on initial data) $\sup_{N,M}\{\bar{H}^N(h^{N,M}(0)) - \bar{\bar{H}}^N\} < \infty$, where $\bar{\bar{H}}^N = \inf_\psi \bar{H}^N(\psi)$, and $\mathscr{H}_N(f^\nabla(0)) \leq CN^d$ with some $C > 0$, see Lemma 2.1 below for the entropy.

Theorem 1.1. *(The hydrodynamic limit) Under Assumption A, we have $\Theta(t) \to 0$ for $t > 0$ as $N, M, K \to \infty$ if it holds for $t = 0$.*

Remark 1.1. If the interaction potential V is strictly convex, the LSI holds for $\mu(\mathrm{d}\phi|\psi)$ with $\rho = CK^{-2}$ and thus Assumption A-(2) is satisfied if $MK^{-3/2} \to \infty$, while Assumption A-(1) follows by applying the method developed in [3]. See Sect. 4 for details.

The rest of this section is some heuristics. Let $\sigma_K(u)$, $u \in \mathbb{R}^d$ be the finite volume surface tension defined by

$$\sigma_K(u) = -\frac{1}{K^d}\log \int_{\mathbb{R}^{\Lambda_K}} \mathrm{e}^{-H_{\Lambda_K,\bar{\phi}_u}(\phi)} \prod_{x\in\Lambda_K} \mathrm{d}\phi(x), \tag{7}$$

where $H_{\Lambda_K,\bar{\phi}_u}(\phi)$ is the Hamiltonian on Λ_K with the boundary condition $\bar{\phi}_u$ on the outer boundary $\partial^+\Lambda_K$ of Λ_K and $\bar{\phi}_u$ is defined by $\bar{\phi}_u(x) = u \cdot x$. Then the measure μ decouples at the mesoscopic level if we remove from H the sum over bonds $b \in \Gamma^*_{N,K,\partial}$, which cross different mesoscopic boxes. The contribution of this sum would be of surface order and moreover, under the condition $\{P\phi = \psi\}$, it is expected that the averaged tilt u over the box $\Lambda_K(i)$ would behave as $\nabla_M \psi(i)$. We could therefore expect that

$$\bar{H}^N(\psi) \sim \frac{1}{M^d} \sum_{i \in \Gamma_M} \sigma_K(\nabla_M \psi(i)).$$

This would imply

$$\bar{D}\bar{H}^N(\psi) \equiv M^d \frac{\partial}{\partial \psi(i)} \bar{H}^N(\psi) \sim -M \sum_{\alpha=1}^d \left\{ \frac{\partial \sigma_K}{\partial u_\alpha}(\nabla_M \psi(i)) - \frac{\partial \sigma_K}{\partial u_\alpha}(\nabla_M \psi(i-e_\alpha)) \right\}.$$

Thus,

$$\begin{aligned} &\left(\bar{D}\bar{H}^N(\psi^1) - \bar{D}\bar{H}^N(\psi^2)\right) \underset{Y}{\cdot} (\psi^1 - \psi^2) \\ &\quad \sim \left(\nabla\sigma_K(\nabla_M \psi^1) - \nabla\sigma_K(\nabla_M \psi^2)\right) \underset{Y}{\cdot} (\nabla_M \psi^1 - \nabla_M \psi^2). \end{aligned}$$

Now Assumption A-(1) is expected to follow from the next remark.

Remark 1.2. Cotar et al. [2] considered a variant $\tilde{\sigma}_K$ of σ_K, which is defined through periodic boundary conditions, and found a class of non-convex V's satisfying the uniform convexity of $\tilde{\sigma}_K$ for K large, that is, there exist $K_0 \geq 1$ and $c_- > 0$ such that

$$\left(\nabla\tilde{\sigma}_K(u) - \nabla\tilde{\sigma}_K(v)\right) \cdot (u - v) \geq c_- |u - v|^2,$$

for every $u, v \in \mathbb{R}^d$ and $K \geq K_0$. We could therefore expect the uniform convexity of σ_K as well. Note that, as we already mentioned, the Gibbs measures have long correlations and therefore the surface tension is actually rather sensitive to the boundary conditions.

Let us assume $\sigma(u) = \lim_{K\to\infty} \sigma_K(u)$ exists. This statement is related to the equivalence of ensembles. Then as $N, M, K \to \infty$, the solution $h^{N,M}(t)$ of the ODE (6) at the mesoscopic level is expected to converge to the solution $h(t)$ of the PDE at the macroscopic level:

$$\frac{\partial h}{\partial t} = \operatorname{div}\left(\nabla\sigma(\nabla h)\right) \left(= \sum_{\alpha=1}^d \frac{\partial}{\partial \theta_\alpha} \left\{ \frac{\partial \sigma}{\partial u_\alpha}(\nabla h(t,\theta)) \right\} \right), \tag{8}$$

for $\theta \in \mathbb{T}^d \equiv (\mathbb{R}/\mathbb{Z})^d$, where $\nabla h(t,\theta) = \{\partial h/\partial\theta_\alpha(t,\theta)\}_{\alpha=1}^d$. Unfortunately, this convergence is left open, especially because the convergence of the surface tension is not obvious due to the character of long correlations in our model.

2 A Priori Estimates

This section gives a priori estimates, which will be used for the proof of the main theorem in the next section. Since the quantities for which we need bounds are mostly related to the $\nabla\phi$-field, instead of the unnormalizable measure μ^N, we work with the probability measure $\mu^{N,\nabla}$ on X_N^∇ for the $\nabla\phi$-field defined by

$$\mu^{N,\nabla}(\mathrm{d}\eta) = \frac{1}{Z_V^{N,\nabla}} \mathrm{e}^{-\frac{1}{2}\sum_{b\in\Gamma_N^*} V(\eta_b)}\, \mathrm{d}\eta_{\Gamma_N}, \tag{9}$$

where $\mathrm{d}\eta_{\Gamma_N}$ is the Lebesgue measure on the affine space $X_N^\nabla := \{\eta \in \mathbb{R}^{\Gamma_N^*}; \eta$ satisfies the plaquette (or loop) condition$\}$ and $Z_V^{N,\nabla}$ is the normalizing constant; see [5], (2.1) for the plaquette condition. The Ginzburg–Landau $\nabla\varphi$ dynamic ϕ_t governed by the SDE (2) determines a dynamic on X_N^∇ by $\eta_t := \nabla\phi_t$. Let $f^\nabla = f^{N,\nabla}(t,\eta)$ be the probability density of the distribution of $\eta_{N^2 t}$ with respect to $\mu^{N,\nabla}$. Then it satisfies the forward equation:

$$\frac{\partial f^\nabla}{\partial t} = N^2 e^H D\cdot(e^{-H} D f^\nabla). \tag{10}$$

Note that the distribution of $\eta = \nabla\phi$ under $f(t,\phi)\mu^N(\mathrm{d}\phi)$ coincides with $f^\nabla(t,\eta)\times\mu^{N,\nabla}(\mathrm{d}\eta)$.

We define the entropy of the probability density $f = f(\eta)$ with respect to $\mu^{N,\nabla}$ as

$$\mathscr{H}_N(f) = \int_{X_N^\nabla} f\log f\, \mathrm{d}\mu^{N,\nabla} \ge 0.$$

Lemma 2.1. *If $\mathscr{H}_N(f^\nabla(0)) \le CN^d$, then we have*

$$\int_0^T \mathrm{d}t \int \frac{1}{f^\nabla(t,\eta)} |Df^\nabla(t,\eta)|_X^2\, \mu^{N,\nabla}(\mathrm{d}\eta)$$
$$= \frac{1}{N^2}\{\mathscr{H}_N(f^\nabla(0)) - \mathscr{H}_N(f^\nabla(T))\} \le CN^{d-2}.$$

Proof. The forward equation (10) shows

$$\frac{\mathrm{d}}{\mathrm{d}t}\mathscr{H}_N(f^\nabla(t)) = -N^2\int \frac{1}{f^\nabla(t,\eta)}|Df^\nabla(t,\eta)|_X^2\, \mu^{N,\nabla}(\mathrm{d}\eta),$$

and this implies the conclusion. □

The following lemma will be used to treat the fluctuation terms $I_{1,f}$ and $I_{2,f}$ in the next section.

Lemma 2.2. *If $\mathscr{H}_N(f^\nabla(0)) \le CN^d$, then we have*

$$\sup_N \sup_{0\le t\le T} E\left[\frac{1}{N^d}\sum_{b\in\Gamma_N^*}|\nabla\phi_{N^2t}(b)|^2\right] < \infty.$$

Proof. We rely on the entropy inequality to show that

$$\frac{1}{N^d}E^{f^\nabla(t)}\left[\sum_{b\in\Gamma_N^*}|\eta_b|^2\right] \le \frac{1}{\epsilon N^d}\log E^{\mu^{N,\nabla}}\left[\mathrm{e}^{\epsilon\sum_{b\in\Gamma_N^*}|\eta_b|^2}\right] + \frac{1}{\epsilon N^d}\mathscr{H}_N(f^\nabla(t)), \tag{11}$$

for every $\epsilon > 0$. However, by means of the partition functions $Z_V^{N,\nabla}$ introduced in (9), the expectation in the right hand side of (11) can be rewritten as

$$E^{\mu^{N,\nabla}}\left[\mathrm{e}^{\epsilon\sum_{b\in\Gamma_N^*}|\eta_b|^2}\right] = \frac{Z^{N,\nabla}_{V-2\epsilon|\eta|^2}}{Z_V^{N,\nabla}}.$$

The subadditivity argument guarantees the existence of the limit (cf. [2], below (1.3)):

$$\sigma_V(0) = -\lim_{N\to\infty}\frac{1}{N^d}\log Z_V^{N,\nabla}$$

at least for small $\epsilon > 0$ (recall that the potential $V(\eta)$ grows at least quadratically as $|\eta|\to\infty$), and this implies that the limit

$$\lim_{N\to\infty}\frac{1}{N^d}\log E^{\mu^{N,\nabla}}\left[\mathrm{e}^{\epsilon\sum_{b\in\Gamma_N^*}|\eta_b|^2}\right] = \sigma_V(0) - \sigma_{V-2\epsilon|\eta|^2}(0),$$

exists and finite. The second term in the right-hand side of (11) is bounded in N and t from Assumption A-(3) by noting that $\mathscr{H}_N(f^\nabla(t))$ is non-increasing in t as we have seen in the proof of Lemma 2.1. The proof is now complete. □

Remark 2.1. This proof is suggested by the Remark in [5], p.20, and does not require the convexity of V. This also simplifies the proof of (45) in [6].

3 Proof of Theorem 1.1

This section gives the proof of Theorem 1.1 under Assumption A.

3.1 Derivative of $\Theta(t)$

Rewriting $\Theta(t)$ as

$$\Theta(t) = \frac{1}{N^{d+2}} \int_{X_N} (\phi - N^{d+2} P^t h(t)) \underset{X}{\cdot} (\phi - N^{d+2} P^t h(t)) f(t,\phi)\, \mu^N(\mathrm{d}\phi),$$

and denoting $\tilde{\phi} \equiv \tilde{\phi}_{N,K,t} = \phi - N^{d+2} P^t h(t)$ for simplicity, we have by (4) for $f(t,\phi)$

$$\begin{aligned}\frac{1}{2}\frac{\mathrm{d}}{\mathrm{d}t}\Theta(t) &= \frac{1}{2N^{d+2}} \int_{X_N} \tilde{\phi} \underset{X}{\cdot} \tilde{\phi}\ N^2 D \cdot (e^{-H} Df)(t,\phi)\, \mathrm{d}\phi \\ &+ \frac{1}{N^{d+2}} \int_{X_N} \left(- N^{d+2} P^t \dot{h}(t)\right) \underset{X}{\cdot} \tilde{\phi}\ f(t,\phi)\, \mu(\mathrm{d}\phi) =: I_1 + I_2.\end{aligned}$$

3.2 The Term I_2

The decomposition of $\tilde{\phi}$ into the sum $\tilde{\phi} = \phi_m + \phi_f$ with $\phi_m = N^{d+2} P^t (P\phi - h(t))$ (the difference at mesoscopic level) and $\phi_f = (id_X - N^{d+2} P^t P)\phi$ (the fluctuation part of ϕ) leads that of I_2:

$$\begin{aligned} I_2 &= -\int_X \dot{h}(t) \underset{Y}{\cdot} P N^{d+2} P^t (P\phi - h(t))\ f(t,\phi)\, \mu(\mathrm{d}\phi) \\ &- \int_X P^t \dot{h}(t) \underset{X}{\cdot} \phi_f\ f(t,\phi)\, \mu(\mathrm{d}\phi) =: I_{2,\mathrm{m}} + I_{2,f}. \end{aligned}$$

The difference between the arguments in [5] and [6], the latter of which is adopted here, actually lies in this decomposition. It will be shown that the fluctuation parts $I_{2,f}$ and $I_{1,f}$, defined below, eventually tend to 0 by an application of the Poincaré inequality combined with Lemma 2.2. On the other hand, the terms $I_{2,\mathrm{m}}$ and $I_{1,\mathrm{m}}$ are objects at mesoscopic level and accordingly the convexity of the coarse-grained Hamiltonian plays a key role to treat these two terms.

3.2.1 The Term $I_{2,f}$

By Schwarz's inequality,

$$|I_{2,f}| \le \left(N^{-d} |N^{d+1} P^t \dot{h}(t)|_X^2\right)^{1/2} \times N^{-1-d/2} \left(\int_X |\phi_f|_X^2\ f(t,\phi)\mu(\mathrm{d}\phi)\right)^{1/2}.$$

For the second factor, by Poincaré inequality, we have that

$$|\phi_f|_X^2 = \sum_{x\in\Gamma_N}\left(\phi(x) - \frac{1}{K^d}\sum_{y\in\Lambda_K([x])}\phi(y)\right)^2 \le \sum_{i\in\Gamma_M} CK^2 \sum_{b\in\Lambda_K^*(i)} |\nabla\phi(b)|^2.$$

Therefore, from Lemma 2.2,

$$\int_X |\phi_f|_X^2\, f(t,\phi)\mu(\mathrm{d}\phi) = O(K^2N^d). \tag{12}$$

On the other hand for the first factor, by the ODE (6), we have

$$\begin{aligned} N^{-d}|N^{d+1}P^t\dot{h}(t)|_X^2 &= N^{-d}\sum_{i\in\Gamma_N}\dot{h}_{[i]}(t)^2 = M^{-d}\sum_{\ell\in\Gamma_M}\dot{h}_\ell(t)^2 \\ &= |\dot{h}(t)|_Y^2 = -\dot{h}(t) \underset{Y}{\cdot} \bar{D}\bar{H}(h(t)) = -\frac{\mathrm{d}}{\mathrm{d}t}\bar{H}(h(t)), \end{aligned}$$

which implies

$$\int_0^T \left(N^{-d}|N^{d+1}P^t\dot{h}(t)|_X^2\right)^{1/2}\mathrm{d}t \le \sqrt{T}\{\bar{H}(h(0)) - \bar{H}(h(T))\}^{1/2} \le C\sqrt{T},$$

from Assumption A-(3). Thus,

$$\int_0^T I_{2,f}(t)\mathrm{d}t = O(N^{-1-d/2}K^2N^d) = O(1/M) \to 0 \quad \text{as } M\to\infty.$$

3.2.2 The Term $I_{2,\mathrm{m}}$

By (1) and (6), we have

$$\begin{aligned} I_{2,\mathrm{m}} &= -\int_X \dot{h}(t) \underset{Y}{\cdot} (P\phi - h(t))\, f(t,\phi)\,\mu(\mathrm{d}\phi) \\ &= \int_Y \bar{D}\bar{H}^N(h(t)) \underset{Y}{\cdot} (\psi - h(t))\, \bar{f}(t,\psi)\,\bar{\mu}(\mathrm{d}\psi), \end{aligned}$$

where $\bar{f}\mathrm{d}\bar{\mu}$ is the image measure of $f\mathrm{d}\mu$ under the map $P : X \to Y$. This term will be treated by combining with the term arising from $I_{1,\mathrm{m};1}$ below.

3.3 The Term I_1

Since $D_x\phi(y) = \delta_{xy}$, by integration by parts, we have

$$I_1 = -\frac{1}{N^d} \int_X \tilde{\phi} \underset{X}{\cdot} Df(t,\phi)\, \mu(\mathrm{d}\phi),$$

and, then by decomposing $\tilde{\phi} = \phi_m + \phi_f$ as above,

$$\begin{aligned} I_1 &= -N^2 \int_X P^t(P\phi - h(t)) \underset{X}{\cdot} Df(t,\phi)\, \mu(\mathrm{d}\phi) \\ &\quad - \frac{1}{N^d} \int_X \phi_f \underset{X}{\cdot} Df(t,\phi)\, \mu(\mathrm{d}\phi) =: I_{1,\mathrm{m}} + I_{1,f}. \end{aligned}$$

3.3.1 The Term $I_{1,f}$

Since ϕ_{f} is a function of the $\nabla\phi$-field, one can replace $Df(t,\phi)\, \mu(\mathrm{d}\phi)$ by $Df^{\nabla}(t,\eta)\times\, \mu^{N,\nabla}(\mathrm{d}\eta)$. Indeed, we may use the integration by parts twice. Then by Schwarz's inequality, we have

$$\begin{aligned} |I_{1,f}| &\le \frac{1}{N^d} \left(\int_X \frac{1}{f^{\nabla}(t,\eta)} |Df^{\nabla}(t,\eta)|_X^2\, \mu^{N,\nabla}(\mathrm{d}\eta) \right)^{1/2} \\ &\quad \times \left(\int_X |\phi_{\mathrm{f}}|_X^2\, f^{\nabla}(t,\eta)\, \mu^{N,\nabla}(\mathrm{d}\eta) \right)^{1/2}. \end{aligned}$$

From Lemma 2.1 and applying Schwarz's inequality, after integrating in t, the first factor behaves as

$$\int_0^T \mathrm{d}t \left(\int_{X_N^{\nabla}} \frac{1}{f^{\nabla}(t,\eta)} |Df^{\nabla}(t,\eta)|_X^2\, \mu^{N,\nabla}(\mathrm{d}\eta) \right)^{1/2} = O\left((N^{d-2})^{1/2}\right).$$

The behavior of the second factor is given by (12) uniformly in $t \in [0,T]$, and thus we have

$$\int_0^T I_{1,f}(t)\, \mathrm{d}t = O\left(\frac{1}{N^d} \left(N^{d-2}\right)^{1/2} (K^2 N^d)^{1/2} \right) = O(1/M) \to 0 \quad \text{as } M \to \infty.$$

3.3.2 The Term $I_{1,\mathrm{m}}$

Lemma 3.1. *For every $f = f(\phi)$ and $\psi \in Y$, we have*

$$\int_X PDf(\phi)\,\mu(\mathrm{d}\phi|\psi) = \frac{1}{N^{d+2}}\bar{D}_M\bar{f}(\psi) + P\mathrm{cov}_{\mu(\mathrm{d}\phi|\psi)}(f, DH).$$

Proof. Except for a slight difference in a scaling, the proof is essentially the same as Lemma 21 of [6]. Indeed, the proof is completed if one can show that the integrations under $\bar{\mu}(\mathrm{d}\psi)$ of the both sides taken inner products with every test function $g = g(\psi) : Y \to Y$ coincide. For the left hand side, we have

$$\begin{aligned}
(L) &\equiv \int_Y \bar{\mu}(\mathrm{d}\psi)\int_X PDf(\phi)\underset{Y}{\cdot} g(\psi)\,\mu(\mathrm{d}\phi|\psi)\\
&= \int_X Df(\phi)\underset{X}{\cdot} P^t g(P\phi)\,\mu(\mathrm{d}\phi)\\
&= \frac{1}{N^{d+1}}\sum_{x\in\Gamma_N}\int_X D_x f(\phi) g_{[x]}(P\phi)\,\mu(\mathrm{d}\phi)\\
&= -\frac{1}{N^{d+1}}\sum_{x\in\Gamma_N}\int_X f(\phi) D_x\big(g_{[x]}(P\phi)e^{-H}\big)\,\mathrm{d}\phi\\
&= -\frac{1}{N^{d+1}}\sum_{x\in\Gamma_N}\int_X f(\phi)\left\{\frac{\partial g_{[x]}}{\partial\psi_{[x]}}(P\phi)\frac{1}{NK^d} - g_{[x]}(P\phi)D_x H\right\}\mu(\mathrm{d}\phi),
\end{aligned}$$

by integration by parts. However, since

$$\sum_{x\in\Gamma_N}\frac{\partial g_{[x]}}{\partial\psi_{[x]}} = K^d\sum_{i\in\Gamma_M}\frac{\partial g_i}{\partial\psi_i} \quad\text{and}\quad \sum_{x\in\Gamma_N} g_{[x]}(\psi)D_x H = N^{d+1}g(\psi)\underset{Y}{\cdot}\mathrm{PDH},$$

we obtain that

$$\begin{aligned}
(L) &= -\frac{1}{N^{d+2}}\int_X\sum_{i\in\Gamma_M}\frac{\partial g_i}{\partial\psi_i}(P\phi)f(\phi)\,\mu(\mathrm{d}\phi) + \int_X g(P\phi)\underset{Y}{\cdot}\mathrm{PDH}f(\phi)\,\mu(\mathrm{d}\phi)\\
&=: (L)_1 + (L)_2.
\end{aligned}$$

On the other hand, for the first term on the right-hand side, we have

$$(R)_1 \equiv \frac{1}{N^{d+2}}\int_Y \bar{D}_M\bar{f}(\psi)\underset{Y}{\cdot} g(\psi)\,\bar{\mu}(\mathrm{d}\psi)$$

$$= -\frac{1}{N^{d+1}} \int_Y \sum_{i\in\Gamma_M} \bar{f}(\psi) \frac{\partial}{\partial\psi_i} \left\{ g_i \mathrm{e}^{-N^d \bar{H}(\psi)} \right\} \mathrm{d}\psi$$

$$= -\frac{1}{N^{d+1}} \int_Y \sum_{i\in\Gamma_M} \frac{\partial g_i}{\partial\psi_i} \bar{f}(\psi)\bar{\mu}(\mathrm{d}\psi) + \frac{1}{N^2} \int_Y \sum_{i\in\Gamma_M} \frac{\partial \bar{H}}{\partial\psi_i} g_i \bar{f}(\psi)\bar{\mu}(\mathrm{d}\psi)$$

$$=: (R)_{1,1} + (R)_{1,2}.$$

For the second term on the right-hand side, we have

$$(R)_2 \equiv \int_Y P\mathrm{cov}_{\mu(\mathrm{d}\phi|\psi)}(f, DH) \underset{Y}{\cdot} g(\psi)\, \bar{\mu}(\mathrm{d}\psi)$$

$$= \int_X g(\psi) \underset{Y}{\cdot} \mathrm{PDH}(\phi) f(\phi)\, \mu(\mathrm{d}\phi) - \int_Y \bar{f}(\psi)\bar{\mu}(\mathrm{d}\psi) g(\psi) \underset{Y}{\cdot} \int_X \mathrm{PDH}(\phi)\, \mu(\mathrm{d}\phi|\psi)$$

$$=: (R)_{2,1} - (R)_{2,2}.$$

In the above computation, we have used that

$$P\mathrm{cov}_{\mu(\mathrm{d}\phi|\psi)}(f, DH) = \mathrm{cov}_{\mu(\mathrm{d}\phi|\psi)}(f, \mathrm{PDH})$$

$$= \int_X f(\phi)\mathrm{PDH}(\phi)\, \mu(\mathrm{d}\phi|\psi) - \bar{f}(\psi) \int_X \mathrm{PDH}(\phi)\, \mu(\mathrm{d}\phi|\psi).$$

Since $(L)_1 = (R)_{1,1}$ and $(L)_2 = (R)_{2,1}$ are obvious, the conclusion follows once we can show that $(R)_{1,2} = (R)_{2,2}$. However, since

$$(R)_{2,2} = \frac{1}{N^{d+1}} \int_Y \bar{f}(\psi)\bar{\mu}(\mathrm{d}\psi) \sum_{i\in\Gamma_M} g_i(\psi) \sum_{x\in\Lambda_K(i)} \int_X D_x H(\phi)\mu(\mathrm{d}\phi|\psi),$$

the conclusion follows from

$$\frac{\partial \bar{H}}{\partial\psi_i}(\psi) = \frac{1}{N^{d-1}} \sum_{x\in\Lambda_K(i)} \int D_x H(\phi)\, \mu(\mathrm{d}\phi|\psi). \tag{13}$$

We finally give the proof of (13). Under the change of variables $\phi = \{\phi(x)\}_{x\in\Gamma_N} \mapsto \{\phi(x) + N\psi_{[x]}\}_{x\in\Gamma_N}$, one can rewrite $\bar{H}(\psi)$ as

$$\bar{H}(\psi) = -\frac{1}{N^d} \log \int_{\{P\phi=0\}} \mathrm{e}^{-H(\phi(\cdot)+N\psi_{[\cdot]})} \mathrm{d}\phi.$$

Taking the derivative in ψ_i and then making the change of variables back to the original expression, we have

$$\frac{\partial \bar{H}}{\partial \psi_i}(\psi) = \frac{1}{N^d} \frac{\int_{\{P\phi=0\}} \sum_{x\in\Lambda_K(i)} N D_x H(\phi(\cdot) + N\psi_{[\cdot]}) e^{-H(\phi(\cdot)+N\psi_{[\cdot]})} d\phi}{\int_{\{P\phi=0\}} e^{-H(\phi(\cdot)+N\psi_{[\cdot]})} d\phi}$$

$$= \frac{1}{N^{d-1}} \int \sum_{x\in\Lambda_K(i)} D_x H(\phi)\, \mu(d\phi|\psi).$$

This proves (13). (One can show it also by multiplying a test function $\tilde{g}(\psi)$ to the both sides of (13) and integrating them by $\bar{\mu}(d\psi)$ as we did before.) □

By Lemma 3.1,

$$\begin{aligned} I_{1,\mathrm{m}} &= -N^2 \int_X (P\phi - h(t)) \underset{Y}{\cdot} PDf(t,\phi)\, \mu(d\phi) \\ &= -\frac{1}{N^d} \int_Y (\psi - h(t)) \underset{Y}{\cdot} \bar{D}_M \bar{f}(t,\psi)\, \bar{\mu}(d\psi) \\ &\quad - N^2 \int_Y (\psi - h(t)) \underset{Y}{\cdot} P\mathrm{cov}_{\mu(d\phi|\psi)}(f, DH)\, \bar{\mu}(d\psi) \\ &=: I_{1,\mathrm{m};1} + I_{1,\mathrm{m};2}. \end{aligned}$$

The integration by parts formula implies that

$$I_{1,\mathrm{m};1} = \frac{M^d}{N^d} - \int_Y (\psi - h(t)) \underset{Y}{\cdot} \bar{D}_M \bar{H}^N(\psi) \bar{f}(t,\psi)\, \bar{\mu}(d\psi).$$

Note that the coarse-grained Hamiltonian appears in the right-hand side due to the operation P.

3.3.3 The Term $I_{1,\mathrm{m};2}$

The only term left is $I_{1,\mathrm{m};2}$. Let $\Delta_M : Y \to Y$ be the discrete Laplace operator defined by

$$(\Delta_M \psi)_i = M^2 \sum_{j\in\Gamma_M:|i-j|=1} (\psi_j - \psi_i), \quad i \in \Gamma_M,\ \psi \in Y.$$

Then, $-\Delta_M : Y_0 \to Y_0$ is one to one, onto and invertible on the space $Y_0 = \{\psi \in Y; \psi \underset{Y}{\cdot} 1 = 0\}$. We now note that $\mathrm{PDH}(\phi) \in Y_0$. Indeed,

$$\mathrm{PDH}(\phi) \underset{Y}{\cdot} 1 = \frac{1}{M^d} \sum_{i \in \Gamma_M} \frac{1}{K^d N} \sum_{x \in \Lambda_K(i)} \sum_{y \in \Gamma_N : |x-y|=1} V'(\phi(x) - \phi(y)) = 0$$

by the symmetry of V. Hence, one can estimate the integrand of $I_{1,\mathrm{m};2}$ as follows:

$$\begin{aligned}
&\left|(\psi - h(t)) \underset{Y}{\cdot} \mathrm{cov}_{\mu(\mathrm{d}\phi|\psi)}(f, \mathrm{PDH})\right| \\
&\quad \le \langle(-\Delta_M)(\psi - h(t)), (\psi - h(t))\rangle_Y^{1/2} \\
&\qquad \times \langle(-\Delta_M)^{-1}\mathrm{cov}_{\mu(\mathrm{d}\phi|\psi)}(f, \mathrm{PDH}), \mathrm{cov}_{\mu(\mathrm{d}\phi|\psi)}(f, \mathrm{PDH})\rangle_Y^{1/2} \\
&\quad =: II_1 \times II_2.
\end{aligned}$$

For the first term on the right-hand side, we have $II_1 = |\nabla_M(\psi - h(t))|_{Y^d}$. On the other hand, for the second term, using the variational representation:

$$\langle(-\Delta_M)^{-1}F, F\rangle_Y = \sup_{\tilde{\psi} \in Y : |\nabla_M \tilde{\psi}|_{Y^d} \le 1} (F \underset{Y}{\cdot} \tilde{\psi})^2, \quad F(\phi) \in Y_0,$$

we have that

$$II_2 = \sup_{\tilde{\psi} \in Y : |\nabla_M \tilde{\psi}|_{Y^d} \le 1} \left|\mathrm{cov}_{\mu(\mathrm{d}\phi|\psi)}(f, \mathrm{PDH} \underset{Y}{\cdot} \tilde{\psi})\right|.$$

At this point, we need the following lemma, which is actually Lemma 22 in [6], and this shows that the LSI implies a nice bound on the covariance:

Lemma 3.2. *Let X be a (general) affine space and assume that a probability measure μ on X is given and satisfies the LSI(ρ), $\rho > 0$, that is*

$$\int_X \Phi(f)\,\mathrm{d}\mu - \Phi\left(\int_X f\,\mathrm{d}\mu\right) \le \frac{1}{2\rho} \int_X \frac{|Df|_X^2}{f}\mathrm{d}\mu,$$

for every $f \in L^1(X, \mu), \ge 0$ and locally Lipschitz. Then we have that

$$\left|\mathrm{cov}_\mu(f, g)\right| \le \frac{1}{\rho}\|Dg\|_{L^\infty(\mu)} \left(\int_X f\,\mathrm{d}\mu\right)^{1/2} \left(\int_X \frac{|Df|_X^2}{f}\mathrm{d}\mu\right)^{1/2}.$$

For $g(\phi) \equiv g_{\tilde{\psi}}(\phi) := \mathrm{PDH} \underset{Y}{\cdot} \tilde{\psi}$, we need the following estimate to apply Lemma 3.2:

Lemma 3.3.

$$\|Dg\|_\infty \leq \frac{C\|V''\|_\infty}{N^{(d+2)/2}\sqrt{K}M} |\nabla_M \tilde{\psi}|_{Y^d}.$$

Proof. Since $g(\phi)$ can be rewritten as

$$g(\phi) = \frac{1}{K^d N} \frac{1}{M^d} \sum_{i \in \Gamma_M} \left\{ \sum_{z \in \Lambda_K(i)} \sum_{y \in \Gamma_N : |z-y|=1} V'(\phi(z) - \phi(y)) \right\} \tilde{\psi}_i,$$

we have

$$D_x g(\phi) = \frac{1}{N^{d+1}} \sum_{y \notin \Lambda_K([x]) : |x-y|=1} V''(\phi(x) - \phi(y)) \big(\tilde{\psi}_{[x]} - \tilde{\psi}_{[y]}\big).$$

This implies that

$$|Dg(\phi)|_X^2 = \sum_{x \in \Gamma_N} |D_x g(\phi)|^2 \leq \left(\frac{\|V''\|_\infty}{N^{d+1}} \right)^2 C K^{d-1} \sum_{i \in \Gamma_M} |\nabla \tilde{\psi}_i|^2.$$

The conclusion follows noting that $\nabla \tilde{\psi}_i = M^{-1} \nabla_M \tilde{\psi}_i$. □

From Assumption A-(2), Lemmas 3.2 and 3.3, we have

$$II_2 \leq \frac{C\|V''\|_\infty}{\rho N^{(d+2)/2}\sqrt{K}M} \left(\int_X f(t,\phi)\, \mu(\mathrm{d}\phi|\psi) \right)^{1/2} \left(\int_X \frac{|Df(t,\phi)|_X^2}{f(t,\phi)}\, \mu(\mathrm{d}\phi|\psi) \right)^{1/2}.$$

Therefore, back to the term $I_{1,\mathrm{m};2}(t)$, we have that

$$\begin{aligned} |I_{1,\mathrm{m};2}(t)| \leq{}& \frac{C\|V''\|_\infty}{\rho N^{(d-2)/2}\sqrt{K}M} \\ &\times \int_Y |\nabla_M(\psi - h(t))|_{Y^d} \bar{f}(t,\psi)^{1/2} \left(\int_X \frac{|Df(t,\phi)|_X^2}{f(t,\phi)} \mu(\mathrm{d}\phi|\psi) \right)^{1/2} \bar{\mu}(\mathrm{d}\psi) \\ \leq{}& \frac{\epsilon}{2} \int_Y |\nabla_M(\psi - h(t))|_{Y^d}^2 \bar{f}(t,\psi)\, \bar{\mu}(\mathrm{d}\psi) \\ &+ \frac{1}{2\epsilon} \left(\frac{C_1\|V''\|_\infty}{\rho N^{(d-2)/2}\sqrt{K}M} \right)^2 \int_X \frac{|Df(t,\phi)|_X^2}{f(t,\phi)} \mu(\mathrm{d}\phi), \end{aligned}$$

for every $\epsilon > 0$. The last bound is simply from $ab \le (a^2 + b^2)/2$ for $a, b \ge 0$. From Lemma 2.1 (which holds also for f in place of f^{∇}), after integrating in t, the second term is bounded by $\{C_2/(\rho\sqrt{K}M)\}^2/(2\epsilon)$.

3.4 Summary and Completion of the Proof of Theorem 1.1

Summarizing all these computations on $\frac{1}{2}\frac{d}{dt}\Theta(t)$ and integrating it in t, we have

$$\frac{1}{2}\Theta(T) \le \frac{1}{2}\Theta(0) + \frac{2CT}{M} + \frac{T}{K^d} + \frac{1}{2\epsilon}\left(\frac{C_2}{\rho\sqrt{K}M}\right)^2$$
$$- \int_0^T dt \int_Y \left(\bar{D}_M \bar{H}^N(\psi) - \bar{D}_M \bar{H}^N(h(t))\right) \underset{Y}{\cdot} (\psi - h(t)) \bar{f}(t, \psi)\, \bar{\mu}(d\psi)$$
$$+ \frac{\epsilon}{2} \int_0^T dt \int_Y |\nabla_M(\psi - h(t))|^2_{Y^d} \bar{f}(t, \psi)\, \bar{\mu}(d\psi).$$

Choosing ϵ in such a way that $0 < \epsilon < 2c_-$, from Assumption A-(1), the sum of the fifth and sixth terms in the right-hand side is non-positive. We thus obtain that $\Theta(T) \to 0$ if $\Theta(0) \to 0$ as $N, K, M \to \infty$.

4 Validity of Assumption A for Convex Potentials

In this section, we show that Assumption A holds for the strictly convex potentials V, that is, those satisfying
Condition V. V is symmetric, C^2 and $V'' \ge c_-$ for some $c_- > 0$.

4.1 Assumption A-(2)

Lemma 4.1. *Assume the Condition V, then the LSI holds with* $\rho = CK^{-2}$. *In particular, Assumption A-(2) is true if* K *and* M *satisfy the condition that* $MK^{-3/2} \to \infty$ *as* $M, K \to \infty$.

Proof. We apply the Bakry–Emery criterion. To this end, it suffices to show that the Hessian of the Hamiltonian $H(\phi)$ considered on the space $X_\psi = \{\phi \in X; P\phi = \psi\}$ satisfies:

$$\mathrm{Hess} H(\phi) \ge CK^{-2} id_X,$$

as quadratic forms. However, for $\xi = \{\xi(x)\}_{x\in\Gamma_N} \in X_0$, the Hessian is given by and estimated from below as:

$$\begin{aligned}
\operatorname{Hess} H(\phi)\xi \underset{X}{\cdot} \xi &= \frac{\mathrm{d}^2}{\mathrm{d}\epsilon^2} H(\phi+\epsilon\xi)\big|_{\epsilon=0} \\
&= \sum_{x,y\in\Gamma_N:|x-y|=1} V''(\phi(x)-\phi(y))(\xi(x)-\xi(y))^2 \\
&\geq c_- \sum_{x,y\in\Gamma_N:|x-y|=1} (\xi(x)-\xi(y))^2 \geq c_- \sum_{i\in\Gamma_M}\sum_{x,y\in\Lambda_K(i)} (\xi(x)-\xi(y))^2 \\
&\geq c_- C K^{-2} \sum_{i\in\Gamma_M}\sum_{x\in\Lambda_K(i)} \xi(x)^2 = c_- C K^{-2} |\xi|_X^2 .
\end{aligned}$$

We have used the Condition V in the third line, then thrown away the sum over the bonds crossing different mesoscopic boxes and applied Poincaré inequality in the last line recalling that $\sum_{x\in\Lambda_K(i)}\xi(x) = 0$ for each $i \in \Gamma_M$. □

4.2 Assumption A-(1)

In the formula (13), since $D_x H(\phi) = \sum_{y\in\Gamma_N:|x-y|=1} V'(\phi(x)-\phi(y))$, the symmetry of V implies that

$$\sum_{x\in\Lambda_K(i)} D_x H(\phi) = \sum_{j\in\Gamma_M:|i-j|=1} \sum_{i,j}^{*} V'(\phi(x)-\phi(y)),$$

where $\sum_{i,j}^{*}$ means the sum over $x \in \Lambda_K(i)$ and $y \in \Lambda_K(j)$ such that $|x-y| = 1$; note that the sum over $y \in \Lambda_K(i)$ cancels and we only have the boundary term. Therefore, from (13), we have that

$$\frac{\partial \bar{H}}{\partial \psi_i}(\psi) = \frac{1}{N^{d-1}} \sum_{j\in\Gamma_M:|i-j|=1} \Phi_{ij}(\psi),$$

where

$$\Phi_{ij}(\psi) = \sum_{i,j}^{*} \int_X V'(\phi(x)-\phi(y))\,\mu(\mathrm{d}\phi|\psi).$$

In the following, for $i, j \in \Gamma_M : |i-j| = 1$ and a function $g : \mathbb{R} \to \mathbb{R}$ (in fact, we will take $g = V'$ or V''), we will write

$$[g]_{ij}(\phi) = \sum_{i,j}^{*} g(\phi(x)-\phi(y)).$$

4.2.1 Computation of (LHS)

Now, let us rewrite (LHS), the left-hand side of the inequality in Assumption A-(1).

Proposition 4.2. *Set*

$$\bar{G} = \sum_{\langle i,j \rangle} \xi_{ij} [V']_{ij}, \tag{14}$$

where $\sum_{\langle i,j \rangle} = \sum_i \sum_{j:j\sim i} (= \sum_{i,j\in\Gamma_M:|i-j|=1})$ *is the sum taken over all directed bonds in* Γ_M, $j \sim i$ *means* $|i-j|=1$, *and*

$$\xi_{ij} = \nabla_M \psi^1(\langle i,j \rangle) - \nabla_M \psi^2(\langle i,j \rangle). \tag{15}$$

Then we have that

$$(LHS) = \frac{1}{2N^{d-1}M} \sum_{\langle i,j \rangle} \xi_{ij}^2 \left\langle [V'']_{ij} \right\rangle(\psi) - \frac{1}{4N^{d-1}M} \mathrm{var}_\psi(\bar{G}), \tag{16}$$

for some $\psi \in Y$, *where* $\langle\ \cdot\ \rangle(\psi)$ *and* var_ψ *represent the mean and the variance under* $\mu(\mathrm{d}\phi|\psi)$, *respectively, in particular,* $\langle \cdot \rangle(\psi) = \int_X \cdot\, \mu(\mathrm{d}\phi|\psi)$.

Proof. We first rewrite (*LHS*) as

$$\begin{aligned}
(LHS) &= \sum_{i\in\Gamma_M} \left(\frac{\partial \bar{H}}{\partial \psi_i}(\psi^1) - \frac{\partial \bar{H}}{\partial \psi_i}(\psi^2) \right) (\psi_i^1 - \psi_i^2) \\
&= \frac{1}{N^{d-1}} \sum_{i,j\in\Gamma_M:|i-j|=1} \left(\Phi_{ij}(\psi^1) - \Phi_{ij}(\psi^2) \right) (\psi_i^1 - \psi_i^2) \\
&= \frac{1}{2N^{d-1}} \sum_{i,j\in\Gamma_M:|i-j|=1} \left(\Phi_{ij}(\psi^1) - \Phi_{ij}(\psi^2) \right) \left\{ (\psi_i^1 - \psi_i^2) - (\psi_j^1 - \psi_j^2) \right\} \\
&= \frac{1}{2N^{d-1}M} \sum_{i,j\in\Gamma_M:|i-j|=1} \left(\Phi_{ij}(\psi^1) - \Phi_{ij}(\psi^2) \right) \xi_{ij},
\end{aligned}$$

where the third equality is from the anti-symmetry of Φ_{ij}: $\Phi_{ji}(\psi) = -\Phi_{ij}(\psi)$, which is shown by the symmetry of V, and the last one is by $\nabla_M \psi(\langle i,j \rangle) = M(\psi_i - \psi_j)$. Then by Taylor's theorem, there exists $a \in (0,1)$ such that

$$(\mathrm{LHS}) = \frac{1}{2N^{d-1}M} \sum_{\langle i,j \rangle} \sum_{k\in\Gamma_M} \frac{\partial \Phi_{ij}}{\partial \psi_k}(\psi(a))(\psi_k^1 - \psi_k^2)\xi_{ij},$$

where $\psi(a) = a\psi^1 + (1-a)\psi^2$. However, similar to the proof of the formula (13) and because of the cancellation due to the symmetry of V as we observed before, we have for $k \in \Gamma_M$

$$\frac{\partial \Phi_{ij}}{\partial \psi_k}(\psi) = M\left\{ \left(1_{\{k=i\}} - 1_{\{k=j\}}\right) \langle [V'']_{ij} \rangle (\psi) - \sum_{\ell \in \Gamma_M : |k-\ell|=1} \mathrm{cov}_\psi\left([V']_{ij}, [V']_{k\ell}\right) \right\}.$$

From this, by noting the anti-symmetry of $[V']_{k\ell}$: $[V']_{\ell k} = -[V']_{k\ell}$, the proof of the proposition is completed. □

Note that $\{\xi_{ij}\}$ is anti-symmetric: $\xi_{ji} = -\xi_{ij}$. What we want to prove is:

$$\text{(LHS)} \geq \frac{c_-}{M^d} \sum_{\langle i,j \rangle} \xi_{ij}^2. \tag{17}$$

4.2.2 Proof of (17) for Convex Potentials

We show that the inequality (17) and therefore Assumption A-(1) hold for the strictly convex potentials V satisfying the Condition V.

Let $\partial \Lambda_K^*(i)$ be the family of all directed bonds $b = \langle x, y \rangle \in \Gamma_N^*$ with $x, y \in \partial \Lambda_K(i)$ and set $\partial \Lambda_{K,N}^* = \cup_{i \in \Gamma_M} \partial \Lambda_K^*(i)$, where $\partial \Lambda (\equiv \partial^- \Lambda) = \{x \in \Lambda; {}^\exists y \in \Gamma_N \setminus \Lambda \text{ such that } |x - y| = 1\}$ stands for the inner boundary of $\Lambda \subset \Gamma_N$. We consider the Dirichlet form defined on the space X_ψ:

$$\mathscr{D}(F, G) = \sum_{b \in \partial \Lambda_{K,N}^*} \int_{X_\psi} \partial_b F \cdot \partial_b G \, \mu(d\phi | \psi), \tag{18}$$

for $F, G : X_\psi \to \mathbb{R}$, where ∂_b is the differential operator defined by

$$\partial_b = \frac{\partial}{\partial \phi(x)} - \frac{\partial}{\partial \phi(y)}, \quad b = \langle x, y \rangle \in \Gamma_N^*.$$

Note that $\{\partial_b\}_{b \in \Lambda_{K,N}^*}$, $\Lambda_{K,N}^* = \cup_{i \in \Gamma_M} \Lambda_K^*(i)$, are all tangent vectors of the space X_ψ so that the dynamics associated with the Dirichlet form (18) preserve the block spin averages ψ. Other Dirichlet forms defined by (18) with the sum taken over $\Lambda_{K,N}^*$ or its any subsets have the same property.

Variational formula for $\mathrm{var}_\psi(G)$. We give the variational formula for $\mathrm{var}_\psi(G) := E^{\mu(\cdot|\psi)}[G; G]$ in terms of $\{\partial_b G\}$. Such formula is useful, since G is a function of V' in the present setting (see (14) below) so that $\partial_b G$ becomes a function of V'' (see Lemma 4.5 below). The variational formula is derived due to the integration by parts formula and its proof is similar to that of Helffer–Sjöstrand representation, see [2] and [3].

Lemma 4.3. (1) *The generator L corresponding to the Dirichlet form $\mathscr{D}$ is given by*

$$LF = e^H \sum_{b\in\partial\Lambda_{K,N}^*} (-\partial_b^*)\big(e^{-H}\partial_b F\big)$$
$$= \sum_{b\in\partial\Lambda_{K,N}^*} \big(\partial_b^2 F - \partial_b H\cdot\partial_b F\big).$$

(2) *The commutator of ∂_b and L is given by*

$$[\partial_b, L] \equiv \partial_b L - L\partial_b = (Q\partial)_b,$$

where the operator $(Q\partial)_b$ is defined by

$$(Q\partial)_b F \equiv (Q\partial F)_b = -\sum_{b'\in\partial\Lambda_{K,N}^*} \partial_b\big(\partial_{b'}H\big)\cdot\partial_{b'}F.$$

The proof of Lemma 4.3 is straightforward. Note that $\partial_b^* = -\partial_b$ and $\partial_b\partial_{b'} = \partial_{b'}\partial_b$. We will denote the expectation $E^{\mu(\cdot|\psi)}[\cdot]$ simply by $E^\psi[\cdot]$.

Proposition 4.4. *(the variational formula) For $G = G(\phi)$, we have that*

$$\mathrm{var}_\psi(G) = \sup_J E^\psi\left[2\sum_{b\in\partial\Lambda_{K,N}^*}\partial_b G\cdot\partial_b J + \sum_{b\in\partial\Lambda_{K,N}^*}\{L\partial_b J + (Q\partial J)_b\}\,\partial_b J\right].$$

If G is $\partial\Lambda_{K,N}$-measurable, then the supremum may be taken over the class of all $\partial\Lambda_{K,N}$-measurable functions $J = J(\phi)$, where $\partial\Lambda_{K,N} = \cup_{i\in\Gamma_M}\partial\Lambda_K(i)$.

Proof. One can assume $E^\psi[G] = 0$ without loss of generality. Let F be the solution of the Poisson equation $-LF = G$. Then by the integration by parts formula, we obtain that

$$\mathrm{var}_\psi(G) = E^\psi[G;G]$$
$$= E^\psi[G(-LF)] = \sum_{b\in\partial\Lambda_{K,N}^*} E^\psi[\partial_b G\cdot\partial_b F].$$

However, we have that $\partial_b F = \{(-L-Q)^{-1}\partial G\}_b$, since

$$-\partial_b G = \partial_b(LF) = L\partial_b F + (Q\partial F)_b =: \{(L+Q)\partial F\}_b,$$

where the second equality is from Lemma 4.3-(2). Thus,

$$\begin{aligned}\mathrm{var}_\psi(G) &= E^\psi\left[\sum_{b\in\partial\Lambda^*_{K,N}} \partial_b G\{(-L-Q)^{-1}\partial G\}_b\right]\\ &= \sup_J E^\psi\left[2\sum_{b\in\partial\Lambda^*_{K,N}} \partial_b G\cdot\partial_b J + \sum_{b\in\partial\Lambda^*_{K,N}} \{(L+Q)\partial J\}_b \partial_b J\right].\end{aligned}$$

The last equality follows from

$$\langle \partial G, A^{-1}\partial G\rangle = \sup_J \{2\langle \partial G, \partial J\rangle - \langle A\partial J, \partial J\rangle\},$$

for (non-negative) operators A, where

$$\langle \partial F, \partial G\rangle = E^\psi\left[\sum_{b\in\partial\Lambda^*_{K,N}} \partial_b F\cdot\partial_b G\right].$$

□

Proof of (17). We return to the proof of (17). In order to apply Proposition 4.4 for the second term of (16), we first compute $\partial_b \bar{G}$:

Lemma 4.5. *For every $b=\langle x_1,x_2\rangle\in\partial\Lambda^*_K(i)$, we have that*

$$\partial_b\bar{G} = 2\xi_{ij}\{V''(\phi(x_1)-\phi(y_1)) - V''(\phi(x_2)-\phi(y_2))\},$$

where $j\in\Gamma_M, j\sim i$, and y_1,y_2 are determined by $y_1,y_2\in\partial\Lambda_K(j), x_1\sim y_1$ and $x_2\sim y_2$. We will denote such i,j,x_1,y_1,x_2,y_2 by $i(b), j(b), x_1(b), y_1(b), x_2(b), y_2(b)$, respectively.

Proof. Since $x_1,x_2\in\partial\Lambda_K(i)$, we have that

$$\partial_b[V']_{ij} = \left(\frac{\partial}{\partial\phi(x_1)} - \frac{\partial}{\partial\phi(x_2)}\right)[V']_{ij} = V''(\phi(x_1)-\phi(y_1)) - V''(\phi(x_2)-\phi(y_2)),$$

with j,y_1,y_2 determined as above. However, the anti-symmetry of $\{\xi_{ij}\}$ and $\{[V']_{ij}\}$: $\xi_{ji}=-\xi_{ij}$ and $[V']_{ji}=-[V']_{ij}$ implies that $\xi_{ji}[V']_{ji}=\xi_{ij}[V']_{ij}$. The sum in (14) is taken over all directed bonds so that each bond $\langle i,j\rangle$ is counted twice. Therefore, the factor 2 appears in the right-hand of $\partial_b\bar{G}$. □

For x,y such that $x\sim y$, $x\in\partial\Lambda_K(i)$, $y\in\partial\Lambda_K(j)$ with $i\neq j$, we set

$$\Psi(x,y;J) = \sum_{z\in\partial\Lambda_K(i):z\sim x} \partial_{\langle x,z\rangle} J - \sum_{w\in\partial\Lambda_K(j):w\sim y} \partial_{\langle y,w\rangle} J. \tag{19}$$

Lemma 4.6. *The three terms in the right-hand side of the variational formula given in Proposition 4.4 are rewritten as follows:*

$$\sum_{b\in\partial\Lambda^*_{K,N}} \partial_b \bar{G}\cdot\partial_b J = 4 \sum_{b=\langle x_1,x_2\rangle\in\partial\Lambda^*_{K,N}} \xi_{i(b)j(b)} V''(\phi(x_1)-\phi(y_1))\partial_b J$$
$$= 4\sum_{\langle x,y\rangle} \xi_{i(x)i(y)} V''(\phi(x)-\phi(y))\Psi(x,y;J), \tag{20}$$

$$E^{\psi}\left[\sum_{b\in\partial\Lambda^*_{K,N}} L\partial_b J\cdot\partial_b J\right] = -\sum_{b\in\partial\Lambda^*_{K,N}} \mathscr{D}(\partial_b J,\partial_b J)\le 0, \tag{21}$$

$$\sum_{b\in\partial\Lambda^*_{K,N}} (Q\partial J)_b\partial_b J = -\sum_{b\in\partial\Lambda^*_{K,N}}\sum_{b'\in\partial\Lambda^*_{K,N}} (\partial_b\partial_{b'}H)\cdot\partial_b J\cdot\partial_{b'}J, \tag{22}$$

where the sum in $\langle x,y\rangle$ *in the right-hand side of* (20) *is taken over all undirected bonds* $\langle x,y\rangle$ *with vertices* $x\in\partial\Lambda_K(i)$ *and* $y\in\partial\Lambda_K(j)$, $i\ne j$, *belonging to different boxes. We denote* $i(x):=i$ *if* $x\in\Lambda_K(i)$.

Proof. By Lemma 4.5, the left-hand side of (20) is equal to

$$2\sum_{b=\langle x_1,x_2\rangle\in\partial\Lambda^*_{K,N}} \xi_{i(b)j(b)}\{V''(\phi(x_1)-\phi(y_1))-V''(\phi(x_2)-\phi(y_2))\}\partial_b J.$$

However, this implies the first equality in (20) by noting that

$$\xi_{i(b)j(b)}V''\big(\phi(x_2(b))-\phi(y_2(b))\big)\partial_b J$$
$$= -\xi_{i(-b)j(-b)}V''\big(\phi(x_1(-b))-\phi(y_1(-b))\big)\partial_{-b}J,$$

which follows from $\partial_{-b}J = -\partial_b J$. To show the second equality in (20), we rearrange the sum in $\langle x,y\rangle = \langle x_1,y_1\rangle$. In fact, for each undirected bond $\langle x,y\rangle$, the summand becomes

$$\xi_{i(x)i(y)}V''(\phi(x)-\phi(y)) \sum_{b=\langle x,x_2\rangle\in\partial\Lambda^*_{K,N}} \partial_b J$$
$$+\,\xi_{i(y)i(x)}V''(\phi(y)-\phi(x)) \sum_{b=\langle y,y_2\rangle\in\partial\Lambda^*_{K,N}} \partial_b J.$$

Thus, the second equality follows by the anti-symmetry of $\{\xi_{ij}\}$ and the symmetry of V''. (21) is immediate, while (22) follows directly from the definition of the operator Q given in Lemma 4.3-(2). □

We now compute $\partial_b\partial_{b'}H$, which appears in the right-hand side of (22).

Lemma 4.7. *We write $b = \langle x, y \rangle$ and $b' = \langle x', y' \rangle$.*

(1) If the distance between two bonds b and b' satisfies $\mathrm{dist}\,(b, b') \geq 2$, *then* $\partial_b \partial_{b'} H = 0$.

(2) If $\mathrm{dist}\,(b, b') = 1$, *$x \sim x'$ and $y \not\sim y'$, then*

$$\partial_b \partial_{b'} H = -V''(\phi(x) - \phi(x')). \tag{23}$$

(3) If $\mathrm{dist}\,(b, b') = 1$, *$x \sim x'$ and $y \sim y'$, then*

$$\partial_b \partial_{b'} H = -V''(\phi(x) - \phi(x')) - V''(\phi(y) - \phi(y')). \tag{24}$$

(4) ($\mathrm{dist}\,(b, b') = 0$) *If $x = x'$ and $y \neq y'$, then*

$$\begin{aligned} \partial_b \partial_{b'} H = & \sum_{z \in \Gamma_N : z \sim x} V''(\phi(x) - \phi(z)) \\ & + V''(\phi(x) - \phi(y)) + V''(\phi(x) - \phi(y')). \end{aligned} \tag{25}$$

(5) ($\mathrm{dist}\,(b, b') = 0$) *If $x = x'$ and $y = y'$ (i.e. $b = b'$), then*

$$\begin{aligned} \partial_b \partial_{b'} H = & \sum_{z \in \Gamma_N : z \sim x} V''(\phi(x) - \phi(z)) \\ & + \sum_{w \in \Gamma_N : w \sim y} V''(\phi(y) - \phi(w)) + 2V''(\phi(x) - \phi(y)). \end{aligned} \tag{26}$$

Proof. From the definition of H given in (3), we have that

$$\partial_{b'} H = \sum_{z \in \Gamma_N : z \sim x'} V'(\phi(x') - \phi(z)) - \sum_{w \in \Gamma_N : w \sim y'} V'(\phi(y') - \phi(w)).$$

Then it is easy to see that $\partial_b \partial_{b'} H$ is given by the formula stated above in each case. □

For b or b' with opposite directions, we can compute $\partial_b \partial_{b'} H$ by this lemma noting that $\partial_{-b} = -\partial_b$ and $\partial_{-b'} = -\partial_{b'}$. We can rewrite the right-hand side of (22) using Lemma 4.7.

Lemma 4.8. *(We assume $d = 2$ for simplicity) The sum in the right-hand side of (22) with the sign changed is rearranged into the sum of the following three terms:*

$$\begin{aligned} \sum_{b \in \partial \Lambda^*_{K,N}} \sum_{b' \in \partial \Lambda^*_{K,N}} (\partial_b \partial_{b'} H) \cdot \partial_b J \cdot \partial_{b'} J = & \sum_{\langle x,y \rangle}^{(1)} I_1(x, y) \\ & + \sum_{\langle x,y \rangle}^{(2)} I_2(x, y) + \sum_{\langle x,y \rangle}^{(3)} I_3(x, y). \end{aligned}$$

Here, $\sum^{(1)}$ *is the sum taken over all undirected bonds* $\langle x, y\rangle$ *with vertices* x, y *belonging to different boxes (i.e.* $x \in \partial\Lambda_K(i)$, $y \in \partial\Lambda_K(j)$ *with* $i \neq j$*),* $\sum^{(2)}$ *is that taken over* $\langle x, y\rangle$ *belonging to an inner boundary of the same box (i.e.* $x, y \in \partial\Lambda_K(i)$*) and* $\sum^{(3)}$ *is that taken over* $\langle x, y\rangle$ *belonging to the same box but with* y *in its inside (i.e.* $x \in \partial\Lambda_K(i)$, $y \in \Lambda_K(i)^\circ \equiv \Lambda_K(i) \setminus \partial\Lambda_K(i)$*). The summands are defined by*

$$I_1(x, y) = 4V''(\phi(x) - \phi(y))\Psi(x, y; J)^2,$$

$$I_2(x, y) = 4V''(\phi(x) - \phi(y)) \left(2\partial_{\langle x,y\rangle} J + \partial_{\langle x,z\rangle} J - \partial_{\langle y,w\rangle} J\right)^2,$$

$$I_3(x, y) = 4V''(\phi(x) - \phi(y)) \sum_{\{z,v,w\}} \left\{\left(\partial_{\langle x,v\rangle} J + \partial_{\langle x,z\rangle} J\right)^2 + \left(\partial_{\langle x,v\rangle} J - \partial_{\langle v,w\rangle} J\right)^2\right\},$$

respectively, where $z, w \in \partial\Lambda_K(i)$ *in the term* I_2 *are determined by the condition "*$z \sim x \sim y \sim w$ *and these four points are different," while* $z, v, w \in \partial\Lambda_K(i)$ *in the term* I_3 *are determined by the condition "*$z \sim x \sim v \sim w$ *and these four points are different" (if* $d = 2$*, such* $\{z, w\}$ *for* I_2 *is uniquely determined, while for* I_3 *there are two ways to choose* $\{z, v, w\}$ *and we take the sum in such two choices).*

Proof. Since Lemma 4.7 shows that $\partial_b\partial_{b'} H$ are linear combinations of $\{V''(\phi(x) - \phi(y))\}_{\langle x,y\rangle}$, we can rearrange the sum in the right-hand side of (22) with the sign changed by identifying their coefficients. First, let us take a bond $\langle x, y\rangle$ satisfying the condition for the sum $\sum^{(1)}$, then the coefficient of $V''(\phi(x) - \phi(y))$ for such bond is given by

$$\begin{aligned} 4\Bigg\{ & -2 \sum_{\substack{z\in\partial\Lambda_K(i):z\sim x\\ w\in\partial\Lambda_K(j):w\sim y\\ z\not\sim w}} \partial_{\langle x,z\rangle} J \cdot \partial_{\langle y,w\rangle} J - 2 \sum_{\substack{z\in\partial\Lambda_K(i):z\sim x\\ w\in\partial\Lambda_K(j):w\sim y\\ z\sim w}} \partial_{\langle x,z\rangle} J \cdot \partial_{\langle y,w\rangle} J \\ & + \sum_{\substack{z\in\partial\Lambda_K(i):z\sim x\\ w\in\partial\Lambda_K(i):w\sim x\\ z\neq w}} \partial_{\langle x,z\rangle} J \cdot \partial_{\langle x,w\rangle} J + \sum_{z\in\partial\Lambda_K(i):z\sim x} \left(\partial_{\langle x,z\rangle} J\right)^2 \\ & + \sum_{\substack{z\in\partial\Lambda_K(j):z\sim y\\ w\in\partial\Lambda_K(j):w\sim y\\ z\neq w}} \partial_{\langle y,z\rangle} J \cdot \partial_{\langle y,w\rangle} J + \sum_{z\in\partial\Lambda_K(j):z\sim y} \left(\partial_{\langle y,z\rangle} J\right)^2 \Bigg\}. \end{aligned} \qquad (27)$$

Indeed, the factor 4 comes from the number of choices of the directions of two bonds b and b', for example in the first and the second sums in (27), we are taking

$b = \langle x, z\rangle$ and $b' = \langle y, w\rangle$. The first and the second sums in (27) come from the terms in (23) and (24) in Lemma 4.7, respectively. The factor 2 comes from the exchangeability of b and b'. The third sum comes from the first term in the right-hand side of (25) taken y for z there (i.e. $b = \langle x, z\rangle, b' = \langle x, w\rangle$ in the present situation), while the fourth sum comes from the first and the second terms in the right-hand side of (26) (taking $b = b' = \langle x, z\rangle$). The fifth and the sixth sums arise for the similar reason. The formula in (27) coincides with $4\Psi(x, y; J)^2$ and therefore we obtain $I_1(x, y)$.

We use the assumption $d = 2$ to compute the two terms I_2 and I_3 (this assumption is just for simplicity; some terms come into $\sum^{(2)}$ from (24) if $d \geq 3$, but no term from that if $d = 2$). Let us take a bond $\langle x, y\rangle$ satisfying the condition for the sum $\sum^{(2)}$, then the coefficient of $V''(\phi(x) - \phi(y))$ for such bond is given by

$$4\Big\{-2\partial_{\langle x,z\rangle}J \cdot \partial_{\langle y,w\rangle}J + 4\big(\partial_{\langle y,x\rangle}J \cdot \partial_{\langle y,w\rangle}J + \partial_{\langle x,y\rangle}J \cdot \partial_{\langle x,z\rangle}J\big) \\ +4\big(\partial_{\langle x,y\rangle}J\big)^2 + \big(\partial_{\langle y,w\rangle}J\big)^2 + \big(\partial_{\langle x,z\rangle}J\big)^2\Big\}. \tag{28}$$

The first term in (28) comes from (23), the second one from (25) and the last three terms from (26). The formula in (28) is equal to $\big(2\partial_{\langle x,y\rangle}J + \partial_{\langle x,z\rangle}J - \partial_{\langle y,w\rangle}J\big)^2$ and thus we obtain $I_2(x, y)$.

Finally, let us take a bond $\langle x, y\rangle$ satisfying the condition for the sum $\sum^{(3)}$, then the coefficient of $V''(\phi(x) - \phi(y))$ for such bond is given by

$$4\sum_{\{z,v,w\}}\Big\{2\partial_{\langle v,x\rangle}J \cdot \partial_{\langle v,w\rangle}J \\ +2\partial_{\langle x,v\rangle}J \cdot \partial_{\langle x,z\rangle}J + 2\big(\partial_{\langle x,v\rangle}J\big)^2 + \big(\partial_{\langle v,w\rangle}J\big)^2 + \big(\partial_{\langle x,z\rangle}J\big)^2\Big\}. \tag{29}$$

The sum of the first two terms in (29) comes from (25), while the sum of the last three terms from (26). This proves the formula for $I_3(x, y)$. □

We are now at the position to complete the proof of (17) under the strict convexity condition on the potential V. Summarizing Proposition 4.4, (16), Lemmas 4.6, 4.8 and omitting the non-negative sums $\sum^{(2)} I_2$ and $\sum^{(3)} I_3$, we have that

$$(\text{LHS}) \geq \frac{1}{4N^{d-1}M} \inf_J E^{\psi}\left[\sum_{\langle x,y\rangle}^{(1)} V''(\phi(x) - \phi(y))I(x, y)\right], \tag{30}$$

where the sum $\sum^{(1)}$ is as in Lemma 4.8 and taken over all undirected bonds $\langle x, y\rangle$ such that $x \in \partial\Lambda_K(i)$ and $y \in \partial\Lambda_K(j)$ with $i \neq j$, and

$$I(x, y) = 4\xi_{ij}^2 - 8\xi_{ij}\Psi(x, y; J) + 4\Psi(x, y; J)^2.$$

To derive (30), we have also used that the first term in the right-hand side of (16) multiplied by $4N^{d-2}M^2$ is equal to

$$2\sum_{\langle i,j\rangle}\xi_{ij}^2\left\langle [V'']_{ij}\right\rangle(\psi)=4\sum_{\langle x,y\rangle}^{(1)}E^{\psi}\left[V''(\phi(x)-\phi(y))\right]\xi_{i(x)i(y)}^2,$$

by recalling that the sum $\sum_{\langle i,j\rangle}$ is taken over directed bonds. Hence, we obtain that

$$\begin{aligned}(LHS)&\geq\frac{1}{N^{d-1}M}\inf_J E^{\psi}\left[\sum_{\langle x,y\rangle}^{(1)}V''(\phi(x)-\phi(y))\left\{\xi_{ij}-\Psi(x,y;J)\right\}^2\right]\\&\geq\frac{c_-}{N^{d-1}M}\inf_J\sum_{\langle x,y\rangle}^{(1)}\left\{\xi_{ij}-\Psi(x,y;J)\right\}^2\\&\geq\frac{c_-}{N^{d-1}M}\sum_{\langle x,y\rangle}^{(1)}\xi_{ij}^2=\frac{c_-}{M^d}\sum_{\langle i,j\rangle}\xi_{ij}^2,\end{aligned}$$

which concludes the proof of (17). We have used the Condition V for the second inequality, while we have expanded the square and used the fact that

$$\sum_{\langle x,y\rangle:x\in\partial\Lambda_K(i),y\in\partial\Lambda_K(j)}\Psi(x,y;J)=0$$

to derive the third inequality. This concludes the proof of (17).

Remark 4.9. The condition $V''\geq c_-$ can be relaxed, since we have used only the fact: $E^{\psi}\left[V''(\phi(x)-\phi(y))\right]\geq c_-$ for every x, y and ψ.

Acknowledgements The author thanks Jean-Dominique Deuschel who suggested the method employed in Sect. 4.2.2.

References

1. Chang, C.-C., Yau, H.-T.: Fluctuations of one-dimensional Ginzburg–Landau models in nonequilibrium. Commun. Math. Phys., **145**, 209–234 (1992)
2. Cotar, C., Deuschel, J.-D., Müller, S.: Strict convexity of the free energy for non-convex gradient models at moderate β, Commun. Math. Phys. **286**, 359–376 (2009)
3. Deuschel, J.-D., Giacomin, G., Ioffe, D.: Large deviations and concentration properties for $\nabla\varphi$ interface models, Probab. Theory Relat. Fields, **117**, 49–111 (2000)
4. Funaki, T.: Stochastic Interface Models, Ecole d'Eté de Probabilités de Saint-Flour XXXIII – 2003, pp. 103–274. Lect. Notes Math., **1869**, Springer, Berlin (2005)

5. Funaki, T., Spohn, H.: Motion by mean curvature from the Ginzburg–Landau $\nabla\varphi$ interface model. Commun. Math. Phys. **185**, 1–36 (1997)
6. Grunewald, N., Otto, F., Villani, C., Westdickenberg, M.G.: A two-scale approach to logarithmic Sobolev inequalities and the hydrodynamic limit. Ann. Inst. Henri Poincaré Probab. Stat., **45**, 302–351 (2009)
7. Guo, M.Z., Papanicolaou, G.C., Varadhan, S.R.S.: Nonlinear diffusion limit for a system with nearest neighbor interactions. Commun. Math. Phys. **118**, 31–59 (1988)

Statistical Mechanics on Isoradial Graphs

Cédric Boutillier and Béatrice de Tilière

Abstract Isoradial graphs are a natural generalization of regular graphs which give, for many models of statistical mechanics, the right framework for studying models at criticality. In this survey paper, we first explain how isoradial graphs naturally arise in two approaches used by physicists: transfer matrices and conformal field theory. This leads us to the fact that isoradial graphs provide a natural setting for discrete complex analysis, to which we dedicate one section. Then we give an overview of explicit results obtained for different models of statistical mechanics defined on such graphs: the critical dimer model when the underlying graph is bipartite, the 2-dimensional critical Ising model, random walk and spanning trees and the q-state Potts model.

1 Introduction

Statistical mechanics aims at describing large-scale properties of physics systems based on models which specify interactions on a microscopic level. In this setting, physics systems are modelled by random configurations of graphs embedded in a d-dimensional space. Since vertices typically represent atoms, the goal is to let the mesh size tend to zero and rigorously understand the limiting behavior of the system.

Although real world suggests that we should focus on the case of 3-dimensional systems, we restrict ourselves to the case of dimension 2. There are two main reasons guiding this choice: first, for models we consider, only very few rigorous results exist in dimension 3, and more importantly, 2-dimensional systems exhibit beautiful and rich behaviors, which are strongly related to this choice of dimension.

C. Boutillier (✉) · B. de Tilière
Laboratoire de Probabilités et Modèles Aléatoires, Université Pierre et Marie Curie, 4 place Jussieu, 75005 Paris, France
e-mail: cedric.boutillier@upmc.fr; beatrice.de_tiliere@upmc.fr

J.-D. Deuschel et al. (eds.), *Probability in Complex Physical Systems*,
Springer Proceedings in Mathematics 11, DOI 10.1007/978-3-642-23811-6_20,

Historically, the most studied graphs are the regular square, triangular, and honeycomb lattices. Since solving statistical mechanics models involves dealing with combinatorial and geometric features of the underlying graph, it has been most convenient to handle the simplest and most regular cases. However in this paper, we deal with models defined on a more general class of graphs, called *isoradial graphs*.

The motivation behind this generalization is to find the most appropriate setting, which exhibits some essential features required to solve the questions addressed, thus allowing for proofs revealing the true nature of the problems, and hopefully a full understanding of the issues at stake. Before going any further, let us recall the definition of a graph, introduce isoradial ones and the corresponding rhombus graphs.

A (simple) *graph* $G = (V, E)$ consists of a set of vertices V and a set E of unordered pairs of distinct elements of V known as edges. A graph G is said to be *planar* if it can be embedded in $\mathbb{R}^2$, meaning that there is a representation of G in $\mathbb{R}^2$ such that points are associated with vertices, simples arcs are associated with edges, and two arcs never intersect except maybe at their extremities. In order to fix ideas, we assume that the graph is finite for the remaining definitions. The connected components of $\mathbb{R}^2 \setminus G$ are called the *faces* of the graph. There is exactly one unbounded face known as the *outer face* and the bounded ones are called the *inner* faces. In the sequel, we will suppose that there is no vertex of degree 1.

Definition 1. A graph is said to be *isoradial* [35], if it is planar and has an embedding in $\mathbb{R}^2$ such that every inner face is a polygon inscribed in a circle of radius 1, and such that all circumcenters of the faces are in the closure of the faces.

From now on, when we speak of the graph G, we mean the graph together with a particular isoradial embedding in the plane. Examples of isoradial graphs are given in Fig. 1 below, in particular the three regular 2-dimensional lattices (square, triangular, and honeycomb) are isoradial.

To such a graph is naturally associated the *diamond graph*, denoted by $G^\diamond$, defined as follows. Vertices of $G^\diamond$ consist in the vertices of G, and the circumcenters of the inner faces of G. Edges of $G^\diamond$ consist of the boundary edges of the outer face of G, and of edges joining the circumcenter of each inner face to the vertices which are on the boundary of this face, see Fig. 2. Since G is isoradial, inner faces of $G^\diamond$

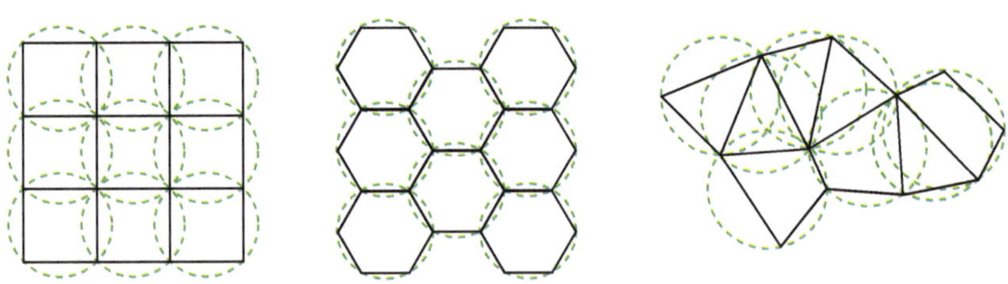

Fig. 1 Examples of isoradial graphs: the square lattice (*left*), the honeycomb lattice (*center*), and a more generic one (*right*). Every inner face is inscribed in a circle of radius 1, represented in *dashed lines*

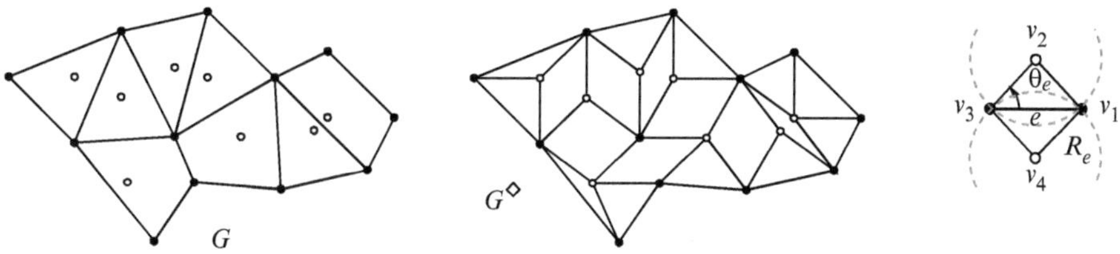

Fig. 2 An isoradial graph (*left*). The white dots are the circumcenters of the inner faces, which are also the vertices of the dual G^*. Its diamond graph is represented in the center. On the *right* is the rhombus R_e and the half-rhombus angle θ_e assigned to an inner edge e

are either side-length-1 rhombi, or half ones when lying on the boundary. Moreover, each edge e of G is the diagonal of exactly one rhombus or half-rhombus R_e of $G^\diamond$; we let θ_e be the half-angle at the vertex it has in common with e. For later purposes, we label vertices of a full rhombus R_e by v_1, v_2, v_3, v_4, as in Fig. 2 (right).

The above definitions naturally generalize to infinite planar graphs, or graphs embedded in the torus obtained as the quotient of $\mathbb{R}^2$ by a periodic lattice. In the case of infinite graphs, we assume that there is some small $\varepsilon > 0$, such that for every $e \in E$, $\varepsilon < \theta_e < \pi/2 - \varepsilon$.

In the past 10 years, surprising results concerning (critical) statistical mechanics models on general isoradial graphs have been obtained by the mathematical community (dimers, spanning trees [35], Ising [13, 14, 17, 18, 47]). The latter are connected to discrete integrability known to physicists for a long time, see for example [8], and yield new insight for models defined on regular graphs.

Isoradial graphs naturally arise in two different approaches to statistical mechanics. This is the topic of the next two sections.

1.1 Transfer Matrices, Star Transformations, and Z-Invariance

The first question addressed when solving a model of statistical mechanics is to compute the *free energy*, which is the growth rate of the *partition function*, counting the weighted number of configurations. An important technique introduced by physicists to solve two-dimensional models is the use of *transfer matrices*, which appears for the first time in the work of Kramers and Wannier [38, 39]. If a model of statistical mechanics with local interactions is defined on a torus of size $n \times m$, then the partition function $Z_{m,n}$ can be expressed as the trace of the n-fold product of a *transfer matrix* T, whose rows and columns are indexed by the configurations of the model in a strip $m \times 1$, and if $\mathscr{C}$ and $\mathscr{C}'$ are two such configurations, then the matrix element $T_{\mathscr{C},\mathscr{C}'}$ is the contribution to the Boltzmann weight of the interactions in $\mathscr{C}$ and $\mathscr{C}'$ in consecutive strips. The free energy of the model f can be expressed in terms of the largest eigenvalue of T denoted by λ_1:

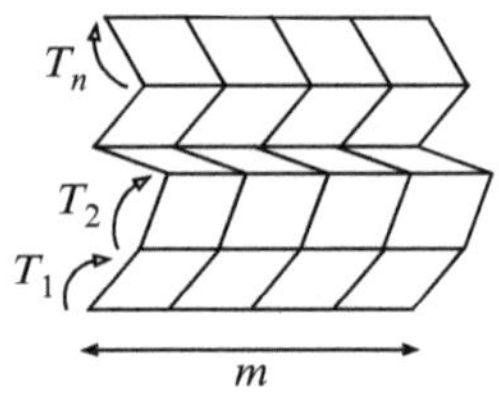

Fig. 3 Transfer matrices for an inhomogeneous model on the $m \times n$ torus

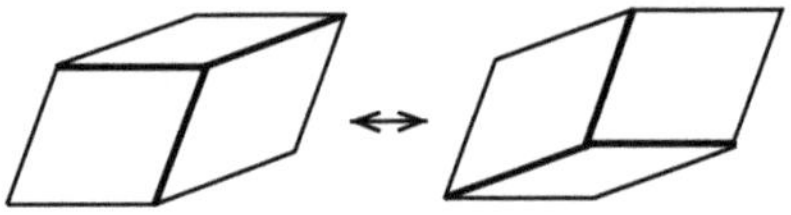

Fig. 4 Star–star transformation (the name *star* referring to the 3-branches stars)

$$f = - \lim_{m,n \to \infty} \frac{1}{mn} \log(Z_{m,n}) = - \lim_{m \to \infty} \frac{1}{m} \log(\lambda_1).$$

If the interaction constants on the graph are not homogeneous, then the transfer matrices used to pass from one strip to another are different (Fig. 3). Nevertheless, if the matrices commute we are able to compute the free energy, as in the homogeneous case, in terms of spectral characteristics of the transfer matrices.

Since the interactions are local, the transfer matrices are themselves sums of local operators R acting on configurations at neighboring sites, and diagrammatically represented by one rhombus of a strip. A sufficient condition to ensure the commutation of transfer matrices is to demand that these R-matrices satisfy some algebraic relation, called the *Yang–Baxter equations*, which can be loosely formulated as follows. Consider the two ways to tile a hexagon with the same three rhombi in Fig. 4. The transformation from one to the other is called a *star–star transformation*. The model satisfies the Yang–Baxter equations if the sum of Boltzmann weights over local configurations of the left-hand side equals those of the right-hand side. This very strong constraint gives a set of equations which needs to be satisfied by Boltzmann weights, see [8,48] for more details. If a nontrivial solution is found, we say that the model is *Z-invariant*.

The condition of Z-invariance yields a natural generalization. Suppose that the graph embedded on the torus of size $m \times n$ is a periodic tiling by rhombi (where the period is independent of m and n). Then by performing a sequence of star–star moves, this graph can be transformed into big pieces of tilted copies of $\mathbb{Z}^2$ [34]. If boundary effects can be neglected, then it suffices to solve the model on each of the copy of pieces of $\mathbb{Z}^2$ using transfer matrices. Summarizing, if the model is Z-invariant, then it can be solved on any periodic graph consisting of rhombi.

Consider an isoradial graph G and recall that a unique rhombus of the underlying diamond graph $G^\diamond$ can be assigned to every edge of G. Then looking at edges rather than rhombi, the star–star transformation becomes a *star-triangle transformation*, see Fig. 5 below.

The definition of Z-invariance naturally extends to this setting, and if the model is Z-invariant the transfer matrix approach can be performed on $G^\diamond$, thus explaining the occurrence of isoradial graphs and Z-invariance in the context of transfer matrices.

Fig. 5 Star-triangle transformation for isoradial graphs

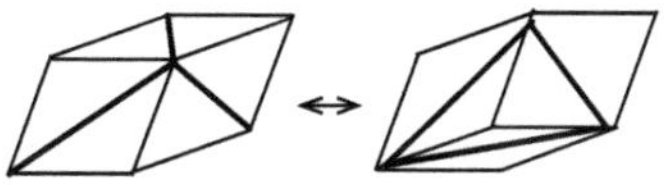

1.2 Conformal Field Theory and Discrete Complex Analysis

Conformal field theory (CFT), introduced by Belavin, Polyakov, and Zamolodchikov [10] (see also [24]), is a theory which aims at describing models at criticality, supposed not to depend on features of the graph, and to be *conformally invariant.* This very strong statement remained largely inaccessible to the mathematics community (except for a few models, such as dimers on the square lattice [33]) until the introduction of the SLE process by Schramm in 1999. The SLE is conformally invariant and conjectured to be the limiting process for many models of statistical mechanics, thus filling a huge gap with CFT. For relations between CFT and SLE, see for example [4, 27]. Several of these conjectures are now solved: loop erased random walk and uniform spanning tree (Lawler, Schramm, Werner [40]), site percolation on the triangular lattice (Smirnov [49]), Ising (Smirnov [50], Chelkak and Smirnov[18]), etc.

It is thus of key importance to have a setting suitable for doing discrete complex analysis, and proving convergence to its continuous counterpart. To this purpose, isoradial graphs are a perfectly suited object. Indeed, when considering an edge and its dual, they consist in the two diagonals of a rhombus, and are thus orthogonal. This allows for a natural discretization of the Cauchy–Riemann equation. Since a lot of developments have happened in this direction, we dedicate it the next section.

2 Discrete Complex Analysis on Isoradial Graphs

As mentioned above, isoradial graphs provide a natural framework for a generalization of the construction in the case of the square lattice. The ideas presented here go back to Duffin [25], and have been developed later by Mercat [47], Kenyon [35], Chelkak and Smirnov [17], Cimasoni [20], and others. There are other possible discretizations of complex analysis on isoradial graphs with remarkable properties. An example using the notion of cross-ratios is given in [12].

2.1 Discrete Holomorphic and Discrete Harmonic Functions

Let us start with the notion of discrete holomorphy. In the continuous setting, the Cauchy–Riemann equations satisfied by a holomorphic function imply that its partial derivatives along two orthogonal unit vectors differ by a factor i. In the setting of isoradial graphs, one takes advantage of the fact that faces of $G^\diamond$ are rhombi, and thus have perpendicular diagonals, to write down finite difference equations which are discretizations of Cauchy–Riemann equations.

Let f be a $\mathbb{C}$-valued function on vertices of $G^\diamond$, let e be an edge of G, and R_e be the corresponding rhombus with the labeling of its vertices v_1, v_2, v_3, v_4, given in Fig. 2. We say that the function f is *discrete holomorphic* at the rhombus R_e, if:

$$\frac{f(v_3)-f(v_1)}{v_3-v_1} = \frac{f(v_4)-f(v_2)}{v_4-v_2}.$$

More generally, we define a discrete operator, $\bar\partial : \mathbb{C}^{V(G^\diamond)} \to \mathbb{C}^{E(G)}$, by:

$$\bar\partial f(e) = i(v_4-v_2)(v_3-v_1)\left[\frac{f(v_4)-f(v_2)}{v_4-v_2} - \frac{f(v_3)-f(v_1)}{v_3-v_1}\right] = \sum_{j=1}^{4} \bar\partial_{e,v_j} f(v_j),$$

where

$$\bar\partial_{e,v_j} = i(v_{j-1}-v_{j+1}), \tag{1}$$

with indices written mod 4. A function f is then said to be *discrete holomorphic on* $G^\diamond$, if $\bar\partial f \equiv 0$. Similarly, we define an operator ∂ by replacing the coefficients $\bar\partial_{e,v_j}$ by their complex conjugate.

The operators ∂ and $\bar\partial$ can be extended to 1-forms (from $\mathbb{C}^{E(G)}$ to $\mathbb{C}^{V(G^\diamond)}$): let g be a $\mathbb{C}$-valued function on edges of G, then for all $v \in V(G^\diamond)$,

$$\partial g(v) = \sum_{e=(v,v')\sim v} \frac{1}{(v'-v)} g(e).$$

The extension of the operator $\bar\partial$ is again obtained by replacing coefficients by their complex conjugate.

Note that the normalization of the operators ∂ and $\bar\partial$ we adopt here differs from that chosen by Mercat, or by Chelkak and Smirnov. One can recover their normalization by multiplying our operators by diagonal operators acting on edges. So these variations do correspond to the same notion of discrete holomorphy.

An notion intimately connected to holomorphy is that of harmonicity. The Laplace operator Δ on G can be written as the restriction of $\partial\bar\partial$ to functions supported on G.[1] Its action on a function f defined on vertices of G is given by

$$\forall v \in G, \quad \Delta f(v) = \partial(\bar\partial f)(v) = \sum_{u\sim v} \tan\theta_{uv}(f(u)-f(v)).$$

A function f is said to be *discrete harmonic* if $\Delta f \equiv 0$. This corresponds to choosing for every edge e, a conductance equal to $\tan\theta_e$.

[1]One can define in a similar way a Laplacian on G^* by restricting the same operator $\partial\bar\partial$ to G^*.

From the factorization $\Delta = \partial\bar{\partial}$, one sees that, like in the continuous case, the restriction of a discrete holomorphic function to G (resp. G^*) is harmonic. Conversely, if H is a harmonic function on G, there exists a harmonic function H^* on G^* such that $H + iH^*$ is discrete holomorphic.

From this choice of Laplacian, one can define the discrete analogues of classical quantities in continuous potential theory (Green function, Poisson kernel, harmonic measure, ...). A surprising amount of estimates have a discrete version, and can be found in [17].

Discrete complex analysis can serve as an approximation of the continuous theory. Mercat [47] proved that the pointwise limit of a sequence of discrete holomorphic functions on isoradial graphs in a domain with a mesh going to zero is holomorphic. See [20] for a generalization of this result to isoradial discretization of compact manifolds. These local convergence results have been supplemented by global convergence theorems [18], implying in particular the uniform convergence for the discrete potential theory objects to their continuous counterparts.

2.2 Discrete Exponential Functions

There is a class of functions playing a special rôle in this theory. These functions, called *discrete exponential functions*, are defined recursively on $G^\diamond$. For any given vertex v_0 of $G^\diamond$ and any $\lambda \in \mathbb{C}$, the function $\mathrm{Exp}_{v_0}(\cdot;\lambda)$ is defined as follows: its value at v_0 is 1, and if v_1 and v_2 are neighbors in $G^\diamond$, let $\mathrm{e}^{\mathrm{i}\alpha} = v_2 - v_1$, then:

$$\mathrm{Exp}_{v_0}(v_2;\lambda) = \mathrm{Exp}_{v_0}(v_1;\lambda)\frac{1+\lambda\mathrm{e}^{\mathrm{i}\alpha}}{1-\lambda\mathrm{e}^{\mathrm{i}\alpha}}.$$

The name comes from the fact that these functions satisfy the following identity: for any pair of neighboring vertices v_1 and v_2 of $G^\diamond$,

$$\mathrm{Exp}_{v_0}(v_2;\lambda) - \mathrm{Exp}_{v_0}(v_1;\lambda) = \lambda\frac{\mathrm{Exp}_{v_0}(v_2;\lambda) + \mathrm{Exp}_{v_0}(v_1;\lambda)}{2}(v_2 - v_1),$$

which is a discrete version of the differential equation

$$\mathrm{d}\exp(\lambda x) = \lambda\exp(\lambda x)\,\mathrm{d}x.$$

satisfied by the usual exponential function. It is straightforward to check from the definition that $\mathrm{Exp}_{v_0}(\cdot;\lambda)$ is discrete holomorphic. In fact, Bobenko, Mercat, and Suris [12] show that any discrete holomorphic function can be written as a (generalized) linear combination of discrete exponential functions, at least under the quasicrystallic assumption (a finite number of slopes $\pm\mathrm{e}^{\mathrm{i}\theta_1},\dots,\pm\mathrm{e}^{\mathrm{i}\theta_d}$ for the rhombi chains): if f is a discrete holomorphic function on $G^\diamond$, then there exists a function $g(\lambda)$ defined on a neighborhood in $\mathbb{C}$ of $\{\pm\mathrm{e}^{\mathrm{i}\theta_1},\dots,\pm\mathrm{e}^{\mathrm{i}\theta_d}\}$ such that

$$\forall v \in G^{\diamond}, \quad f(v) = \oint_{\Gamma} \mathrm{Exp}_{v_0}(v;\lambda) g(\lambda)\, \mathrm{d}\lambda,$$

where Γ is a collection of disjoint small positively oriented contours around the possible pôles $\pm \mathrm{e}^{\mathrm{i}\theta_j}$ of the integrand.

2.3 Geometric Integrability of Discrete Cauchy–Riemann Equations

An important feature of the discrete Cauchy–Riemann is the property of *3D consistency* [12]. Let f be a discrete holomorphic function on $G^{\diamond}$. Given its value on three vertices of a rhombus, the other can be computed using the discrete Cauchy–Riemann equation.

If a *star–star* transformation is realized on the graph $G^{\diamond}$ (see Fig. 4), can we determine the value of the function f at the new vertex in the center, so that f is again discrete holomorphic? A priori, there are three different ways to compute a possible value: one for each rhombus. It turns out that the three values obtained are equal.

Under the quasicrystallic assumption, $G^{\diamond}$ can be seen as a monotone surface in $\mathbb{Z}^d$, projected back to a properly chose plane. The star–star move is a local displacement of this surface along the faces of a cube. If given two monotonic surfaces Σ_1 and Σ_2 in $\mathbb{Z}^d$ that can be deduced one from the other by a sequence of star–star moves, then there is a canonical way to extend a discrete holomorphic function f defined on Σ_1 to Σ_2 by "pushing" its values along the deformation making use of the Cauchy–Riemann equations. In this sense, this 3D consistency can be considered as a notion of *integrability* [11]. This property is very much related to the Z-*invariance* in statistical mechanics.

2.4 Generalization of the Operator $\bar{\partial}$

The double graph G^D of G is a planar bipartite graph constructed as follows, see also Fig. 6 below. Vertices of G^D are decomposed into two classes: black vertices, which are vertices of G and of G^*, and white vertices, which are the centers of the rhombi of $G^{\diamond}$, seen either as edges of G or as those of G^*. A black vertex b and a white vertex w are connected by an edge in G^D if the vertex corresponding to b is incident to the edge corresponding to w in either G or G^*. That is, edges of G^D correspond to half diagonals of rhombi of $G^{\diamond}$. It turns out that G^D has an isoradial embedding (with rhombi of side-length $\frac{1}{2}$) obtained by splitting each rhombus of $G^{\diamond}$ into four identical smaller rhombi.

Fig. 6 The double graph G^D of the graph on the right of Fig. 1: in *thick full lines* is the original graph G, in *full lines* is the dual graph G^*, in *dotted lines* is the underlying $G^\diamond$ graph

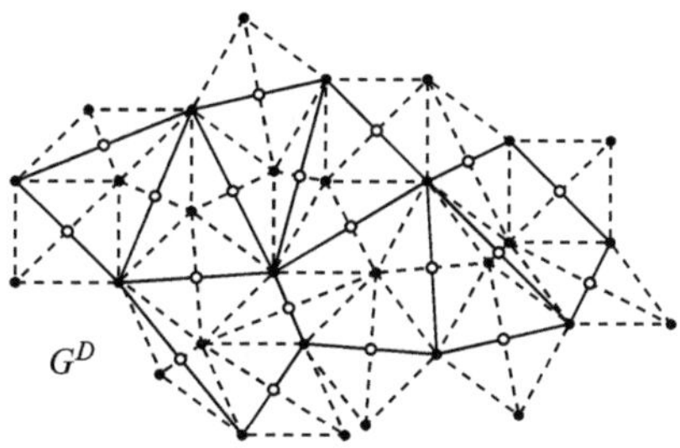

The operator $\bar{\partial} : \mathbb{C}^{V(G^\diamond)} \to \mathbb{C}^{E(G)}$ of Sect. 2.1 can be interpreted now as an operator $\bar{\partial} : \mathbb{C}^{B(G^D)} \to \mathbb{C}^{W(G^D)}$, where $B(G^D)$ (resp. $W(G^D)$) denotes the set of black (resp. white) vertices of G^D. Let f be a $\mathbb{C}$-valued function defined on black vertices of G^D, then for every white vertex of G^D:

$$\bar{\partial} f(w) = \sum_{b \sim w} \bar{\partial}_{w,b} f(b).$$

Let w, x, b, y be the vertices of a rhombus enumerated in cclw order, so that $w \in W(G^D)$ and $b \in B(G^D)$. Then by Formula (1),

$$\bar{\partial}_{w,b} = i(x - y). \tag{2}$$

This definition can be extended to any bipartite isoradial graph, see Sect. 3.1.

For a detailed study of this operator on isoradial graphs on compact surfaces and its connections with discretization of geometric structures (especially spin structures), see [20].

3 Dimer Model

The *dimer model* represents the adsorption of diatomic molecules on the surface of a crystal. The surface of the crystal is modeled by a planar graph $G = (V, E)$, which we assume to be finite for the moment. A *dimer configuration* of G is a *perfect matching* of G, that is a subset of edges M such that every vertex of G is incident to a unique edge of M. Denote by $\mathcal{M}(G)$ the set of dimer configurations of G. An example of dimer configuration is given in Fig. 8. Assume that a positive *weight function* ν is assigned to edges of G, that is every edge e of G has weight ν_e. Then when the graph G is finite, the probability of occurrence of a dimer configuration M is given by the *dimer Boltzmann measure*:

$$P_{\text{dimer}}(M) = \frac{\prod_{e \in M} \nu_e}{Z_{\text{dimer}}},$$

where $Z_{\text{dimer}} = \sum_{M\in\mathcal{M}(G)}\prod_{e\in M}\nu_e$ is the normalizing constant known as the *partition function*.

The dimer model has the attractive feature of being exactly solvable, meaning that there is an explicit expression for the partition function. This fundamental result is due to Kasteleyn [30] and independently to Temperley and Fisher [51]. It relies on the Kasteleyn matrix, which we now define. A *Kasteleyn orientation of the graph* G is an orientation of the edges such that every face is *clockwise odd* meaning that, when traveling clockwise around edges of a face, an odd number of them are co-oriented. The corresponding *Kasteleyn matrix* K is the associated signed, weighted, adjacency matrix of the graph: lines and columns of K are indexed by vertices of G, and for any two vertices x, y of G, the coefficient $K_{x,y}$ is:

$$K_{x,y} = \begin{cases} \nu_{xy} & \text{if } x \sim y \text{ and the edge } xy \text{ is oriented from } x \text{ to } y, \\ -\nu_{xy} & \text{if } x \sim y \text{ and the edge } xy \text{ is oriented from } y \text{ to } x, \\ 0 & \text{otherwise.} \end{cases}$$

Theorem 1 ([30, 51]). *The dimer partition function is,*

$$Z_{\text{dimer}} = |\operatorname{Pf}(K)| = \sqrt{\det(K)}.$$

Remark 1. The Pfaffian, denoted by Pf, of a skew-symmetric matrix is a polynomial in the entries of the matrix, which is a square root of the determinant.

Proof (sketch). Refer to [31] for details. When writing out the Pfaffian of an adjacency matrix as a sum over permutations, there is exactly one term per dimer configuration. The issue is that each term comes with a signature. The purpose of the Kasteleyn orientation is to compensate the signature with signs of coefficients.

As a consequence of Theorem 1, Kenyon derives an explicit expression for the dimer Boltzmann measure.

Theorem 2 ([32]). *Let $E = \{e_1 = x_1y_1, \ldots, e_k = x_ky_k\}$ be a subset of edges of the graph G. Then, the probability that these edges are in a dimer configuration chosen with respect to the dimer Boltzmann measure is;*

$$P_{\text{dimer}}(\{e_1, \ldots, e_k\}) = \left|\left(\prod_{i=1}^{k} K_{x_i,y_i}\right)\operatorname{Pf}((K^{-1})_E)\right|,$$

where $(K^{-1})_E$ is the submatrix of K^{-1} whose lines and columns are indexed by vertices of E.

Proof. Since the proof is very short, we repeat it here. The probability of dimer configurations containing edges of E is given by the weighted number of these configurations divided by the weighted number of all configurations. By expanding the Pfaffian along lines and columns, this yields:

$$P_{\text{dimer}}(\{e_1,\ldots,e_k\}) = \left|\left(\prod_{i=1}^{k} K_{x_i,y_i}\right)\frac{\operatorname{Pf}(K_{E^c})}{\operatorname{Pf}(K)}\right|.$$

Using Jacobi's formula for Pfaffians [29], $|\operatorname{Pf}(K_{E^c})| = |\operatorname{Pf}(K)\operatorname{Pf}((K^{-1})_E)|$ yields the result.

In Sect. 3.1, we state Kenyon's results for the $\bar{\partial}$ operator on infinite, bipartite, isoradial graphs. Then, in Sect. 3.2, we relate them to the corresponding dimer model.

3.1 Dirac Operator and Its Inverse

As mentioned in Sect. 2.4, the $\bar{\partial}$ operator, also referred to as the *Dirac operator* in [35], can be extended to infinite bipartite isoradial graphs. Let G be such a graph, the set of vertices can be divided into two subsets $B \cup W$, where vertices in B (the black ones) are only incident to vertices in W (the white ones). Then $\bar{\partial}$ maps $\mathbb{C}^B$ to $\mathbb{C}^W$: let f be a $\mathbb{C}$-valued function on black vertices, then

$$(\bar{\partial} f)(w) = \sum_{b\sim w} \bar{\partial}_{w,b} f(b),$$

where, if w, x, b, y, are the vertices of the rhombus R_{wb} in cclw order, $\bar{\partial}_{w,b} = i\,(x-y)$ is given by Formula (2).

One of the main results of the paper [35] is an explicit expression for the coefficients of the inverse $\bar{\partial}$, as a contour integral of an integrand, which only depends on a path joining the two vertices. In order to state the result, let us introduce the following functions defined on vertices of G. Fix a reference white vertex w_0 and let λ be a complex parameter. Let v be a vertex of G, and $w_0 = v_0, v_1, \ldots, v_k = v$ be a path in $G^\diamond$ from w_0 to v. Each edge $v_j v_{j+1}$ has exactly one edge of G (the other is an edge of G^*). Direct it away from this vertex if it is white, and toward it if it is black, and let $\mathrm{e}^{\mathrm{i}\alpha_j}$ be the corresponding vector. The function f_v is defined inductively along the path, starting from $f_{v_0} \equiv 1$, and

$$f_{v_{j+1}}(\lambda) = \begin{cases} f_{v_j}(\lambda)(\lambda - \mathrm{e}^{\mathrm{i}\alpha_j}) & \text{if the edge } v_j v_{j+1} \text{ leads away from a black or towards a white vertex,}\\ \frac{f_{v_j}(\lambda)}{(\lambda - \mathrm{e}^{\mathrm{i}\alpha_j})} & \text{otherwise.}\end{cases}$$

It is easy to see that the function is well defined, i.e. independent of the choice of path from w_0 to v. An important point is that, by using a branched cover of the plane over w_0, the angles α_j are defined in $\mathbb{R}$ and not $[0, 2\pi[$. Then, Kenyon has the following theorem.

Theorem 3. *Coefficients of the inverse Dirac operator can explicitly be expressed as:*

$$(\bar{\partial}^{-1})_{b,w_0} = \frac{1}{4\pi^2 i} \oint_C f_b(\lambda) \log(\lambda)\, \mathrm{d}\lambda, \qquad (3)$$

where C is a closed contour oriented counterclockwise, containing all poles of the integrand and with the origin in its exterior.

Remark 2. The remarkable features of this theorem are the following. Explicit computations become tractable, since they only involve residues of rational functions. Moreover, the Formula (3) has the surprising feature of being "local", meaning that if the graph is changed away from the vertices w_0, b and a path joining them, the value of the inverse Dirac operator stays the same.

3.2 Dirac Operator and Dimer Model

The Dirac operator $\bar{\partial}$ can be represented by an infinite matrix K, whose lines (resp. columns) are indexed by white (resp. black) vertices of G, and the coefficient $K_{w,b}$ is given by:

$$K_{w,b} = \bar{\partial}_{w,b} = i(x-y) = 2\sin(\theta_{wb})\mathrm{e}^{\mathrm{i}\alpha_{wb}},$$

where θ_{wb} is the half-angle of the rhombus R_{wb}, and $\mathrm{e}^{\mathrm{i}\alpha_{wb}}$ is the unit-vector in the direction from w to b. The matrix K resembles a Kasteleyn matrix of the graph G, but differs in two instances.

1. Rows and columns are indexed by only "half" the vertices. To overcome this, one can define a weighted adjacency matrix $\tilde{K}$ whose lines and columns are indexed by all vertices of the graph: $\tilde{K} = \begin{pmatrix} 0 & K \\ -K^T & 0 \end{pmatrix}$. Then, we would have $\mathrm{Pf}(\tilde{K}) = \pm\sqrt{\det(\tilde{K})} = \pm|\det(K)|$.
2. The weights of the edges have an extra complex factor of modulus one, instead of a sign given by a Kasteleyn orientation. In [35], Kenyon shows that the matrix is *Kasteleyn flat*, meaning that for each face of G whose vertices are labeled by $u_1, v_1, \ldots, u_m, v_m$, the coefficients of the matrix K satisfy:

$$\arg[K_{u_1,v_1} \cdots K_{u_m,v_m}] = \arg[(-1)^{m-1} K_{v_1,u_2} \cdots K_{v_m,u_1}].$$

   Using a result of Kuperberg [37], this implies that when the graph is finite and planar, K behaves as a usual Kasteleyn matrix, thus yielding an explicit formula for the partition function.

One of the main conjectures of the paper [35] is to show that the inverse $\bar{\partial}$ operator, which we write K^{-1} using the matrix notation, is related to the dimer model on the infinite, bipartite, isoradial, graph G. Its relation to the Gibbs measure is given in the following.

Theorem 4 ([22]). *There is a unique Gibbs measure P_{dimer} on dimer configurations of G, such that the probability of occurrence of a subset of edges $E\{e_1 = w_1b_1, \dots, e_k = w_kb_k\}$ in a dimer configuration of G chosen with respect to the Gibbs measure P_{dimer} is:*

$$P_{\text{dimer}}(\{e_1,\dots,e_k\}) = \left(\prod_{i=1}^{k} K_{w_i,b_i}\right) \det_{1\le i,j\le k}[(K^{-1})_{b_i,w_j}], \tag{4}$$

where $(K^{-1})_{b_i,w_j} = (\bar{\partial}^{-1})_{b_i,w_j}$ is given by (3).

Proof (sketch). Fix the edge set E, and cut out a finite piece of the rhombus graph $G^\diamond$ containing E and paths joining vertices of E. Use a result of [22] by which any finite piece of the graph $G^\diamond$ can be filled with rhombi in order to become part of a periodic rhombus tiling of the plane. Define the probability of the edge set E as the weak limit of the Boltzmann measures on the natural toroidal exhaustion of the periodic graph. Use the uniqueness of the inverse Dirac operator and its locality property to deduce that this expression coincides with (4). Use Kolmogorov's extension theorem to show existence of a unique measure on G which specifies as (4) on cylinder events.

3.3 Other Results

3.3.1 Free Energy

Assume that the infinite bipartite isoradial graph is also periodic, and let $G_1 = G/\mathbb{Z}^2$. In [35], Kenyon proves an explicit expression for the "logarithm of the normalized log of the Dirac operator", which has the surprising feature of only depending on rhombus angles of G_1:

$$\log(\det{}_1 \bar{\partial}) = \frac{1}{|V(G_1)|} \sum_{e\in E(G_1)} \left(\frac{1}{\pi} L(\theta) + \frac{\theta}{\pi}\log(2\sin\theta_e)\right).$$

Kenyon conjectures it to be related to the *free energy* of the corresponding dimer model, result that we prove this conjecture in [21].

3.3.2 Gaussian Free Field

There is a natural way of assigning a height function to dimer configurations of G. In [23], we show that when dimer configurations are chosen with respect to the Gibbs measure of Theorem 4, the fluctuations of the height function are described by a Gaussian free field. Note that this proof holds for all dimer models defined on bipartite isoradial graphs, in particular for $\mathbb{Z}^2$ and the honeycomb lattice.

4 Ising Model

The Ising model, first considered by Lenz [42], has been introduced in the physics literature as a model for ferromagnetism. The vertices of a graph $G = (V, E)$ represent atoms in a crystal with magnetic moment reduced to one component that can take two values ± 1. A *spin configuration* is thus a function σ on V, with values in $\{+1, -1\}$. Magnets with opposite moments tend to repel each other, which has a cost in terms of energy. From these considerations, the *energy* of a configuration σ on a finite graph is defined by:

$$\mathscr{E}(\sigma) = -\sum_{e=uv\in E} J_e \sigma_u \sigma_v,$$

where the positive numbers (J_e) are called *interaction constants*. The probability of a occurrence of a spin-configuration is then defined using the *Ising Boltzmann measure*:

$$P_{\text{Ising}}(\sigma) = \frac{\exp(-\mathscr{E}(\sigma))}{Z_{\text{Ising}}}.$$

The normalizing factor, $Z_{\text{Ising}} = \sum_\sigma \exp(-\mathscr{E}(\sigma))$ is the *Ising partition function* on G.

If the graph is drawn on an orientable surface with no boundary,[2] then the partition function can be written as a combinatorial sum over contours. There are in fact two well-known such expansions: the so-called *high-temperature expansion* (contours on G) and the *low-temperature expansion* (contours on G^*). The contours in the low-temperature expansion have a nice interpretation in terms of the dual closed curves separating zones of different spins.

The most studied case is the Ising model on (pieces of) regular lattices where the interaction constants are taken to be constant equal to β, the inverse temperature. Kramers and Wannier [38] discovered a duality for the Ising model on the square lattice: the measure on contours on G^* coming from the low-temperature expansion is equal to that obtained by considering the high-temperature expansion of the Ising model on G^* at another temperature β^* satisfying $\sinh(2\beta)\sinh(2\beta^*) = 1$. This showed that the self-dual temperature for which $\beta = \beta^* = \log\sqrt{1+\sqrt{2}}$ is the critical point if this one is unique. Later, Lebowitz and Martin-Löf [41] proved that this is indeed the case.

Note that the case of $\mathbb{Z}^2$ is singular since the graph is isomorphic to its dual. This duality can be generalized to any graph and its dual. For example, the low-temperature contour measure for the Ising model on the honeycomb lattice with inverse temperature β is the same as the high-temperature measure for the Ising model on its dual, which is the triangular lattice. But using star-triangle

[2]The correspondence can be extended to surfaces with boundary by including in addition to the closed contours a certain number of paths connected to boundary.

transformations, one can go back to the Ising model on the honeycomb lattice. This generalizes the duality we had for the square lattice.

On isoradial graphs, instead of taking the same interaction constants for all edges, it is natural to take J_e to be a function of the angle θ_e. In order to ensure some integrability at the discrete level, we can impose that the Boltzmann weights satisfy the star-triangle relations, yielding a one-parameter family of Z-invariant interaction constants. There is a generalized duality relation: the high-temperature expansion on G corresponds to the low-temperature expansion on G^* for a dual value of parameter. We qualify as *critical* the interaction constants corresponding to the self-dual value of the parameter, given by the following formula,

$$J(\theta_e) = \frac{1}{2}\log\left(\frac{1+\sin\theta_e}{\cos\theta_e}\right),$$

and refer to the corresponding Ising model as *critical Z-invariant*. See [8], Chap. 7.13 for the parameterization of the star-triangle relation. They coincide with the critical inverse temperatures of the homogeneous triangular, square, and honeycomb lattices for θ_e identically equal to $\frac{\pi}{6}, \frac{\pi}{4}, \frac{\pi}{3}$ respectively. The Ising model on two-dimensional graphs is in correspondence with other well-known models of statistical mechanics: the dimer model and the q-random cluster model with $q = 2$. We differ the discussion of the correspondence with dimers to Sect. 4.2.

There are essentially two ways of taking the large graph limit of the Ising model. Either you let the mesh tend to zero at the same time as the number of vertices of the graph goes to infinity, in order to get continuous objects in a bounded region of the plane (macroscopic level); or you let the graph tend to infinity, keeping the mesh size fixed, to get a model of statistical mechanics on an infinite graph.

The first approach has been adopted by Chelkak and Smirnov [18] to prove conformal invariance of the Ising model on bounded domains of the plane with Dobrushin boundary conditions. The second approach was used by the authors in [13, 14] to construct a Gibbs measure for the critical Ising model on infinite isoradial graphs.

Before explaining these two approaches, let us mention some works by physicists on the Z-invariant model at and off criticality by Au-Yang and Perk [1–3] and by Martinez [44, 45].

4.1 Conformal Invariance

Let Ω be a bounded domain of $\mathbb{R}^2$, and fix two points a and b on the boundary. For every ε, let $G^{\diamond}_{\varepsilon}$ be an approximation of Ω by a rhombus-graph[3] with rhombi of side

[3] On the boundary, we put only half-rhombi such that only "black" vertices are exposed on the boundary.

length ε. Let also a_ε and b_ε be approximations of a and b, located at the center of half-rhombi on the boundary.

Consider the Ising model on the isoradial graph G_ε corresponding to black vertices of $G_\varepsilon^\diamond$, with Dobrushin boundary conditions, *i.e* fixed spins $+1$ on one arc from a_ε to b_ε, and -1 on the other. Introduce, following Chelkak and Smirnov [18], a twisted version of the partition function, considered as a function on edges of G_ε, or equivalently on inner rhombi of $G_\varepsilon^\diamond$. Let z denote a generic rhombus of $G_\varepsilon^\diamond$,

$$\tilde{Z}_{G_\varepsilon,a_\varepsilon,z} = \tilde{Z}_\varepsilon(z) = \left(\sin\frac{\theta_z}{2}\right)^{-1} \sum_{\substack{\text{closed contours}\\ +\text{ curve } \gamma: a_\varepsilon \rightsquigarrow z}} \left(\prod_{e:\text{ piece of contour}} \tan\frac{\theta_e}{2}\right) \mathrm{e}^{-\frac{\mathrm{i}}{2}\mathrm{wind}(\gamma)},$$

where $\mathrm{wind}(\gamma)$ is the winding of the curve γ from a to z. If we remove the prefactor and the contribution of the winding, this would be for $z = b_\varepsilon$ the low-temperature expansion of the partition function of the Ising model on Ω_ε with Dobrushin boundary conditions between a_ε and b_ε.

Now define for all inner rhombi z of $G_\varepsilon^\diamond$,

$$F_{G_\varepsilon,a_\varepsilon,b_\varepsilon}(z) = F_\varepsilon(z) \propto \frac{\tilde{Z}_\varepsilon(z)}{\tilde{Z}_\varepsilon(b_\varepsilon)},$$

up to a multiplicative factor, introduced for technical reasons.

Chelkak and Smirnov prove that the function F_ε and is discrete holomorphic by comparing configurations differing by local arrangements near z, and that it solves some discrete Riemann–Hilbert boundary value problem. They then deduce using approximation results [17] that, as $\varepsilon \to 0$, the function F_ε converges to the function solving the analog continuous Riemann–Hilbert boundary value problem, which is conformally covariant.

Moreover, they prove that this observable satisfies a martingale property with respect to the growth of the curve from a_ε to b_ε: given the first n steps of the curve $(\gamma_0 = a_\varepsilon, \gamma_1, \dots, \gamma_n)$, the expected value of the observable $F_{G_\varepsilon\setminus[\gamma_0,\gamma_{n+1}],\gamma_{n+1},b_\varepsilon}(z)$ is equal to $F_{G_\varepsilon\setminus[\gamma_0,\gamma_n],\gamma_n,b_\varepsilon}(z)$.

This martingale property together with the convergence to a conformally covariant object is sufficient to imply the convergence of the interface between a and b to a chordal SLE, with parameter $\kappa = 3$.

In the same paper, Chelkak and Smirnov construct another discrete holomorphic observable using not the loops separating clusters of spins, but those separating the clusters in the corresponding random cluster model[4] with $q = 2$. This observable is a direct generalization of the one introduced by Smirnov in [50] for the square lattice. Again, this observable is discrete holomorphic and satisfies a martingale property. In this case, the scaling limit of the interface is a chordal SLE(16/3).

---

[4] Also known as the Fortuin–Kasteleyn percolation.

4.2 The Two-Dimensional Ising Model as a Dimer Model

It turns out that a general Ising model on a graph drawn on a surface without boundary can be reformulated, through its contour expansion, as a dimer model on a decorated graph. This correspondence is due to Fisher [26]. Since a lot of exact computations can be carried out on the dimer model, this correspondence has been proven to be a useful tool to study the Ising model [31, 46]

The idea is to construct a *decorated version* $\mathcal{G}$ of the graph G with a measure preserving mapping between polygonal contours of G and dimer configurations of $\mathcal{G}$. This is an example of *holographic reduction* [52]. We now present the version used by the authors in [13, 14], which has the advantage of using decoration with cyclic symmetry. For other versions of Fisher's correspondence, see [19, 26].

The decorated graph, on which the dimer configurations live, is constructed from G as follows. Every vertex of degree k of G is replaced by a *decoration* consisting of $3k$ vertices: a triangle is attached to every edge incident to this vertex, and these triangles are linked by edges in a circular way, see Fig. 7 below. This new graph, denoted by $\mathcal{G}$, is also embedded on the surface without boundary and has vertices of degree 3. It is referred to as the *Fisher graph* of G.

Here comes the correspondence: to any contour configuration C coming from the high-temperature expansion of the Ising model on G, we associate $2^{|V(G)|}$ dimer configurations on $\mathcal{G}$: edges present (resp. absent) in C are absent (resp. present) in the corresponding dimer configuration of $\mathcal{G}$. Once the state of these edges is fixed, there are, for every decorated vertex, exactly two ways to complete the configuration into a dimer configuration. Figure 8 below gives an example in the case, where G is the square lattice $\mathbb{Z}^2$.

In order to have a measure preserving mapping with dimers, the dimer weight function ν is given for the special case of the Z-invariant critical Ising model by

$$\begin{cases} \nu_e = \cot\frac{\theta}{2} & \text{if } e \text{ comes from a rhombus of } G \text{ with half-angle } \theta, \\ \nu_e = 1 & \text{if } e \text{ belongs to a decoration.} \end{cases}$$

A Kasteleyn matrix K is constructed on $\mathcal{G}$. One of the main results of [14] is that its inverse has a "local" expression in the spirit of that obtained by Kenyon in [35] for the inverse Dirac operator, see also Theorem 3.

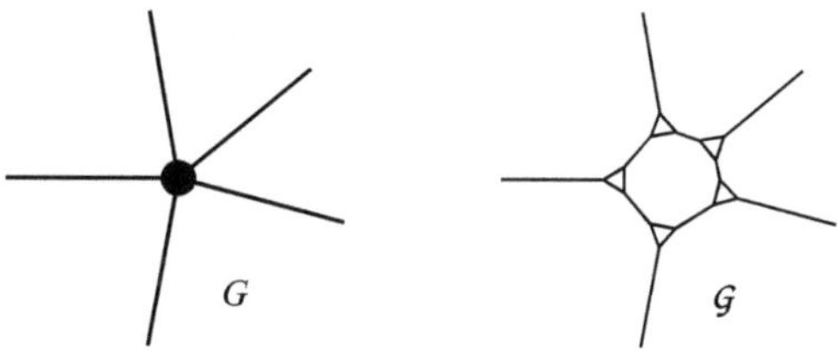

Fig. 7 *Left*: a vertex of G with its incoming edges. *Right*: corresponding decoration in $\mathcal{G}$

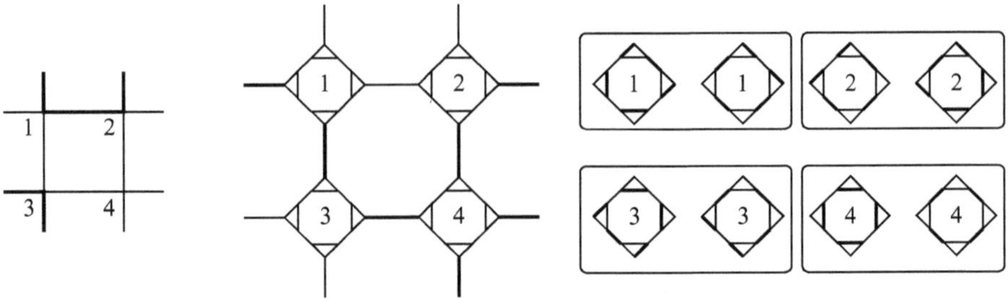

Fig. 8 Polygonal contour of $\mathbb{Z}^2$, and corresponding dimer configurations of the Fisher graph

Theorem 5 ([14]). *If x (resp. y) is a vertex of $\mathcal{G}$ and belongs to the decoration corresponding to the vertex $\mathbf{x}$ (resp. $\mathbf{y}$) of G, then $(K^{-1})_{x,y}$ has the following expression:*

$$(K^{-1})_{x,y} = \frac{1}{4\pi^2} \oint_{C_{x,y}} g_x(\lambda) g_y(-\lambda) \operatorname{Exp}_{\mathbf{x}}(\mathbf{y}; \lambda) \log(\lambda) \, d\lambda,$$

where g_x (resp. g_y) is a simple rational fraction of λ that depends only on the geometry of the decorated graph in an immediate neighborhood of x (resp. of y), and the contour of integration $C_{x,y}$ is a simple closed curve containing all poles of the integrand, and avoiding the half-line starting from x in the direction of y.

Then, as for the dimer model, we obtain an explicit expression for a Gibbs measure, and recover the explicit formula for the partition function, obtained by Baxter [6], at the critical point, see [14] for precise statements.

5 Other Models

We briefly discuss now some aspects of other models of statistical mechanics: first some models related to the Laplacian, then the q-Potts model and its random cluster representation, and finally the 6-vertex and 8-vertex models.

5.1 Random Walk and the Green Function

We already mentioned the Laplace operator on functions of vertices, corresponding to conductances on edges given by $c(e) = c(\theta_e) = \tan \theta_e$. For this particular choice of conductances, the associated random walk is a martingale and the covariance matrix associated with a step is scalar. Up to a time reparameterization, its scaling limit is a standard two-dimensional Brownian motion [9, 17]. Kenyon [35] proves a *local formula* for the inverse of the Laplacian, the Green function Δ^{-1}, on an infinite isoradial graph:

$$\forall x, y \in V(G), \quad \Delta^{-1}_{x,y} = -\frac{1}{8i\pi^2}\oint_C \mathrm{Exp}_x(y;\lambda)\frac{\log\lambda}{\lambda}\,\mathrm{d}\lambda,$$

where C is a positively oriented contour containing in its interior all the poles of the discrete exponential function, and the cut of the determination of the logarithm. He also proves precise asymptotics of G, improved in [15]. This Green function gives information on the random walk on G with conductances $\tan\theta$, but can also be used to gather properties of other models from statistical mechanics with an interpretation in terms of electric network, such as random spanning trees.

A *spanning tree* of G is an acyclic connected subgraph of G whose vertex set contains all vertices of G. On a finite graph, if the weight of a spanning tree equals the product of the conductances of its edges, the partition function is given, via Kirchhoff's matrix-tree theorem [36], by the determinant of any principal minor of the Laplacian. One can construct a measure on spanning trees of G as limit of measures on finite graphs where a spanning tree would have a probability proportional to its weight. The statistics of edges present in the random spanning tree are given by a determinantal process on edges: the probability that edges $\{e_1 = v_1w_1, \dots, e_k = v_kw_k\}$ are present is given by

$$P_{\text{tree}}(\{e_1, .., e_k\}) = \left(\prod_{i=1}^{k}\tan\theta_i\right)\det_{1\le i,j\le k}\big(H(e_i, e_j)\big),$$

where H is the impedance transfer matrix [16] defined by

$$H(e_i, e_j) = \Delta^{-1}(v_i, v_j) - \Delta^{-1}(v_i, w_j) - \Delta^{-1}(w_i, v_j) + \Delta^{-1}(w_i, w_j).$$

5.2 *q*-Potts Models and the Random Cluster Model

A natural generalization of the Ising model is the Potts model with q states (or colors). Spins at neighboring vertices interact only if they have different colors. The energy of a spin configuration σ is $\mathscr{E}(\sigma) = -\sum_{e=uv\in E} J_e \mathbb{I}_{\sigma_u\neq\sigma_v}$. For $q = 2$, we recover the Ising model. This model can be mapped to the (inhomogeneous) random cluster model [28] as follows:

- Spins of different colors are in different percolation clusters,
- Neighboring spins sharing an edge e of the same color are connected with probability, $p_e = 1 - \mathrm{e}^{-2J_e}$.

Once again, we want the interaction constants to be functions of the half-angle of the corresponding rhombus. It is possible to take them to satisfy the Yang–Baxter equation. Moreover, if we impose a generalized self-duality, then there is a unique choice for the function p [34]:

$$p(\theta) = \frac{\sqrt{q}\sin\left(\frac{2r}{\pi}\theta\right)}{\sin\left(\frac{2r}{\pi}\left(\frac{\pi}{2}-\theta\right)\right)+\sqrt{q}\sin\left(\frac{2r}{\pi}\theta\right)}, \quad \text{with } r = \cos^{-1}\left(\sqrt{q}/2\right).$$

Although a formula for the free energy of that model is known [8], little is known except for the case $q = 2$ corresponding to the Ising model.

5.3 6-Vertex and 8-Vertex Models

To conclude, let us briefly mention that the 6- and 8-vertex models, introduced originally on the square lattice, have natural generalization on isoradial graphs, or more precisely on the dual of its rhombic graph $G^{\diamond *}$, which is a 4-regular graph. A configuration of the 6-vertex model (resp. of the 8-vertex model) is an orientation of the edges of $G^{\diamond *}$ such that the number of ingoing edge (and thus of outgoing edges) at each vertex is equal to 2 (resp. is even). These models can be solved through Bethe ansatz on the square lattice (in the sense that their free energy can be computed [5, 43]), and possess a very rich algebraic structure. There are some studies on their Z-invariant generalizations [6, 7], but many questions still need to be solved.

References

1. Au-Yang, H., Perk, J.H.H.: Critical correlations in a Z-invariant inhomogeneous Ising model. Phys. A **144**(1), 44–104 (1987)
2. Au-Yang, H., Perk, J.H.H.: Q-dependent susceptibilities in ferromagnetic quasiperiodic Z-invariant Ising models. J. Stat. Phys. **127**, 265–286 (2007)
3. Au-Yang, H., Perk, J.H.H.: Q-dependent susceptibility in Z-invariant pentagrid Ising model. J. Stat. Phys. **127**, 221–264 (2007)
4. Bauer, M., Bernard, D.: CFTs of SLEs: The radial case. Phys. Lett. B **583**(3–4), 324–330 (2004)
5. Baxter, R.J.: Eight-vertex model in lattice statistics. Phys. Rev. Lett. **26**(14), 832–833 (1971)
6. Baxter, R.J.: Free-fermion, checkerboard and Z-invariant lattice models in statistical mechanics. Proc. Roy. Soc. Lond. Ser. A **404**(1826), 1–33 (1986)
7. Baxter, R.J.: Perimeter Bethe Ansatz. J. Phys. A **20**(9), 2557 (1987)
8. Baxter, R.J.: Exactly solved models in statistical mechanics. Academic, London [Harcourt Brace Jovanovich Publishers] (1989) Reprint of the 1982 original
9. Beffara, V.: Is critical 2D percolation universal? In: In and out of equilibrium. 2, vol. 60 of Progr. Probab., pp. 31–58. Birkhäuser, Basel (2008)
10. Belavin, A.A., Polyakov, A.M., Zamolodchikov, A.B.: Infinite conformal symmetry in two-dimensional quantum field theory. Nuclear Phys. B **241**(2), 333–380 (1984)
11. Bobenko, A.I., Suris, Y.B.: *Discrete differential geometry*, vol. 98 of Graduate Studies in Mathematics. American Mathematical Society, Providence, RI (2008); Integrable structure

12. Bobenko, A.I., Mercat, C., Suris, Y.B.: Linear and nonlinear theories of discrete analytic functions.. Integrable structure and isomonodromic Green's function. J. Reine Angew. Math. **583**, 117–161 (2005)
13. Boutillier, C., de Tilière, B.: The critical Z-invariant Ising model via dimers: The periodic case. Probab. Theory Relat. Fields **147**, 379–413 (2010) 10.1007/s00440-009-0210-1
14. Boutillier, C., de Tilière, B.: The critical Z-invariant Ising model via dimers: locality properties. Comm. Math. Phys. **301**, 473–516 (2011) 10.1007/s00220-010-1151-3
15. Bücking, U.: Approximation of conformal mappings by circle patterns. Geom. Dedicata **137**, 163–197 (2008) 10.1007/s10711-008-9292-7
16. Burton, R., Pemantle, R.: Local characteristics, entropy and limit theorems for spanning trees and domino tilings via transfer-impedances. Ann. Probab. **21**(3), 1329–1371 (1993)
17. Chelkak, D., Smirnov, S.: Discrete complex analysis on isoradial graphs. Adv. in Math. **228** 1590–1630 (2011)
18. Chelkak, D., Smirnov, S.: Universality in the 2D Ising model and conformal invariance of fermionic observables. Inv. Math.
19. Cimasoni, D.: A generalized Kac-Ward formula. J. Stat. Mech. Theory Exp. **2010**(07), P07023 (2010)
20. Cimasoni, D.: Discrete Dirac operators on Riemann surfaces and Kasteleyn matrices. J. Eur. Math. Soc. to appear
21. de Tilière, B.: Partition function of periodic isoradial dimer models. Probab. Theory Relat. Fields **138**(3–4), 451–462 (2007)
22. de Tilière, B.: Quadri-tilings of the plane. Probab. Theory Relat. Fields **137**(3–4), 487–518 (2007)
23. de Tilière, B.: Scaling limit of isoradial dimer models and the case of triangular quadri-tilings. Ann. I. H. Poincare B **43**(6), 729–750 (2007)
24. Di Francesco, P., Mathieu, P., Sénéchal, D.: Conformal field theory. Graduate Texts in Contemporary Physics. Springer, New York (1997)
25. Duffin, R.J.: Potential theory on a rhombic lattice. J. Combin. Theory **5**, 258–272 (1968)
26. Fisher, M.E.: On the dimer solution of planar Ising models. J. Math. Phys. **7**, 1776–1781 (1966)
27. Friedrich, R., Werner, W.: Conformal restriction, highest-weight representations and SLE. Comm. Math. Phys. **243**(1), 105–122 (2003)
28. Grimmett, G.: The random-cluster model, vol. 333 of Grundlehren der Mathematischen Wissenschaften [Fundamental Principles of Mathematical Sciences]. Springer, Berlin (2006)
29. Ishikawa, M., Wakayama, M.: Minor summation formulas of Pfaffians, survey and a new identity. In: Combinatorial methods in representation theory (Kyoto, 1998), vol. 28 of Adv. Stud. Pure Math., pp. 133–142. Kinokuniya, Tokyo (2000)
30. Kasteleyn, P.W.: The statistics of dimers on a lattice: I. the number of dimer arrangements on a quadratic lattice. Physica **27**, 1209–1225 (1961)
31. Kasteleyn, P.W.: Graph theory and crystal physics. In: Graph Theory and Theoretical Physics, pp. 43–110. Academic, London (1967)
32. Kenyon, R.: Local statistics of lattice dimers. Ann. Inst. H. Poincaré Probab. Statist. **33**(5), 591–618 (1997)
33. Kenyon, R.: Conformal invariance of domino tiling. Ann. Probab. **28**(2), 759–795 (2000)
34. Kenyon, R.: An introduction to the dimer model. Lecture notes ICTP (2002) ICTP Lecture note series volume XVII (ISBN 92-95003-25-X) May 2004
35. Kenyon, R.: The Laplacian and Dirac operators on critical planar graphs. Invent. Math. **150**(2), 409–439 (2002)
36. Kirchhoff, G.: Über die Auflösung der Gleichungen, auf welche man bei der Untersuchung der linearen Vertheilung galvanischer Ströme geführt wird. Ann. Phys. **148**, 497–508 (1847)
37. Kirchhoff, G.: An exploration of the permanent-determinant method. Electron. J. Combin. **5**(46), 34 (1998)
38. Kramers, H.A., Wannier, G.H.: Statistics of the two-dimensional ferromagnet. Part I. Phys. Rev. **60**(3), 252–262 (1941)

39. Kramers, H.A., Wannier, G.H.: Statistics of the two-dimensional ferromagnet. Part II. Phys. Rev. **60**(3), 263–276 (1941)
40. Lawler, G.F., Schramm, O., Werner, W.: Conformal invariance of planar loop-erased random walks and uniform spanning trees. Ann. Probab. **32**(1B), 939–995 (2004)
41. Lebowitz, J.L., Martin-Löf, A.: On the uniqueness of the equilibrium state for Ising spin systems. Comm. Math. Phys. **25**, 276–282 (1972)
42. Lenz, W.: Beitrag zum Verständnis der magnetischen Erscheinungen in festen Körpern. Physik. Z. **21**, 613–615 (1920)
43. Lieb, E.H.: Exact solution of the problem of the entropy of two-dimensional ice. Phys. Rev. Lett. **18**, 692–694 (1967)
44. Martìnez, J.R.R.: Correlation functions for the Z-invariant Ising model. Phys. Lett. A **227** (3–4), 203–208 (1997)
45. Martìnez, J.R.R.: Multi-spin correlation functions for the Z-invariant Ising model. Phys. A **256**(3–4), 463–484 (1998)
46. McCoy, B., Wu, T.T.: The Two-Dimensional Ising model. Harvard University Press, Cambridge (1973)
47. Mercat, C.: Discrete Riemann surfaces and the Ising model. Comm. Math. Phys. **218**(1), 177–216 (2001)
48. Perk, J.H.H., Au-Yang, H.: Yang-Baxter equations. In: Françoise, J.P., Naber, G.L., Tsun, T.S. (eds.) Encyclopedia of Mathematical Physics, pp. 465–473. Academic, Oxford (2006)
49. Smirnov, S.: Critical percolation in the plane: Conformal invariance, Cardy's formula, scaling limits. C. R. Acad. Sci. Paris Sér. I Math. **333**(3), 239–244 (2001)
50. Smirnov, S.: Conformal invariance in random cluster models. I. holomorphic fermions in the Ising model. Ann. Math. **172**(2), 1435–1467 (2010)
51. Temperley, H.N.V., Fisher, M.E.: Dimer problem in statistical mechanics-an exact result. Philos. Mag. **6**(68), 1061–1063 (1961)
52. Valiant, L.G.: Holographic algorithms. SIAM J. Comput. **37**(5), 1565–1594 (2008)

MIX
Papier aus verantwortungsvollen Quellen
Paper from responsible sources
FSC® C105338

If you have any concerns about our products,
you can contact us on
ProductSafety@springernature.com

In case Publisher is established outside the EU,
the EU authorized representative is:
Springer Nature Customer Service Center GmbH
Europaplatz 3, 69115 Heidelberg, Germany

Printed by Libri Plureos GmbH
in Hamburg, Germany